Handbook of
Communications
Systems
Management

Second Edition

Handbook of
Communications
Systems
Management

Second Edition

JAMES W. CONARD, EDITOR

AUERBACH PUBLISHERS
Boston and New York

Auerbach Publishers
210 South Street
Boston MA 02111 USA

Contributors

UYLESS BLACK, *President, Information Engineering Inc, Falls Church VA*

LAYNE C. BRADLEY, *Assistant Vice-President–Information Systems, Coventry Corp, Arlington TX*

NORMAN CARTER, *President/CEO, Development Systems International, Studio City CA*

DOROTHY CERNI, *Management Standards Coordinator, Digital Equipment Corp, Littleton MA*

PETER CLUCK, *National Audit Consultant, Pansophic Systems Inc, Conshohocken PA*

WILLIAM COLLINS, *Consulting Engineer, Codex Corp, Mansfield MA*

PHILIP CROSS, *Vice-President MIS, Group Health Inc, New York*

EDWARD J. DELP, *Professor of Electrical Engineering, Purdue University, West Lafayette IN*

PHILLIP EVANS, *Director of Telecommunications, FMC Corp, Dallas*

LOIS FRAMPTON, *Manager, Distributed Processing Architecture, Digital Equipment Corp, Littleton MA*

WALTER J. GILL, *Vice-President and Chief Scientist, Network Equipment Technologies, Redwood City, CA*

JERRY GITOMER, *Manager UNIX Development, NPRI, Alexandria VA*

GIL E. GORDON, *President, Gil Gordon Associates, Monmouth Junction NJ*

JOHN GUDGEL, *Senior Network Design Engineer, Wang Information Services Corp, Lowell MA*

C. STEPHEN GUYNES, *Regents Professor, University of North Texas, Denton TX*

JAN L. GUYNES, *Associate Professor, University of Texas at Arlington, Arlington TX*

JAMES HANSEN, *Professor of Information Systems, Brigham Young University, Provo UT*

HUGH HARVARD, *Principal, Harvard Consulting Services, Dallas*

WILLIAM H. HEFLIN, *President, William H. Heflin Associates, Albuquerque NM*

GILBERT HELD, *Director, 4-Degree Consulting, Macon GA*

GERALD ISAACSON, *President, Information Security Services, Northborough MA*

PHILIP N. JAMES, *Lecturer, Department of Information Systems, California State University, Long Beach CA*

Contributors

VINCENT JONES, *Consultant in Computer Networking, Tenafly NJ*

CELIA JOSEPH, *Consultant (Independent), Concord MA*

ROBERT KLENK, *Senior Systems Analyst, Medical Center of Delaware, New Castle DE*

ELLEN KOSKINEN-DODGSON, *Consultant, Telecommunications Management Consultants Inc, Vancouver, British Columbia*

GREG KOSKINEN-DODGSON, *Consultant, Telecommunications Management Consultants Inc, Vancouver, British Columbia*

DAVID P. LEVIN, *President, NetComm Inc, New York*

L. MICHAEL LUMPKIN, *Marketing Director, Bytex Corp, Southborough MA*

JOHN MAZZAFERRO, *Director of Marketing, IN-NET Corp, San Diego*

JAMES F. MOLLENAUER, *President, Technical Strategy Associates, Newton MA*

NATHAN J. MULLER, *Manager, Consultant Relations, General DataComm Inc, Middlebury CT*

KURUDI H. MURALIDHAR, *Principal Scientist, Industrial Technology Institute, Ann Arbor MI*

KENNETH NOWLAN, *Director, GNA Consulting Group Ltd, Vancouver, British Columbia, Canada*

ROBERT L. PERRY, *Senior Market Research Consultant, Arnold MD*

HEIDI PETERSON, *Image Processing, IBM Thomas J. Watson Research Center, Yorktown Heights NY*

ROBERT A. PRICHARD, *Network Planning Manager, Public Service Co of Colorado, Denver*

ANAND V. RAO, *Owner, Rao Communications, New York*

JEFFREY SCHRIESHEIM, *Technical Director PC Integration, Digital Equipment Corp, Littleton MA*

OTTO I. SZENTESI, *Vice-President and Director, Siecor Corp, Hickory NC*

JOSEPH TARDO, *Consulting Engineer, Digital Equipment Corp, Littleton MA*

RON G. THORN, *Associate Professor, University of Texas Pan American, Edinburg TX*

JON W. TOIGO, *Independent Writer and Trainer, Toigo Productions, Dunedin FL*

JEFFREY EARL TYRE, *Regional Engineering Manager, CMC, Rockwell International Co, Santa Barbara CA*

MICHAEL VARRASSI, *Consultant, V&M International, Richardson TX*

FLOYD WILDER, *Consultant, Wilder Associates, Anaheim CA*

JOHNNY S.K. WONG, *Assistant Professor, Iowa State University, Ames IA*

Contents

INTRODUCTION.   xi

PART I   PLANNING FOR COMMUNICATIONS SYSTEMS.   1

I-1   **Strategic Planning for Communications Systems**   5

I-1-1   Strategic and Long-Range Information Resource
        Planning   7
I-1-2   Network Planning Guidelines   23
I-1-3   Planning for Distributed Processing.   31
I-1-4   Local Networks and Strategic Planning   43
I-1-5   Incorporating LANs into Long-Range Plans   57

I-2   **Communications System Design Planning**   67

I-2-1   Selecting Network Design Software.   69
I-2-2   Selecting a Wide Area Network Design Tool.   79
I-2-3   Graph Theory and Network Design   101
I-2-4   Graph Theory and Multidrop Line Routing.   109
I-2-5   A Systematic Approach to Network Optimization.   119
I-2-6   Examining Integrated Communications Cabling
        Alternatives   133
I-2-7   Implementing a Micro-Mainframe Link.   145

I-3   **Security Planning**.   167

I-3-1   Communications Security Concepts   169
I-3-2   Security in Open Communications Architectures   185
I-3-3   Communications Security Standards.   195
I-3-4   Security in Financial and Messaging Applications.   215
I-3-5   A LAN Security Review.   225
I-3-6   An Overview of Computer Viruses.   237
I-3-7   Protection of Communications Ports and Lines   249

PART II   COMMUNICATIONS NETWORKING   265

II-1   **Architecture and Standards**.   267

II-1-1   Standards Organizations and Their Procedures   269

Contents

II-1-2   Functional OSI Profiles............................   283
II-1-3   OSI Systems Management...........................   303
II-1-4   Internet and TCP/IP...............................   315
II-1-5   Communications Standards in CAD/CAM..............   333
II-1-6   ETSI: A European Standards Organization   347

II-2   Networking Services   357

II-2-1   X.25 Facilities...................................   359
II-2-2   Recommendation X.25: Changes for the
         1988–1992 Cycle..................................   389
II-2-3   T1 and Beyond....................................   415
II-2-4   An Update on ISDN................................   429

II-3   Technology   437

II-3-1   Advances in Modem Modulation Techniques   439
II-3-2   The Practical Side of Voice and Data Integration   455
II-3-3   Communications Protocols.........................   465
II-3-4   Electronic Mail Systems   475
II-3-5   Trends in Fiber-Optic Technology   485
II-3-6   An Overview of Digital Image Bandwidth Compression..   501
II-3-7   The PBX in Perspective   517

PART III   LOCAL AND METROPOLITAN NETWORKS.............   539

III-1   Local Network Methodologies.....................   541

III-1-1   The Manager's Guide to Selecting a Local Area
          Network...   543
III-1-2   Expanding Token-Ring Networks   553
III-1-3   Failure Analysis in Token-Ring LANs   559

III-2   Metropolitan Area Standards and Technology.......   567

III-2-1   The IEEE 802.6 Metropolitan Area Network
          Distributed-Queue, Dual-Bus Protocol   569
III-2-2   Metropolitan Area Network Standards from
          IEEE 802.6   597
III-2-3   Performance Evaluation of Metropolitan Area
          Networks..   611
III-2-4   An Overview of FDDI..............................   627

PART IV   COMMUNICATIONS SYSTEMS MANAGEMENT..........   649

IV-1   Selection and Procurement Considerations   651

IV-1-1   Modem Selection Factors..........................   653
IV-1-2   Selecting and Evaluating Statistical Multiplexers   671

IV-1-3   Reducing the Costs of Equipment and Outside Services   683

IV-1-4   How to Choose a Systems Integrator   693

IV-2   **Implementation Guidelines**   705

IV-2-1   Data Network Design Fundamentals.   707

IV-2-2   System Acceptance Testing   719

IV-2-3   An Integrated Communications Cabling Network   731

IV-2-4   The Manufacturing Automation Protocol and the
Technical and Office Protocol   763

IV-3   **System Operations**   773

IV-3-1   A Manager's Perspective of Network Management   775

IV-3-2   Managing the Building of a Network Control Center   787

IV-3-3   Evaluating the Effectiveness of Data Communications
Network Controls   815

IV-3-4   The Third-Party Maintenance Alternative   831

PART V   **COMMUNICATIONS SYSTEM ISSUES AND TRENDS**   839

V-1   **Supplementary Topics.**   841

V-1-1   Maintaining Accurate Documentation   843

V-1-2   Guidelines for Using Consulting Services   853

V-1-3   Training Communications Professionals   867

V-1-4   Evaluating Training Methods and Vendors   875

V-1-5   Weighing the Costs and Benefits of Standards
Involvement   887

V-1-6   Protecting the Network Through Regulatory
Involvement   891

V-2   **Future Trends in Communications**   897

V-2-1   Managing the Transition to Electronic Data Interchange . .   899

V-2-2   ISDN, OSI, and Signaling System #7   917

V-2-3   Transborder Data Flow   923

INDEX   931

Introduction

IN SELECTING MATERIAL for this second edition of the *Handbook of Communications Systems Management*, I have sought articles that are interesting, informative, timely, and of lasting value over the near-term period of one to five years. These criteria support the objective of creating a desk reference that the communications systems manager can turn to as a source of useful information.

The focus on the near term requires the sacrifice of both very short-term and very long-term information. This trade-off, however, is acceptable because the communications manager relies heavily on the trade press and periodicals for current information. In addition, to include longer-term information would require prediction of the future, which produces notoriously unreliable results. Prediction is an activity more appropriate for soothsayers and gurus than editors.

Nevertheless, the selection process forced me to examine current developments and to try to see how these developments will effect communications systems over the near term. In other words, I had to look for trends.

COMMUNICATIONS AND SOCIETY

The longer-term trends in communications probably cannot be separated from those in information systems and, for that matter, from those of human culture. Human society is moving toward what McLuhan called "the global village," which others have characterized as the information age, and which will be supported by what I call the globalization of communications.

Microcomputers make it possible to provide an individual, rather than an enterprise or an organization or a town, with immense computational power. This resource is available to the individual regardless of geographical location or proximity to other people. This means that it should now be possible to construct a decentralized society, with individuals contributing within their community of interest and cooperating with other groups through a global communications system.

One result of this movement toward global information systems is the decreasing need for isolated information systems of any kind. In a globally distributed information system the individual components begin to disappear as separate entities and instead merge into a global utility. This could, for example, mean the disappearance of local networking as a separate identifiable discipline. From the user's point of view, what is needed is not a local network but local availability of a globally distributed information resource. Still another example of the move toward global information systems is the expansion of the definition of information from the traditional data that was relevant to digital computers to voice, images, video, and graphics.

There is little doubt that communications is the key to the global information utility. Cooperative multivendor, transnational movement of information will require increasingly effective communications. Communications systems will have to have greatly increased capacity at much higher speeds and at relatively low cost.

THE TECHNOLOGY OF COMMUNICATIONS

The technologies most likely to support global networking are switching, transmission, routing mechanisms, signaling and control, and high bandwidth services. Significant advances are being made in each of these.

Switching

Switching is the direction of information through a multinodal network from source to destination. Because it is impractical to build a global, fully meshed network, switching technology is critical to cost-effective and responsive networking. For economic and performance reasons, digital rather than analog switching is essential. The importance of digital switching technology has certainly been recognized by the telephone industry. Since 1980, more than 40% of the switching offices worldwide have been equipped with programmable digital switches, replacing electromechanical and analog switches, and by the year 2000 90% of the worldwide switching capacity will be digital.

Another aspect of switching technology is the emergence of fast packet switching. Fast packet is an efficient technology because it allows variable, on-demand channel capacity. It is being developed primarily for front-end and PBX applications and in support of frame relay services. Fast packet is similar to X.25 packet switching but at much higher rates and much lower transit delays. Fast packet targets are for delays on the order of 50 msec and data rates in the megabits-per-second range. With digital end-to-end connectivity, fast packet will permit voice as well as data and images to be packet switched.

The high speeds being proposed for fast packet will require the deployment of new switching fabrics (fabric is the term used to describe the design and structure of switching activity) that support on-the-fly switching. Switching stages are implemented as very fast hardware made up of silicon gates. Routing through the gates is triggered by a header field or tag that identifies a virtual channel. (A three-stage switching node, for example, requires a three-bit tag.) A packet arriving on any input and carrying the same tag will be directed to the same output line. This approach is much faster and more efficient than software switching supported by routing tables.

Transmission

The changes in transmission technology will involve a rapid conversion toward all-digital transmission facilities. Telephone networks still provide most of the world's services for information exchange, and these networks, especially the long-haul trunks interconnecting switching offices, are now about 80% digital. A significant portion of these trunks are fiber based. Trends point to an all-fiber long-haul network to be in place shortly after the turn of the century; this network is called the Synchronous Optical Network (SONET). This structure is intended to provide standard fiber-optic interfaces, its capacity rising in modular increments from 51.84M bps to 2.5G bps.

The integration of the digital transmission facilities with the digital switching technologies discussed earlier has been termed the Integrated Digital Network, or IDN. Extension of this digital capability to the users' premises is the basis for the Integrated Services Digital Network (ISDN). ISDN is intended to make bandwidth on demand available to users for any application. The digital capability makes it possible to integrate all sorts of mixed media services on single digital access lines. This can result in significant cost savings, economies of scale, and much more flexibility for the user.

Routing

The deployment of global high-speed digital networks will undoubtedly require the development of new protocols that will work efficiently in the megabit-to-gigabit data rate environments. One such new protocol, CCITT V.120, adapts the ISDN signaling protocol to efficiently carry user information on the higher-speed channels.

Another example of protocol development in support of high-speed digital transmission is the asynchronous transfer mode, a packet-oriented transfer mode that uses asynchronous time-division multiplexing. In this form of multiplexing, the users' information is organized into fixed-size cells. A cell includes user information and a transport header. The header carries a virtual channel identifier. This permits cells belonging to the same user stream, even though they could follow different routes, to be directed to the proper destination. The routing mechanism can be supported by the new switching fabrics discussed earlier. Asynchronous transfer mode cells are assigned on demand. Sequence integrity can be preserved by using connection-oriented techniques.

Signaling and Control

The transition to all-digital networks will be accompanied by the design and deployment of new signaling and control technology; this will support the global interconnection of digital networks into a true information utility. Current activity is concentrated on the development of a common channel (common to all signaling functions and paths) signaling system called signaling system #7. Although it was initially deployed to support digital telephone networks and ISDN, this new signaling system has enormous potential in private networks as well as in the provision of transaction-based services. This is because the signaling network is actually a high-speed, reliable, transaction-oriented packet-switched network.

High-Bandwidth Services

Another technological trend that will profoundly influence the globalization of communications is broadband ISDN (B-ISDN). B-ISDN will probably be the next significant communications development after ISDN. Projected for implementation during the latter half of this decade, and on into the next century, B-ISDN will support user interfaces at 600M bps and beyond. At these data rates, B-ISDN can accommodate a range of bandwidth-intensive interactive and distributed services such as video telephony, archive retrieval, and medical imaging as well as applications for entertainment, publishing, and education.

THE REST OF THE PICTURE

Other trends and issues can be expected to effect global networking. One of the more intriguing is cellular transmission. Cellular radio technology, which permits wireless networks, is now gaining momentum. Wireless networks have local area, packet radio wide area, and personal communication applications. The latter is a highly significant development because it decouples the user from the physical location of the communicating device, another step toward the true global communications tool discussed earlier.

For trends to actually predict the future, they must be linear. Many influences, however, can distort the trend line. The success or failure of open communications and information processing systems architectures is an example: the failure of the concept of openness would retard the development of the global network. Other examples of issues that can bend trend lines are regulation, legal constraints, security, and, of course, politics and economics.

Whether future communications systems will take the form suggested here is, of course, unknowable. What is certain is that the communications system manager must remain constantly aware of the sometimes subtle changes occurring now that could have immense impact on the system of tomorrow.

ORGANIZATION

In this second edition of the Handbook we have maintained the structural organization of the original. The five major parts are divided into thirteen sections, the subjects ranging from initial planning through trends. I sincerely hope that the volume will prove a useful guide and reference as you pursue your management tasks, and I will be very happy if the Handbook remains within arm's reach and becomes dog-eared from use.

ACKNOWLEDGMENTS

My utmost appreciation goes to the many contributors to this book. Without them there could be no Handbook. My appreciation goes also to the staff at Auerbach Publishers, whose skill helped to place this volume in your hands. A special thanks to Susan McDermott, Paul Berk, and Kim Horan Kelly, all of whom encouraged and supported my efforts on this project.

JAMES W. CONARD

Part I
Planning for Communications Systems

PLANNING IS A VERY IMPORTANT—some would say the most important—part of the communications systems manager's job. If management activities were to be compared to traveling along a road, then a plan is, in effect, a milestone on the road. Information in the plan helps to define a location. At the same time, the plan defines the overall direction of the road ahead. Certainly, the plan does not detail all of the twists and turns nor all of the potholes and boulders. It does, however, provide a large-scale road map into the future.

Communications systems planning is an important task that consists of many phases, each flowing into the next and with continuous feedback and alteration processes in operation. The major phases include assessing user requirements, translating user requirements into objectives, documenting the goals and objectives (i.e., writing the plan), and finally, executing the plan.

Assessing user needs is one of the keys to a successful plan. It is also one that often requires more than a little educated guesswork. The usual way of determining network user requirements is to ask questions. This may involve both a formal approach (as with a questionnaire) or a more informal approach (e.g., through meetings) or a combination of both. The trouble with this is that although some useful information is obtained, the users frequently do not know what their requirements are. The planner then must supplement the questionnaire results with such actual measurements as traffic loading and geographic dispersion.

An important but often overlooked aspect of communications network planning is an assessment of the organization's plans for the future. These plans must be evaluated to determine the impact on the network plan. If, for example, the organization is planning expansion of operations over the life of the network, this expansion needs to be quantified into users, terminals, and connection requirements. A planned merger or divestiture can similarly affect the plan. A manager comfortably implementing an IBM SNA network may suddenly discover that a new subsidiary is a DECnet user. A planned reduction of personnel is equally important to the planner.

The difficulty in this aspect of planning is that the communications systems manager may not be privy to this type of information, which is often confidential. To do a proper job, the network planner must at least attempt to obtain a sense of the organization's direction.

These difficulties point out the importance of another phase of the network planning process. The network planner should gather all pertinent information into an architectural or network objectives plan. Whatever the name, the intent is to document the goals and objectives of the network deployment project. Each requirement is specified. The aggregate requirements lead to sizing and other specifications.

The plan must cover all aspects of the network. All assumptions must be listed. Projected costs, schedule, and equipment and software requirements must be specified. Environmental considerations must be addressed. Both short-term needs and long-term projections must be defined. Every aspect of the new or upgraded system must be examined.

Recognizing that the only certainty in a plan is change, the network planner must establish a control mechanism for modifying the plan as implementation proceeds. Applications change, new ones are added, old equipment is retired, and new devices are acquired. Departments are reorganized, moves are made, and buildings are constructed or leased. All require modifications to a plan that was gospel only months ago. If each change is not documented, along with its cost and its effects on the schedule, the communications systems manager may have to accept the responsibility for implementation problems.

When the plan is drafted, the most interesting part begins: getting approval signatures. The plan really represents a contract between the planner and the organization. The communications systems manager uses the plan document to tell the organization's senior management what the plan is, how much it will cost, and how long it will take and to ask senior management for approval. It is amazing what happens when an individual is asked to sign a plan. The oral approval (which is much easier to obtain) becomes a commitment, and the individual is much more aware of the implications. This awareness is often translated into a second look at the situation, and this in turn translates into additional requirements or a demand for more accurate projections. The system planner must be willing to gamble a little because the signature process may take a while. The planner may want to initiate the implementation phase before all signatures are obtained.

Once senior management has signed the plan or at least is in the process of approving it, the system planner begins to manage the early stage of the implementation phase. This involves evaluation of alternatives, make-or-buy decisions, translation of the objectives document into hardware and software design plans, and generation of requests for proposals.

Implementation is the subject of a later portion of this handbook. In Section I-1, strategic and long-term planning are explored. Section I-2 examines several practical aspects of system design planning. Finally, in Section I-3, some of the many details of security planning are discussed.

Section I-1
Strategic Planning for Communications Systems

MOST OF US PROBABLY KNOW from experience that the maxim "Plans never work, but planning always does" applies to all aspects of communications systems planning. This section is aimed at providing the communications systems manager with some useful advice and guidelines on the subject of strategic planning for communications systems. The chapters describe the complex interrelationships among the many components of the distributed information system. The contemporary business enterprise typically is supported by both computational and communications resources. These resources are of both tactical and strategic importance. Their provisioning, deployment, and operation is probably critical to the success of the enterprise. This success can be achieved only through the planning process.

The planning process is the subject of Chapter I-1-1, "Strategic and Long-Range Information Resource Planning." This chapter defines the strategic planning problem, the process, and the various levels of planning. It includes a thorough discussion of the elements of strategic planning and provides an excellent set of guidelines for the planner. This chapter is general enough to be applicable to any computer or communications system yet detailed enough to guide the creation of an organization-specific plan.

The communications system has become an important business tool. It is essential, therefore, that strategic plans for communications networks address business as well as technical issues. Chapter I-1-2, "Network Planning Guidelines," offers planning and implementation guidelines for the development of network strategic plans. The chapter emphasizes the business aspects of planning, including management requirements, constraints, planning team organization, training, and support considerations.

Chapter I-1-3, "Planning for Distributed Processing," addresses the considerations that are unique to the distributed processing system as contrasted with local or centralized processing systems. It reviews those issues that must be considered when planning a distributed resource information system. Application structures for the distributed environment are discussed. Communications standards that become important in a distributed multivendor environment are identified and reviewed.

The last two chapters of this section explore the impact of local networks on the strategic planning process. Chapter I-1-4, "Local Networks and Strategic Planning," examines the principles of planning using a planning model. The model takes a top-down approach to the elements of planning, beginning with an unconstrained logical model of an ideal system. The next step is to analyze the current situation with regard to existing system technology and other factors. This analysis is then compared with the ideal system. The result is a deficiency list, which forms the basis of a plan specifying actual projects to be executed in order to approach the objectives of the ideal system.

Chapter I-1-5, "Incorporating LANs into Long-Range Plans," focuses on the key long- and short-term operational issues that must be addressed in fitting local networks into the corporate strategic plan. The thrust of this chapter is to explore specific architectural questions that will lead to the development of a workable action plan.

Chapter I-1-1
Strategic and Long-Range Information Resource Planning

Philip N. James

THE ROLE OF COMPUTER-BASED information systems in business has changed significantly during the past 25 years. Today, most organizations depend on complex, increasingly interdependent, online multidepartmental systems to succeed and indeed survive. Recent advances in minicomputers and microcomputers, which seem to reduce this complexity through greater local autonomy, actually create more problems by increasing the difficulty of maintaining data and processing consistency and by introducing new complexities in communications. Furthermore, senior managers are aware that the information resource should provide a competitive advantage, and they expect their technical managers to tell them how. In this environment, new approaches to strategic information resource planning must be found to supplement traditional planning methodologies.

Traditional long-range information systems planning and capacity planning continue to ensure the bread-and-butter role of the information resource—to support the internal operations of the organization—as well as to provide the foundation on which new planning approaches must be built. These new approaches must be based on strategic business considerations—the nature of the organization and the business environment in which it operates. These new approaches should produce business plans that include strategic uses of the information resource in conjunction with the other organizational resources (e.g., finances, personnel, facilities, and equipment) to capture business opportunities and achieve business objectives.

THE PLANNING PROCESS

Planning is the process that creates results, confirms or determines the organization's direction, explains the reason for the direction, and commits to the support of that direction. The written plans provide interim guidelines and are subject to continual review and updating. It is the evolving understanding and team spirit fostered by the planning process that keeps the organization effective and responsive.

As Exhibit I-1-1 shows, planning is a cyclical process, directing actions that cause results and learning from the results. This exhibit illustrates the important role that an understanding of the organization's business environment plays in the formulation of plans.

Exhibit I-1-2 shows that the ultimate product of the planning process is a set of action plans for each manager that supports the mission and goals of the organization. The planning process takes place at every level and in every unit of the organization; in effective planning-driven organizations, all of these processes are articulated so that everyone is working toward the same set of organizationwide goals and objectives.

Exhibit I-1-3 shows how the basic planning questions affect the development of action plans. The organization's mission and its businesses (i.e., products and services) define its customers and vice versa. The organization seeks to serve its customers by taking advantage of its strength, its competitors' weaknesses, and opportunities in the business environment while avoiding threats, all with respect to each of its businesses. Because

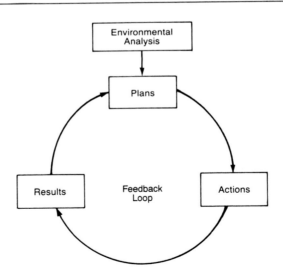

Exhibit I-1-1. The Planning Process

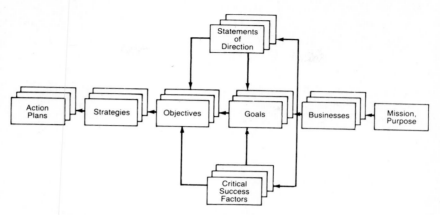

Exhibit I-1-2. The Objectives of Strategic Planning

these elements change over time, planning must be a process that responds to the changes as they occur.

LEVELS OF PLANNING

Strategic Planning

The strategic level of planning requires a sophisticated understanding of the organization, its mission, the businesses it will be in and why, its customers and their needs, and its competitors. Managers can make their greatest impact in this area. If they understand these concepts and can identify uses of information in pursuit of objectives at this level, they can help their organizations use the information resource to gain a competitive advantage. However, this must be a synergistic process using the experience, expertise, and perspective of each senior manager. From this interaction, solid strategies can evolve.

There are two critical factors to effective strategic planning. The first is having strategic vision, which serves as a touchstone for employees at every level in the organization in their daily decision making. The second critical factor of strategic planning is a thorough understanding of the business environment and its movement (e.g., customer needs and perceptions, competitor effectiveness in satisfying those needs and perceptions, and opportunities and threats derived from social, political, and economic changes). The information in the environment relevant to the organization's business should be monitored regularly to detect changes that may signal the beginning of a trend. Many business failures can be attributed to lack of detailed or current understanding of the business environment.

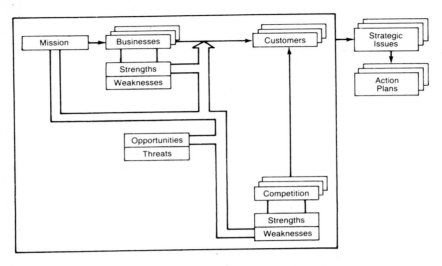

Exhibit I-1-3. Strategic Business Planning

Documentation of the strategic plan should be clear and concise and directed at senior management. It should be prepared in a timely and responsive fashion and frequently revised as a snapshot of the current state of the process. When major changes in the business require major revisions to the plan, a "skunk works" or "tiger team" approach may be best. All other planning efforts flow from the strategic plan and are directed by it.

Tactical Planning

Tactical planning is concerned with developing action plans for the acquisition and deployment of the organization's resources to carry out the strategies identified in strategic planning. Budgets, plans for specific projects, and master manufacturing plans are examples of tactical plans. Long-range information resource plans often are more tactical in approach than strategic plans, despite their longer time frames.

Operational Planning

Operational planning deals with the daily operations that are necessary for accomplishing the action plans. Shop floor scheduling in manufacturing and production scheduling in an information systems shop are examples. At a level below operational planning are practices and procedures, desk or workstation instructions, and other detailed plans that ensure the execution of operational plans.

Long-Range Planning

Often confused with strategic planning, the term *long-range planning* applies only to the time frame of the planning horizon. Usually, any plan covering more than a two-year period should be considered long range. A long-range plan can also be considered strategic in nature if it addresses strategic issues as previously discussed. A strategic plan, however, is not necessarily long range. A major change in organizational mission or direction may require a plan that at first cannot be extended beyond a year or two, at which point experience will determine a clearer future.

The Elements of Planning

Planning can also be categorized by requirement. A comprehensive information resource plan, for example, would address the following elements:

- The information needs of the organization (i.e., strategic, tactical, and operational) are often expressed as an information architecture.
- The organizational resources required to supply acquisition, deployment, stewardship, and disposal needs, including:
  —Human. Managerial, professional and technical, operational, recruitment, skills and training, the human support infrastructure, career planning, and management succession.
  —Physical. Computer, office automation, and nonautomated equipment, other equipment and supplies, and facilities.
  —Financial. Operating and capital expenditures.
  —Information. That which is handled by information technology as well as by libraries, mail rooms, file cabinets, conversations, and people's memories.
- The organizational plans for acquiring, developing, deploying, and controlling the other resources.

THE REASON FOR PLANNING

A formal planning process forces managers to articulate their objectives, priorities, and action plans. The advantages of a formal, documented information resource plan include:

- Better information that is more accessible and in a usable form.
- Better budget justification for information technology projects, ensuring that resources will be available when needed.
- Better use of the funds spent on information technology.
- Smoother running of all components in a complex information technology environment.
- Reduction of costly maintenance and corrective work.

- In a distributed environment, identification of cost-saving and cost-sharing opportunities.
- More effective responses to unexpected changes in the business environment, thus avoiding crisis management.

ORGANIZATIONWIDE STRATEGIC INFORMATION RESOURCE PLANNING

Many planning principles, if properly translated, are applicable at any level. For example, the strategic planning process can be conducted by the organization's CEO or on a lesser level by the organizational unit doing the planning. Tactical and operational planning may be delegated, whereas strategic planning may not.

As discussed, the strategic planning process is important, not the products. It is the interaction leading to shared understanding and a shared direction that produces results, not the prescriptive steps of an action plan. Furthermore, the success of the process depends on whether the right people are present and on their degree of participation. A planning methodology can ensure only that most of the right questions are asked; the answers must come from those who know them or can participate in developing them.

The role of the professional planner is to catalyze, simplify, and support the process, usually through a methodology that ensures that the process occurs continuously, involves the right people in the appropriate manner, and addresses the right issues. Among other things, this includes focusing on what is important: the 20% of an organization's activities that determine 80% of an organization's effectiveness. To whatever degree is reasonable within the culture of the organization, strategic information resource planning should be an integral part of strategic business planning.

Enterprise Analysis

The critical first step in information resource planning is to acquire an understanding of the organization being supported. Most organizations consist of a single enterprise—a flow of material and information used to deliver a family of products and services to meet a need. The available human and physical resources determine the effectiveness of this flow. In the private sector, the ordering of needs is carried out in the market by determining the enterprise's ability to earn a profit from the sale of its products and services. In the public sector, political considerations provide the basis for the ranking of needs.

The basic business processes of most enterprises are remarkably stable. However, an organization's method of executing a process changes frequently. The details of each process, then, must change to reflect a changing product mix and market. Nevertheless, the basic structure of information and material flow is surprisingly constant.

It is important to develop a model of the enterprise that shows the relationships among the business processes. The model's level of detail depends on the organization's particular needs. For strategic planning, the focus is on issues, directions, and strategies, and a simple model that concentrates on the activities affecting critical success factors usually suffices and is a good starting point in any organization. For more detailed long-range tactical and operational planning, the entire enterprise should be modeled. The business processes may be dissected into their component parts (i.e., the entities involved in information and material flow, and the relationships among them). This entity-relationship approach can lead directly to the design of corporate data bases, often through the use of automated tools.

Once the enterprise model is complete, it is used to derive an information architecture, which illustrates the flow of information through the enterprise. (Furthermore, a material flow architecture that provides the basis for planning for other enterprise resources can be derived.) This information architecture becomes the basis for applying the information technologies that support the organization. These technologies include the portfolio of data processing applications, office technologies (including decision support and end-user computing), and data communications. This approach to information resource planning ensures that it is directed by organizational business needs and not by the information technologies and their advocates.

For optimal effectiveness, both the enterprise model and the architecture derived from it must be subjected to rigorous configuration management to ensure that both remain valid and current as the business evolves. It is important to choose a methodology that fits well with the enterprise culture. Some of the more powerful modern methodologies require a commitment of resources and a level of understanding beyond the willingness of some organizations to attain, even when the payoff is high.

Environmental Analysis

Effective environmental analysis is usually the most important critical success factor in strategic information resource planning. An organization's information resource is embedded in a complex structure that includes the enterprise itself; the business environment and market in which it operates; the social, cultural, political, and economic environment within which it conducts business; and the information resource management profession, including the dynamic world of information technology. Successful information resource management depends on understanding and being aware of this environment to detect change, assess its impact, and trigger corresponding changes in the strategic plan (see Exhibit I-1-4).

In addition to the organization's strategic business plan, the evolving environment of information resource planning includes changes in:

- Ideas about the nature and use of the information resource and cultural attitudes toward these changes.
- Technologies for managing the information resource, including changing cost-performance relationships.
- Attitudes toward technology.
- Characteristics, perceptions, and availability of information systems professionals.
- Characteristics and perceptions of information resource users.

However, environmental analysis is rarely included in information resource planning. As a result, information systems departments have often failed to anticipate change. Examples of such failures include the pressure for distributed processing and the introduction of microcomputers.

The environment can be studied in many ways. Among them are business planning techniques, which focus on customers (i.e., users), competitors (i.e., vendors and other information services suppliers), strengths and weaknesses, and opportunities and threats. In addition, futures research methodologies seek to identify the most probable range of alternative futures in order to develop effective strategies for dealing with them. Informa-

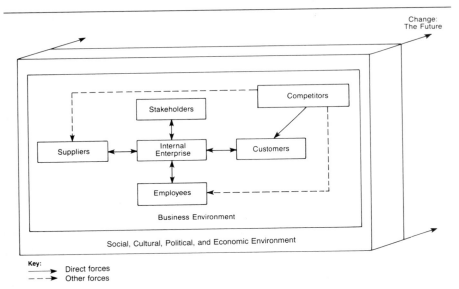

Exhibit I-1-4. Planning Environments

tion about futures research techniques, which include Delphi, inferential scanning, and cross-impact analysis, can be obtained from academic centers for futures research.

Strategic Synthesis

Building the strategic synthesis is the primary step. Here, the information collected in enterprise and environmental analyses is blended into a strategic plan. This step requires the active participation and leadership of the top corporate executives supplemented by the organization's most strategic thinkers, regardless of their role in the organization.

In many organizations, the strategic synthesis still traditionally exists in the CEO's thoughts but is not documented. Increased participatory management, organizational and environmental complexity, and accelerating change, however, make it preferable to develop a documented plan through a group process involving top executives. This approach builds team commitment to the plan by providing a sense of ownership. The result is to focus the energies of the group toward achieving the planned results. Properly disseminated throughout the entire enterprise, the plan (or at least its basic strategic vision and directions) can act as a powerful catalyst for getting the entire organization on board.

Opportunities for the use of the information technologies as competitive weapons are also determined here. The application of value-chain analysis (see Exhibit I-1-5) or the customer resource life cycle (based on an expansion of IBM BSP's requirements, acquisition, stewardship, and disposal analysis) in an environment that encourages synergy between business thinking and information technology thinking provides such opportunities with an excellent chance of becoming visible.

ELEMENTS OF STRATEGIC PLANNING

Enterprise analysis, environmental analysis, and strategic synthesis must address all elements of strategic planning if the resulting plan is to be effective in guiding the organization toward the future. These elements are discussed in the following sections.

Market. In addition to the organization it has traditionally supported, the information systems market increasingly extends to the organization's stakeholders, including customers, competitors, and suppliers (see Exhibit I-1-4). Departments that sell their services as one of the businesses of the organization will have additional markets.

Mission. The many missions of information resource management can sometimes conflict. Many of the information needs of the organization are still best satisfied with large applications running on mainframes; this tradi-

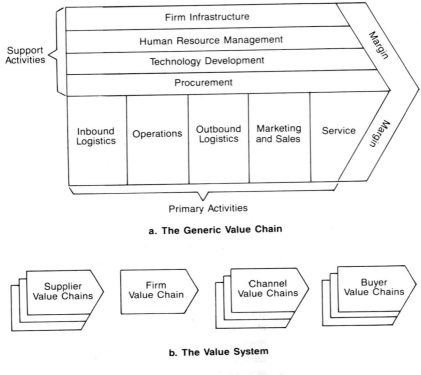

a. The Generic Value Chain

b. The Value System

Exhibit I-1-5. Value-Chain Analysis

tional role is the backbone of most information resource management organizations. Today, most of these organizations also need to embrace new approaches. Two examples include end-user computing, in which support, education, and encouragement are more important than operating efficiency; and competitive systems, in which imagination and an understanding of the business are critical in identifying opportunities. The mission statement must be carefully crafted for and endorsed by the entire organization.

Market Share.   For each of the businesses, or missions, of information resource management, there will be a market, customers, and competitors. The CIO needs to know the organization's market share and that of each of the principal competitors and how all of these are changing.

Strategic Vision.   Strategic vision is a statement of the department's philosophy and of the future character it seeks to achieve.

Goals and Objectives.   Goals are long-term aims toward which the organization directs its efforts. Strategic directions point toward goals. The strategic vision is a synthesis based on a set of goals. Objectives are specific, quantifiable short- and mid-term aims that are realized en route to goals, toward which progress can be measured. (Some organizations reverse these definitions.) The department's goals and objectives must be specified and ranked.

Milestones.   Milestones are specific, scheduled events or deliverables representing progress toward objectives.

Roadblocks.   Roadblocks are any actual or potential obstacles to achieving an objective. As many as can be anticipated at any given time should be identified, and strategies for minimizing their impact should be developed; there will be an ample number that cannot be anticipated.

Customers, Competitors, Stakeholders.   Customers who buy services or products from the department and competitors who offer competing services and products to those customers are both stakeholders because both have a strong interest in the department, what it does, how it does it, and how well it is doing. Other stakeholders are employees, enterprise managers, stockholders, and regulatory agencies.

In information systems, the nature of customers and other stakeholders is often unclear. In a payroll system, for example, there is the customer for the system (the vice-president of finance), the customers for the services provided by the system (the paymasters), and the stakeholders affected by the system (employees to be paid). There is ample evidence that an in-depth understanding of customers and other chief stakeholders is important.

Strengths and Weaknesses.   To compete effectively, the department must have a clear understanding of its strengths and weaknesses as well as those of each of its competitors as they relate to the needs and perceptions of its customers. Organizations must be willing and able to analyze their weaknesses and develop strategies for overcoming or avoiding them in order to survive.

Opportunities and Threats.   A critical factor in the competitive environment is the environment itself. This includes the state of the economy, the state of technology, and anything else that could affect the department's ability to compete. From this standpoint, the department must identify opportunities and threats in order to capitalize on the former and avoid the latter.

Issues.   Many issues surface and are resolved during the planning process; those that remain unresolved become strategic issues. Most are resolved through compromise.

Critical Success Factors.   These indicators determine what factors are critical to the success of the department. They are leading indicators, not trailing indicators like accounting data.

Scenarios.   Although scenarios are most effective when done graphically, narrative descriptions can provide an awareness of detail that may be overlooked in other representations. The assumptions used in building scenarios must be documented carefully.

Action Plans.   The planning process is not complete until action plans for each manager are in place that ensure the completion of milestones and the achievement of objectives.

STRATEGIC AND LONG-RANGE PLANNING GUIDELINES

Hierarchy

The tangible product of the planning process is a hierarchy of documented plans. The strategic plan states the organization's mission, strategic vision, goals and objectives, strategies, and issues in the context of an understanding of the enterprise and its environment. The long-range plan supports the strategic plan and is a road map for implementing the strategies and achieving the objectives through the effective deployment of resources. It also provides historical information so that both program continuity and discontinuity are visible. The budget is a detailed statement of actions, including the use of resources, for accomplishing the tasks scheduled for the next year or two. The action plans, which include management plans for each project, functional area, and organizational unit, are the plans through which work is actually accomplished.

This hierarchy of plans is the product of a hierarchy of processes. They provide snapshot views of the integrated planning process as well as documentation. Among their primary purposes is the documentation of scheduled needs for financial, physical, human, and information resources. Perhaps the principal value of documented plans is their use in the redeployment of resources in response to change.

Plan Contents

In addition to an executive summary and introduction, a comprehensive plan should contain the sections described in the following paragraphs. It is best to organize the plan so that the strategic plan, the long-range plan, the action plans, and the bulk of the supporting data are contained in a set of volumes carefully constructed in terms of their relationships and uses.

Philosophy, Mission, and Direction.   This section summarizes the organization's current concept of the information resource, its management, and its use in business; how those concepts are likely to change; and the primary catalysts for change during the planning period.

The Enterprise.   This section documents the department's understanding of the enterprise, its information requirements, how those requirements are being met, and how this situation is expected to evolve during the planning period.

The Business Environment.   This section documents the nature of the information resource management environment and how it is likely to change during the planning period.

The Nature and Scope of the Information Resource.   This section should discuss the various components of the information resource, explain how they are managed and coordinated in the enterprise to ensure effective support, and describe how they and their relationships are expected to evolve during the planning period. Representations of the organization's telecommunications network, applications portfolio, systems software environment, and capacity for information storage and processing should be included. Quantitative data on the size, structure, and value of the information resource and the use of other resources to manage it should be summarized here and detailed in an appendix.

Special attention should be given to changing requirements for availability and reliability and plans for meeting these changes. Accessibility, including user friendliness, training, the support infrastructure following training, the provision of easier-to-use tools, and the general improvement of computer literacy should all be addressed.

Goals, Objectives, Strategies, and Milestones.   These should summarize action plans in place; the action plans themselves may be included in an appendix.

Support of Business Plans.   This section delineates how the information resource will be used to improve competitive advantage, increase income, improve productivity and effectiveness, reduce costs, and otherwise achieve business objectives.

Benefits.   Information systems managers have generally been unsuccessful in convincing general management of the value and benefits of information technology support. This section should provide information that improves the situation, in quantitative form when possible.

Security and Disaster Recovery.   The organization should have plans for the physical and logical security of its information and for the tools that process it, including plans for recovery in case of a disaster. The strategic directions and summaries of these plans should be included here; the plans themselves may be included as an appendix.

Issues. This section discusses the important issues that prevent achieving the plan or any of its elements. The impact of each issue and alternative and recommended strategies for resolving it should be discussed.

Assumptions. This valuable section gathers all the assumptions used in developing all elements of the plan. This section needs careful review as changes in the environment are detected, and it should point toward specific plan revisions required by such changes. When a manager is seeking to monitor change, this section can trigger useful ideas.

Figures and Schedules. The quantitative information that supports long-range and strategic planning serves several purposes in the enterprise. The most effective quantitative information puts the current situation and future plans into a historical perspective so that discontinuities are evident and understood and the validity of strategic directions can be assessed. Past performance against plans should also be documented and targets established for improved planning effectiveness. The information should be displayed in pictorial and graphic form whenever possible, with tables providing specific information as needed. A typical planning horizon is 5 years; ideally, the historical perspective should cover at least 10 years. Organizations should plan their data collection and retention policies with this historical perspective in mind.

It is important to recognize that many of the figures and schedules that form the quantitative part of a long-range and strategic plan are maintained separately as part of the regular management processes of an organization. Occasionally, it may be desirable to modify or combine existing management tools into a form that provides a longer-term perspective—both future and past—and to maintain them in that form. If this is done, a snapshot of them can be included in any planning documents that are produced.

ACTION PLAN

Putting these principles and ideas into practice is not likely to be easy. The following guidelines may help the manager get started:

1. Assessing the current planning processes, especially with respect to the following questions:
   —Is it a continuous process that can detect and respond to change?
   —Are the right people involved? Substantively?
   —Have the organization and its information requirements been documented effectively? If not, can they be documented?
   —Have the main forces in the organizational environment affected the planning process? Can the organization's support of the planning process be documented?

2. When deficiencies occur, developing objectives, strategies, and action plans for repairing them.
3. If the planning process appears too complex or is subject to information overload, stepping back and focusing on critical success factors— A thorough understanding of the mission and the expectations of senior management, customers, and other stakeholders is necessary.
4. Structuring planning documents so that the critical factors of the plan stand out and are persuasive to senior management.
5. Considering adopting an annual reporting process that sets forth in lay terms the accomplishments versus the plans for the current year, the benefits the organization realized from them, and the plans and their projected benefits for next year.

Chapter I-1-2
Network Planning Guidelines

Layne C. Bradley

TECHNOLOGICAL ADVANCES DURING the past few years have made communications networks a major information management tool. However, it is the business and strategic, rather than the technical, perspective that propels most organizations into larger, more complex networks. The major factors that drive this network approach include productivity issues, such as processing costs, effective resource management (e.g., sharing work loads among multiple computing sites), and personnel costs associated with developing and running computer systems.

Other concerns are the growth of microcomputers and their integration into processing networks and the trend toward unattended operations (i.e., networks of large host processors connected to smaller remote processors that can operate almost automatically, with little or no intervention by operations personnel). Communications systems managers also recognize the need for an effective disaster recovery plan to ensure that processing will continue even if one or more processing centers become inoperable.

Although organizations differ in structure, goals, and hardware and software used, their reasons for implementing networks are similar:

- To achieve load leveling across resources or to balance the work load among several processing centers (see Exhibit I-1-6)—The desire to optimize resources is a major force behind this trend. In the configuration depicted in Exhibit I-1-6, distributed computer networks are used to balance work loads among multiple CPUs.
- To provide remote support or to use computer networks to support additional processing capacity at remote sites that are not necessarily major standalone processing centers (see Exhibit I-1-7)—Remote unattended operations are associated with this category. In the configuration depicted in Exhibit I-1-7, distributed computer networks

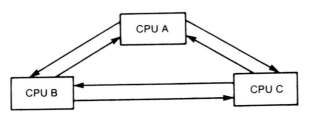

Exhibit I-1-6. Load Sharing

are used to support additional processing capacity to and from remote sites as required. In some cases, CPUs A, B, and C could represent unattended processing sites.

- To integrate microcomputers into mainframe networks—Exhibit I-1-8 depicts an integrated network, in which microcomputers can operate in a standalone mode as well as interface with a host mainframe for such activities as downloading portions of data bases.
- To provide realistic disaster recovery capabilities—For example, in the configuration depicted in Exhibit I-1-9, an alternative processor (CPU C) can be used for continued network processing in the case of extended downtime on CPU A.

Networks are the lifeblood of many organizations. Effective network management, therefore, is critical to the long-term success of the organization. This chapter reviews a network planning approach that considers business and technical issues and presents guidelines for its implementation.

BUSINESS CONSIDERATIONS

The application of networks in the business environment has become a major force. Consequently, communications systems managers, who are responsible for implementing the network in a technical sense, must also realize that they are implementing a major business tool. Therefore, the

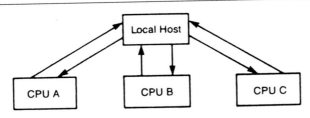

Exhibit I-1-7. Remote Support

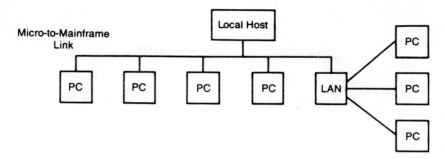

Exhibit I-1-8. Microcomputer Integration

manager must clearly understand the business of the organization and the role of the network within it.

Senior management must provide substantial input during the planning phases of the network to ensure that the network responds well to business needs. When developing the network plan, the communications systems manager must work closely with senior management to address the following basic business concerns:

- The organization's strategic business objectives.
- Return on investment goals.
- Organizational structure.
- Centralized versus decentralized operations.
- Management styles and philosophy.

In addition to the business issues mentioned, the following technologies must be considered:

- Microcomputers.
- Local area networks (LANs).
- Departmental computing.

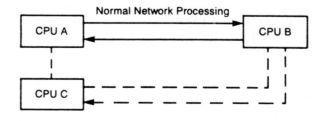

Exhibit I-1-9. Disaster Recovery

- Distributed data bases.
- Unattended computer operations.
- Office automation systems.

NETWORK PLANNING

Planning for the implementation of a network is a complex task. Breaking the process down, however, makes network planning more manageable. The communications systems manager must remember that thorough planning is the key to success.

Business Considerations

As a result of significant strides in technology, the technical aspect of networking is today better defined than before, though it is still complex. The communications systems manager must realize, however, that technical planning alone is insufficient. Business considerations play a major role in implementing a network.

Functional objectives concern network performance and thus involve considerable technical consideration. Service-level network agreements are derived from this process. Functional objectives basically determine the functions for which the network is designed and the computer capacity that is required to perform them.

Determining the Scope of the Network

Determining the scope of the network requires both short- and long-term planning. For example, the initial short-term objective of the network may be to link two of the organization's data centers to allow for load sharing; the long-term strategy can include the acquisition of several companies in the following three to five years.

In the case of the long-term strategy, the primary question is how to handle the information-processing needs of these companies. Because such specific knowledge about companies usually is not known far in advance, management should develop a broad plan that addresses the possible information needs of new companies. The plan should consider two basic types of companies: a company with a computer system of its own that must be integrated into the network and a company without a computer system.

The following information should be obtained before defining the scope of the network:

- Organizational objectives regarding network use.
- The type of distributed computer network to be established initially (e.g., a LAN versus a large host-to-host remote processing network).

- The type and volume of work expected to be processed on the network.
- Whether the network should initially be limited to specified users or applications.
- The short- and long-term impact of the network on the organization and staffing.

Defining Management Requirements and Constraints

Defining management requirements and constraints will help the communications systems manager identify the management parameters of the network. Specifically, budgets are established (both start-up and ongoing for at least three years) and time frames for successful network implementation are identified. Project control must be established within the organization. For example, it must be determined whether the organizational steering committee or communications systems management has complete control over network implementation.

Risk analysis from a managerial perspective is also important. Two sides to this process must be considered: the risks that might arise as the project progresses (e.g., excessive costs, turnover of key personnel, technical limitations, and organizational constraints) and the risks of not implementing the network. The latter might include loss of competitive edge, inability to pursue corporate acquisitions effectively, and inability to expand markets rapidly as a result of slow distribution of information to geographically dispersed divisions. Proper risk assessment requires that the manager:

- Identify the risks.
- Identify the worst-case results relating to those risks.
- Establish a set of management guidelines to follow in each case.

Determining Organizational Capabilities

In addition to considering the general impact of a network on an organization, the communications systems manager must address the network's detailed effects to determine the organization's ability to initiate such a complex project. If not, the organization must explore longer-term measures for initiating the project as well as implementing and managing the network.

IMPLEMENTING THE NETWORK

Establishing a Project Team

The first step in implementing the network is to establish a project team. If the manager follows the planning process described, several staff members will be involved in the decision to actually implement the network. This group would serve as part of the project team.

The most important decision is whether the project team should be permanent or temporary—that is, whether the manager should view the task as any other new development task and use the standard organization or whether a special team should be established. This decision is based primarily on the planning analysis conducted earlier. However, because the implementation of a distributed computer network is not a typical task for the organization, a special team consisting of the most skilled personnel is usually preferable. If a special team is established, its primary tasks will include developing the required procedures and policies and, if necessary, recommending organizational changes to support the network on an ongoing basis.

Because the idea of distributed computer networks may be new to the project team, training must be included in the overall planning effort. This training will affect the expected schedule for implementing the network.

After the project team has been established, standard project management controls and procedures can be followed. However, some revision might be necessary as the project progresses if network planning is a new area for the organization.

Management of the project will be even more difficult if the network involves geographically dispersed data centers and the project team is made up of representatives of all of these centers. Effective coordination of activities and centralization of authority in a single project leader rather than a committee are absolutely critical to success in such an environment.

Establishing Network Control

Two major areas of network control are change control and problem tracking and resolution. Change control is not new in communications systems. Production applications are usually protected from new or revised processing changes through a formal change control mechanism. Although both areas are similar, change control for a distributed computer network can become critical. Depending on the use of the network (e.g., distributed point-of-sale operations in a retail environment), network failure caused by unplanned or uncontrolled changes could be disastrous from a business point of view. The criticality of the network to the actual business of the organization may demand a change control mechanism separate from others used in the organization.

Effective tracking of problems and their timely resolution are also mandatory. Procedures must be implemented to quickly identify and then track problems through to resolution and provide follow-up reports to determine what happened and why, the action taken, and preventive measures to adopt. Several vendor-supplied systems address this area and should be investigated by the project team and implemented if not already in use.

Training Operations and User Personnel

Effective training of all personnel is essential to the overall success of the project. However, determining the type and level of training to offer is difficult because of the varied levels of expertise required by operations and user personnel.

For operations personnel, three levels of training must be accomplished: managerial, functional, and technical. Managerial training involves familiarizing all appropriate managers with the basic concepts, capabilities, limitations, objectives, controls, and procedures necessary for the successful operation of the network.

Functional training, which is much more detailed, is given to those personnel who will be directly involved in the ongoing operation of the network. The network communications control group and some computer operations personnel usually receive this training.

Finally, detailed technical training is required for the personnel supporting the network (e.g., the technical software support group and any in-house hardware maintenance personnel).

As with any project, the depth of training required depends on the expertise of assigned personnel. Training requirements can be assessed through a skills analysis effort, in which required skills are matched against the expertise of personnel. Gaps indicate where training must be developed and applied.

On the user side, training is primarily at the management and functional levels. Management training is for user managers and relates the basic concepts, reporting procedures, problem tracking and resolution, and service-level agreements as they apply to the network. Functional training teaches users how to use the network on an operational basis and how to report problems with network facilities.

Establishing Ongoing Support

The final step in implementing a distributed computer network is to establish a support and maintenance system for the network. If the project team comprises personnel from several areas of the information resources department, those personnel will probably return to their previous responsibilities when the project is completed. Therefore, a permanent support group must be established.

This unit must be planned and staffed early in the project's development so that personnel can be trained and assigned. The support group must be in contact with the network users. This can be done through a network control center that the user can contact for quick resolution of any problems. The support group must be adequately staffed and trained to ensure successful operation of the network.

ACTION PLAN

The implementation of a distributed computer network involves a complex interaction of hardware, software, security, personnel, and procedures. This complexity is compounded by the need for interaction between communications systems personnel and users at remote locations.

From the communications systems manager's viewpoint, the following points should help ensure the success of a distributed computer network:

- An extensive analysis and planning effort must precede the decision to implement the network.
- The decision to implement the network must be based on sound business reasoning—The network should not be implemented simply because it is technically feasible and appealing.
- The structure of the network must complement the organization's business strategy.
- The complexity of the task must not be underestimated.
- Understaffing must be avoided—Lack of appropriate expertise can lead to failure.
- Senior management must understand the magnitude of the task, the short- and long-term costs, and the risks—Its commitment must be obtained at the outset.
- The structure of both the project team and the ongoing support group must be carefully considered.
- Training shortcuts must be avoided—Learning through mistakes can be disastrous in such a project.

Although this is not a comprehensive list, it highlights the major management considerations. Failing to address them properly will threaten the outcome of the project.

In a project of such magnitude, the tendency is to focus on the short-term objective—getting the network implemented and running—while allowing the future to work itself out. This reaction is a typical one and is usually caused by pressure from senior management to see a quick return on investment.

Network planning, however, must become an ongoing, critical part of the communications systems manager's responsibilities. Although all the details for the following five years cannot be projected, the communications systems manager must have some understanding of the future information needs of the organization, how the network will support those needs, and what actions must be taken to ensure that the network will continue to provide the required service.

Chapter I-1-3
Planning for Distributed Processing

Lois Frampton
Jeffrey Schriesheim

A DISTRIBUTED PROCESSING SYSTEM is a set of interconnected cooperating information-processing systems, which may include mixed architectures and capabilities. Each component of a distributed processing system is autonomous, providing a distinct function or set of functions. The key to distributed processing is to analyze the different functions required in a system to identify the necessary processing and support requirements and to place the appropriate processing capabilities at the locations where they are needed.

A distributed processing system can include all types of computers, from microcomputers to mainframes. The exact processing requirements selected should support the functions required at different geographic or local locations. Distributed processing includes the processing of voice, video, and graphics as well as alphanumeric data. Some distributed processing systems are homogeneous in the sense that the same operating system is used on all the CPUs in the system. The larger a distributed processing system is, however, the more likely it is to be heterogeneous—composed of different hardware configurations and different operating systems.

The components of a distributed processing system are separate—that is, they can be relocated, new components can be added, and existing components can be reallocated. A distributed processing system can be expanded in small increments, on an as-needed basis, allowing systems to be tailored to suit the exact needs of a particular requirement. Incremental growth leads to reduced cost as well as to risk containment; applications can span technology changes and proprietary product boundaries.

Applications that run in a distributed processing environment must be composed of clearly separable modules. These modules communicate with one another by explicitly passing data among themselves or by sharing

31

access to files or data bases. Modules running on different computers cannot share an address space, so communication among modules using global storage is not possible.

Partitioning an application into separate modules permits the distribution of processing among computers in a building, on a university campus, or across the world. Once communications between modules is explicit, mechanisms can be designed that work whether the modules are on the same system, on the same local area network, or separated by a wide area network. Using such protocols as a remote procedure call (discussed in a later section) eliminates the need to make changes to application programs to allow them to execute in these different environments.

Some software modules provide a well-defined service to other modules. These service providers, called servers, are the distributed versions of library subroutines that are used in many different applications. Requesting modules, known as clients of the services, access services by issuing requests to server modules. Service actions are initiated when a client sends a request to a server and are completed when the client has received a response. The roles of client and server are relative to a particular service being performed; a server may itself act as a client of another server to accomplish part of the computation it was asked to perform. A typical distributed processing system might contain servers to process electronic mail, query data bases, access a remote file system, or perform a long computation on a fast CPU. In a distributed processing system, clients and servers cooperate to perform an application. This is an extension of older networked systems whose interactions were limited to messages and file transfers.

EVOLUTION OF DISTRIBUTED PROCESSING

During the early 1970s, most information processing was performed on large, centrally located computers. With the advent of smaller, less expensive computers, enterprises with multiple sites began to use small processors in branch offices and then transfer the data to the home office for corporatewide use. This type of processing was usually executed in batch mode without stringent deadlines, but batch processing was slow and time had to be allowed for retransmission of data in case of failures.

ARPAnet was developed during the mid-1970s by researchers for the Defense Advanced Research Project Agency; it provided store-and-forward messaging, file transfer, and remote execution capabilities in a single large interenterprise network of heterogeneous systems. The ARPAnet protocols are still used today in a series of networks that are themselves networked together to provide worldwide communications for government-related activities among universities, government agencies, and industry. ARPAnet protocols are also used in commercially available network systems.

Third-generation distributed processing is made possible by recent advances in data base technology, low-cost processors, and storage, and the
large high-speed reliable networks now being built. Distributed systems can
be configured that are fast enough and reliable enough to support the text
processing, information interchange, and data processing requirements of
enterprises with a single network or a connected set of networks.

Distributed processing systems are used in many different situations. For
example, in a manufacturing environment, a single distributed processing
system could be used to enter an order from sales personnel in the field,
issue work orders to manufacture the requested item, track the manufacturing process, and when the item is available, issue an invoice along with
the details to dispatch it to the customer. The same system could also be
used for information systems development work, accounting, personnel,
payroll, and intercompany electronic mail. The location of the various functions is immaterial; what is important is that the information on the network
is available to the functions where and when it is required.

IMPLEMENTATION CONSIDERATIONS

Factors that play little or no role in the implementation of an application
running on a single computer become critical in the design of a distributed
application and in the installation of the hardware and software that is to be
used for a set of distributed applications.

Capacity Planning

Distributed processing is an acceptable alternative to local processing
only if its performance is adequate, especially if the network is used for
interactive or time-dependent applications. Planning for distributed processing involves determining not only the CPU and storage requirements of
a set of applications but also the communications requirements. When the
system is designed, the individual functions that require access to computer
resources should be identified. These functions should then be placed
where they are most frequently accessed and used (e.g., a module that
frequently accesses a data base should be located close to that data base).
When the individual functions have been identified and located, the capacity of the connections and processors can be calculated on the basis of the
rate of data exchange between the separate functions. This capacity should
support the query rate and the volume of responses to these queries.

Reliability

Because of the modular approach of distributed processing, the reliability
of an application can be increased. In some situations, reliability is essential; for example, in Paris, printed telephone directories have now been

33

replaced by small video terminals located in homes as well as businesses and public telephone sites. The availability of this system must be continuous for almost all sites. Systems with such stringent reliability requirements achieve them by using redundant components. Such redundancy can be cheaper in a distributed processing environment. When two or more systems are used for a given application, if one system fails, the entire work load can be processed by the remaining systems with little or no manual intervention.

Digital Equipment Corp's networking product set, DECnet, uses these concepts to support the network itself. If a system on the network's route to a given destination fails, a new route is computed dynamically and remaining network traffic for functioning systems proceeds without operator intervention and often without perceptible performance degradation.

Management

Communications hardware and software make the job of managing the computing environment more important and more complex. Distributed processing networks often span departments within a company, and different operational styles and goals within departments can present problems. Software components often vary to suit different departmental computing styles. Although this is generally an asset, it can become a problem if lax security in some departments permits unauthorized access to sensitive data in other departments.

Large distributed processing systems are feasible only if they can be well managed. Managing such an environment implies that many tasks must be automated so that human intervention is required only to tailor systems for the applications they process. Components within a distributed processing system should be designed for remote management. Small distributed processing systems can be managed centrally; large networks and geographically separated systems require distributed management. It is not feasible to expect computer operator coverage for every component of a distributed system; in fact, many components are now designed for remote management and are built without consoles. Nevertheless, some central management is necessary to plan the physical connections between sites and to establish corporate security guidelines.

As the use of distributed processing systems increases, it is becoming difficult to distinguish between local system management and network management. Users who do not want to learn to manage their microcomputers or workstations need to have their systems managed remotely. For example, DECnet includes a remote system management capability that allows systems managers to generate a prototype system and download it onto several workstations, eliminating the need for users to become systems managers.

Security

Additional security precautions are necessary in a distributed processing system because it is no longer possible to lock the system up and limit access to a few trusted personnel. When access to a system could be limited to a single building, it was not very difficult to secure it by mechanical means. Enclosed environments, however, are too restrictive for applications that require users to access data over telephone lines or over networks between sites.

Security considerations can be broken down into three subcategories:

- Authentication—Demonstrating that a system or user is who it claims to be.
- Access control—Controlling who is allowed different types of access to system resources.
- Data encryption—Preventing data from being interpreted by persons without knowledge of or access to the keys used to encrypt it.

Authentication.   When a credit card is used to access an automated teller machine (ATM), the combination of the information encoded on the card and the password supplied by the card holder authenticates the identity of the holder to the banking system. Both the card and the password are necessary to prove that the person accessing the ATM is authorized and that the card has not been stolen. Authentication between distributed systems uses similar principles.

Access Control.   Access control is used to limit the type of access that a person or program has to a computer resource. This type of control is not new; only systems managers are allowed to perform certain system operations and only certain programs are allowed to read certain files. What is new for a distributed system is that access control must be defined for all users of the system, not just for users with local accounts on the system containing the resource.

Data Encryption.   Data encryption is a very old technique used to encode data so that it can not be interpreted by someone without access to the encryption keys or encoding rules. Data encryption has increased in sophistication in recent years as its use becomes more necessary: sophisticated network listening devices and eavesdropping techniques have been developed, and machines are now so fast that it is possible to break many encoding schemes by simple trial-and-error methods.

Shared Data

The various components of a distributed application might require access to the same data. There are two ways to accomplish this: the data can be transferred from one part of the application to another using a file transfer or a document transfer protocol, or the data can be stored at a central site and accessed from other systems as needed. Most distributed processing systems provide both methods of data sharing.

Several years ago, Sun Microsystems Inc introduced Network File System, which is now commonly used in UNIX. The Distributed File Service is a similar service available on DECnet. These services differ from the file transfer services formerly offered in networks because they provide remote access to an entire file system, not just to a single file at a time. For example, these new file services allow a user to obtain a listing of all the files in a directory or delete groups of files with a single command. A file service allows microcomputer and small workstation users to increase the amount of file storage available to them, and also allows files to be shared between users and applications.

Access to Printers

Recent advances in document-publishing products and printer technology have outstripped the support provided by traditional print systems. In a distributed environment, users can create documents with complicated formats, sophisticated fonts, and embedded graphics on an inexpensive microcomputer and then print these documents on an expensive printer shared by many users.

Load Balancing

Distributed processing can potentially be used for load balancing; if multiple servers provide the same service, it is reasonable to try to improve their use by balancing the load among them. Load balancing in a distributed environment is more difficult than in a local environment because the cost of communications and user accessibility must be considered before the decision on where to place a piece of equipment or perform a task is made. For example, jobs should be printed where users can conveniently pick up the output, but the nearest or most convenient printer may not be the fastest one available. Several research projects concerned with load balancing are under way, and significant progress can be expected in the next few years.

THE ROLE OF STANDARDS IN DISTRIBUTED PROCESSING

Interoperation between heterogeneous systems requires the adaption of standards, either de facto or officially sponsored. Two different systems can communicate only if both systems speak the same language—that is, share common protocols. Even when systems share the same underlying communications protocols, to exchange more than just simple unstructured messages they need higher-level applications protocols to perform such operations as formatting mail messages, transferring files, and accessing remote data bases.

Most networks now support digital data, with graphics and image processing becoming increasingly common. Networks will also soon include voice and video as integrated components. Standards must be developed quickly to address all forms of information processing in a distributed processing environment.

Existing Standards and Standards Activity

Most heterogeneous distributed processing systems in operation today rely on protocols whose concepts originated in ARPAnet; these protocols are therefore limited in the size of the network they can support. Since 1978, however, the International Standards Organization (ISO) and the International Telephone and Telegraph Consultative Committee (CCITT) have been developing a family of communications protocol standards that provide the services specified in the ISO basic reference model for open systems interconnection (OSI).

Standards that are compatible with the reference model are called OSI standards; most of these deal with defining the nature of the communications between systems (e.g., how an Ethernet operates). Relatively few current standardization activities are focused on such general-purpose services as electronic mail, which provide direct support to user applications. Application-specific standards that operate over OSI-conforming networks are also under development for office systems, banking, and business applications.

In 1987 it became apparent that full-function distributed processing required additional standards to permit coordination among systems that share the processing of an application. The ISO has begun a new project to develop a basic reference model for open distributed processing. (The term *open* indicates that participation in the network or distributed processing system is open to any vendor's components as long as those components conform to the appropriate ISO standards.)

Security

The ISO has produced a Security Architecture (ISO 7498 Part 2) that provides a general description of security services and related mechanisms and defines the places where these services may be provided. The ISO is working on standards for authentication and has produced one standard pertaining to encryption, Data Encipherment—Physical Layer Interoperability Requirements (ISO 9160). The ISO has decided that it will not standardize specific encryption algorithms because of the political implications involved in such standardization. One such standard is the American National Standard Data Encryption Algorithm (ANSI X3.92-1981).

Access control depends on an organization's security requirements as well as the particular application in question. The technology already exists for limiting access to systems by using smart cards that have data about the owner encoded on them. These cards are similar to the cards used to access ATMs, and their standardization will most likely follow a route similar to ATM cards. There is also a need for at least minimal controls on access to data bases in an open distributed environment.

Network Management

If a network is composed of equipment from a variety of vendors, no one vendor's proprietary management software will support the entire distributed system unless standards are used by all the various components. Unfortunately, ISO standards for network management are behind schedule, but the ISO is developing the common management information protocol to pass information between management modules in separate systems. The common management information services protocol allows network management applications to communicate with each other to obtain and set management attributes, report events, and invoke other services. Standards for configuration management, fault management, performance management, and accounting will be emerging in the future.

Mail

Interest in electronic mail and messaging is high. The world's postal, telegraph, and telephone organizations are actively working on standardization through the CCITT; computer vendors and others interested in message handling are working with the ISO on the joint ISO/CCITT project. The technology is available to type mail into a word processor, transfer it through a private network to a public network gateway, and then send it over the public network to a destination private network. CCITT X.400 defines the standards that support message handling and transfer; some of these standards are already approved and others are near approval. Many vendors, including AT&T, Data General, Digital Equipment Corp, Hewlett-Packard, IBM (Europe only), Telecom Canada, Telenet, and Western Union intend to provide X.400-related services.

Names and Addresses

An international standard for paper mail addresses has been in use for many years. This standard allows post offices to forward mail to most locations on the globe. Now that networks are becoming a reality, a similar scheme must be developed for the names and addresses of machines, the people who use them, and the applications that run on them. These names must be unique; there cannot be two different systems with the same name and address in a network.

Machines communicate with each other using addresses that resemble international telephone numbers. Machines have no trouble keeping track of long strings of numbers, but people prefer to use names that they can remember. A directory service is the component of a distributed processing system that provides a mapping between a name and an address. Given the name of an object, the directory service can provide the address and, in some cases, other attributes of the name.

In the ISO, the standard for the directory service (ISO DIS 9594 Parts 1–8) is in the approval process, and its use as a directory for the names and addresses used for message handling has been defined. Using the service for other types of names and addresses (e.g., the names of servers and their network addresses) is under study.

Names in a large network require some structure, so that individuals can create names without first confirming that the name is unique. The scheme being considered by the ISO is hierarchical; the following is an example:

ISO.USA.ACME_CORP.engineering.wheels.tires.smith.mystuff

Names for objects that might be referred to internationally start with either ISO or CCITT, followed by the name of the country and the name of the enterprise. A registry must be established within each country to ensure that the strings used to represent enterprises are unique within that country. The ISO is in the process of selecting a format for the names and addresses that it will use and determining how they will be registered. Registering by country, however, will not be convenient for large multinational corporations.

Remote Procedure Calls

Many interactions between parts of an application involve a request and a response; for example, an application requests a file server to read a record and the server responds with the record. Historically, distributed applications have used such verbs as SEND to make a request and RECEIVE to access the response. Such programs are not transparent to the communications systems being used and can often be debugged only in a distributed environment.

During the early 1980s, a more transparent concept—remote procedure call—was developed. Modular programming techniques advocate the partitioning of a program into a number of procedures, so that each procedure can be partially debugged separately and then combined to form the program. Data is passed between the procedures using argument lists.

A call to a remote procedure looks like a call to a local procedure in the program. The difference is that the two procedures are not linked together; in fact, they need not reside on the same system. Any procedure can be made into a remote procedure provided that it does not share global storage with other procedures. The remote procedure call facility takes care of converting between data formats when different operating systems or programming languages are used. Most remote procedure call facilities can transparently locate the program being called through the use of a directory service.

There are a number of remote procedure call implementations running in university distributed processing environments. Apollo Computer Division of Hewlett-Packard offers remote procedure call in its Network Computing System. Most of these facilities are, however, incompatible with OSI. OSI standardization activities have begun with European Computer Manufacturer's Association Standard 127, "Basic Remote Procedure Call Using OSI Remote Operations," which will be submitted to ISO for processing as an international standard.

Shared Data

The ISO is working on standards for data sharing by transferring files and for data sharing by access to data at a remote site. File transfer, access, and management (FTAM) was approved by the ISO as IS 8571 in 1988. This protocol is more powerful than earlier standards that permitted the transfer of files on magnetic tape because, for some frequently occurring types of files, it also provides for conversion of data formats between systems. The document transfer protocols being developed within the ISO will encode documents in a standard format so that they can be understood by all conforming word processing and editing systems.

There is also an ISO group working on remote data base access. This protocol will use Structured Query Language (SQL) to send requests for data base access between systems. To share access to a relational data base, a server for the data base will be defined by placing the name and address of the server in the directory service. Applications requiring the use of the data base will consult the directory to find the data base server and then access the server using the remote data base access protocol.

Transaction Processing

Many applications fall into the category of transaction processing. A transaction is a series of actions that must all be completed or none of them can be completed. The most common way to achieve this all-or-nothing effect is by using a protocol called two-phase commit. In a distributed processing environment, a transaction may require two or more systems to be completed; therefore, it is necessary to roll back the transaction not only when the application determines that it must do so, but also if one of the systems or some part of the network fails.

The ISO is developing the commitment, concurrency, and recovery protocol for coordinating the actions of the systems involved in a transaction. The ISO is also developing a protocol for transaction processing that will use the commitment, concurrency, and recovery protocol in addition to other OSI protocols. The transaction processing standard will allow the use of such protocols as remote data base access and remote procedure call within the scope of a transaction.

Distributed Printing

To solve the problems of the varying sophistication of printers in a distributed system, the ISO has been working on the definition of a distributed print standard. The resulting standard will support the distribution of users of a print system while providing equal access to the shared resources offered by the system. The print system protocols will also support current printing features as well as many future features.

Data Interchange

In a homogeneous set of systems, data interchange is usually accomplished without conversions, although on some systems data representation depends on the programming language used. When the OSI reference model was written, it was recognized that most large networks would be heterogeneous, making data conversion a fact of life, so OSI includes the concept of a presentation layer.

The purpose of the presentation protocol is to negotiate data formats that both the sender and receiver can support. Most OSI implementations will support the representation of data in Abstract Syntax Notation One (ASN.1), ISO 8824-8825. If two different systems do not share a more convenient format, the sender can convert from its native format to ASN.1 and the receiver can then convert from ASN.1 to its native format. Further work must be done on ASN.1 to permit more efficient encoding of long lists and large matrices of elements when they all have the same characteristics.

SUMMARY

Although distributed processing in commercial environments is still in its infancy, there is little doubt that its use is spreading rapidly and will grow even more quickly when implementations of more standards are available from vendors. Commercial applications usually have long lifetimes because of the high cost of implementation. Therefore, even if an application is not initially intended to be distributed, it is important to design it so that it can be adapted for distribution later. The best way to do this is to design an application as a set of modules performing separate functions. If a vendor separates calculations from user interactions, these two functions can be performed on different systems, so that if the mode of interacting with the user changes to take advantage of new input/output devices, there is no need to change the calculations.

The job of distributing an existing application is much easier when the application is written in a standard programming language and avoids using features specific to a particular operating system. Similarly, using packaged software that is available in multiple environments or using software that is provided by manufacturers committed to developing software that adheres to existing and new standards aids the distribution of applications.

In some application areas (e.g., the Manufacturing Automation Protocol initiated by General Motors), products for open distributed processing are already installed and running. The availability of messaging and file transfer products makes some applications possible today. The distribution of other applications must await the availability of remote data base access and transaction processing and products conforming to remote procedure call standards. In the meantime, as the number of installed distributed processing systems increases, system designers are discovering new techniques and creating requirements for further standardization in the area of open distributed processing.

Chapter I-1-4
Local Networks and Strategic Planning

Robert Prichard

DURING THE LAST FIVE YEARS, requirements for communications within functional units of an organization have accelerated, while transparent communications between functional units have had to be maintained. These demands have driven network users to seek ease of connection and less reliance on proprietary operations. Gone are the days when separate networks were used to meet individual requirements within an organization. Today, a single intelligent network must integrate the individual unit requirements and facilitate transparent communications.

This chapter can help the communications systems manager select local area network (LAN) solutions to meet the organization's current needs while successfully integrating the LAN into the overall communications network architecture. This process requires a top-down planning sequence that defines the specific LAN solution in the context of the strategic direction of the entire communications network.

STRATEGIC PLANNING

The strategic planning process is composed of three basic tasks:
- Projecting communications system requirements into a three- to five-year time frame.
- Establishing a flexible systems development process capable of responding to internal and external change.
- Defining expected system results to ensure that the system design produces them within the specified time.

Strategic systems planning is very different from conducting a technical system analysis to determine the performance of a particular system that will be used as the corporate network standard. If operational and performance issues are allowed to dominate the LAN selection process, the result will be a bottom-up model that meets technical criteria but that most likely does not fit the existing network. Technical system analysis is vitally important, but analysis is only a subset of strategic systems planning. Once the strategic system issues are in place, a strict technical analysis can be performed with confidence that the solution will meet short- and long-term goals.

Modeling's Role in Strategic Planning

Building a model to describe a corporate communications system has two parts:
- The selection of a starting point.
- The definition of the process used to describe the system.

The process for describing the system can follow inductive or deductive reasoning. With the inductive (or bottom-up) approach, the designer moves from the specific to the general. In other words, the designer starts with an application and fits the network to it. With the deductive (or top-down) approach, the designer moves from the general to the specific. This approach starts with the network and fits the application to it.

Two factors affecting the selection of a communications system modeling and planning approach are the corporation's organizational structure and its technological maturity.

The communications system plays a specific role in the organization's ability to deliver the goods and services that keep it in business. The perceived importance of this role determines where the communications function resides in the organizational structure and also determines management's response to capital expenditure requests for system improvements. Budget items are generally evaluated on the basis of four types of requirements:
- Legal requirements—New legislation or amendments to existing legislation affecting the way the corporation does business. These legislative changes may require modification of the communications systems.
- Operational requirements—Requirements that come from system users (e.g., system reliability, security, performance, and interconnection with other service or supply companies).
- Economic requirements—Requirements that will result in cost avoidance or reduced expenses. These requirements are evaluated with economic analysis methods to determine whether they will improve the profitability of business operations.
- Customer service requirements—Requirements necessary to improve the corporation's image among its customers. Management views

these requirements as critical to the corporation's strategic positioning in a competitive marketplace. Internal communications systems generally require some modification to take advantage of new interface options with customers.

The second factor affecting the modeling approach is technological maturity, which is defined as the level of communications system sophistication that can be supported internally by the corporation. It can be measured by observing how the communications system design and operations functions are performed. If the design function is predominantly vendor controlled or if the maintenance and operations functions are predominantly controlled by an outside source, technological maturity is low.

A high level of technological maturity is present when an organization uses a standards-based network development plan that is not vendor specific nor driven by vendor product change. This type of network design allows for multiple vendor equipment and does not require hardware-intensive solutions or nonstandard software fixes for compatibility.

In operations, technological maturity is demonstrated by using internal personnel to cost-effectively provide operations and maintenance functions. Contractors or supplemental help is used at the discretion of the corporation during peak periods, not routinely.

In actual communications systems development, a staff's degree of technological maturity will fall at some intermediate point between total vendor control and total internal control, as shown in Exhibit I-1-10. The modeling approach used to plan the long-range development of the communications system will determine the direction of movement along this continuum. Bottom-up modeling is generally used when the trend is toward vendor or outside control. The top-down approach is generally used when the trend is toward internal control.

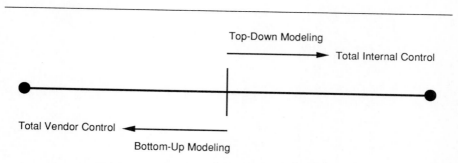

Exhibit I-1-10. The Control Continuum

Bottom-Up Modeling.   Bottom-up modeling views the network through the application, as illustrated in Exhibit I-1-11. The application, not the network, is the corporate resource. The user application, based on the installed (and mostly vendor-specific) hardware, drives the selection of equipment and technology. Because the majority of local network applications required little outside connectivity in the past, the premise worked well and provided the required computing resources and communications at the facility or campus level.

Bottom-up design usually leads to the development of multiple, separate LANs, voice, and other data networks. The exception is when a single vendor provides all services. In the current environment, a single vendor cannot supply all the resources required to operate a major corporate

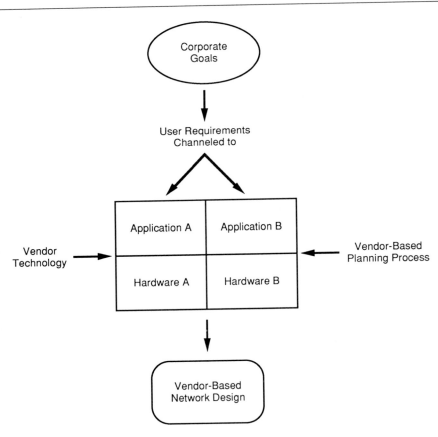

Exhibit I-1-11. The Bottom-Up Model

communications network. Furthermore, connectivity requirements for LANs are no longer isolated at the facility or campus level. With bottom-up systems planning, communications functions are likely to be fragmented and integrated systems transparency virtually impossible.

Top-Down Modeling.   Top-down modeling views the network as a corporate resource (see Exhibit I-1-12); individual devices or local networks are resources to be accessed by the entire organization. The network must be able to transport all types of data. The particular application remains important; however, the local network is designed within the context of the larger corporate network. The network provides the connectivity and the hardware required for protocol-independent data, voice, and image transmission.

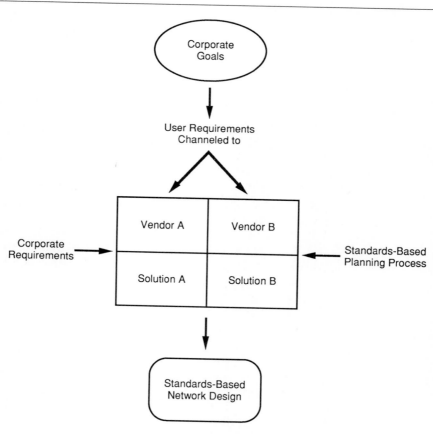

Exhibit I-1-12. The Top-Down Model

Top-down modeling supports a corporation's strategic planning efforts. The network is a corporate resource, and it should be designed with a standards-based architecture such as that provided by the open systems interconnection (OSI) model. An OSI-based architecture will have several strategic advantages. One is transparency of operation among the different types of systems on the network. This means that LANs can be selected to meet user needs without compromising access to the entire communications network. Another is the division of the OSI model into lower layers, which provide end-to-end data transfer resources, and upper layers, which provide applications-oriented information transfer resources. This allows corporate responsibilities for these functions to be divided along the lines of the OSI model. The result will be smoother operation of the corporation's communications services.

The elements composing the top-down strategic planning process are detailed in Exhibit I-1-13. Each element in the model is described in greater detail in the following sections.

The Ends Plan

The ends plan defines the parameters of the communications system as if it were installed today and a current system did not exist. The design is not constrained by cost, organizational structure, or available hardware. The ends plan identifies a set of logical models rather than a special physical system. The term *logical model* is defined as a set of specified interfaces and information flows, not the physical equipment and media required to implement the network.

This set of logical models is then structured into a network architecture, which is defined as the logical assembly of the communications system's functional components. The architecture provides a basic interconnection plan showing how the network will carry information and interface to required services. It also provides a standards-based topology to interconnect strategic communications functions and provide information and resource access regardless of location. Other related requirements include:

- Connectivity among systems, hardware, and procedural elements.
- A single system image to any user or vendor subsystem, regardless of its function or geographic location.

There are four categories of input to the ends plan:

- User communications requirements—These are the resource requirements expressed by the working units that maintain the customer interface or the organization's production of goods and services. They are often based on vendor suggestions and may not reflect an understanding of the strategic needs of the organization. User communications requirements usually are the basis for budget item requests.

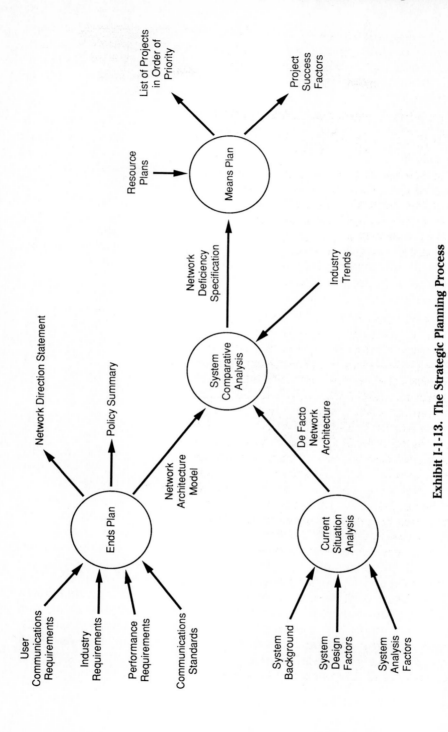

Exhibit 1-1-13. The Strategic Planning Process

- Industry requirements—These are specific to the corporation's business (e.g., the use of Manufacturing Automation Protocol/Technical Office Protocol (MAP/TOP) in the automotive industry, or the migration to the proposed Utility Communications Architecture in the power utility industry).
- Performance requirements—These are driven by operational or customer service issues. Generally, they are technical items relating to response time, throughput, graphics support, nodal delay, or related performance issues. Customer service requirements may be such items as call routing, voice response, direct data base access, or the average response speed.
- Communications standards—These are driven by operational, industry, or economic issues. They allow seamless interfaces, feature and function transparency, and system integration. Standards requirements generally promote operational efficiencies within the communications network; they also can assist in the development and execution of disaster recovery procedures.

The output from the ends plan process becomes input for the system comparative analysis. The primary output is the following written material:

- The network direction statement—This document provides the basic principles that guide network development. Such areas as network structure, network composition, network capacity, and risk evaluation are addressed in this document.
- The policy summary—This document outlines corporate guidelines for communications system responsibility. The system design, installation, and maintenance functions are described in the policy summary.
- The network architecture model—This is the set of models and the supporting documentation that define the logical assembly of the communications system. The network architecture model includes the architecture plan showing how the network will carry information and interface to the required services.

The Current Situation Analysis

Usually, a company has a communications system already in place, which may be as simple as leased telephone services from the local common carrier or as complex as a multinode, nationwide voice and data network. This existing system and the basic services that the business must provide are the starting points for the analysis of the current situation.

The communications manager requires three types of information to conduct a thorough situation analysis: system background, system design factors, and system analysis factors.

System Background.   The system background provides specific knowledge about why certain equipment is in place and how it was selected. The decisions made and approved by management in the past are an indication of company policy.

System Design Factors.   These are the constraints existing at the time of the system design, including the state of technology, corporate business direction, public communications policy and regulation, and general economic climate. The following is a more detailed examination of each of these factors.

State of Technology.   Changes in technology most directly affect system design. For example, new standards and larger function sets are continually introduced for LANs. Examples of the new standards include the IEEE 802.6 (metropolitan area networks) and 802.9 (integrated voice/data LANs) as well as the ANSI standard on the fiber distributed data interface (FDDI).

Corporate Business Direction.   The business supported by the communications system may take various directions over time. Changes may result from changing markets in which the primary product or service is obsolete or changed, or the corporation may be involved in mergers and acquisitions in which new subsidiaries are being acquired or sold. Most often, changes occur with the natural turnover of management personnel. For example, a conservative management will hold system levels at the status quo, concentrating mainly on operation and maintenance issues. On the contrary, an aggressive management will demand that the system design accommodate the merging of new technologies and new systems of companies acquired in the course of business. The particular mode of the company has a major impact on system design and strategic planning.

Public Communications Policy and Regulation.   Common carriers (local and long distance), services, and tariffs represent real costs for operating the communications network and heavily affect the design of the network architecture. In the past, common carriers have had little impact on LANs. Today, however, twisted-pair wiring is an option for LAN traffic in the 10M-bps range. Therefore, the cost of and jurisdiction over local wiring in the plant have become strategic issues.

The communications manager may also need support from the local or long-distance carrier if the distributed processing environment requires a combination of private and public network facilities. Although LAN technological advances allow for the use of bridges, routers, and gateways to create communications access, the actual communications media may span tens or hundreds of miles. Most companies must lease these types of transmission facilities from a common carrier. And even when the company has a large internal network that supports LAN connectivity and distributed processing, common carriers still can have a significant effect on the company's strategic plan for disaster recovery. The communications

What is the current pace of ecomonic activity in the company's industry?

Are demand factors pushing production of a particular set of goods or services?

Are profits increasing or decreasing?

Are additional facilities being added to compensate for the growth pattern?

Are consolidations and reorganizations occurring?

If the market is surging, is the observed growth actual or just a swell?

Can a slowdown that will leave an excess of facilities be predicted?

Do communications requests advance linearly with new service requirements, or is there some lag between the two?

How much lead time is required for new communications services?

What types of services may be required for related systems issues (e.g., safety, reliability, and spare capacity)?

Exhibit I-1-14. Questions for Economic Evaluation of System Expansion

---

manager must design a network architecture that permits the company's system to interface directly with the common carrier at multiple points to provide prompt service restoral using carrier transmission facilities.

General Economic Climate. The economy affects the pace of communications growth. In robust economic times, the growth rate accelerates; conversely, in depressed times, it slows. The economic climate combined with the phase of communications system maturity determines the corporate communications network growth and may be used to estimate future expansion. The communications manager can evaluate the economic climate and its impact on the company's network by addressing the questions listed in Exhibit I-1-14.

System Analysis Factors. System analysis factors are used to compare the network and its performance with management's expectations. The communications manager must provide answers to the following questions:

- Technical—How is the network configured today? Who is connected to it? What services are provided to users, maintenance and operations personnel, and network designers? What situations are supported well, marginally, or not at all? Where are the supported locations? How are they connected? What types of services are required at each location, and why is each location important to the company? When will new service requests be issued and for which locations? What is driving the changes?
- Cost—How much of the overall enterprise budget is communications expenditures? Is that amount increasing? If so, at what rate?
- Performance—What is management's view of the services provided by the communications group? How do they compare with the costs,

reliability, and response times of services contracted from common carriers or other outside sources?

- Profitability—How would the corporation's profitability be affected if services remained fixed at the current level? Could the current grade of service continue to be provided with no capital expansion?
- Capacity—How does the current system's capacity compare with projected future requirements as outlined in the company's strategic business plan? (This comparison should be based on quantifiable items that directly relate to performance requirements.) When will the current system be unable to meet the company's performance requirements?

The System Comparative Analysis

The output of the current situation analysis process is a de facto network architecture. This architecture and supporting documentation are used as input to the system comparative analysis, which is a process that analyzes the differences between the de facto network architecture and the network architecture model developed in the ends plan. The output of the process is a network deficiency specification.

There are three kinds of input to the system comparative analysis: the ends plan network architecture model, the de facto network architecture, and the evaluation of current industry trends. The ends plan network architecture model and the de facto network architecture have been addressed already. The communications manager must evaluate the following three categories of current industry trends:

- Standards—Standards are the doorway to the future. General knowledge of the types of standards and their applications is more important than global technical knowledge of specific standards. Because standards help define the interfaces and protocols required to design or maintain the network architecture, understanding them will help the communications manager gain advantage in competitive bidding situations, clarify vendor claims, and identify potential implementation problems.
- Technology developments—New technology developments must be monitored by the communications manager. Reading current materials, obtaining information directly from vendors, and attending trade shows and industry conferences are all methods for staying informed of new developments.
- Product evaluations—Product evaluations are a direct source of information because they can provide specific benefits and performance as well as functional and interface information. In the process, the communications manager will get to know the vendors and the local marketing groups and learn about product support and customer

service. To properly perform the evaluation process requires a considerable amount of the communications manager's time.

A properly designed corporate network architecture should not be designed around one vendor's product or product line. The best network for the corporation is one designed with flexible, standards-based interfaces and protocol transport facilities that will provide the widest range of options possible.

The output from the system comparative analysis procedure is the network deficiency specification. This document identifies the deficiencies between the de facto network architecture and the ends plan network architecture model, with consideration given to the business and evolving industry trends. The deficiencies are summarized in descending order of importance as follows:

- System architecture deficiencies.
- System standards deficiencies.
- System technology deficiencies.
- Transmission system deficiencies.

The Means Plan

The means plan process creates a document that specifies projects to be implemented and the period of time for implementation. The goal is to balance network deficiencies and corporate resources to provide optimum user, application, or technology solutions while maintaining alignment with the strategic or long-term communications plans.

The network deficiency specification and the resource plans are the input to the means plan. In strategic communications system planning, the communications manager will use the resource plans to schedule time, personnel, and money.

A project schedule must be prepared, incorporating time for the following activities:

- Documentation preparation, which includes system design, technical specification development, and financial justification preparation.
- Management approval.
- Vendor selection and product evaluation.
- Procurement and equipment production.
- Personnel training, equipment installation, test and acceptance, and production cutover.
- Project finalization, including closing out all related project jobs or tasks, finalizing all vendor payments, checking red-lined construction drawings (i.e., drawings that conform to the actual installation) for accuracy, ensuring that all planned services are cut over to the new system and that previous services are terminated, checking all as-built drawings for accuracy and ensuring that they are posted to the master

drawing file, and providing for the turnover of system responsibility to operations and maintenance personnel.

Personnel also must be scheduled. Generally, engineering and operations personnel have direct responsibilities, with approval and support roles being played by management, finance, and procurement personnel. Engineering personnel provide system design, system specifications, and system documentation packages. Operations and maintenance personnel provide new system training, installation, testing and acceptance, and cutover. Contingency personnel planning is required to ensure that skilled replacement personnel are available should personnel substitution be necessary.

The size of the project budget determines whether the project must be capitalized or expensed. Generally, capitalized items are part of the budget cycle, whereas expensed items are charged against a single budget line item. For capitalized items, funding approval must be obtained from management; the funding approval documentation often contains a lease-versus-buy comparison. Disbursement schedules also must be prepared to allow for actual cash flows during the year.

The output from the means planning process is a document containing a ranked list of projects and a list of individual project success factors. The first list sequences projects according to their strategic importance. Each project targets an identified network deficiency (or multiple deficiencies) and through implementation moves the de facto architecture toward the ends plan architecture model.

The list of project success factors identifies measurable performance indicators. Each project is measured by how much it narrows the distance between the de facto network architecture and the ends plan architecture model. Other internal or external factors, however, can influence the project's success in reducing the gap between the current system and the desired system. The gap may be even wider after the project is implemented, which is the result of circumstances that have changed in a dynamic environment. The project success factors measure whether the new system accomplished its specified design criteria independent of other influencing factors.

SUMMARY

The strategic application of LAN technology to communications challenges within an organization can best be served by the top-down planning approach described in this chapter. The benefits of this approach—which requires creating a top-down model of the corporate communications system, an ends plan, a current situation analysis, a system comparative analysis, and a means plan—will be a data communications system that is transparent to corporate users and fully compatible with the business goals and plans of the corporation.

Chapter I-1-5
Incorporating LANs into Long-Range Plans

Robert Prichard

CHAPTER I-1-4 DISCUSSES the application of strategic planning to local area networks (LANs). This chapter discusses how LANs can be incorporated into the overall communications system architecture of the organization. The discussion focuses on the key strategic, tactical, and operational questions and includes an action plan to help communications managers make long-range plans for including LANs in their organization's communications system.

SELECTING LAN TECHNOLOGY

Today's communications environment requires that LANs maintain communications systems integrity and present a single system image to all users. Communications system integrity allows any resource a communications path to any other compatible resource in the network. To the application, the network appears as a single broad highway with no contention for communications resources. LAN technology, therefore, must be integrated into the network architecture using standard interfaces and protocols.

Using the top-down strategic planning approach, the communications manager can be sure that LAN operating features will function correctly after the LAN is connected to other internal or external facilities. Top-down planning also allows for modular network growth, economic use of existing facilities in technology transition, less vendor control, and individual project cost justification.

Network Layers

Networks, as defined in strategic plans, can be separated into the layers illustrated in Exhibit I-1-15: facility or campus networks, hubs and hub networks, metropolitan area networks (MANs), and wide area networks

(WANs). Each of these network layers uses a different combination of the three basic types of LAN technologies: PBX, baseband, and broadband.

Facility and Campus Networks.   A facility can represent any organizational unit. It may be a single building or part of a building or multiple buildings in a contiguous geographic location, organized around a common function.

PBX networks are used in the campus or facility for local low-speed connections to asynchronous or synchronous hosts. The highest concentration of LANs is in the campus or facility, because this is where user groups cluster. Baseband LANs are used most frequently; broadband LAN use is less frequent but can be extremely effective in some campus environments and high-rise or multiple-building complexes.

Hubs and Hub Networks.   A hub is a facility with access to the next level of network services or higher-speed transmission links. The same combinations of LANs may be found in the hub as in the facility or campus environment. A hub network is the group of facilities or campus networks supported by a hub. The hub is the service gateway for passing traffic from facilities to network resources and vice versa. The classification of a particular location as a hub or a facility is based on the types of services provided, switching capabilities, and media access. The focus for the communications manager is on functions, not physical hardware, though generally the two are parallel.

Metropolitan Area Networks.   The MAN connects the hub nodes. It can be a T1, DS3, or higher-speed network and is usually transported by optical fiber or digital microwave. The IEEE Standard 802.6 for MANs and the ANSI standards for the fiber distributed data interface (FDDI) are two examples of proposed industry standards in this area. Both baseband and broadband technologies are used for MANs. The common carriers are looking at a robust high-speed MAN, called switched multimegabit data service (SMDS), to transport both voice and data traffic.

The private MAN does not fit every application. Some organizations are not large enough to support a private network with multiple hub locations; other networks use common carrier facilities to connect hub nodes and use LANs primarily as internal communications networks within facilities or campuses.

Wide Area Networks.   The WAN is used to transport data over large geographic areas. These systems may be public or private and may contain elements of digital or analog microwave, terrestrial land line, optical fiber, or satellite carriers. The WAN has no impact on a LAN unless the LAN belongs to a network that has a distributed architecture and wide geo-

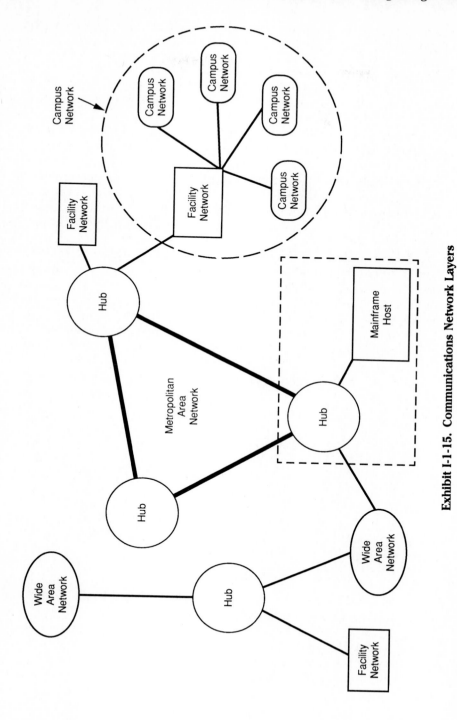

Exhibit 1-1-15. Communications Network Layers

graphic range. In that case, transport of data may be required from one LAN to a second LAN using the WAN. The communications manager should look at this contingency when planning, even if the application is not a current requirement.

System Justification

The technical alternatives for LAN service should be evaluated as a system justification analysis. System justification is composed of technical and economic justification.

Technical justification is a written document that focuses on whether a LAN is appropriate for servicing needs and on which particular LAN configuration provides the best technical solution. Answers to the following questions provide an information framework for a complete technical analysis:

- User profile—Who are the users, and what are their job descriptions? What departments will use the LAN, and which group will use it the most? How many people will use the network?
- Distribution profile—What kind of geographic distribution of services is required? Are users located in the same facility? If so, how are communications services distributed? What transmission resources are available within the facility? How difficult are new installations?

    If the users are not in the same facility, are they in a campus environment, or distributed differently? What communications media are available between the locations? Are there any related restrictions or factors (e.g., unavoidable costs, tariffs, or local regulations) that affect the distribution of services?
- Traffic profile—What volume of traffic must be supported? How much of it is internal and how much external? When are the primary and secondary hours of operation, and when are the typical peak periods? What is the distribution of transaction types (e.g., file transfers, print jobs, communications, common applications access, and data base inquiries) on a regular day and during peak times? What is the maximum acceptable response time for the individual user? How many I/O transactions will the file server process? Will local printers attached to terminals or network printers be used? How much memory is required to support network operations? How many and what size of records will be stored? What retention and access guidelines will be used? How much system downtime can be tolerated?
- System profile—Which technology best meets both user and network requirements? What hardware is required to implement the system? Is it deliverable today? Are the external communications services delivered through a communications server or by a gateway device

attached to a terminal node? What external host access is required? How will nodes be attached to or deleted from the network? What hardware is required for each new node installation? Does the network configuration allow for easy troubleshooting and diagnostics?

Economic justification is the balance to the technical justification. The process of system selection should not be driven by finding the most expensive system in today's dollars; the system should be selected on the basis of the best combination of cost and performance.

Objective evaluation is not always easy. Soft dollar savings from a LAN (e.g., productivity improvements) are difficult to measure in real terms. For example, costs that are avoided by installing a new system may be masked by additional service requests. The new service requests combined with the up-front cash outlays may in fact cause the system costs to rise for a period of time. If management has approved the project on the basis of cost reduction statistics alone, the communications manager's recommendation may be criticized. Therefore, the communications manager must understand the company's financial planning methodology and the manner in which cost analyses are presented to management to represent new system costs appropriately.

The following are basic cost factors to be considered when preparing an economic analysis:

- Hardware costs—The type and cost of the supporting hardware, including file servers, print servers, communications servers, peripherals (e.g., printers, plotters, digitizers, and high-resolution monitors), and storage devices must be documented.
- Software costs—The costs of the supporting software must be documented, including the cost and availability of site licenses for applications software.
- Communications and networking costs—Costs to be accounted for include bridges, routers, gateways, cabling, transceivers, transceiver cables, modems, user and network interfaces, concentrators, and cross-connect hardware.
- Maintenance costs—This area is often overlooked when system costs are calculated. Up-front costs are incurred in training maintenance personnel in LAN operation and troubleshooting. In addition, test equipment and troubleshooting and analysis software must be purchased.

  Furthermore, there are ongoing maintenance costs. How much labor is required to move, add, or change a terminal within the network? How many people will be needed to maintain network performance levels? What other duties and responsibilities do the maintenance personnel have? Will such equipment as bridges, routers, or gateways be required to add communications functions? Is special cabling required for every installation? Is the addition or deletion of a network node disruptive to the entire network?

- Management costs—LAN management is an often overlooked cost. LAN managers are responsible for resource and functional operations management. Resource management includes creating and maintaining records for locations, types, and numbers of devices; software revision levels; global addressing schemes; physical configurations; network problems; and other associated functions. Functional operations management includes the addition and deletion of workstations, network hardware and software, file server setup, access identifications, and user assistance. Depending on the network's size and type and on the relative sophistication of the users, 20% to 100% of one person's time will be required for network management. More time will be required during the initial installation, with requirements decreasing as the users become more familiar with network operation.

Installation and Integration Requirements

Much of the information used to prepare an installation and integration plan is compiled in the process of completing the technical system justification. The five component parts are a technical plan, a procurement plan, an implementation plan, a cutover plan, and postcutover operations.

The technical plan lists technical specifications defining the product to be procured for the selected network. Exhibit I-1-16 provides an example of the information required in the technical plan.

The technical plan should also address whether the installation will be turnkey with contracted maintenance, turnkey with internal maintenance, or an internal installation with internal maintenance.

The procurement plan is the documentation required to meet the company's purchasing policies. When preparing the procurement plan, the communications manager should:

- Make vendor requests—Vendor requests will be one of three types: a request for information (RFI), a request for proposal (RFP), or a request for quotation (RFQ). The RFI is used as a technical and financial screening document to narrow the field of possible products. The RFP is a more formal document requiring detailed system configuration information and firm system pricing. The RFQ is the document used when a system has been selected but additional hardware is needed. This document is used to compare pricing from different vendors with the same product.
- Compare vendor packages—The vendor bids should be compared on price/performance ratios, not just on dollar amounts. Most important, the comparisons should be done in the context of the whole com-

LAN Type
Broadband, baseband, or PBX
Access method
OSI levels 1 and 2 supported

Network Architecture
Architecture type
Control protocol
Topology
Transmission medium
Data rate
Maximum network length
Maximum number of workstations supported
Types of workstations and emulation packages supported

Communications Capabilities
Gateways, bridges, routers, and brouters supported
Data communications equipment supported
Electrical and mechanical interfaces supported
Communications link support (speeds and interfaces)
OSI levels 3 and 4 protocols supported
Microcomputers supported
Mainframe link

Operating System Functions
Operating system platform
Internal functions:
• File locking
• Record locking
• Print spooler

Communications Media
Backbone cable type
Topology of work area
Cable lengths
Wiring access closets
Distribution wiring to terminals
Connectors required
User interfaces required
Communications and interface equipment

Additional Information
Vendor information
Product history
Test and troubleshooting capabilities
Configuration aids
Security and LAN management provisions
Training for maintenance personnel and LAN manager
Documentation:
• Users manual
• Technical reference manual

Exhibit I-1-16. Technical Planning Information for LAN Installation

munications system. If all of the RFPs represent different products and vendors, the comparison process may be complete at this point. A second round of comparisons on the basis of an RFQ may be required to competitively bid the same product.

- Select a vendor and finalize negotiations—The communications manager, along with a group of support personnel, should choose a vendor and forward the documentation to management for required approvals. Once approval is given, final delivery, support, and payment schedules are negotiated with the vendor.

The implementation plan is the project's preinstallation action plan; its primary focus is physical site preparation and coordination. Dates are established for all interdepartmental functions, vendor activities, and contracted services. Key elements of the implementation plan include:

- Facilities preparation.
- Internal departmental coordination.
- Detailed wiring plans and installation.
- Equipment configurations.
- Circuit ordering, installation, and testing.
- Equipment installation and testing.
- Customer and user training.

The cutover plan provides detailed installation procedures and a timetable for bringing the system online. Vendor and user responsibilities during cutover will vary depending on the installation and maintenance agreement.

The postcutover operations plan is used to describe network test and acceptance procedures and how to install live traffic on the network. Acceptance may come after traffic is running on the network and all of the network functions have been demonstrated. Once the acceptance agreement is signed, payment should be made to the vendor.

LAN Management Requirements

The LAN will change over time. To manage it properly, the communications manager must control its growth and ensure that its design remains consistent with the long-range plans for the entire communications network. The areas of network administration, network expansion, and network security all must be managed.

The network administration function seeks to maintain consistency between the LAN protocols and communications network protocols. Its goal is to provide transparency for all applications. The communications manager must monitor the protocol stacks used in the LAN applications to ensure that these same protocol stacks are supported by the communications network's transport mechanism. Most applications will require access to another system or to a remote resource at some time, and if the communications path must be patched together with nonstandard interfaces, the network's integrity is compromised.

As the user community requests additional LAN applications, the communications manager has the responsibility to strictly enforce technical design parameters in network expansion interconnection. Specifically, the physical interfaces (both electrical and mechanical), the transmission media, and the protocols must be consistent. Sloppy electrical and mechanical interfaces create noise generators that corrupt data and reduce throughput. Improperly selected transmission media will have the same type of limiting effects, though the symptoms may be different and more difficult to diagnose. The critical task for the communications manager is to maintain consistency in network expansion so that the maximum performance is obtained from the whole communications network.

In today's distributed communications environment, security is a key issue. The connectivity that supports multiple network access can be used to gain unauthorized access to any network in the communications system. Therefore, multiple levels of security are required in the network. The communications manager is responsible for implementing network security within a consistent framework.

ACTION PLAN

To incorporate LANs into long-range or strategic plans, the communications manager should:

- Communicate to corporate management the need for long-range, strategic communications system planning—Management support must be obtained for a project to develop a strategic communications system plan.
- Develop a framework for top-down modeling—If no strategic plan exists, the bottom-up modeling process is probably in place. Tough issues must be addressed to change the systems modeling process.
- Communicate the importance of a standards-based communications network architecture for integrating LANs into long-range planning.

Section I-2
Communications System Design Planning

THE DESIGN OF COMMUNICATIONS SYSTEMS and networks is a mixture of science, art, and black magic. The manager moving from the planning to the design phase with a network project will need to make use of every available support structure. Experience will help. Knowledge of available tools will help. Guidelines can be valuable starting points. Only the manager can provide the experience. We can, however, offer descriptions of useful methodologies and tools that can be applied to the system design process.

Chapters I-2-1 and I-2-2, "Selecting Network Design Software" and "Selecting a Wide Area Network Design Tool," constitute a two-part exploration of computerized network design tools. Such tools are intended to simplify and support the task of achieving that elusive goal: an optimum network design. The first chapter offers a rationale for the acquisition of network design software tools; this discussion is followed by a list of selection criteria and associated considerations. Chapter I-2-2 is more specific in addressing the attributes and parameters associated with centralized, distributed, and hybrid network configurations.

In a similar approach, Chapters I-2-3 and I-2-4 combine to review the theory and practice underlying the application of graph theory to communications network design. Chapter I-2-3, "Graph Theory and Network Design," discusses the relationship of network topology to graphs, examines the properties of graphs, and then uses this baseline to consider connection matrices describing network modes and interconnecting links. An example of the application of this approach to determining the ideal location for concentrator equipment concludes this chapter. Chapter I-2-4, "Graph Theory and Multidrop Line Routing," focuses attention on an example of the application of graph theory to a commonly encountered multidrop network topology. Also included is a useful automated routing decision matrix using BASIC.

Continuing the discussion of design tools, Chapter I-2-5, "A Systematic Approach to Network Optimization," provides suggestions for a very organized approach to the problem of network design. The process is divided into exploration, modeling, and integration phases. Each phase is broken into component parts with useful suggestions and evaluation guidelines provided for each step in the process.

In Chapter I-2-6, "Examining Integrated Communications Cabling Alternatives," the design emphasis switches from the external wide area environment to the local distribution and external access environment. Integrated cabling can offer many benefits to the communications system manager. An organized, well-planned approach to handling voice, data, and image on a single distribution and access system is far more economical than a haphazard vendor-by-vendor approach. The chapter discusses planning methods and cabling requirements. Case studies for cost comparison are included.

The final chapter of this section returns to the most persistent design issue: the micro-mainframe link. In Chapter I-2-7, "Implementing a Micro-Mainframe Link," the author explores recent advances in technology that allow lower cost implementation of this important aspect of networking. The major components of such a link—software on the host, the interconnecting communication link, and software on the microcomputer—are reviewed.

Chapter I-2-1
Selecting Network Design Software
John E. Gudgel

ONE WAY THAT COMMUNICATIONS SYSTEM MANAGERS can control their network's communications costs and prepare for future expansion (or consolidation) is through the use of automated network design tools. There are many different tools available, ranging considerably in both price and capabilities; this chapter describes the benefits of such tools and suggests some questions that should be asked before one is purchased.

THE PROBLEM OF OPTIMAL DESIGN

General Problem

In most communications systems departments today, engineers still design and optimize corporate networks by scratching ideas on a pad of paper, pounding figures through calculators, and scanning traffic reports or telephone bills. Typically, this design activity is not an uninterrupted process; the designer may be responsible for other projects or may constantly have to fight real or imaginary communications system fires. Through sheer perseverance, the designer may eventually develop an "optimum" network configuration. Unfortunately, by the time it is finished months may have passed and the optimum design may no longer be optimum.

In general, the major challenge to ongoing engineering support and planning for communications networks is how to organize and process all of the complex factors that need to be considered for a time-sensitive design, including:
- Current and potential circuit and equipment costs.
- Network traffic patterns.
- Bandwidth and equipment requirements.
- Network resiliency and robustness.

- User performance needs.
- Anticipated changes and growth.

Although all of these factors must be considered in virtually every communications network design, their individual importance and complexity may vary depending on the size of the corporation and the extent of its communications needs.

The Problem in Small Companies

In small companies, often only a few employees are involved in planning, engineering, administering, and operating the various corporate wide area networks (WANs). Although the size of these networks may be small, they still can be vital to the company's sales and marketing, manufacturing, or research and development activities.

Communications system managers often have difficulty deciding how to use their limited staffs to the best advantage. There may be only one or two people responsible for all voice and data communications in the company. On the voice side, managers may have to manage PBX or key system installations, analyze wide area telecommunications service (WATS) versus direct distance dialing (DDD) proposals, order and administer phone lines, handle branch or home office phone problems, and try to control overall company voice communications costs.

On the data side, managers may have to select, install, and support data communications equipment and lines. They may also have to serve as their organization's contact with public data network providers, be familiar with multiple protocols, and understand the many factors that can affect data communications performance. These same individuals might also be responsible for their organization's facsimile, telex, or local area network (LAN) facilities.

The Problem in Large Corporations

The communications system departments in large corporations can consist of hundreds of individuals. Here, the major problem is the number and magnitude of the networks involved.

Typically, companies that transform themselves from successful small firms to multinational corporations undergo periods of virtually uncontrolled communications growth, during which network planning and cost control are subordinated to satisfying networking requirements and during which individual groups frequently act independently to satisfy their communications needs. When growth slows and management begins to look more closely at expenses, they often find that the corporation is spending large sums for a hodgepodge of separate, incompatible networks, abandoned or forgotten dial and private lines, and obsolete equipment.

The voice network may consist of hundreds of key systems and PBXs purchased from a variety of vendors; serviced by a multitude of central office trunks, WATS, and foreign exchange lines; and interconnected by a vast web of private tie lines. The corporate headquarters phone room may be comparable in size to a large metropolitan telephone company central office, with staffs dedicated to installing phones, engineering phone-company or privately owned transmission facilities, and servicing one or more large PBXs.

Data communications may consist of many groups providing engineering and support to the many centralized and distributed WANs and LANs. One or more IBM shops, for example, may handle all of the dial-up and leased-line SNA (System Network Architecture) networks used to access the corporate mainframe. Another group may specialize in operating the X.25 packet-switching or message network that interconnects thousands of minicomputer or microcomputer users worldwide. And every building in the company may have one or more groups, each of which maintains its own on-site LAN.

In this crowded and confusing communications environment, many large corporations have attempted to restore order by consolidating communications activities under one central communications or network services organization, by looking at integrating various networks onto a single platform, and by acquiring the personnel and the tools needed to plan and engineer their networks to meet future needs.

One tool used by both large and small companies over the past five years to help organize communications information, control costs, and plan for network changes is network design software. Specific design tool requirements must be formulated before purchase, because the software is sold by several vendors, whose products vary in terms of:
- One-time licensing fees.
- Recurring maintenance costs.
- Types of computers on which they run.
- Size and kinds of networks they can design.
- Overall performance and capabilities.

GENERAL NETWORK DESIGN TOOL SELECTION CONSIDERATIONS

There are certain general considerations that should be analyzed before the organization purchases any type of network design software. These considerations include:
- User network design applications.
- System requirements.
- The user interface.

- Help aids and vendor support.
- One-time and recurring costs.

The User's Network Design Requirements

The first question that should be asked before beginning the search for a network design tool is what it will be used for. To answer this question, the type and size of the network to be designed should be considered.

The Type of Network to Be Designed.   Most network design software is programmed to configure and optimize specific types of networks. For example, several tools analyze centralized data networks. These programs typically allow the user to take into account protocol characteristics, queuing delays, special equipment parameters, and other variables commonly associated with IBM SNA and other similarly focused networks.

There are also tools available for the design of voice, distributed data, and local area networks. Furthermore, several vendors have used common platforms as the basis for design tools in different product lines. Some of these share data bases, allowing the design of hybrid centralized and distributed and integrated voice and data networks.

The Size of Network to Be Designed.   There are limits to the size of the networks the programs can design. Examples of such limits include the maximum number of terminals and controllers allowed in a centralized network design, switching and terminal nodes in a distributed data network topology, or PBXs in a tandem voice network. These limits may be inherent in the software itself or may be determined by the speed or memory capacity of the computer on which it runs.

The existence of these limits generally means it is wise to clearly define the size of the desired networks before beginning the search for a design tool. Vendors should be asked to specify their product's design size limitations and the factors that affect it. Furthermore, the buyer should test the software under production conditions to ensure that it performs as advertised.

System Requirements

Network design software is available to run on mainframes, minicomputers, and microcomputers. Some vendors also sell time-share access to their design tools. Because the cost differences between the purchase of a mainframe or minicomputer and a personal computer are substantial, the type of system required to run the software is a major factor in choosing a design tool.

In general, mainframe- and minicomputer-based design tools are more powerful than the microcomputer-based tools. They typically can handle larger networks, produce designs faster, and have more capabilities. Not surprisingly, they are also almost always more expensive than microcomputer-based design programs.

The vast majority of network design software available today runs on IBM-compatible personal computers, whether a PC XT or an 80286- or 80386-class PC AT or PS/2. Because network designs often involve complicated algorithms and large amounts of data, most design tool vendors will recommend that the most powerful compatible microcomputer available be used.

The tool may require that the microcomputer be equipped with specific software or hardware. A typical microcomputer-based design tool might require the following:

- DOS version 3.0 or greater.
- Microsoft Windows.
- 640K bytes or more of RAM.
- 10M bytes or more of free disk memory.
- Video graphics array (VGA) or an enhanced graphics adapter (EGA).
- A serial or bus mouse.
- A printer or plotter.

Similarly, mainframe- or minicomputer-based tools may also require that the system be equipped with specific hardware and software. The expense of acquiring this hardware and software must be considered during the overall tool acquisition cost/benefit analysis.

The User Interface

Ease of use is another important consideration in the selection of a network design tool. In general, more powerful (and expensive) design tools have more functions but are harder to use. Many tools, however, have special features and graphics capabilities that make them easier to use. These features and capabilities include the use of a mouse and graphics icons, graphics capabilities, mapping, and output reports. These factors are discussed in the following paragraphs.

The Use of a Mouse and Graphics Icons.   Many tools use special software (e.g., Microsoft Windows) that allows the entry of commands through the use of a mouse and on-screen icons. With this interface, users can follow the icon-driven menus step by step through the design process without having to enter text commands that are difficult to remember.

Graphics Capabilities.   The typical network that would be designed using a network design tool consists of a series of nodes (e.g., terminal and switching nodes) interconnected by communications links (e.g., telephone company circuits and inside wiring). As network designs grow larger and more complicated, it is helpful to be able to view the network topology on the computer screen. Many tools have sophisticated color graphics depicting equipment and line types using a variety of colors and symbols. A mouse enables the user to zoom into one portion of a network design, move a node, add or delete a link, and make other topological changes.

Mapping.   The ability to plot node locations on a map of the US is particularly useful for domestic WAN designs. Design tools with mapping capabilities usually plot locations using the first six digits of the node's telephone number (NPA-NXX), which it automatically translates into vertical and horizontal (VH) map coordinates. Nodes are then displayed on the map in true geographic relation to one another. Some tools also show state and local access and transport area (LATA) boundaries, allowing the user to anticipate the telephone company circuit tariffs for the connection of the network's nodes.

Output Reports.   Once design information has been entered and processed, it is important that the tool produce meaningful reports for analysis by the user. Such reports might include detailed or summary information on the network topology, bandwidth allocation, node or link performance, equipment and line costs, and other components of the design. Preferably, the user should be able to define the format in which this information is presented and, if desired, be able to download it to a spreadsheet or other software package so that it could be analyzed in more detail or incorporated into a word processing document.

Help Aids and Vendor Support

Even if a network design tool is reputed to be user-friendly and possesses all of the user interface attributes mentioned in the previous section, it is almost inevitable that troubles will occasionally be encountered when the tool is run. There should be help aids and support mechanisms available to help solve such problems. Some of the help aids and vendor support commitments one might want to ask about when evaluating tools are discussed in the following paragraphs.

Training.   Because the network design process can be complicated and involve numerous steps, it is advisable to attend a formal training session to become familiar with the tool and perform a few practice designs. During the evaluation process, questions should be asked about the availability of training and whether it is included in the purchase price. A customer may want to have several people trained. Inquiries should be made about the location of training and what the additional cost might be for extra people. For these customers, some vendors offer on-site training or package deals for groups of company employees.

Error Messages and Online Help.   One useful feature is online help, which allows the tool to assist in diagnosing design problems through either error messages or menus and screens. After hours of entering data and waiting for a preliminary design, nothing is more frustrating than to have an error occur that terminates processing but to not know why the failure occurred. Some, but not all, design tools produce detailed error messages. These help to quickly pinpoint flaws in data input or design parameters that might cause a processing failure. Other tools may have a help key that explains possible reasons for a particular problem and recommends ways to resolve it. This online help may also be part of an overall tool tutorial that can be accessed and read by new users.

Thirty-Day Trial.   It is difficult to evaluate the utility of a design package without first being trained on it and then having the opportunity to use it on a meaningful design. For this reason, most design tool vendors allow prospective customers to use the software free of charge for 30 days to determine whether it meets their requirements.

Usually, such trials are set up with mutually agreed-on acceptance criteria. From the buyer's side, these criteria might stipulate that the tool perform as advertised or that it be able to do specific design tasks required by the user. The vendor's expectation is that if the tool passes the acceptance test, the customer will buy the software. The actual process is rarely so straightforward. Small problems may occur, for example, that would not cause the buyer to reject the software outright, or nontechnical (i.e., political or economic) reasons may prevent a customer from fulfilling its obligations at the end of the trial. For this reason, both parties may wish to add contingency clauses to the trial agreement specifying escrow monies and cancellation penalties to cover such possibilities.

Vendor Guarantees and Ongoing Support.   Because the purchase of network design software can be a major expense, the buyer should expect the vendor to guarantee its product. Unlike hardware guarantees, which may simply state that if the equipment breaks during a specified period it will be fixed or replaced, software guarantees may obligate the vendor to supply revisions or upgrades as they occur, provide consulting, or fix identified bugs in a timely manner.

After network design software has been purchased and installed, the vendor should make available trained telephone support personnel who can help users to solve problems. Such support might include the ability to dial into the customer's computer in order to run diagnostics. In some instances, it may be necessary for the vendor to visit the customer site, in which case it would be helpful to know ahead of time what the response time and cost of this service call may be. All of these vendor guarantees and support commitments should be clearly defined in the purchase agreement.

One-Time and Recurring Costs

The one-time cost for network design software can vary considerably, from a few thousand to hundreds of thousands of dollars. Furthermore, there may be other recurring fees for keeping the software current and supportable. Some of the factors that affect these one-time and recurring costs include the type of software license, maintenance requirements, and tool modularity. These factors are discussed in the following paragraphs.

The Type of Software License. Particularly for design tools that run on personal computers, the type of software license purchased can have an important impact on the tool's recurring and nonrecurring costs. The type of software license that a company should acquire when purchasing a network design tool depends on the number of employees that are expected to use the tool and their location. If only one or two people at one location are going to use the tool, the company should purchase a single copy of the software. If many people at one location need to access the tool simultaneously, the company can buy either multiple individual licenses (often volume discounted) or a site license that includes sufficient copies of the software to satisfy the local need. Finally, if a company intends to have many employees at many corporate sites use the tool, a corporate license may be the best option. Such a license allows employees around the country or around the world to transport and run the software in an unrestricted manner, as long as it is used for company business.

Site and corporate licenses have the added advantage of giving the customer more clout with the vendor when requesting software enhancements or changes. Network design tool vendors are often willing to customize their software to satisfy specific company needs, especially if it means additional licensing or special development revenue.

Maintenance Requirements. Different tools have different software maintenance requirements. Because, for example, telephone company line costs are an important design parameter for WANs, the accuracy and the timeliness of the circuit tariff data base is critical. Keeping a circuit tariff data base current is a costly and time-consuming task because the vendor has to track hundreds of federal- and state-tariffed and nonregulated telephone company costs. For this reason, the recurring fee for WAN design tools can be expensive, in some cases costing tens of thousands of dollars per year.

Tool Modularity.   Many vendors have developed software modules to meet a variety of network design requirements. For example, one vendor might have separate tools to design voice, centralized or distributed wide area data, local area, and integrated voice and data networks. This vendor can sell each module individually or combine them into a complete package. The recurring and nonrecurring costs that customers therefore have to pay depend on their specific requirements, the capabilities inherent in each design module, and the way the vendors decide to package and price their product.

SUMMARY

These are some of the general factors that should be considered before choosing a network design tool. Chapter I-2-2 describes some of the WAN-specific attributes that should be evaluated before deciding which design tool to buy. Exhibit I-2-1 contains a list of names, addresses, and telephone numbers for a few of the vendors of communications network design software.

Aries Group Inc
1500 Research Blvd, Suite 320
Rockville MD 20850
(301) 762-5500

BGS Systems
128 Technology Center
Waltham MA 02254
(617) 891-0000

Connections Telecommunications Inc
15 Christy's Dr
Brockton MA 02401
(508) 584-8885

DMW Commercial Systems
2020 Hogback Rd
Ann Arbor MI 48104
(313) 971-5234

General Network Corp
25 Science Park
New Haven CT 06511
(203) 786-5140

HTL Telemanagement Ltd
3901 National Dr
Burtonsville MD 20866
(301) 236-0782

John Bridges & Associates
1600 S Stemmons, Suite 450
Lewisville TX 75067
(214) 436-8334

Network Synergies Inc
1215 Potter Dr
West Lafayette IN 47906
(317) 742-9000

NMI Network Management Inc
130 Steamboat Rd
Great Neck NY 11024
(516) 829-5900

Quintessential Solutions
1335 Hotel Circle South, Suite 206
San Diego CA 92108
(619) 692-9464

Technetronic Inc
7927 Jones Branch Dr, Suite 400
McLean VA 22102
(703) 749-1471

Telco Research
1207 17th Ave South
Nashville TN 37212
(615) 320-6176

Exhibit I-2-1. Network Design Software Vendors

Chapter I-2-2
Selecting a Wide Area Network Design Tool

John E. Gudgel

CHAPTER I-2-1 SUGGESTS SEVERAL REASONS for companies of all sizes to consider acquiring network design software: to better organize and expedite the network design process, to identify potential network line and equipment savings, to quickly produce time-sensitive network redesigns, to improve design productivity of small engineering staffs, and to generate proposals for the sale of networks. It is also recommended that the potential buyer consider several general criteria before selecting network design software. These general considerations involve: the type of network to be designed; the size of network to be modeled; the design tool's system requirements; the way the user will interface with the software; the tool's graphics, mapping, and reporting capabilities; the help aids and other vendor support mechanisms available; and the one-time and recurring costs for necessary hardware and software.

These basic considerations are the criteria that the organization should consider first when evaluating network design software. Once all the general questions have been answered, however, and the customer has a clear picture of the types of wide area networks (WANs) it wishes to use the software to design, there are numerous network-specific criteria that should be examined to verify the tool's ultimate utility and worth. This chapter describes some of these network-specific attributes and suggests some questions that should be asked before a design tool purchase decision is made.

WIDE AREA DATA NETWORK DESIGN TOOL CONSIDERATIONS

In this chapter, WANs are defined as communications systems, either voice or data, that are composed of geographically dispersed nodes interconnected by transmission links. Their size, performance capabilities, and the types of circuits and equipment they are composed of distinguish them from local area networks (LANs), which are generally located in one building, and metropolitan area networks (MANs), which are usually confined to a single campus or city.

Wide Area Data Network Topologies and Applications

Wide area data networks can be subdivided into several categories on the basis of their general layout (topology) and the applications they support.

Centralized Data Networks. In a typical centralized data network (see Exhibit I-2-2), terminals located all over the country or world exchange information with a single host, usually a mainframe or large minicomputer, located in a main computing center. Ordinarily, the terminals are connected by local cables or point-to-point circuits to regional or district cluster controllers, which are in turn connected to the central computer by point-

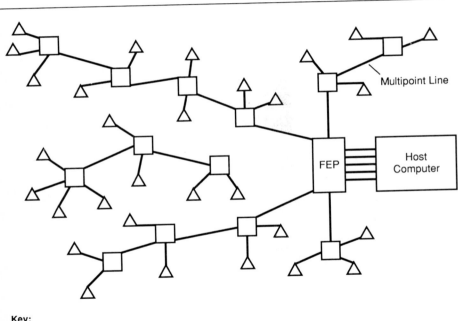

Key:
☐ Cluster controller
△ CRT

Note:
FEP   Front-end processor

Exhibit I-2-2. Typical Centralized Data Network Topology

to-point or multipoint telephone company lines. All of the controller data circuits are connected, in a star configuration, to a central hub where the host is located.

All traffic flows from the terminals to the hosts (or vice versa) over a single path. Usually, there is no alternative routing available, which means that if a failure occurs in any of the data equipment or along the circuit between the terminal and the host, communications are disrupted. The best example of a centralized data network would be a typical IBM Systems Network Architecture (SNA) WAN.

Distributed Data Networks.   Distributed data networks (see Exhibit I-2-3) are the second type of wide area data network. They differ from centralized data networks primarily in their traffic flows, robustness, and overall topology. In a distributed data network, all terminals and hosts are generally peers and can communicate with one another.

In a typical configuration, widely distributed microcomputers or minicomputers are connected by point-to-point tail circuits to regional switching nodes, which are in turn interconnected by high-speed backbone

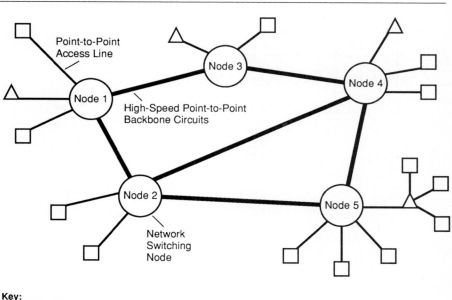

Exhibit I-2-3. **Typical Distributed Data Network Topology**

trunks. Usually, each switching node is connected to at least two other switching nodes, providing alternative routing in case of a failure in the primary route equipment or line. The best examples of distributed data WANs are the public packet-switching networks (e.g., Telenet, Tymnet, and WangPac).

Hybrid Networks. In this chapter, a hybrid network is a catchall category encompassing WANs composed of multiple network elements. This can include networks that have both centralized and distributed components and public and private elements or that integrate voice and data (see Exhibit I-2-4).

A design model of such a network can be extremely complicated, involving very high bandwidth facilities, multiple protocols, and diverse traffic patterns and routing. A serious financial commitment, in both dollars and resources, is also involved in the design of such a network. Software that will model these networks is becoming available; because this networking environment is changing so rapidly, however, much of the software is incomplete.

Anyone can design a WAN, given unlimited bandwidth or financial resources. The simplest solution is just to interconnect every network point to form a complete mesh. The real challenge in network design is to create a topology that meets the user's application, performance, and growth requirements for the least cost. The two principal components associated with wide area data network costs are the telephone company circuits and equipment.

WAN Circuit Pricing

It is extremely important that WAN design tools accurately price various types of telephone circuits and incorporate these costs into a design. The monthly telephone bill for private circuits in a large corporate network can run into the hundreds of thousands of dollars. Communications systems managers are often hesitant to make major changes to a functioning network, and once a redesign decision is made, it can take months to order and install new lines. Therefore, an initial design error that incorporates unnecessary circuits into a topology can cost an organization millions of dollars over the life of the network.

The challenge to design tool software vendors is to provide and maintain a circuit pricer that can handle the variety of different facility types, tariffs, carriers, and special pricing options that must be considered in precisely determining the one-time and monthly costs for telephone company lines.

The prospective buyer of a WAN design tool should look very closely at the circuit pricing functions, the most important of which are discussed in the following sections.

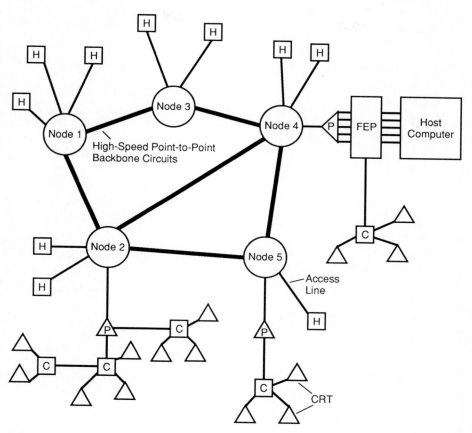

Notes:
C Controller
H X.25 host
P Protocol converter

a. Centralized-and-Distributed Hybrid Topology

Exhibit I-2-4. Hybrid Network Topologies

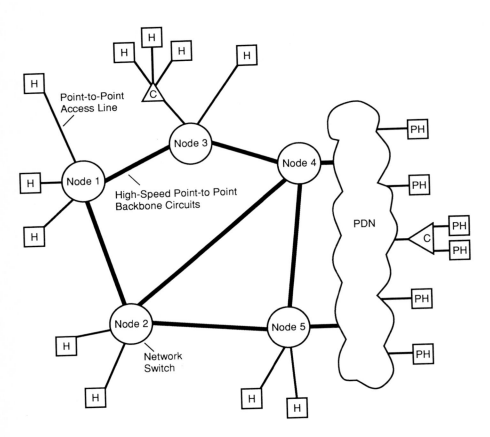

Notes:
C   Concentrator
H   Private network host
PDN Public data network
PH   Public network host

b. Public-and-Private Hybrid Topology

Exhibit I-2-4. Hybrid Network Topologies *(Cont)*

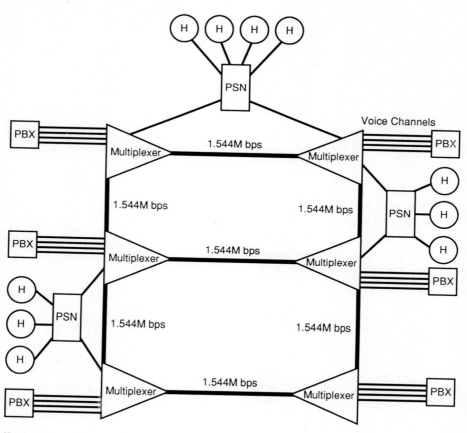

Notes:
H   Distributed network host
PBX   Public branch exchange
PSN   Packet-switching node

c. Integrated Voice-and-Data Hybrid Topology

Exhibit I-2-4. Hybrid Network Topologies *(Cont)*

Pricing by NPA-NXX. The user should be able to define a domestic circuit to be priced by entering the first six digits of the phone number (NPA-NXX) at the end-point locations (for point-to-point lines) or intermediate points and end-points (for multipoint lines). The pricer should then convert the NPA-NXX into vertical and horizontal (VH) coordinates and determine the proximity of this point to the nearest telephone company serving office. This function is important because without it, the user would have to manually determine a location's VH position and the distance to the telephone company point of presence. In a large network, this would be a monumental task. A phone number can usually be much more easily obtained and entered.

Pricing of Multiple Facility Types. Telephone company facilities used in WANs can be divided into three categories:
- Switched (i.e., dial) facilities.
- Nonswitched (i.e., dedicated) facilities.
- Special facilities.

Within these three categories, carriers can provide a variety of line types and services, and the design tool should be able to price most of them.

Switched facilities include direct distance dialing (DDD), foreign exchange (FX), and wide area telecommunications service (WATS). They are commonly used in voice networks as well as dial-up data networks. The cost for such facilities typically includes a fixed monthly amount for access as well as a use fee that is based on a number of variables, including the call distance, the time of day, and the length of the call.

The difference between DDD, FX, and WATS from the user's perspective is mainly one of price. Each facility type is groomed to meet specific user needs and calling patterns. DDD is generally used at locations with low-volume, two-way calling, and FX by companies with moderate-volume, one-way traffic to a particular city or town.

There are two basic types of WATS: INWATS (800 service), and OUTWATS. INWATS is used by companies with moderate to large volumes of calls coming from widely dispersed outside sources going to a specific location. OUTWATS is used by companies that have moderate to large volumes of traffic going from a particular location to widely dispersed sites. WATS used to be a straight-banded service with users ordering individual circuits to handle calls coming from or going to a particular region. Now there are virtually banded WATS, WATS with multitiered pricing, and WATS with T1 access. Pricing all these services is very difficult.

A pricing program should be able to price switched calls of a specified duration made at a designated time of day between two specified locations. Some network design tools will process this information and display the cost break-even points for switching from DDD to WATS or from WATS to private lines.

Telephone company nonswitched (i.e., dedicated) facilities encompass all the varieties of private lines (i.e., analog, low-speed digital, and high-speed T1 and T3 circuits).

Analog private lines can be either two-wire or four-wire, point-to-point, or multipoint. The most common analog lines used in private networks are the four-wire voice-grade 3002 type. Many design tools price only this type of line. There are other types of analog lines, such as those used in teletype or high-security defense networks. If non-3002–type analog circuits will be used in a network, it is probably wise to check with the design tool vendor to determine whether these circuits can be priced.

Many companies are gradually converting their private networks from analog to digital facilities, and a variety of digital facilities can also be ordered. First, there are the traditional Dataphone Digital Services (DDS), which has been offered by AT&T and the Bell operating companies for several years. DDS is currently offered at data rates of 2.4K, 4.8K, 9.6K, and 56K bps. AT&T now offers DDS with a secondary channel, providing customers with improved monitoring and control capabilities. Several local exchange carriers have begun to offer both 19.2K-bps DDS and variable-rate DDS to customers, and several of the interexchange carriers have announced fractional T1 or DS0 digital services, which use the digital access and cross-connect system in the central office to supply digital circuits running at 64K bps to 1.544M bps (e.g., AT&T's Accunet Spectrum of Digital Services [ASDS]). Eventually, traditional DDS services may be replaced by the newer DS0 offerings; until this occurs, however, design tool vendors will have to try to incorporate pricing for all of these digital services in the tariff data bases.

The final category of dedicated facilities typically used in network designs includes the 1.544M-bps T1, 2.048M-bps E1 (European T1), and 45M-bps T3 lines. These lines are mainly used by large companies for local bypass circuits and in integrated voice and data networks. Pricing is usually straightforward; however, because E1 service is not available in the US and because T3 is not widely used, pricing for these two services is rarely found in the tariff data bases of network design tools. If such circuits are to be included in a network design, the pricing program may need to be customized by the vendor.

The special facilities category includes telephone company virtual private networks and non–telephone company transport services, such as the public data networks. Both of these transport types make use of switched and nonswitched telephone company facilities; the cost of these services is computed very differently.

The telephone companies have devised a method to give their larger virtual private network customers the advantages of private networks (i.e., fixed costs) without their operating problems. This is accomplished through hardware and software that, from the user's point of view, dedicates a chunk of the public network to the customer's exclusive use. The charge for virtual private network service can typically include fixed costs for network access and a specified maximum volume of traffic. In some cases, the telephone company charges for connect time, but this is usually at a tariffed rate substantially lower than the equivalent tariffed DDD or WATS use fees.

Public data networks are in an operational sense similar to telephone company virtual private networks. Customers access the public data network by either buying a leased-line connection or dialing into a local packet assembler/disassembler. For each call, the user is assigned a virtual circuit dedicated to its use for the duration of the session. Unlike virtual private networks, however, most public data networks are operated by nonregulated companies, and therefore their rates are not tariffed. These rates typically include fixed charges for access and use-sensitive costs for both connect time and traffic (in kilocharacters or kilosegments). They may also charge for special value-added services (e.g., protocol conversion). Rates are published, and some network design tool vendors have begun to include them in their circuit pricers.

Extensive Carrier Tariff Coverage. In the US, telecommunications common carriers—and their services, organized according to federally created local access and transport areas (LATAs)—are regulated by both federal and state governments. There are separate tariffs for intraLATA, intrastate-interLATA, and interstate interexchange channels, and there are hundreds of common carriers; this means that the tariff data base required to price switched calls and private lines must be extensive. No pricer has every single tariff; however, interstate interexchange carriers, intrastate-interLATA, and intraLATA tariffs should most definitely be included.

Interstate interexchange carrier tariffs include Federal Communications Commission (FCC) Tariffs 9 and 11, which govern the charges AT&T can assess for various interstate services, as well as the various other common carrier tariffs, which cover MCI, US Sprint, and other common carriers' rates.

Each state public utilities commission has the power to establish the rates for telecommunications activities within its borders. Intrastate-interLATA DDD, WATS, and private-line services are supplied by the interchange companies (e.g., AT&T and MCI). The state may decide to adopt the federal interstate tariff established by the FCC for each interexchange carrier or may establish its own tariff for each carrier.

Within the confines of a LATA, basic telecommunications (i.e., intra-LATA) services must be provided by the local exchange carrier. The local exchange carrier can be one of the 23 Bell Operating Companies or any of the approximately 1,700 independent telephone companies (e.g., GTE). It is important for a pricer to be equipped with as many local exchange carrier tariffs as possible, because they are used to price both pure intraLATA circuits as well as the access portion of intrastate-interLATA and interstate lines.

Supply Recurring and Nonrecurring Charges for Complete Service. There are five basic rate elements that are defined in federal and state tariffs and that are used to price various telecommunications services. These rate elements are the usage surcharge, the interexchange carrier charge, the channel termination charge, the service termination charge, and the access coordination charge.

The use surcharge is assessed by carriers for using a line and is based on call minutes, packets, transactions, or message units. This charge is usually associated with switched facilities (e.g., DDD or WATS).

The interexchange carrier charge is the tariffed monthly service and one-time installation charge the interexchange carriers can assess for the portion of a circuit between their rate centers. This charge, which usually applies only to dedicated circuits, is itself composed of three subelements. The first is a distance-sensitive interoffice channel fee that is based on the airline mileage between the carrier's rate centers. The second element is a central office connection fee for attaching the interoffice channel to the local access line. The last element comprises channel options, which include such alternatives as line conditioning and signaling.

The channel termination charge is the tariffed monthly service and one-time installation charge assessed at each end of an interexchange carrier channel to extend it to the local exchange carrier's rate center. These portions of a circuit are priced in a similar manner to the interexchange charge, consisting of channel mileage, connection, and options fees.

The service termination charge is the charge assessed by the local carrier to connect a user's premise to the local exchange carrier's rate center. For dial users, this is the one-time installation and monthly service fee required to give them access to the switched telephone network. For private-line users, this is the one-time installation and monthly charge for extending their private line at each end from the local exchange carrier's rate center to their premises. It includes all of the rate subelements associated with the channel termination and service termination charges (i.e., mileage, connection, and options fees) as well as optional charges for premise inside wiring.

The access coordination charge has been implemented because inter-LATA circuits are composed of numerous segments supplied by different carriers. The users have the option of either ordering each segment themselves or asking the carrier to order the entire circuit for them. For an access coordination charge most carriers not only will coordinate the installation of the various segments but will act as a single point of contact for the entire circuit in case of problems and will combine the individual segment costs onto a single monthly bill.

The use surcharge and service termination charge are used to price switched facilities. The interexchange carrier, channel termination, service termination, and access coordination charges are used in various combinations to price dedicated lines. A network design tool pricer should be able to take information supplied by the user and apply these tariff rate elements to price a circuit end-to-end.

WAN Equipment Selection and Deployment

WAN equipment includes such hardware as modems, data service units, line drivers, assemblers/disassemblers, concentrators, controllers, packet switches, matrix switches, multiplexers, and PBXs. Various vendors' makes and models of this equipment can be differentiated from one another on the basis of not only their functions but also such characteristics as performance capabilities, port capacity, and protocol support.

Many network design tools allow the user to define the types of equipment that can be used in a design and the conditions when it should be included. A design, for example, may indicate that a 9.6K-bps line is required between two points. Some design tools will not only price the circuit but assign a pair of 9.6K-bps modems to the circuit, assuming that one has been defined in the equipment data base.

Wide area data network design involves a trade-off between the use of expensive network equipment (i.e., high capital costs) and circuits with expensive monthly charges (i.e., costs that continue for the lifetime of the network). Many network design tools allow the user to assign either a one-time purchase or a monthly lease cost to each network device, and some tools will even incorporate these costs into the overall design process. An example is a WAN design tool that includes concentration site locators to highlight sites where concentrators, multiplexers, switches, or controllers should be deployed according to the trade-off between additional equipment and extended tail circuit costs. Finally, as part of the reporting process, most tools will display an inventory of the equipment used in a design along with the one-time and monthly costs that might be incurred if the network were deployed.

WAN Performance Modeling

One function that is clearly important to the WAN design process is the ability to model performance and allocate bandwidth on the basis of user application response time and availability requirements. To model performance, the network design tool must be able to take numerous factors into account. These factors include:

- Protocols used.
- Traffic flows.
- Projected growth.
- Network delay.
- Response time.
- Network reliability and availability.

These factors are discussed in the following sections.

Protocols Used. In virtually all data networks, information is transmitted from one network device to another using mutually agreed-on rules and procedures. The rules and procedures for communications system interaction, which can either be proprietary to the system or conform to established standards, are known as network protocols.

Protocols are very important to network designs because they are the framework for network operation. The network functions governed by protocols include message segmentation and reassembly, flow and error control, synchronization, sequencing, addressing, routing, and multiplexing. Furthermore, protocols themselves can be a major source of network traffic because they add overhead to every message. This overhead has been known to constitute 40% or more of the traffic in some networks. Many network design tools allow the user to define the network protocols that will be used. The tool can then use this information to predict how the protocol overhead will affect traffic flows and ultimately the amount of bandwidth that will have to be allocated.

Traffic Flows. Real data traffic patterns are another important factor affecting network performance. Such traffic patterns vary considerably on the basis of the type of network. In a centralized data network, for example, most traffic moves from a terminal to a central host or vice versa; in a distributed data network, peer devices all communicate with one another.

Many tools allow the user to define the traffic patterns found in a network by entering this information into a traffic data base grid. Traffic in these grids is expressed in such units as transactions or kilocharacters during the busy hour. These traffic patterns, influenced by protocol overhead and preset response time requirements, are then analyzed by the design tool to determine what equipment should be deployed and how much bandwidth should be allocated.

Projected Growth.   Data communications engineers should design networks not only to meet current needs but to satisfy projected growth over a reasonable time frame (e.g., one to two years). Growth in network traffic will change the traffic flows, which means that performance models will also be affected. One useful feature that many tools offer is the ability to multiply traffic volumes by a growth factor either on a node-by-node or a networkwide basis. This makes it possible for the user to see the impact of growth and adjust the design accordingly.

Network Delay.   Another factor that must be included in design tool performance modeling is the network transmission delay. Each piece of network equipment introduces some delay, attributable to the throughput limitations of individual circuits, polling cycle time, and the processing capacity of host computers.

Typically, network design software lets the user enter equipment delay factors when setting up the network hardware data base. These delay factors are often stated in the vendor specifications accompanying the equipment. Line propagation and host delay factors are usually entered at a later stage of the design process. In addition, some tools permit users to incorporate special delays (e.g., those encountered in satellite transmissions) in their designs.

Response Time.   The definition of response time really depends on the type of WAN. In an interactive centralized data network, response time is often defined as the sum of the round-trip delays, from the time a message leaves a terminal until it is processed by the host and a reply is returned to that terminal. In a typical store-and-forward distributed network, response time is the sum of the one-way delays a message experiences during its transmission from one host to another.

Whatever the definition, response time is an important measure of network performance. For this reason, most network design tools allow users to specify minimum response time standards; the software incorporates this information into a design. For traffic during the busy hour, most tools will design the network so that 90% or more of the messages will experience less transmission delay than the amount that the user has specified as acceptable.

Network Reliability and Availability.   Network reliability and availability is another way in which engineers measure network performance.

Some communications systems managers require that their networks be configured with redundant equipment and alternative circuit routing so that there is no possible single point of failure. In such networks, if a failure occurs, all traffic can be redirected over reciprocal network facilities.

What this means to the design process is that maximum use levels must be defined for both circuits and equipment to ensure that these components can handle the extra traffic loads associated with rerouting when a failure occurs. Tool users should be able to preset these maximum use levels for each node and link in the network. They also should be able to specify equipment redundancy and alternative routing requirements so that they can incorporate this information into a design.

Bandwidth Allocation and Network Configuration

The programming procedure furnishes the network design tool with user response time and availability requirements, equipment specifications, delay factors, protocol definitions, traffic volume estimates, and projected growth. If the software can use the circuit pricer and hardware data bases to compute network costs, the next step is for the tool to allocate bandwidth and design a network configuration that meets user needs at the least cost.

The actual design process can take from several minutes to several hours, depending on the network's size. During this time, a design tool examines numerous configurations to verify that they conform to all of the established user standards. The tool usually begins, however, with a fully meshed topology, then looks for possible concentration points and unnecessary lines. The tool eventually reaches a point at which no further changes can take place without violating one of the constraints. At this point, the design is complete and the tool reports its findings. This report can include a graphic display of the network topology and detailed or summary sheets outlining such items as the types of equipment or lines assigned, performance information, and node-by-node costs.

NETWORK-BY-NETWORK DESIGN TOOL SPECIFICATIONS

Once all of the capabilities of various network design tools are understood, the network designer needs to determine the capabilities required to handle specific network design needs. The following is a summary of the most important functions, arranged according to network type.

Centralized Data Network Design Tool Specifications

The following are some criteria for centralized data networks that should be examined before a design tool is chosen.

Circuit Pricing.   A centralized data network design tool should be able to price all of the circuit types previously mentioned. Special attention should be paid to the ability of the tool to accurately price multipoint circuits. In particular, the tool should be able to compare various multipoint pricing options, including the use of either interexchange or local exchange carrier bridging.

Equipment Selection and Deployment.   Along with the typical networking equipment (e.g., modems, data service units, and multiplexers) used in all of the WAN types, there is certain equipment that is mainly associated with centralized data networks. This equipment includes special terminal types, cluster controllers, and central hosts and front-end processors.

Special terminal types include IBM physical unit (PU) type 2.0 and 2.1 devices, such as 3270 CRTs and 3780 printers. The quantities and types of these terminals should be defined in the equipment data base along with any unique characteristics.

Examples of cluster controllers are those belonging to the IBM 5250 family. The user should be able to specify the numbers and types of terminal each controller can support and the devices' buffer sizes and port capacities.

For most network designs, the brand of central host (e.g., IBM System 3090) or front-end processor (e.g., IBM 3705/3725) is not important; what counts is the data traffic generated by the device and its effect on overall network performance. In particular, it is important that the user be able to specify the host–to–front-end-processor turnaround time because this affects transaction response time.

Protocol Definition.   The most common line protocols used in centralized data networks are IBM 3270 binary synchronous communications (BSC) and IBM 3274/76 synchronous data link control (SDLC). It is helpful for a tool to come equipped with these protocols already predefined. However, various users may implement IBM protocols slightly differently, and there are a number of other non-IBM protocols, both proprietary (e.g., DEC DDCMP) and standardized (e.g., CCITT link access protocol B), that can be used in centralized data networks. For this reason, a centralized network design tool should allow users to define their own protocols.

Traffic Flow.   The typical traffic flow in a centralized data network consists of a terminal sending input messages to a central host and then receiving an output reply. Transactions can be interactive or batch file transfers. The design tool user should be able to define each transaction type on the basis of its size (in characters), frequency (transactions during the busiest hour), and priority.

Growth.   The user should be able to readily change the frequency and type of transactions and to add terminals and controllers to see what impact growth would have on a network design.

Network Delay and Response Time.   Centralized data networks experience all of the equipment and propagation delays mentioned previously. However, there are several additional delays that centralized network design tools should recognize when computing overall round-trip network response time. These delays include those related to modem polling, input and output queuing in the controllers and front-end processors, and host turnaround time. The tool should be able to take all the expected delays into account and compute the response times for various drops and even for specific transaction types on a particular line.

Bandwidth Allocation and Network Configuration.   If the user has entered all of the necessary location, equipment, protocol, traffic, and response time information, the design tool should be able to produce an optimum least-cost line topology. This topology should specify the proper combination of terminals and controllers that can be attached to each multipoint line and how these lines should be laid out to maximize user performance while minimizing network cost. Once a centralized data network topology is produced, the engineer should be able to alter selected parameters to see the effects on network cost and performance of changes to line protocols, traffic flows, and line speeds.

Distributed Data Network Design Tool Specifications

The following are some criteria for distributed data networks that should be examined before a design tool is chosen.

Circuit Pricing.   All types of switched and dedicated facilities can be used in distributed data networks. The predominant types, however, are point-to-point analog and digital. The tail circuits in such networks are often very short, so it is crucial that the tool be able to comprehensively price local intraLATA lines. In addition, distributed data networks can support a substantial number of foreign hosts, so it is useful to be able to price both international and domestic private lines. Finally, because the alternative to a private distributed data network is a public data network, the user should have the power to price public data network services and compare them with proposed private configurations.

Equipment Selection and Deployment.   Along with the modems, data service units, and multiplexers found in other types of WANs, there are special devices that are usually found only in distributed data networks.

These devices include data switches, packet assemblers/disassemblers, and hosts and concentrators.

Most distributed data networks use data switches (i.e., packet, message, or circuit switches) to cost-effectively move data between peer hosts. Typically, these switches are interconnected by high-speed backbone trunks in a topology that allows alternative routing in case of a primary path failure. The user should be able to define in the equipment data base the various switch types that can be used in a design and their specific characteristics, including port capacity, throughput, and line speed support.

Many X.25 networks have the ability to support multiple network protocols. This is done through the use of packet assemblers/disassemblers, which can convert messages in asynchronus, bisynchronous, SDLC, or other proprietary protocols to X.25 so that they can be transported across the network. Distributed data network design tools should allow the user to define these devices in the equipment data base, specifying the hardware's unique protocol and performance characteristics.

Each network switching node in the distributed network usually supports a number of remote end nodes that are connected to the switch by point-to-point tail circuits. The end node equipment can be a single host or a series of hosts connected to a concentrator. As with packet assemblers/disassemblers and switches, the user of a distributed data network tool should be able to define these devices.

Protocol Definition. The line protocol most frequently used in distributed data networks is high-level data link control or its internationally standardized equivalent, CCITT link access protocol B (LAP-B). This is the line protocol used in most X.25 packet-switching networks. However, there are other line protocols used in distributed data networks (e.g., the Transmission Control Protocol and Internet Protocol), and virtually every public data network implements X.25 slightly differently. For this reason, tool users should be able to define the specific protocols that will be used and incorporate them into the design.

Traffic Flows. Because each host in a distributed data network may wish to communicate with every other host, the traffic matrix in these networks can be very large. Therefore, it is very important that the user verify that a tool can handle the anticipated traffic flows.

Traffic can include real-time interactive applications though the predominant traffic type is usually store and forward. Messages can travel along multiple paths, which can be predetermined (as in a permanent virtual circuit type of packet network) or dynamically selected by the switches (as in a datagram type of packet network). The user should ask how a tool handles traffic routing to ensure that message switching as done in the model parallels message switching as done in the real network.

Growth. As in centralized data networks, growth and change are an everyday occurrence in distributed data networks. Growth can include adding new end or switching nodes or increasing traffic volumes. The user should be able to add a traffic growth factor to the design equations or add new nodes to existing designs to see what impact this growth will have on the overall network configuration.

Network Delay and Response Time. The importance of network delay and response time in a distributed data network really depends on the applications being supported. For interactive applications, response time is a very important design factor because users will rarely tolerate delays of more than a few seconds. However, for store-and-forward applications (e.g., electronic mail), response time is not as critical. Such messages are usually transmitted to users on other network hosts, and the senders generally understand that there will be some lag time before the recipients see, read, and respond. For these applications, delays of several minutes or even hours can be acceptable.

To improve response time for interactive messages, most networks are set up to allow a host or individual users to assign a priority to each transaction. Likewise, distributed network design tools should allow users to assign priorities to messages according to application type so that the design can more closely simulate real network operation. In addition, these tools should make it possible for engineers to enter special delay factors commonly associated with distributed data networks, including packet processing time, line and switch use, and hop count.

Network Reliability and Availability. One of the characteristics of distributed data networks that differentiates them from centralized data networks is that the former generally make greater use of alternative routing and therefore have, on average, better reliability and availability. To include alternative routing in a distributed network design, users should have the option of specifying minimum connectivity and maximum line use so that the tool can model traffic flows under various failure scenarios.

Bandwidth Allocation and Network Configuration. As with design tools for centralized data networks, distributed network design tools should be able to produce an efficient, least-cost topology conforming to the design parameters. The tools may first identify possible sites for network switches or concentrators on the basis of traffic flows and circuit cost savings. Once the user has approved or disapproved these sites, the tool will connect each end node to the nearest switch and lay out backbone circuits to handle the internodal traffic. During this process, the tool may select hardware that has sufficient port capacity and power to handle the anticipated traffic load. Once a final topology is completed, the tool should produce both a graphic display of the network and multiple reports that describe bandwidth allocation, equipment deployment and use, and costs.

Hybrid Networks

As previously described, hybrid networks include communications systems that have both centralized and distributed components, that combine public and private elements, or that integrate voice and data. Design of these specialized WANs requires that a tool be equipped with some additional capabilities, which are discussed in the following sections.

Circuit Pricing. All three of the hybrid network types make use of point-to-point and multipoint private lines. In public-and-private hybrids, however, it is also important that the tool be able to price both access and traffic costs for various public data network services. For integrated voice-and-data networks, it is important that the tool be able to price DS0 or fractional T1, 1.544M-bps T1, and 45M-bps T3 lines.

Equipment Selection and Deployment. The combination of services that is the trademark of hybrid networks often requires the use of special equipment. In the case of mixed centralized-and-distributed configurations, some protocol conversion is typically performed by either a packet assembler/disassembler or other variety of black box. In integrated networks, T1 or T3 multiplexers or a digital access and cross connect system may be required. Because most networks can be classified at one time or another as hybrid, it is helpful to make a list of all the possible equipment one might want to use in a present or future design and to verify that all of this hardware can be defined in a tool's equipment data base.

Protocol Definition. In theory, all types of protocols might be used in a hybrid network. Therefore, it is particularly important that the tool selected be flexible enough to allow the definition of many protocols.

Traffic Flow. The direction and amount of traffic flow in a hybrid network really depends on the applications being supported. Because both centralized-and-distributed and integrated voice-and-data hybrids in a sense encompass multiple networks, the traffic matrix can be enormous. It is important to verify that a design tool purchased to model hybrid networks can deal accurately with these large amounts of traffic.

In public-and-private hybrids, traffic is divided between two or more networks, so the actual traffic flowing over any one entity is less than it would be over a single, unified private network. Here, the principal design consideration might involve ways to minimize use-sensitive costs for traffic flowing over the public network.

Growth.   One useful feature of hybrid networks is that they generally are more adaptable to growth or change. Adding circuits or applications can be both cheaper and quicker in hybrid networks than in the more rigid centralized or distributed topologies. For example, in integrated networks, there frequently is spare capacity on the T1 or T3 spans to support additional channels. This ability to increase network traffic inexpensively should be a factor in a network design, especially if change is anticipated. A tool used for hybrid designs should keep track of spare bandwidth and show how it might be allocated as the network expands.

Network Delay and Response Time.   Network delays in a hybrid network can vary. In an integrated voice-and-data network, for example, if high-speed T1 or T3 lines are being used in a circuit-switching configuration, the network delays are usually small. On the other hand, the delays caused by protocol conversion in a centralized-and-distributed hybrid can be substantial. Likewise, substantial delays sometimes are experienced in public-and-private hybrids depending on the paths messages take as they transit various networks. For a design tool to accurately model response time in a hybrid network, the user must know the numerous delay factors that might be encountered and be able to enter this data for consideration during the design process.

Network Reliability and Availability.   Especially in integrated voice-and-data networks, so much information is moving on each circuit that many companies feel that it is crucial to configure them with either alternative routing or emergency backup lines. In some cases, this may mean that a company has an active spare T1 that is used only in an emergency. Or a company can take advantage of the ability of some multiplexing and switching equipment to automatically reallocate bandwidth to critical applications during a failure. Whatever the method, it is definitely desirable to be able to model the impact of alternative routing schemes on network performance. A design tool with this capability can help to identify problems before they happen.

Bandwidth Allocation and Network Configuration.   The appearance of the final graphic representation of a hybrid network can be remarkably similar to either centralized or distributed designs: there will be terminals, hosts, concentrators, and private lines. There will also be some major differences, however. The bandwidth required by integrated voice-and-data hybrids can be considerable, and there typically is greater use of multiplexing. Public-and-private hybrids incorporate a public network half, whose topology and characteristics may be unknown.

The reports generated by the tool should describe in detail how bandwidth is allocated and traffic is routed. If T1 or T3 lines are recommended, the tool should give some cost justification for these circuits. For public-and-private hybrids, the tool should show how much traffic is being routed over the public network and what the use charges will be.

ACTION PLAN

This chapter has described in detail some criteria to consider before the organization acquires a WAN design tool. These criteria can vary considerably depending on the types of networks to be designed. Because of this variability, it is recommended that the potential design tool owners follow these steps to help make their search a successful one:

- Knowing the user network design needs.
- Gathering information on potential tools.
- Writing a detailed specification.
- Arranging to get demonstrations of various tools.
- Mailing a request for proposal to selected vendors.
- Evaluating proposals to determine whether they meet the specifications.
- Negotiating a 30-day trial to verify tool performance.
- Securing all vendor support guarantees in writing.
- Making sure all users receive proper training.

Chapter I-2-3
Graph Theory and Network Design

Gilbert Held

A COMMUNICATIONS NETWORK can be viewed as a series of transmission paths that interconnect different devices, including terminals, multiplexers, concentrators, port selectors, and similar equipment. If transmission paths are viewed as branches and equipment clustered at a common location are viewed as nodes, a network can be redesigned in graph form consisting of branches that are formally referred to as links and nodes that are connected to one another by one or more links (see Exhibit I-2-5).

When the network schematic at the top of Exhibit I-2-5 is converted to its graphic form, the host computer and the two modems at the central site location are considered an entity and denoted as node 1. The terminal and modem at remote sites A and B are considered an entity for nodes 2 and 3. Finally, the modem, control unit, and set of terminals at remote location C are considered an entity representing node 4 when the network is redrawn as a graph.

GRAPH PROPERTIES

Prior organizational requirements and constraints resulted in the original network design that the graph in Exhibit I-2-5 represents. Rather than consider requirements and constraints at this time, this discussion begins without prior conditions by examining the representation of a potential network in which four remote locations are to be connected to a host computer. Exhibit I-2-6 illustrates a graph that contains five nodes and will be used both to discuss the properties of graphs and to develop and examine the properties of a connection matrix resulting from the topology of the graph. The versatility of using graphs to represent networks can be envisioned by considering a few of the possible network configurations Exhibit I-2-6 could represent:

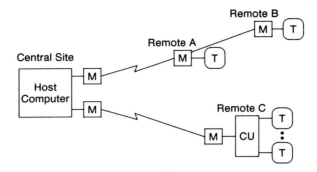

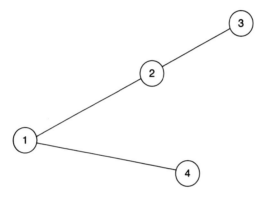

a. Simple Communications Network

b. Network in Graph Format

Notes:
CU   Control unit
M   Modem
T   Terminal

Exhibit I-2-5. Network and Graph Relationship

- A central computer site and four remote sites that must communicate directly with the central site.
- The possible routes allowed for the construction of a single multidrop line connecting each remote site to the central computer.
- Five remote locations clustered within a geographic area that must communicate with a distant computer system (e.g., five cities in a state east of the Mississippi that must communicate with a computer located in Denver).

The five-node graph illustrated in Exhibit I-2-6 contains nine links, with each link representing a line or branch connecting two nodes. Each link can be defined by a set of node pairs (e.g., [1,2], [1,3], [1,4], or [1,5]). If it is assumed that the flow of communication represented by a graph is bi-directional, there is no need to distinguish between link (i, j) and link (j,i), where i and j represent node numbers. Therefore, link (i, j) can be defined as link (j,i). Furthermore, the graph representing a set of N nodes and L links can be defined as $G(N,L)$.

If i and j are distinct nodes in the graph $G(N,L)$, the term *route* can be defined as an ordered list of links (e.g., $[i,i_1], [i_1,i_2], \ldots [i_n, j]$) so that i appears only at the end of the first link and j appears only at the end of the last link. With the exception of nodes i and j, all other nodes in the route appear exactly twice, because they represent an entrance into a node from one link and an exit of the node by a link to another node. On the basis of the preceding route properties, the route will not contain a loop or represent the retracing of a link.

Two other properties of graphs that warrant attention are cycles and trees. A cycle is a route that has the same starting and ending nodes. Therefore, the route (1,2), (2,4), (4,5) in Exhibit I-2-6 can be converted into the cycle (1,2), (2,4), (4,5), (5,1) by using link (5,1) to return to node 1. Link (5,1) can be replaced by links (5,3) and (3,1) to obtain a different cycle that also returns to node 1. A tree is a collection of links that connect all nodes in a graph and contain no cycles.

Although a comparison of a route and a tree may appear trivial, the difference between the two is distinct and has a key applicability to one specific area of network design. Unlike a route, which may or may not connect all nodes in a graph, a tree connects all nodes. This means that an analysis of the different tree lengths constructed from a common graph can

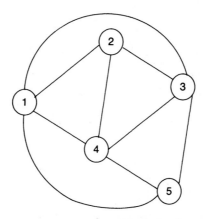

Exhibit I-2-6. Five-Node Graph

be used to determine the optimum route for a multidrop line that inter-connects network locations represented as nodes on a graph.

THE BASIC CONNECTION MATRIX

A graph $G(N,L)$ that contains N nodes and L links can be described by an N-by-N matrix. In this matrix, the value of each element is based on whether or not a link connects node i to node j. Therefore, if X is the connection matrix, the value of each element X_{ij} is as follows:

$X_{ij} = 0$ if no link connects i to j
$X_{ij} = 1$ if there is a link between i and j

The variable i represents the ith row of the connection matrix, and the variable j represents the jth column of the connection matrix. Exhibit I-2-7 represents the connection matrix for the graph illustrated in Exhibit I-2-6.

CONSIDERING GRAPH WEIGHTS

Although a basic connection matrix illustrates the relationship between links and nodes, that relationship is expressed as a binary relationship. In reality, the assignment of values to links can represent a link's length, cost, transmission capacity, or a similar value, as long as the values for all links are expressed in the same term. The assignment of values to links results in a real number known as a weight being placed on, above, or below each link on a graph. If the weight of a link between nodes i and j is denoted as W_{ij}, it is assumed that $W_{ij} = W_{ji}$, because there is no distinction made between links (i, j) and (j,i). Similarly, if there is no link between nodes i and j, W_{ij} is assigned the value 0.

To illustrate the use of graph weights, for a five-node graph in which each node is connected to another node by a link, the link distances between nodes are as indicated in Exhibit I-2-8 and can be used to construct a weighted connection matrix. The resulting matrix illustrated in Exhibit I-2-9 might represent a geographic area consisting of a city and suburbs in which five offices are located. A typical network design problem in this example

$$X = \begin{bmatrix} 0 & 1 & 1 & 1 & 1 \\ 1 & 0 & 1 & 1 & 0 \\ 1 & 1 & 0 & 1 & 1 \\ 1 & 1 & 1 & 0 & 1 \\ 1 & 0 & 1 & 1 & 0 \end{bmatrix}$$

Exhibit I-2-7. Basic Connection Matrix

Link	Distance in Miles
(1,2)	9
(1,3)	11
(1,4)	8
(1,5)	13
(2,3)	7
(2,4)	14
(2,5)	12
(3,4)	8
(3,5)	6
(4,5)	6

Exhibit I-2-8. Link Distances

---

is determining the optimum location for the installation of a control unit among the five offices. With this type of problem, each location must be considered as a potential installation site for the control unit. Because a terminal in each of the other offices would be connected to the control unit by the use of a leased line, the installation location selected should minimize the distance to all of the other offices.

USING A WEIGHTED CONNECTION MATRIX

Although the weighted connection matrix listed in Exhibit I-2-9 can be easily analyzed, this would be more complex if there were 20, 30, or even 50 offices within the geographic area. In such instances, it is preferable to develop a computer program that analyzes the connection matrix to determine an optimum installation location for a control unit or another type of data concentration device. A program for analyzing the five-by-five connection matrix in Exhibit I-2-9 can be expanded to solve more complex problems involving the selection of a data concentration site within a geographic area containing hundreds of offices.

Exhibit I-2-10 contains a BASIC language program listing that computes the route distance in a graph from each node to every other node, providing the user with the information necessary to select an optimum node location from a five-node graph. The first line in the program contains a dimension

---

$$X = \begin{bmatrix} 0 & 9 & 11 & 8 & 13 \\ 9 & 0 & 7 & 14 & 12 \\ 11 & 7 & 0 & 8 & 6 \\ 8 & 14 & 8 & 0 & 6 \\ 13 & 12 & 6 & 6 & 0 \end{bmatrix}$$

Exhibit I-2-9. Weighted Connection Matrix

(DIM) statement that allocates space for a two-dimensional five-by-five (25) element array labeled X, which represents the connection matrix and the five-element array labeled DISTANCE. The first pair of FOR-NEXT loops initializes the element values of the connection matrix at the READ statement by assigning a value from a DATA statement to X_{ij} on the basis of the values of I and J in the FOR-NEXT loops. The values in the DATA statements correspond to the row values of the connection matrix listed in Exhibit I-2-9.

The second pair of FOR-NEXT loops begins by initializing each element of the one-dimensional DISTANCE array to 0. Next, the route distance from node i to all other nodes is computed as J varies from 1 to 5. The last FOR-NEXT loop in the program displays the results of the route distance calculations.

Exhibit I-2-11 contains the results of the execution of the program listed in Exhibit I-2-10. Although the program execution simply lists the route distance by node and leaves it to the user to visually identify the minimum route distance, the following section discusses how the program can be modified for more practical use.

```
DIM X(5,5), DISTANCE(5)
FOR I = 1 TO 5
FOR J = 1 TO 5
READ X(I,J)
NEXT J
NEXT I
DATA 0,9,11,8,13
DATA 9,0,7,14,12
DATA 11,7,0,8,6
DATA 8,14,8,0,6
DATA 13,12,6,6,0
REM FIND MINIMUM ROUTE DISTANCE
FOR I = 1 TO 5
DISTANCE(I) = 0
FOR J = 1 TO 5
DISTANCE(I) = DISTANCE(I) + X(I,J)
NEXT J
NEXT I
REM PRINT RESULTS
FOR I = 1 TO 5
PRINT ''NODE''; I; ''ROUTE DISTANCE ='';
 DISTANCE(I)
NEXT I
```

Exhibit I-2-10. BASIC Language Program

```
NODE 1 ROUTE DISTANCE = 41
NODE 2 ROUTE DISTANCE = 42
NODE 3 ROUTE DISTANCE = 32
NODE 4 ROUTE DISTANCE = 36
NODE 5 ROUTE DISTANCE = 37
```

Exhibit I-2-11. Program Execution

PROGRAM MODIFICATION

To simplify computing the cost of leased lines on the basis of the distance between rate centers, the Bell System developed a vertical and horizontal (VH) coordinate grid system that overlays all of the continental US and a large portion of Canada. The VH coordinate grid system is used by most communications carriers and several independent firms as the foundation for developing programs that compute location to place a data concentration node within a geographic area, the path that represents a minimum-route multidrop circuit, and other network design functions.

Each rate center corresponds to an area code and three-digit telephone prefix, and software is available that allows users to enter the six digits that represent office locations to obtain their VH coordinate location. Some vendors sell tables listing VH coordinates of thousands of cities in the US and Canada; others market a data base of VH coordinates that can be incorporated into user-developed software. Assuming access to one or more methods of obtaining VH coordinates is available, the previously described program can be modified to automatically compute link distances in the following manner.

The grid formed to define the VH boundaries between two locations represents a two-dimensional plane. Because of this, a slightly modified version of the Pythagorean theorem can be used to calculate the distance between pairs of VH coordinates representing two locations. The Pythagorean theorem states that the hypotenuse of a right triangle is equal to the square root of the sum of the squares of the other two sides. Therefore, if C is the hypotenuse and A and B are the other two sides of the triangle, the length of the hypotenuse is:

$$C = \sqrt{A^2 + B^2}$$

In the VH coordinate grid system, A and B can represent the difference between pairs of VH coordinates. Therefore, A can be replaced by V_1—V_2 and B by H_1—H_2; the subscripts 1 and 2 represent locations 1 and 2. Finally, to convert VH coordinate points to mileage, the resulting sum of the squares must be divided by 10 before taking the square root. The formula for calculating the distance (D) in miles between two locations expressed as VH coordinates is:

$$D = \text{INT}\left(\sqrt{\frac{(V_1 - V_2)^2 + (H_1 - H_2)^2}{10}} + 0.5\right)$$

In the equation, 0.5 is added to the result of the computation before the integer (INT) is taken because a communications carrier is permitted by tariff to round the computed mileage to the next higher mile when performing cost calculations.

To illustrate the VH coordinate system, the mileage between Denver and Macon GA is computed. The VH coordinates of Denver are (7,501;5,899), and Macon's coordinates are (7,364;1,865). Therefore, the distance between those two locations can be calculated as follows:

$$D = \text{INT}\left(\sqrt{\frac{(7,501 - 7,364)^2 + (5,899 - 1,865)^2}{10}} + 0.5\right) = 1,277 \text{ miles}$$

On the basis of the preceding description of the VH coordinate grid system, the program listed in Exhibit I-2-9 can be easily modifed to compute distances between nodes rather than work with predefined link distances. VH coordinates or the area code and three-digit prefix of the telephone exchange are entered for each location. For the area code and exchange, a commercially available data base can be integrated into a program to allow VH coordinates to be retrieved. Once the VH coordinates for each node are determined, the distance equation can be used to determine the link distances between nodes. Thereafter, the program would operate as previously described to compute the total route distance from each node to all of the other nodes in the network. Finally, for networks that have a large number of nodes, the program can be modified to sort and list the route distances in ascending order by node. This would allow users to easily consider alternative data concentration locations if the better location is not suitable for installing the required equipment.

SUMMARY

Now that the properties of graphs and the use of a connection matrix to determine the optimum location to place data concentration equipment have been detailed, the examination of the use of graph theory to solve another common network design problem will continue: where a multidrop line should be routed to minimize its distance. Chapter I-2-4 examines graph use to determine an optimum multidrop line route.

Chapter I-2-4
Graph Theory and Multidrop Line Routing
Gilbert Held

PRIM'S ALGORITHM, named for its developer, was first published in 1957 in the *Bell System Technical Journal*. Use of this algorithm leads to the development of network trees in which the link distances between nodes have been reduced to a minimum. Therefore, a tree structure of this type is known as a minimum-spanning tree, and the algorithm is referred to as a minimum-spanning-tree algorithm.

The application of Prim's algorithm requires the use of a network diagram and its conversion into a graph. Exhibit I-2-12 illustrates a typical multidrop routing problem. Four remote sites must be connected to a central computer site by a common multidrop line routed from the central site to each of the remote sites.

The network illustrated in Exhibit I-2-12 can be converted into a graph, as indicated in Exhibit I-2-13. The central site computer becomes node 1, and remote sites B, C, D, and E become nodes 2 through 5.

Now that a graph representing potential network structures has been developed, the next step to be performed before Prim's algorithm is applied is to assign link weights. A multidrop line is a leased line whose cost is usually proportional to its length, so link weights are typically based on the distance between nodes. Because of the variability of tariffs, this may not always be a correct assumption. Some commercially available programs assign values to link weights on the basis of the actual cost of a potential line routed between two nodes. In this chapter, link weights are based on distances between nodes.

Assume for the sake of illustration that the distances between the nodes shown in Exhibit I-2-13 correspond to the entries in Exhibit I-2-14. Those distances can then be used as the link weight assignments. In Exhibit I-2-15, all link weight assignments are shown on the graph.

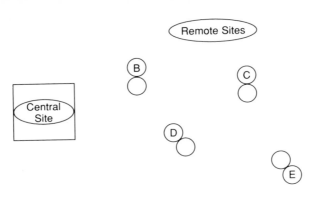

Exhibit I-2-12. Terminal Locations to Be Serviced

To use Prim's algorithm to construct a minimum-spanning tree, a node must first be selected. A two-node subgraph $G(N_iL_i)$ (where i equals two elements in the set 1 to n) is then constructed by connecting the first node to its nearest neighboring node. This is, in effect, selecting the minimum link distance between the first node and all other nodes in the graph. Next, the previously constructed two-node subgraph is expanded into a three-node subgraph by connecting one of the nodes in the two-node subgraph to the nearest node not contained in the subgraph. This process is repeated until all of the nodes in the graph are connected to one another.

To illustrate the use of Prim's algorithm, it can be applied to the graph shown in Exhibit I-2-15. The link weights represent mileage between nodes. Exhibit I-2-16 lists the steps involved in developing a minimum-spanning

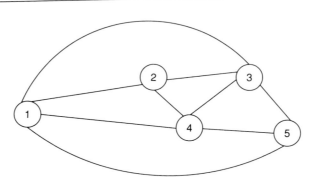

Exhibit I-2-13. Conversion of Network Locations into a Graph

Link	Node Distance (miles)
1,2	30
1,3	60
1,4	60
1,5	90
2,3	50
2,4	40
3,4	20
3,5	50
4,5	30

Exhibit I-2-14. Node Distance Relationships

tree based on the creation and expansion of subgraphs. The steps listed correspond to the creation of the minimum-spanning tree illustrated in Exhibit I-2-17; the resulting tree connects all nodes and has a total link weight of 120.

The effect of the dual links connected to node 4 should be noted. A modem installed at node 4 must have a fan-out feature to transmit data received from link (2,4) onto links (4,3) and (4,5). The fan-out feature is also required when the direction of data flow is reversed, toward the central site computer located on node 1, if the origin of the data is one terminal located on either node 3 or node 5.

If modems with the fan-out feature are not available, a minimum-spanning-tree–derived configuration cannot be used, either. Then the use of a modified minimum-spanning-tree technique must be considered. A tree with the minimum link length but with only one link routed from each node is needed.

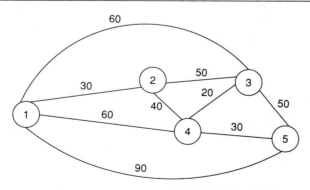

Exhibit I-2-15. Assignment of Link Weights

COMMUNICATIONS SYSTEM DESIGN PLANNING

Step	Nodes in Subgraph	Link Addition
1	1	(1,2)
2	1,2	(2,4)
3	1,2,4	(4,3)
4	1,2,3,4	(4,5)
5	1,2,3,4,5	—

Exhibit I-2-16. Constructing the Minimum-Spanning Tree

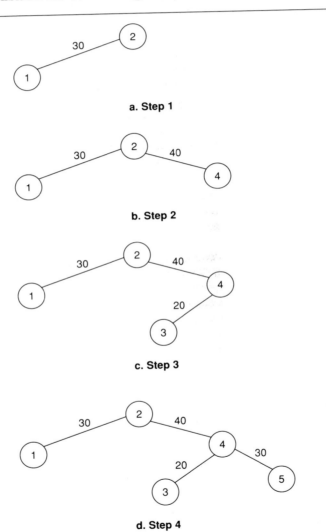

a. Step 1

b. Step 2

c. Step 3

d. Step 4

Exhibit I-2-17. Expansion of a Subgraph into a Minimum-Spanning Tree

Step	Nodes in Subgraph	Link Addition
1	5	(5,4)
2	5,4	(4,3)
3	5,4,3	(3,2)
4	5,4,3,2	(2,1)
5	5,4,3,2,1	—

Exhibit I-2-18. Constructing a Modified Minimum-Spanning Tree

MODIFIED MINIMUM-SPANNING-TREE TECHNIQUE

To develop a modified minimum-spanning tree, the location farthest from the node connected to the computer must be selected first. In the example, this is node 5. Next, subgraphs are formed. However, the following constraints must be followed when the subgraphs are developed. First, if a node has two links attached, it cannot be used to form any more subgraphs; second, the destination link should be selected last.

Exhibits I-2-18 and I-2-19 illustrate the steps needed to develop a network configuration using a modifed minimum-spanning tree. The two constraints must be followed, but the resulting network configuration will not require the use of fan-out modems. Exhibit I-2-18 lists the steps in the development of a modified minimum-spanning tree. Exhibit I-2-19 illustrates the tree, in which all nodes are limited to a maximum of two links.

A comparison of the tree illustrated in Exhibit I-2-19 with the minimum-spanning tree constructed in Exhibit I-2-17 shows that the total link weight of the tree increased by 10 miles, from 120 to 130 miles. This is one of the trade-offs network designers may have to consider—extended distance versus the elimination of equipment with a fan-out capability. In this example, a network manager could determine the extra cost associated with obtaining modems with a fan-out capability and compare that with the additional mileage cost required for their elimination.

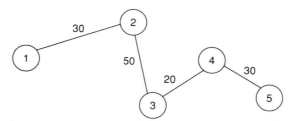

Exhibit I-2-19. Modified Minimum-Spanning Tree

USING A CONNECTION MATRIX

A computer program can be written to automate the development of a minimum-spanning tree and a modified minimum-spanning tree. It requires the use of a connection matrix, a two-dimensional matrix in which each element X_{ij} is assigned the value of the link distance between nodes i and j. Each column represents the links from a single node (e.g., column 1 contains the elements for the links to node 1).

Using the values from the graph in Exhibit I-2-15 produces the five-by-five element connection matrix shown in Exhibit I-2-20. To keep track of the nodes as the connection links are formed, a second, one-dimensional array is used. This array contains an element N for each node, the value of N equal to the numeric designation of the node. The one-dimensional matrix for Exhibit I-2-15 contains the five elements 1, 2, 3, 4, 5 for nodes 1 through 5.

Because Prim's algorithm allows any node to be the initial node, solution of the matrix can begin on any column of the two-dimensional matrix. For the sake of this example, the column for node 1, the central site, is considered first. The column is searched for the element representing the shortest link connected to node 1. This is element X_{21}, the lowest value in that column, equal to 30.

The selection of element X_{21} indicates that node 1 must be connected to node 2 in developing the first subgraph. Because nodes 1 and 2 are now connected, elements N_1 and N_2 in the one-dimensional N array are assigned the value 0 to denote that they are in the subgraph. Similarly, because $X_{ij} = X_{ji}$, X_{12} in the second column of the two-dimensional matrix is assigned the value 0 to preclude the possibility of selecting a reverse loop when another column is searched. Exhibit I-2-21 illustrates the composition of the node and the connection matrixes after these operations have been performed.

Because node 2 has been connected to node 1, the connected node number can be used as a column index for searching the revised two-dimensional array illustrated in Exhibit I-2-21. This results in the selection of element X_{42} because its value of 40 represents the minimum link distance from node 2 to all other nodes not currently contained in the subgraph. The

$$
X = \begin{bmatrix} X_{11} & X_{12} & X_{13} & X_{14} & X_{15} \\ X_{21} & X_{22} & X_{23} & X_{24} & X_{25} \\ X_{31} & X_{32} & X_{33} & X_{34} & X_{35} \\ X_{41} & X_{42} & X_{43} & X_{44} & X_{45} \\ X_{51} & X_{52} & X_{53} & X_{54} & X_{55} \end{bmatrix} = \begin{bmatrix} 0 & 30 & 60 & 60 & 90 \\ 30 & 0 & 50 & 40 & 0 \\ 60 & 50 & 0 & 20 & 50 \\ 60 & 40 & 20 & 0 & 30 \\ 90 & 0 & 50 & 30 & 0 \end{bmatrix}
$$

Exhibit I-2-20. Connection Matrix for Graph in Exhibit I-2-15

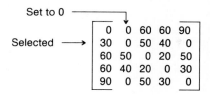

a. Node Matrix

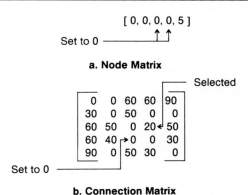

b. Connection Matrix

Exhibit I-2-21. Node and Connection Matrix Values

---

selection of element X_{42} indicates that link (2,4) is added to the developing minimum-spanning tree as well as the next column to be searched—in this instance, column 4. A search of column 4 in Exhibit I-2-20 results in the selection of element X_{34}, whose value is 20, because that element's value is the minimum link distance from node 4 to all other nodes. After link (4,3) is added to the subgraph, elements N_3 and N_4 in the N node matrix are set to 0. In addition, because $X_{ij} = X_{ji}$, element X_{43} in the two-dimensional connection matrix is set to 0 to preclude its selection when column 3 is searched. Exhibit I-2-22 illustrates the revised node and connection matrixes after these operations have been performed.

[0, 0, 0, 0, 5]

Set to 0 ⎯⎯⎯⎯⎯⎯ ↑ ↑

a. Node Matrix

⎯⎯ Selected

0	0	60	60	90
30	0	50	0	0
60	50	0	20◄	50
60	40 ►0	0	30
90	0	50	30	0

Set to 0 ⎯⎯⎯⎯⎯⎯

b. Connection Matrix

Exhibit I-2-22. Revised Node and Connection Matrix Values

115

In the node matrix illustrated in Exhibit I-2-22, only element N_5 (whose value is 5) is not equal to 0. This indicates that the minimum-spanning tree can be completed by connecting node 5 to the nearest node. Searching column 5 results in the selection of element X_{45}, which has the lowest nonzero value, indicating that the addition of link (4,5) is all that remains to complete the development of the minimum-spanning tree.

```
REM CONNECTION MATRIX FOR MODIFIED MINIMUM SPANNING TREE - NO FANOUT
CLS
PRINT "LINK SELECTION SUMMARY"
DIM n(5), x(5, 5)
REM initialize node matrix element values
FOR i = 1 TO 5
        n(i) = i
        NEXT i
REM initialize connection matrix element values
FOR j = 1 TO 5
FOR i = 1 TO 5
        READ x(i, j)
        IF x(i, j) = 0 THEN x(i, j) = 9999 'make missing links into large #
        NEXT i, j
        DATA 0,1,3,2,4,1,0,6,2,8,3,6,0,3,5,2,2,3,0,2,4,8,5,2,0
WEIGHT = 0'link weight counter
REM start in column 1
j = 1 'j is the column pointer into connection matrix
1 IF n(j) > 0 GOTO 2'node selected not connected
        FOR k = 1 TO 5
        IF n(k) > 0 THEN j = n(k)'pick first unconnected node
        NEXT k
2       i = 1'pick a legal value in row
4        x = x(i, j)
        IF n(i) > 0 GOTO 3
        i = i + 1
        GOTO 4
3            FOR i = 1 TO 5 'search column j for smallest value
                IF n(i) = 0 THEN x(i, j) = 9999
                IF x(i, j) ≤ x AND n(j) > 0 THEN
                x = x(i, j)
                indexi = i   'i index of smallest value in column
                indexj = j   'j index of smallest value in column
            END IF
            NEXT i
            IF x = 9999 GOTO 20'when smallest value=9999 we end
            WEIGHT = WEIGHT + x
REM segment selected is link connecting nodes j to i
PRINT "LINK CONNECTS NODES"; indexj; " TO "; indexi; " LINK WEIGHT="; x
            n(indexj) = 0'set node element to 0 to represent connection
            REM test if all nodes connected
            FOR i = 1 TO 5
            IF n(i) > 0 THEN 10 'if a node element >0 all nodes not connected
            NEXT i
            END
10          REM since Xij=Xji, set Xji to high value to preclude its use
            x(indexj, indexi) = 9999
REM search column denoted by indexi which represents node just connected
            j = indexi
            GOTO 1
20 PRINT "TOTAL LINK WEIGHT =", WEIGHT
```

Exhibit I-2-23. BASIC Program Listing

$$\begin{bmatrix} 0 & 1 & 3 & 2 & 4 \\ 1 & 0 & 6 & 2 & 8 \\ 3 & 6 & 0 & 3 & 5 \\ 2 & 2 & 3 & 0 & 2 \\ 4 & 8 & 5 & 2 & 0 \end{bmatrix}$$

a. Connection Matrix Used

```
LINK SELECTION SUMMARY
LINK CONNECTS NODES 1  TO 2  LINK WEIGHT = 1
LINK CONNECTS NODES 2  TO 4  LINK WEIGHT = 2
LINK CONNECTS NODES 4  TO 5  LINK WEIGHT = 2
LINK CONNECTS NODES 5  TO 3  LINK WEIGHT = 5
TOTAL LINK WEIGHT # 10
```

b. Program Execution

Exhibit I-2-24. Operation of BASIC Program

This solution of the node and connection matrixes to create a minimum-spanning tree can be performed by computer. There are several commercial programs available, but it is possible to write one to perform the same function. Regardless of the program's origin, the key to its effective operation with a large number of nodes is a fast search routine that quickly finds the lowest value in each column of the connection matrix.

Exhibit I-2-23 shows a BASIC program written to solve for a modified minimum-spanning tree, one in which the final configuration will not require fan-out modems. In this program, which creates a node matrix (n) and a connection matrix (x), nonexistent links are represented by the value 9999. This ensures that nonexistent links will not be selected in the searches of the connection matrix columns. For similar reasons, whenever link X_{ij} is selected, the value of link X_{ji} is set to 9999 so that it will not be selected in subsequent searches.

One other item in the program that warrants elaboration is the statement IF n(i) = 0 THEN x(i,j) = 9999. This statement says that if a node has been connected, element i in column j that corresponds to the row cannot be selected and its value is set to 9999.

The upper portion of Exhibit I-2-24 is the initial connection matrix that the program in Exhibit I-2-23 operated on. The lower portion of Exhibit I-2-24 shows the execution of the program. It produced a link weight of 10. If node 3 were attached to node 4 instead of node 5, which would result in a fan-out, the link weight would be reduced to 8. Therefore, the execution of this program illustrates the trade-off between a true minimum-spanning tree and the modified tree for which no fan-out modems are required.

Chapter I-2-5
A Systematic Approach to Network Optimization

Nathan J. Muller

COMMUNICATIONS IS A VERY UNSTABLE and unpredictable industry. Just when it seems as though the optimal network design has been achieved, new services, new products, or price wars force the company to start all over. One result is that the technology currently relied on suddenly becomes obsolete, replaced by something faster and more efficient.

A comprehensive network analysis is extremely complex because network topology, traffic load, facility costs, equipment types, and communications protocols must be considered simultaneously with such performance parameters as queuing, blocking, and reliability. In addition, response time and throughput objectives must be considered.

The designer must also consider the type of traffic that the network must support: voice, data, image, full-motion video, or a mix of these. Most currently available microcomputer-based design tools address only a few aspects of the optimization process, taking into account various technical and economic factors but overlooking strategic implications.

Conversely, the input requirements of some high-end design tools can be quite demanding. A data base of system design parameters must be developed to compute the various performance attributes of the network (e.g., actual response times or end-to-end grade of service). The designer might need to gather such information as user port speed, host port speed, input message length, output message length, host processing time, maximum link speed, and the frequency with which various applications are run during the peak busy hour at each location. Equipment and facility performance parameters (e.g., delays) must be provided.

Every aspect of the network's architecture must be completely understood before the appropriate analytical tools can be applied with any degree of confidence. All this information, however, can be quite voluminous. Rarely does an organization have the luxury of waiting the usual three to six months to implement a detailed or lengthy data collection process and then waiting for a follow-up analysis before making decisions about upgrading or expanding its network. Rarely do designers appreciate the amount of work necessary for data collection and review (the process of ensuring that all of the data used in the design process is correct, relevant, and in the proper format). In fact, requiring such details of the designer can drain energy and cause frustration, which can adversely affect performance.

These difficulties can be avoided by breaking down the design effort into discrete, manageable phases that successively eliminate alternative designs until the most feasible solution emerges. The design process can be divided into three discrete phases: exploration, modeling, and integration.

EXPLORATION

The objective of exploration is to ascertain the general design approach and to determine the various line speeds that are required so that the force of the design effort can be narrowed to specific types of equipment. During this phase, graphic displays should be used instead of computer printouts, which are better used for documenting the final selection.

The general approaches to networking include time-division multiplexing, statistical time-division multiplexing, and X.25 packet switching (see Exhibit I-2-25). Mixing one approach with another is also possible, as in the case of consolidating lower-speed traffic by statistical time-division multiplexing to a hub, where traffic from multiple sources is consolidated by time-division multiplexing for high-speed transport over the T1 backbone between hubs (see Exhibit I-2-26). Sites with unusual traffic patterns or locations in which inexpensive and abundant fiber bandwidth is available may be considered separately from the rest of the design.

Protocols

A variety of equipment may be in use on the network, including modems, terminals, microcomputers, workstations, concentrators, front-end processors, and host computers. This means that multiple communications protocols are in use on the network, which may pose problems.

Gathering protocol-related information from second-hand sources may not be sufficient. During the exploration phase, the communications protocols of these devices must be verified to ensure the efficient flow of data throughout the network. Performing a pilot test may be the only way to

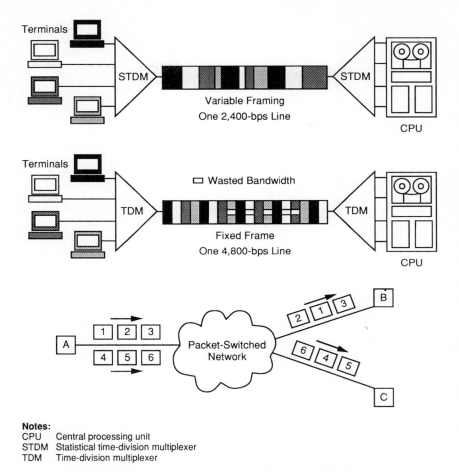

Terminals

STDM | STDM

Variable Framing
One 2,400-bps Line

CPU

Terminals

☐ Wasted Bandwidth

TDM | TDM

Fixed Frame
One 4,800-bps Line

CPU

A [1] [2] [3] [4] [5] [6] → Packet-Switched Network → [2] [1] [3] B, [6] [4] [5] C

Notes:

CPU Central processing unit
STDM Statistical time-division multiplexer
TDM Time-division multiplexer

Exhibit I-2-25. General Approaches to Networking

verify the accuracy of this information. This entails monitoring the equipment on the existing network under a realistic traffic load to verify what protocols are used at the link and network levels. In this way, information is gathered on such variables as bit error rates, buffer overflows, retransmissions, mean time between failures, and responsiveness to control keys.

Because various devices may support different protocols, the designer has several options available for dealing with them. One is the use of appropriate protocol converters—classic black-box solutions that allow computers operating with different protocols to communicate with each other.

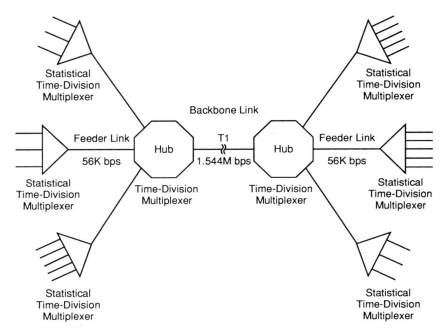

Exhibit I-2-26. Complementary Relationship of Time-Division Multiplexers and Statistical Time-Division Multiplexers

Protocol conversion is the process of reformatting or converting one protocol to another. For example, a protocol converter can be used to reformat asynchronous data for transmission over a synchronous data link or vice versa. The growing popularity of packet networks has created the need for X.25 protocol converters to allow synchronous and asynchronous devices to communicate over X.25 networks. These devices are commonly known as packet assemblers/disassemblers (PADs) and can be obtained as an adapter card, which is inserted into a microcomputer as a packet assembler/disassembler, or as a standalone device.

Another solution is to isolate different protocols to prevent applications from sharing lines. This can be confusing in wide area networking because protocols can be run together over the backbone but may or may not be run together over certain feeder links. For example, statistical time-division multiplexers are typically used on feeder links but differ in the type of protocols they support. They analyze the data that each channel is supplying to determine what protocol is being used and provide such features as priority scanning and channel-based error checking, which can limit throughput. Time-division multiplexers, on the other hand, are totally transparent to the data that each channel is supplying and are deployed as network backbones to ensure the free flow of data between major nodes.

The most efficient way to handle this problem is to treat different protocol applications as traffic layers, so that they can be selectively routed over separate lines or share bandwidth on other portions of a route, depending on the technology employed on each end of the links. Some high-end network optimization tools provide a color-coded display of bandwidth and traffic by protocol to help the designer work this out.

Switching Technology

The choice of switching technology depends on the opportunities available for sharing bandwidth. The extent to which bandwidth is used determines whether traffic may be interleaved dynamically onto a smaller bandwidth or merely aggregated statically in dedicated fashion onto a larger bandwidth.

Low-bandwidth use with bursty, interactive traffic is suitable for such dynamic bandwidth allocation technologies as polled multidrop, statistical multiplexing, and packet switching. High-bandwidth use with full-duty cycle batch, print, and file-transfer traffic is better suited to such static bandwidth technologies as frequency-division multiplexing or time-division multiplexing, both of which dedicate a full-time path to each user.

A topological display—by location—of terminal line speeds, number of terminals, required bandwidth (obtained by the product of line speeds multiplied by the number of terminals), and estimated peak-hour load in bits per second is helpful in determining bandwidth use. Aggregating line bandwidth by geography is useful in evaluating the potential for regional hubs and in sizing the speeds of backbone trunks.

The sources and destinations of traffic must be identified, including the location of host computers. Decentralized patterns of communications to several computer centers or in distributed data base environments lend themselves to such switching technologies as time-division multiplexing, multiple-aggregate statistical multiplexing, packet switching, or data circuit switching.

Traffic Types

Traffic should be characterized by type (e.g., interactive, batch, message size, file transfer, print, or CRT screen). In addition, the duty cycle during peak hours must also be determined. Usually, heavy-duty batch and print applications are deferred to off-peak hours to alleviate throughput loads during peak hours, thereby saving money on dial-up connections through the public network. In this case, such applications may be ignored for the purpose of peak-hour line sizing.

Management Reports

Detailed call, use, line-error rate, and uptime reports are helpful in ensuring the accuracy and completeness of network information. If no performance history is available because the company has recently decided to implement a private network, a simple model can be developed to show the peak-hour call intensities in terms of call arrival and duration at each location. From this information, a percentage breakdown can be developed for intralocation and interlocation traffic and the interlocation traffic can be further broken down for each of the locations. Alternatively, a typical day and hour may be chosen. For each location, a profile is developed that consists of calling and called numbers, start time, and duration for the chosen period.

Because front-end processors and statistical multiplexers may slow systems down, reports about buffer use are helpful. These reports reveal how often buffer thresholds are exceeded and identify what those thresholds are. When those levels have been exceeded, input/output problems may occur and system throughput can suffer. For this reason, the lengths of buffer queues must be reviewed during the exploration phase.

Traffic Load

Traffic load (i.e., voice and data calls that actually get through the switch and on the network) can be estimated by dividing monthly traffic by the number of business days in the month and multiplying that figure by a peak factor for the busy hour in the day, typically between 13% and 20%. If traffic is calculated in terms of characters, the result is multiplied by the number of bits per character and then divided by 3,600 seconds per hour to obtain the number of bits per second during the peak hour.

Traffic load can be estimated from historical use logs or monitors. Virtually all systems perform functions that relate to the logging of messages and the gathering of statistics regarding their own operation; the MIS department may therefore have historical information available in a usable format. Traffic load can also be estimated by multiplying the number of messages sent in the busiest hour by the size of the messages in bits and dividing that by 3,600 seconds to obtain the peak load in bits per second.

Another estimate can be obtained by multiplying the terminal line speed by the number of terminals at a particular location, applying a percentage duty-cycle factor for use during the busy hour, and dividing the result by concentration ratios of eight to 16 for interactive asynchronous traffic, two for bisynchronous traffic, and one for batch or print traffic. Growth plans— whether at existing sites, new sites, or both—should be included in this analysis.

Only traffic actually sent across the communications lines, however, should be considered. Some terminals require the remote host computer to echo back input characters and others do not. Some applications send change characters only when they transmit screens, compress blanks, evoke locally stored screen displays by transmitting short codes, or otherwise optimize screen communication dialogues by as much as 40% with compression techniques.

Low-bandwidth applications suggest the use of a local bridging or multidrop configuration to move traffic onto spare line capacity and to advance service to distant points from service already provided by shorter lines at closer points. Multidrop or bridging operations usually require all terminals to conform to a common communications line polling protocol. To ensure flexibility in this area, the applications that the terminals must support should be checked for conformance to the line-polling protocol.

Criteria for satisfactory response-time goals should be established to evaluate future performance. Simple computerized response time and throughput queuing models (see Exhibit I-2-27) may be used to demonstrate throughput and response time payoffs for higher line speeds and more expensive, higher-speed equipment. The maximum number of terminals or users that can be supported by a line of given speed before response time deteriorates exponentially must be determined. Such comparisons must come from the operating region that originates the traffic.

If bandwidth use is low, statistical multiplexing or packet switching can be used for aggregating traffic at a hub or access node. Protocol support of the statistical multiplexers must be determined; if support for different protocols is required and the equipment provides that support, a mixed set of protocols can share the same aggregate bandwidth. That is not possible on multidrops. Statistical time-division multiplexers, PADs, and packet switches conserve bandwidth by dynamically allocating it to active terminals only. On lines with low-bandwidth use, this dynamic allocation results in substantial improvements in bandwidth efficiency.

If line bandwidth use is 70% or more, it is difficult to fill the remaining bandwidth. Many of these lines, however, can be consolidated at an access node for long-haul transport over a higher-speed backbone.

Frequency-division multiplexing can also be used to combine lower-speed channels onto a higher-speed backbone, often by using the same type of physical facility for the aggregate. This is implemented by running the line at a higher speed, at the cost of slightly more expensive terminating equipment.

It is also possible to use time-division multiplexing to pool demand for bandwidth onto more economical, higher-speed bulk transport facilities (e.g., T1 at 1.544M bps or T3 at 45M bps). With time-division multiplexing, individual bandwidth channels are still dedicated to particular lines, permitting full protocol transparency. This is especially useful when a variety of obscure protocols must be supported or when high throughput is important.

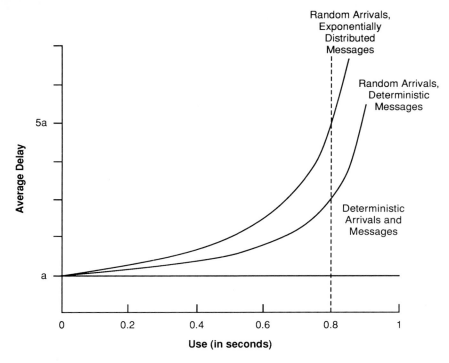

Exhibit I-2-27. Response Time Curves

Data circuit switching allows terminals to be used in a multipurpose manner for flexible connections to alternative host computers. This offers substantial cost savings, because the number of front-end processor and host computer ports can be minimized.

Sometimes a hybrid approach to networking may prove more efficient and economical. This entails the use of several statistical multiplexers to concentrate low-speed interactive terminals operating at 2,400 bps to 9,600 bps onto higher-speed 56K-bps lines for high line occupancy, batch, or printer applications at 9.6K or 19.2K bps. In addition, compressed voice at 32K bps or 16K bps may be combined for transport over the T1 backbone. Polled multidrop tail circuits can collect traffic from isolated locations into a multiplexer at a hub.

The public network may be used to carry data traffic if traffic patterns and session durations permit. This is particularly worthwhile for low-traffic volumes that cannot cost-justify dedicated leased lines.

Multidrop Configurations

Multiplexers can be multidropped to allow traffic to be leapfrogged, portions of traffic to be dropped and inserted, and multiplexers to be switched through intermediate pass-through nodes. The public-switched network may be used as a backup to a private leased-line network. As an interim measure, the public-switched network can even be used exclusively until private leased lines are installed and put into service.

The economic benefits of using a multidrop configuration to piggyback and leapfrog traffic on a single line instead of many individual lines is appreciated by most communications managers. The concentration function of frequency-division multiplexers, statistical multiplexers, PADs, and packet switches that combine many channels of a given speed onto one or more aggregates also provides clear economic benefits. Because the same amount of traffic may be carried over fewer lines, substantial cost saving on facilities and a faster payback on equipment may be realized.

Multiplexers and packet switches offer further potential economic benefits because low-speed channels can be concentrated on higher-speed network aggregates. The economic comparison is not as obvious because the higher-speed lines usually cost more than lower-speed lines. Economic evaluation usually adheres to crossover charts (see Exhibit I-2-28), which show the number of low-speed lines for particular circuit lengths that a single higher-speed line must replace for line costs to break even.

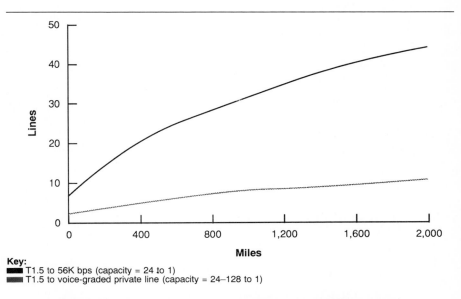

Key:
▬ T1.5 to 56K bps (capacity = 24 to 1)
▬ T1.5 to voice-graded private line (capacity = 24–128 to 1)

Exhibit I-2-28. Cost Ratio Crossover Chart for Tariffs 9 and 11

COMMUNICATIONS SYSTEM DESIGN PLANNING

Sensitivity Analysis

Another aspect of the exploration phase of network design is sensitivity analysis. The effect of significant deviations in traffic load estimates on the choice of line speeds and on associated costs must be examined. When speed choices are between 2,400 bps and 4,800 bps or between 4,800 bps and 9,600 bps, differing by a ratio of 1:2, a factor-of-two increase in the traffic estimate should make the decision easy, except in very low traffic cases. When, however, the choice is 56K bps versus 1,544M bps (a ratio of 1:28) or 64K bps versus 2.048M bps (a ratio of 1:32), it is unlikely that the decision would be affected by a factor-of-two increase in traffic estimates, except in marginal cases.

The economies of scale for some networks may allow traffic to be redistributed over fractional T1 links at a relatively small additional cost. These relatively small costs add fault tolerance to the network, provide a hedge against congestion, and build in a cushion for future growth. In some cases, substantial saving in annual line costs may accrue, despite the addition of fractional T1 links.

Transmission Variables

The analysis of the reliability and quality of various facilities must take into consideration the fact that noise levels, signal power levels, losses, phase jitter, delays, error rates, downtimes, and restoral times are often specified by the carrier. Digital lines usually provide higher-quality and more reliable service than analog lines.

When digital lines are not available, however, higher-quality conditioned analog lines can be leased at a slightly higher cost than unconditioned lines (see Exhibit I-2-29). In such cases, the actual measured reliability should be tracked for exceptions, trends, and comparisons with the stated reliability measures provided by the carrier.

Disaster Recovery

When traffic is concentrated onto a single backbone facility, there is a risk of stranding locations if certain links go down. At isolated sites, it is

Service	Installation Charge per IOC	Monthly Charge
D1	$249.00	$22.30
D6	$700.00	$50.00

Note:
IOC   Interoffice channel

Exhibit I-2-29. Cost Comparison: D1 Versus D6 Conditioning Installation and Service Charges

128

beneficial to add fault tolerance to the network with economical fractional T1 links to alternative locations. If the primary link goes down, critical communications can continue to pass until it can be restored.

Over high-density routes, traffic can be divided between two competing fiber carriers. This way, disaster recovery becomes an inherent property of the network, rather than a service that must be subscribed to at extra cost. Other safeguards that merit consideration include alternative routing, back-up dial-up to the public network, and speed fallback.

Protection against catastrophic failure must also be considered (e.g., deploying equipment with built-in redundancy and using surge protectors and uninterruptible power supplies). The resources spent for such prepara-tions should be weighed against the value to the organization of unin-terrupted computer availability. The implementation of exotic disaster re-covery solutions could negate the dollars saved through multiplexing. Attention should also be given to environmental factors (e.g., temperature, humidity, air quality, and flooding potential), because they also contribute to network failures.

Lease Versus Buy

Equipment lease or purchase costs amortized over a payback period need to be analyzed. Recurring line-cost savings extend into the indefinite future, which can cause one-time equipment purchases to be ignored in the exploration phase.

The real challenge in performing a lease-versus-buy study is to include all of the costs associated with the network and not just the initial cost of the investment. Annual expenses associated with each type of investment must be included (e.g., maintenance, utilities, space, and insurance).

User Requirements

Finally, the exploration phase must account for special features required by different classes of users. These features include: centralized network control (with capabilities for remote diagnostics, restoral, and speed fall-back), backup dial-up, priority access, alternative routing, speed changes, remote reconfiguration, time-of-day reconfiguration, parity, error and flow control, encryption, specialized interfaces, and support for specific char-acter sets and graphics applications.

MODELING

At the completion of the exploration phase, one or two alternative net-work configurations and ranges of line speeds have been determined. Compatibility problems—particularly with regard to protocols and special features—between applications and specific equipment types indicated by the proposed configurations have been investigated and reconciled.

Specific equipment types have been selected. Reference checks, demonstrations, and pilot tests have been conducted with vendors.

Now, the network must be laid out and priced. Computerized network design tools allow an analyst to evaluate an existing network topology or a proposed network topology. For a point-to-point segment—with or without drops—the pricing tool used in the modeling phase evaluates the cost of incremental additions or localized changes to the network. It is also used to evaluate service alternatives between a given pair of end points, with the current network as a baseline for comparison.

For time-division multiplexers, aggregate bandwidth limitations are applied. For statistical multiplexers, PADs, and packet switches, concentration ratios for various types of traffic and traffic protocols are applied. The numbers and locations of concentrators can be either automatically optimized according to the trade-off of line costs for equipment costs or manually configured.

In this phase, multidrops can be reconfigured using rules of thumb about the maximum number of drops or terminals permitted per line, partly on the basis of the throughput and response time queuing analysis of the exploration phase. This does not entail the collection of location-by-location transaction detail typically required by most commercially available network design tools. Such design tools demand much more detail about message sizes, use frequencies, protocol details, and equipment delays than is really required during this phase. An estimate is all that is required to demonstrate the economic benefits of proposed changes based on a high-level macro routing analysis.

INTEGRATION

Design tools for integration use standard network optimization algorithms to size line and route traffic to meet response time objectives and throughput requirements, while trading off line costs and equipment costs in an integrated fashion.

The design tool's graphics capability should include: location specification by area code and exchange or by city and state; display labels associated with nodes and links; line highlighting and differentiation by color, dashes, or line widths; and easy information archival and retrieval.

Its spreadsheet capabilities should include: listing of ports, lines, costs, and equipment counts by location; providing tools for reasonableness and completeness checks; scaling or converting traffic statistics; documenting scenario assumptions in the data base; modifying and editing without destroying the original data base; and easy information archival and retrieval.

The pricing module of the design tool should include up-to-date tariffs of all appropriate jurisdictions, carriers, technologies, and grades of service—

including those for analog, conditioned lines, digital, packet, and high-capacity digital. This is important because even a small error in the tariff data base can lead to a large error in design. This module should include data bases for determining service availability, monthly recurring charges, and fixed installation charges.

The same module should allow the designer to update and maintain the tariff data base, identify faulty scenarios in which services are not available, and specify a preferred carrier as well as a backup carrier when service from the preferred carrier is not available. This module should also permit the analyst to identify bridging options and to retrieve equipment pricing.

Throughput and response-time modeling should take the following into account:

- Use frequencies or intensities at various locations.
- Applications (e.g., batch, interactive, message, print, CRT screen, and file transfer) and their categorization by message size.
- Priorities and foreground and background status of various applications.
- Average and worst-case response times.
- Equipment protocols, speeds, capabilities, capacities, delays, and access restrictions.

Finally, a report of line use and traffic in bits per second must be provided for reasonableness and validation checking.

Circuit-holding time analysis produces estimates of the number of host computer ports required to meet call-blocking requirements. This type of analysis applies constraints on the number of simultaneous calls or virtual circuits that each node can support. The designer should also be able to obtain a report of erlangs (i.e., a measurement of telephone conversation traffic that equals 3,600 seconds of phone conversation) for reasonableness and validation checking.

The optimization module should route multidrop configurations on the basis of least cost and determine optimal numbers and placement of concentrators, including both equipment costs and line costs in optimization decisions. The analyst can manually override placement of certain equipment and lines.

STRATEGIC THINKING

In network design, it is easy to take into account various technical and economic factors and completely overlook the strategic implications. For example, with T1 lines, to qualify for a monthly rate of $9.50 or less per mile, a three- or five-year commitment with AT&T must be made. (This assumes that monthly charges for long-distance calls meet the minimum requirement to qualify for one of AT&T's volume pricing plans.)

Much can happen during that time, however, that could make an organization regret such long-term commitments. As the monthly per-mile rates for T1 continue to plummet, what started out as the most economic network may quickly turn into the most expensive one. It might be wiser to pay slightly more the first year and reap more savings in successive years as T1 rates drop.

Strategic thinking is also crucial when comparing the various offerings of carriers. For example, rate differentials of 78% to 88% exist between the dial-up 56K-bps services of MCI and US Sprint to AT&T's current Accunet Switched 56 Service. From an economic perspective, it would make more sense to select MCI's 56K-bps Switched Digital Service or US Sprint's VPN-56.

From a reliability standpoint, however, these services may not measure up to that offered by AT&T. AT&T's DDS averages 99.5% error-free seconds for 56K-bps transmissions, delivering error-free performance 99% of the time. The relatively new 56K-bps services of MCI and US Sprint do not have service records comparable to those of AT&T. Although MCI and US Sprint claim that their services support 99.8% error-free seconds at that speed, neither carrier guarantees this level of performance. Before making long-term commitments to any carrier for switched 56K-bps service, the organization may wish to try all three and monitor performance.

SUMMARY

As carriers continue to phase out analog lines in favor of digital lines while trying to recoup heavy investments in fiber networks, competition will only intensify. This means pricing changes will become more frequent and discount formulas more complex, necessitating a continual evaluation of what is available to ensure an optimally configured network.

In the process of using design tools to optimally configure their networks, communications departments not only can realize efficiencies and economies but can gain new insight into the relationship of various configuration approaches and their ability to support organizational objectives. In understanding how networks can support organizational objectives, the role of the communications manager takes on a strategic dimension.

Chapter I-2-6
Examining Integrated Communications Cabling Alternatives

David Levin

AN INTEGRATED COMMUNICATIONS cabling network is a custom-designed network capable of handling a variety of communications (e.g., voice, data, video, access control, facsimile, and imaging). Installation of such a network is important to any organization with more than 25 staff members that is relocating to new office space, because it is less expensive than having individual vendors install a dedicated cabling network for their own communications equipment.

In addition, when properly designed, this type of cabling network can support additional communications equipment at no additional cabling cost. In short, it offers a company the ability to change its communications technology in response to its business needs without being limited by the cabling network.

THE PLANNING METHODOLOGY

Designing an integrated communications cabling network is much easier than the wide variety of existing technologies might lead the user to believe. The planning methodology begins by listing current local and remote communications requirements and then extrapolating these requirements over the period of the lease. Exhibit I-2-30 lists the local and remote communications requirements for a representative 200-employee service organization over a 10-year span.

These requirements have been divided into five categories: voice communications, local data communications, remote data communications, local video communications, and other miscellaneous requirements.

Communications Requirements	Required at Time of Move	Required in 2 Years	Required in 10 Years
Voice Communications Requirements			
Station to local station	X	X	X
Station to remote station	X	X	X
Voice mail		X	X
Automated attendant			X
Local Data Communications Requirements			
Single-channel token-ring LAN	X	X	X
Multiple-channel token-ring LAN		X	X
Single-channel Ethernet LAN	X	X	X
Multiple-channel Ethernet LAN			X
MS-DOS microcomputers to value-added network		X	X
MS-DOS microcomputers to TWX-telex network		X	X
TWX-telex terminal to access line	X	X	X
Remote Data Communications Requirements			
Local microcomputer LAN to remote LAN		X	X
Local microcomputers to remote at 1,200 bps to 2,400 bps	X	X	X
Local asynchronous terminal to remote host		X	X
Local Video Communications Requirements			
Video training			X
Video teleconferencing			X
Security surveillance	X	X	X
Other Communications Requirements			
Facsimile transceiver to access line	X	X	X
Copier use collector to microcomputers			X
Access control devices	X	X	X
Imaging terminals		X	X

Note:
Requirements are for a typical 200-person service organization.

Exhibit I-2-30. Typical Local and Remote Communications Requirements

---

Voice communications encompasses local telephone stations communicating among themselves or with remote stations, voice mail communications, and automated attendant communications applications. The organization described in Exhibit I-2-30 expects to employ voice mail services within two years of business occupancy and automated attendant services within 10 years.

Local data communications requirements are divided into the following: single- and multiple-channel token-ring local area networks (LANs) for microcomputers using MS-DOS software; single- and multiple-channel Ethernet LANs; microcomputers connecting to value-added networks; microcomputers connecting to teletypewriter exchange (TWX)-telex networks and TWX-telex terminals connecting to an access line.

At move-in time, the local data communications requirement is for only a single-channel LAN for MS-DOS microcomputers—either a token-ring LAN or an Ethernet LAN.

Within two years, the organization will require either multiple-channel token-ring or Ethernet LANs, resulting from growth in the number of devices, increased communications traffic, and the need to divide different work groups. In addition, the organization will begin using microcomputers and dedicated TWX-telex terminals to access value-added networks as well as the TWX-telex networks. Within 10 years, as the number of file servers increases, multiple-channel LANs will be required. As LAN traffic increases, performance considerations necessitate that the organization be divided into smaller, logical work groups.

Remote data communications requirements are considerably more limited than local communications requirements. At move-in time, the organization has microcomputers communicating with remote information services with data rates from 1,200 bps to 2,400 bps. Within two years, microcomputer LANs will have to communicate with remote LANs, and microcomputers will have to emulate asynchronous terminals to communicate with remote host computers.

Although many organizations dismiss local video communications as being too futuristic or too costly, for the organization described in Exhibit I-2-30, video requirements are clearly on the 10-year horizon. Many organizations require security surveillance at move-in time but usually prefer to install dedicated cabling for this purpose. Most organizations will want to take advantage of video training and video teleconferencing applications within the next 10 years, but those same organizations incorrectly believe that a separate cabling network is required for this purpose.

A rapidly growing number of facsimile transceivers (i.e., a transmitter and receiver, more commonly referred to as a fax machine, or, within microcomputers, fax cards) are being installed in offices along with access control networks to permit staff members easy access at all times. Devices that track copier use are also beginning to appear in organizations that can obtain reimbursement for these expenses from their clients. Copier-tracking devices connect to a microcomputer that collects the use information. These are being networked within offices that usually have more than one copier.

Finally, the most important growth in communications technology within the next 10 years is based on image handling. These devices will most likely consist of microcomputers connected to each other as well as to external large-screen video monitors.

CABLING SUPPORT FOR TYPICAL COMMUNICATIONS REQUIREMENTS

Although numerous cabling alternatives exist, the choices are not as unwieldy as most people would assume. For example, an integrated communications cabling network always includes some form of twisted-pair cabling. Twisted-pair cabling is usually supplemented by additional twisted-pair cabling, coaxial cabling (either baseband or broadband), or fiber-optic cabling.

Exhibit I-2-31 presents three of the most popular integrated communications cabling network alternatives. The twisted-pair-only alternative typically uses six or eight pairs of solid copper cable with a braid or foil shield, terminated in two or three connectors. IBM Type-2 cable, which consists of two pairs of 22-AWG (American Wire Gauge) cable for data and four pairs of 24-AWG cable for voice, is the most popular example. Contrary to IBM specifications, these cables may be terminated in nonproprietary connectors without compromising the technical integrity of the network.

The eight-conductor RJ45, designated as the ISDN connector, is frequently used. Eight cable conductors can carry a variety of signals, including an RS-232 interface with conductors for ground, transmit data, receive data, data set ready, data terminal ready, carrier detect, clear-to-send, and request-to-send signals. The asynchronous interface unit boundary between an Ethernet terminal server and a terminal also uses eight conductors.

With two RJ45 connectors (one for voice and one for data), a total of 16 conductors (eight pairs) per station outlet can be connected. The voice and data twisted pairs either share the same sheath from the communications closet to the station outlet or are separately sheathed.

AT&T insists on separate sheaths in their premise distribution system cabling. The premise distribution system specifies two four-pair cables from station outlet to communications closet. The AT&T premise distribution system also specifies that feeder cables in individual sheaths be used between communications closets, thereby separating all voice communications from all data communications traffic.

Separate sheaths for voice and data between closets are a wise choice technically, with few additional costs. Separate sheaths from the outlet to the communications closet are costly (because of the requirement for dual-cable runs) yet offer little technical benefit. In most instances, digital and analog telephone sets coexist easily with digital data. The biggest problems arise when digital data is mixed with the coin signaling of a pay telephone—a rare occurrence in today's office environment.

The eight-pair cable for the RJ45 connector should have either 24-gauge or 22-gauge conductors. To maximize design flexibility, however, it is better when all eight pairs are made out of the heavier 22-gauge conductors. The

	Cabling Alternatives		
	A	**B**	**C**
Cable Components			
Twisted-pair cabling	Yes	Yes	Yes
Baseband coaxial cabling	No	No	Yes
Broadband coaxial cabling	No	Yes	No
Local Data Communications Requirements			
Single-channel token-ring LAN	Y-F	Y-F	Y-F
Multiple-channel token-ring LAN	N	Y-F	N
Single-channel Ethernet LAN	Y-F	Y-F	Y-F
Multiple-channel Ethernet LAN	N	Y-F	N
MS-DOS microcomputers to value-added network	Y-F	Y-F	Y-F
MS-DOS microcomputers to TWX-telex network	Y-F	Y-F	Y-F
TWX-telex terminal to access line	Y-F	Y-F	Y-F
3270 CRT to 3270 controller	Y-L	Y-F	Y-L
Local Voice Communications Requirements			
PBX station to PBX switch	Y-F	Y-F	Y-F
Auxiliary station to telephone company demarcation	Y-F	Y-F	Y-F
Local Video Communications Requirements			
Video training	N	Y-F	N
Video teleconferencing	Y-L	Y-F	Y-L
Security surveillance	N	Y-F	N
Other Types of Communications Requirements			
Facsimile transceiver to access line	Y-F	Y-F	Y-F
Copier use collector to microcomputers	Y-F	Y-F	Y-F
Access control devices	Y-F	Y-F	Y-F
Imaging terminals	Y-L	Y-F	Y-L

Notes:
A   Twisted-pair cable
B   Twisted-pair-and-broadband cabling hybrid
C   Baseband cabling hybrid
N   No support
Y-F   Yes, with full-function support
Y-L   Yes, with limited-function support

Exhibit I-2-31. Cabling Support for Typical Communications Requirements

---

incremental cost of using 22-gauge instead of 24-gauge conductors is approximately 30% for materials, with no additional cost for labor.

The eight-pair station cable needs either a braid or foil shield to minimize interference from external sources (e.g., fluorescent lights and electric motors). Although the AT&T premise distribution system specifications allow both voice and data pairs to be unshielded, the 10% incremental cost of foil shielding provides inexpensive insurance. Even the 30% incremental cost of braid sheathing an eight-pair single cable represents, because of the cable's reduced susceptibility to outside noise, a better investment than the incremental labor cost of unshielded dual cabling.

Customized eight-pair cable is commonly used when twisted-pair cabling is combined with other cabling media (e.g., baseband coaxial, broadband coaxial, or fiber-optic cable). The use of two of these cabling hybrids is compared in Exhibit I-2-31.

Alternative B combines twisted-pair cable with broadband coaxial cabling. Alternative C combines twisted-pair cable with Ethernet baseband coaxial cabling. Alternative A is the twisted-pair cabling by itself. Although Ethernet devices can operate on twisted-pair cable, this comparison is based on the installation of thickwire Ethernet backbone cabling between floors and thinwire Ethernet cable looping through the office space, with a maximum of 30 stations per thinwire loop.

When these three cabling alternatives are compared using the criteria depicted in Exhibit I-2-31, there are some surprising results. All three alternatives provide full-function support for single-channel token-ring LANs, single-channel Ethernet LANs, and microcomputers connecting to value-added and TWX-telex networks. Token-ring and Ethernet LANs use baseband signaling, a scheme in which only one signal can exist on the cable during information transfer. Broadband LANs support frequency division multiplexing, which permits simultaneous transfer of information on multiple channels. Therefore, only cabling alternative B, which combines twisted-pair and broadband coaxial, supports full-function multiple-channel LANs. IBM's broadband equivalent to its token-ring system is its PC Network product; these share the same software environment.

The twisted-pair and twisted-pair-and-baseband hybrid types offer limited-function support for 3270 terminal-to-controller communications. In both, the twisted-pair cabling uses baluns to connect 3270 terminals with the controllers. The solution is acceptable when the population of 3270 devices is less than 100 but unacceptable when the population of 3270 devices is more than 100.

The broadband environment supports network interface units that contend for controller ports—a situation that results in significantly more terminals than controller ports. The broadband network interface units also support the attachment of asynchronous ASCII terminals and printers as well as Centronics-compatible microcomputer printers. The broadband environment is technically generic, allowing different products to coexist on different channels.

Although all three cabling alternatives provide full-function support for voice communications, the twisted-pair and twisted-pair-and-baseband hybrid provide little support for local video. The twisted-pair and twisted-pair-and-baseband cabling hybrid do not support video surveillance or video training requirements. Their support for video teleconferencing is limited because a codec (i.e., coder-decoder), costing at least $25,000, is required to convert the video image to a digital data stream of less than 1.6M bps. The broadband hybrid, however, easily supports video communications on dedicated channels with codecs that cost less than $1,500 each.

All three cabling alternatives support facsimile devices, copier use collectors, and card access devices. The broadband cabling hybrid provides full-function support for imaging terminals, as opposed to limited support from the twisted-pair and twisted-pair-and-baseband cabling hybrids caused by single-channel limitations.

TWO COST COMPARISON CASE STUDIES

A leading service organization recently relocated its European headquarters and closely examined three cabling alternatives for its 200-outlet, multifloor office space: IBM Type-2 cable; Ethernet with one eight-pair, 22-gauge twisted-pair cable per outlet; and broadband with one eight-pair, 22-gauge twisted-pair cable per outlet. The costs of the IBM Type-2 cabling and the broadband-hybrid cabling are within 5% of each other. The cost of the Ethernet hybrid cabling network is 60% higher because, with one transceiver per drop, all wiring must be run in cabling trenches.

The chosen design included less expensive generic outlet connectors and closet terminations, IBM Type-2 cable, and a broadband coaxial network. Management expressed a desire to be well insulated from cabling issues in their new office space.

Another leading professional service organization recently relocated its Manhattan headquarters and closely examined three cabling alternatives for its 1,600-outlet, multifloor office space: IBM Type-2 cable with broadband; broadband with a single 22-gauge, eight-pair cable per outlet; and the AT&T premise distribution system. In this design, four connectors were recommended in each station outlet plate for maximum flexibility. In the cost analysis shown in Exhibit I-2-32, LAN components were included to make 1,200 outlets ready for connection to microcomputers.

The IBM Type-2 cabling network (alternative B) is the most expensive alternative because of the high cost of the cable and termination components. The second most expensive alternative was the broadband hybrid network (alternative C). Surprisingly, the cost differential between the AT&T premise distribution system network (alternative A) and the broadband hybrid network is insignificant. The broadband and twisted-pair hybrid cabling network, however, is considerably more robust in its functional capability than the AT&T premise distribution system twisted-pair cabling environment.

Premise distribution system advocates point out that this cabling scheme also includes fiber-optic cabling. Unfortunately, the premise distribution

Cabling Alternative	A	B	C
Twisted-pair cable type	Two 4-pair	6-pair	8-pair
Broadband coaxial cable	No	No	Yes
Gauge/shielding	24-AWG/no	24/22/yes	22-AWG/yes
Closet terminations	AT&T 110	IBM rack	M66 block
Cabling Materials			
Station cable	$ 135	$ 330	$ 175
Feeder cable	80	105	100
Outlet plates	45	55	45
Closet terminations	40	50	25
Patch cords	90	60	10
Twisted-pair cabling supplies	15	20	10
Broadband backbone cable	0	0	30
Broadband drop cable	0	0	45
Amplifiers and passives	0	0	20
Connectors and broadband supplies	0	0	20
Translators	0	0	10
Total Cabling Materials	$ 405	$ 620	$ 490
Cabling Labor			
Pull twisted-pair feeder cables	$ 85	$ 85	$ 85
Pull twisted-pair station cable	220	160	120
Terminate feeder cables	55	50	40
Terminate stations in closet	160	120	95
Terminate outlet plates	75	85	75
Voice cross connections	15	15	15
Data cross connections	25	25	10
Install broadband backbone	0	0	120
Install broadband drop cables	0	0	60
Certify the cabling network	30	20	40
Field conditions allowance	75	75	75
Total Cabling Labor	$ 740	$ 635	$ 735
Grand Total (materials and labor)	$1,145	$1,255	$1,225

Notes:
A AT&T-IS Premise Distribution System
B IBM Type 2 cabling network
C Broadband hybrid network

Exhibit I-2-32. Integrated Communications Cabling Network Alternative Costs (based on 1,600 outlets with 1,200 LAN-ready using IBEW Manhattan contractor)

---

system fiber-optic cable provides only an 8M-bps baseband-backbone connection between communications closets. The high bandwidth of this medium cannot be used with only twisted-pair cabling at the outlet plate.

Even IBM's inclusion of fiber in connections to the desktop has received poor acceptance in the marketplace. Because only a few applications justify the cost of adding fiber to the desktop, organizations are cynical about installing the correct type of fiber-optic cable. When an application

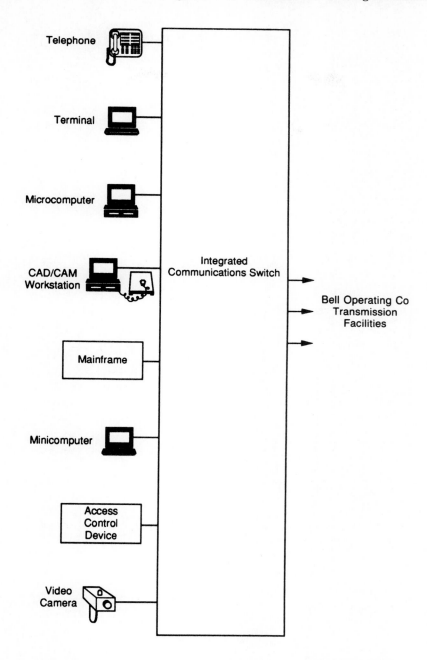

Exhibit I-2-33. The Projected Integrated Local Communications Environment

COMMUNICATIONS SYSTEM DESIGN PLANNING

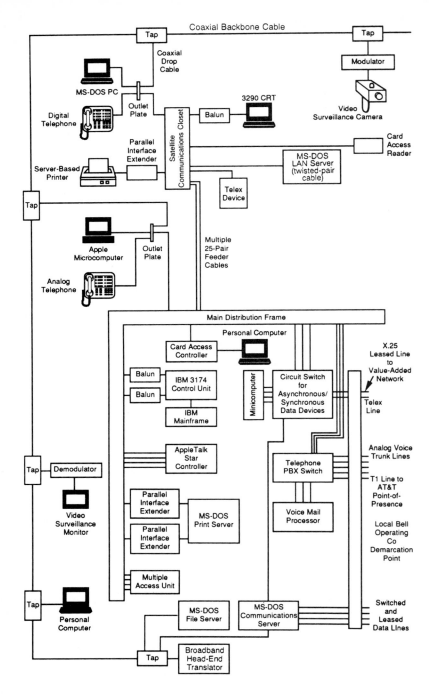

Exhibit I-2-34. Today's Actual Integrated Local Communications Environment

that justifies desktop fiber from a business viewpoint does arise, it is highly probable that the equipment will require optical fiber different from the one originally installed.

THE KEY TO AN INTEGRATED LOCAL COMMUNICATIONS ENVIRONMENT

Exhibit I-2-33 shows the typical integrated local communications environment of the future. Most industry authorities have been promising an integrated communications switch for the last 15 years, and examples of such a switch abound: the switching multiplexer, the integrated voice and data private branch exchange, the combination packet and circuit switch, the combination T1 multiplexer and matrix switch, and the digital cross-connect switch.

Because most of these products overlap in both function and market, all contend for the same budget dollars. In reality, most of these products satisfy needs that are poorly addressed by the other types of products. The majority of these products have specific cabling requirements that vendors would like the organization to believe are so different that they need a dedicated cabling network. In reality, that is not the case. Most desktop products operate within a certain spectrum of cabling alternatives. Therefore, the organization must maximize the flexibility of the integrated communications cabling network to support as much of this cabling spectrum as economically possible.

SUMMARY

Unfortunately, the integrated communications switch of the future is likely to remain more of an academic concept than an economically competitive product offering. The integrated local communications environment of the future exists today, however, and the key to building this environment is the cabling network.

Exhibit I-2-34 shows an actual integrated local communications environment that exists in both the US and Europe. This network has become a de facto standard among autonomous offices of an international professional services organization.

In today's actual integrated local communications environment, a multitude of communications technologies coexist, including a multiple-channel LAN for MS-DOS microcomputers and a video surveillance network, both operating on the coaxial broadband network. Twisted-pair cabling supports access control, telephone, voice mail, a single-channel Appletalk network, IBM 3270-type terminal equipment, printer interface extenders, facsimile transceivers, telex terminal equipment, and asynchronous terminal equipment.

Chapter I-2-7
Implementing a Micro-Mainframe Link

Robert Klenk

MICRO-MAINFRAME LINKS are an integral component of an organization's overall information processing system. These links were important because they allowed microcomputers to access some of the power of a mainframe. The introduction of the Intel i486 microprocessor, however, has produced microcomputers whose processing power rivals that of mainframes. Today, the emphasis has shifted to the sharing of data over the micro-mainframe link.

A micro-mainframe link or connection enables microcomputers to share both processing power and data with a mainframe system. This distribution of computing power throughout a computer network gives users greater flexibility. In addition, today's microcomputers have faster processors, larger capacity hard drives, and better backup devices than their predecessors. The latest technology includes CD-ROM (compact disk read-only memory) disks and CD-WORM (compact disk write-once/read-many) disks. Tape drives can now back up 2.2G bytes on a single tape cartridge. An IBM Personal System/2 (PS/2) Model 80 may incorporate a 20-MHz 80386 processor with a 314M-byte hard drive and 16M bytes of RAM. A microcomputer this powerful may be atypical, but most microcomputers are attached to local area networks (LANs) that furnish the average user with greatly increased computer power.

Significant improvements have been made not only in microcomputer hardware technology but in the software needed to complete the micro-mainframe link. Terminal emulation and file transfer programs permit a wider variety of connections and substantially improve ease of use. Hardware and software developments also permit microcomputers to function more efficiently in a LAN in which microcomputers share data and peripherals.

Improvements in microcomputer software contribute to the increased interest in processing mainframe data on microcomputers. Packages capable of organizing and processing large stores of data are now available, combining spreadsheet, data base management, and graphics functions. Other microcomputer improvements include updates to such programming languages as C and Pascal, and newer versions of BASIC. Data base packages for the LAN environment have improved, and record and file locking problems are being addressed. For example, Paradox from Borland International Inc allows a record to be read by one user and updated by another, with the update appearing on the first user's screen automatically. In addition, because microcomputer software is relatively easy to use, many user departments acquire it to perform tasks that were once restricted to the mainframe.

Although microcomputers help increase productivity, their use can create implementation problems. They are sometimes incompatible with the mainframe and other microcomputers in the organization; as a result, it is difficult to maintain integration standards. The micro-mainframe link can help alleviate this problem, depending on the organization's needs and its hardware and software. This chapter discusses the current and potential applications of the link and reviews some of the problems facing its implementation.

APPLICATIONS FOR THE MICRO-MAINFRAME LINK

Hardware and software vendors continually market new products designed to establish interdepartmental and mainframe communications. The following are current uses for the micro-mainframe link:

- Terminal emulation.
- Cluster controller emulation.
- Accessing internal and external data bases.
- Downloading data and programs from the mainframe.
- Collecting data on the microcomputer for transmission to the mainframe (i.e., uploading).
- Internal data communications networking (e.g., electronic mail).
- Using the mainframe as a file server for microcomputers.
- Programming on microcomputers and transferring programs to the mainframe.

The following sections discuss each of these uses in detail.

Terminal Emulation

Microcomputers can be used to emulate a variety of popular terminals in both synchronous and asynchronous modes, as illustrated in Exhibit I-2-35.

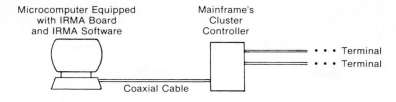

a. **Synchronous Connection**

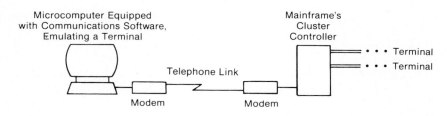

b. **Asynchronous Connection**

Exhibit I-2-35. Examples of Microcomputer-Mainframe Connections

Synchronous terminal emulations use a product such as IRMA, from Digital Communications Associates (DCA), and are directly connected through coaxial cable to a cluster controller. The cable is connected to the IRMA board in the microcomputer, and the microcomputer runs IRMA software as if any of IBM's 3270 terminals were connected to the mainframe.

Many microcomputer communications packages make asynchronous connections through the combination of a modem at the microcomputer, a telephone line, and a modem at the mainframe's cluster controller (e.g., an IBM 3274 or 3276). The microcomputer software emulates an asynchronous terminal such as an IBM 3101, DEC VT-52, DEC VT-100, or DEC VT-220. When the microcomputer is used for terminal emulation, the user has the added benefit of local storage capability and such standalone options as word processing, financial analysis, and graphics capability.

IBM's 3270 personal computer, introduced in 1983, was a milestone product affecting the micro-mainframe link. It could run seven windows simultaneously: four for mainframe host sessions, two for user notepads, and one for PC-DOS software. The bulk of data conversion, uploaded and downloaded, was the user's responsibility, but the 3270 provided the basic tool for data transfer. The 3270 personal computer supported the Distributed Function Terminal mode, which allows the 3270 emulator to have multiple System Network Architecture (SNA) host sessions concurrently.

The Macintosh MacTerminal is a newer communications tool that uses a graphic user interface. It can set up communications using pull-down menus and dialog boxes. It can be used to emulate DEC VT terminals, a simple TTY terminal, and an IBM 3278 terminal.

Cluster Controller Emulation

The cluster controller receives data from the host computer, interprets the switching information, and sends the data to the proper terminal. The controller multiplexes the input from all the terminals and sends it through a single line to the host.

A microcomputer can also emulate a controller (e.g., IBM's 3274) that connects terminals to the mainframe. A microcomputer equipped with the appropriate hardware and software can be used as a gateway to the mainframe. Individual microcomputers or those in a LAN can thus be connected to the mainframe.

The cluster controller can be emulated by using either synchronous or asynchronous connections. A microcomputer can emulate a cluster controller either by directly connecting to the terminals or by using a modem-to-telephone-to-modem structure similar to that shown in Exhibit I-2-35.

Apple Computer Inc offers a cluster controller that emulates an IBM 3276 or 3271 cluster controller but allows asynchronous terminals or microcomputers to be connected. The Apple cluster controller communicates, either through a synchronous modem or coaxial cable, with the IBM 37X5 controller and supports the binary synchronous communications (BSC) protocol and the synchronous data link control (SDLC) protocol associated with IBM's SNA. When a microcomputer is used as a cluster controller emulator, the connection must be made to a front-end processor, which may have few ports.

Accessing Data Bases

Using a modem and communications software, microcomputers can access internal data bases and such external services as CompuServe, GEnie, The Source, and Dow Jones News/Retrieval. Advances in modem technology, communications software, and telephone service have made access to data easy, reliable, and rapid. For example, it is routine for communications to occur at a rate of 9,600 bps over switched lines.

Another example of data base access software is dBASE/Answer (comarketed by Informatics General Corp and Ashton-Tate). It links IBM PCs to IBM's data base, IDMS (Computer Associates International Inc), or ADABAS (Software AG of North America).

Downloading

Because end users need access to data that traditionally resides on the mainframe (e.g., inventory, work orders, labor costs, and billing information), downloading has become one of the most requested uses for the micro-mainframe link.

Uploading

This capability supplements key-to-disk and key-to-tape data entry systems and enhances front-end data editing. For example, managers throughout an organization can prepare individual budget projections on microcomputers, the information can be uploaded to the mainframe, and the data can be consolidated by mainframe software for reports.

Files can be transferred (i.e., uploaded and downloaded) either synchronously or asynchronously. One problem with early attempts at asynchronous transmission was the slow speed of data transfer. Software such as BLAST (BLocked ASynchronous Transmission), from BLAST Communications Research Group in Baton Rouge LA, addresses this problem. By blocking the asynchronous transmission, file transfer occurs at a much faster rate.

An IBM PS/2 can be fitted with a 3270 emulation board and 3270 emulation software to make the micro-mainframe connection. A file transfer option exists, as long as the mainframe software IND$FILE is available. The combination can exchange data between the microcomputer and the mainframe in either direction, with some limitations. The uploaded file cannot go directly into a generation data group (GDG) file, which would be helpful for creating a history of the upload. If a GDG is needed, a separate job would have to be run. Some limitations result because the file transfer uses the same communications link as the terminal emulation.

Internal Data Communications Networking

This capability allows mainframes to communicate with individual microcomputers or with microcomputers connected in a LAN (see Exhibit I-2-36). In addition to an internal mail system, electronic mail (E-mail) is popular and can sometimes extend to other organizations. Transend Corp (Portola Valley CA) offers a product, PC COMplete, that allows mail messages and files to be exchanged. A microcomputer user can dial a mail system, log on, upload and download information, then log off.

Another example of internal networking is a mainframe system that handles the large-volume processing of a payroll for the organization. Microcomputers with data entry software collect the data from workstations throughout the organization; the data is consolidated at a single LAN server

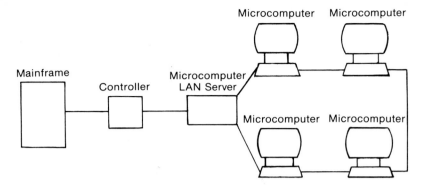

Exhibit I-2-36. Micro-LAN Link

and uploaded to the mainframe. The LAN server upload procedure contains controls that ensure that no data is lost in the transfer and that input batches equal the final received by the mainframe. Response time is better on the edits at the microcomputer end, so some cost savings can be realized by using microcomputers.

Using the Mainframe as a File Server for Microcomputers

Users who perform their work on microcomputers and then send programs and data to the mainframe have greater control over the information. The regular backup and recovery performed by mainframe operations protect user data, and the user has access to the mainframe's much larger file storage capacity. When each user department develops its own style of data storage, however, an undesirable duplication of data may occur. To alleviate this problem, many MIS departments keep standardized common data in a central data base.

Developing Programs on Microcomputers for Transfer to the Mainframe

This capability reduces the mainframe's work load, gives programmers computer time for development and testing work, and provides access to a computer when the mainframe is down. Micro Focus Workbench from Micro Focus Inc (Palo Alto CA) can even emulate IMS data bases on the microcomputer. In conjunction with its microcomputer implementation of COBOL, it provides for development and support of the entire life cycle of a project, from its data base design to programming.

The micro-mainframe link can be the means whereby data is transferred between unlike computer systems. For example, a microcomputer may be emulating a terminal for the Texas Instruments 990 minicomputer. Establishing a link between the minicomputer and an IBM mainframe directly may prove to be time-consuming, but if the 990 is already capable of downloading a file to the microcomputer, that file can then be uploaded to an IBM mainframe using an in-place micro-mainframe link. Exhibit I-2-37 shows how the transfer between two unlike computer systems takes place.

HOW THE LINK IS ESTABLISHED

The micro-mainframe connection consists of three distinct components:
- The software used on the mainframe.
- The communications device that connects the mainframe and the microcomputer.
- The software used on the microcomputer.

Each component requires some customization because the hardware, software, and operating systems of most organizations are rather diverse. Exhibit I-2-35 illustrates these components, which are described in detail in the following sections.

Mainframe Software

Mainframe systems run several operating systems and have even more diverse methods for file structuring (e.g., data base systems, sequential files, indexed files). Data can be stored in sequential files (e.g., ISAM, VSAM, and BDAM files). These numerous structures make the micro-mainframe link more difficult for vendors to accomplish. Before data can be transmitted to the microcomputer, it must be extracted from the mainframe data structure and prepared for the transfer. The micro-mainframe system

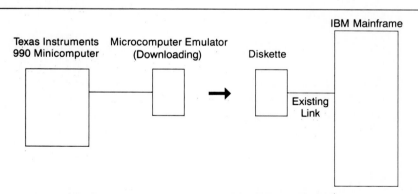

Exhibit I-2-37. Data Transfer Between Unlike Systems

vendor usually provides the mainframe software needed to accomplish this task.

Communications Devices

When the data is extracted from the mainframe, there must be a communications connection to the microcomputer. The most common link uses the communications lines from the mainframe's controller. The lines run to a device, usually a board, installed in the microcomputer. DCA's IRMA board and accompanying software allow the microcomputer to emulate IBM 3270 terminals and transfer files to and from the mainframe. Some of the mainframe's ports can be configured to communicate with the microcomputer through modems located at each end, but this would require further customization within the link.

At some point during the file transfer process, a conversion must occur. The IBM mainframe uses EBCDIC to interpret character bits, whereas most microcomputers use ASCII. The conversion of EBCDIC to ASCII, and vice versa, can occur at almost any point in the link; a black box on the communications line is the most common tool for conversion.

The microcomputer might need to communicate with the mainframe using synchronous rather than asynchronous transmission protocol (the microcomputer's native method of data transfer). As in the method of conversion for interchange codes, a device inserted at some point in the line can convert one protocol to another.

Microcomputer Software

The third component of the micro-mainframe link is the microcomputer software that will receive and initiate data. This software prepares the inquiries directed to the mainframe, sends and receives the data, and, if necessary, emulates terminals. The file transfer software often places data from the mainframe in a standard DOS text file (sometimes called a print file); with some links, however, the data can be placed directly in the application's file format.

The Application Connection (T-A-C), from Lotus Development Corp, which is discussed in detail in a later section, is one example of a product that downloads data from the mainframe directly into the microcomputer's application software. Mainframe data from various data bases and file structures can be downloaded directly into Lotus 1-2-3, Symphony, or dBASE files on the microcomputer.

Handshaking

Products currently offered for the micro-mainframe link require that software reside on both the microcomputer and mainframe so that terminal emulation and file transfer can occur. This software ensures that proper handshaking requirements are met. The two computer systems agree that communication will occur, that the format is compatible, and that the mode of communications (e.g., simplex, half duplex, full duplex) is decided.

Although some vendors can provide the mainframe software, an organization may still need to customize it, depending on file structure and the method for accessing data in the mainframe. For example, a microcomputer cannot directly access a VSAM file, so the mainframe software must convert the VSAM file into one the microcomputer can read. If the vendor cannot provide this capability, the organization must use a COBOL program to create the file the microcomputer can read.

Most microcomputers communicate with the host through a terminal protocol. In the SNA-driven architecture, all connections are under control of the master computer. Even so, SNA has precisely defined the network so that communications are transparent to the user. The trend is now toward peer-to-peer relationships, as occurs in IBM's System Application Architecture (SAA). The benefit of SAA as an alternative to an SNA network environment is that processing can be distributed rather than centralized. Resources are readily available to end users and the data for various applications can be kept on more than one computer, ensuring data backup. This approach to networking can improve the computer-to-computer transfer needed by today's businesses.

DIFFICULTIES ENCOUNTERED WITH THE LINK

Organizations implementing a micro-mainframe link often encounter problems in the following areas:

- Protocol conversion.
- Conversion between ASCII and EBCDIC.
- Incompatible hardware.
- Slow transfer rate.
- Untested products.
- Processing capacity of microcomputers.
- Installation.
- Security and data integrity.

These problems are discussed in detail in the following sections.

Protocol Conversion. Most protocol converters emulate an IBM 3270 on one end and some variety of hardware and software combination on the other. A single synchronous line connects the converter to the host, the terminals are connected to the converter by asynchronous lines, and the configuration functions as a controller with intelligent terminals attached. The converter does the work of storing all keystrokes from the attached devices and, when necessary, transmits a complete screen from each device or terminal. This is necessary to create a 3270 data stream that the mainframe can understand. Because of the various protocols used in the communications industry, data transferred from microcomputer to mainframe must undergo changes. Transport Control Protocol/Interconnect Protocol (TCP/IP) was developed in an attempt to standardize protocol not only for the micro-mainframe link but for computer-to-computer transmission. It is an industry-standard protocol that allows noncompatible systems from different vendors to exchange data.

Users can communicate with machines using TCP/IP either remotely or through a direct cable connection. TCP/IP is a set of basic networking functions that are supported by approximately 50 network and computer vendors and is the basis for some micro-mainframe links. Even with this standard protocol, however, black boxes may still be needed at points in the network to convert one protocol to another.

To change to TCP/IP, a company will need to invest in network bridges, network adapter cards, and network management tools. The network bridges allow connection of LANs, connection of remote LANs over T1 communications lines, and connection to a mainframe network. The benefits of such a change include ease of connectivity with organizations already using TCP/IP.

ASCII-EBCDIC Conversion.   Because EBCDIC is the IBM standard, communications with IBM mainframes require that EBCDIC character codes be used. Most microcomputers, however, use ASCII code.

Incompatible Hardware.   Even when modems can transmit data between one type of microcomputer and a mainframe, hardware or operating system requirements might prevent microcomputers made by different manufacturers from using the same modem. With the wide variety of microcomputers found in user departments, the confusion caused by the need for separate modems and communications packages for each of them highlights the need for standardization before connections are established.

Slow Transfer Rate.   Depending on the product, file transfer between microcomputers and mainframes is usually a slow process and often consumes the resources of both systems. The introduction of OS/2 for the microcomputer, which allows the link to operate in the background, has reduced some of these problems. (OS/2 provides multitasking for the microcomputer.)

Untested Products.   The market for micro-mainframe link products is continuing to expand, and in their eagerness to capture market share, some vendors announce new products before they are available. Therefore, managers should insist on a thorough demonstration or references for specific products.

Microcomputer Processing Capacity. Although the newest micro-computers have processing speeds of 33 MHz, can address up to 16M bytes of RAM, and can store 800M bytes on optical disk, most microcomputers do not have these capabilities. Most have 10M- to 20M-byte hard disks, which can be used up quickly in mainframe inquiries and the microcomputer's usual processing. A LAN, which enables many microcomputers to access data downloaded to a file server, helps alleviate this problem.

Installation. Individual data centers must work with the data communications department to explore communications hardware and software configurations.

Security and Data Integrity. Data base administrators responsible for securing data must implement controls to limit the increased accessibility of mainframe files.

ORGANIZATIONAL ISSUES

One significant obstacle in the establishment of a micro-mainframe link has been the absence of a business focus in designing the connection, evaluating products, and installing the systems. The connection is often made simply as a technological accomplishment. In the selection of a link, data communications has the opportunity to educate and direct the user.

By and large, the technical problems of the micro-mainframe link are solved as users ask for more or different types of data. Maintenance of the link utility and transfer procedures are ever more time-consuming and complex. In addition, not everyone gets the same results from data that is ostensibly from the same source. For example, two managers obtain data from the same data base and download it into a spreadsheet on the micro-computer, yet their models produce different results. A mechanism has to be in place to determine whether each manager captured data at the same time from the same place. One of the problems with downloaded data is that it becomes static after it is downloaded.

In general, users cannot determine the resources needed to accomplish the link or the technical obstacles involved. The data communications department must design the process with the user community in mind.

Implementation Steps

The micro-mainframe link should be an extension of data processing, not a replacement or short-term solution to a backlog problem. The following steps are suggested to help the communications systems manager alleviate problems with the link:

- Maintaining a business focus—The person assigned the job of implementing the link should preferably be an analyst rather than a technician.
- Developing an understanding of the behavior, habits, and interests of users.
- Creating a block diagram—General information flow is more helpful than a technical file-name-by-file-name flowchart.
- Developing a clear picture of how the link will be used.
- Building in flexibility.
- Obtaining management commitment.
- Starting out small—After a few projects produce results, other links can be incorporated throughout the organization.

It is also important to ask the following questions:

- What data is to be exchanged?
- Where is the physical data located (e.g., in the data base, in the sequential file)?
- Does the problem require a general or specific solution?
- What transfer rate is needed for the volume of data?
- Will the data be shared?
- When will the data be transferred?

Standards and Policies

Organizations also need to develop standards and policies regarding data transfer by micro-mainframe links. For example, for hardware acquisition, a standard procedure would be not to buy and install a link without the involvement of the data communications department. Data security policies can be improved as well.

Advanced Technology

In advanced systems, a microcomputer will run a portion of the processing while the mainframe runs another portion and communicates the results back to the microcomputer. All this activity will be transparent to the user. As technology advances, MIS managers will need to address the question of software development (e.g., who writes software for the mainframe and who writes software for the microcomputer?) and data resource management issues (e.g., who is in charge of the information for the organization?). They must also decide which new packages to put on the network.

SECURITY AND CONTROL

As more microcomputers and micro-mainframe links are incorporated throughout an organization, more users have potential access to data. More important, microcomputer users can often circumvent the security restrictions and controls imposed on the mainframe.

Currently, access is usually restricted only by the use of microcomputer passwords and ID numbers. In addition, data cannot be controlled after it has been downloaded to the microcomputer. Because diskettes are easily transported, data can be easily distributed to unauthorized personnel.

Data integrity must also be safeguarded. For example, a novice user accessing spreadsheets could accidentally write over valuable data. Information that is downloaded and uploaded is easily subject to modification on the microcomputer, and the usual mainframe audit trail might be lost in the microcomputer environment. User departments should therefore implement controls that restrict users to their level of authorization. Like mainframe data controls, microcomputer controls can be placed over files to maintain different levels of access (e.g., read only, read and update, and privileged user).

Remote communications over standard dial-up lines are susceptible to violation from computer hackers. With a minimal amount of hardware and a telephone, hackers can access mainframes in the same way authorized users do: they dial the number of the data center computer and enter a correct password and identification number. In some cases, the microcomputer is programmed to dial numbers until a carrier-ready signal is reached, indicating that a mainframe is answering the call. The hacker attempting to guess a correct password or ID number is sometimes clued by friendly messages designed to help legitimate users log on to the system. Hackers also share information on computer bulletin boards and in newsletters for hackers.

Although most hackers seek only the personal satisfaction of breaching security, some have seriously damaged essential data files. Unauthorized penetration of private data bases can be minimized by sophisticated entry protection, but by their nature, remote communications can make any data base more vulnerable.

Moreover, a new security risk has been identified as a result of the micro-mainframe link: a microcomputer may be used as an unauthorized source of remote access to the mainframe system (see Exhibit I-2-38). A microcomputer may be attached to the mainframe through an emulation board and software, and it may also be equipped with a modem for communications. The microcomputer can be programmed to receive a call from a remote site. The user calls the microcomputer remotely and thus

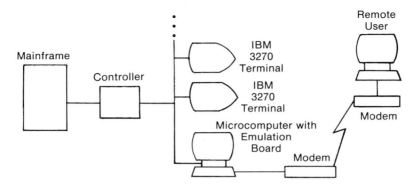

Exhibit I-2-38. A Security Risk in the Micro-Mainframe Link

can use the mainframe's resources as though the terminal were located on site.

Although most organizations have strict policies authorizing remote access to the system, the use of a microcomputer with an emulation board can circumvent the normal procedures and is difficult to detect. The mainframe only detects that another terminal is attached. On the newer notebook and laptop computers, modems are generally internal, so the link is not readily detected.

Security Proposals

The following controls, which are well known in the mainframe environment, are being proposed for use in the micro-mainframe environment:

- Passwords and ID numbers.
- Software and hardware controls.
- Front-end checking for valid data.
- Enforced policies and procedures for downloading data.

These security controls are discussed in detail in the following sections.

Passwords and ID Numbers. When an authorized microcomputer user requests a program or file from the mainframe, the proper password should be supplied automatically. Building in the passwords and procedures to access the mainframe poses another problem, however. For example, the user of a portable computer may store the telephone numbers, log-on procedures, and account numbers for access to CompuServe and organizational files; to log on to the service, the user performs only a few simple keystrokes. Mere access to the hardware provides automatic access to a wide range of data. Therefore, microcomputers should be closely guarded in the end-user environment.

Software and Hardware Controls.   Because hackers can obtain passwords and identification numbers, one of the most secure, though expensive, solutions is data encryption. The authorized microcomputer user must be issued a special decrypting mechanism that interprets the encrypted data. The mainframe should record every attempted log on, and this record should be reviewed regularly.

Another security measure is to implement communications equipment that double-checks the remote user's authorization. This protective device answers the remote call, requests a password, and then searches a telephone directory of authorized users to find the user's number and call the user back. The connection to the mainframe is thus initiated by the mainframe. (This solution, however, restricts authorized remote users to certain locations and can be circumvented by various telephone configurations.)

Front-End Checking for Valid Data.   Although data integrity is usually ensured by mainframe software, microcomputers can perform this function, thereby saving valuable mainframe time and reducing system response time.

Enforced Policies and Procedures for Downloading Data.   The security of the microcomputer link requires responsible end users who adhere to well-defined policies that state who has access to which data. Because data can now be transported easily on diskettes, there is an even greater need to enforce the standards that control access to both hardware and software.

Because users are rarely confronted by log-on and log-off software at the microcomputer level, and because microcomputers exist in a relatively unsecured environment (especially if many users have access to the same microcomputer), vendors have introduced access control software and hardware. Such products as Watchdog (Fischer International Systems Corp, Naples FL) and PC/Data Access Control System (Pyramid Development Corp, West Hartford CT) provide microcomputers with the same security as mainframes. Identification numbers and passwords can be issued to users, and the degree of access allowed each user can be restricted. The following restrictions can be applied:

- No booting from a diskette.
- Limited access to files and directories.
- File encryption.
- Limited access to communications ports.
- Limited access to any diskette (i.e., the user can gain access only to the hard disk, not to the diskette or, in certain cases, not even to the printer.)

By using this security system, data downloaded from a mainframe can be protected from unauthorized access. Access control should be required if the microcomputer contains mainframe data and is equipped with a modem for remote access; if no control is imposed, an unauthorized user can call the microcomputer and access the information.

MICRO-MAINFRAME LINK PRODUCTS

Lotus Development Corp provides a micro-mainframe link that is a companion product to its popular spreadsheet packages (i.e., Lotus 1-2-3 and Symphony). The purpose of this product, called The Application Connection (T-A-C), is to enable microcomputer users to access mainframe data so that the details of the transfer are transparent. T-A-C has two components of the micro-mainframe link: software on the mainframe that allows it to connect to such data bases on DB2, SQL/DS, and SAS as well as microcomputer software that allows transfer to such microcomputer applications as Lotus 1-2-3, Symphony, and dBASE.

T-A-C users can work with a menu similar to the spreadsheet's own menu system. Requests for data transfer can be submitted to the mainframe through virtually any communications link. T-A-C can, for example, access IMS, IDMS, and VSAM through a Focus connection. The software on the mainframe receives the request, translates the request into instructions for the data structure, and then extracts the data. The data is transferred directly into the microcomputer application file.

T-A-C can also be used to upload information to, as well as extract it from, the mainframe. Some organizations have placed restrictions on data that is to be uploaded; however, one of T-A-C's features allows the file request to access only data that the user is ordinarily authorized to obtain. It does this by connecting to the security systems often used in the mainframe environment, namely RACF or ACF2.

T-A-C is an important advance toward simplifying the micro-mainframe link and should lead to greater use of the link. The user is presented with a familiar system: microcomputer menus. In addition, T-A-C's capability to upload and download data will help integrate the link into an overall information processing plan.

Another product permitting broad access to the resources of the mainframe is Tempus-Access from Micro Tempus Inc. It allows microcomputers to extract information from a variety of mainframe systems. The mainframe software that extracts and formats the data for transmission to the microcomputer is DYL-280 from Sterling Software Inc. DYL-280 can read sequential, ISAM, BDAM, and VSAM files as well as such data base systems as IMS, IDMS and IDMS/R, and CA-Datacom/DB. DYL-280 extracts data into a format that can be received by most microcomputer applications (e.g., fields that are separated by commas). Programmers' Workshop, a module that facilitates access to a greater extent, can store in a few keystrokes the commands needed to access mainframe data. Tempus-Access is available in versions supporting MVS/TSO, MVS/CICS, and DOS/VSE.

Tempus-Access was developed for end users who are not technically oriented. The product attempts to present the user with a common view of the data across several mainframe file structures. Users specify the fields they wish to access by filling in blanks on the microcomputer screen. Users enter selection parameters and sorting order, and the request is submitted to the mainframe. The DYL-270 on the mainframe processes the request and extracts the information from the proper file. Users do not need to know the details of the file's access method. This capability is most beneficial when trying to extract data from several file structures and is even more important when a DBMS such as IMS is involved. The extract is downloaded through a communications line to a microcomputer (e.g., Tempus-Access works with the IRMA board), from which the microcomputer application package must import the data. Many microcomputer software packages allow data import only if the field and record separators are uniform.

FUTURE DEVELOPMENTS

As more organizations implement some type of micro-mainframe connection, several developments are likely to occur. These are described in the following sections.

End-User Computing. End-user computing is one trend that has contributed to the implementation of the micro-mainframe link. When designing and developing their own applications, end users often use fourth-generation languages (4GLs), which permit them to develop applications without an indepth knowledge of the details of a programming language. The details of file structure (e.g., file access methods or data base structures) are transparent to users.

True end-user computing has not caught on very quickly, because a great deal of computer and systems analysis knowledge is demanded of the end user. The average end user is not experienced enough in systems analysis to put together a well-designed data processing system.

There is also the problem of security within the system. Users cannot usually determine how to protect the integrity of data. They also have difficulty figuring out the most effective way to get the data and have little idea of the effect their request will have on the mainframe or microcomputer. Users still need an analyst to determine what data is needed where and the best method for getting the data to the end user.

Easier-to-Use Links.   Currently, the link is technical in nature. Users do not know how data is stored in DB2, IMS, or any other data base management system in use on the mainframe. They may know the microcomputer DBMS, but they need more technical skills to make the transfer of files from microcomputer to mainframe. As the user base of micro-mainframe links expands, users will find it easier to access mainframe data and send data to the mainframe. Details of the link will be transparent to the user, and the commands needed to perform transfers will become less complicated.

Greater Emphasis on Security.   The mainframe environment provides a secure system for accessing information. The link enables users to access a greater amount of information, and because most microcomputers or LANs place a lesser degree of importance on security and control, managers implementing the link must educate users on security controls and procedures. An organization must also formulate policies regarding information obtained from the mainframe.

One security measure is to download the security attributes of the mainframe file along with the file itself. Whenever the microcomputer user wishes to access the data, the rules are checked and applied to the request. This approach is used only on the most sensitive files (e.g., payroll) but is effective in closing the security gap between microcomputer and mainframe. Without some type of security system on the microcomputer, files remain unprotected.

Microcomputers Incorporated into the Overall Information Processing Plan.   Microcomputers seem to have created an end-user revolution rather than an evolution. Some organizations are merely reacting to users' demands for information access rather than carefully planning how the microcomputer can fit into the entire MIS process. Hastily constructed links may solve short-term problems. To ensure that management's goals are achieved, however, the long-range impact on the organization must be considered.

Cost Reductions.   Advances in technology are reducing the cost of implementing the micro-mainframe link. Increasing competition among vendors will further reduce costs. Microcomputer advances include the development of systems using the Intel i486 microprocessor. Dell Computer Corp offers a 486-based computer running at 33 MHz, with up to 64M bytes of RAM and up to 650M bytes of hard disk storage. This speed and power greatly increases the capability of microcomputers to handle the size of files that are downloaded from a mainframe.

Networking Improvements.   In many microcomputers and LANs, problems are not discovered until there is a mishap. These problems usually require a great deal of manual intervention, which can cause difficulties in a communications network. Features that allow portions of the micro-mainframe link to test the system periodically, however, can be added. The self-test would automate the problem-solving process ordinarily performed by users, which will free the user troubleshooters to handle other tasks. Intelligent products can report the status to the network as a whole, thus increasing the efficiency of the network.

More Widespread Use of the LAN-to-Mainframe Link. As more organizations implement LANs, there are clear advantages to such a link. First, a LAN-to-mainframe connection is cost-efficient. Second, LAN users can share the downloaded data and thus reduce the number of inquiries to the mainframe.

Often, custom-written software is needed to extract information from mainframe data structures (e.g., data bases, VSAM files) and transmit it to a usable form on the LAN. For example, an organization extracts information from a mainframe IMS data base to a sequential file, then downloads the file to a LAN. Software on the LAN then has the capability to format the data directly into common microcomputer software formats. A menu-driven system allows the user to select a format (e.g., dBASE application, a Lotus 1-2-3 cell format, or a comma-delimited format). With a mainframe file downloaded from the LAN, it is important to carry down the timing of the file and, if needed, any security protection that exists from the mainframe.

Advanced Technology. The first micro-mainframe links seem primitive compared with current technology, and even more advances are being made. One such advance is the introduction of the link on a series of chips rather than on a microcomputer board. In addition, the industry can expect increases in the speed at which files are transferred, and the physical space required for the hardware will continue to decrease.

Improved Standards. Micro-mainframe links are often deficient when working with a variety of microcomputer applications, a variety of mainframe data structures and operating systems, or virtually any communications technology. This problem is a result of the proprietary nature of most products. Products that can overcome these obstacles will be available to facilitate implementation of the micro-mainframe link. These advances will decrease the importance of the problems encountered in the link.

Manufacturers of data communications equipment define protocols to establish connectivity among the range of their products or product lines. Each vendor's equipment, however, does not necessarily connect to another manufacturer's equipment. A lack of adherence to industrywide protocols has led to a situation in which each manufacturer has designed proprietary means for communication. Thus, a product such as a micro-mainframe link might work well with one vendor's equipment but have no connectivity to other manufacturers' products or to such services as packet-switching networks.

Several organizations define industry standards for data communications within the US and internationally. The Institute of Electrical and Electronics Engineers (IEEE), the American National Standards Institute (ANSI), and the Electronic Industries Association (EIA) set industry standards in the US. The International Telecommunication Union (ITU), the International Telephone and Telegraph Consultative Committee (CCITT), and the International Standards Organization (ISO) set international standards.

All of these organizations, however, do not work together and can produce different protocols. The protocols defined by each organization are identified by its initials and the number of the recommended standard. For example, the most basic connection, the physical link between data terminating equipment and data communications equipment is defined by EIA RS-232.

Committees are still working on standards, but apart from some agreement on electronic data interchange (EDI), they are still not industrywide. Although many standards have been proposed, each manufacturer chooses whether or not it will adhere to them. For example, IBM's SNA adopts many of the protocols defined by the ISO's open systems interconnection (OSI) model for networked systems but does not comply with the standard in every situation. In general, to capture a wider market share, manufacturers make products that comply with all standards.

One standard that will directly affect micro-mainframe processing is the Application Program Interface (API), which allows a microcomputer applications program to simulate the keystrokes of a human operator and communicate with a mainframe application. An extension of API, developed by IBM, is the High Level Language Application Program Interface, which allows microcomputer programs written in BASIC or COBOL to communicate directly with the mainframe application.

ACTION PLAN

Communications hardware and software capabilities are expanding as new products are announced. Communications systems managers seeking to implement a micro-mainframe link should keep up to date with the latest developments in microcomputers as more powerful systems become available, with data communications standards proposed by standards-setting organizations, and with the availability of link products. In addition, managers should consider the tactical and strategic position of using microcomputers for overall MIS needs and should determine the type of link desired (e.g., a standalone microcomputer emulating a terminal or a connection to a LAN).

Communications systems managers should take the following steps to ensure successful implementation of a micro-mainframe link:

- Contacting the manufacturer of a micro-mainframe link and determining whether the product meets the organization's needs.
- Observing the link in operation at other installations.
- Coordinating the efforts of MIS and user departments in setting up the link.
- Implementing the link in tactical and strategic steps.
- Considering requirements for securing data at each point of the link; sensitive data can be accessed by the microcomputers and can reside on relatively uncontrolled microcomputers.
- Scheduling training sessions for nontechnical users of the micro-mainframe link.

To achieve the greatest benefit from the micro-mainframe link, communications systems managers must help organizations integrate microcomputers into overall MIS goals.

Section I-3
Security Planning

THIS SECTION DISCUSSES AN ISSUE of both immediate, daily concern and continuing, long-term concern: security. It provides the communications system manager with answers to such questions as, what exactly does *secure* mean with reference to a communications network? How much security is enough? What standards are available? Does open architecture imply insecure architecture? And are computer viral infections a serious cause for concern?

Chapter I-3-1, "Communications Security Concepts," establishes a baseline by reviewing the fundamental design and operational requirements for a secure communications system. The exhibits in the chapter list the security factors for each component of a communications network.

With the baseline established, the section moves to a discussion, in Chapter I-3-2, of "Security in Open Communications Architectures." This chapter reviews security concepts and discusses security services and mechanisms with respect to the International Standards Organization (ISO) reference model for open systems interconnection (OSI). Such techniques as cryptography, with its subsets of public and secret key encryption, are explained.

Chapter I-3-3, "Communications Security Standards," focuses on the standards approach to system security. The chapter delineates organizations involved in standards development and the current status of national and international security standards and concludes with a list of the probable future developments in this important work.

Chapter I-3-4, "Security in Financial and Messaging Applications," extends the theme of Chapter I-3-3. It discusses the security features that have been included in American National Standards Institute (ANSI), International Telephone and Telegraph Consultative Committee (CCITT), and ISO standards. Examples are key distributor and confidentiality features in ANSI X.9, authentication features in X.12 for electronic document interchange, and similar features in CCITT and ISO versions of X.400 electronic messaging and X.500 directory services.

Local networks present unique security problems because of the distributed nature of the control mechanisms. Chapter I-3-5, "A LAN Security

Review," offers a step-by-step plan for conducting a security review of a local area network. The chapter includes a discussion of techniques and principles unique to the local networking environment.

Computer viruses are a threat to every computer resource. Viruses frequently enter the computer system through the communications network. For this reason, viral infection is of concern to the communications systems manager. Chapter I-3-6, "An Overview of Computer Viruses," examines the history of computer viruses, categorizes types of viruses, and suggests measures for mitigating this threat.

Most organizations' communications systems are connected to the telecommunications facilities provided by common carriers. These access ports provide a vulnerable point of attack for anyone attempting to disrupt the system. Chapter I-3-7, "Protection of Communications Ports and Lines," suggests access control as the main line of defense in this situation. Hardware can be used to augment existing security features. The benefits of this approach, a discussion of the types of hardware solutions, and a set of useful appendixes are the subjects of this section's concluding chapter.

Chapter I-3-1
Communications Security Concepts

Gerald I. Isaacson

THE IMPORTANCE OF DATA COMMUNICATIONS SECURITY increases in relation to the boundaries of the total system and the demands imposed by traffic and its value. Secure and reliable systems begin with deliberate intent and evolve through the serious consideration of each system component's security and reliability during design, implementation, and daily operation.

Although reliability and accuracy are the primary concerns of communications systems design, consideration should be given to providing flexible safeguards to which additional security controls can be attached without major network redesign. Such safeguards can reduce a system's vulnerability to unauthorized external attacks, increase reliability, and facilitate recovery in the event of a disaster or user error.

Communications systems, with the exception of those used in classified defense and intelligence operations, were developed to operate in nonhostile environments. However, in today's environment of multiconnected public switching networks in which microcomputers can emulate host computer responses and users are increasingly aware and capable, control must be integrated into network design.

Ensuring data communications security involves many considerations, from the structure of the total organization to the selection of personnel and equipment. This article highlights factors that play a significant role in attaining the degree of security needed to guarantee a required level and class of service.

REQUIREMENTS ANALYSIS

In a secure communications system, transactions are always delivered to the correct recipients. The system must deliver a message with the same content that was entered; unauthorized persons cannot learn of the transaction or delay it during transmission. Authorization must be established for the originator and the recipient, for both the transaction type and its sending or receiving station.

A security violation is a breach of the network between the originator and the recipient. A violator can access a message for unauthorized monitoring, acceleration, deferment, or modification or can introduce data into or remove it from the network.

Goals for a Secure Communications System

Although a totally secure system is unobtainable, it is possible to provide adequate security that is consistent with the risks involved and within the constraints imposed by personnel, physical, technical, and financial resources.

An adequately secure system depends on a sufficient level of reliability, or operational readiness. Reliability refers not only to the continued operational status of the equipment and programs but also to the system as an entity, including personnel, procedures, logistical requirements, and maintainability. The goals for a secure and reliable communications system are:

- Data accepted by the system will never be lost, misrouted, modified, disclosed, duplicated, accelerated, or delayed without authorization.
- The system will never be completely out of service, though it may occasionally provide less service than usual.
- Failure of a single network node or transmission path will not cut off users of that node or path from the network—Individual users may be cut off if the transmission path to a single user fails.
- All errors in message transmission, reception, content format, or processing will be detected, causing prompt notification of operations personnel responsible for the system.

The first step in obtaining operational readiness in communications networks is to determine the level of equipment reliability required and the number and extent of failures that can be permitted without jeopardizing the system's objectives. System planners must then design the network and its components to achieve the specified goals.

The Relationship of Exposure to Effect

The protection a secure communications system provides is not a function of the cumulative levels of security of all points in the transfer of data or messages but is related to the security level at each point of the passage.

Although the risk of a security violation decreases as traffic flows from individual stations toward the destination and is concentrated in the data pipeline, the seriousness of a violation or system malfunction increases.

The transmission of unauthorized messages from an endpoint is a serious violation but may be a minor problem compared with the disruption of service at an exchange point or a switching center. A major security breach or system failure at a concentrator or switch within the network could have far-reaching consequences.

The data security administrator should keep in mind that the cost of a protective program increases rapidly once adequate levels of protection are achieved and absolute security is attempted. Both the effect and threat of a violation must be realistically appraised.

The development of standards for communications networking has become critical by the sheer number of standards that have been adopted by international standards bodies. The open system interconnection (OSI) model, which defines a communications environment in terms of functional layers, is an example of the coordination of standards toward a common end. Standards for reliability and security are needed to ensure adequate protection across network boundaries and within private networks. Such standards are being addressed in various layers of the OSI model. As the number of nodes and gateways between networks continues to grow, enabling users to establish worldwide connectivity across multiple networks, the possibility that the end user's identity cannot be verified increases. The International Standards Organization (ISO), the Institute of Electrical and Electronics Engineers (IEEE), and the American National Standards Institute (ANSI) continue to address these issues.

The proper legal safeguards must also be established to protect network users from problems outside their responsibility and jurisdiction (e.g., unauthorized data entered into the network).

Security and reliability in data communications systems can be examined from two viewpoints. The first concerns the requirements that must be imposed during system design and implementation, and the second focuses on the controls needed to ensure the system's safe daily operation.

DESIGN REQUIREMENTS FOR SECURE DATA COMMUNICATIONS

Several factors must be considered at each level in the design and implementation of a communications system to ensure adequate security and reliability. These factors include system requirements, system components, messages (data), and accountability and reconciliation features.

Requirements for System Security

The four basic considerations for security in a communications environment are: network access, detection of intrusions, personnel, and security costs.

Preventing Physical Access. The most obvious step in establishing a secure communications environment is to prevent unauthorized access to the installation and system components, including input and output terminals, communications lines and switching points, computer centers, and data storage files. Protection of the computer center should include:

- Limited access to the communications facility through guard-controlled entries.
- Automated access control systems, such as manual and electronic keys, magnetic or other electronic card systems, and biometric access controls (e.g., retinal scan, signature dynamics, voice recognition, and geometry or fingerprint analysis).
- Secured communications frames and cable rooms, including alarm systems and intrusion detection.

These controls are usually specified during physical site planning.

Security measures for the protection of stored data begin with limited physical access to the storage facility. The facility's vulnerability to fire and other dangers can be reduced by physically separating it from the computers. The installation of shielding between the storage facility and public or unguarded areas protects the files from magnetic interference. Because damage occurs only if files are close to the source of the field, positioning the storage units away from the external walls may suffice.

Overt or accidental radio frequency interference may also cause computer and communications malfunctions if the computer center is close to high-power radar transmitters or high-voltage power lines. Although adequate site and network planning helps prevent these problems, shielding may be required in some cases. The threat of emanations to communications systems may grow as traditional vulnerabilities are reduced by means of better controls. In classified environments, equipment that meets the Department of Defense TEMPEST requirements is used to reduce this exposure. This level of protection is generally unwarranted for commercial environments but may become necessary in the future.

Access to transmission facilities is harder to prevent because such facilities are outside the protected environment of the terminal or computer center. Even in a private leased network, part of the message transmission path may consist of public lines. As the messages enter and leave a common carrier facility, they are susceptible to intrusion. In such cases, methods other than physically securing the lines must be used to prevent unauthorized monitoring of data transmissions. Alternatives include encrypting transmissions and designing protective features into each message.

Preventing Operational Access. If prevention from unauthorized physical access to the installation fails, the next level of security must prevent operational access to the system. The following safety features should be considered during the design of the data communications network:

- Computer-controlled terminals.
- Stringent sign-on and sign-off procedures.
- Terminal and operator identity verification by computer.
- Input and output message control (sequence numbering).

Because an intruder can still gain access to the main computer or control center, second-level safeguards must be in place, including:

- Physical separation of computer control from network control.
- Operator password for computer access.
- Restrictions on the types of data that can be sent or received at the control centers.

Operational access to stored data can be restricted through the use of key and password file protection for tapes and disks. In some cases, data can also be encrypted.

Protection from Unauthorized Monitoring. Data communications systems are monitored for two reasons: to scan for a particular transaction or to gather data (e.g., about overall operation, volume, or financial statistics). Operational or financial data is usually available later in its organized form (i.e., after it has been collated and analyzed). Monitoring lines to accumulate raw data would be much more difficult than obtaining it from personnel with access to the data in its final form.

Protection against scanning for a particular message lies in making the use of monitoring equipment uneconomical compared with other, more easily countered methods or making the data unusable in its transmitted form (e.g., by using data encryption). The data transmission path is unprotected outside the restricted physical environment of the switching center and is vulnerable to monitoring at any point. Because an intruder may access the transmission lines, the data itself must be protected. Design considerations for the protection of data include the use of:

- Message encryption.
- Multiplexed transmission lines.
- Synchronous, continuous data streams.
- The highest-speed transmission facilities feasible or available.
- Alternate paths and rotary line configurations.
- Packet switching technology.
- Satellite transmission.

Data encryption provides the highest level of security. The use of physical hardware devices on each line, cryptographic programs in the computers, or a combination of these methods involves trade-offs. Each alternative must be examined in terms of the level of security required.

The other safeguards are easier to breach than encryption. However, they limit accidental monitoring or casual scanning of the data being transmitted by requiring more complex equipment and techniques. Although safeguards can be chosen on the basis of technical and economic factors, communications lines and procedures play an integral part in maintaining network security and operational readiness.

More than 70% of all voice and data transmission in the US is accomplished by radio transmission, including terrestrial microwave and satellite. With microwave transmission, the exposure area may encompass several miles. The area in which a satellite broadcast can be received encompasses thousands of miles. Because of the growing availability of receiving equipment and the near impossibility of discovering eavesdroppers, it is increasingly important to encrypt radio transmission to prevent disclosure of sensitive information.

Requirements for Component Security and Reliability. The basic physical components of a data communications system are the computers, the network, and the terminal facilities. Each component is vulnerable (to a varying degree) to operational failures or security violations. The factors listed in Exhibits I-3-1, I-3-2, and I-3-3 must be considered during system design if the vulnerability of system components is to be reduced.

Physical Design Requirements	Operational Readiness of Equipment	Computer System Architectural Design
• Location • Physical construction • Entry facilities • Fire and other safety requirements • Equipment layout • Air-conditioning requirements • Site security (e.g., guards, fences) • Electrical supply (protection against power and irregularities)	• Alternate hardware requirements • Emergency power, contingency air conditioning, backup voice and data communications capability • Maintenance schedules and availability • Equipment reliability history • Access to spare parts, both on site and at a remote parts depot • Diagnostic test equipment and personnel availability • Disaster recovery plans	• Redundancy of peripherals and mainframes • Types of system architecture (shared-load or fully equipped standby designs) • Data storage medium • Encryption devices for data storage • Operator and maintenance console security (e.g., magnetic card sign-on devices, key locks)

Exhibit I-3-1. Major Factors in Computer Center Security

Transmission Path Characteristics

- Network selection: wide area network, metropolitan area network, local area network, public or private packet switched, private microwave, local bypass, cable TV Type of transmission media: twisted-pair cable, coaxial cable, fiber-optic cable, terrestrial microwave, satellite transmission
- Contingency circuits and alternate routing
- Error monitoring and network control
- Leased-line or public switched dial-up circuits
- AT&T or other common-carrier circuits

System Design Characteristics

- Network protocol: asynchronous, bisynchronous, X.25, X.32, carrier sense multiple access with collision detection (CSMA/CD), token passing
- Transmission speeds: 300-, 1,200-, 2,400-, and 9,600-bps, T1
- Topology: star, bus, hierarchical, ring, distributed
- Use of concentrators, multiplexers, PBXs, front-end processors
- System requirements: store and forward, online transaction–oriented, remote job entry, dial up
- Use of dial-back capabilities, encryption, message authentication (Financial Institute Message Authentication Standard [FIMAS])

Exhibit I-3-2. Factors in Network Facility Security

Message-Oriented Security and Reliability

Many factors that influence communications security and reliability are inherent in the design of the messages or traffic carried through the system. The first design requirement must be the selection of an adequate standard code set with appropriate parity-checking or error-correcting codes. Other key requirements that must be considered during system design include message authorization, validation, and delivery as well as the protection of message data and message accountability.

Message Authorization. The authorization process begins with the checks and approvals needed before data is entered into the system. This preparation must be considered part of the total communications security program. At this point, an authorized message in machine-readable form is presented to the system for transmission. The programmed security procedures must include, at least, the following steps in the authorization process:

- Validation of the originating terminal with respect to ownership and the correct station for the line.
- Verification that the station is authorized to transmit at that time.
- Confirmation of the operator sign-on.
- Validation of message format.
- Verification of operator (station) authority to transmit the message type.

Equipment Type	Reliability Considerations	Physical Protection
• Telegraphic	• Contingency equipment	• Terminal location
• Programmable	• Maintenance and spare	• Terminal access
• Computer	parts	• Terminal control
• Encryption devices	• Alternate power supply	mechanisms
	• Software requirements	
	• Error-detection capability	

Exhibit I-3-3. Factors in Terminal Security

- Validation of message-numbering sequence.
- Testing for correct authorization codes embedded in the message.

The procedures that should be followed if a message does not pass all authorization steps must also be considered during system design. Depending on the severity or frequency of the occurrence, the procedures can include:

- Rejecting the message.
- Rejecting the message and notifying supervisors.
- Disconnecting the line and preventing further transmission over it.
- Switching the line to monitor-only status, discontinuing message processing, and notifying supervisors.

Because any errors that the system detects can be part of a pattern indicating an attempt to breach system security, the design must include the capability to detect and log any incidents and provide exception reports to highlight them.

Message Authentication.   Once an authorized originator has entered a message into the system, the procedures to be followed should address the parameters for message authentication. These specific parameters indicating message form and content must be standardized and adhered to strictly. At a minimum, automated authenticity tests for messages should include:

- Positional edits for correct control characters, address and data fields, and line and format constraints.
- Data validation for routing numbers, addresses, type code, and user-specific, content-oriented information.
- Authorization checks for coded data, test words, and other security tests (e.g., multiple identical currency fields).

The financial industry has adopted the ANSI standard X.9.17, which specifies a method for authenticating a transaction's contents. The method involves passing a message appended to the transaction through an encryption process using the Data Encryption Standard (DES) algorithm and then transmitting the block with the message as a control element. The transaction itself is not encrypted for transmission. Upon receipt, the appended message is decrypted using the DES key. The result is compared with the original unencrypted message; if any change occurred, the results will differ, indicating a modification of the message during transmission.

The authentication process increases the level of system security, helps ensure that the data required for delivery processing is valid, and provides an opportune time in the processing cycle to capture data for message accountability. The automated procedures should include the return of a positive message-acceptance indication to the originator once a message has been authenticated.

Message Delivery.   Secure message delivery implies that message routing is verified and checked for validity and authentication and that unauthorized line monitoring does not occur. (Monitored data is unusable.) It also implies that all messages have been delivered and are accounted for and that no message has been altered, duplicated, accelerated, or delayed.

The routing verification procedures should be designed to ensure that:
- The destination is a valid point in the network and is authorized to receive the type of traffic involved.
- A positive connection is made with the station and that the connection is validated before and after message transmission.
- Verification of message acceptance is received from the terminal on delivery, with terminal identification included in the acknowledgment.
- Unbroken, sequential output numbers are transmitted as part of the message.
- A log of all messages transmitted is kept.
- The queuing and routing algorithms enable traffic to be processed efficiently, preventing undue delay of messages in transit.

An additional method for ensuring secure delivery may include facilities for proof of delivery.

In systems that use packet-switching technology, a message can be segmented into more than one packet. Each packet can follow a different route to the reassembly point for transmission to the user or recipient. Because routing within the network varies, the reassembly and alternative routing characteristics are of primary concern.

The use of unique system reference numbers is an important aspect of message protection and identification. A system reference number, however, does not enable a receiving node in a network to know whether it has received all messages addressed to it within a time period that is short enough to permit possible action. The loss of a message will be discovered (possibly days later during the course of business or accounting), but not by means of communications procedures.

To solve this problem, the sending station should enter a number (in sequence, according to recipient) with the message. If a number is received out of sequence, the recipient is alerted that intervening messages are delayed. Although it may be cumbersome, this procedure alerts the operator to the type of failure when a computer switch receives messages and does not detect this type of occurrence.

Sequence validation at the point of receipt can be done online or may be incorporated into close-of-business procedures. Because the individual components of a transmission may arrive out of sequence at the final assembly point (e.g., as in packet networks), the system's ability to reassemble the packets and handle error conditions is of primary importance. Although logical message sequencing and numbering (e.g., as in packet transmission) can still be carried out, even full messages may arrive out of sequence because of the possibility of different routing.

Protection of Message Data.   In communications environments, there is little protection against unauthorized monitoring of network lines. To provide the required degree of security, the system design must include various methods of protecting the data in the message. The use of code words in the message or encryption devices and processes at various points in the system limits the chance for unauthorized modification or reading of messages.

Encryption can be implemented in various ways and by using different encryption algorithms. The best-known method for commercial, nonclassified environments is the use of the DES algorithm, which is approved by the National Institute of Standards and Technology (NIST). The DES is the basis for the Financial Institution Message Authentication Standard (FIMAS) as well.

Public-key cryptosystems, another encryption alternative, do not require the transmission of a secret key, a major drawback with DES and traditional encryption systems. However, public-key cryptosystems are slow compared with DES substitution/transportation methods. Some encryption systems use public keys for DES key transmission and DES for data transmission.

When messages and data are multiplexed or packet-switched, it is more difficult to detect a single message using low-technology devices. However, the equipment and knowledge needed to violate such systems are accessible, and the security of such transmissions should not be overrated.

In an ideal environment free from line interference, a high degree of line monitoring can be used in the system to detect errors caused by potential intruders. In this way, monitoring or interception devices attached to the network can be detected quickly when they disturb the synchronization of the transmission and cause errors.

Message Accountability. Communications system security must guarantee message accountability; once a message has been accepted by the system, it must be delivered (as received) to the proper recipient, and no message may be lost, delayed, or accelerated. The system must be designed so that every message is safely stored on a permanent device from which it can be restored to the active system when required. Multiple copies of messages must be stored to ensure accountability if a device fails. In distributed environments, it is advisable to retain copies of all traffic at all centers. Internal system controls, such as numbering systems or other internal addressing mechanisms, must be designed to permit message data to be retrieved as required. Internal system controls must contain audit trail data that, at minimum, includes:

- Input- and output-station and line identification and sequential number.
- Delivery date and time.
- Number of copies delivered.
- Message status (normal, duplicated).

The system must be designed to require acceptance or rejection of all messages.

The receiving station should confirm identification before and after a message is delivered and should acknowledge delivery automatically. The switching system files should be checked for the previous day's traffic at least once every 24 hours. The system must therefore be designed to accommodate file-aging mechanisms.

System recovery programs must be designed to account for all in-transit messages after a failure and must also be able to restore the active message file and continue service without modifying message priority. Retrieval programs must be designed to ensure that only authorized personnel access delivered messages. Retrieved messages should also be identified as not originals and should be independent of all accounting processes.

The design of a message-trace system should also be considered so that messages inside and outside the system can be traced. The internal trace can be done through intermediate message logging during the switching process; the external trace can be facilitated by the addition of a retrieval capability based on the sender's reference data.

Offline programs should be designed to allow a complete analysis of message logs, which may involve edited printouts of message logs and message-search programs. If security restrictions do not allow network operating personnel to access message data, the programs should be designed to use only control information and message routing data.

Recovery from Communications System Failure

A complete communications system or its various components may fail during normal operations. In the early stages, failures can result from software deficiencies, program malfunctions, or software modifications. Failures can also be caused by hardware malfunctions in mainframes or peripheral devices. Depending on the system design, it may be necessary to shut the system down for a short time during the operating day if changes must be made in major system parameters such as communications line tables or account structures. Recovery from such failures or shutdowns must be included in the system's operational requirements.

The recovery procedures should address the accountability of messages in transit (including those acknowledged by the switching center). Historical data (e.g., retrieval files, message logs) should be safeguarded. Recipients should be informed of the current or potential status of all transmissions following recovery. All messages must therefore be identified as possible duplications, unless the transmitting center has received verification of the messages' delivery and acceptance.

Other critical data (e.g., the current network configuration) must be safely stored if the system is to be rapidly and accurately recovered. Recovery to normal status involves regaining control of in-transit message files, network status, and error statistics; resynchronizing communications lines; reopening magnetic tape logs; and other housekeeping required for an orderly restart. The physical status and temporary routing status of the terminals should also be retained.

As long as an invalid message is in the system, failure will recur. The invalid message, along with other relevant information, should be transferred to the appropriate output device, allowing the system to return to normal. Communications personnel can then examine the data to discover the cause of the malfunction.

OPERATIONAL REQUIREMENTS FOR SECURITY AND RELIABILITY

For a communications system to operate daily in a secure and reliable environment, controls, procedures, and safeguards must be designed into the system. The next step is to ensure that these tools are used effectively. It is important to address the physical aspects of system operations and traffic and message control.

Operational Communications Security

The overall data security program should extend to the data communications facilities. An effective security program provides a realistic, obtainable level of protection. Control measures that are too stringent will not be observed, and this disregard will soon extend to all security precautions. The primary goals should be to achieve a level of control that can be practiced and to observe it strictly. Identity badges should be used and checked, visitors should be escorted through restricted areas, files and materials should be locked, and passwords and codes should be safeguarded.

Fire drills should be scheduled, and the staff should be advised immediately of new emergency procedures. With respect to identification, particular attention should be paid to vendor and service personnel who have access to critical equipment and sensitive operational areas. Security mechanisms (e.g., automatic door locks) should be in working order at all times and should not be disabled for convenience. Controlling system documentation is as important as controlling equipment. Detailed program listings and specifications as well as code books and operations manuals should be secured when not in use.

Operational Reliability

The use, scheduling, and monitoring of daily and periodic maintenance programs is vital to a communications system's reliability. During the design stage, equipment reliability should be a primary consideration; other reliability factors correspond to equipment use and maintenance. The daily operational plans must provide for a scheduled maintenance program.

The operational staff should ensure that there are spare units on site or an extensive spare parts inventory at a remote depot, particularly if the system is heavily loaded. The staff should also frequently check system performance, use, downtime, and repair time and monitor input power and environmental conditions, such as temperature and humidity. Maintenance programs and parts inventory for remote locations must be readily available. The test equipment required by the maintenance personnel is as important as the personnel are.

Because of the tremendous increase in the use of communications in business, disaster recovery or contingency planning must be addressed from both an operational readiness and a design perspective. When systems are developed, contingency alternatives must be specified in the original concept. Because communications circuits may be expensive to back up, system design may require nontransmission capabilities (e.g., moving magnetic tapes) to provide emergency backup in the event of a disaster.

Communications System Operations

Program changes, a major factor in the daily operation of a system, significantly influence system security and reliability. Several factors contribute to providing the necessary levels of procedure and control:

- Complete control of all program change requests with multiple authorizations.
- Full documentation in accordance with published standards for all programs and procedures.
- Independent software to test changes and provide for operational and staff acceptance before changes are introduced into the system.
- A procedure for handling emergency situations in a secure and efficient manner.

Daily responsibilities of the system's operational personnel include:

- Investigating all error messages, reports, and alarms.
- Monitoring communications lines for failures and problems.
- Monitoring network status for operational, out-of-service, or alternatively routed stations and lines.
- Monitoring traffic queues for congestion.
- Controlling tapes, disks, and other system materials to ensure proper labeling and retention.
- Maintaining backup for programs, tapes, and other material.
- Examining system printouts, program dumps, and recovery printouts.
- Controlling changes to the system's software, devices, terminals, or circuits to ensure that they have been authorized and properly logged.
- Monitoring vendor and maintenance personnel.
- Controlling testing during operational hours.
- Ensuring that all simultaneous changes to hardware and software are necessary.

Traffic and Message Control

A security and reliability program must also address traffic and message control during daily operations, including all aspects of file retention, recovery, and system housekeeping. One daily procedure involves scanning all active message files for undelivered traffic. Although some of these messages can be intentionally deferred for subsequent delivery, others can inadvertently remain in the system as a result of system software problems or an unnoticed fault in a terminal. The procedure provides a list of all traffic pending delivery since the last search and indicates its status and length of time in the file. Operations personnel are directed to take the appropriate action.

Another important aspect of security and reliability in daily operations is the maintenance of off-premises and secure on-premises storage of vital programs and message files. This maintenance procedure involves determining the retention criteria for message logs and the use of encrypted files. Backup files and programs should be reviewed and used regularly to guarantee their readiness in an emergency.

In most systems, it is advantageous to limit the time during which data must remain confidential. Six months after a transaction occurs, access to the data may no longer be limited. The same may be true only one hour after a financial transaction has been settled and all accounting has been completed. Time limits can provide managers with a more informed view of the best method for providing security.

In message control, all message errors should be reconciled daily, as should traffic delivery figures for messages sent and received. The use of offline reconciliation programs to scan message logs and balance traffic received against traffic delivered can aid in this task.

SUMMARY

A secure and reliable communications system is possible only if a view of the total system environment is maintained. The security effort should be distributed and realistic.

The design of the communications system plays a significant role in providing the operational tools needed to maintain a secure environment. If security controls are not implemented during the design of a system, it will be extremely costly or even technically impossible or infeasible to add them later. Communications system security depends on an early, well-considered commitment to its requirements.

Chapter I-3-2
Security in Open Communications Architectures

A HIGH SCHOOL STUDENT GAINS illegal access to a service bureau and corrupts the files of its client users. A worker intercepts financial transmissions and diverts millions of dollars into a fictitious account. A seemingly innocent Christmas message, broadcast by electronic mail, is retransmitted, in turn, to the mailing lists of each recipient, clogging the system and denying service to users. Only the more sensational examples of breaches of computer security are reported in the press, but they amply illustrate a problem of increasing concern to both the consumers and suppliers of computer services: protecting information that is processed and stored in—and exchanged between—computers.

The need for information security has long been recognized and addressed with respect to national security and other sensitive applications. The preceding examples underscore the vulnerability of information and its related resources in the context of common business applications. As the use of computers, terminals, and data communications has grown, so has the risk. Part of this increased vulnerability can be attributed to the dramatic rise in the volume of information being processed. A second, and probably more significant reason, concerns recent changes in the nature and extent of information processing.

Processing that was once done at a central site on expensive machines to which physical access could be easily controlled is now, more often than not, widely distributed, taking place on relatively inexpensive equipment. Not only is access to these machines often uncontrolled, but they, together with their data, can be stolen quite easily. The extent to which data transmission has become commonplace, even for inexpensive home computers, makes unauthorized access to remote business machines a not-insurmountable challenge. Furthermore, the accelerating trend toward global communications and worldwide connectivity has heightened interest in measures to protect these resources.

This chapter is chiefly concerned with security in computer networks and open communications architectures and in particular with the open systems interconnection (OSI) reference model, an architectural framework that is receiving worldwide recognition and support. The intent is to define security in this context and to review the fundamental concepts, highlighting those of particular concern to the OSI model. Although the focus here is security in data communications, network security cannot be divorced from end-system security. Unless this interdependence is understood, security measures invoked in one area may be rendered ineffective by a lack of complementary measures in the other. Reference is therefore made to security measures that are external to, but still support, communications security.

NETWORK SECURITY CONCEPTS

Assets, Threats, and Vulnerabilities

The security concepts and techniques described in this chapter apply to processing environments that are not constrained by national security requirements (although, in many cases, the methods employed are similar). The objective of the security effort is to minimize the risk of exposure of assets to certain threats or to reduce the probability of known threats exploiting certain vulnerabilities.

In a computerized environment, assets consist of data (in transit or in stored form) and physical equipment—computers, modems, lines, switching equipment, and software, for example—as well as data communications and data processing services. In addition to these tangible assets, security protection can encompass the availability and survivability of a system, ensuring its ability to continue a predefined level of processing in the event of serious system failure or a natural catastrophe.

The various threats to a system can be categorized as follows:
- Unauthorized destruction of an asset.
- Unauthorized modification or corruption of an asset.
- Loss of an asset.
- Unauthorized disclosure of information.
- Disruption of service.

Threats may be deliberate or accidental, and they may be active (i.e., resulting in a change to the system) or passive (i.e., no change to the system). An example of a passive threat involves browsing through confidential files with the intent of copying information without modifying its content.

Certain threats originate within the organization, typically when a legitimate user behaves irregularly or exceeds a prescribed level of authorized system use. Most computer-related crime, for example, is believed to rely on inside help. Inside attacks often involve modification of the system, although passive attacks, such as gaining unauthorized access to confidential records, are common.

Threats that originate outside the organization are more likely to involve eavesdropping on data transmissions, other nondestructive interference with data flow, and obtaining information or system access by means of false credentials.

A system is considered vulnerable if it is susceptible to any of the preceding threats. Communications systems and their media are subject to reliability problems, to degradation, and to environmental disruptions in addition to the threat of unauthorized use. Each of these is considered a specific vulnerability against which a system must be adequately protected. Defining this adequacy usually entails tradeoffs that yield something less than full protection.

Concerns in a Data Communications Environment

In many cases, the very fact that data is being transmitted is what renders it vulnerable. Any time data is sent over a public network—or over a private network if part of that network extends beyond the user's premises—it must traverse a series of links and carrier equipment, each of which is vulnerable to certain threats. Local area networks, which are usually wholly under user control, also present security problems because gaining access to the network is often an easy matter. Routine network maintenance and monitoring and the use of such management tools as line monitors and recorders can provide easy cover for the unauthorized observation of network traffic and collection of data.

Much valuable information can be obtained simply by monitoring transmitted data. In most cases, monitoring data passing through communications media is a relatively easy task (fiber-optic cable being a notable exception). The use of microwave and radio-relay stations makes long-distance transmission vulnerable to radio interception. Furthermore, all electromechanical devices, including disk drives, printers, terminals, and telephone switches, give off emanations from which the data stream often can be reconstructed.

Although eavesdropping is generally considered a passive threat, it can provide the basis for attacks whose purpose is to redirect information or to obtain authorization for unauthorized activity. For example, by recording and subsequently replaying a valid authentication sequence, it may be possible to gain access to a system. Impersonation—masquerading as a legitimate user—usually entails some form of either replay or substitution of valid data.

Disruption of service can take several forms: preventing receipt of messages at a specific destination, blocking the systemwide transmission of messages, and diverting messages to a different destination are examples. Techniques for disrupting service include flooding the system with spurious traffic and interfering with network relay functions.

Trojan horses and trapdoors are two decidedly devious threats that lie within the system. They can be written by insiders or implanted by outsiders as a result of hardware and software purchases or routine maintenance.

A Trojan horse causes an authorized function to effect an unauthorized activity—for example, forwarding a message to an unauthorized destination as well as to its legitimate destination. A trapdoor sets off an unauthorized activity when a certain sequence of events is detected. A disgruntled employee, for example, may arrange for the deletion of vital files should the employee's name disappear from the payroll.

Threat Assessment and Risk Analysis

The application of security measures to counter specific threats includes an evaluation of the risk involved and the cost of protection. Cost must be weighed against the value of the asset to be protected. In general, it is not desirable to spend more than the asset is worth in protecting it. Security measures should make the cost of breaching security greater than the benefit. (This assumes that the intruder is motivated by financial gain rather than by, for example, revenge or personal satisfaction.)

Security can be thought of as a kind of insurance. Risk analysis and threat assessment are used to determine the extent of risk and the level of protection that can be justified. Indeed, the purchase of insurance may prove a valid response to some threats.

NETWORK SECURITY TECHNIQUES

Protection against threats to systems and their data is achieved by the application of security mechanisms. Countering a specific threat may require one or perhaps a combination of mechanisms. The objective of protection may be to prevent a breach of security or simply to signal when a violation has occurred. Other mechanisms are used to recover in the wake of a security violation. Security mechanisms thus fall into one of three camps: prevention, detection, and recovery.

Cryptographic Techniques

Cryptographic techniques constitute an important class of security mechanisms. Cryptography can be applied to conceal the content—or even the presence—of data, either in storage or during transmission. Cryptographic techniques can play a significant role in maintaining data confidentiality, integrity, and authentication.

Confidentiality of data—that is, its concealment from those not authorized to see it—is often achieved by cryptographic means. This technique involves applying a cryptographic function, or algorithm, to the data, which in its unencrypted form is said to be in cleartext. The output of the encryption process, referred to as ciphertext, cannot be readily processed by an unauthorized user, nor can the meaning of the data easily be uncovered. The algorithm selected for the encryption process must be robust enough so that the cleartext cannot easily be recovered—that is, deciphered—from the ciphertext without knowledge of the cryptographic parameters. The processes of encryption and decryption are illustrated in Exhibit I-3-4.

Two types of encryption are commonly used for encoding information in commercial applications that require such protection: private-key (symmetric) encryption and public-key (asymmetric) encryption.

Private-key encryption transforms data by means of a cryptographic algorithm (one that need not be kept secret) and a binary number, called a key, that is known only to the sender and receiver of the data. The best known of the private-key algorithms is the Data Encryption Standard (DES), which has been designated a federal information processing standard. DES uses a 64-bit key, 56 bits of which are for encryption and the remaining 8 bits for

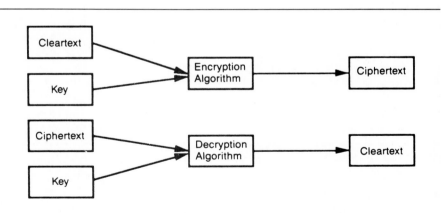

Exhibit I-3-4. The Encryption and Decryption Process

error detection. The output of the process is determined by the key used. Because only holders of the key can decipher encrypted data, the key must be protected. Exhibit I-3-5 illustrates the process.

Public-key encryption, in contrast, uses two keys: a private key for decryption and a public (i.e., nonprivate) key for encryption. The private key is computed from the product of two prime numbers; the encryption key is computed from the same numbers, using a publicly available formula. Public-key encryption works because although the encryption key may be widely known, it is not feasible to derive the decryption key from this information. The best-known algorithm for public encryption is the RSA algorithm, developed by Rivest, Shamir, and Adleman in 1977 at the Massachusetts Institute of Technology. Exhibit I-3-6 illustrates the public-key encryption process.

Clearly, keys play a critical role in the cryptographic process, and, for this reason, management of the keys is central to the success of any protective measure that relies on encryption. Key management entails the generation and distribution of keys as well as the control of their use. Some limitation generally is imposed on the lifetime of a key.

Data can be said to have integrity if it can be shown that the data has been neither altered nor destroyed without authorization. To verify the integrity of transmitted data, the source and destination of the data both calculate a cryptographic checkvalue. Similar to a checksum or a redundancy check, this value is computed through the use of confidential cryptographic parameters. Any modification or loss of data during the transmission is detected because the checkvalue computed by the receiver will not correspond with that computed by the sender and appended to the data.

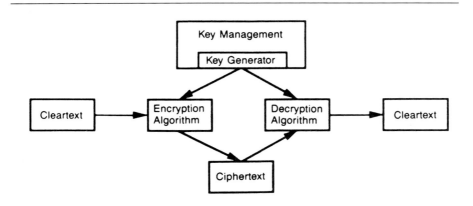

Exhibit I-3-5. Secret Key Encryption

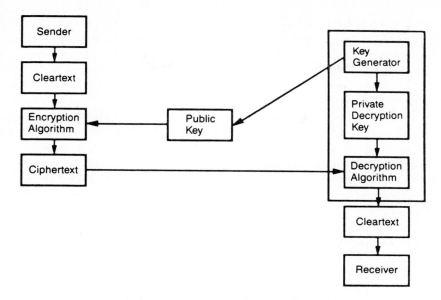

Exhibit I-3-6. Public-Key Encryption

Even data protected by encryption and cryptographic checkvalues is vulnerable. By observing the nature of the data stream and its traffic patterns—frequency, volume, and direction, for example—an intruder can deduce useful information, such as the location and content of certain fields, from which authentication information can be derived in the attempt to gain access to the system. In many cases, therefore, the protection of information on traffic flow is desirable.

Traffic-flow security mechanisms involve encryption not only of the data stream but of the gaps between data. Spurious traffic is inserted in these gaps before encryption, thus concealing not just the data but its volume, frequency, and pattern.

Full traffic-flow security can be implemented only with end-to-end encryption. It is not achievable in a datagram environment, in which messages are sent individually, or in networks in which the addressing information must appear in cleartext at each relay. One word of caution: the use of traffic-flow security measures on networks whose charges are based on the volume of data transmitted (e.g., by block or packet) can prove very expensive, because of the cost of transmitting the spurious traffic between data.

Verification Techniques

Certain security mechanisms are designed to verify the source of a message and its receipt. A digital signature, for example, involves signing a data unit with a private code or key. This signature cannot be created by anyone other than the keyholder. The receiver can thus confirm the source of the data; correspondingly, the sender cannot later deny having sent it.

Notarization mechanisms, for their part, rely on a trusted third party (e.g., an arbitrator or notary) to attest to the accuracy of certain properties concerning the data, such as the time of sending or receipt or its origin.

Access control mechanisms restrict resource access to those who are so authorized. Account codes, passwords, and tokens are among the commonly used techniques. Authorization guidelines generally are established in the security policy, and the basis for authorization is considered the distinguishing factor between different approaches to policy. For example, one type of policy, called identity based, permits access on the basis of the need to know; another, called rule based, grants access based on the sensitivity of data and the clearance level of the individual seeking access.

Authentication techniques seek to verify the identity of the partners in a data communications exchange. The specific techniques vary with the type of exchange and the protection required but generally include passwords, cryptographic techniques, time stamping, handshaking techniques, digital signatures, and notarization.

Using passwords is considered a weak authentication technique. Although it provides protection against accidental error, password protection is vulnerable to deliberate attacks, such as replay attempts. Passwords usually are used for authentication only when both the communicating partners and the communications facilities can be trusted. Protection against active attacks requires the use of encryption for authentication. Replay attacks can be detected, though not prevented, by the use of time stamping and synchronized clocks. Three-way handshakes can also be used to provide mutual authentication and to protect against replay. Nonrepudiation and digital signatures are appropriate authentication techniques when the communicating partner is not trusted.

The notion of trust entails confidence that hardware and software will function correctly. The demonstration of trust is a complex process not unlike proving the reliability of a computer program and involves the use of formal proof techniques and verification and validation procedures.

It can be important to ensure that the channel over which data is to be sent is trustworthy. If routing controls are implemented, the specification by the sender of certain routing caveats can verify that the paths over which the data is transmitted possess the degree of protection required by the sensitivity of the data.

Emanations Protection

Physically tapping into a communications circuit poses a greater risk of detection than does intercepting radio and microwave transmissions or electromagnetic emanations. As noted, all electromechanical devices emit detectable emanations.

To fully protect radio and microwave transmissions from interception, either the data must be encrypted or transmission must be restricted to secure communications media. Encryption, however, is not a practical solution to the problem of emanations; rather, the only effective way of dealing with emanations is to reduce or eliminate them. This can be done by enclosing all electromechanical devices in a shielded room, an expensive and inconvenient step. A second approach is to purchase equipment that is designed specifically to suppress or prevent emanations. Such equipment is generally identified as conforming to a Department of Defense TEMPEST profile.

COMPLEMENTARY MEASURES

Although network security mechanisms are, for the most part, specific to communications technology, the additional support of broad security measures is essential. Effective network security relies as much on physical security, administrative security, and personnel security as it does on its own specific mechanisms. In many cases, for example, lack of proper personnel and physical security procedures negates the purpose of technical mechanisms. Although a full range of security measures is not required (nor can it be justified) in every instance involving sensitive information, an awareness of these interdependencies is crucial to selecting the most appropriate mix of protective measures.

Thus, some applications may require full cryptographic protection along with supporting physical, personnel, and data security measures. In other cases, it may be sufficient simply to restrict access to the computer or lock up diskettes. In general, as illustrated in Exhibit I-3-7, there is a hierarchy of protective measures. The application of security measures to networks, as in most other areas, involves recognizing the threats to and vulnerabilities of the system, knowing the range of protective measures available to counter the threat, and then making a judgment regarding the degree of protection to be implemented, bearing in mind the risks and costs involved.

Configuration and network management play key roles in ensuring that management has an up-to-date and accurate picture of network activity (e.g., what lines terminate where, what changes are being made to the system, and who authorized them). Audit trails complement this aspect of security control by facilitating the collection of information on network and system activity and thus permitting postincident analysis.

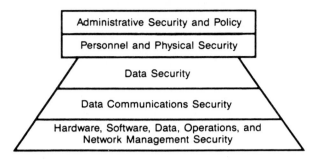

Exhibit I-3-7. The Hierarchy of Security Measures

SUMMARY

Network security should be of concern to all involved in the transmission of data. Awareness of network security concepts—and particularly of the vulnerabilities of the specific system—by those who use and manage the network can help minimize risk. By understanding the application of security concepts to communications architectures, systems designers can incorporate relevant protective mechanisms. Finally, all installations are vulnerable to some extent; but particularly in those that process sensitive data, at least one senior member of the technical staff should be well versed in information processing and network security theory and practice.

Chapter I-3-3
Communications Security Standards

Joseph J. Tardo

DATA COMMUNICATIONS STANDARDS PROMOTE the widespread interconnection of large numbers of geographically dispersed, heterogeneous computing elements. This ability to access information and resources, however, can introduce major security problems and risks if it is not suitably controlled. It is therefore important for data communications standards to include appropriate security features and mechanisms.

The need for security is becoming increasingly acute as larger networks become commonplace. Early standardization forestalls the proliferation of proprietary approaches to security that tend to negate the benefits of a standards-based solution.

Security problems are generally complex and are best approached methodically. High-level requirements and goals should be expressed in terms of security policies that indicate criteria for identification of and access to sensitive or critical information and resources. These policies should also detail circumstances under which individuals can be held accountable for their actions and specific procedures (e.g., dual control) that are designed to institute a set of continuous internal checks or audits.

To plan for and eventually configure a secure system on the basis of standards, a communications systems manager needs to know which security features are necessary to support security policies, which standards exist (or are likely to exist soon) that specifically address security, and which standards have (or are likely to have) the needed security components or extensions. This chapter offers guidelines for assessing specific security needs and focuses on security in general communications facilities, including open systems interconnection (OSI). The standards field is constantly changing as new standards are approved and new groups are formed. The emphasis in this chapter is therefore on approved standards, with a brief mention of ongoing developments.

STANDARDS BODIES ACTIVE IN SECURITY

Almost all standards bodies that are developing data communications standards have some security activity or interest. Some standards bodies, however, exist primarily for the purpose of producing security standards. Representatives of computer manufacturers and communications service providers are the major contributors in these committees, although a number of user groups are also represented.

Within the US, the American National Standards Institute (ANSI) Accredited Standards Committee X3T5.7—Information Processing Security—and the Institute of Electrical and Electronics Engineers (IEEE) standards committee 802.10—Interoperable LAN Security—focus primarily on encryption. Internationally, the International Standards Organization/International Electrotechnical Commission Joint Technical Committee 1, subcommittee 20 (ISO/IEC JTC 1 SC20)—Data Cryptographic Techniques, for which ANSI X3T1 is the US Technical Advisory Group (TAG)—has developed standards for encryption algorithms and for the standardized application of these algorithms.

ISO/IEC JTC 1 SC20 developed procedures for the registration of cryptographic algorithms and standards for both physical link encryption and modes of operation for 64-bit block ciphers. ANSI X3T1—Data Encryption—developed the standard for the data encryption algorithm (DEA), which is essentially the same as the Data Encryption Standard (DES) algorithm published by the US government. Maintenance responsibilities for this algorithm are currently being reassigned within ANSI. The ISO, mostly for political reasons, has elected to publish standards for applications of encryption only, not for actual cryptographic algorithms.

ISO/IEC JTC 1 SC21 Working Group 1—Security Rapporteur Group—focuses primarily on communications security architectural questions relating to OSI and other communications models. This committee was responsible for drafting the OSI security architecture, ISO 7498/2, and is actively working on other security models and frameworks. The US technical advisory group is ANSI committee X3T5.7.

ISO/IEC JTC 1 SC21 Working Group 6—Security Rapporteur Group—is primarily focused on OSI upper-layer security issues. This committee drafted the authentication amendment to the association control application service element (ACSE) standard, ISO 8649/AM 1, and is working on other upper-layer security components, including the security exchange application service element (SE-ASE). The US TAG is also ANSI committee X3T5.7.

ISO/IEC JTC 1 SC21 Working Group 4—Management Security Rapporteur Group—focuses primarily on security issues in the management of OSI networks. The US technical advisory group is the security management ad hoc group within ANSI committee X3T5.4.

JTC 1 security work was reorganized in 1989. A new subcommittee, ISO/IEC JTC 1 SC27, is expected to assume the work program of SC20 but with a broadened scope, permitting it to address some security issues beyond cryptography. This group should provide a focus for coordinating all the security standards development work within JTC 1.

The European Computer Manufacturers Association (ECMA) technical committee (TC) 32 technical group (TG) 9 considers security in distributed open systems and has produced a report on a security framework for open systems. It has also published ECMA 138, Data Elements and Service Definitions. TC32 TG9 is developing a distributed system security architecture.

The International Telephone and Telegraph Consultative Committee (CCITT) Study Group VIII considered placement of security in the telematic services during the study period that ended in 1988. CCITT Study Group VII drafted the directory authentication framework and the specifications for security teleservices in directory systems and message transfer systems included in the 1988 X.500 and X.400 recommendations. During the 1992 study period, security standards are being developed jointly with ISO by the distributed applications security group.

Most standards interpretive bodies have elements dealing with security as well. Examples of these subgroups include the Security Special Interest Group of the National Institute for Science and Technology (NIST) Implementor's Workshop, the security working group within the Manufacturing Automation Protocol (MAP) and Technical Office Protocol (TOP), and the European Open and Secure Information Systems (OASIS) committee.

A table of international standards organizations with their respective responsibilities is given in Exhibit I-3-8.

OSI SECURITY ARCHITECTURE (ISO 7498/2)

Any discussion of data communications security standards invariably begins with the OSI security architecture. Familiarity with this standard is recommended for anyone in the security field; in particular, its extensive list of security definitions is very useful. This section provides a brief overview of ISO 7498/2 and assumes a familiarity with the seven-layer OSI reference model (ISO 7498) shown in Exhibit I-3-9.

General Description

ISO 7498 provides a reference model architecture that serves as the conceptual basis for the development of standards for open systems. ISO 7498/2 extends this basic model to include security-related concepts and architectural elements for circumstances in which communications must

Standards Organization	Committee	Subcommittee	Responsibility
ISO	JTC 1	SC6	Transport, network encryption, and lower-layer security model
		SC18	Secure messaging
		SC21	OSI security architecture, presentation encryption, application authentication, higher-layer security model, and frameworks
		SC20	Cryptographic techniques, digital signatures, and physical-layer encryption
	TC68		Financial message authentication, wholesale, retail, and bank cards
ECMA	TC32	TG9	Open systems security
CCITT	SGVII SGVIII		Directory and messaging security in telemetric services
IEEE	802.10		LAN encryption

Exhibit I-3-8. Principal International Standards Organizations Involved in Security Standards

be protected against threats to security. It identifies basic security services and mechanisms and describes where each should be placed within the seven-layer OSI model. It also provides guidelines for and identifies constraints on the use of these services and mechanisms in an attempt to provide a consistent security approach within OSI.

Basic Security Goals

The communications systems manager must be concerned with controlling access, protecting data, and providing individual accountability and auditing throughout the information processing environment. OSI focuses only on the visible aspects of communication paths, not on end-system measures. The basic communications security goals of the OSI security architecture, therefore, include:

- The prevention of unauthorized disclosure of information.
- The prevention or detection of unauthorized tampering with or modification of information or services, including the substitution,

Layer 7	Application
Layer 6	Presentation
Layer 5	Session
Layer 4	Transfer
Layer 3	Network
Layer 2	Data Link
Layer 1	Physical

Exhibit I-3-9. The OSI Basic Reference Model

destruction, or corruption of information or services (e.g., masquerading) or the reordering or replaying of messages.
- The prevention or detection of the unauthorized use of resources or services.
- The prevention or detection of unauthorized denial of service or interruption of service attempts.

The precise definition of authorization in all of these goals depends on the particular security policy in effect. Two forms of authorization policy are identified in the OSI security architecture:
- Identity-based authorization—Permission is based ultimately on the identity of the requestor, including his or her membership in a group.
- Rule-based authorization—Permission is based on matching security labels and attributes associated with the requestor and the resource.

Identity-based policies are usually discretionary in nature; the granting of access rights is generally under the control of the owner or creator of the resource rather than the system administration. The familiar access controls used to determine individual, group, and organizationwide privileges are an example of a mechanism that can be used in conjunction with an identity-based authorization policy. Rule-based policies, which are generally mandatory and imposed by the system administration, operate so that security-access attributes are always determined by the system according to the rules. Military classifications and clearances are an example of a mandatory, rule-based policy mechanism.

Security Services and Mechanisms

The OSI security architecture identifies a number of security services and mechanisms. These are discussed in the following sections.

Authentication. There are two forms of authentication services. The first, peer-entity authentication, applies only to connection-oriented services and peer entities named within OSI. When service is provided by layer n, it provides corroboration to the next higher layer ($n+1$) that its peer entity

($n+1$) in a particular connection or association is the one claimed. For example, the transport layer can vouch for the identities of session layer entities communicating through a transport connection because these entities are uniquely determined by transport service access points.

The second form of authentication service, data-origin authentication, is similar to the first except that it applies on a connectionless (i.e., single data unit) transfer basis. A distinction is made between these two forms of service because more effective mechanisms can sometimes be employed for peer-entity authentication.

Authentication mechanisms rely on an exchange of information or credentials to prove or substantiate a claimed identity. One familiar mechanism uses secret passwords, in which both ends know the password and one proves its identity to the other by producing the shared secret. The secret is lost if a third party is able to eavesdrop or if the password is presented to the wrong challenger.

A more robust authentication mechanism uses a shared secret function, in which one entity challenges another by presenting data that only the correct remote entity could properly transform. Cryptographic techniques, such as data encryption or digital signatures, can be used in place of a password to demonstrate the knowledge of a secret key without disclosing the secret. It is important to note that each challenge could be made effectively unique so that every response would match exactly one challenge. These responses could not then be replayed, as can encrypted passwords or even encrypted time vectors.

The bidirectional data exchange inherent in connections and associations permits authentication by means of challenges and responses, whereas the single one-way transfer inherent in connectionless data does not. At best, data-origin authentication permits an assumed challenge, such as a time-varying quantity or sequence number, to be used to approximate an identifier that can only be used once. Peer-entity authentication can also provide resistance to the unauthorized replaying of authentication exchanges, whereas data-origin authentication provides only limited indications of misuse.

Access Control.   Access to systems or applications is usually granted on the basis of the authenticated identity of the remote system or application. The OSI security architecture provides access control services at the application, transport, and network layers; some of the security mechanisms available include access control lists, capabilities, and security labels.

Data Confidentiality.   The data confidentiality service protects against the inadvertent or unauthorized disclosure of data in transit. A number of forms of the basic service are identified in the OSI security architecture.

Connection-oriented and connectionless services may be appropriately provided at any layer except the session layer, depending on the particular configuration, because no one layer can provide protection for all contingencies. For example, the most effective protection against an external threat on a single link is usually link encryption. For point-to-point links, encryption at the physical layer usually suffices. Encryption at the physical layer is also the basis of the traffic-flow confidentiality service, which protects against the possibility that an intruder might observe such patterns in communications traffic as the length of messages.

For multiparty links and LANs, or LANs interconnected through bridges, encryption at the data link layer may be sufficient. When relatively secure LANs are interconnected using public packet-switched data networks, network layer encryption between points of subnetwork attachment may be more appropriate. In other circumstances (e.g., for end-to-end encryption over multiple subnetwork hops), confidentiality should be established at the transport layer or above. Selective encryption of fields within a message (rather than the encryption of entire associations or connections) can be provided only by a combination of mechanisms at the presentation and application layers.

Confidentiality can also be provided by mechanisms other than encryption. In the OSI security architecture, the only alternative specifically discussed is routing control. A measure of traffic-flow confidentiality can also be realized by traffic padding—either within messages or with dummy messages transmitted at regular intervals—at the network and application layers.

Data Integrity.   The data integrity service protects against the unauthorized alteration of data in transit. As with the data confidentiality service, the basic service takes several forms.

The connection-oriented integrity service may be applied at the network, transport, and application layers. In addition, connection-oriented data integrity may include automatic recovery (i.e., retry) at the transport and application layers. The connectionless integrity service may be provided, without recovery, at the network, transport, and application layers. As with data confidentiality, data integrity can also be provided for selected data fields by a combination of mechanisms in the application and presentation layers. Specific data integrity mechanisms can include the use of a cryptographic checkvalue, possibly with security sequence numbering or time stamps.

Nonrepudiation.   The nonrepudiation service provides protection against a denial that communication took place. This service may be provided on behalf of the recipient (ie., proof of origin) or the sender (i.e., proof of delivery). Because it depends on particular application semantics, it is provided only in the application layer. Nonrepudiation may be mechanized using a combination of data integrity and digital signature techniques in conjunction with key management (e.g., the archiving of keys). Implementing nonrepudiation will most likely involve the use of at least one trusted third-party notary.

Management. The OSI security architecture also provides a section outlining security management concepts. Primary among these is the security management information base, a conceptual aggregate of all relevant security information necessary to provide OSI security services and mechanisms. The security management information base is a logical concept whose elements may be implemented in various ways in real open systems; it is not itself a specified component. Another concept introduced in the management section is that of the security domain. Although not defined precisely, a security domain is a collection of entities that are under a common security administration and subject to the same security policy.

The OSI security architecture identifies several specific systems security management activities, including security event-handling management, security audit management, and security recovery management. It also describes a number of security mechanism management functions, including the management of cryptographic key material, access control, authentication, and routing control.

Impact and Acceptance

The influence of the OSI security architecture standard was felt even before it was formally published. The standard is considered by many to be a landmark public document on network security. It provides an extensive list of definitions, but more important, it also provides a basis for other work in the area, most notably the Department of Defense (DoD) Secure Data Network Study (SDNS) and OASIS, as well as continuing work in security frameworks and models within ISO.

ECMA AND STANDARDS FOR SECURITY IN OPEN SYSTEMS

As the ISO was developing the OSI security architecture, ECMA published technical report TR/46 on security in open systems. TR/46 was developed by ECMA task group TG9 on security under ECMA technical committee TC32, whose overall area of responsibility is communications, networks, and system interconnection.

TR/46 identifies the following functional requirements for security:
- Access control, including authentication, authorization, and proxy access (permitting entities to act on one's behalf).
- Resource protection.

- Information protection in processing and storage as well as during an actual interchange.
- Management, including administration, auditing, and recovery.

TR/46 focuses on security functions necessary for building secure distributed open systems applications. In particular, its security model is based on the approach of subjects accessing objects, with access mediated by security facilities according to user and application policy requirements. Conceptually, it identifies 10 security facilities:

- The subject sponsor—The processing element or elements acting directly on behalf of the user.
- Authentication.
- Association management.
- Security state.
- Security attributes management.
- Authorization.
- Interdomain.
- Security audit.
- Security recovery.
- Cryptographic support.

The report also identifies supportive security applications as possible mechanisms for developing components within a security architecture. Supportive security applications are application processes that follow the client-server model described in TR/42's framework for distributed office applications. These processes may be used to provide or manage security functions within a security domain or to provide support for secure interactions between domains.

This work is in ECMA-138, which describes data elements and services used to implement the concepts in TR/46. It provides a much more detailed model based on security policies and domains, including security services interfaces and protocol security attributed data structures. These are cdde-scribed using abstract syntax notation, in ASN.1, per ISO 8824. The authentication service has six primitive operations (i.e., AA-Authenticate, AA-Change-Authentication, AA-TerminateAuthentication, AA-CheckID, AA-Recovery, and AA-Management) that use security attributes such as privilege attribute certificates and integrity class.

Relation to ISO 7498/2

The ECMA work builds on the ISO work by providing a conceptual basis for addressing security in the upper layers as well as specific protocol definitions to support this model. It also deals with distributed (multiparty) applications. Many felt that the OSI security architecture was limited by the scope of the reference model, which stresses the pairwise aspects of the interconnection of open systems.

Communications systems managers should consider TR/46 as containing mostly valuable tutorial information for understanding the problem and a potential solution methodology for eventual distributed application security. The definitions in ECMA-138 will eventually be candidates for inclusion by reference in other applications protocol standards.

CRYPTOGRAPHIC TECHNIQUES

The first published standards dealing with security techniques defined a particular encipherment algorithm, the data encryption algorithm (DEA). DEA (ANSI standard X3.92-1981) was essentially the ANSI version of the US government's Data Encryption Standard (DES), published by NIST; subsequent ANSI standards were initially based on this particular algorithm. Once work on cryptographic techniques moved into the international arena, however, the focus shifted to multiple algorithms.

In addition to encryption algorithms, integrity mechanisms, digital signatures, authentication exchanges, encryption in communications protocols, and key management have been the subject of standardization work.

DEA and DES

Considerable controversy has surrounded the Data Encryption Standard and data encryption algorithm since their inception. Detractors have claimed that the algorithm was deliberately weakened during its review process so that the length of the key is shorter than originally proposed and that it has a trapdoor that allows knowledgeable parties to easily decipher transcripts. No hard evidence has emerged, however, that the DES is vulnerable to any form of known cryptanalytic technique other than the brute force method of trying all possible keys. Furthermore, the US Department of State, which controls the export of all encryption equipment under the International Traffic in Arms Regulations, tends to impose tighter constraints on the export of the DES than on other proprietary algorithms. NIST continues to reaffirm the DES whenever it comes up for periodic review; it is published as Federal Information Processing Standard 46 (FIPS-46) and is available from the National Technical Information Service.

The general concept of standardized, public encryption algorithms itself has been controversial. Proponents argue that because a standardized algorithm is used by vast numbers of experts, it is more likely that any weaknesses will be discovered early on. Opponents, pointing to the experience of the Allies in World War II, claim that knowing the algorithm gives the cryptanalyst a major advantage. Furthermore, opponents insist that it is ill-advised to rely solely on one algorithm given the increasing processing power of today's computers. For the DES in particular, the possibility of winning a brute force attack against its 56-bit key becomes more realistic with each advance in supercomputer technology.

Within the international standards community, the opponents of the DES have prevailed; political considerations intervened when ISO JTC 1 SC20 attempted to publish the DES as ISO 8227. Efforts to publish the draft international standard, although approved through the usual technical balloting process, resulted in an ISO Central Secretariat decision not to standardize cryptographic algorithms. Even without an official ISO document, however, the DES has been incorporated into a number of subsequent standards, in particular for financial applications.

Modes of Operation

The DES is a block cipher that provides a functional one-to-one mapping for each key from 64-bit vectors onto 64-bit vectors. Problems arise when larger blocks of data are encrypted because repeated patterns in the data that span 64-bit blocks (e.g., sequences of null or blank characters) produce repeated ciphertext patterns.

The modes-of-operation standards extend the usefulness of DES to such situations as data communications streams. Modes of operation are essentially procedures for combining cleartext and ciphertext using simpler operations, such as bitwise conditional vector complementation (exclusive-OR). NIST initially published three modes of operation in FIPS-81, and these were carried over intact to ANSI X3.106-1983. These modes are electronic codebook (ECB)—obvious block-by-block encryption—and two chaining modes that mask repeated data patterns—cipher-block chaining (CBC) and cipher feedback (CFB).

In cipher-block-chaining mode, the previous ciphertext block is combined with the current cleartext block using bitwise complementation (exclusive-OR) before the DES block encryption function is applied. The decryption process requires the two reverse steps: the ciphertext blocks are first passed through the DES and then undergo bitwise complementation with the immediately preceding ciphertext block. Cipher-block chaining can be applied only to messages that are a multiple of the 64-bit block size. To encrypt arbitrary-length messages, a form of passing is generally used. To process the first block of a message, a value known as an initialization vector is used in place of the ciphertext. In CBC mode, either the initialization vector can be sent encrypted in ECB mode as the first block of ciphertext or the same initialization vector can be used for all messages to save transmission. In the first case, a new initialization vector could be chosen for each message, thereby masking repeated transmissions of the exact same message. In the second case, the initialization vector is generally considered key material and is protected as if it were the actual DES key, although its effect is mainly on the first 64-bit block of text.

The CFB mode uses two similar steps but applies them in a different order. Cipher feedback first passes part or all of the previous ciphertext block through the DES and then combines the result with the next cleartext block using exclusive-OR. A new initialization vector is used to initialize each message cipher stream and is usually added to the beginning of each message. Cipher feedback can be applied on data that is blocked into sizes smaller than a DES 64-bit block, as is required for cipher-block chaining. In fact, because CFB block size can be as small as 1 bit, cipher feedback is often preferred for certain kinds of communications applications—for example, those that require encrypting 1 bit at a time.

The ECB, CBC, and CFB modes of operation were extended into the international standards arena after they were published by NIST and were published as ISO 8372 after they were generalized to apply to 64-bit block cipher algorithms. Standards for modes of operation for general bit-length block ciphers are described in ISO draft-proposed standard 10116.

Registry of Cryptographic Algorithms

Following the ISO Central Secretariat's decision not to standardize cryptographic algorithms, SC20 introduced a work item to develop a registry of these algorithms, published as ISO/IEC 9979:1990. The registry provides formal ISO names for registered algorithms, and although it does not require that details of the algorithms be divulged, it does require that the intended application area, interface requirements (i.e., whether it is a block or stream cipher), and a battery of test words be supplied. The registry is intended to provide an alternative to algorithm standardization by encouraging the development of algorithm-independent standards. Appointment of the National Computer Center in the UK as registration authority is being finalized.

Integrity Mechanisms and Digital Signatures

A block cipher such as the DES can be used to compute a cryptographic checkvalue analogous to the way addition and accumulation are used to compute a checksum. ISO 9797:1989 describes such a procedure using an n-bit block cipher as the function block and exclusive-OR for accumulation, much like the CBC mode of operation. Essentially the same procedure is used in ANSI X9.9-1987 for computing checkvalues on electronic funds transfer messages, but X9.9 specifically requires the use of the DES algorithm.

The ISO Central Secretariat ruling also prevented public-key algorithms, such as the Rivest-Shamir-Adleman (RSA) algorithm, from becoming standardized. ISO distinguishes public-key algorithms from such conventional algorithms as the DES by referring to them as asymmetric rather than symmetric. Asymmetric algorithms have two keys—an encryption key and a decryption key—whereas symmetric ciphers have essentially one key.

Asymmetric algorithms are useful for digital signatures. For example, the RSA algorithm permits encryption and decryption to be performed in either order. Assuming that the decryption key is kept secret, anyone with the corresponding public-encryption key can verify the source of a decrypted data buffer.

ISO is currently drafting two digital-signature standards. Proposed standard DP 9796 computes a digital signature with message recovery in which the text to be signed is completely included in the signature and can be recovered during the signature verification procedure. The original text is encoded in a redundant form, making it unlikely that a random string of bits would be considered a valid signature. DP 9796 is self-contained in that it includes the description of the cryptographic procedures for signature computation and does not refer to any other document.

Digital signature with message recovery can be used to sign only data that is no longer than half the block length of the public-key algorithm. Another proposed standard computes a digital signature in which a hash or compressed encoding is signed rather than the actual data, thereby permitting arbitrarily long data (e.g., entire files) to be signed. To verify such a signature, the alleged original text is used to compute the hash, and then the public key is used to recover the hash as originally computed by the signer. The ISO is preparing committee draft 10118 to provide one or more standardized hash functions suitable for use in signature with shadow and other standards.

Peer-Entity Authentication Mechanisms

Several draft-proposed standards are under development for cryptographic peer-entity authentication mechanisms. DP 9798 uses an n-bit symmetric key algorithm and permits a challenge-response exchange. By encrypting a received challenge with its secret encryption key, one entity can demonstrate that it knows the commonly held key without divulging that key directly, as would be the case if a simple password scheme were used. Furthermore, if the challenge is chosen at random so it cannot be reused, the response cannot be recorded and then replayed later. DP 9798 also accommodates the use of a third-party authentication server. DP 9799 is similar to DP 9798 but uses asymmetric algorithms.

Choice of Encryption Algorithm

It may be necessary to decide whether to specify a standard algorithm (e.g., DES) or a proprietary algorithm. Despite continued controversy, DES remains cryptographically strong, and files and communications encrypted using DES are unlikely to yield to direct cryptanalysis. Usually it is the keys that are compromised or subverted, not the cryptographic system itself.

The decision rests on two main issues: interoperability and export. Although the use of DES cannot guarantee interoperability, use of nonstandard algorithms effectively precludes it. Export issues are of greater concern because, except for financial applications, proprietary algorithms are often easier to obtain outside the US and may often be exported or imported with fewer restrictions.

LINK ENCRYPTION

Many well-established standards cover encryption at the lowest two layers of OSI. These are primarily limited to providing the data confidentiality service. Physical-link encryption represents the most generally available form of communications encryption today, with equipment readily available from several vendors. Many such devices comply with existing standards to some extent, providing a useful gauge of their quality, but conformance to standards does not necessarily guarantee interoperability.

For example, none of the standards described in this section address key management procedures (e.g., the secure exchange of working keys). In general, easy key management tends to be the strongest motivation for selecting a particular manufacturer's offering over another, but reliance on proprietary provisions works against multivendor interoperability goals.

FIPS 139 (Federal Standard 1026)

FIPS 139, formerly Federal Standard (FED-STD) 1026, applies to the physical layer only and specifies the use of DES in 1-bit CFB mode to provide data confidentiality. FED-STD-1026 was originally developed by the National Communications System and published by the General Services Administration. As with other federal standards, FED-STD-1026 was developed primarily to meet the needs of government purchasing. It has since been taken over by NIST and republished as a Federal Information Processing Standard.

Essentially, FIPS 139 covers the minimal requirements for interoperability with security during actual data exchange. In particular, it dictates lengths and formats for initialization vectors and describes when to send the vector to resynchronize cryptographically after reestablishing communications. This standard, though not developed for this purpose, can provide confidentiality with data integrity if a higher-layer protocol provides suitable error detection. For example, if the high-level data link control protocol—described in ISO 4335 and ISO 3309—is used, its cyclic redundancy check might provide the necessary integrity mechanism.

FIPS 139 applies to simple point-to-point environments. It includes provisions for both synchronous and asynchronous (i.e., start-stop character) framed transmission, including specifications for partitioning the initialization vector when various asynchronous transmission character lengths are used. To an external observer, synchronous physical-layer encryption provides a continuous stream of random bits, masking the presence of data on the line.

ANSI X3.105-1983

The ANSI link-encryption standard, X3.105-1983, applies to both the physical and data link layers. Like FIPS 139, it provides only the data confidentiality service. Most of the physical-layer provisions are compatible with FIPS 139 but differ in certain details. For example, the ANSI standard places fewer restrictions on the choice of initialization-vector lengths and permits the use of 8-bit cipher feedback, which can be more quickly implemented than a 1-bit cipher feedback.

X3.105 introduces the concept of data encrypting equipment (DEE) situated between the data terminal equipment (DTE) and the network interface data communications equipment (DCE). In most cases, the DCE is connected to a local modem; X3.105 specifies how the DEE should interact with Electronic Industries Association (EIA) RS-449 and RS-232 interchange circuits.

X3.105 also covers encryption with data link protocols. At the data link layer, encryption is on a per-message basis, which permits multipoint physical links in which multiple stations share the same physical link and transmit at different times, using different cryptographic key relationships. Physical-layer encryption alone does not support multipoint links. Data link protocols may be preferable to physical link protocols for certain line conditioning techniques (e.g., statistical multiplexing) or when only a portion of the traffic needs to be protected. The major drawback to the use of data link encryption is that it leaves message pattern and header information exposed; physical-layer encryption provides better protection against traffic analysis.

X3.105-1983 accommodates various options for encryption at the data link layer. The preferred approach simply replaces the contents of the data information field (the I-field) with encrypted data, leaving header fields in the clear. The encrypted data includes a control byte and initialization vector in addition to the actual encrypted data; the I-field is handled as if it were to be transmitted transparently by whatever link protocol is in use, and the data enciphering equipment computes a new frame check code trailer. No integrity check code is performed. This form was advocated by manufacturers who had designed their DEE products to receive data according to the link protocol, encrypt it, and forward the encrypted data to the decrypting device in what amounted to a cascaded data link connection.

An alternative protocol-sensitive approach to data link–layer encryption is full-frame encryption. In this approach, encryption begins at a specified point in the frame and continues to the end. The encrypted portion can include the block check code, which can then function as a form of integrity check. This type of operation is more complicated than the preferred approach because in certain circumstances it can generate bit patterns that can confuse data link equipment. The standard does not specify a solution to this problem, and individual manufacturers of data encrypting equipment have provided their own. Full-frame encryption was advocated by manufacturers who had taken a bit or byte cut-through approach to the design of data encrypting equipment.

ISO 9160

The ISO link-encryption standard, ISO 9160, published in 1987, is essentially consistent with FIPS 139 and the physical-layer components of ANSI X3.105. The ISO standard goes into great detail about the interface between DEE devices and the operation of the data communications equipment and the interface between DTE devices and DEE devices relative to internationally standardized protocols and CCITT-recommended protocols (e.g., V.24). The ISO document also discusses additional initialization-vector signaling methods.

ISO 9160 is algorithm independent, reflecting the ISO Central Secretariat's decision against standardizing encryption algorithms, and is therefore perhaps overly generalized in places. The annex to the standard, however, refers to the operation of the standard with the ANSI DES algorithm and contains most of the provisions of FIPS 139.

FIPS 140 (FED-STD-1027)

FIPS 140, formerly FED-STD-1027, covers the physical security of implementations of link-encryption equipment using the DES. For example, it describes the types of indicators that must be provided and how they should be controlled by a physical lock and key. It also addresses such things as emanation limits, antitampering requirements, battery-backup operations, continuous self-test procedures, and requirements for key protection, including conditions under which key material is to be zeroized.

FIPS 140 is essentially the same as FED-STD-1027, which was originally developed by the Communications Security Organization of the National Security Agency and published by the General Services Administration. The

National Security Agency used FED-STD-1027 as a basis for certifying commercial-grade DES algorithm link-encryption devices as suitable for use by such government agencies as the Treasury Department or the Federal Bureau of Investigation.

Starting in 1988, the National Security Agency shared responsibility for certifying commercial-grade encryption equipment with NIST. NIST is in the process of revising and updating FIPS 140. Much of the National Security Agency's earlier certification process was transferred to its commercial communications security endorsement program, to which a new set of documents applies. The program applies mainly to high-grade products that are capable of protecting both government-sensitive and DoD national security–classified information using proprietary, classified algorithms of the National Security Agency other than the DES. Documents produced by the program will not be published as standards but will, instead, be made available to qualified vendors on an as-needed basis by the National Security Agency.

FIPS 140 remains perhaps the only document of its type and is often cited in discussions concerning the physical security of encryption devices. Because it primarily addresses standalone DEE devices, its provisions tend to be overly restrictive for other types of devices (e.g., embedded option module and board-level devices).

DEVELOPMENTS TO WATCH

OSI Frameworks and Models

ISO/IEC JTC 1 SC21 has embarked on an ambitious program of security frameworks and models that are intended to provide the basis for a uniform, consistent implementation of security mechanisms in layer standards. As currently defined, frameworks describe a general service available in many layers that can be distributed among many systems. Work has begun on frameworks for authentication, access control, nonrepudiation, confidentiality, integrity, and security auditing.

Models, on the other hand, address both the implementation of services within particular layers and the interaction of these layer services. The upper-layer architecture security model, as its name implies, details how such security mechanisms as encryption, authentication, and key management are to work in the upper three OSI layers (i.e., the application, presentation, and session layers). The model will address such areas as when and how authentication procedures are to be invoked in setting up and maintaining associations and will define the security relationships between various system and subsystem entities. An analogous model is under development for the lower layers.

End-to-End Encryption

Activities are planned or are already under way that define end-to-end encryption protocols at the OSI presentation, transport, and network layers. These are discussed in the following paragraphs.

Presentation Encryption. Presentation encryption services are expected to be characterized as under the control of a particular application and will apply to selected data fields rather than to entire connections. SC21 WG6 is studying the issue of how to specify the type of protection to apply to elements in the data stream and how to specify the associated cryptographic keying parameters in each instance.

Transport Encryption. Transport encryption services are expected to apply to entire connections on an end system–to–end system basis. The minimum level of protection would be specified at the service boundary using arguments to service primitives. Administration-imposed protection policies and keying, however, will probably be implemented by means of network management functions.

Of the three layers for which encryption protocol extensions are planned, standards for the transport layer will probably be available first. Two ISO committees—SC20 WG3 and SC6 WG4—are addressing transport encryption; the work in SC6 will be conducted under an approved joint work item involving CCITT as well. The SP4 protocol, published as the result of the National Security Agency–sponsored secure data network study, is a likely candidate for this work.

Network Encryption. Network-layer encryption can protect all communications between a pair of systems, whether between end systems or intermediate systems (i.e., routers or gateways). Network encryption, therefore, can be applied between entire subnetworks. When applied between end systems, network-layer encryption is much like transport-layer encryption: it uses a system-to-system keying relationship, as does SP4, rather than a new cryptographic key for each connection. Protocols for other applications of network-layer encryption—the National Security Agency's SP3 is an example—tend to be more complicated. SC6 is considering drafts for network-layer encryption.

IEEE committee 802.10 is developing a standard for confidentiality, integrity, and access control services for LANs. The standard will most likely operate between the logical link control and media access control sublayers. Compatibility with the suite of 802 protocols is expected. The standard will also include key and layer management provisions.

Security Management

ISO committee SC21 WG4 has an active rapporteur group that is in the advanced stages of developing part 7 of the management information services definition dealing with security management. This document addresses a number of security-related management services, including security event logging and audit management, security recovery reporting, key management, access-control management, routing-control management, and data confidentiality and integrity protection management. The document also details the concept of the security management information base.

Kerberos Authentication

As part of the trend toward open operating system architectures, many organizations have been formed with the goal of standardizing existing implementations of the UNIX operating system. Prominent examples of such organizations are X/Open, the Open Software Foundation (OSF), UNIX International, and IEEE POSIX. These groups generally prefer to incorporate existing designs rather than develop new ones, and as they become more interested in security mechanisms for distributed applications, it seems likely that they will consider Kerberos, the network-authentication and key-distribution service developed under the Massachusetts Institute of Technology's project Athena.

Kerberos requires that a client user first contact the trusted, online Kerberos key-distribution service (KKDC) to obtain a ticket for a designated service. This ticket contains a DES key that is generated by the service for this session, encrypted under the secret key of the service, and bound to additional information, including the identity of the client and the time of the request. The service also returns the session key encrypted under the client's master key. On Athena workstations, the client key is obtained as a function of the user's password, thereby granting a granularity of authentication to the individual user rather than to a system. This key is subsequently used by the client to create an authenticator that, together with the ticket, authenticates the client when the client accesses the remote service.

SUMMARY

Except for physical-layer link-encryption devices, few if any products covered by existing security standards are yet available. Consequently, at least in the short term, a company committed to a standards-based solution to its security problems may have to install nonstandard security products or extensions. This is especially true in the area of authentication, in which a plethora of expensive and incompatible devices is currently being marketed. A company's policy statements should make it clear that such solutions represent temporary solutions to immediate, acute security problems. The emerging security features in standards should be carefully considered when planning long-term solutions.

Chapter I-3-4
Security in Financial and Messaging Applications
Joseph J. Tardo

THIS CHAPTER DESCRIBES SECURITY FEATURES in two important application areas: financial services and messaging and directory systems.

FINANCIAL SERVICES STANDARDS

The standards discussed in this section were written specifically for financial services applications, not for general applications. These documents, however, have often been cited in a wider context. Financial services standards are developed in the US by ANSI committee X9 and internationally by technical committee TC-68, which is at the same level in the International Standards Organization (ISO) hierarchy as JTC 1.

ANSI X9.9: Message Authentication Code

ANSI X9.9, the most widely known financial services standard, is used for computing a Data Encryption Standard (DES)–based cryptographic checkvalue on wholesale (e.g., interbank) electronic funds transfer messages. This checkvalue is known as a message authentication code. X9.9-1987 is often cited as the source for the definition of the message authentication code, even though it also includes quite a few operations specific to electronic funds transfer (EFT) processing.

In the standard, electronic funds transfers are specified—in a form analogous to an abstract syntax (i.e., independent of representation)—as strings of uppercase characters, numerals, and certain punctuation marks. These characters can be sent coded in many ways (e.g., by teletype), and a single EFT might be coded according to several schemes as it is relayed through

215

an interbank network. It is still recoverable, however, as a string of characters, with special characters delimiting the different message fields.

X9.9 contains editing rules that specify how fields are to be selected and encoded into a buffer of bits for the message authentication code computation. It includes a table that provides an 8-bit ASCII encoding for each character. The actual message authentication code computation uses the DES in cipher-block chaining (CBC) mode with a zero initialization vector and performs all the encryption steps, discarding all but the final ciphertext block. In effect, computing the message authentication code is analogous to computing a checksum except that the DES algorithm is used as the function instead of addition and exclusive-OR is used to accumulate the running checkvalue.

The ISO version of X9.9, ISO 8731/1, reflects an earlier version of the standard. Both ANSI X9.9 and ISO 8731/1 provide for data origin authentication with single datagram data integrity (in ISO 7498/2 terminology) but without confidentiality.

X9.17: Key Distribution

ANSI X9.17-1985 defines key distribution procedures and protocols that allow wholesale interbank electronic funds transfer systems to support X9.9 and companion standards. As with X9.9, X9.17 is often cited outside of the financial services context as an example of a general approach to key distribution and has as a result occasionally been required in circumstances in which not all provisions of X9.17 were desirable or appropriate.

Because key-distribution messages must use the same communications links as electronic funds transfers, they are specified in the same abstract syntax form. The standard defines message-exchange types for three environments:

- Point-to-point—A message exchange in which one terminal requests a data encryption or authentication key from a peer with which it already shares a master key-encrypting key. The peer generates the working key and then returns it encrypted under the shared key-encrypting key.
- Key-distribution center—A message exchange in which peer terminals do not share a key-encrypting key with each other but share their own key-encrypting key with a third-party service, the key-distribution center. On receiving a key request, a terminal requests that the key-distribution center generate a working key and return it encrypted twice, once under its own key-encrypting key and once under the initiating requestor's key-encrypting key. The terminal decrypts its part to recover the working key and forwards the other part to the initiator.
- Key-translation center—A message exchange in which the receiving peer terminal does not share a master key with the initiator but gen-

erates the working key and sends it encrypted under a key shared with the key-translation center. The key-translation center, which knows the initiator's master key, reencrypts it for the initiator and then relays it back to the peer.

X9.17 also defines ancillary procedures that support key management. These include the use of double-length keys, triple encryption, key notarization, and the use of key counters and offsets.

ANSI X9.23: Message Confidentiality

ANSI X9.23 provides for confidentiality, without integrity or origin authentication, for an electronic funds transfer communications environment. Essentially, X9.23 provides rules for encoding characters into ASCII, padding messages to multiples of 8 bytes, and encrypting using the DES algorithm in CBC mode. The procedures apply either to entire messages or to selected, delimited fields. Of particular importance in this standard are its filtering rules for the transparent transmission of 8-bit quantities over traditional interbank, character-oriented media (e.g., teletype).

ANSI X9.26: Access Management

ANSI X9.26 standardized the use of the DES algorithm for node authentication and for encoding personal authentication information such as account numbers and personal identification numbers (PIN). An international version, DIS 10126-1, is being prepared.

ANSI X12.42: Electronic Document Interchange Authentication

This document was jointly produced by the X9 committee and the X12 committee on electronic document interchange (EDI). It provides editing rules for X12-standardized EDI message formats that are analogous to those of X9.9 and X9.23. It defines ways to include message-authentication codes in EDI messages, for encrypting designated fields, and for including X9.17 cryptographic-service messages in electronic documents. The standard provides for mechanizing message protection—which may be applied at the functional group level, the transaction set level, or some combination of the two—using security header and trailer segments defined in X12.42 (which refers to X9.9, X9.23, and X9.17 for service and mechanism definitions).

CCITT RECOMMENDATIONS

The International Telephone and Telegraph Consultative Committee (CCITT) is empowered, by international treaties, to produce documents that member countries are committed to follow, whereas compliance with

ISO standards is purely voluntary. In practice, however, when CCITT makes recommendations, compliance may often be negotiable, and many countries adopt ISO documents rather than develop their own compulsory national standards—as, for example, in government purchasing. ISO and CCITT have agreed to avoid duplication of work and to avoid issuing technically incompatible standards when their respective standards and recommendations overlap in scope. Achieving this goal, however, has sometimes been complicated by differences in the organizational structure and work programs of the two organizations.

CCITT gave serious attention to security considerations when formulating its recommendations. In particular, CCITT added extensive security services and mechanisms to its recommendation on directories and message-handling systems. It also added cryptographic-authentication exchange to a protocol for accessing public packet-switching data networks.

In certain areas, CCITT has taken the lead in developing standards, although with full ISO participation. However, because the two organizations operate under very different procedures—CCITT is on a four-year cycle, whereas ISO uses a ballot-by-ballot procedure with different member body organizations authorized to approve items at different levels—the texts of the corresponding ISO and CCITT documents are not always exactly equivalent.

CCITT X.500 and ISO 9594: Directory Systems

The CCITT X.500 series of standards for directory services is also issued as ISO 9594. Optional security services are specified in X.500 through X.507. A framework for authentication is presented in X.509 (also ISO 9594-8).

The goal of the X.500 standards is to make possible a large-scale, globally distributed data base of information to facilitate communication. Part of the goal is to provide a basis for developing a worldwide, automated directory assistance containing network terminal addresses for sending electronic mail.

X.500 directories have two major architectural components: directory agents and a directory information base. To access information, the end user makes requests to a directory user agent. The directory user agent functions as a client accessing a directory system agent, which has access to the actual information. The address information is intended to be distributed among many directory system agents, and the standards define protocols permitting a variety of ways for the agents to process requests.

Information in the directory information base is indexed using a hierarchical naming scheme. Each entry is a collection of attributes, certain of which are distinguished as the key fields for accessing a data base record. The directory information tree conceptually includes all such distinguished attributes, so a set of distinguished names of entries can be derived by following the vertices of the tree. In practice, the directory should contain one global name hierarchy that branches from a common root into different country names, then through registration authorities and subauthorities to leaf entries that are the actual objects being accessed.

X.509 augments X.500 by adding cryptographically derived authentication information as optional attributes of the object in the directory. These attributes take the form of certificates—collections of information about the actual object named in the entry (e.g., the values of certain attributes) that have been bound together in a certified way using a digital signature. Certificate values are included as attributes of entries. For example, it can be shown that a particular certificate correctly represents the binding of a distinguished name to a public key by virtue of the fact that it has been digitally signed by a mutually trusted administrative entity. X.509 introduces the concept of a certification authority to represent such an administrative entity. Directory user agents and directory system agents include procedures for assembling one or more certificates as needed to support authentication.

Optional Security Services.   The optional security services specified in X.500 through X.507 include:
- Control of access to directory entries.
- Authentication of users accessing the directory.
- Authentication of retrieved values to users.

The standards are written to allow these services to be used for applications other than directory applications. In particular, the CCITT standards for message-handling systems—the X.400 series—rely heavily on these same services for security features.

Although access control is specified as a service, X.500 stops short of defining a mechanism for accomplishing it, leaving this implementation to the local system. Protocols for the exchange of access control information will be standardized, but not the access control mechanisms themselves. Early X.500 implementations will provide only minimal support for access control to directory entries.

X.500 accommodates digital signatures to provide authentication of data origin and to ensure data integrity. The standard allows users to request digital signatures for retrieved values in order to verify that they indeed come from legitimate service components. Similarly, users can sign local directory service agent requests to enable the service agent that is the actual custodian of the requested information to enforce its access controls. This is possible because X.500 permits directory system agents to obtain information from a chain of other directory system agents rather than directly. X.500 does not provide for confidentiality of information, nor does it provide for authentication of error returns or referrals (as when a directory system agent that does not have the information returns a hint about its location rather than passing the request on to another directory system agent).

CCITT X.509 and ISO 9594-8: Authentication Framework. X.509 is one of the few standards that applies public-key encryption technology to the problems of distributing security-related attributes for authentication and controlling the distribution of cryptographic keys in a very large user community. X.509 defines two kinds of authentication: simple, password-based and strong, cryptographic-based. Simple authentication, though much less secure, was included in X.509 because it can be implemented in the short term. Its use is not, for example, affected by export control laws. Password-based systems, however, are vulnerable to a variety of attacks. A system could, for example, misrepresent itself to trick unsuspecting users into revealing their passwords.

For simple authentication, the directory does not maintain the cleartext password; rather, it supports a version transformed by a one-way hash function. The hash function, known by any verifying entity, can be applied so that the password would appear in cleartext only on the local system where it was entered.

Because a well-chosen hash should be irreversible in any practical sense, it should prevent untrusted components in the directory from recovering the password, thereby preventing an intruder from gaining service on a local system. However, there is little or nothing to prevent a determined entity from masquerading as a legitimate directory service component and merely replacing the hashed password.

Strong authentication is resistant to such attacks, permitting identities to be verified without their learning enough to later masquerade as a different entity. In strong authentication, a challenge that need never be reused is selected by the object entity and sent to the subject entity requesting authentication, encrypted under the subject entity's key. This key will be accurate because it appears in a certificate digitally signed by a trusted certification authority. Each entity can look up the other's certificate attribute in the directory information base, or each can exchange certificates directly with the other. An entity is considered authentic if it can prove it knows the private encryption key that will reverse the transform performed with the public key, thereby returning the decrypted challenge. In addition to the two-message challenge/response exchange, X.509 defines one-way and two-way authentication procedures that use time stamps and random number values to prevent the replay of authentication information.

X.509 contains both informal and formal (i.e., abstract syntax) descriptions of certificates and of the service parameters that affect simple and strong authentication. Certificates contain effective dates and expiration dates, algorithm identifications, the issuer's name, and the subject's name and public key. Certificates holding certificate authority or issuer certificates are associated with the names of certificate authorities.

To verify a certificate, an entity needs the public key associated with the certificate authority that issued it. If both the subject and object entities subscribe to the same authority, verification is immediate. In other cases, X.509 provides algorithms to establish an unbroken chain of trust—a chain of certificates that begins with a trusted key and continues with each key verifying the next public key in the chain until a trusted version of the certificate authority key for the sender's certificate is obtained.

Certification authority entries also have a blacklist attribute for listing all the revoked certificates, much the way a credit card company lists invalid or stolen cards. User and certification authority certificates that have been revoked are listed on separate attributes. No foolproof method of revocation exists, however; the approach adopted in the standard represents a compromise. An intruder could, for example, prevent access to the blacklist with an effective denial-of-service attack. The NIST-OSI Implementer's Workshop has provided additional definitions for hash functions and digital signatures suitable for use with X.509.

Although X.509 is intended to be algorithm-independent, only one public-key encryption algorithm is actually known to supply the described services—the Rivest-Shamir-Adleman (RSA) algorithm, which is computationally intensive. Annex C of X.509 describes the use of the RSA algorithm with various key parameters, one of which is a short, common, public-verification key component that reduces the computing required for signature verification.

CCITT X.400 and ISO 10021: Message-Handling Systems

The CCITT X.400 recommendations for message-handling systems include a rather extensive set of optional security services. Message-handling systems provide a facility for users to exchange messages on a store-and-forward basis. The originating user interacts with the message-transfer system by submitting and retrieving messages through a local user agent, which in turn acts as a client to a message-transfer agent server. A message-transfer agent can store and forward messages on behalf of users. The message-transfer system consists of a set of message-transfer agents that cooperate to eventually deliver messages to the message-transfer agent that services the addressed end user's user agent. User agents, in addition, can interact with the message-transfer system through a message store, the primary purpose of which is to accept delivery of messages on behalf of a particular end user for later retrieval by that end user's user agent.

User agents can request that information be returned from the message-transfer system in the form of a receipt—or report—indicating whether or when the message was actually delivered. Users can also submit probes, or dummy messages, simply to obtain a report.

When servicing a request, message-transfer agents can expand distribution lists, split requests (i.e., submit a probe or message to multiple recipients), redirect or reroute messages, and convert to alternative encodings.

The X.400 Security Services Model. The X.400 security services rely in large part on X.509 security definitions and mechanisms. Secure-access management services provide support for controlling access within the message-transfer system. Peer-entity authentication services can be applied on user-agent-to-message-transfer-agent or message-transfer-agent-to-message-transfer-agent links using either X.509 simple or strong authentication and are used for access-control decisions. The security-context service matches security labels on messages and agents to restrict the flow of information to unauthorized or untrusted transfer or user agents.

X.400 provides data-origin authentication services for messages, probes, and reports; these are mechanized using digital signatures as specified in X.509. During X.400 message-origin authentication, the originator of a message or report supplies a certificate along with a token that contains such items as security labels, integrity-check fields, proof-of-delivery requests, and content-confidentiality keys. Some fields of the token are encrypted, whereas others are integrity protected. The originator digitally signs the messages when submitting them.

Proof of submission and proof of origin are included as part of the origin-authentication services; they take the form of reports originating from within the message-transfer system. These reports include such items supplied by the message-transfer system as the originating message-transfer agent's certificate and a token containing a digitally signed message-submission identifier, the time of submission, and the arguments of the submitted message. The user can use the token and certificate to verify the report.

X.400 includes three data confidentiality services: the connection-confidentiality service, the content-confidentiality service, and the message-flow confidentiality service. The standard does not, however, specify mechanisms for implementing these services.

The connection-confidentiality service protects protocol exchanges between message-transfer system components and is provided by the underlying communications service. The content-confidentiality service is applied on an end-to-end basis between originator and final recipient. Message-transfer systems support this service by providing the parameters for transferring key information securely and by providing indications that the data needs to be decrypted on receipt. The message-flow confidentiality service essentially provides double enveloping and message padding to hide address and size information from untrusted components of the message-transfer system.

Three data integrity services—the connection-integrity service, the content-integrity service, and the message-flow integrity service—function similarly to their data confidentiality service counterparts. A fourth data integrity service—the message-sequence integrity service—assures recipients that the messages they receive have not been reordered or replayed.

The data integrity and data confidentiality services are mechanized either in the user agents or in the underlying communications subsystem, not in the message-transfer system.

Nonrepudiation services provide information that can be used to prove to a third party that a particular form of communications took place. These services are directly provided from within the message-transfer system by trusted components. The service protects against three threats: denial by the originator that the message was sent, denial by the recipient that the message was received, and denial by the message-transfer system that the message was ever submitted for delivery. The implementation of nonrepudiation is perhaps the least understood of all of the X.400 services and involves the potential addition of trusted entities (e.g., third-party notaries).

X.400 also identifies a message-security labeling service, intended to be used with the security-context service, that permits messages to be individually labeled on a limited basis. The security-management services primarily support labeling and context services but also permit entities to change their authentication credentials. The register security-management service allows the security administration authority to set allowable user labels in the message-transfer agent; an analogous service is provided for the message store. A change-credentials service is also provided.

Security Aspects of CCITT X.32

CCITT recommendation X.32 defines a protocol by which data terminal equipment (DTE) establishes access to data communications equipment (DCE) on public packet-switched data networks. The motivation for X.32 was to provide dial-up X.25 service. It includes an optional cryptographic challenge/response exchange using public-key encryption, which is mechanized in both the layer 2 high-level data link control (HDLC) exchange information (XID) frame and in the layer 3 X.35 protocols. The appendix of the recommendation includes a description of the algorithm and certificate format, which is not necessarily compatible with X.509.

The intention of the X.32 security recommendation appears to be that the service providers managing DCE devices should provide certificates to subscribers. The burden of computation for authentication rests with the subscriber in that the DTE device must encode a digital signature whereas the DCE device performs a much simpler verification computation.

It is not clear to what extent the X.32 authentication procedures will be supported in the basic product offerings for private X.25 networks.

Privacy-Enhanced Mail

The Internet Engineering Task Force (IETF) issued three documents that defined privacy-enhanced mail protocol elements and procedures for the Internet community. RFC1113, RFC1114, and RFC1115 provide for message authentication and encryption, primarily using the DES algorithm. They also include public key–based key distribution provisions that use the X.509 syntax for public key certificates.

SUMMARY

Achieving a secure system—one that adheres to and enforces stated security policies—requires that policy requirements be taken into account in such diverse areas as physical plant planning and personnel as well as in the selection of communications protocols and services. The use of available security features and mechanisms provided in communications standards must similarly be motivated by policy requirements.

Standardized security mechanisms provide at best only the necessary components within an overall security strategy. Such features do not, merely by their incorporation, guarantee security, nor can one expect standards developers to derive a simple, complete, nonredundant set of security mechanisms sufficient for all or even a majority of circumstances.

Chapter I-3-5
A LAN Security Review

Robert Klenk
Peter Cluck

THE INTRODUCTION OF A LOCAL AREA NETWORK (LAN) into a computing environment introduces some security and reliability concerns. In the single-user microcomputer environment that usually precedes a LAN installation, the protection of data is simpler: one user controls a microcomputer, including that computer's physical security, data backup, and the reliability of the information in the system. In a LAN environment, resources are shared. The system may have a central file server that contains such applications as spreadsheets, word processing, and data base management as well as critical and sensitive data. A LAN turns each system location into a miniature data center, complete with data center security concerns.

In this chapter, based on the suggestions from the risk assessment performed, examples of the control measures as implemented by the Novell Inc LAN operating system called NetWare are discussed. General examples apply to most LAN operating systems, and implementations specific to NetWare are pointed out.

A review of security and control in the LAN estimates the level of:
- Confidentiality of information on the LAN.
- Reliability of information on the LAN.
- Reliability of the network (e.g., is the data backed up periodically as needed?).
- Performance of the network.

THE LAN ENVIRONMENT

The LAN hardware includes the cables and circuit boards that physically connect the microcomputers or mainframes to the LAN and carry messages around the system. The software is the LAN's operating system.

Most LANs transmit data over coaxial cable or twisted-pair wires. Fiber-optic technology has great potential for very high speed data transmission and combined data and video service, but it is more expensive. Optical fiber has a higher bandwidth, better security, no electrical interference, and fewer installation problems.

The network topology is the physical design or structure of the system. In general, topologies can be described as trees, stars, or rings, but in large networks, a combination of these architectures may be used. A star LAN has a central hub and a point-to-point communications circuit to every other device on the network. If the central controller stops working, the entire network stops. This architecture has high overhead cost and vulnerability. Ring LANs pass messages from node to node until they reach their destinations. Each element plays an active role in transmission, so failure of two or more elements can isolate sections of the ring. The tree architecture, also called a bus, arranges elements like leaves on a tree, and the devices share the main transmission medium. Messages are transmitted along the bus to other attached devices. Each topology presents different control considerations.

One role of the LAN software is to provide services directly to the user; another role is to support the applications that run on the network. NetWare designates one machine in the network as the file server. It contains the shared hard disk and runs the network operating system. This ensures data integrity for the network and allows for proper control and management of all network resources.

NetWare's fault tolerance increases the dependability of the LAN by safeguarding against failure in critical parts of the network hardware. For example, NetWare makes two copies of the file allocation tables (FATs) and directory entries and stores them on different disk cylinders. If a failure occurs in a directory sector, NetWare automatically switches to the duplicate directory. When the system is turned on, NetWare automatically performs a complete self-consistency check on each duplicate directory and FAT. Every time data is written to the network disk, the network automatically performs read-after-write verification to guarantee that the data is legible.

LAN SECURITY REVIEW

The first step in performing a security review is to obtain an understanding of the LAN and its applications. NetWare structure and commands are described in several volumes of product documentation.

NetWare is similar to MS-DOS in that both provide commands that allow manipulation of stored data. MS-DOS manages data stored locally in the microcomputer and supports the application running on the microcomputer; NetWare manages network access and data stored on the network hard disk. NetWare does not replace the operating system of a workstation that is still needed for running applications.

LAN software has three major components: the operating system (DOS in this case), the shell, and the file server. The shell and file server are NetWare components in this case. DOS and the shell are present at most workstations. The file server software manages the network resources and synchronizes disk access so that there are no conflicting file requests. The network server runs the file server software; workstations run only the shell and DOS.

A virtual console function, called FCONSOLE, lets authorized users perform file server console operations from any workstation on the network. The user must have the security level equivalent to supervisor to execute commands that affect more than one workstation. The activities performed by FCONSOLE include controlling and shutting down the file server, viewing the server's status, monitoring file-locking activities, broadcasting console messages, purging salvageable files, viewing software product information, viewing a station's console privileges, restricting access of management features to certain users, listing a connection's open files, listing connections using an open file, and viewing the system mapping table, physical disk stats, and disk channel stats.

User Rights and Access

The NetWare SETLOGIN utility is an important tool for customizing the LAN environment. It allows drive mapping to be set up, programs to be run, messages to be displayed, and other functions to be performed automatically. The functions executed with SETLOGIN determine the chain of events that take place during log-in. SETLOGIN can be used to customize the network log-in procedure for each user.

NetWare permits use of the MAP command to map a particular drive letter to particular directories during log-in (e.g., the Lotus 1-2-3 directory can be mapped to drive C, or the WordPerfect directory can be mapped to drive D). Mapping also sets up the search-drive facility, which is similar in function to the DOS PATH command.

The NetWare commands are usually in SYS:PUBLIC; most default scripts include this as a search drive. Only the system supervisor should have access to all NetWare commands, and only those commands that are needed by a user should be available to that user.

Certain files can be executed during the SETLOGIN script. Those files are executed during log-in (e.g., #WP will execute WordPerfect). Similar to executing a .BAT file in DOS, this can be done using the INCLUDE command from SETLOGIN. All or part of a log-in script can be stored in an ASCII text file. This is usually done to facilitate standard log-ins for certain groups of users.

Additional variables in the SETLOGIN script are based on three elements: user names, workstation attributes, and the date and time. The script can be customized to display the date and time and a user's name. Tests can be performed on the workstation or user ID using the script's if-then programming features.

Restricted Commands

Each network operating system has certain commands that are so powerful they must be restricted to authorized individuals. Such commands:
- Add, change, or delete users.
- Establish connection to other LANs.
- Powerfully affect files across the networks (e.g., hide files).
- Powerfully affect the security of the LAN (e.g., alter read or write access in directories).

In NetWare, these commands typically are restricted:
- SYSCON—This enables a system supervisor to modify a file server's directory and security structure and add users, groups, and directories to a server.
- HIDEFILE—Hides a specified file so that it does not appear in a directory search and cannot be deleted or copied.

Exposures and Threats

During a LAN security review, the data security administrator should classify the exposures and threats to the LAN. A major exposure is a network user who is not properly identified. The most common means of identifying users is to issue a user ID and password to each user. Controlling user IDs (e.g., requiring periodic password change requirements) is necessary. In addition, restrictions can be placed on user log-ins, as shown in Exhibit I-3-10.

NetWare provides a specific security function—the SECURITY command, which lets the system supervisor check for possible holes in the network security, identify who has been given the security level of supervisor, and check conformance to password rules. Occasionally, this utility can supply useful and surprising results. The command must be executed by someone with a supervisor security level.

Unauthorized Devices. To prevent an unauthorized device from transmitting on the LAN, the configuration should be verified and as many of the locations visited as possible.

NetWare System Configuration V2.12   Monday June 8, 1992 11:06 am
User CLUCK__P On File Server PENN5

Account Restrictions For User CLUCK__P

	Account Disabled:	No	
	Account Has Expiration Date:	No	
AL	Date Account Expires:		ns
AR	Limit Concurrent Connections:	Yes	
BA	Maximum Connections:	1	
BI	Allow User To Change Password:	Yes	
BI	Require Password:	Yes	
BR	Minimum Password Length:	6	
BR	Force Periodic Password Changes:	Yes	ces
CA	Days Between Forced Changes:	60	ns
CA	Date Password Expires:	August 11, 1992	
CL	Limit Grace Logins:	Yes	s
CL	Grace Logins Allowed:	1	
CO	Remaining Grace Logins:	1	
▼ CO	Require Unique Passwords:	Yes	
	Limit Disk Space:	Yes	
	Maximum Disk Space (in kilobytes):	5,000	

Exhibit I-3-10. User Log-In Setup Screen

Unauthorized Commands.   Unauthorized commands can potentially be stored on the microcomputers that are used for workstations, and some of these workstations may have hard disks with other copies of the NetWare commands. Nevertheless, the user's overall system security level still indicates what that user can do. For example, if the user has a SYSCON-security level, that user can access files only at that level.

Unauthorized File Sharing.   A LAN might allow unauthorized file sharing or resource sharing if the network is not secure. Some users may not have the proper security level assigned to them, or they may have update capabilities for files that should be restricted to read-only access. The available NetWare rights include READ, WRITE, OPEN, CREATE, DELETE, CHANGE (directory rights), MAKE (new subdirectories), ERASE (existing subdirectories), SEARCH (directories), and MODIFY (file status flags).

Applications

LAN applications include those usually run on a microcomputer (i.e., word processing, data base management, spreadsheet, graphics, and communications). Communications applications may include dial-up access to outside services and a gateway to mainframe sessions.

NetWare provides application program interfaces (APIs) for network applications. For example, there is a resource accounting API that allows applications to use information generated by the accounting portion of the operating system. A network management utility that bills for resource use is one example. The queue management API is used to manage print jobs for a print server or a script for an archive or for a batch job server. The network diagnostics API creates information for applications that depend on statistical information from the network. File server function calls are accessible from the virtual console API. Any applications that are designed to monitor system security can use the security API. This interface permits the review of unauthorized access.

NETWORK ADMINISTRATION

The NetWare system has the ability to limit a user's access to designated directories, to specific workstations, or to certain times of day. The number of resources that one user can tie up during a period of time can also be limited. The system uses nonreversible password encryption to store passwords, and an optional capability requires users to change their passwords after a specified period, not reuse previous passwords, and use passwords of a minimum length. In addition, the system has a parameter to specify the number of invalid log-in attempts before the system locks the user out. These optional controls are highly recommended.

Physical Security

Safeguarding a LAN starts with physical protection of the LAN hardware. The degree of physical security depends on the risk analysis that identifies threats and exposures. The main LAN server requires special protection because it can act as a system console where commands can affect the entire network. The LAN workstations might require a limited-access area if, for instance, the data accessible to its users is sensitive. Components of the LAN and their locations and intended uses should be evaluated carefully.

Media for carrying the LAN communications include:
- Copper cable.
- Fiber-optic cable.
- Wireless transmission equipment (i.e., microwave, radio, and infrared transmission).

If the information on the LAN is sensitive, the most secure medium to use is fiber-optic cable; the least secure is wireless transmission. Physical protection from wiretaps should be installed with sensitive data. For example, communications involving national security are often sent over wires. The pipes carrying these wires are filled with a liquid that, if the pipe is cut, sets off a pressure-sensitive alarm.

Ordinary LAN installations will not require this extreme physical protection. However, the degree of physical protection of a network from wiretapping depends largely on the medium used to carry the data transmission. It is easy to passively wiretap (i.e., just listen to) a copper cable. It is much more difficult to listen in on fiber-optic cable. Regardless of the transmission medium, few problem-solving technicians have devices to listen to data as it crosses the network.

Access Control

User names and passwords are part of the LAN log-in procedure. During log-in, NetWare asks for a user name; when the name has been accepted as an authorized user of the network, NetWare asks for a password. (Passwords are optional but are recommended. Longer passwords and frequent password changes are additional recommended options.) When a password has been accepted, the user is recognized as a particular profile. According to the profile, the user is given access to certain applications, data files, commands, and network resources and is given specific rights to the applications and data files on the appropriate access list. In NetWare, such rights are specifically defined as read, write, open, create, delete, parental, search, and modify.

Password security controls access to the file server. Trustee rights specify the degree of access a user has to files in a directory. Directory security determines what a user can do within a directory. File attributes determine what access is available to individual files.

Security Limitations

Many LANs have components that are outside the immediate physical premises. To verify the components and connections of a network, the LAN reviewer might have to visit several locations. To view all the security parameters, the reviewer must have a supervisor security level.

Audit Trails

Audit trails are the record of LAN activities. Because the record of which user did what may be crucial to security, the audit trail should include invalid log-in attempts and attempts to access data outside a user's restrictions. These logged items require follow-up to determine whether security policies are being violated.

Recent versions of NetWare have the ACCOUNTING option, which enables the system to record log-ins and system use. Other NetWare products, such as LTAUDIT, are available to provide more flexible audit trails of system use.

Backup and Disaster Recovery Planning

A disaster recovery plan should ensure that essential applications are restored as quickly as possible after a disaster. System backup is always a necessity.

External devices (e.g., tape drives, removable disks, cassettes, parallel drives, cartridges, and optical disks) can be used to provide backup capability for files. It is up to the data security administrator to consider the criticality of the system. Factors to be considered are uninterruptible power supplies, redundant processors, inventory of additional equipment, and problem-detection tools.

A review of the LAN should include a check to see that backups are performed regularly. Evidence should substantiate this, and the data security administrator might want to test the backup tapes or disks on another system.

Concurrency

LANs introduce the problem of record concurrency (i.e., when two or more users try to access the same file or record simultaneously). In Net-Ware, there is a system utility called HOLDON/HOLDOFF, which can hold a network file open so that it cannot be accessed by two or more users simultaneously. Most LANs have a comparable utility.

SPECIAL RISKS AND CONCERNS

Special security concerns are introduced with the LAN environment. Some LANs permit an open port on the file server, where portable or laptop computers can link to them. This should be eliminated if it is not acceptable for a particular situation.

In addition, all LANs are not standalone networks. Other networks may be accessed from a LAN, or a LAN may tie into others. The LAN security review should determine what other networks are accessible and whether they access the LAN in a secure manner. Asynchronous bridge software is included with the NetWare operating system. This allows bridges to take LANs beyond local-distance boundaries. It makes connecting with another LAN as easy as changing a computer's disk drive.

Another special risk is server-to-server communications. This should be reviewed in the same manner as network-to-network communications.

A link to a mainframe always raises the question of information integrity. The token-ring attachment enables a LAN supporting IBM's token-ring topology to include a 3174 mainframe with the token-ring support option as if it were another workstation on the LAN. Some early LAN gateways did not properly disconnect a user from a mainframe session. That is, when a user logged out of a mainframe session, the LAN did not disconnect the host session. The LAN then accepted the next user requesting a mainframe session as the first user, which created a security problem.

Dial-up access presents another concern. There are many products that enable a remote user to dial up and gain access to a LAN. Once this is done, a remote user may have the same capabilities as a local user. Dial-up access should be treated as another application, and who can use it and what characteristics those users have once they are connected to the LAN should be reviewed.

ACTION PLAN

A step-by-step approach to the LAN security review must be tailored to the specific needs of the organization. A detailed understanding of the hardware and software that is running on the LAN should be gained before the review. System documentation manuals are the best source for this information. The review should include steps to:

- Identify the users with supervisor system privileges.
- List all users on the system and what they can access.
- Evaluate the exposures (e.g., mainframe access, system utilities, directory sharing, and trustee rights).
- Run the SECURITY (or equivalent) utility to identify holes in the system—This will help identify any users with supervisor privileges, who do not require passwords and are not required to change them.
- Identify the file server on the LAN and review the overall physical security.
- Review the system for external connections (i.e., ports and modems) and draw a schematic diagram of the LAN.
- Use SYSCON (or an equivalent utility) to review the existing controls.

Performing the Review

The LAN security review should include steps to:

- Determine the scope of the system, including what external devices the LAN can use—The NETGEN (or equivalent) file should be reviewed.

- Use the SYSCON command to identify who the system users are—The groups existing in the system should be identified and the account restrictions for users listed.
- Identify the critical resources on the system—SYSCON can be used to determine who has access and whether it is authorized.
- Use SYSCON to determine what other LANs can be attached—If possible, the users of the other LANs should be determined.
- Have the SECURITY utility executed—The output from the screen should be printed and reviewed for any holes in security.
- Identify the system administrator and review how users are given access and granted rights—System defaults should be verified.
- Review any disaster recovery plans that exist for the LAN (assuming that it is critical enough to warrant one)—The availability of backup devices should be considered.
- Review the backup of files and software, determine its location, evaluate its environmental security, and verify that the proper files are being retained.
- Observe the physical security of the file server, cabling, modems, and location of any external devices.
- Review the logical process of connecting to mainframes (dial-up or gateway)—These connections can support file transfer with CICS, VM/CMS, or MVS/TSO environments.
- Review some of the log-in scripts to determine whether certain users have automatically signed on to applications during their log-in process.
- Review the list of network users who are currently logged in, using the USERLIST command-line utility.

Interpreting the Results

The results of the review should be presented to management, who determines the action to be taken on outstanding security and control issues. Based on the criticality of the applications running on the LAN, the following recommendations can be made:

- Users should have only the access levels required to perform their routine job functions.
- All users should be required to use log-in passwords and to have them changed regularly—Passwords should be a minimum of five characters and should not be reused.
- System software and critical files should be backed up regularly and stored in a secure location.

- Additional critical hardware devices should be obtained to provide a secure level of redundancy in case of primary device failure.
- There should be a formal procedure to add users and resources to the LAN.
- Because of the critical nature of the file server, the access to this component should be properly controlled.

Chapter I-3-6
An Overview of Computer Viruses
Jon William Toigo

COMMUNICATIONS SYSTEMS MANAGERS are rightly concerned by the computer virus threat to established security and disaster recovery techniques. Virus programs are becoming more sophisticated; they are more difficult to anticipate, prevent, and detect. To complicate the problem, each new type of virus must be fully analyzed before the development of a reliable prevention tool. Viruses have become an issue of concern to insurers, many of whom have disclaimed policy coverage for virus-related data loss. The recommendations offered in this chapter are responses only to known virus programs. Unfortunately, experts expect that new generations of such programs will be able to circumvent current preventive measures, necessitating continued awareness and preventive efforts.

A BRIEF HISTORY

The computer virus was first unveiled several years ago by its creator, Fred Cohen of the University of Southern California. Cohen called this security threat a computer virus because of its infective behavior, similar to biological viruses. Cohen's computer virus could remain undetected in a computer for many years, then become active and infect vital programs and data files by modifying them to include some version of itself. According to Cohen, these viruses could be introduced surreptitiously, leave few traces in most systems, act effectively against modern security controls, and require only a minimum of specialized knowledge to implement.

Viruses have since become a growing concern for communications systems managers and professionals; numerous professional journal articles, conference workshops, and even a Computer Virus Industry Association (primarily for antivirus software and computer security product vendors and services) have resulted.

SCOPE OF THE PROBLEM

Since Cohen's initial examination of the computer virus, several university computer centers, government and business computers connected to electronic mail networks, and private microcomputers (especially those accessing public electronic bulletin boards) have fallen prey to viruses. The Internet virus, which infected at least 6,200 UNIX-based computers and disrupted the work of more than 8 million programmers and other MIS personnel, is the most damaging incident reported to date.

Viruses target these types of networked systems for several reasons. University computer environments, electronic networks, and bulletin board systems are relatively unprotected and often allow user anonymity, so computer criminals who wish to propagate viruses can access these networks easily. These environments are sometimes referred to as social because of the ease of access to shared information. After a virus is planted in such an environment, virus-contaminated software is unintentionally disseminated by the network's users, spreading infection in a geometric progression.

Inevitably, viruses have reached business systems. Reports have begun to circulate regarding virus-related damage in nearly all sectors of the business world. Few companies, however, have publicly admitted suffering a virus-related disaster. Understandably, in the corporate environment, there is incentive to keep such matters quiet. If an organization publicizes the vulnerabilities of its system, this may lead to additional attacks as well as to customer dissatisfaction and fear.

Several cases have been reported that help to define the scope of businesses' vulnerability to computer virus attacks. One victim was the Miami branch of the Bank of South Carolina: a disgruntled employee placed a virus that was intended to slow the recording of deposits from .5 to 3 seconds per transaction in the bank computer system. The virus could have placed the bank one month behind in bookkeeping within one week had it not been purged after being discovered in the course of a routine audit.

A Fort Worth case, which became public with the prosecution of the virus author, involved a fired employee of USPA & IRA. The employee hacked into the insurance company's IBM System/38 and planted a virus program that deleted 168,000 employee records.

In another case, the Providence Journal Co was infected by a virus that spread to several microcomputers, destroying nearly $10,000 worth of data and equipment by rendering hard disk drives unusable. That experience pointed to another dimension of the virus dilemma: the degree of insurance coverage for virus-related losses. Insurance is often viewed as the stopgap between failed disaster prevention and total disaster, yet existing insurance policies did not cover the losses to Providence Journal. Accrued dollar losses totaled less than the company's MIS insurance deductible. Furthermore, many insurers are reconsidering whether they will cover virus-related damages at all. Insurance vendors question whether the loss of data can be reasonably valued and insured.

Government computers have also been infected by viruses. The Navy Regional Data Automation Center in Jacksonville FL was infected by a virus disguised as a public domain data archiving program that was downloaded from a bulletin board system. Operating the software caused the data on the hard disk of the host microcomputer to be irreparably damaged. Other agencies, including the Environmental Protection Agency, NASA, the Department of Defense, and the CIA, have been beset by viruses that have destroyed data on IBM and Macintosh microcomputers and VAX/VMS, Unisys, and IBM mainframes.

Some observers attribute the great damage potential of computer viruses to the decentralization of computing and the associated lack of centralized management and monitoring; the vast majority of cases have involved microcomputers, lax security, and a lack of computer use standards. In most cases, viruses were transmitted with public domain software or were deliberately introduced with malicious intention. There are confirmed examples of virus programs debilitating systems on most hardware platforms, on networks, and even within microprocessors.

TYPES OF VIRUSES

The targets of computer viruses are typically nonspecific. Often, a virus infecting an organization's computer is not originally intended for that computer, though viruses can be directed and effective. The following sections describe three of the most common types of computer viruses encountered by MIS professionals.

Boot Infectors

Boot infectors are viruses that attach themselves to a diskette or hard disk at sector 0, the boot sector (containing the bootstrap loader program). The virus is activated when the system is booted and often will infect any diskette that is used with the infected system, even when a warm boot, or reinitialization, is performed. The virus writes itself to the boot sectors of any diskettes that are used with an infected system.

The most notorious boot infector is the Pakistani Brain virus. The brain virus is believed to have reached the US from Pakistan by way of universities in the UK. In fact, the names and addresses of the virus authors in Lahore, Pakistan appear at the beginning of the program together with the recommendation that victims contact them for a cure. Originally conceived as a protection scheme, the virus code was designed to debilitate unlicensed duplicates of application software developed by the authors. Somewhere along its journey from Asia to the US, however, the virus code was altered to become the malicious and highly infectious virus it is today.

This particular virus damages a disk by embedding itself in the boot sector, then scrambling data. It creates three successive clusters in six sectors of a DOS-formatted disk and classes them as bad sectors. The code then copies the file allocation tables and jumbles numbers, destroying the boot sector and the file allocation tables and rendering the data on the disk unusable. The Pakistani Brain virus received its name from a signature it left on an infected disk, which appeared as the volume label of the disk.

Systems Infectors

A second type of virus (as classified by the Computer Virus Industry Association) is identified by its ability to attach itself to system files and remain memory resident. A characteristic of these viruses is that they make subtle changes to the infected system over a period of time (e.g., gradually increase transaction processing times or change system error or informational messages). Sometimes, they are designed to monitor system variables (e.g., the system clock) and activate their destructive mechanisms when a certain condition is met (e.g., a date or time).

A virus can be programmed to create havoc at a specific time (e.g., the Friday the 13th virus, which was introduced at Hebrew University in Israel). Such virus time bombs are particularly challenging for traditional data protection methodologies. If virus-infected programs are backed up, there is every chance that reloading the backup software following a virus-related system failure will result in a repetition of the failure. Virus programs can remain dormant for as long as their authors wish, potentially infecting many generations of backup tapes and disks.

Executable-File Infectors

According to the Computer Virus Industry Association, viruses infecting executable files (with the file name extensions .COM or .EXE) are the most dangerous. They infect other systems by attaching duplicates of themselves to specific executable files loaded on any disk that comes into contact with the virus-infected host. Infection may occur during direct exposure (e.g., when using a diskette) or through a network connection.

How viruses spread from one disk to another, across networks, or over data communications lines is a matter of considerable concern. The Lehigh University virus illustrates one common method of virus transmission. This virus was contained in the stack spaces of the COMMAND.COM program on an infected diskette. Because the virus used empty but allocated storage space, the size of the COMMAND.COM file was not increased. By reading the COMMAND.COM program each time a DOS command was executed against another disk, the virus spread to the subject disk.

The Lehigh virus destroyed disks by erasing them. Erasure did not occur immediately after infection; a counter in the virus program recorded the number of DOS command calls, and when the counter reached 4, the virus erased the disk on which it resided. The newly infected disks similarly self-destructed after they infected four more disks, and so on. (The selection of 4 as the counter value appears to have been arbitrary. It could have been 40 or 400.)

A VIRUS PROGRAMMING PRIMER

Although there are many virus types that each use different hiding places, time delay functions, and sophisticated camouflage techniques, a highly infectious executable-file virus basically operates in the following manner (see Exhibit I-3-11). The virus program code replaces the first program instruction of a host program with an instruction to jump to the memory location following the last instruction in the host program. The virus program code is inserted at that memory location. Next, the virus simulates the instruction replaced by the jump. The virus program then jumps back to the second instruction in the host program, and the balance of the host program is executed. In this way, every time the host program is run, the virus will infect another program, then run the host program.

This is the most common method by which a virus spreads, and it does so at phenomenal speed. In tests at the University of Southern California, Cohen introduced viruses into computer systems. None of the viruses took more than an hour to infect the entire system, and some took as little as five minutes. Later tests, directed against a system equipped with security software that restricted data flow and program access according to user access levels, allowed Cohen (as a user with a low security clearance) to infect all programs and data on the system—even those files restricted to users with high-level security clearances.

With many viruses, damage to systems and data is not the direct result of virus transmission but of the activation of some part of the virus program that is designed to debilitate the system. After it has infected files on the disk, a virus usually responds to some trigger (e.g., a counter or a clock), or it goes immediately to work performing its programmed function (e.g., file deletion or cluster reallocation). In this way, the computer virus resembles a Trojan horse program—but one with a sophisticated means for penetrating and propagating within a computer before debilitating it.

241

Step 1

The virus code replaces first program instruction with one to jump to the memory location following the last program instruction.

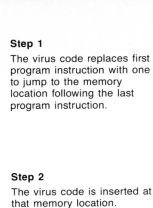

HOST PROGRAM
00000
00010 FIRST INSTRUCTION

00B12 LAST INSTRUCTION

Step 2

The virus code is inserted at that memory location.

HOST PROGRAM
00000
00010 FIRST INSTRUCTION

00B12 LAST INSTRUCTION VIRUS

Step 3

The virus simulates the host program instruction replaced by the jump.

HOST PROGRAM
00000
00010 FIRST INSTRUCTION

00B12 LAST INSTRUCTION
VIRUS CODE

Step 4

The virus jumps back to the host's second program instruction.

HOST PROGRAM
00000
00010 FIRST INSTRUCTION
00020 SECOND INSTRUCTION

00B12 LAST INSTRUCTION
VIRUS CODE

Step 5

Every time the host program is run, the virus will infect another program, then run the host program.

HOST PROGRAM
00000
00010 FIRST INSTRUCTION
00020 SECOND INSTRUCTION

00B12 LAST INSTRUCTION
VIRUS CODE

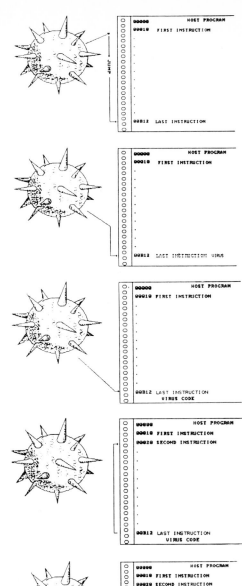

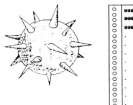

Exhibit I-3-11. How Viruses Work

242

COUNTERACTING VIRUSES

Many businesses and government organizations have instituted standards and procedures to prevent computer viruses from infecting their systems. In the community of medium-sized and small businesses, however, there is still a dearth of information about computer viruses, how to prevent them, and how to detect them. Two organizations that disseminate information to concerned managers are:

Software Development Council
PO Box 61031
Palo Alto CA 94306
(415) 854-7219

Computer Virus Industry Association
4423 Cheeney St
Santa Clara CA 95054
(408) 727-4559

Together with professional computer journals, these resources can help to keep managers apprised of developments in virus prevention.

Virus prevention begins with the analysis of the current computer systems with which a company is working. These systems need to be verified as virus free before any controls can be established that will help prevent and detect a newly introduced virus.

Detecting computer viruses can be complex and time consuming. Often, viruses hide where their presence cannot be detected except by bit-level file comparison. Some indications that a virus may have contaminated a system are:

- Programs take longer to load.
- Disk access time seems excessive for even simple tasks.
- Unusual error messages occur with increasing frequency.
- Indicators are lit on devices that are not being accessed.
- Less memory is available than usual; some larger programs cannot load.
- Program or data files are erased.
- Disk space is less than seems reasonable.
- Executable files (i.e., those with .EXE or .COM file name extensions) have changed size.
- Unfamiliar hidden files have appeared.

Such indicators as longer-than-usual program-loading delays or excessive disk access may only be perceptible to those who use the system regularly, so the users' impressions are extremely important.

Some computer virus detection products also provide canary programs (named for the canaries that miners used to send into shafts to check for poisonous gas). Canary programs can be loaded, run, and then examined for signs of virus contamination. Although not completely reliable for all types of viruses, these programs can be useful indicators of contamination.

After a system has been examined and, with reasonable certainty, deemed virus free, a global snapshot, or system status log, may be created. (Several antivirus software products provide this capability.) This snapshot includes the boot sectors of the disk, system interrupt vector addresses and code, system files, and all .EXE and .COM files. Because it is unusual for these items to be updated frequently, a periodic comparison of a new global snapshot to the original should reveal the presence of a virus. Several antivirus products provide utilities for automatically performing such a comparison during boot-up and even following a restart. Exhibit I-3-12 provides a partial listing of available antivirus software vendors and products.

Some of these products can detect viruses before they cause damage. Virus programmers constantly upgrade the elements of their programs to conceal their viruses and elude detection, however, and these programmers are usually aware of which preventive measures are implemented on a targeted system.

In addition to the methods already described, communications systems managers can take commonsense steps to help guard systems against viruses. Many are drawn from established security practices but are restated in the following sections to emphasize their effectiveness against the computer virus threat.

Microcomputer Virus Countermeasures

The following is a list of techniques that can help prevent viruses in a microcomputer environment:

- The use of public domain or shareware programs should be restricted or eliminated, or every such program should be thoroughly validated and tested before use.
- Production software should be subjected to nondestructive testing before use—This should include shrink-wrapped software and all updates and revisions.
- If possible, microcomputers should be isolated and delegated to specific users—Software sharing should be restricted.
- The use of modems should be controlled and communication with bulletin boards should be confined to a single, standalone microcomputer—All shareware should be tested before use.
- Regular examination of disk boot records should be performed—Viruses often reveal themselves through odd changes in byte numbers or quizzical messages.

Product Name	Environment	Vendor	Function
Antidote	PC and PS/2	Quaid Software Ltd 45 Charles St Toronto Ontario Canada M4Y 1S2	Examines files for alternation during boot-up
Antigen	PC	Digital Dispatch 1580 Rice Creek Rd Minneapolis MN 55432	Attaches itself to programs, detects modifications, and stops virus program execution once detected
C-4/Retro V	PC	Interpath 4423 Cheeny St Santa Clara CA 95054	Claims to filter out all 39 known virus programs
Data	PC	Digital Dispatch 1580 Rice Creek Rd Minneapolis MN 55432	Blocks viruses during Physician downloading; monitors write-to-disk commands. User can set up list of protected files and directories that will be monitored for changes
Disk Watcher	PC	RG Software Systems 2300 Computer Ave Suite 1-51 Willow Grove PA 19090	Release 2.0 contains virus protection software
Dr Panda Utilities	PC	Panda Systems 801 Wilson Rd Wilmington DE 19803	Three programs: Physical, Monitor, and Labtest protect against reformatting of the hard disk, rewriting the boot sector, writing the stack space of FATs, and formatting of diskettes and hard disks
Electronic Filter	IBM Information Network	IBM PO Box 30021 Tampa FL 33630	Prohibits the transfer of programs within an IBM system
Flu Shot Plus	PC	Ross M. Greenberg Software Concepts Design 594 Third Ave New York NY 10016	Filters out virus program code
Immunize	PC through mainframe	Remote Technologies 3612 Cleveland Ave St Louis MO 63110	A low-level antiviral program
Interferon	PC Apple	Sirtech 10 Spruce La Ithaca NY 14850	Detects and recognizes signals that viruses give off when they are present; code at the end of an application or changes in common system files
Mace Vaccine	PC	Paul Mace Software 400 Williamson Way Ashland OR 97520	Protects system files, hard disk boot sector, and partition tables from virus infection
NoVirus	PC	Digital Dispatch 1580 Rice Creek Rd Minneapolis MN 55432	Memory-resident virus checker
TCELL	VAX/VMS UNIX	Secure Transmission A.B. Sweden	Monitors file seals randomly; uses backup to fix

Exhibit I-3-12. Antivirus Software Vendors and Products

Product Name	Environment	Vendor	Function
Vaccinate	PC	Sophco PO Box 7430 Boulder CO 80306	Vaccinate is itself a virus that uses a program called SYRINGE to innoculate programs and a program called CANARY for sacrificial use with suspect programs
Vaccine	Bulletin Board System Operators	CE Software Co 1854 Fuller Rd West Des Moines IA 50265	Alerts BBS sysops to virus activity
VirALARM	PC	Lasertrieve Inc 395 Main St Metuchen NJ 08840	Protects programs by erecting a barrier
Virusafe	PC	ComNet Co 29 Olcott Sq Bernardsville NJ 07924	Memory-resident program that checks programs for infections
Virus RX	Apple	c/o Applelink or BBS 20525 Mariani Ave Cupertino CA 95014	Informs the user of any infections; user reloads from backup software

Exhibit I-3-12. *(Cont)*

- Critical and sensitive files and programs should be write protected.
- A schedule of backups and removal of backups from the facility where they originated should be strictly enforced—Monthly backups should be retained for at least one year.
- The user community should have access to information on computer viruses and related security issues—Users should be aware of the signs of possible contamination.
- In-house programmers should be encouraged to attend training seminars that teach techniques for developing virus-resistant program codes.
- If a boot diskette is used, the boot should always be done with an approved boot diskette.
- Volume labels on all fixed disks and diskettes should be meaningful and should be inspected regularly.
- When a second microcomputer is used for printing, the disk containing the file to be printed should not be bootable and should not contain system files.

Network Virus Countermeasures

A set of recommendations for combating viruses in LAN environments follows:

- The countermeasures listed for microcomputer environments should also be observed with network nodes and workstations.

- A network should be disconnected if a virus is detected—All work-stations should be tested for viruses before reconnection.
- Network security procedures should be enforced to ensure that un-authorized access is not possible—There is considerable information to suggest that the Internet virus would not have been so successful if proper user ID and password measures had been enforced.
- Shareware or bulletin board downloads should not be filed in com-mon file-server directories where they can be accessed by all nodes.

Mainframe Virus Countermeasures

The following preventive measures are effective against viruses in main-frame environments:

- Interfaces should be designed so that direct access is restricted to mainframes from communications networks.
- Users should be prevented from performing direct terminal-to-main-frame updates—Batch updates should be performed when possible.
- Terminal emulation software should be stored in a separate sub-directory—Executable files should not be stored in the same directory as emulator software.

In addition to these countermeasures, standard security procedures still apply. Exposure of computer systems to viruses can be greatly reduced by following the general-use procedures described.

Recovery Techniques

When a virus has contaminated a system, recovering from the infection with minimum data loss is a difficult undertaking. A main obstacle to recovery is the identification of the virus.

Viruses often destroy themselves when they become active, and though remnants of the virus may still be present, they can be difficult to collect from damaged media. The virus remnant must be taken apart and ana-lyzed. In the Internet case, Lawrence Livermore Laboratories required 16 hours to develop a fix program. They detected the virus, dropped connec-tion with the Internet network, eliminated the virus from their systems, re-connected to the network, and were promptly reinfected. Finally, they dropped the connection and developed a fix program to prevent reinfec-tion.

The Computer Virus Industry Association recommends the following steps for recovering from a virus infection in a microcomputer or LAN environment:

1. Power down equipment.
2. Seek professional help.
3. As an alternative to the use of professional help, reboot from original (write-protected) system diskette.

4. Back up all nonexecutable files.
5. Perform a low-level format on the disk.
6. Replace system and executable files.
7. Restore data from backup.

A more detailed procedure for recovering from an infection is available from the association.

The computer virus is a threatening reality. The threat posed by viruses is, however, inversely proportional to the security awareness of the end-user and programming population and to the quality of preventive programs that a company employs. The reported statistics on viruses, often cited by those who want to minimize the threat of the virus, indicate that 94% of reported virus events are not actually caused by viruses. Of the remaining 6%, most are nonclassifiable, presumably because the virus has destroyed all traces of itself. Many times, a virus can be mistaken for a hardware or software bug; however, there are clear-cut indications of computer viruses in about 3% of the reported incidents. The Internet incident was one such case and incurred more than $97 million in damage. In light of the potential magnitude of such a disastrous event, the importance of undertaking a program of prevention is well worth the small cost of a reliable antivirus software package and the attention of the communications systems manager and staff.

Chapter I-3-7
Protection of Communications Ports and Lines

COMPUTER HARDWARE SUPPORTS various physical and logical ports that are used to connect terminals and other external devices. In the simplest sense, a port is a socket into which a dedicated terminal or modem is plugged so that it may communicate with the host machine. There is generally no special hardware protection for these ports. The simplest form of port security might seem to be to simply unplug, or switch off, a communications line when it is not in use. This is seldom done, however, because dial-up access usually is not a scheduled event and thus requires that the port be accessible at all times.

Another vulnerability inherent in dial-up communications is the fact that dial-up modems are not treated differently from direct terminal connections in terms of system access. The hardware often provides no way to inform the operating system that an incoming user has gained access by way of a dial-up modem instead of a dedicated circuit. Although all users, regardless of whether dial-up or direct-access mode is used, are viewed by the operating system as being equal for access purposes, the probability of unauthorized access is far greater in dial-up mode.

There are four basic requirements for dial-up security; if they are not adequately met by the host itself, additional protection may be needed to supplement the host's security capability. The basic requirements are described in the following sections.

User Identification and Authentication. A user ID and password process is the first, and often the only, access control mechanism that a direct-access or dial-up user encounters when entering a system. If this mechanism is unavailable or weak, several other access control techniques can fulfill the same purpose. Most of the external dial-up protection devices described in this chapter are designed to strengthen system identification and authentication.

Security Event Logging. All dial-up activity between a host and a user must be monitored in order to uncover intrusion attempts or successes. System logging, which can also help gauge system use and identify user difficulties, should be used for this. If adequate system journaling is not possible, as is the case with many small or unsophisticated systems, several dial-up security devices with communications logging capability can be used to perform this function as part of an access-control strategy.

Limiting the Attacks. A system must be able to counter guessing attacks. An intruder who does not know the correct access codes makes many guessing attempts, often by running a program that generates and rapidly tries a series of passwords. Mechanisms that restrict the number or speed of log-on attempts per connection are useful.

Concealment of Information. It may be necessary to protect information communicated over dial-up lines from disclosure or interception, especially if the information is confidential or susceptible to fraud. Hardware that encrypts the information on the line can prevent disclosure, and other mechanisms can provide a cryptographic seal to authenticate the message contents.

After discussing the benefits of add-on hardware security devices, this chapter describes the various hardware alternatives available. It also includes a section presenting factors to be considered during an evaluation of security needs.

THE BENEFITS OF SECURING DIAL-UP LINKS

Although there are many trade-offs in implementing hardware security devices, there are also significant advantages. The primary advantage is that hardware security devices reduce dependence on software or procedural security mechanisms, many of which are not particularly strong in practice or are not readily available for all computer systems. Two other notable benefits can be gained by applying hardware protection to the communications link:

- Separation of functions.
- Additional layers of protection.

Separation of Functions. With hardware security devices, separation of functions is gained through:

- Externalization—A set of security functions is located outside the machine, physically and logically separated from the host. This separation reduces the degree of dependence on the software and procedural controls present in the system.
- Kernelization—A portion of the security function is assigned to a single dedicated mechanism for reduced and controlled access by way of communications. This separation can be further enhanced by assigning responsibility for these functions to communications or security personnel, instead of to systems programmers, who typically administer the technical aspects of operating system security.

Additional Layers of Protection.　Hardware security devices on a system's communications links provide formal protection of the network itself, which is a new concept for commercial and unclassified systems. Most hardware protection is designed to control authorization to a single system object, the communications port. Other software and procedural security mechanisms should still be used to reduce logical exposure to the remainder of the system.

THE SIX TYPES OF HARDWARE

To protect any set of dial-up communications ports, two basic approaches can be taken, each of which involves adding hardware-protective devices to the dial-up circuit. These approaches are referred to as the one-end and two-end solutions, depending on the placement and configuration of the protective hardware.

The One-End Solution

This solution calls for a separate password on the communications link itself and uses hardware to protect only one end of the communications link. Two types of devices are available, one for installation on the host computer and the other on the user's terminal. These devices perform a basic user authentication screening function, generally without requiring users to obtain extra equipment.

The Two-End Solution

Tighter security is gained by using a matched set of hardware-protective devices for both ends of the dial-up circuit (computer and terminal). These devices can communicate with each other in various ways to perform their security functions. The four types of equipment are classified by function. Three perform authentication functions, of the user, the user's terminal or location, or the message or data transmitted on the circuit. The fourth type

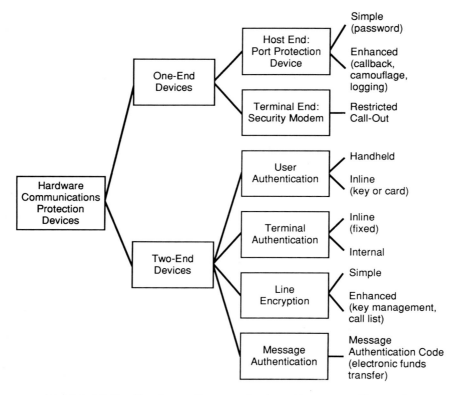

Exhibit I-3-13. Hardware Communications Protection Alternatives

is line encryption, which conceals transmitted data and, in addition, authenticates the user or originating terminal by exchanging encryption keys.

Currently available hardware communications protection devices are summarized in Exhibit I-3-13.

ONE-END PROTECTION: STRATEGIES AND FEATURES

One-end protection devices improve user access control by performing a preliminary call-screening or authentication function. Typically, such a device is independent of the computer. Most versions of one-end protection devices are installed at the host computer end, but some newer multi-function devices are connected to the user's terminal.

The following discussion separates these devices into two categories: those that can be placed on the host end of the circuit (port protection devices, or PPDs) and a newer and more flexible type of device that can be placed at the terminal end (a controlled-access security modem).

Host Port Protection Devices

A PPD, an external device fitted to a communications port of a host computer, authorizes user access to the port. It is separate from and independent of the computer's built-in access control functions, and it is designed specifically to control dial-up access.

PPDs can operate between the host and modem (digital side) or between the modem and telephone set (analog side); some modems include PPD functions in a single unit. Placement depends on system configuration and security needs. Once connection and user validation take place, the PPD becomes passive in the circuit. Exhibit I-3-14 depicts a dial-up circuit with a PPD installed.

The four primary features of PPDs are described in the following sections.

Password Tables. All PPDs require the user to enter a separate authenticator, or password, in order to access the computer's dial-up ports. This is the primary protection offered by PPDs. All PPDs also limit the number of sign-on attempts per telephone connection.

Callback to Call Originator. Some users erroneously describe all PPDs as callback devices, when most PPDs do not have, or need, that capability. Callback is a second level of user authentication beyond the standard PPD password table. A PPD with callback ordinarily requires that the user enter a PPD table password. The PPD then disconnects the line and locates the telephone number that is associated with the password given. The PPD then calls the user at that number, and host connection is made.

Hiding the Port's Existence. All PPDs camouflage the computer's dial-up ports so that the computer cannot be identified by an unauthorized caller. Although this is commonly a side effect of some password entry methods, it can be engineered separately. Some PPDs located on the analog side use a synthesized human voice to hide the modem tone on initial connection. Digital-side PPDs can send their own screen displays by way of the modem to the user's terminal, which masks the kind of computer they are protecting.

Attack Signaling or Journaling. Most PPDs warn of or record a dial-up attack. Some models use front-panel display lights; others maintain internal logs in RAM, and the most expensive models use the disk storage of dedicated microcomputers to record many types of communications activity information. Systems that use the callback approach may need a micro-

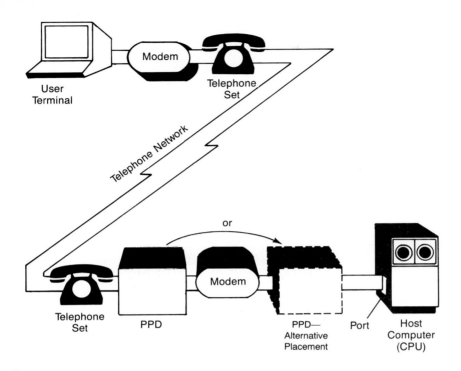

Note:
PPD   Port protection device

Exhibit I-3-14. Dial-Up Circuit with Port Protection Device

computer to record information for generating telephone use bills to system users.

Controlled-Access Security Modems

Controlled-access security modems, installed on user terminals, are single-user modems that incorporate a set of outbound call-screening security functions to control access to the host from the user end.

With a security modem, a dial-out connection is made only after the user enters a password. Inside the modem, the password is matched in a secured table with a dial-out telephone number sequence needed to connect the user to the specified host computer. The table can also contain a complete log-on sequence for transmission to the host once connection is made; it is advisable, however, not to include the log-on password in this

254

sequence. Users generally have no control over the connection information stored in the security modems, and the security administrator can telephone those units and change that information as needed.

TWO-END PROTECTION: STRATEGIES AND FEATURES

In high-security systems, screening of the port by password protection may still seem inadequate, and a more positive identification of the specific terminal or user attempting access may be desired. Additional resistance to tapping or tampering with communications traffic may also be needed. The two-end approach uses a security device at the user terminal end that matches a device or special software at the host computer.

Two-end protection can significantly raise the level of communications security and can increase user convenience. However, these benefits are often accompanied by a substantial increase in cost as well as by other drawbacks. In addition, the extent of the risk may not justify installing such a level of security in a given system. These issues must be considered before a purchase decision is made.

Matching Algorithms for Additional Security.   Most two-end security techniques use complex algorithms that are uniquely associated with specific terminals or users to create one-time passwords. The hardware or software at the host computer end knows which algorithm is associated with each user or terminal and can use that algorithm to perform a mathematical computation and then challenge the user or terminal device to do the same. If the user's response matches that generated by the host end, the host has authenticated the identity of the communicating party with a high degree of certainty. This challenge-response approach does not require the user to remember anything that may be written down or given to another person. The authenticator devices are constructed so that the algorithm cannot be copied. However, the devices are still subject to being loaned, lost, or stolen.

Trade-Offs in Cost and Flexibility.   The two-end approach requires that each dial-up user or terminal possess a particular type of authentication device and that the host computer be equipped with a matching device or special software—both of which substantially increase the cost of securing a dial-up network. Costs vary widely according to the level of security provided and other features. For example, costs can be as high as $6,000 per user-host link if sophisticated traffic concealment is needed in addition to access control. Most user or terminal authentication devices cost between $50 and $100 per user, plus the equipment or software required at the host end.

Two-end security devices can be separated into the challenge-response types that provide user or terminal authentication (access control) and those that offer concealment safeguards against eavesdropping (encryption) or tampering (message authentication). The second type also provides strong access controls. The purchaser must determine whether the concealment function is necessary.

Devices in the two-end category generally are easier to use than one-end devices, primarily because passwords do not need to be memorized and connection delays can be shorter. Nevertheless, the approach is more complex because there are more items to install, maintain, break, and misplace.

User Authentication Tokens: One-Time Passwords

Two-end devices that perform highly secure user authentication are based on the concept of a unique token that is used like a mechanical password. A token is a small item (e.g., a plastic card) that is required in order to access the system. Each token has a special algorithm or some other unique and noncopyable identifier embedded in it. The host computer identifies the user uniquely by means of the token's distinctive characteristics.

Most varieties of user authentication tokens are handheld and require no terminal attachments. Examples are a calculator with special circuitry, a plastic card that continuously displays a time-based authenticator, and a light-sensitive wand designed to read and interpret special terminal challenge displays sent by the host.

For most tokens, the user must enter challenge information sent by the host into the token. A liquid crystal display (LCD) on the token then shows the computed result of the challenge. After the user enters that authentication information through the terminal, the host then reads the authentication information, compares it with the correct answer previously generated, and decides whether to approve access.

Terminal Authentication

The second two-end device authenticates specific user terminals. Terminal authentication takes the following forms:
- Add-on terminal authentication devices.
- Embedded terminal authentication capability.

Add-On Terminal Authentication Devices. Terminal authentication devices work much like user authenticators. They use matching pairs of devices inserted in the communications circuit, with one device placed between the terminal and modem and the other attached to the host computer's port. A typical product includes a four-port unit for the host end that can generate challenges to the small portable units that connect to the terminals. Each terminal is uniquely encoded for identification by the host unit.

Hybrid versions of terminal authenticators are also available; some can authenticate many users at the same time. For example, a newer version of the four-port unit described above has a slot in which users insert any prevalidated magnetic-stripped card (e.g., a bank or charge card). Another popular product takes a similar approach, requiring each user to insert a personal plastic card embedded with identification circuitry into the unique terminal unit. Both of these products automatically accept the challenge from the host, use the algorithm or data in the user's token to perform the required calculations, and then transmit the results to the host for verification.

Embedded Terminal Authentication Capability. Many standard terminals or workstations already have internal circuitry that supports the assignment of unique terminal identifiers. This capability is also called answerback memory. These identifiers are either fixed and preassigned (hardwired) or, more commonly, special memory locations in firmware that can be changed to the desired code sequence during terminal setup. It is usually possible to conceal this code once it is entered so that it cannot be read or copied by the user.

The host system can use this feature by sending the standard ASCII code ENQUIRE as a challenge to the terminal, which responds with the answerback memory contents for authentication. Some commercial software communications packages for microcomputers as well as some modems also have this feature.

Line Encryption Devices

Encryption is the process of scrambling information in a predetermined way so that it is unintelligible to anyone who does not know how to unscramble it. This process has been used by governments for centuries to protect secrets during transmission, but it has seldom been used elsewhere. Because attempts are constantly made to break codes, sophisticated encryption methods have been invented. These newer encryption methods can be used efficiently only by computers or special microcircuitry.

The Data Encryption Standard (DES), which was developed under the sponsorship of the National Institute of Standards and Technology for use within the federal government and elsewhere, uses a highly complex and effective algorithm. The DES requires the entry of a 64-bit key sequence, 56 bits of which are used for encryption and decryption. Because each bit can be either on or off, an extremely large number of keys is possible—which is the source of the DES's strength. It is difficult to discover the key used to encrypt a message with the DES, even using sophisticated computerized techniques.

Because it has several attributes that cover most communications security needs, encryption is the best way to ensure data privacy in dial-up communications. The primary rationale for using encryption is that it conceals information passing over the communications link, protecting it from disclosure. In addition, in some modes, encryption can ensure the integrity of a message so that tampering or transmission errors can be identified. (It should be noted, however, that the process of message authentication, which is discussed below, is more effective for ensuring message integrity.) Finally, the uniqueness of the encryption key that is shared by sender and receiver enforces a high degree of user authentication. If both sender and receiver share a single key, they must have exchanged it or been assigned it by a third party.

Encryption devices can take two forms. In the more traditional form, the circuitry is enclosed in a small box that is connected in series between the port and the modem, on either end of the communications circuit. In the newer form, designed for microcomputers, all circuitry is contained on a single circuit board that is plugged into one of the standard slots inside the computer. With the latter form, the circuit board generally can be used for encryption of internal disk files as well as for communications. With either form, the host's communications ports are fully protected from intruders.

There is one common problem with communications encryption. If the key used by the sender and receiver is the only real security, the security surrounding the procedure used to exchange the key is extremely important. Most current encryption systems rely on users to transfer keys manually, which may or may not be secure—an intruder may have an opportunity to intercept the key while it is in transit. Newer and more sophisticated encryption devices have been developed that can be linked together so that they automatically identify each other and exchange encryption keys without human intervention.

Message and Data Authentication

Message authentication is a two-end approach that was originally designed for electronic funds transfer (EFT). It can easily be used to verify the integrity of any collection of data being transmitted or stored and to ensure that alterations are not made without being detected. In EFT, it is important to verify that the contents of a message have not been changed, because those messages are actually electronic checks that are subject to fraud or embezzlement.

The banking industry, in conjunction with the National Institute of Standards and Technology and the American National Standards Institute (ANSI), has developed ANSI Standard X9.9 for message authentication in EFT. This standard uses the DES with a prespecified key to authenticate selected fields in an EFT message, or an entire data message, to ensure that the message is not altered in transit. A message authentication code is calculated as a cryptographic function of the cleartext message. The message authentication code is then appended to the cleartext message to serve as a cryptographic authenticator. The message authentication code can then be checked by the recipient by duplicating the original message authentication code generation process with an identical device and the DES key.

The same process of generating a verifiable seal against tampering could be used effectively in various business applications. Communications links protected full time by message authentication devices would be highly resistant to intruders.

EVALUATING SECURITY NEEDS

Determining dial-up security needs can be a very complex process. The following set of questions should help focus the decision process and aid the systems manager in deciding on a course of action.

Is Better Dial-Up Security Needed?

The following criteria can help determine whether a computer system needs supplemental dial-up communications security devices.

Defining Security Requirements for Information Flowing on Dial-Up Circuits.   Three factors can be used to determine security requirements for collections of information or the systems that process them. The first is sensitivity to disclosure, which is the negative impact that could occur if the information in the system were disclosed to unauthorized persons. The second factor is availability, the impact on the organization if the information or processing system is not available within a specified period of time. The third security measurement factor is integrity. If the information must be highly accurate to be useful or if it may be the target of fraudulent modification, this factor becomes important.

Characteristics of a Dial-Up Circuit Needing Communications Security.   If the current resistance of the host system's operating system to outside penetration is low, the potential exposure by way of dial-up communications networks may be high. This is particularly true if transmitted information is very sensitive. If intruders can gain access to the system to change, tap, or interfere with communications, additional security protection is probably needed.

A dial-up circuit needing strong communications security is one with one or more of the following characteristics:
- It handles data that must not be modified or disclosed.
- It supports processes that are very time sensitive.
- It permits easy access to fragile data bases or files.

If Improved Security Is Needed, Is One-End or Two-End Best?

If management determines that dial-up security devices are needed to bolster communications security capability, it must decide on the general type of device. The following criteria can help determine whether the one-end (host or terminal PPDs) or the two-end mechanism will best meet the computer system's security needs.

Integrity and Sensitivity to Disclosure. When the information that may be accessed by dial-up is very sensitive to disclosure or fraudulent modification, one of the two-end approaches that involves encryption should be used. For information that is low to moderate in sensitivity, a one-end approach that provides added ability to screen out intruders through access control barriers may be appropriate.

User Resistance to Remembering More Passwords. When users are highly resistant to remembering extra passwords for access control, one of the two-end approaches that performs user or terminal authentication by way of a token or an add-on box may be appropriate. Possession of the token is functionally identical to remembering a password.

User Resistance to Connection Delays. When higher levels of user authentication are required, but users are resistant to delays in connecting to the system, one of the two-end devices—a terminal security modem or a PPD without callback—may be appropriate. None of the two-end devices uses the time-consuming callback approach, but some induce their own form of user connection delays by requiring the user to receive a challenge, process it with the token, and then enter the result on the keyboard.

If PPDs or Security Modems Are Desired, What Features Are Needed?

When additional security should be in the form of low to moderate improvement in user access control (identification and authentication), PPDs or security modems may be needed. The following criteria are useful for selecting and implementing PPDs.

Access Security Versus Password Entry Methods. There are three basic methods for entering the password into a PPD, each with its own security or convenience considerations. Some units require the user to respond with voice to challenges in such a way that a numeric password is formed. This is time consuming and is not appropriate for users of direct-connect modems instead of telephone sets. Similar units require the user to enter a numeric password on the telephone keypad. The problems with this approach are that some terminals may not have keypads, and more important, the numeric password does not have enough possible variations to be highly secure. Nonetheless, the voice and keypad methods do hide the host's modem tone from intruders.

The third method of password entry is by means of the user's terminal keyboard. This approach permits the creation of far stronger passwords, because any character of the password can be any one of the 128 characters in the ASCII character set. Even terminals with direct-connect modems can use this method. The host port's modem tone can be heard on connection, but the password strength and the ability of this type of PPD to camouflage the type of host computer being accessed should be sufficient to thwart penetration attempts, although it may not eliminate them.

Security Evaluation of Various Features.   Two PPD features merit special discussion. An important standard feature that all units share is the procedure for changing security tables. Low-security PPDs permit this to be done either manually or by way of a connected terminal with no special external security controls. Higher-security devices require a special password plus a physical key to enter the device into supervisory mode for table maintenance.

One controversial optional feature of many PPDs that gives additional protection but has numerous drawbacks is callback. Once almost synonymous with PPDs, callback can serve as a second password hurdle, but in many systems the user may call in from several different telephone numbers. Furthermore, if the first PPD password procedure is strong, a callback system may not be needed unless management wants to control the locations from which dial-up users may call. Major drawbacks include user connection delays, reversal of toll charges, and increased security problems with table administration. A further potential problem is that hackers have identified a strategy for penetrating certain PPDs by exploiting the way in which these devices perform the callback process. It is useful to note that all of the newer high-end PPDs either do not use callback or make its use optional.

If Two-End Security Is Needed, Which Approach Is Best?

When the user authentication features of the PPD or security modem do not meet the security requirements of the dial-up communications network, one of the four two-end security device approaches may be appropriate. The following factors should be considered in the selection of a particular two-end device.

Information Sensitivity. If the information transmitted on the dial-up network is so sensitive to disclosure that it should be protected against wiretaps, the best solution is a form of line encryption.

Information Integrity. If information must be communicated through dial-up lines without modification, the best solution is to use message or data authentication through a hardware device that performs the MAC generation process.

Terminal Location. If it is important to know that a specific terminal is being used or that the communications come from a specific location, the best solution is to use the existing terminal authentication capability, if available, or a terminal authentication device. However, if all that is needed is a check on the originating location of the call, a PPD with callback will also do the same job, possibly at a lower cost.

User Identification. If it is necessary to ensure that a specific individual is accessing the system, one of the various user authentication token devices should be implemented. Line encryption can also help if the user is required to enter an encryption key in order to use the device.

What Are the Trade-Offs in Adding Dial-Up Security Devices?

The prospective buyer of hardware for communications protection should carefully consider the adverse impact of installing these devices. This impact can arise in user convenience, systems management effectiveness, and cost.

User Convenience. Users may resist additional passwords for PPDs or security modems. The typical user may perceive authentication tokens (e.g., cards or wands) as a nuisance. The set of administrative procedures associated with maintaining some manual forms of encryption key management is even more onerous. Imposing these additional requirements for the sake of security is unnecessarily burdensome unless they are clearly needed because of system risks.

Similarly, connection delays caused by security measures are often unwelcome. Some delays result from PPD callback procedures. Other procedures, such as the manual entry of an identification string generated by a handheld authenticator token, also generate connection delays of a minute or so.

Systems Management Effectiveness. When system security weaknesses are examined closely, the most common problems discovered are administrative. In other words, more security potential is typically available in a system than is realized. This is especially true of the user account name and password scheme. The problem is a human one, and imposing hardware protective devices will not solve it. In fact, it may exacerbate it. For example, when an organization decides to install PPDs on the numerous dial-in lines attached to its primary computer, a new set of problems will immediately surface. Perhaps the most obvious of these is the problem of managing an additional password system, separate from that used by the host computer. The procedures for assigning and changing passwords for PPDs must be rigorous, or the real protection they can offer will be reduced. This usually means that more people will be needed to administer the system, especially if the organization takes the opportunity to separate out the communications security function from the computer security function.

Cost. Communications protection devices typically cost several hundred dollars per line. The bare minimum cost per port to install hardware protection is approximately $200 and can range into the thousands of dollars, depending on the approach and level of security desired. Along with this initial capital cost is the recurring cost of maintaining and repairing the devices. Other direct and indirect dollar costs imposed by these devices may include the following:

- User inefficiency (one minute per connection multiplied by many connections per year adds up quickly in terms of salary).
- Computer processing delays while user or terminal authentication takes place.
- Increased host computer telephone costs because of callback procedures that require that session connections originate at the host end.

All of the costs involved must be identified and estimated to determine the true cost of installing additional dial-up security protection. This final cost should then be compared with an estimate of the current risk from damage resulting from dial-up intrusion, to evaluate whether the new devices are warranted.

SUMMARY

Various alternatives for improving dial-up security with add-on devices have been presented here. It is important to determine which, if any, of the devices can help an organization sufficiently to warrant their purchase. Each device provides enhanced dial-up security at some cost, in real dollars or in efficiency.

Part II
Communications
Networking

COMMUNICATIONS SYSTEMS MANAGERS ARE, in a very real sense, engaged in the value-added resale business. The business product is information delivery services. The services are characterized in terms of technical attributes such as delivery rate, error performance and other integrity features, geographic coverage, and timeliness. In order to "sell" these services to the using "customers," the manager must provide an infrastructure of communications facilities on which the delivery services can be based. Because it would be prohibitively expensive for the enterprise to construct a network connecting all possible sources and destinations, the manager will probably choose to lease services from a carrier and supplement these with, perhaps, some private facilities and certainly with privately owned equipment and software. The manager has thus taken a purchased network service; added value to it in the form of enhanced services, accessibility, software features, and equipment; and resold it to organizational users.

That this organizational networking is a very significant and lucrative business can be seen from the fact that the 1989 market for networking services purchased from carriers was approximately $200 billion. As might be expected, the largest portion of this expenditure, approximately $150 billion, was for voice services. Voice is, of course, a very real information source and is increasingly being brought under the aegis of a common information services management structure.

The balance of $50 billion in carrier revenues was derived from data services, image delivery services, and leasing of facilities for private networks. The $200 billion revenue figure is expected to grow to around $700 billion by the end of the decade. If the estimated $95 billion spent for data communications products alone is added to the carrier revenue, the importance of managing these networking resources is clearly shown.

This value-added networking business has several important underlying structural elements. Each must be considered when the communications staff is designing, assembling, and deploying the network.

Architectural considerations are, or certainly should be, foremost in the network plan. The architecture affects each and every decision made about the network. The topological arrangement of nodes, hosts, controllers, and terminals is covered by the architectural plan. The architecture provides the basis for selection of facilities. It identifies the need for and the location of protocol drivers and, perhaps, protocol converters. The architectural approach governs equipment and software procurement, aids in identification and resolution of compatibility problems, and shows the way toward migration to newly emerging technologies.

A second consideration in communications networking is the judicious and timely application of recognized communications standards. Application is judicious in the sense that standards must be carefully selected and carefully applied to the network. Application of standards for the sake of standardization is not the goal. The goal is to enhance the performance and reduce the cost of the network services by the practical application of standards. Timely application of communications standards is also important. If the network manager attempts to apply the standard too early, there is risk of rapid change in the technology rendering the equipment obsolete. It is always dangerous to be sitting on the leading edge of technology. On the other hand, waiting too long can mean missing the benefits of standard approaches and being bypassed by the technological change.

A third element to be considered is the availability of services. As stated earlier, it makes sense to buy, rather than build, many of the services required as the baseline for the effective network. The manager must establish that the desired service is available in the locations where it is required. Many services, especially the newer ones, are available but are a long way from being ubiquitous.

A final consideration is technology. The rapid pace of technological change shows no signs of abating. The communications manager must somehow keep abreast of technology so that an opportunity for substantial improvement will not be missed. This must be done while coping with the daily problems of managing the existing network.

Part II is devoted to an exploration of these aspects of communications. Each plays a major role in the business of networking.

Section II-1
Architecture and Standards

THE MANAGER-ARCHITECT MUST DEVOTE a great deal of energy to construction of a network framework that can support the organization's information processing requirements. The results, however, are manifested only in the abstract: the network cannot be entered or observed; only the results of the network operation can be seen. As with any successful architecture, the framework is invisible but the integrity of the network structure depends on it.

As it is with architecture, so it is with standards. Only the effects can be observed. Improper application of standards can be a disaster. Proper choice and application can have considerable benefit in cost and performance.

In this section, we examine several architecture and standards issues. In Chapter II-1-1, "Standards Organizations and Their Procedures," the author explains the national and international standards bodies' structures, membership, and procedures. This is useful background in determining the role to be played by standards in the network.

Chapter II-1-2, "Functional OSI Profiles," looks at the practical application of the open system concept. A profile is a selection of open systems interconnection (OSI) tools directed at a specific application (e.g., manufacturing). The chapter explains the development of the procedures for international standard profile definition.

Systems management is an obviously important facet of communications networking. Chapter II-1-3, "OSI Systems Management," describes the open networking approach to management services and protocols.

In the interim preceding wide-scale implementation of OSI protocols, the Transmission Control Protocol and Internet Protocol (TCP/IP) suite is fulfilling many requirements. This approach is the subject of Chapter II-1-4, "The DARPA Internet Project and TCP/IP Protocols." Beginning with the historical perspective, the chapter provides descriptions of the architectural design and each of the networking mechanisms that make up TCP/IP.

Computer-aided design and manufacturing relies heavily on communications. Chapter II-1-5, "Communications Standards in CAD/CAM," examines the role of local networks in linking design workstations in the office and factory environments.

The final chapter of this section, Chapter II-1-6, looks at "ETSI: A European Standards Organization." The move in the European Community toward open markets beginning in 1992 is creating the need for communitywide standards. The European Telecommunications Standards Institute is leading this effort.

Chapter II-1-1
Standards Organizations and Their Procedures
Michael Varrassi

STANDARDS WORK IN THE US is directed by several organizations. An understanding of the particular role played by each of these groups is essential to the successful development of national and international standards. This chapter reviews these standards organizations and their international counterparts and examines how a proposed standard is submitted for consideration.

NATIONAL STANDARDS ORGANIZATIONS

The American National Standards Institute (ANSI)

The American National Standards Institute (ANSI) was founded in 1918 as the American Engineering Standards Committee with the objective of reducing economic losses and, consequently, conserving national resources. In 1928, the organization became the American Standards Association and remained so until it received its current name during the late 1960s.

ANSI's mission is the development of consensus standards to serve constituencies in government and industry as well as consumers, professional societies, and associations. This goal is accomplished by coordinating private sector activities in the development of national standards and identifying areas in which standards are needed. Among the services that ANSI provides are the verification of due process and the supervision, designation, publication, and maintenance of approved standards. ANSI membership is open to all who are directly and materially affected by a proposed standard. Individuals are permitted to attend meetings as observers, though they cannot vote on the proposals.

ANSI thus functions as a coordinator of voluntary standards work in the US for both the private sector and the government. ANSI does not develop standards; rather, its function is to ensure that its committees follow due process in the approval of proposed standards as an American National Standard. This is done by observing the following concepts:

- Openness—Providing adequate notice of meetings and ensuring that no unreasonable financial requirements or conditions are set.
- Fair representation of interest—Preventing dominance by a single group.
- Diverse membership categories—Including producers, users, and general-interest groups.
- Written procedures—Formulating documented guidelines and making these available to any interested party.
- Specific appeals procedures—Developing an appeals process that is unambiguous, realistic, responsive, impartial, and substantive.
- Listing of proposals—Publicizing proposals that require balloting in ANSI's Standards Actions.
- Consideration of all views and objections—Putting mechanisms in place to resolve conflicting opinions.
- Consideration of standards proposals—Ensuring that new proposals as well as revisions are evaluated.
- Maintenance of accurate records—Tracking and retaining all drafts, amendments, and resulting actions, with supporting data and documentation on the disposition of objections.

The Committee on Information Processing Systems. Several ANSI committees are involved in the definition, development, and approval of communications and information processing standards. Of these, the most significant is perhaps the American Standards Committee X3, which is responsible for standards affecting computing systems and peripheral equipment. The technical subcommittees that make up X3 are X3A on recognition media, X3B on storage media, X3H&J on languages, X3I on data representation, X3K on documentation, X3S on communications, and X3T&V on systems technology. Specific task groups of interest are X3V1 on office systems and X3T5 on Open Systems Interconnection (OSI).

The T1 Committee on Telecommunications. The T1 committee was formed as a nonprofit organization in 1983 to represent exchange carriers with an interest in standards and related technical fields and to function as the secretariat for work on independent interconnection standards. Before divestiture, the Bell system provided telephone service for the vast majority of the US. Together with its manufacturing organization, Western Electric, and its research facility, Bell Laboratories, the Bell system thus provided the primary impetus for developing US communications policy, and its technical views became de facto standards. The T1 Committee on Telecommunications was formed to meet the postdivestiture need for developing consensus communications standards that formerly were the province of AT&T.

The mission of the T1 committee is to develop technical standards and reports on interfaces to US networks, develop positions on subjects related to international standards bodies, review the interconnection and interoperability of interfaces with end-user systems, and provide carrier information and enhanced service, including switching, signaling, transmission performance operation, administration, and maintenance. Currently, six technical subcommittees carry out the standards development work of T1 as follows:

- T1E1—Carrier-to-customer installation interfaces.
- T1M1—Internetwork operations, administration, maintenance, and provisioning.
- T1Q1—Performance.
- T1S1—Services, architectures, and signaling.
- T1X1—Digital hierarchy and synchronization.
- T1Y1—Specialized subjects.

The Electronic Industries Association (EIA)

The EIA is a nonprofit organization that represents the interests of manufacturers. Its focus concerns areas of legitimate public interest in order to stimulate public interest in electronics, national defense, communications, education and entertainment, and the evolution of technology. The EIA provides a public forum for national laws and policies.

The Institute of Electrical and Electronics Engineers (IEEE)

The IEEE was formed by the merger of the Institute of Radio Engineers and the American Institute of Electrical Engineers. Its objective is to provide a basis for ratings in order to determine performance specifications for equipment and materials. Proposed standards are intended as guidelines for gaining practical field experience. The IEEE has published more than 500 standards and oversees 400 independent and ongoing standards projects.

The National Institute for Standards and Technology (NIST)

NIST was founded in 1901 as the National Bureau of Standards to provide a basis for the orderly conduct of industry and commerce by ensuring the compatibility of measurement standards needed by industry, consumers, scientists, and the government. It acts as a catalyst for the development of advanced technology, coordinates voluntary product standards, sponsors network sounding boards, and performs research for government and private agencies. In addition, NIST verifies that physical and chemical measurements derive from a consistent set of standards, develops certain data and improved measurement techniques for the engineering community, and conducts research in engineering and applied science.

NIST became involved in the development of computer and communications standards with the formation of the Institute for Computer Science and Technology (ICST). Its areas of concentration include software, data, and equipment standards; networking; and software management. By the mid-1970s, NIST, through the ICST, had managed the development of 44 standards and guidelines and participated in the development of 79 national and 86 international voluntary standards.

Underwriters Laboratories

Underwriters Laboratories emerged as a nonprofit organization in 1894 to ascertain, define, and publish standards, classifications, and specifications for materials, devices, products, equipment, constructions, methods, and systems, in addition to providing other information that reduces or prevents bodily injury, loss of life, and other damage from those. The organization has approved more than 500 Standards for Safety, more than 70% of which are recognized by ANSI.

INTERNATIONAL STANDARDS ORGANIZATIONS

The International Telecommunications Union (ITU)

International standards work formally began with the creation of the ITU in 1865, culminating nearly 20 years of effort to establish telegraph service between a score of European nations, including Russia and Turkey. The factors that influenced this effort include the expanding role of consumers and users in standards development, the increased interest in standards as a means of improving industrial and product safety, and new technologies using interdisciplinary solutions. In the intervening century, the development of international standards promoted an expansion of international trade, facilitated the development of new markets, and laid the groundwork for the onset of the information age.

Recently, the need to develop international standards as a precursor to national standards has become the primary objective of the ITU. The intent of this approach is to avoid the restrictive conflicts that arise when national standards precede these international counterparts. In the communications industry, for example, the rapid growth of international markets has accentuated the need for standards that can transform the notion of a global marketplace into a reality.

Today, the work of the international standards forum is increasingly extending beyond mere refinement of technology and its application and is actually expanding technological frontiers by defining new capabilities. Integrated services digital network (ISDN) and messaging directories are but two examples of this change in role. Other areas of emphasis for ITU are improving the efficiency of international communications and information exchange, integrating voice and data communications, and integrating diverse information technologies. Focal points in the integration of diverse information technologies include text preparation and interchange, local area networks, text processing (e.g., videotex, teletex, and facsimile), certification, ergonomics, office equipment and supplies, product safety, and programming languages. The need for such effort is clearly underscored by the rate of growth of the communications market: an estimated 200% every six years.

With its headquarters in Geneva, Switzerland, the ITU comprises four agencies: the International Telephone and Telegraph Consultative Committee (CCITT), the International Radio Consultative Committee (CCIR), the International Frequency Registration Board (IFRB), and the Secretariat (see Exhibit II-1-1). As their titles suggest, the CCITT is concerned with communications standards activities, the CCIR with radio-related issues, and the IFRB with allocation of the electromagnetic spectrum. The Secretariat

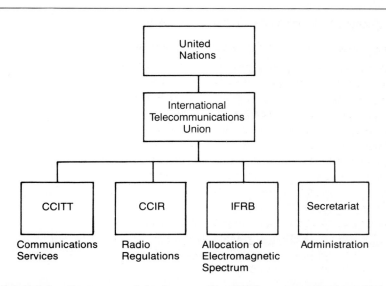

Exhibit II-1-1. Structure of the International Telecommunications Union

undertakes the administrative work necessary to support the three standards and regulatory arms of the ITU.

The International Telephone and Telegraph Consultative Committee. The CCITT was formalized as a permanent component of the ITU in 1956, with a directive to regulate, plan, and coordinate the standardization of international communications for all users. This mission translates into three broad objectives: to maintain and extend international cooperation among all members of the union, to promote the development of technical facilities to their most efficient operation, and to harmonize the actions of nations in the attainment of their goals. Clearly, the goals of the CCITT vary according to the target user group. For example, the goals in developing standards for the industrialized nations are to bring order to the industrialization process and to encourage innovation. The same standards become tools to facilitate the transfer of technology for the developing nations of the world.

Work is accomplished through various study groups (see Exhibit II-1-2 for a listing) and a laboratory dedicated to subjective and objective telephonometric problems. The CCITT produces two series of documents: the telegraph and telephone regulations, which are imperative, and the CCITT recommendations, which are flexible standards that relate to technical and operational concerns, tariffs, administrative directives, and terminology.

Study Group	Area of Interest
Study Group I	Services
Study Group II	Network Operations
Study Group III	Tariff and Accounting Principles
Study Group IV	Maintenance
Study Group V	Protection Against Electromagnetic Effects
Study Group VI	Outside Plant
Study Group VII	Data Communications Networks
Study Group VIII	Terminals and Telematic Services
Study Group IX	Telegraph Network and Telegraph Terminal Equipment
Study Group X	Language for Telecommunication Applications
Study Group XI	Switching and Signaling
Study Group XII	Transmission Performance of Telecommunications Networks and Terminals
Study Group XV	Transmission Systems and Equipment
Study Group XVII	Data Transmission over the Telephone Network
Study Group XVIII	Integrated Services Digital Network (ISDN)

Exhibit II-1-2. CCITT Study Groups, 1984–1992 Study Period

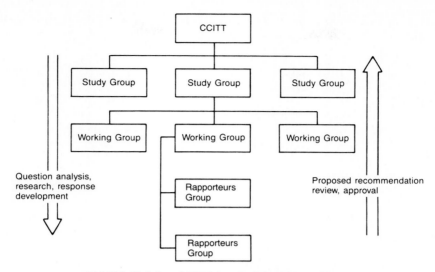

Exhibit II-1-3.   CCITT Study Group Structure

Although they are not binding, the recommendations form the basis of many international agreements, and their intent is to ensure quality of service for interconnection of networks. In some countries, they are referenced as part of formal interconnection agreements; in these cases, compliance becomes mandatory.

The structure of a CCITT study group varies with the topic and the specific requirements of the group. In most cases, however, a study group takes its work assignments from questions assigned for study during the CCITT plenary period. Individual questions are assigned to subsets of the study group, called working parties. Each working party can compromise smaller groups, whose purpose is to address specific aspects of the assigned problem; these groups, in turn, are known as rapporteurs groups. Exhibit II-1-3 illustrates the typical structure of the CCITT study group.

The investigative questions themselves are formulated during the later stages of a plenary period; consensus is reached at the final plenary assembly. Work assignments are then issued to specific study groups and their respective working parties. As work during the next plenary period proceeds toward the resolution of these specific questions, the results are passed from the rapporteurs group to the working party for review and from there to the study group. Each review usually yields comments, concerns, and suggestions, which are documented as new contributions that define changes requested in the existing work. Comments serve to continually refine the development of the answers to the questions posted at the outset of the plenary period. These resolutions eventually are documented as new or revised recommendations.

Two points require emphasis. The first is the iterative nature of all standards work: the creation, alteration, and review of a draft standard is a continual process. Second, although this approach can be construed as unnecessarily arduous, it is precisely this process of refinement that results in the creation of an effective global communications environment.

The CCITT is made up of the member countries of the United Nations. Participation is also permitted for recognized private operating agencies and scientific and industrial organizations, provided these secure the consent and sponsorship of their respective governments.

The Development of CCITT Recommendations. As noted, recommendations take root in study questions that are agreed to at the outset of a plenary period. These can consist of entirely new questions or of topics left unresolved in the previous period. The study is based on contributions from the various participants engaged in the process. These contributions are reviewed during regularly scheduled sessions held during the study period. Recommendations are adopted or rejected during a plenipotentiary assembly, a gathering of representatives from each member nation that is convened at the close of each four-year study period. Exhibit II-1-4 presents an overview of the process.

Accelerated procedures can be invoked when the need for rapid progress is expressed prior to a formal plenary assembly. Accelerated procedures require unanimous approval and become formal recommendations when voted on at the plenary session. Considerable care is taken to ensure that contributions that are considered for the accelerated process have undergone significant review to avoid the need to revise subsequent recommendations.

US Participation in the CCITT. The US, through its participation in the UN, is a voting member of the CCITT. The US position on standards issues is developed by the US National CCITT Committee. Membership in this committee, which is administered by the US State Department, is open to any interested individual or organization. Its purpose is to promote US interest in the CCITT; provide guidance on policy or positions in preparation for study group, working party, and plenary meetings; and advise on both the disposition of proposed contributions and the resolution of administrative and procedural problems related to the US National Committee.

The US National Committee is composed of four study groups in addition to a joint working party on ISDN. The four study groups focus on specific areas of interest that correspond to the CCITT study groups as follows:

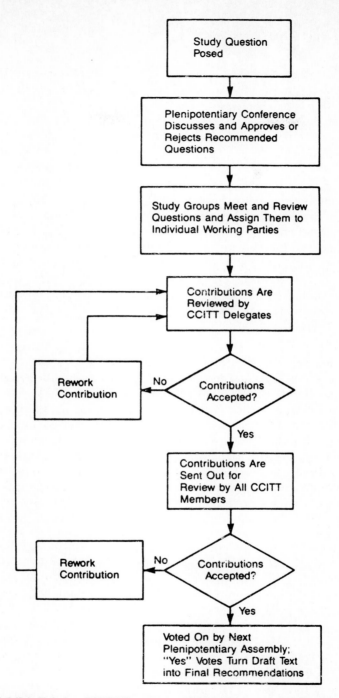

Exhibit II-1-4. CCITT Recommendation Process: An Overview

US Study Group	*CCITT Study Group*
A	I, II, III, and IX
B	XI and XVIII
C	IV, V, VI, X, XII, and XV
D	VII, VIII, and XVII
ISDN Joint Working Party	Representatives from all CCITT study groups

Contributions to the standards effort can be developed and submitted by any interested individual or group. All US contributions submitted to the CCITT must first pass through the US National CCITT Committee to ensure that a proposed contribution does not conflict with any US standards position. Although the US government provides the administrative framework for the development of standards positions, the work itself represents a collaborative effort among representatives of the public and private sectors involved in such committees as T1 and X3. Positions are communicated to the CCITT by a formal US delegation made up of these informed parties from both sectors.

The International Standards Organization (ISO)

The ISO was established in 1947 to achieve worldwide agreement on international standards. More specifically, its objective is to provide standards at an international level with a view to improving international collaboration and communication and to promoting smooth and equitable growth of international trade throughout the world.

The ISO today consists of 90 member countries. Any national body with a formal and recognized standards organization is permitted membership; those without such an organization can attend as observers, or correspondents. Correspondents are granted regular membership as their standards organizations mature. Although they are kept informed of ISO activities, correspondents cannot vote.

The scope of the ISO's work encompasses any activity that crosses national boundaries. The organization itself consists of approximately 2,400 technical bodies and more than 1,500 working parties whose charge is to facilitate world trade and contribute to public safety. The ISO has published more than 6,400 International Standards.

The US representative to the ISO is ANSI. This liaison is maintained through the close coordination of work among the various ANSI committees, such as X3, and their ISO counterparts, such as, in this case, TC97. The purpose of this alignment is to complement one another's standards work. In a similar fashion, the CCITT and the ISO are drawing closer, symbolic of the merging of data processing and communications.

The International Electrotechnical Commission (IEC)

The IEC is a voluntary standards organization that was established in 1904 to promote safety, compatibility, interchangeability, and acceptability. Its goal is to prevent such divergences of international electrical standards as the use of 110 volts in North America and 220 volts in Europe. It also focuses on the development of a universal technical language for terminology, universal electrical and electronic signals, and test methods. New projects are proposed by members to the committee of action, which then assigns work to the appropriate technical committee. Drafts of proposed standards are circulated by central offices and national committees.

The IEC has a structure similar to that of the ISO and likewise has its headquarters in Geneva. Its general assembly, which meets annually, is represented by 40 nations working through 82 technical committees and 700 working groups. As with the ISO, ANSI is the US representative to the IEC.

SUBMITTING A PROPOSED STANDARD

National Standards

Standards generally begin as contributions from an individual or corporation. A contribution contains a title that indicates its substance and a background section that identifies the environment being addressed and places limits on the use of the contributions (standards may well end up as part of a government regulation). The body of a contribution then enumerates the technical details of the proposed recommendation. The level of detail present should enable an individual reviewing the contribution to gain a complete and unambiguous understanding of the proposed standard. The ANSI style manual describes the preparation of contributions.

In many cases, a specific contribution addresses only one of several issues related to a proposed or existing recommendation. Once it is prepared, the contribution must be presented to the appropriate review organization. At this point, the writer of the standard must address the question of scope. Is the goal to produce a national standard (i.e., an American National Standard), an international standard, or both? The following example presumes that the intent is to submit the proposal to the ANSI process (see Exhibit II-1-5). For a standard to become an American National Standard, it must be reviewed and accepted by one of three standards organizations: the ISO, the IEC, or the ITU.

A communications standard is submitted to the T1 Committee on Telecommunications or through the X3 Committee on Information Processing Systems. Working through the chairperson of the appropriate technical

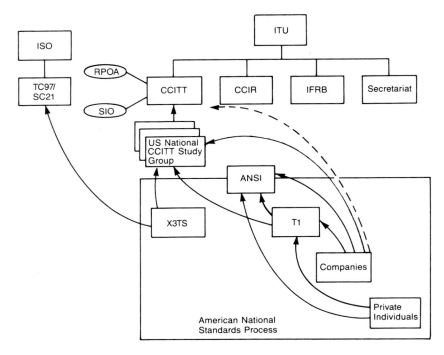

Exhibit II-1-5.   International Communications Standards Process

subcommittee, the originator of the contribution routes it to these groups for consideration. Because proposals often encompass several technical areas, the contribution can be routed to more than one technical subcommittee or ad hoc group for consideration.

It is customary for the originator of the contribution to present the contribution to the appropriate review group. This presentation consists of a brief overview of the proposal and of any areas of general concern. A discussion of the merits of the proposal ensues. The goal is to produce a consensus of support for the contribution. Usually, there are three potential outcomes of a presentation. First, the proposal produces the desired consensus. In this case, the proposal is forwarded to the technical subcommittee plenary with a recommendation for approval. Second, the discussion results in suggestions for changes to the proposal. If the originator is willing to make the recommended changes, the contribution can then be forwarded, with those changes, to the plenary for approval. Although this process appears to be fairly expeditious, several iterations may be necessary before the proposal is deemed ready for approval. The reality of standards work is a slow, methodical evolution from what exists to what is desired. In most cases, numerous positions initially exist, only one of which can prevail. This is the challenge of the standards development process.

Finally, if the originator is unwilling to make the changes or if the proposal is rejected outright, this information is forwarded to the plenary.

International Standards

If a contribution is regarded as a potential international standard, the T1-approved proposal, in addition to being sent to ANSI, is forwarded to the appropriate US study group for review. If the proposed contribution has undergone the scrutiny for which T1 is known, it generally will be approved expeditiously.

As suggested in Exhibit II-1-5, the originator of a contribution can submit the proposal directly to the appropriate US National CCITT study group for consideration. This is usually done when there is substantial knowledge of the proposal or when there are minor modifications to existing positions.

If a significant amount of technical analysis is deemed necessary (as occurs frequently in a technical subcommittee of T1), it is generally avoided at the US National CCITT study level. If this type of review is needed, the proposal may be rejected with the recommendation that a group of inter-ested parties meet to work out a suitable alternative. Conversely, if a pro-posal has been reviewed in the open forum of the T1 communications environment, it usually receives matter-of-fact approval for consideration as a US position. In addition, US study groups A and D maintain their own ad hoc working parties that develop contributions to be forwarded to the CCITT.

The consensus process turns out to be both a strength and, in some cases, a weakness of contributions generated as US positions. Because a broad range of discussion occurs before a proposal is accepted as a US position, these positions carry significant weight in the international arena. In fact, because the US standards process is open to all parties with direct and material interest in the development of a standard, an increasing num-ber of representatives of other countries participate in the process.

As might be imagined, this creates some unusual challenges in develop-ing lobbying positions for the international standards process. Standards are inherently a combination of technical and political views that result from compromises reached by means of negotiation. Thus, fallback positions must be developed to achieve desired results. This process can become very difficult if the person or group with whom the negotiations will take place knows the bottom-line position before negotiations begin. There have also been cases in which foreign interests have attempted to block pro-posals through the US consensus process, as in the videotext standards development of the early 1980s.

In most nations, government involvement in the international standards development community is natural, because communications generally is handled by the government or by a postal telephone and telegraph entity owned or operated by the government. This is not true in the US. One reason for the US National CCITT structure is to give all interested parties in the US an opportunity to contribute to the development of US policy and positions on telecommunications. These positions are communicated in the manner similar to that in which proposed contributions are developed. When study group or working party sessions are to be convened, the US State Department takes nominations for members of the US delegation. Those appointed to the delegation become specialized employees of the US government for the duration of the meetings. This is significant because of the principles on which this process is founded. When US positions are being developed through the standards process, standards originators represent their company or interest group. Once positions are agreed to at the US National CCITT Committee level, they become US positions; specific company or interest group identifications are no longer considered. Those who are selected as US delegates to the CCITT are expected to support and articulate the positions that have been adopted.

Chapter II-1-2
Functional OSI Profiles
Dorothy M. Cerni

SUCCESSFUL WORLDWIDE COMMUNICATIONS networks of the future will provide information exchange by supporting the interoperability of heterogeneous computer and communication products. An internationally standardized solution to the difficulties associated with developing an integrated multivendor networked system is based on the seven-layer open systems interconnection (OSI) reference model. The reference model classifies the total communications task into seven smaller, functionally separate pieces, or layers, and provides an overall framework for the developent of OSI standards. Many international standards, both completed and in draft form, are currently in place for each of the seven OSI layers. It is doubtful, however, that a full complement of international OSI protocols will ever be defined, because new applications and new communications technology will always require new standards.

For OSI to be realized practically, communicating systems must do more than simply conform to one or more OSI standards as written. Overall, OSI standards offer too much flexibility and contain too many optional classes, subsets, and parameters to provide practical interoperability as written, and using implementations of these standards, without prior agreements about the choices within these options, may lead to suboptimum performance and excessive implementation costs. Consequently, manufacturers and users interested in OSI products have united to reach agreements on precisely stated subsets or combinations of OSI standards as well as selected options within each standard that are considered necessary to support a given user function (e.g., such applications as electronic mail or such environments as packet-switched data networks).

Documents that specify the agreements reached in regional or topic-oriented groups have been called by various names: profiles, functional standards, implementation agreements, or implementor specifications. This chapter refers to these agreements as profiles and, for international activity, International Standardized Profiles (ISPs); the layer standards that make up a profile are distinguished from the profile itself. (As with most new endeavors, the profile concept preceded an agreed-on terminology. The Manufacturing Automation Protocol [MAP] version 2.1 is generally recognized as the first protocol.)

Conformance testing of systems that implement a specific profile, in turn, provides another measure of assurance that the implementing systems will be compatible. In general, conformance testing will determine whether a specific implementation adheres to the relevant OSI protocol standards by testing both the functional capabilities and the dynamic behavior of the implementation. The specifics regarding conformance testing and interoperability of OSI implementations are beyond the scope of this chapter.

OSI STANDARDS DEVELOPMENT

The international OSI standardization effort was initiated by the International Standards Organization (ISO). The US member of the ISO is the American National Standards Institute (ANSI), which is responsible for coordinating and managing US participation in the 164 technical committees and more than 2,000 subcommittees and working groups of ISO. The standardization work of ISO covers almost all areas of human commerce except electrotechnical issues, which are the responsibility of the International Electrotechnical Commission (IEC), and communications issues, which are the responsibility of the International Telephone and Telegraph Consultative Committee (CCITT). An international standard is the result of agreements worked out in technical committees and subcommittees by the member bodies. There are currently more than 5,000 ISO standards.

The international OSI effort, which assumed a major role in ISO during the past decade, was centered in the ISO's Technical Committee 97 (TC97), Information Processing Systems. Prodigious efforts have resulted in almost 200 OSI international standards distributed among the seven layers. Each of these standards has taken more than four years to develop, passing through the ISO-regulated stages of new work item, draft proposal, and draft international standard, before reaching the stable level of international standard. Because international standards are public documents, the user is assured that any changes made to the standards will be done in an open, predictable way, thereby protecting any investment in products and systems.

In 1987, ISO/TC97 underwent a historic reorganization: the ISO and the IEC formed a joint committee—Joint Technical Committee 1 (JTC 1), Information Technology—which combined the 14 subcommittees of ISO/TC97 with two relevant IEC committees (see Exhibit II-1-6). The merger guarantees that the ISO and the IEC will collaborate on complementary information technology work. The IEC, like the ISO, is a nontreaty, voluntary international standards organization. Formed in 1904, the IEC's standardization work includes electrical work of all kinds. The United States is rep-

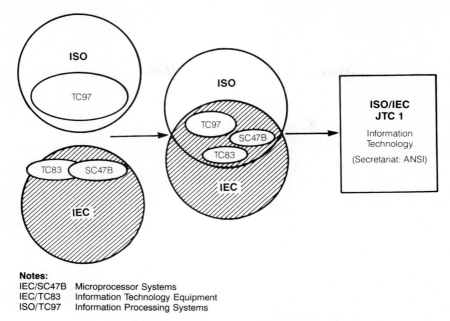

Notes:
IEC/SC47B Microprocessor Systems
IEC/TC83 Information Technology Equipment
ISO/TC97 Information Processing Systems

Exhibit II-1-6. Formation of ISO/IEC JTC 1

---

resented in the IEC by the US National Committee, which is a part of ANSI. Exhibit II-1-7 lists the ISO/IEC JTC 1 subcommittees that have a major role in the development of OSI standards.

In the US, work on information technology standards is done in several committees. The most significant of these is Accredited Standards Committee X3 (ASC X3), Information Processing Systems. Because ASC X3 operates under its open procedures, ANSI has accredited ASC X3 to present standards for balloting as US national standards. The X3 administrative committees and technical subcommittees are under the jurisdiction of the X3 Secretariat—the Computer and Business Equipment Manufacturers Association (CBEMA). The scope of the work done in the technical committees of ASC X3 corresponds to that done in JTC 1.

The CCITT is a permanent study committee of the International Telecommunications Union (ITU), founded in 1865, now a specialized agency of the United Nations. The ITU, through the work of its 157 member nations, has responsibility for regulating, planning, coordinating, and standardizing international communications of all kinds.

In contrast to the ISO and the IEC, whose basic purposes are to develop international standards, the ITU has several purposes, only one of which is

ISO/IEC JTC 1/SC 6:   Telecommunications and Information Exchange Between Systems
  Working Group 1:   Data Link Layer
  Working Group 2:   Network Layer
  Working Group 3:   Physical Layer
  Working Group 4:   Transport Layer

ISO/IEC JTC 1/SC 18:   Text and Office Systems
  Working Group 1:   User Requirements and SC 18 Management Support
  Working Group 3:   Document Architecture
  Working Group 4:   Procedures for Text Interchange
  Working Group 5:   Content Architectures
  Working Group 8:   Text Description and Processing Languages
  Working Group 9:   User-Systems Interfaces and Symbols

ISO/IEC JTC 1/SC 20:   Data Cryptographic Techniques
  Working Group 1:   Secret Key Algorithms and Applications
  Working Group 2:   Public Key Crypto-Systems and Modes of Use
  Working Group 3:   Use of Encipherment Techniques in Communication
                       Architectures

ISO/IEC JTC 1/SC 21:   Information Retrieval, Transfer, and Management for
                       Open Systems Interconnection
  Working Group 1:   OSI Architecture
  Working Group 3:   Data base
  Working Group 4:   OSI Systems Management
  Working Group 5:   Specific Application Services
  Working Group 6:   OSI Session, Presentation, and Common Application Services
  Working Group 7:   Open Distributed Processing

ISO/IEC JTC 1/SGFS: Special Group on Functional Standardization
  Working Group: Taxonomy of International Standardized Profiles

Exhibit II-1-7. ISO/IEC JTC 1 Subcommittees and Working Groups

---

the establishment of communications standards (called recommendations by the CCITT). The US member of the ITU, and thus of the CCITT, is the Department of State, and the US standards committee most involved with CCITT matters is ASC T1, Telecommunications, which has the same relationship to ANSI as does X3. The secretariat of T1 is the Exchange Carriers Standards Association (ECSA).

The CCITT is also involved in the OSI effort and, when possible, has developed recommendations that conform to the technical content of ISO standards in areas of joint interest. Similarly, some ISO standards are based on already developed CCITT recommendations. Certain OSI profiles include both CCITT recommendations (e.g., X.25) and ISO standards. The extensive liaison between the ISO and the CCITT during the past decade reflects the rapid convergence of computer and telecommunication technologies.

In 1980, the Institute of Electrical and Electronics Engineers (IEEE) formed the Project 802 Committee, which began work on local area network (LAN) standards. The IEEE LAN standards, now also finding a home in the ISO family of standards, form the foundation for the OSI profiles defining the transport of data over LANs.

OSI FUNCTIONAL PROFILE DEVELOPMENT

The international organizations involved in OSI-related functional profile development and publication are of three types:
- Open OSI implementors' workshops.
- Industry-specific organizations.
- Government groups.

A description of the detailed technical contents of the various profiles is beyond the scope of this chapter; further information can be obtained by contacting the appropriate profile organization. Addresses are provided in Exhibit II-1-8.

Corporation for Open Systems
1750 Old Meadow Rd
Suite 400
McLean VA 22120

COSAC
Treasury Board Secretariat
140 O'Connor St
Ottawa, K1A OR5
Canada

European Workshop for Open Systems
Rue Brederode,2
B-1000 Bruxelles
Belgium

Guide to the Use of Standards
Elsevier Science Publishing Company Inc
52 Vanderbilt Ave
New York NY 10017

ISO/IEC JTC 1/SGFS SECRETARY
Nederlands Normalizatie-Instituut
Postbus 5059
2600 GB Delft
The Netherlands

US MAP/TOP User Group
One SME Dr
PO Box 930
Dearborn MI 48121

NIST Workshop for Implementors of OSI
Building 225, Room B-217
National Institute of Standards
  and Technology
Gaithersburg MD 20899

OSI Asia and Oceania Workshop
c/o Interoperability Technology Association
  for Information Processing, Japan
Sumitomo Galen Bldg 3F
24-Dalyko-cha
Shinjuku-ku, Tokyo 160, Japan

POSI
Kikai-Shinko Bldg
Room 313
5-8 3-Chome
Shibakoen, Minaato-ku
Tokyo 105, Japan

SPAG SERVICES SA
1-2 Ave des Arts, Bte 11,
B-1040 Brussels
Belgium

UK GOSIP
CT2 A4 Network Systems Branch
Central Computer and
  Telecommunications Agency
Riverwalk House
157/161 Millbank
London SW1P 4RT
UK

US GOSIP
The US Government's OSI User
  Committee
Building 225, Room B-217
National Institute of Standards
  and Technology
Gaithersburg MD 20899

Exhibit II-1-8.   Profile Organization Addresses

OSI Implementors' Workshops

There are three OSI workshops currently operating: the National Institute of Standards and Technology (NIST) Workshop for Implementors of OSI, the European Workshop for Open Systems, and the OSI Asia and Oceania Workshop.

NIST Workshop for Implementors of OSI. The NIST Workshop for Implementors of OSI was formed in 1983 by US industry to serve as an open public forum for future users and potential suppliers of OSI products. The workshop, which is hosted by the Institute for Computer Science and Technology (ICST), publishes stable OSI implementation agreements representing widespread technical consensus.

The workshop organizes its work around topic-specific special interest groups (SIGs) that prepare the technical documentation describing their agreements. The NIST workshop SIGs are: directory services; file transfer, access, and management (FTAM); lower layers; management; open document architecture; OSINet; security; transaction processing; upper layers; virtual terminal; and message-handling systems (X.400). An executive committee of SIG chairpersons administers the workshop. A plenary assembly, consisting of all workshop delegates, considers SIG motions and other workshop business (e.g., approving the publication of the agreements). The SIGs are encouraged to work with standards organizations and user groups and to seek widespread technical consensus on implementation agreements through international discussions and liaison activities. The resultant agreements specify options and parameters of OSI standards that implementors can use to develop specifications for product development. Participation in all aspects of the workshop is open to all interested persons worldwide.

The workshop takes specifications of OSI protocol standards and produces agreements on implementation and testing particulars of functionally related groups of these standards. The first publication on stable agreements was "Stable Implementation Agreements for Open Systems Interconnection Protocols," Version 1, Edition 1, December 1987 (NBS Special Publication 500-150). This document defines agreements as stable if they meet the following criteria:

- The agreements are based on final standards (e.g., ISO international standards or CCITT recommendations) or nearly final standards (e.g., an ISO draft international standard with no significant changes expected).
- The agreements are approved by the NIST workshop plenary for progression from the ongoing agreements document to a draft of the stable agreements document and, after a period of review, have been passed to the final stable agreements—The only changes allowed

after approval will be clarifications, errata, and the correction of omissions discovered during implementation of the agreements.

The most significant and far-reaching agreements based on the NIST workshop output are: the US Government Open Systems Interconnection Profile (US GOSIP), the Manufacturing Automation Protocol (MAP), the Technical and Office Protocols (TOP), and the OSI Protocol Platform of the Corporation for Open Systems (COS). Each of these is discussed later in this chapter.

European Workshop for Open Systems.   The European Workshop for Open Systems was founded in 1987 with the cooperative efforts of nine European associations working to define the requirements for information technology interworking. The workshop is defined as a platform on which such organizations as users, vendors, postal, telephone, and telegraph organizations, standardization bodies, and academia can meet and agree on the policy and detailed definitions covering true market needs. Its objectives are:

- To provide a focal point in Europe for the study and development of the technical contents of OSI profiles and corresponding test specifications.
- To provide an open European forum for all experts willing to join and to contribute.
- To cooperate with workshops in other regions to achieve globally harmonized results and to make its output available for rapid standardization worldwide and in Europe.
- To ensure cooperation with the Conference of European Posts and Telegraphs and the European Telecom Standards Institute.

The European Workshop for Open Systems functions with autonomous management and budget but is legally part of the European committee for standardization, within the overall structure of the joint European standards institution, which includes the European Committee for Electrical Standardizations. The development of profiles in the workshop takes place in expert groups that are equivalent in function to the NIST SIGs. Both the expert groups and the oversight body, the technical assembly, are open to those interested and able to contribute. The steering committee is open to Europe-wide, primarily nonprofit, organizations. The workshop intends that its output documents be freely available and directly usable as input to the international ISO/IEC ISP process as well as to the parallel program of European functional standardization.

OSI Asia-Oceania Workshop.   In direct response to the need to harmonize functional profiles on a worldwide basis, as required by the ISO/IEC ISP process, Japan's Promotion Conference for OSI (POSI) requested that the Interoperability Technology Association for Information Processing establish an OSI workshop. The major responsibility of the workshop is to review proposed draft ISPs for the Asia-Pacific Ocean area. The OSI Asia-Oceania Workshop's purpose may be summarized as follows:

- The workshop is to be a facility for open discussions about OSI implementors' specifications.
- The workshop is to be open to all researchers and engineers involved in this field in the Asia-Oceania area.
- In cooperation with the corresponding workshops in Europe and the US (i.e., the European Workshop for Open Systems and the NIST workshop), the workshop is to incorporate the results of the work done in the ISO toward the development of ISPs.

The workshop is directed by a steering committee and has a structure similar to that of the NIST workshop: a plenary assembly (called a plenary board) and individual-subject SIGs. The workshop secretariat is the Interoperability Technology Association in Information Processing, Japan.

Industry-Specific Groups

The following five industry groups have a specific interest in functional profile development, although not all of them are in the business of developing profiles:

- MAP User Group.
- TOP User Group.
- Standards Promotion and Application Group (SPAG).
- COS.
- POSI.

MAP and TOP User Groups. Although MAP and TOP are often used together (i.e., MAP/TOP), each represents an independent effort to specify a set of OSI options and services to meet specific application problems. MAP, designed for the factory floor, and TOP, designed for the technical office, have no required relationship but do interact in many ways. In general, the differences between the two specifications occur in OSI layers 1 and 2 and in overlapping selections at layer 7. MAP and TOP user groups have grown to the World Federation of MAP/TOP Users, which has its own committee and meeting structures. The World Federation has four regions, each of which may decide to involve MAP only or MAP and TOP.

In the US, both MAP and TOP user groups have a technical subcommittee structure that concentrates on specific functional areas. A major subcommittee responsibility is to study user requirements and specify the protocol solutions that meet those requirements. This work is done in close cooperation with the NIST OSI Implementors' Workshop.

MAP. MAP is a seven-layer, broadband, token-bus-based communications specification for the factory environment. MAP was initiated during the early 1980s by General Motors (GM) to permit full-plant automation requiring communications between heterogeneous computers and programmable devices. From the beginning, GM recognized that, ultimately, only the use of nonproprietary international OSI standards would ensure universal user support for their program, which would in turn activate product availability. The technical agreements expressed in the successive versions of MAP have been largely reached in appropriate NIST workshop SIGs. MAP includes certain capabilities (e.g., network management and directory services) for which ISO standards are not yet available.

MAP version 3.0 illustrates a problem common to all OSI-related implementations: determining when a profile is stable enough to warrant an implementation that will be in force long enough to protect the user's investment in hardware and application code. The MAP work completed by 1988 reached the point of development that permits a strategy or policy aimed at encouraging implementation yet protecting investment as future specifications are added that provide new functions. Accordingly, the US MAP/TOP User Group, with the support of the World Federation, has stated that version 3.0 provides a stability baseline and that all releases for at least six years must include mechanisms for interoperability with MAP version 3.0. Compatible additions will be reviewed for inclusion in future releases of the MAP specification on an annual basis. The only changes that will be allowed to the MAP 3.0 specification itself are errata changes.

TOP. TOP, the seven-layer specification based on OSI and including the carrier sense multiple access/collision detection (CSMA/CD) LAN IEEE 802.3 specification, was initiated by the Boeing Company to provide standards for data communications between the company's mainframe, minicomputers, and microcomputers. In version 3.0, TOP provides data interchange solutions for business, engineering, and publishing environments that need to exchange information in a multivendor environment. It is applicable to users who need to communicate with other organizations, both inside and outside their company.

Two purposes of the TOP specification are to:
- Promote the design and testing of TOP products that meet the requirements of a significant market segment.
- Act as a procurement tool for users.

TOP specifies the standards needed to perform document exchange, graphics exchange, product data exchange, electronic mail, file transfer, printer/plotter services, and general network connectivity. TOP does not limit users to those functional areas explicitly defined but allows users to take advantage of a consistent set of communications architectures and protocols to develop solutions to their specific problems. TOP networks will be able to transfer data to and accept data from any of the MAP networks, thus linking the engineering and office environments to the factory floor. TOP is also fully compatible with US GOSIP.

SPAG. Originally composed of 12 leading European manufacturers of information technology products, SPAG was founded in 1983 at the request of the Commission of the European Communities. The commission's initial objective was to receive guidance from SPAG in the promulgation of a European information technology standards policy for providing a homogeneous and broader European market for information technology products. The resultant European policy, which is also followed by the European Free Trade Area, has targeted OSI implementation as the basis for public procurement for 17 European countries. According to this plan, all member countries will eventually adopt the functional standards published as European standards, which will be implemented by the publication, in each country, of national standards identical in presentation and content to the European standards.

The goal of this effort is the elimination of differences in national standards, thereby eliminating technical barriers to commerce in information systems and data communications products in European markets. To assist in the achievement of this policy, the major joint European standards institutions in conjunction with the Conference of European Posts and Telegraph and Telephone, formed a joint information Technology Steering Committee. A major input to this process has been the extensive compilations of profiles developed by the SPAG Technical Committee and expert groups and published in "The Guide to the Use of Standards," from which the European functional standards have been developed and published as European prestandards.

In 1986, a separate commercial entity, consisting of 8 of the 12 SPAG companies, was formed as SPAG Services. Its purpose was to offer development, testing, and validation services for products conforming to European functional standards. Recently, SPAG and SPAG Services became integrated into SPAG Services SA, and the new company now has an expanded shareholder base as well as associate members.

COS. COS International was established in 1986 with the long-term goal of providing an international vehicle for accelerating the introduction of interoperable, multivendor products and services operating under agreed-to OSI, integrated services digital network (ISDN), and related international standards to ensure the acceptance of an open communications environment in world markets. The 65-plus vendors, users, and carriers who are members of COS participate by means of both an administrative board–

executive committee structure, as well as a technical strategy forum–steering committee/technical committee structure. The COS staff provides a range of testing services for OSI products alleging compliance with COS-selected functional profiles and will implement a product and service certification scheme.

The functional profiles tested by COS are found in the COS profile specification document, which is produced by the strategy forum. The document consists almost entirely of references to the relevant sections of the NIST OSI Workshop Implementors' Agreements and the appropriate standards themselves.

One goal of the COS testing program is to confer a COS mark on products or services that meet COS requirements. The COS-mark program has elements of formal certification, in that it is an official acknowledgment of successfully negotiated conformance testing. In addition to being a visual and symbolic indicator, however, the COS mark implies certain commitments by the vendors and COS to achieve successful communication between COS-marked products.

POSI. POSI was founded by six major computer manufacturers and Nippon Telephone and Telegraph. The purpose of POSI is to promote OSI standards in manufacturing by exchanging information among member companies and by cooperating with other groups worldwide. In particular, POSI objectives are:

- Corporate-level policy decisions and information exchange aimed at promoting OSI.
- Corporate-level policy decisions regarding the development of OSI subsets and functional standards.
- Dissemination of results achieved.
- Information exchange and cooperation between European and US computer manufacturers.

The conference itself consists of executives of the member companies. The Japanese system for promoting OSI is complicated, and POSI plays a leading role. Most of the technical work is done by the Interoperability Technology Association for Information Processing, Japan, which is responsible for both OSI profile development and conformance testing. The chairs of the technical committees of this association, however, are most often selected from POSI member companies.

Government Documents

Many government organizations worldwide, both national and regional, have already developed information technology policies that make mandatory, or recommend strongly, the eventual procurement of OSI computer products for public sector use. In addition to the European community organizations discussed, these government entities include the US, Japan, Canada, and Australia. Three government-sponsored documents written to assist agencies and departments in understanding and implementing OSI functional profiles, according to the specific needs of a particular government, are: UK GOSIP, US GOSIP, and the Canadian Open Systems Application Criteria (COSAC).

UK GOSIP.   The UK GOSIP document is the responsibility of the Central Computer and Telecommunications Agency of HM Treasury. It precisely reflects the agency's policy commitment to OSI, originally promulgated in 1984. GOSIP is a user application profile, intended to provide specific technical guidance to both government departments and equipment suppliers. GOSIP is complementary to such other agency documents as management guides and surveys of major OSI product suppliers. It also provides the user with the agency's view of worldwide activities in conformance testing. The GOSIP specification as such is not mandatory for government departments but is expected to have wide application (beyond that of the public sector) and the agency encourages other major users to adopt identical or equivalent specifications. It is the stated intent of UK GOSIP to be as compatible as possible with other major profile efforts, including MAP, TOP, and US GOSIP. In keeping with this philosophy, a European procurement handbook for open systems, incorporating version 3 of UK GOSIP, is being developed for potential implementation across the European economic community.

US GOSIP.   US GOSIP, developed by the US government's OSI Users Committee, is based on NIST workshop agreements. GOSIP addresses the federal government's need to move immediately to OSI multivendor interconnectivity without sacrificing the essential functions already implemented in critical networking systems. GOSIP is consistent with and complementary to MAP and TOP specifications. A modified version of this document, issued as a federal information processing standard in 1988, is mandatory for government users (although exceptions still apply). The standard specifies applicability and implementation procedures. US government agencies have a grace period until 1990 to comply; it is expected that vendors will have brought many more OSI-conformant products to the marketplace by then.

COSAC.   COSAC reflects the 1987 government policy that endorses OSI as a preferred information technology strategy and requires that departments and agencies state a clear preference for OSI-based products and services. COSAC is a dynamic guide to the identification of particular profiles, providing general guidance for the application of OSI in government. Its main objective is to assist government groups in migrating from their current information processing systems to systems that conform with OSI standards and to identify for Canadian suppliers the related technology requirements. As such, the document will reflect evolving standards and profiles, test results, and user experiences. COSAC will attempt to reference any relevant existing profiles and will clearly identify areas of compatibility and incompatibility.

Specific governmental approaches, as embodied in UK GOSIP, US GO-SIP, and COSAC, will inevitably retain a unique national focus. Even so, each organization expresses the intent to continue the close liaisons already established with other public and private efforts; the goal is to achieve the maximum compatibility possible without losing insight of individual government requirements.

FUNCTIONAL PROFILE HARMONIZATION

The underlying theme of all these discussions regarding workshops, industry groups, and government profile documents is the accelerating effort and drive toward the development of OSI profiles and the implementation of these profiles in commercial products. None of these efforts is being made in isolation; extensive cooperative initiatives have been pursued beyond those described in this chapter.

The MAP and TOP programs represent industry-specific, user-led initiatives that were in operation before any organized, open forum (e.g., the NIST workshop) was established; both became identified with the workshop efforts. SPAG, a vendor-only group, pursued the actual development of profiles, which were then inserted into the general European profile development and acceptance work and will now be absorbed into the larger European workshop programs. COS, an international not-for-profit vendor and user organization, does not develop profiles but rather bases its agreements on the work of such open, recognized groups as the NIST workshops. COS will perform conformance tests on implementations. POSI, a group with limited vendor participation, is intended to be an OSI public-relations group, exchanging information, supporting the work on functional profiles done by the Japanese interoperability technology association, and interacting with the international community. COS, POSI, SPAG, and the MAP/TOP User Group have already forged numerous bilateral and multilateral harmonization efforts. In addition, such public activities as the 1988 Enterprise Networking Event International, cosponsored by the US MAP/TOP User Group and COS, make the various developments in OSI network implementation and testing highly visible. Efforts are being made by these groups to promote international coordination of conformance tests and test methods to achieve globally accepted tests and test results and to accelerate the availability of test facilities.

It is commonplace to assert that profiles are developed to provide the basis for OSI-product development and conformance testing with the ultimate goal of system interoperability. Therefore, if profiles implemented worldwide differ, the resultant products will not be interoperable even if they all seemingly conform to OSI standards. There is an inevitable tension between the needs and requirements of individual organizations on one hand and the advantages of a universal market of OSI products based on identical implementation profiles on the other. ISO has chosen to step into this paradoxical world to participate in the effort toward convergence. The activities of interested ISO members and the four industry-specific groups— MAP/TOP user groups, SPAG, COS, and POSI—indicate that the differences in organizational intent and function have been eclipsed by a common interest in worldwide interoperability.

ISO/IEC JTC 1 Special Group on Functional Standardization

In 1985, TC87 began to define its role in the international effort to group information technology standards into functional stacks of standards. Some experts felt that standards implementation was beyond the realm of the ISO; others felt that the intrinsic interrelationship between the profiles and the standards made it imperative for the ISO to become involved at some level. Therefore, at the TC97 plenary in 1986, a special working group on functional standardization was formed to study the issue. Consequently, the Special Group on Functional Standardization was chartered with the following action items:

- Definition of functional standardization and functional standards.
- Development of a catalog of functional standards with appropriate classification.
- Definition of a methodology for achieving functional standardization.
- Development of a set of operating procedures and assessment of resources.
- Execution of the review of proposed draft functional standards.
- Consideration of requirements for functional standards on conformance and maintenance.
- Development of an expeditious publication procedure.

The Special Group on Functional Standardization is now the ISO/IEC JTC 1/SGFS. This group has the status of an ISO subcommittee and reports directly to JTC 1.

Definition of an ISP

ISO and IEC jointly agreed to a new publication, the ISP, which has the status of an international standard in the international community. The ISP has been defined as an internationally agreed-to, harmonized document that identifies a group of standards together with the options and parameters necessary to accomplish a function or set of functions.

The Liaisons to the ISO: SPAG, COS, POSI, and the World Federation of MAP/TOP Users

Concurrent with the study of the ISO's role in functional profile work, the ISO Executive Board and Council were examining the role of organizations worldwide (i.e., SPAG, POSI, COS, and the World Federation of MAP/TOP Users) that were recognized as having a special interest in convergent OSI functional profile development. Even though these four organizations had no special relationship with the ISO, representatives from each were invited to brief the ISO board on activities regarding the ISO process, particularly relating to the development of standards needed for profile establishment. These four organizations provide the ISO with priorities of needed standards development and harmonized profiles that can potentially feed into the ISO-based process of recognition, approval, and publication of functional profiles. There were two direct consequences of this action on the part of the ISO General Secretary, each of which marked a first in ISO history:

- A joint meeting was held between the chairs of TC97 and the OSI-related technical subcommittees and representatives from the four organizations to review the ongoing work of TC97 and compare it with the priority requirements for standards identified by the four organizations.
- The ISO recognized these four groups as special liaisons to the Special Group on Functional Standardization—This special-liaison status permits the groups to participate directly, rather than through a national member body, in the special group activities but does not confer voting privileges.

Taxonomy Framework and Directory of Profiles

The JTC 1 Special Group on Functional Standards and its Taxonomy Group have produced a draft technical report, DTR 10,000, "Taxonomy Framework and Directory of Profiles," which defines profiles as they are documented in ISPs and gives guidance to organizations making proposals for draft ISPs on the nature and content of the documents being produced. Although applicable to all areas of competence within JTC 1, priority is given at this initial stage of development to profiles in the OSI area—that is, those that specify OSI base standards and those expected to be used in conjunction with these base standards.

According to DTR 10,000, base standards specify single-layer or multilayer procedures and formats that facilitate the exchange of information between systems. They provide options—anticipating the needs of a variety of applications and taking into account the different capabilities of real systems and networks.

DTR 10,000 is largely concerned with the concept of a profile, which is defined as a set of one or more base standards—and, if applicable, the identification of chosen classes, subsets, options, and parameters of those base standards—necessary for accomplishing a particular function. One or more profiles, with accompanying text detailing the format and structure, constitutes an ISP. The document expresses a basic intent that profiles should not contradict base standards but should specify permitted options for each base standard and suitable values for those parameters left unspecified in the base standards.

The profiles define combinations of base standards for the purpose of:

- Identifying the base standards as well as the appropriate classes, subsets, options, and parameters necessary to accomplish identified functions for such purposes as interoperability.
- Providing a system of referencing the various uses of base standards that is meaningful to both users and suppliers.
- Providing a means to enhance the availability of consistent implementations of functionally defined groups of base standards, which are expected to be major application systems components.
- Promoting uniformity in the development of tests for systems that implement the functions associated with the profiles.

Accordingly, a profile should leave open as few options as possible in the base standards to which it refers in order to maximize the probability that the implementations can interoperate. The profile cannot specify any requirements that contradict or cause nonconformance to the base standards to which it refers. It may, however, contain conformance requirements that are more specific and limited in scope than those of the base standards to which it refers. The capabilities and behavior specified in the profile will always be valid in terms of the base standards, but the profile may exclude some valid optional capabilities and optional behavior permitted in those base standards.

In summary, then, the basic principles for an OSI profile are that it specifies the application of one or more OSI base standards in support of a specific requirement for communications between systems. It does not require any departure from the structure defined by the OSI reference model, but it does define only relevant parts of the function, not the systems as a whole.

The profile taxonomy is the structure and classification within which profiles fit. It gives a first-level specification of profiles, including any determined technical constraints. It both classifies individual profiles and specifies several relationships among them.

The following classes of OSI profiles have been recognized:

- T—Connection-mode transport profiles, related to subnetwork type.
- U—Connectionless-mode transport profiles, related to subnetwork type.
- A—Application profiles using connection-mode transport service.
- B—Application profiles using connectionless-mode transport service.
- F—Interchange format and representation profiles.
- R—Relay functions between T or U profiles.

Process for Submission of Proposed Draft ISPs to the JTC 1 Special Group on Functional Standardization

The process inherent in the submission of a preliminary draft ISP to ISO for approval as an ISP by a JTC 1–recognized organization differs from the traditional standards development process in several ways. It is important to examine these differences to have a clear understanding of the ISP process under the ISO umbrella. Although there are similar (and even identical) aspects to the traditional standardization approval process, the ISP work is a new activity for ISO, done in new ways and leading to new documents. Exhibit II-1-9 summarizes some of the differences.

The first trial run of the process involves the submission of three preliminary draft ISPs. The three drafts are detailed in Exhibit II-1-10.

The six stages in the idealized procedural mechanism for developing an ISP are:

1. The submitting organization announces to the Special Group on Func-

ISO STANDARD	ISPs
Completed documents, ready for approval at any stage (preliminary draft, draft ISP, or ISP), are developed (or reworked) in committee.	Documents may be prepared outside of the ISO process.
Consensus is required for each step of the process within the rules established by ISO.	The global harmonization is the responsibility of the submitter.
The process is guaranteed to be open (e.g., membership is broad-based and balanced, procedures are fair and regulated, reviews consider all comments).	The assurance of the openness of development is also the responsibility of the submitter.
Each proposed standard has at least two technical committee votes (for draft ISP and ISP status).	The draft IBP is submitted to only one vote of the technical committee (for ISP status).

Exhibit II-1-9.   Traditional Versus ISP Standardization Approval

1. Preliminary Draft ISP TAann:[1]   "Connection-mode transport service over connectionless network service" (COS and the World Federation of MAP/TOP Users[2])

    Profile a)   TA11[3]:   In public-switched data networks—permanent access (COS)

    Profile b)   TA51:   In local area networks with CSMA/CD ISO 8802.3 (World Federation of MAP/TOP Users)

(The number of profiles could become lengthy)

2. Preliminary Draft ISP TB/TC/TD/TE: "Connection-mode transport service over connection-mode network service in public-switched data networks—permanent access" (POSI)

    Profile a)   TB111:   Transport protocol classes 0 + 2 + 4

    Profile b)   TC111:   Transport protocol classes 0 + 2

    Profile c)   TD111:   Transport protocol class 0

    Profile d)   TE111:   Transport protocol class 2

3. Preliminary Draft ISP AFT11:   "FTAM: Simple file transfer" (SPAG)

Notes:
1. Preliminary Draft ISP numbers and titles are preliminary place holders.
2. The submitting organization is given in parentheses.
3. Profile numbers are those contained in the taxonomy.

Exhibit II-1-10.   Preliminary Draft International Standards Profiles

---

tional Standardization its intent to develop and submit a globally harmonized document to the ISO process of review and approval.

2. The special group appoints an expert review team to examine the preliminary draft ISP and accompanying materials.

3. The preliminary draft ISP and explanatory report are submitted to the special group—The target time for ISP publication by the ISO is 7 to 10 months.

4. The review team prepares and distributes a review report.

5. The preliminary draft ISP advances to the draft ISP level and a ballot takes place according to JTC 1 procedures.

6. Upon publication of the ISP, the submitting organization assumes responsibility as the maintenance authority for the ISP.

SUMMARY

In the confusing international arena of information technology implementation, the JTC 1 SGFS has carved out an ISP pathway connecting the world of profile developers to that of profile testers. The first preliminary draft ISPs gaining access to this path test the usefulness of the signposts, the appropriate direction and length of the path, and the worth of the journey. After emerging as ISO-approved ISPs, the documents will be available for public use and will provide a basis for the development of uniform, internationally recognized test systems. The achievement of such a goal will prove to be the acid test of the usefulness of maintaining such a pathway within the ISO world. It is not an exaggeration to state that the decision by the ISO and the IEC to found the Special Group on Functional Standardization has established a new venture of such magnitude that all the consequences of this decision will not be played out for many years to come.

Chapter II-1-3
OSI Systems Management
William Collins

INTEREST IN OPEN SYSTEMS INTERCONNECTION (OSI) has increased dramatically over the past several years. Progress on standards for OSI management is one of the main reasons behind the growing interest.

A communications manager's task is to facilitate the exchange of information between all the offices in a company's network, regardless of the type of computer system installed at each office. Connecting computer systems from different vendors has often been an extraordinarily difficult task; developments in OSI are helping to resolve this situation.

OSI will permit a communications manager to design a flexible communications system that will connect systems productively and efficiently throughout the corporate structure. The design will require the use of the comprehensive basic reference model of OSI, OSI management services and protocols, and OSI management applications. OSI leaves room for system growth and change in an elegant way: it standardizes the protocols between functions but not the design for implementing those functions, thus ensuring compatibility among different systems while encouraging innovative engineering.

BASIC REFERENCE MODEL

The OSI basic reference model provides substantial flexibility in designing communications systems. It defines seven layers of functions involved in data communications and the services and protocols required to perform these functions.

The basic reference model defines OSI systems management in terms of management applications that communicate with one another through application protocols. These management applications provide the following capabilities:

- Mechanisms for monitoring, controlling, and coordinating all managed objects within open systems.
- The ability to manage objects related to single or multiple layers.
- A framework for connecting open systems that allows suppliers to construct their individual systems in unique ways but still allows manageability across all open systems in the network.

Although the uniformity of many aspects of OSI simplifies the communications manager's job, it leaves many issues that must be resolved on a case-by-case basis. Managers have the responsibility not only for implementing the OSI layer services and protocols but for deciding which layers of the OSI reference model are applicable to a given problem, which OSI system management service and protocol is best suited for a specific implementation, and whether these services and protocols can be properly translated to a specific implementation.

ISSUES OF OSI SYSTEMS MANAGEMENT

OSI systems management provides a mechanism for the exchange of information about monitoring, controlling, and coordinating communications resources in open systems. The resources may be both real pieces of equipment and logical representations of real equipment. The OSI management framework uses the term *managed objects* to describe these resources and to distinguish the objects themselves from management information about them.

Most management information exchanged between open systems requires the same communications services as those for other application layer exchanges. Systems management communications is effected, therefore, through application layer protocols. The scope of OSI systems management includes the following:

- Defining mechanisms for monitoring, controlling, and coordinating all OSI resources across open systems.
- Modeling systems management application processes, which are the management functions in an open system.
- Specifying systems management application entities, which are the aspects of systems management application processes that concern OSI communications.
- Using application layer services and protocols for the exchange of management information and control.

Exhibit II-1-11 illustrates the relationship among these aspects of OSI systems management in the context of a connection between two open systems.

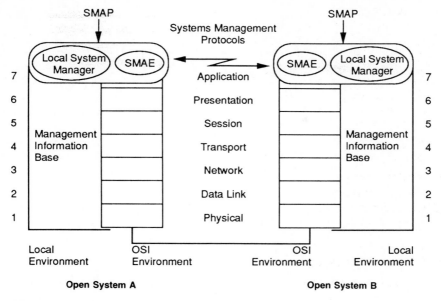

SMAP
SMAP
Systems Management
Protocols

Local System Manager · SMAE · Application · SMAE · Local System Manager

7		7
6	Presentation	6
5	Session	5
4	Transport	4
3	Network	3
2	Data Link	2
1	Physical	1

Management Information Base

Management Information Base

Local Environment · OSI Environment · OSI Environment · Local Environment

Open System A · **Open System B**

Notes:
SMAE   Systems management application entity
SMAP   Systems management application process

Exhibit II-1-11.   Systems Management Information Exchange

SYSTEMS MANAGEMENT APPLICATION PROCESS

A systems management application process is an application process that performs management functions. It consists of a systems management application entity to carry out communications with other systems management application processes, a management information base, and possibly one or more managers that provide various functions.

A management information base is a conceptual repository containing the OSI systems management data in an open system that is available to the OSI environment. The management information base data may be provided by a local systems manager through a systems management application process or by a remote open system through either system management protocols by way of the systems management application entity or through layer management protocols. The management information base data is available to all of these sources. It is organized according to the OSI systems management standard on the structure of management information, which defines the format used to identify the OSI management data.

SYSTEMS MANAGEMENT APPLICATION ENTITIES

In the application layer of the OSI reference model, application entities represent the communications aspects of an application process. Systems management application entities contain a number of service elements, including common management information service elements, remote operation service elements, and association control service elements. The relationship among these service elements is illustrated in Exhibit II-1-12.

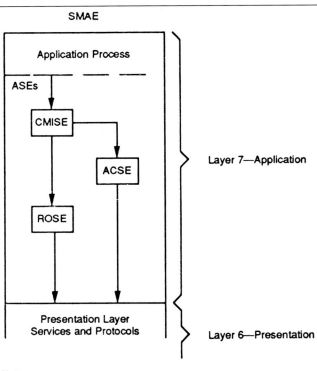

Notes:
ACSE   Association control service element
ASE   Application service element
CMISE Common management information service element
ROSE   Remote operation service element
SMAE Systems management application entity

Exhibit II-1-12. Systems Management Application Entities

Services	Type
M-INITIALIZE	Confirmed
M-TERMINATE	Confirmed
M-ABORT	Nonconfirmed
M-EVENT-REPORT	Confirmed/nonconfirmed
M-GET	Confirmed
M-SET	Confirmed/nonconfirmed
M-ACTION	Confirmed/nonconfirmed
M-CREATE	Confirmed
M-DELETE	Confirmed

Exhibit II-1-13. Summary of Common Management Information Services

---

Common Management Information Service Element

The common management information service element provides OSI common management information services through the use of the OSI common management information protocol.

The definition of OSI common management information services is currently under development within Working Group 4 of the International Standards Organization and International Electrotechnical Commission Joint Technical Committee 1 Subcommittee 21 (ISO/IEC JTC1/SC 21) and, in the US, American National Standards Institute (ANSI) X3T5.4. The services—listed in Exhibit II-1-13—allow OSI systems to exchange information between systems management applications processes. Confirmed services require a response and nonconfirmed services do not; some services require both.

M-INITIALIZE.   The M-INITIALIZE service, a confirmed service, is used by a common management information service element user to establish an association with a peer common management information service element user and forms the first phase of management information service activity. This service is used only to create an association and may not be issued on an established association. It is mapped to the association control service element.

M-TERMINATE.   The M-TERMINATE service, a confirmed service, is used by a common management information service element user to cause a normal release of an association with a peer common management information service element user. It is mapped to the association control service element.

M-ABORT.   The M-ABORT service, a nonconfirmed service, is used by a common management information service element user to cause an abrupt release of an association with a peer user. It is mapped to the association control service element.

M-EVENT-REPORT.　The M-EVENT-REPORT service is used by a common management information service element user to report an event to a peer common management information service element user. It is defined as both a confirmed and a nonconfirmed service.

M-GET.　The M-GET service, a confirmed service, is used by a common management information service element user to retrieve management information values from a peer user.

M-SET.　The M-SET service is used by an invoking common management information service element user to request the modification of attribute values by a peer user. It is defined as both a confirmed and a nonconfirmed service.

M-ACTION.　The M-ACTION service is used by a common management information service element user to request a peer user to perform an action on a managed object. It is defined as both a confirmed and a nonconfirmed service.

M-CREATE.　The M-CREATE service, a confirmed service, is used by an invoking common management information service element user to request a peer common management information service element user to create a representation of a new managed object instance and simultaneously to register its identification. The representation should include the object's identification and the values of its associated management information.

M-DELETE.　The M-DELETE service, a confirmed service, is used by an invoking common management information service element user to request a peer user to delete a representation of a managed object instance and to deregister its identification.

This limited set of common management information services permits a variety of management communications.

Common Management Information Protocol

The common management information protocol provides the common management information services. Although a detailed discussion of these protocols is beyond the scope of this chapter, it is important to note that they use the connection-oriented remote operation service as an underlying service. This service is itself being modified by the ISO to relax restrictions that do not allow responders to invoke operations and to allow an alignment between the ISO and the International Telephone and Telegraph Consultative Committee (CCITT) versions of this service.

Although the common management information protocols specification requires connection-oriented remote operation services, it is not desirable to use it for system management exchanges that do not require a transport connection when the traffic for common management information protocols operations is light. In this situation, a connectionless remote operation service is preferable. For example, when a systems management application entity is reporting events randomly, the high overhead would make it costly to maintain a transport connection. Although OSI has traditionally been connection oriented, work has begun in the ISO to specify connectionless services for certain applications.

Remote Operation Service Element

The remote operation service element provides remote operation services through the use of the remote operation protocol. Application entities that are interactive may require interactive protocols. For example, if application process A requests that an operation be performed by application process B, application process B will attempt to perform the operation and then report the outcome of the attempt back to application process A. This interaction is illustrated in Exhibit II-1-14.

The specification and implementation of the interaction between application processes A and B is achieved by a remote operations service element. Remote operation service elements provide a uniform mechanism for defining and representing operations and their outcomes. They can support operations that always report results, operations that never report results, operations that report only success, and operations that report only failure.

One application process can perform operations provided by another by entering into an exchange of operation protocol data units. These units,

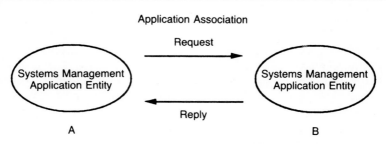

Exhibit II-1-14.   Remote Operations Model

defined by such remote operation services as Invoke, ReturnResult, ReturnError, and Reject, can be described as follows:

- Invoke—Requests that an operation be performed and carries the description of the operation to be performed.
- ReturnResult—Reports the successful completion of an operation.
- ReturnError—Reports the unsuccessful completion of an operation.
- Reject—Reports the receipt and rejection of an invalid unit.

Because remote operation service elements provide an environment for interactive protocols, the developer of an interactive application process that uses an interactive protocol must specify the nature of the operation, the results, and the errors that are specific to the application.

Association Control Service Element

The association control service element provides association control services through the association control protocol. It is concerned with providing a service and a protocol that support the establishment and release of application associations, the identification of the application contexts applicable to the association, the selection of an initial application context, and the transfer of user information between peer application entities. It also provides the means for identifying the presentation and session layer requirements for supporting the application association.

ISO/IEC JTC1/SC 21 Working Group 6 is currently defining a context management facility that supports the definition of more than one application context, the deletion of an existing defined context, and the selection of a current application context from a set of defined contexts. The management of application context and presentation context, however, requires further study.

Application Context Management

Some ISO/IEC JTC1/SC 21 standards committees consider an application association to involve the defined application context list and the current application context. Each application association has one defined application context list that contains the names of all application contexts agreed upon by both communication application entities. Association control specifies three services for managing the defined application context list: A-ASSOCIATE, A-CONTEXT-DEFINE, and A-CONTEXT-DELETE.

Each transmission of the protocol data unit of an application association has exactly one current application context, which must be a member of the defined application context list. The A-CONTEXT-SELECT service specifies the current application context for transmission from the initiator of the service to its recipient peer application entity.

In an application environment with a single application association, context management is not required. In this case, the application control service elements are considered to be part of the application context of the user application and do not form a separate application context. Context is acquired through the A-ASSOCIATE service, the context can be relinquished through the A-RELEASE service, and the context can be aborted through either the A-U-ABORT or A-P-ABORT services.

In an application environment with more than one application service element, context management may be needed. In this case, the service elements used to acquire, relinquish, and abort the context depend on the circumstances. Context-acquiring services include A-ASSOCIATE, A-CONTEXT-DEFINE, and A-CONTEXT-SWITCH. Context-relinquishing services include A-U-ABORT and A-P-ABORT.

SPECIFIC MANAGEMENT FUNCTIONAL AREAS

The *OSI Management Framework* DIS 7498 includes standards for specific management functional areas that define the procedures used to accomplish management tasks. These include configuration management, fault management, security management, performance management, and accounting management. Exhibit II-1-15 illustrates these specific management functional areas as service users of common management information service elements.

Configuration Management. Configuration management facilities control, identify, and collect data from and provide data to OSI resources to

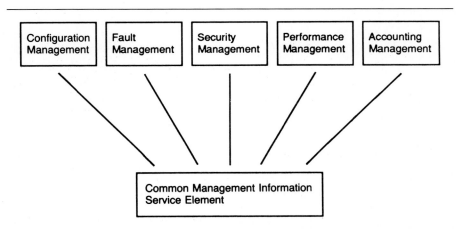

Exhibit II-1-15.   Specific Management Functional Areas and Common Management Information Service Element

permit continuous operation of interconnected devices. The facilities provide for setting the open system parameters, initializing and closing down OSI resources, collecting data about the open system state both on a routine basis and in response to a significant change of state, and providing specified data to open systems on request.

Fault Management. Fault management facilities permit the detection, isolation, and correction of abnormal operations in the OSI environment. Faults, which may be persistent or transient, cause open systems to fail to meet their operational objectives. Faults manifest themselves as errors in the operation of an open system; error detection provides the mechanism for recognizing faults. The facilities provide for maintaining and examining error logs, accepting and acting on error detection notifications, tracing faults, carrying out a sequence of diagnostic tests, and correcting faults.

Security Management. Security management facilities protect OSI resources. The facilities provide for authorization, access control, encryption and key management, authentication, and the maintenance and examination of security logs. An application service element might use the security management facilities to request authentication of a communication partner from a trusted third party. An implementation might request security facilities to provide the audit trails needed by fault management.

Performance Management. Performance management facilities evaluate the behavior of OSI resources and the effectiveness of communications activities. The facilities provide for gathering statistical data for the purposes of planning and analysis and for maintaining and examining the logs of system state histories.

Accounting Management. Accounting management facilities make it possible to set charges and identify costs for the use of system resources. The facilities provide for informing users of costs incurred or resources consumed, establishing accounting limits for the use of OSI resources, and permitting costs to be combined when multiple OSI resources are used to achieve a given communications objective.

STRUCTURE OF MANAGEMENT INFORMATION

The structure of management information refers to the logical structure of OSI management information. According to the *OSI Management Framework* and the *Management Information Service Overview*, this information is structured in terms of managed objects, their attributes, the operations that may be performed on them, and the notifications that they may issue. The set of managed objects in a system, together with its attributes, constitutes that system's management information base.

The structure of management information defines the concept of managed objects and the principles for naming managed objects and their attributes so that they can be identified in management protocols. It also defines a number of subobject and attribute types that are, in principle, applicable to all classes of managed objects. These include the common semantics of the object or attribute types, the operations that may be performed on them, and the notifications that they may issue. The structure of management information also defines the relationships that may exist between the various object types.

SUMMARY

OSI systems management services and protocols are a result of an effort to design concrete solutions to real problems. They permit communications managers to turn an ad hoc communications network system into a uniformly managed network, easily changing the configuration of a communications device or computer peripheral many miles away. Access to each open system in a network requires the use of the comprehensive structure of the OSI management services and protocols and their subcomponents. Attention to the intricacies of the various service elements can result in clear benefits, most notably a standard way to manage communications between network components.

Chapter II-1-4
Internet and TCP/IP

Jeffrey Earl Tyre

TCP/IP STANDS FOR a pair of networking protocols, the Transmission Control Protocol (TCP) and Internet Protocol (IP). These protocols are part of a family of protocols that have evolved from the Internet project, a national research effort initiated by the Department of Defense (DoD) that today networks thousands of government, commercial, and educational computers around the world. TCP/IP has become synonymous with the Internet project, because these two protocols provide the major network communications procedures within this family of protocols. The two most familiar application protocols of the TCP/IP protocol family are the file transfer protocol (FTP) and Telnet, a terminal access protocol.

Exhibit II-1-16 illustrates some typical components of the Internet environment. These devices, physically attached to various networks, internetwork through the support of the TCP/IP protocol family. Specifically, this exhibit shows devices connected to two local area networks (LANs) that are bridged together that have wide-area access, by means of a gateway, to the Defense Data Network (DDN) packet-switching network. Each computer with access to the Internet can be networked to any other computer on the Internet through the use of a unique address assigned to each computer. The networks and packet routing techniques employed make it possible to readily address computers within a building or across the nation.

The TCP/IP protocols have been readily embraced by the industry as a standardized thread connecting many heterogeneous computing environments. Because of the maturity and nonproprietary nature of these protocols, the TCP/IP protocol family is widely used outside the scope of the Internet project and its administrative jurisdiction, as a networking platform between any given group of computers for which protocol software and complementary network interface hardware are available. To help distinguish between a particular Internet project (the national networking project) and a generic TCP/IP internet project (which could be used either within or outside the context of the Internet), this chapter refers to the first with an initial capital letter and the second with a lowercase initial letter.

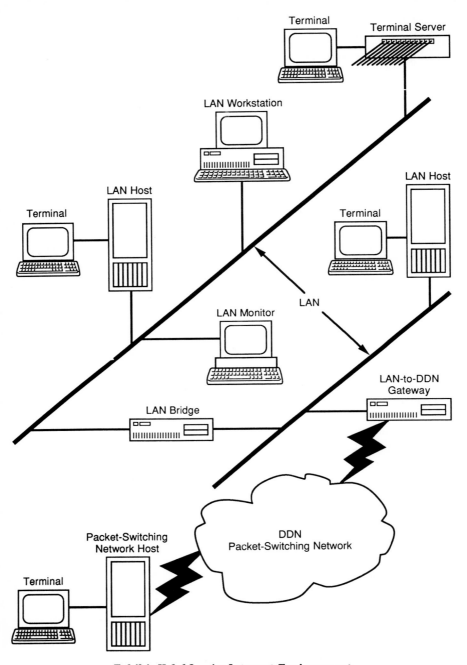

Exhibit II-1-16.   An Internet Environment

A major focus of this chapter is to review the history and administration of the Internet project. In addition, it discusses individual protocols that are part of the TCP/IP family and illustrates how the protocols provide a layered set of rules by which computer processes communicate.

HISTORY

The Internet project began in research supported by the DoD. The Advanced Research Project Agency Network (ARPAnet) is a DoD-funded project that was established in 1969 as an experimental packet-switching network, the forerunner of today's X.25 wide area networks. The ARPAnet led to the development of protocols that work on several administratively separate yet interconnected networks.

These networks provide communications services for hundreds of computers, primarily in the defense industry and scientific research communities. For example, the Military Network (MILnet) was splintered off from the ARPAnet to provide a separate military network facility. Along with several other classified networks, the ARPAnet and the MILnet come under the authority of the DDN.

One of the fastest growing segments of the Internet community is the NSFnet, a project primarily funded through the National Science Foundation to provide networking services to academic and industrial research groups. The internet backbone of the NFSnet is actually being used to replace sections of the ARPAnet (which is limited to telephone lines and modems) with T1 and T3 transmission capabilities. Data traffic rates are increasing in some regional areas of the NSFnet at the rate of 20% to 30% per month. Such a growth rate indicates that the NSFnet will become an important national data network.

The TCP/IP protocol family also has become a de facto standard for business and industry, where it is often used in non-Internet situations. A principal reason for this is that the International Standards Organization (ISO) set of networking protocols, the open systems interconnection (OSI) reference model, has not been comprehensively formalized to the point of full implementation by vendors or their customers. In comparison, the TCP/IP protocols are one of the most mature and widely supported sets of networking protocols.

For example, TCP/IP has been programmed as an integral part of the University of California at Berkeley's BSD version of the UNIX operating system. Support of these networking protocols within the UNIX operating system has facilitated its use by such companies as Amdahl Corp, Data General Corp, Hewlett-Packard Co, Sun Microsystems Inc, and Unisys Corp. Other computing environments also enhanced with the TCP/IP protocol family include the Apple Macintosh, IBM PC-DOS and VM, Digital Equipment Corp VMS and Micro-VMS, the Santa Cruz Operation Inc XENIX, and Wang Laboratories Inc VS. Much of this can be attributed to the open forum within which the TCP/IP protocol family has evolved.

PROTOCOL STANDARDIZATION

Internet activities are coordinated by the DDN Network Information Center (NIC), which is part of SRI International (Menlo Park CA). This organization oversees the evolution of Internet standards through the distribution and review of technical reports called requests for comments (RFCs). RFCs can be formalized into official protocols (e.g., TCP and IP), which are standards issued for compliance by those groups using the Internet. Furthermore, these protocols may be adopted by the DoD as a military standard (MIL-STD) specification. For example, TCP and IP are respectively listed as RFC 793 and RFC 791 as well as being MIL-STD-1778 and MIL-STD-1777 specifications. Important RFCs continue to emerge (e.g., IBM's NetBIOS, which is being implemented on top of the TCP/IP networking functions). In general, it is hoped that by openly publishing RFCs, a uniform set of standards will be available to all who use them.

NETWORK ADMINISTRATION

The administration of any TCP/IP network depends on whether a connection to the Internet is to be provided, as opposed to use of the TCP/IP protocol family without a requirement for access to the Internet. If an Internet connection is desired, a network number must be obtained from the DDN NIC. Essentially, internetworking between TCP/IP networks is supported by the assignment of network numbers unique to each network. All devices attached to the same network have the same network number, with individual devices on a network having a different host address. The local administrator has the authority to assign a host address to the NIC network number, and the two in combination form a complete IP address for each network device.

This process identifies any device that has access to the Internet. If no access to the Internet is required, administrators are free to assign both network numbers and host addresses, as long as the basic rules regarding network and host addressing are followed in the manner previously described.

In reality, an IP address consists of four parts, commonly represented by the variable $a.b.c.d$, with each part consisting of 8 bits (decimal values actually range from 0 to 255) and yielding IP addresses that look something like 121.10.4.1. The four-part format yields three classes of networks, with each class effectively defining the number of hosts that can be identified through the use of this addressing convention. With n representing the network component and *h* representing the host component, the four-part format yields three possible addressing schemes, which are discussed in the following sections.

Class A: *n.h.h.h.* In this class, the first byte represents the network number and the remaining three bytes represent the host address. This class is used for networks with a large number of hosts.

Class B: *n.n.h.h.* In this class, the first and second bytes represent the network number. The third and fourth bytes are used for the host address. This class is used for networks with an intermediate number of hosts.

Class C: *n.n.n.h.* This class has many networks, indicated by use of the first, second, and third bytes in combination to represent the network number. Only the fourth number is used to assign a host address. This class is used for networks with a small number of hosts.

ARCHITECTURAL DESIGN

As with many network protocol architectures, the TCP/IP protocols are functionally grouped into layers. The principal benefit of a layered architectural design is to provide protocols that specialize in network services and that work with processes in adjacent layers to form a comprehensive set of network communications mechanisms. Exhibit II-1-17 presents both

ISO Open Systems Interconnection	DoD Internet Project
Application	Application
Presentation	
Session	
Transport	Transport
Network	Internet
Data Link	Data Link
Physical	Physical

Exhibit II-1-17. Architectural Comparison

an architectural model for the TCP/IP protocols and the seven-layered OSI standard reference model. As can be seen from the exhibit, the OSI and TCP/IP architectural models share major structural elements. However, an important aspect of the evolutioin of the TCP/IP protocols has been their integration with communications specifications from a variety of underlying network interfaces. Specifications for these network interfaces are often derived from standards set by committees outside the Internet forum, such as X.25 from the International Telephone and Telegraph Consultative Committee (CCITT) and the 802 committee standards from the Institute of Electrical and Electronics Engineers (IEEE).

This chapter distinguishes between those protocols that are administered by the DDN NIC and those standards set by other organizations. It does not detail the standards set by the other organizations but does introduce the essential elements of these network interfaces.

Internet Protocols

The core of the TCP/IP family of protocols provides the essential application, host-to-host (i.e., endpoint) transport, and internetworking (i.e., intermediary-point) protocols that have been implemented on a variety of network interfaces.

Application layer protocols of the TCP/IP model are designed to provide services similar to the protocols that are segregated between the upper three layers of the OSI model, including application user and programming interfaces, data and file type declaration, and network session initiation requests to the transport layer protocols.

Transport layer protocols from both models provide similar functions, in general involving addressing, network connection establishment, data transmission flow control, and network connection release. The transport layer is considered to support host-to-host communications, because it is the lowest layer involving peer-to-peer processing on the source and destination host computers and not on an intermediary device. A transport-layer network session dialogue between the source and destination hosts is transparently supported by the internetworking elements of the TCP/IP protocol family.

Internet layer protocols provide a common link between the various network interface specifications (e.g., through the IP addressing scheme). Internetworking is achieved by network routing mechanisms that act on the flow of data between two host endpoints.

Network Interface Protocols

The network interface protocols are the collection of separate specifications from the various underlying networks with which the TCP/IP protocols have been integrated. Examples of the networks supported include satellite and radio packet, X.25, Ethernet, and token ring.

Data link layer protocols are defined by the specifications of the various network interface standards. These protocols initiate, control, and terminate the flow of the communications link across the physical network. This includes error checking for reliable data transmission across the physical media. Some networks implement their own addressing scheme for network device identification with address specifications at this layer.

Physical layer protocols from networking standards supported in both architectural models define reliable, bit-level data transmission on network media. Primary concerns include mechanical, electrical, and procedural functions of the interface between the network device and the physical network.

NETWORKING DEVICES AND UNDERLYING MECHANISMS

Three general types of entities are involved in network communications: software processes, network devices, and physical networks. A process is a software program that is executing on a computer. All executing computer programs consist of one or more processes. The networking protocols discussed are implemented as executable computer programs and therefore consist of computer processes.

Network devices are computer units physically attached to the network (e.g., host computers, terminal servers, network monitors, gateways, routers, and bridges). Network devices have software processes executing on them that support network communications.

The term *network*, in a strict sense, applies to the media involved in network communications. This chapter applies the term *physical network* to distinguish the network media entity from the more generalized concept of a network, which commonly refers to all three types of entities involved in network communications.

Exhibit II-1-18 illustrates the concept of process-to-process communications, within the context of the TCP/IP layered architecture presented in Exhibit II-1-17. Two major components constitute process-to-process communications over a TCP/IP network. Dialogue between protocols of corresponding layers on physically separated devices is termed peer-to-peer communication and is represented by the horizontal arrows. The vertical arrows depict the flow of data within a device, between adjacent protocol layers, which is commonly referred to as layer-to-layer communication processing.

Several major network components are represented in Exhibit II-1-18, along with the conceptual features mentioned previously. Applications are represented by the letters X, Y, and Z to illustrate paired application layer

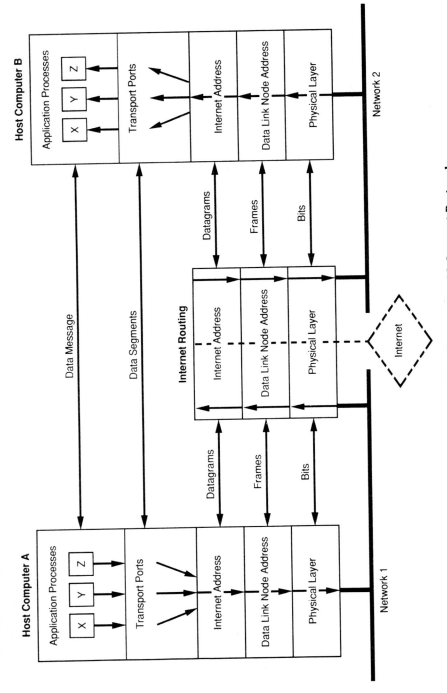

Exhibit II-1-18. Peer-to-Peer Processing with Internet Protocols

processes on networked host computers. These can be either the standard TCP/IP protocol family applications (e.g., Telnet and FTP) or other programs that make programming calls as part of an application programming interface (API). An example of the latter is the BSD UNIX socket library of programming function calls, which interface to the lower-layer (i.e., transport layer and below) protocols.

Host computers are designated A and B in the exhibit. These are the computational platforms for the source (A) and destination (B) networking processes. These devices can range from personal computers to supermainframes. As long as the device designates a TCP/IP networking endpoint, including such specialized devices as terminal servers, it is considered to be a host.

Physical networks are designated network 1 and network 2 in the exhibit to represent the local or packet-switching network to which a host computer is attached. The TCP/IP protocol family is supported by a range of network interfaces (e.g., Ethernet and X.25) and media types (e.g., coaxial and twisted-pair cable and optical fiber) that these interfaces use.

Internet gateways, referred to in the exhibit as internet routing, are network devices that reside between network hosts to provide internetwork routing mechanisms. Gateways either can be statically preset or can dynamically determine the optimal datagram path across networks, with the topology of the internetwork retained in memory on each gateway. Information is exchanged between gateways to determine current network traffic loads across specific networks and to adjust accordingly to lighten traffic loads.

Underlying Mechanisms

Exhibit II-1-19 extends the concepts introduced in Exhibit II-1-18 by illustrating the transformation of application data into network packets, packet transmission and routing, and the reassembling of data packets. Beginning from the application layer of the source host, the data unit that is handled by application processes can be thought of as a data message. The term *data* is used in Exhibit II-1-19 to designate the original application layer data message. Depending on the size of the data message, it may be necessary to transmit data over the network in the form of smaller data units than are being handled at the application layer. Therefore, an initial modification of the data message may be to divide it into smaller pieces, termed data segments. This mechanism is called segmentation and is performed at the transport layer, along with the countermechanism of reassembly of the data message at the destination host computer.

Once segmentation has occurred, further processing is performed to supplement each data segment with control information, a mechanism

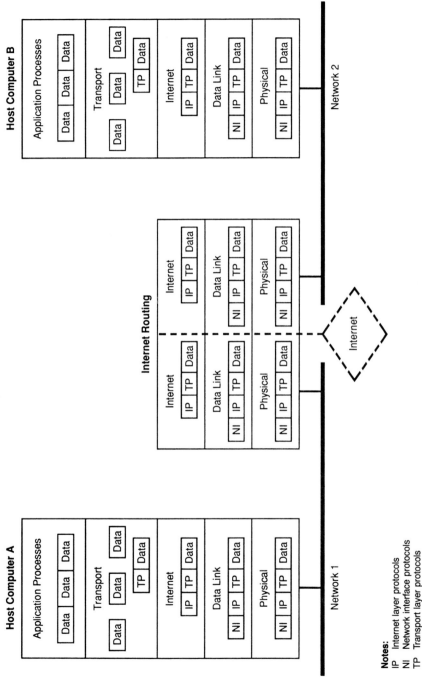

Exhibit II-1-19.   Data Segmentation, Encapsulation, and Reassembly

Notes:
IP   Internet layer protocols
NI   Network interface protocols
TP   Transport layer protocols

324

referred to as encapsulation. Control information includes features such as addressing (i.e., identification), destination receipt acknowledgment, and error correction.

Encapsulation can occur at each subsequent architectural layer, beginning at the transport layer, as information is added to each data segment during processing by each protocol that handles it. This is designated in Exhibit II-1-19 by TP for transport layer protocols, IP for internet layer protocols, and NI for the network interface protocols. The control information added to the data segments is often referred to as a header, to distinguish it from the original application layer data. Data units that have undergone encapsulation at the internet layer are referred to as internet datagrams. In addition, the terms *packet, frame*, and *bit* refer to data units that are applicable to the protocol format for the various network interfaces supporting the TCP/IP protocol family.

A major component of the control information is addressing—that is, identification of network entities. Earlier in this chapter, the identification of network devices (i.e., the host address) and physical networks (i.e., the network number) was included in a discussion of network administration. Identification of each process endpoint active over the Internet must also be available to provide true process-to-process communications. Endpoints are data structures that are under the control of the computer processes that constitute the transport layer protocols and are commonly referred to as ports. Coresident endpoints on a host are distinguished from each other by a port number. In summary, to specify an endpoint (e.g., TCP port) with which to communicate on a specific destination host, a three-part address is used that consists of the port number and an Internet protocol address, the host, and the physical network.

Intermediary Processing. This discussion concentrates on networking mechanisms that are involved at the communications endpoints, in the source and destination hosts. Intermediary processing during internetworking can involve a series of IP datagram routing events on gateways and networks that are located between the source and destination hosts. Exhibits II-1-18 and II-1-19 symbolically represent all possible routing events under the label internet routing. Each internetworking transmission uses the Internet layer protocols to perform routing decisions based on the Internet address (i.e., the packet's ultimate destination) and routing control information (i.e., the best way to get it there at the time). As with network hosts, each Internet gateway has a unique Internet address that is used not only for communications between gateways but for communications between the source host and the local gateway when it has been determined that the destination host does not reside locally. Internetwork routing relies on the Internet layer and does not use the facilities of the transport layer, as illustrated in Exhibits II-1-18 and II-1-19.

The reverse of the mechanisms described occurs at the destination or intermediary network device, as data is processed from lower- to upper-layer protocols and the header information pertinent to each layer is examined. In the case of intermediary devices, the processes involved in going from upper to lower layers will be invoked again as data is being forwarded to the destination host. As the network packets successively reach the transport layer of the destination host, they are fully reassembled into the original data message that is usable at the application layer.

OVERVIEW OF THE TCP/IP PROTOCOLS

Many discussions of networking protocol architectures begin with the lowest layer, the one defining physical attachment to the network, and work their way up to the application layer. This chapter emphasizes the user's perspective—essentially, how user data that resides on a local host computer is transformed into data packets, transmitted across the network, and reassembled into user data on a remote host computer. The architectural layers of the TCP/IP protocols have been designed to provide the networking services discussed in the following sections. Although not specific to the TCP/IP protocols, overviews of the network interface layers (i.e., data link and physical) are included as part of a complete architectural discussion.

The Application Layer. This layer consists of programs that interface users or host processes with the lower-layer networking protocols. These programs essentially are starting points for the support of basic host computer applications (e.g., terminal access, file copy, and electronic mail) beyond the local host computer. Application layer programs initiate network communications requests to the protocols of the transport layer, which are then active in establishing, controlling, and closing communications process endpoints. TCP/IP includes the following application layer protocols:

- Telnet (RFC 854 and RFC 855/MIL-STD-1782)—This virtual terminal service provides network access to destination hosts from terminals attached to source hosts, which include terminal servers. In many applications, destination host computers use a device driver, called the network virtual terminal, to negotiate sessions with a wide variety of terminal types.
- File transfer protocol (RFC 959/MIL-STD-1780)—This file transfer protocol (FTP) is used between hosts. One FTP session requires at least

two TCP port connections: one that facilitates a command mode and one or more others that provide the connection used for the actual data transfer. In command mode, such activities as file directory administration and file parameter settings (e.g., ASCII versus binary) are performed.

- Simple mail transfer protocol (RFC 821 and RFC 822/MIL-STD-1781)— This electronic mail protocol is intended for use between existing host mail systems. In other words, the user interfaces with a local host mail program, which uses the local simple mail transfer protocol (SMTP) sender module to communicate with an SMTP receiver module on the remote host system. The receiving agent of the SMTP protocol interfaces with the reciprocal electronic mail service on the remote host system.
- Domain name service (RFC 822 and RFC 823)—This protocol provides a distributed name service that is accessible by system administrators and programs from host sites. Users can query a domain name service (DNS) for the Internet address by providing the host name. Because the DNS is a data base, information other than the Internet address can be maintained, including records of the types of services provided by different hosts.
- NetBIOS (RFC 1001 and RFC 1002)—Implementation on the TCP/IP protocols is a recent RFC development that has been introduced to make use of this widely supported IBM PC standard. NetBIOS provides an application program interface that provides a front end to networking services similar to those found in the transport layer. An inherent NetBIOS limitation is that it does not define internetworking capabilities. Combining NetBIOS with the TCP/IP protocols is mutually beneficial to both sets of protocols because it provides TCP/IP with NetBIOS's broad personal computer hardware and software support and supplements NetBIOS with TCP/IP's inherent support of internetworking services.

The Transport Layer. The transport layer contains the protocols that are responsible for the establishment, control, and termination of network connection between the data structure endpoints that reside on the source and destination hosts. Transport protocols service communication initiation requests from application layer processes while servicing requests from the internet layer below by identifying the destination port of incoming data. The TCP/IP transport layer includes the following protocols:

- Transmission Control Protocol (RFC 793/MIL-STD-1780)—This protocol provides reliable, bidirectional network communications between peer processes. TCP is reliable because it maintains and controls the state of the connection and ensures that all data has been received by the destination host. TCP breaks an application layer data message into segments and adds a TCP header to the beginning of each segment that contains control information. Part of this control information is segment order, which is maintained through use of a sequence number field. Data flow control is provided through the exchange of current and expected sequence numbers between the source and destination ports. Each data segment is addressed with source and destination port numbers. In addition, a checksum field is used for bit-level error checking of both the TCP header and the data.

- User datagram protocol (RFC 768)—Like TCP, the user datagram protocol (UDP) provides source and destination port numbering within a header. However, UDP provides unreliable network communications between peer processes; it does not provide destination receipt acknowledgment and thus cannot guarantee delivery of data to the destination host. In addition, UDP does not implement mechanisms for data flow control, because segmentation is not supported (all data must fit into one segment). The principal gain achieved by lack of data flow control and error checking is that UDP provides faster data throughput compared with TCP.

The Internet Layer.   The internet layer protocols implement device and network addressing that supports communications between network devices and provides routing services for transmitting data across an internet. By providing internetworkwide addressing and routing, these protocols handle the underlying data transmission and switching mechanisms in support of the transport layer protocols. The internetwork route that the data takes is transparent to the transport layer protocols. The internet includes the following protocols:

- Internet Protocol (RFC 791/MIL-STD-1777)—This protocol provides the Internet address to identify network devices and physical networks. A second set of important mechanisms is IP fragmentation and reassembly. During routing, datagrams may need to travel across a variety of network types (e.g., X.25, Ethernet, and phone lines). Because each type of network defines a different maximum packet size, IP must provide a way to resolve size conflicts. If a maximum packet size is less than the datagram size, fragmentation must be applied to divide the datagram into smaller units. IP is designed to maintain the integrity of the datagrams through a reassembly process at their destination. An IP header is attached to the outgoing transport layer segments, forming an IP datagram. Major elements within the IP header are the Internet source address, Internet destination address, and IP header checksum fields.
- Internetwork control message protocol (RFC 792)—This protocol

provides a mechanism for IP devices to communicate control information about the network. The information usually consists of error messages sent back to the datagram source when an IP device determines that a problem has occurred in the transmission of a datagram that it has received.

- Address resolution protocol (RFC 826)—This protocol provides a method for translating between a 32-bit IP address and a 48-bit Ethernet address. The Ethernet address is actually implemented at the data link layer. Both of these addresses can be used to reference a network device, but only the Internet address is usable across all subnetworks. This is a critical step in that translation is between an Internet address and the node address of the underlying network interface.

The Network Interface

The various network interfaces to which the Internet protocols have been ported contain specifications that are defined by other standards organizations. Primary examples of these underlying networks are X.25 and Ethernet. The most common features of these interfaces can be discussed in general within the context of a data link layer and a physical layer.

Data Link Layer. Data link control protocols to which the TCP/IP protocols have been implemented include the ISO high-level data link control (HDLC), the CCITT link access procedure (LAP) and link access procedure—balanced (LAP—B), and the IEEE 802.2 logical link control (LLC). Networking specifications are concerned at this layer with two general considerations: reliable data transmission across the physical media, including initiation of data transmission, error checking of the active link, and termination of data transmission; and addressing network devices with formats and procedures governed by the underlying physical network's specifications.

Framing is a data link layer mechanism performed by many networking specifications to surround the data unit with both a header and trailer, creating a data unit referred to as a frame. This is the final encapsulation before the data is transmitted on the network. A frame error control mechanism used by many data link protocols is based on a cyclic redundancy check method and incorporated as an inherent part of framing, usually termed the frame check sequence. Framing involves a specific format for each type of network standard involved. For example, RFC 894 is a TCP/IP specification that defines how Internet datagrams are to be framed for transmission on an Ethernet.

A link-level node address, identifying the physically attached devices, has the Internet address mapped to it, as was discussed with address resolution protocol. The addressing mechanism involved at this layer is dependent on the specifications of the underlying network. For example, HDLC provides a frame address field. It is used to identify secondary devices dependent on a primary device that is providing the principal data communications link. This type of multiple link is classified as a tiered-multipoint network. In comparison, the Ethernet standard demonstrates how a node is addressed on a peer-multipoint network, in which each node has an equivalent addressing status.

The Physical Layer. The physical layer of each networking standard defines reliable bit-level data transmission on a network medium. Major functions include mechanical properties (i.e., the size and physical configuration of the device interface) and electrical properties (i.e., changes in voltage represent bits) as well as procedural concerns of transmitting data. A wide variety of physical interface standards have been implemented for the TCP/IP protocols, including the EIA RS-232 and RS-449, IEEE 802.3, Ethernet, DDN 1822, and CCITT V.35.

SUMMARY

Support of the TCP/IP protocol family has grown significantly. The TCP/IP protocol family offers a well-documented and mature set of rules for networking. TCP/IP's technical merits, as well as its documentation and maturity factors, contribute to decisions to select these networking protocols for connectivity to numerous types of hosts. Many of these decisions are based on the fact that there is a void between the need for networking and the broad implementation of the OSI standards. TCP/IP has served the needs of a diverse set of groups, ranging from those that require a nonproprietary R&D networking environment to commercial interests who could not wait for OSI.

There is little doubt that OSI will eventually be the most widely supported networking standard. Two of the most interesting questions that need to be asked are how soon this will occur and what the real driving force behind the transition will be. If OSI technically and functionally offered little advantage over TCP/IP networking, OSI replacement of TCP/IP and other networking protocols would be a lengthy process because of the large number of hosts currently implementing TCP/IP networking.

However, there are distinct improvements in OSI's protocols in the area of applications. It is here that a robust functional capability—such as that offered with file transfer, access, and management (FTAM), message handling system (X.400), directory (X.500), and the common management information protocol (CMIP)—currently provides advantages that are distinctly appealing to users. This user appeal, especially in the commercial sector, may provide widespread support of OSI. Many estimates indicate that support of OSI will increase during most of the 1990s in much the same way that use of TCP/IP grew during the 1980s. The federal government has clearly stated that OSI, through the government open systems interconnection profile (GOSIP), will be the networking protocol of choice and will be mandatory in network product and service acquisitions. This is the most definitive indication, at least in the US, of the transition from TCP/IP to OSI.

Chapter II-1-5
Communications Standards in CAD/CAM

James F. Mollenauer

DURING THE LAST FEW YEARS, data communications has emerged as a critical component of computer-aided design and manufacturing (CAD/CAM). Its growing importance represents both a response to changes in computer technology and an attempt to increase the application of CAD/CAM to the production process.

The first generation of CAD systems typically involved multiple users sharing a mainframe or minicomputer. Finished designs were printed out on plotters, and the resulting plots were handled in a manner similar to manually generated engineering drawings. They could be copied on blueprint machines and marked up in red pencil, but the design data remained on the CAD computer (except for its storage on a reel of magnetic tape).

For automated production, the CAD computer could be equipped with a paper-tape punch. The design could be converted to tool-path instructions on the tape, which would then be hand-carried to a numerically controlled milling machine. This level of automation in manufacturing is not at all new. The weaving of cloth under the control of punched cards was introduced on the Jacquard loom during the 1830s. The most significant changes in product design have concerned not the application of computers but their rapid proliferation.

The availability of inexpensive microprocessor computing power has changed the development environment significantly over the past several years. Intelligent workstations have replaced terminals that provided only a keyboard, screen, and printer for design work. In many cases, the computing power of these new workstations substantially exceeds that of the shared computers of the previous generation. Nevertheless, large projects require the cooperative efforts of many designers, and for this the workstations must be interconnected.

In tandem with the need to share data on large projects is the desirability of centralized program storage. A CAD system comprehensive enough to support the design of an airliner, together with the applications built on it, might involve tens or even hundreds of megabytes of software. Efficient networking can obviate the need to supply each workstation with the disk space to store this amount of data.

In fact, networking allows for a further partitioning of the CAD process, separating the task into the CAD operations performed on the data representing the part and support of the user interface, including graphics and a variety of I/O devices. Although microprocessor speed can double each year, speed does not come at a constant price. Assigning the computer-intensive and data-intensive aspects of the design process to a centralized server makes sense when the performance of the server is 5 to 20 times that of a Digital Equipment Corp VAX 11/780, the flagship superminicomputer of 10 years ago.

The result of this partitioning is a clustering of workstations around servers. Small local area networks (LANs) provide communications links between servers and their client workstations; the servers, in turn, are interconnected by a backbone network. An example of such a configuration is the Prime/Computervision CADDSnetwork system, shown in Exhibit II-1-20, which features CADDServers and CADDStation clients (CADDS is an acronym for Computer-Aided Design and Drafting System). The CADDS 4X program is partitioned; one copy of the main task is in the server, and copies of the graphics and I/O support software are distributed to each workstation. Both the server and its workstations use a UNIX-based operating system. Local storage is provided on the client stations for frequently used files, although diskless stations are also supported.

Other steps in the overall CAD/CAM process also depend on networking. Specifically, networking provides the necessary links to analysis machines, engineering and manufacturing data bases, and the factory floor. Design analysis for thermal and mechanical stress, aerodynamics, and other computing-intensive properties typically takes place on a specialized analysis computer. Depending on the application, this machine may range from a superminicomputer (e.g., a VAX) to a very large supercomputer (e.g., a Cray). Although the recent introduction of workstations capable of running at a rate of more than 10 MIPS may alter the distribution of work, the need for specialized crunch nodes is unlikely to disappear. Networking is necessary to convey the data to these machines.

Although the design process is being decentralized to workstations, centralized control of data remains essential. Both work-in-process and finished designs must be archived, and access and modification controls must be implemented to prevent simultaneous design modification by two or more engineers.

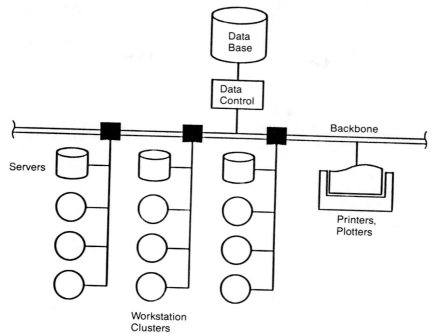

Note:
Servers provide file storage and perform computationally intensive CAD operations. Clusters are internally connected through Ethernet; an Ethernet backbone interconnects the clusters and attaches to the data management system and shared peripherals.

Exhibit II-1-20.   Workstation-Based CAD Network

The solution to this control issue is a centralized data base. These data management systems differ from typical data bases that are optimized for retrieval of small records. The CAD/CAM data base is optimized for retrieval of very large items, and it is concerned with authorization level, concurrent access, and the associativity of components in assemblies.

From the engineering data base, the product design is sent to the factory. Numerically controlled tools and robotic assembly equipment can run directly from CAD data with little human intervention. In most cases, however, the data does not come directly from the engineering data base. For reasons of geography or efficiency, the factory may well run its own computer system. Finished designs are sent to this system along a communications path that need not be as fast as the links between the workstations and the engineering data base; only completed designs are transmitted across the link. Within the factory, the need for speed resurfaces as data is

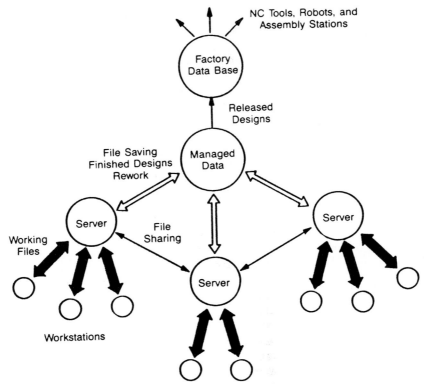

Exhibit II-1-21. Design Data Flow in a CAD/CAM System

fed repetitively to the manufacturing machines. This arrangement is illustrated in Exhibit II-1-21.

The passage of data from a concept to a finished product generally involves a succession of computers; rarely does one vendor supply the best solution for all stages of the process. Thus, computers from many vendors are found in the typical manufacturing corporation. As a result, standards are needed to move the data from one computer to another and to specify the meaning of the data once it arrives at its destination.

STANDARDS FOR DISTRIBUTED APPLICATIONS

The dispersal of design activities to individual workstations has created a demand for data communications both among workstations and between workstations and servers. Whether the servers store files, provide access to shared peripherals, or carry out computation, high-speed communications geared for bursts of data is imperative. CAD files often comprise many megabytes of data and generally must be transmitted within a minute or so. High-resolution graphics usually involve transfers of a megabyte or less, but the time requirements are even more demanding, generally on the order of just a few seconds. Local area networks are able to meet both these demands.

The Ethernet LAN

The preferred LAN in computer-aided design is Ethernet. Ethernet features a peak speed of 10M bps and a cable-sharing mechanism that is highly efficient at low to moderate loads. Response can degrade under high load factors, but these are rare in practice. Most Ethernet networks in production environments run at load factors of 3% or less, and peaks rarely exceed 10%. The ability of current workstations and minicomputers to obtain data from a file system, process the higher-level protocols, and then put the data on an Ethernet network limits the contribution from any one data source. If the load grows substantially, partitioning the network into clusters of workstations interconnected by a backbone cable can keep the load on any one Ethernet at a moderate and workable level.

The importance of proprietary LAN designs in the workstation area is declining. These networks, however, are still popular in the microcomputer market, with its lower speed requirements and greater cost sensitivity. A peak speed of 64K bps, for example, though much too slow for a CAD workstation, would suffice for many microcomputer file-transfer applications. If a group of microcomputers is being used for CAD purposes, however, it would probably pay to install Ethernet to take advantage of such facilities as central file storage, archiving, and access control, which are now offered by the major CAD vendors.

The introduction of Ethernet on existing telephone wiring makes it cost-effective with any kind of desktop computer, including IBM PCs and compatibles, Macintoshes, or UNIX-based workstations. The cost of cabling in the original Ethernet cable is significant: the Teflon jacket version required in return-air plenums is close to $3.00 per foot, not including labor.

The TCP/IP Protocol Set

The higher network protocol layers are now entering a period of change. The current preferred protocol set is Transmission Control Protocol and Internet Protocol (TCP/IP), which was developed by DARPA, an agency of the Department of Defense. This protocol set likely will give way during the next five or so years to international standards, but for now it is supported by most CAD/CAM networks.

The IP layer provides internet routing, moving data from one Ethernet LAN to another. It is generally used to route data from a cluster over the backbone to another workstation or server, either on the backbone or on another subnetwork. Data can also be routed over wide area networks, in which remote systems or networks are connected to form a coherent whole.

An increasingly popular alternative to IP-based routing is the use of a bridge. A bridge interconnects two LANs, in some cases by means of an intermediate wide area link, through the use of the link-level address rather than an explicit internet address. Because no assumptions are made about the network protocols above Ethernet, this approach provides a high degree of flexibility.

Bridges can usually be classified into two types. Preloaded bridges contain a list of stations on the home network; other destinations are assumed to be located on the other side of the bridge. Learning bridges determine the location of stations from the source addresses in the data packets. A learning bridge may forward some packets unnecessarily, but it adapts automatically to changes in the network configuration, sparing the administrative burden of maintaining duplicate tables in several systems.

TCP, for its part, integrates a multiple-hop journey into a reliable transmission. The protocol checks sequencing, sets timers to detect lost or discarded packets, and runs an end-to-end checksum. Because these functions consume overhead, most TCP/IP implementations also provide, for less-demanding applications, a simpler, nonconnected protocol known as the user datagram protocol. Applications that use this can perform their own checks on data sequencing and integrity. These applications include file access, remote log-in, and graphics transfer.

In addition to these transport protocols, there are two application protocols. The file transfer protocol is used to move files from one system to another. Because it transfers files in their entirety, the file transfer protocol is not the protocol of choice for accessing large CAD files. Telnet, on the other hand, provides a mechanism that enables a terminal to access a remote computer. It works well for text but has no graphics capability; remote CAD terminals must therefore use another access method.

Remote File Access

In the case of remote file access, proprietary or de facto standards provide the capability missing in the TCP/IP package. The two systems that have achieved the widest implementation are Sun Microsystems' Network File System (NFS) for workstations and Novell's NetWare for microcomputers. Both allow files located on remote servers to be treated at the application level as if they were located on the local machine. Files are accessed piecemeal: only the requested records are sent over the network.

An important distinction between these systems lies in their degree of openness. Sun has actively encouraged adoption of the NFS standard by other manufacturers; more than 100 firms have licensed it, although not all have products on the market yet. Because it includes a protocol—called XDR—for treating the diverse data formats of different machines, NFS works well in a heterogeneous environment. NetWare, on the other hand, involves a proprietary operating system on a microcomputer-based server and addresses only microcomputers as clients.

Other LAN Standards

Other LAN standards find application in the CAD/CAM environment, although to a lesser degree. Even in the case of Ethernet, the version used most commonly is not the official Institute of Electrical and Electronics Engineers (IEEE) 802.3 version, but the earlier proprietary version promulgated by Digital Equipment, Intel, and Xerox. In fact, Digital Equipment alone accounts for about 42% of the installed Ethernet base. The IEEE standard differs relatively little from the proprietary version in most respects, but it does add a logical-link sublayer that provides most of the link services of the wide area X.25 protocol set as an option. Because TCP/IP predates these services, it does not use them, and it is usually teamed with Ethernet Version 1 or 2 rather than with IEEE 802.3.

This situation, however, is certain to change in the future. The federal government is starting to require compliance with its Government Open Systems Interconnection Profile (GOSIP) stack, and this will require the use of not only 802.3 for Ethernet installations but the International Standards Organization (ISO) standards for higher layers.

Token-based standards have not yet made inroads in the CAD/CAM environment. IBM's token-passing ring is just coming into use, primarily to interconnect PCs. However, the new 16M-bps version of the token ring may attract users for such performance-sensitive applications as CAD. Most IBM CAD installations use IBM's 5080 workstation, which is directly connected to a mainframe I/O channel without an intervening LAN. Digital Equipment workstations and mainframes, on the other hand, are usually interconnected through Ethernet, using the proprietary DECnet protocols for the higher layers.

Among the standards organizations there is consensus that the token-passing bus (IEEE 802.4) is the appropriate standard for the factory environment. It offers bounded delay time and a rigid cabling system, based on cable TV technology, that is resistant to electrical noise. Network buyers, however, have not shared this unanimity. Thousands of Ethernets have been installed in manufacturing companies without a significant incidence of problems. Although Ethernet delays can become considerable under heavy load conditions, such conditions are far from the norm. Extremely critical applications (e.g., emergency plant shutdown) can justify dedicated wiring in any case. Because the Ethernet cabling has four ground wraps, its noise immunity is very effective: at least two orders of magnitude better than the levels of radio-frequency noise found around heavy electrical machinery.

The success of Ethernet in the CAM environment can be attributed to its earlier availability. Token-bus standards and associated equipment have come on the market only within the last two years. Whether Ethernet will maintain its momentum remains to be seen; as long as it is backed by such vendors as Digital Equipment, its presence is unlikely to diminish soon.

Future LAN Technology

As processing power increases, so too does the need for increased speed in network connections. The current generation of LANs, with speeds as high as 10M bps, handles existing CAD requirements quite well. Shortly, however, CAD applications will demand higher-speed networks. This need will be met by fiber-optic systems.

The first fiber-optic standard to emerge is the Fiber Distributed Data Interface (FDDI). This standard is being developed by American Standard Committee X3T9.5. FDDI is a ring-based system that runs at 100M bps, using a token-passing protocol that is a modified version of the one used in IEEE 802.5. Originally conceived as a computer room interconnection, FDDI is considerably less distance-sensitive than many other LAN protocols, and as a result, it achieves a total loop length of almost 200 kilometers (about 125 miles). FDDI maintains compatibility with the IEEE 802 family of standards by running under the logical link control specified by IEEE 802.2.

A second thrust of development is metropolitan area networks (MANs), which are being standardized by IEEE working group 802.6. The 802.6 technology is a dual-fiber bus installed in the form of a ring with a capacity of 310M bps. Because the majority of the bits transferred between corporate locations are digital voice, the MAN effort has been optimized to support voice as well as data communications.

The 802.6 standard has been developed with the participation of the communications industry, and it can be expected to be compatible both with the other IEEE 802 standards through the logical link control layer as well as with the broadband integrated services digital network (ISDN) standards in the international communications community.

The 802.6 standard is the basis for the switched multimegabit data service being planned by the telephone companies. This system will provide the data service of 802.6 but not the isochronous capability designed for voice. It will provide a bandwidth guarantee to the user (in amounts less than the total capacity of 45M bits) as well as such services as address screening and group addressing.

THE ROLE OF CENTRALIZED CONTROL

The proliferation of computing power, although it has broadened the engineer's design capabilities, has created some significant management challenges. When data is distributed over many workstations, it becomes more difficult to organize projects. File backup becomes more cumbersome, and the likelihood of concurrent modification increases. Clearly, some degree of centralized control must be imposed.

Managing CAD Data

The high-speed LAN provides the solution to the control problem: a managed central file store. This is not simply a file system with transparent access, as is the case with Sun Microsystems' NFS, but an organized data base that relates to the structure of both the corporate organization and the data itself.

A local area network moves data in and out of a data base quickly. Older data communications technologies would prove too slow in transmitting CAD files, which can easily run several megabytes in size. With a LAN, a designer who needs to work on a specific part requests copies of the relevant files, which are then sent over the network to the workstation. The original data remains in the data base, available to others on a read-only basis until the user returns it, modified or not.

Permission to access and modify the data is set up to correspond to the design organization. Hierarchical levels of authority grant users the ability to read, modify, and approve designs. These levels can become complex, reflecting the real working practices of engineering departments. For example, certain individuals may be allowed to mark up a design but not to alter any elements. The design originator retains control over modifications and uses the notations to guide rework.

A given part can be represented by a multiplicity of files. One may contain the basic geometry, or model, of the part; others may contain drawings from diverse perspectives derived from the first file. Still another file may contain finite-element meshes superimposed on the part as an aid to simulation studies. Different permission schemes may apply to each of these related files. For example, in the Prime/Computervision PDM data base product, 99 levels of authority can be specified to accommodate this multiplicity.

Managing such complexity requires state-of-the-art data base technology. Relational data bases can provide the needed access control, but they are slow in coping with the large volumes of data generated in CAD. Hence, as a compromise, access control is handled by a relational data base, but the data itself resides outside the data base, in the file system of the host computer. The relational data base contains indexes that point to the location of the data.

Standard Interfaces to the Data Base

Standards are equally important in dealing with the data base. IBM's Structured Query Language (SQL), as a de facto standard, makes it possible to provide equivalent facilities on a variety of platforms—both hardware and data base—from different vendors. Although SQL is not an official standard, it enjoys wide acceptance, and it is available from a variety of manufacturers of mainframes to microcomputers.

Data management systems designed for CAD lend themselves to other purposes as well. The content of the managed files need not necessarily be understood by the management system. Some applications may make use of the data content when, for example, reconstituting drawings for display or plotting; however, in general, files of any type can be managed. This means that CAD files from diverse vendors can be accommodated; such utilities as the file transfer protocol are used to move the files into the managed environment.

Software components can be handled in a similar manner. Source code entails the same management concerns as mechanical design: access and concurrency control and associativity among the modules that make up a complete software system. For example, Boeing applies a CAD data management system to software components.

DATA EXCHANGE

The exchange of data among different CAD systems has long been a primary goal. Although not yet fully realized, the capability does exist for many kinds of parts and graphic entities. A limitation on the exchange of data between two systems is—and always will be—that both must support the same entities. If not, data must somehow be approximated or simulated on the system to which it is sent; as a result, a round trip may entail a loss of information.

In a multiple-system environment, the number of translations is of the order n-squared when individual source-to-target translations are used. Through the use of an intermediate format, however, translations into and out of that format require only $2n$ translations. Although two translation steps are needed, the low cost of computing power makes this a judicious move. Of course, the intermediate format must represent a superset of all the individual system's capabilities; otherwise, a loss of information will occur. Although currently a limiting factor in data exchange, the incompatibilities here are being reduced.

The first and still most prevalent data exchange standard is the Initial Graphics Exchange Specification (IGES). Drafted in 1979, it was officially adopted by the American National Standards Institute in 1981. The current version of IGES incorporates such features as electrical and printed circuit board support as well as data compression.

Although not yet widely implemented, data compression is valuable in relation to IGES because of its cumbersome format. IGES relies on punched-card images, in which information is arranged in specific columns or character positions. As a result, IGES files are almost always much larger than the native formats of vendor systems. Of course, general-purpose compression techniques (e.g., Huffman coding) can be applied to the file, but then the file must be decompressed before each use.

A spin-off of IGES has recently become popular in Europe. SET, an exchange standard originated by Aerospatiale in France, is conceptually rooted in IGES but it addresses two of IGES's drawbacks: the bulkiness of its implementation and its less-than-complete coverage of all vendors' features. SET attempts to represent fully any vendor's features; whether those features can be translated out of SET and into another vendor's system is still not guaranteed. Compactness of representation has been achieved with files one-fifth to one-third the size of uncompressed IGES files.

Beyond IGES, the adoption of the newer Product Data Exchange Standard (PDES) is likely. The PDES is a reworking of IGES that better meets international requirements. The standard relies on a constructive representation of solids instead of wideframe boundaries. Although PDES data formats are not upwardly compatible with IGES, it will be possible to translate IGES files to PDES when PDES gains wider acceptance.

THE ERA OF INTERNATIONAL STANDARDS: MAP, TOP, AND STEP

In networking as well as in data exchange, international standards being developed under the auspices of the ISO are certain to play a dominant role in the future.

The ISO protocol layering scheme is the open systems interconnection (OSI). Although this set of standards is just now reaching completion, the conceptual division of the communications process on which it is based has already become the norm. For example, TCP/IP, although not an ISO standard, largely conforms to the OSI reference model. ISO counterparts of the TCP/IP protocols and its higher-level utilities have been developed, as shown in Exhibit II-1-22. The ISO approach adds an explicit presentation layer to control page and screen formatting and character sets (although not content) as well as a separate session layer. Functions are either similar to or a superset of the corresponding TCP/IP function.

One of the drawbacks of the renewed focus on standards today is that there simply are too many of them. Interoperability of diverse equipment is, in fact, hindered by the presence of multiple options at each of the seven

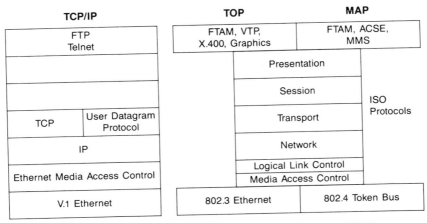

Exhibit II-1-22.   A Comparison of TCP/IP, MAP, and TOP

levels of the protocol. As a result, it has become necessary to create a profile, or second-order standard, to indicate which standard to use. Two such profiles are of major interest in CAD/CAM: Manufacturing Automation Protocol (MAP) and Technical and Office Protocol (TOP).

MAP for the Factory

MAP, sponsored by General Motors, is a means of specifying data transport and semantics for factory-floor automation. MAP endorses the IEEE 802.4 token-passing bus as the preferred LAN; added to this are the standard ISO network, transport, session, and presentation layers. At the application layer, the protocol specifies the file transfer, access, and management (FTAM) for file transfer, the manufacturing message system (MMS) for the content of messages bound for robotic equipment, and the association control for service elements (ACSE) for program-to-program communications.

This set of protocols should make communications possible between heterogeneous devices, not only in moving data from point to point but in communicating the meaning of that data. The MMS, based on the Electronic Industries Association's EIA-511 standard, provides the coding for controlling the equipment that machines parts and assembles them into a finished product.

Problems of compatibility have arisen, however—not among vendors using MAP but between versions of the standard itself. Version 3.0, issued in 1988, is not upwardly compatible with version 2.1, issued in 1985. FTAM has been retained and enhanced—it can now support accessing parts of a remote file rather than the entire file—but the MMS and the ACSE are not compatible with their predecessors from version 2.1. As a result, large-scale implementation of MAP has been delayed pending the availability of products using version 3.0. A significant amount of momentum has been lost in this delay.

Meanwhile, variations of MAP have already appeared. One that has found considerable acceptance is an abbreviated protocol stack known as enhanced performance architecture (EPA), which is also known as mini-MAP. This version eliminates the network, transport, session, and presentation layers on single-hop networks. It has been supported by the process control industry and by a European consortium known as Communications Networks for Manufacturing Applications (CNMA).

CNMA has lobbied for inclusion of Ethernet—the IEEE 802.3 version—as an alternative to the token-bus approach for the underlying network. This proposal, however, has encountered resistance on the grounds that many manufacturers will support only one medium, and interoperability will suffer if two standards proliferate.

TOP for the Office

A companion protocol set to MAP is TOP. MAP is optimized for the factory floor, TOP for the office environment. As shown in Exhibit II-1-22, its middle protocol layers are identical to those of MAP, but it diverges at the top and bottom layers.

The lower layers are based on the 802.3 version of Ethernet. Although token-bus cabling is readily installed in open-bay factories, Ethernet cabling is better suited to individual offices. In addition, the perceived token-bus advantages of noise immunity and bounded worst-case delay are not critical in office work.

At the top of the protocol stack, TOP retains FTAM and the ACSE and adds electronic mail (X.400) and virtual terminal facilities. The MMS is not needed.

Because CAD is essentially an office activity and because CAM extends to the factory floor, support of both protocols is desirable in CAD/CAM installations. The natural dividing point between the two activities is at the data base level, in the separation of the engineering and factory data bases. CAD workstations will embrace TOP (at least, when potential performance problems with ISO protocols are resolved), and NC tools and controllers will gravitate toward MAP.

Originally proposed by Boeing, TOP today is benefiting from a boost by the federal government. Government contracts are stipulating the use of the POSIX protocol set, which is essentially identical to TOP; TCP/IP is being allowed only as a temporary alternative.

STEP: International Data Exchange

Paralleling the work on international standards for data transport is a corresponding effort in data exchange representations. Work on PDES has been expanded in scope and has now joined forces with the ISO effort in a program known as STEP. The situation indicates the eventual acceptance of the IEEE 802 local area network standards as ISO standards.

When it is released during the 1990s, STEP—the French acronym for Standard for the Exchange of Product Data—will complete the evolution of CAD/CAM standards from proprietary conventions through national standards to international standards. As production of goods takes place on an increasingly international scale, the standards necessary to expedite the process should be in place.

Chapter II-1-6
ETSI: A European Standards Organization
Michael Varrassi

THE GEOPOLITICAL NATURE OF EUROPE has had a profound influence on the need for communications standards there. Comprising a large number of sovereign nations in a relatively small geographic area, Europe, notwithstanding the relatively short distances involved, is the site of many individual national communications standards and numerous and distinct operating agreements. To sell manufactured goods or provide communications services in Europe has required as many as 21 individual operating agreements. Interoperability problems plague communications among European and non-European countries. These incompatible standards have made it difficult to prepare for the open European market, a problem that affects European companies and governments and non-European companies wishing to do business there. The governments of Europe were finally compelled to recognize the need for standardization and to form the European Telecommunications Standards Institute (ETSI).

From its inception, the ETSI was intended to establish common positions and combine individual views for subsequent introduction to international standards forums. The idea was that ETSI activities would be similar to the standards activities under way in the US, the goal being to speed the development of international standards. Exhibit II-1-23 describes this process.

Regional standards organizations can resolve conflicting views within the membership before the proposals reach the international standards forums. ETSI could help expedite the formalization of European positions before the establishment of international standards positions, speed the production of worldwide standards through a preapproval process, and anticipate worldwide standards, preventing the premature introduction of national standards within individual European countries.

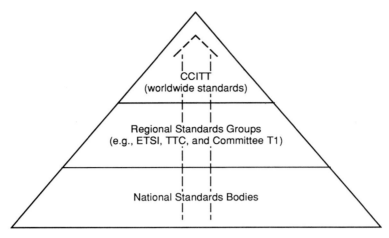

Exhibit II-1-23.   The New International Standards Process

ETSI is a young organization. In 1987, discussions were held in the Conference of European Postal and Telecommunications Administrations (CEPT) concerning the creation of a standards organization to unify the European communications standards development process. These discussions led to the development and approval of a memorandum of understanding in countries comprising the CEPT, signed in January 1988. The first general assembly meetings were held in March 1988, followed closely by the first technical assembly that same month.

STRUCTURE OF ETSI

There are two main governing bodies within ETSI: the technical assembly, which addresses specific technical work functions; and the general assembly, which is an overall governing or management board and secretariat.

The technical assembly consists of a hierarchy of subgroups. Individual project teams are the most fundamental level. These teams work in closed groups on specific topics; they produce the technical draft text that ultimately becomes the basis for a standard. Individuals who participate in these teams work full time on solving a particular question, which means, in general, doing the work needed to create a draft standard. When completed, the work of the project team is passed to the governing subcommittee that is responsible for the overall work area.

Subcommittees perform their work and provide guidance and technical input to the project teams by holding periodic group meetings at various locations throughout Europe. The subcommittees take the work of project teams and prepare draft standards. At this level, technical details drive the decision process. The work of the subcommittee is passed to a technical committee.

The technical committee's purpose is to obtain agreements from the members representing the ETSI countries and determine whether a given draft standard should move forward in the approval process. This is the first point at which politics becomes a significant factor in what may result. Once an agreement has been reached, the draft standard is passed to the technical assembly.

The technical assembly has two functions. The first is to approve the draft standard and recommend the approval to the general assembly. (The general assembly then passes the standard to participating countries for a period of public comment.) The second function is to develop and approve the technical work program for the upcoming year. The information flow among these groups is depicted in Exhibit II-1-24.

Funding ETSI's Work

ETSI activities are of two types: annual and multiannual, and within these two categories are either costed or voluntary. Currently, ETSI finds that most work programs are accommodated during a single annual project cycle. Costed work programs are funded by ETSI through the contributions of its

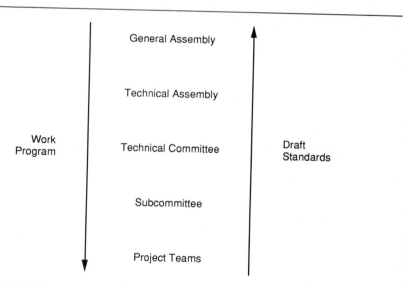

Exhibit II-1-24. Technical Work Flow Through the ETSI

membership. Voluntary programs are of interest to only one or a few members; therefore, the funds necessary to support voluntary work come directly from the interested members.

This system works well. In fact, from inception through completion and approval, a standard takes an average of 300 days, an astonishingly short period of time when compared with the results of other regional and international standards bodies. A large investment in personnel makes possible these results. There were more than 100 full-time experts involved in ETSI standards work in 1989. The technical assembly had the participation of over 1,000 technical experts.

VOLUNTARY STANDARDS

ETSI produces three types of output: voluntary standards, which are composed of interim European Technical Standards (I.ETSs), European Technical Standards (ETSs), and technical reports.

After the ETSI technical and general assemblies have developed and approved a particular standard, the text of the standard goes to the appointed national standards organization within the participant country for public notice and comment. This process is similar to the American National Standards Institute (ANSI) public notice procedure used within the US. During this period, generally 90 days, all comments received are returned to ETSI. ETSI reviews the comments and, unlike ANSI, may at its sole discretion decide whether to let the original text stand, modify the text as may be indicated in the received comments, or begin the process again.

Currently, no mechanism exists for the resolution of comments by multiple review. One additional iteration would double the time required to approve the standards, and ETSI feels this cost is too high.

MANDATORY STANDARDS

There has been much discussion within the US regarding the mandatory nature of the ETSI standards. In fact, ETSI does not produce, through the procedure just described, mandatory standards. A political action that follows ETSI's technical work produces mandatory standards known as Normal European Telecommunication Standards (NETSs); this action is taken by the European Economic Council and is outside the jurisdiction of ETSI.

ETSI usually knows when a specific work area is likely to produce a NETS. This occurs when a CEPT committee notifies ETSI that work in a particular area will be submitted to the formalized process necessary to produce the mandatory standard. Examples of this type of notification include the digital telephone standard and group 3 facsimile work. When these decisions were being made, formal notification was given to non-European companies that were thought to have an interest. The US companies were sent draft copies of the standards and were asked to provide their input. Although this was done, there is no mechanism within ETSI to provide feedback to let those companies know whether the comments are to be incorporated into the standards.

ETSI AND GOVERNMENT CONTROL

There is no direct government control of ETSI—it enjoys complete autonomy from the 21 countries for which it develops standards. Governments have a liaison to the management group through the various national standards organizations, the CEPT, and other government agencies. The role of the national standards organizations is to make sure that public notice is given during the development of standards. Once a standard is approved, the national standards organizations are responsible for the introduction of the standards within their country. The national standards organization also is responsible for withdrawing from use within that country any conflicting standards.

Exhibit II-1-25 describes the steps necessary to prepare a mandatory standard or an NETS. The ETSI technical assembly prepares the proposed standard and the standard is passed to the ETSI management committee for final approval (1). Once approved at this level, it is passed to the national standards organizations for public notice (2). Comments are submitted to

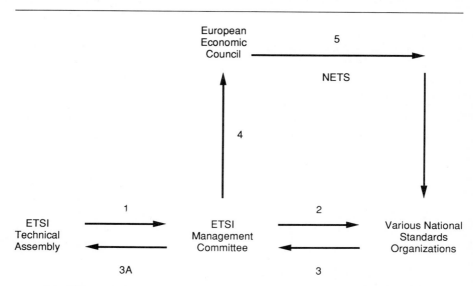

Exhibit II-1-25. Turning Voluntary Standards into Mandatory NETSs

the ETSI management committee (3), and then back to the technical groups, if necessary (3A). The European Economic Council (4) then votes whether the ETS should become a NETS. Final NETSs are then returned to the national standards organizations for incorporation (5).

COMPOSITION OF ETSI

As mentioned, ETSI was intended from its inception to work independently of the governments for which it produces standards. Currently, there are 21 governments that support and participate through the general assembly in the ETSI process. Technical and financial support comes from ETSI members and participant companies.

Membership in ETSI is broken down into five categories: research bodies, public network operators, manufacturers, administrations, and users and private network providers. Recent membership in ETSI was 183 organizations. The composition of the membership is shown in Exhibit II-1-26.

The distinction between administrations and public network providers should be noted. In some European countries, the administration and the public network providers are the same entity. An example of this is the French administration, whose public network provider is French Telecom. In other countries, however, they may be distinctly different organizations. This generally occurs where there is more than one provider of public network services. Some examples of this situation include the UK, where the Department of Trade and Industry is the administration and where British Telecom and Mercury provide public network services. The US is also an example, where the US Department of State functions as the administration and there are a large number of public network providers.

Other Participation

Observers, special guests, and counselors also participate in ETSI. Observers, as the name implies, are permitted to attend general assembly and technical assembly sessions but are not allowed to participate in the proceedings. There is no provision for the participation of organizations outside

Category	Number of Members
Manufacturers	112
Administrations	26
Public Network Providers	21
Users and Private Service Providers	19
Research Bodies	5

Exhibit II-1-26.   ETSI Membership by Category, November 1989

Europe in the lower working levels of ETSI. There are, however, negotiations under way that might provide this type of liaison. There are 25 observers of ETSI activities.

Special invited guests consist of organizations directly invited to ETSI assemblies, and represent groups from outside Europe. The majority of these groups are formal standards organizations. Among them, from the US, are ANSI, the Telecommunications Industry Association (TIA), and Committee T1. The final group, counselors, consists of two governmental bodies: the European Economic Council and the European Free Trade Association (EFTA). The EFTA is the organization that represents countries not part of the European Economic Council.

Areas of Activity

Currently, there are three major areas of ongoing activity within ETSI: communications, information and technology, and broadcasting. The second two areas, information and technology and broadcasting, overlap standardization areas previously the responsibility of other European groups. Work on information and technology standards now is performed by ETSI instead of the European Standardization Committee and the European Committee for Electrotechnical Standards. Broadcasting (including cable activities) covers standards that were the responsibility of the European Broadcasting Union.

Within the fields of technical study, there have been specific areas of focus. Legal issues, particularly intellectual property rights, are being addressed by the intellectual property rights technical committee. The strategic review committee examines areas for future ETSI work. Technical work has been focused on integrated services digital network (ISDN) and mobile communications, with an emphasis on terminals and intelligent networks. Technical standards efforts are coordinated through the activities of 12 technical subcommittees: network aspects, business telecommunications, signaling protocols and switching, transmission and multiplexing, terminal equipment, radio equipment and systems, the special mobile group, paging systems, satellite earth stations, equipment engineering, advanced test methods, and human factors.

LIAISON WITH ETSI

Three levels of formal liaison are recognized by ETSI, but these apply only to other European organizations. These levels are differentiated by the amount of information and the coordination required between ETSI and the liaison body and are described in Exhibit II-1-27. Liaisons with individual companies do not occur.

Type of Activity	Amount of Coordination
Level 1: Exchange of information	Low
Level 2: Coordination of work	Medium
Level 3: Cooperation in new work ventures	High

Exhibit II-1-27.   Liaison Between ETSI and Outside Organizations

---

The general assembly has set specific guidelines for the future exchange of information. They are as follows:

- The exchange of information can be carried out only with formal associations representing manufacturers, users, government bodies, or standards organizations—There is no plan for an exchange of information with organizations outside Europe.
- Exchange of documentation must be made in strict accordance with negotiated agreements between ETSI and the organization.
- Dissemination of the ETS, I.ETSs, and technical reports is permitted.
- Some types of working documents may be distributed, with the prior approval of the general assembly—These document types include the annual work program, multiannual work program, and draft standards.
- The first draft of a document may be distributed if the approval of the chairman of the technical committee is given.

The general assembly does not authorize unilateral release of information and looks for reciprocity in any information release agreement.

SUMMARY

One question that continues to be asked is how interested companies can contribute to the ETSI process. Currently, annual workshops are designed to provide input for the development of the annual work program. For voluntary standards that are destined to become Normal European Technical Standards, input is solicited from various international manufacturing and service providers. As mentioned earlier, some providers and manufacturers in the US were given copies of proposed standards that were to become mandatory and were invited to comment on them. Unfortunately, there is no formal way to learn the results of this input and there are no iterative means of resolving conflicts between contributions. The reason for this refusal is the amount of time it would take—600 days on the average.

To provide individuals and companies outside Europe with information on ETSI and its work, ETSI has expanded its sources of information. Newsletters, notes on emerging technologies, and basic handouts and status reports are issued. Additional information about ETSI is available from the following individuals:

Armin Silberhorn
Senior Manager, Head of Department of Technical Organization
Route des Lucioles Sophia-Antipolis
B.P. 152-06561
France
Telephone: +33.92.94.42.19
Facsimile: +33.93.65.47.16
Telex: 470.040F

Professor Gagliardi
Director, ETSI
Route des Lucioles Sophia-Antipolis
B.P. 152-06561
France
Telephone: +33.92.94.42.00
Facsimile: +33.93.65.47.16
Telex: 470.040F

Section II-2
Networking Services

NETWORKING SERVICES, ESPECIALLY THOSE PROCURED from common and value-added carriers, provide the transmission infrastructure for contemporary communications networks. Except for leased facilities, the most ubiquitous services today are those based on the packet-switching technology exemplified by X.25. In the realm of private leased lines, the trend is toward increased use of digital facilities of the T1 genre. These will migrate, albeit slowly, toward the integrated services digital network (ISDN). These aspects of networking services are the subject of this section.

Chapter II-2-1, "X.25 Facilities," takes a very comprehensive look at the facilities provided by packet-switching networks. Facilities are what some might call options. Most are subscription choices that are then invoked on a call-by-call basis.

X.25 has evolved continuously since the publication of the first meaningful standard in 1972. Keeping up with changes can be a problem for both network providers and users. Chapter II-2-2, "Recommendation X.25: Changes for the 1988–1992 Cycle," provides detailed explanations of the important differences between the 1984 and 1988 versions of X.25.

T1 has become a widely used and a very cost-effective service for application communication networks. It is often the choice for high-speed digital backbone corporate networks and is useful for integrating voice and data. Chapter II-2-3, "T1 and Beyond," examines the characteristics and applications of T1 as well as providing an introduction to SONET and T3 networks.

Chapter II-2-4, "An Update on ISDN," closes this section. The chapter discusses the current state of the latest digital technology to be deployed—integrated services digital networks.

Chapter II-2-1
X.25 Facilities
Uyless Black

A WEALTH OF INFORMATION is available on the CCITT X.25 standard for packet-switching networks. Certain topics are rarely explained in this literature, however, perhaps because the X.25 standard itself is rather terse in its explanation of them. The quality-of-service features, which are also called facilities, are an example. Because the X.25 standard is written as a reference and not as a tutorial, the description of each facility is restricted to a few sentences or paragraphs. The facilities are the foundation of many X.25 services, however, and this brief treatment does not reflect their importance. This chapter, which provides a detailed tutorial on the X.25 facilities, is intended to supply some of the missing information.

X.25 OVERVIEW

This section consists of a brief overview of the structure of X.25 to provide a context for the descriptions of the facilities that follow.

The formal title of the X.25 standard is *Interface Between Data Terminal Equipment and Data Circuit Terminating Equipment for Terminals Operating in the Packet Mode on Public Data Networks*. As Exhibit II-2-1 illustrates, X.25 defines the procedures for the exchange of data between a user device, or data terminal equipment (DTE), and a network node, or data communications equipment (DCE). What the standard defines as a DCE is more accurately called a data switching exchange (DSE). For consistency with the standard, however, this chapter uses DCE.

Because X.25 defines the procedure by which a user DTE device and a packet network node (DCE) establish a session and exchange data, it also defines, in effect, the procedures by which two packet-mode DTE devices communicate through a network. The procedures include those for identifying the packets of specific user terminals or computers with logical channel numbers, for acknowledging or rejecting packets, for error recovery, and for flow control.

X.25 provides for sequence numbers in packets to allow the DTE and DCE to keep packets in proper order and to provide a way for them to acknowl-

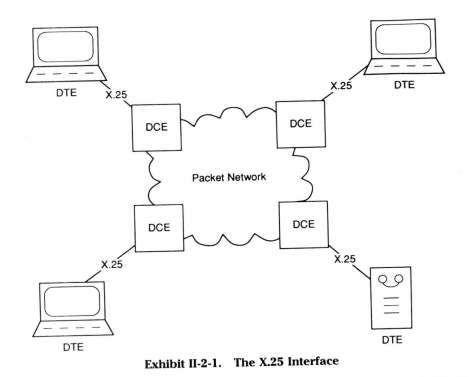

DTE · X.25 · DCE · X.25 · DTE · DCE · Packet Network · DCE · DCE · X.25 · X.25 · DTE · DTE

Exhibit II-2-1.   The X.25 Interface

edge their successful receipt. The P(S) field is used to order packets that are being sent. The P(R) field is used to acknowledge received packets. The default values for both the P(S) and P(R) fields are 0 through 7.

X.25 also allows DTE devices and DCE nodes to specify the size of transmit and receive windows that determine the number of packets they can send or receive without flow control restrictions. For example, if a DCE had a receive window of 2 (the X.25 default value), the DTE device could send it two packets without flow control restrictions and without waiting for an acknowledgment. This parameter permits the DTE and DCE to control the rate at which they receive incoming packets. The sequence number values and the transmit window values do not have to be the same.

X.25 provides two ways for DTE devices to establish and maintain communications:

- Permanent virtual circuits.
- Virtual calls.

A permanent virtual circuit, shown in Exhibit II-2-2, is analogous to a leased line in a telephone network; it ensures a connection through the packet network between the transmitting and receiving DTE devices. Permanent virtual circuits are established by prior agreement among the users and the packet network carrier. The carrier then receives a logical channel number for the users. Thereafter, when a transmitting DTE device sends a packet into the packet network, the identifying logical channel number in the packet indicates that the requesting DTE device has a permanent virtual circuit connection to the receiving DTE device. As a result, services are provided by the network and the receiving DTE device without further session negotiation. A permanent virtual circuit requires no call-setup or clearing procedures, and the logical channel is continually in data transfer state.

A virtual call (also called a switched virtual call) resembles a telephone dial-up call in that it requires setup and breakdown procedures, as Exhibit II-2-3 indicates. The calling DTE device issues a special X.25 packet, called a call-request packet, with a logical channel number and the address of the called DTE device to the network. The network uses the address to route the call-request packet to the DCE that is to support the call at the remote end. This DCE then sends an incoming-call packet to the proper DTE device. The receiving DTE device, if it chooses to acknowledge and accept the call request, transmits a call-accepted packet to the network. The network then transports this packet to the requesting DTE device in the form of a call-connected packet. The channel then enters a data transfer state, creating an end-to-end virtual circuit.

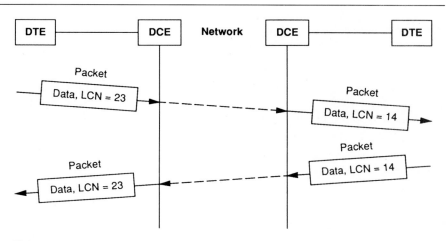

Note:
LCN   Logical channel number

Exhibit II-2-2.   A Permanent Virtual Circuit

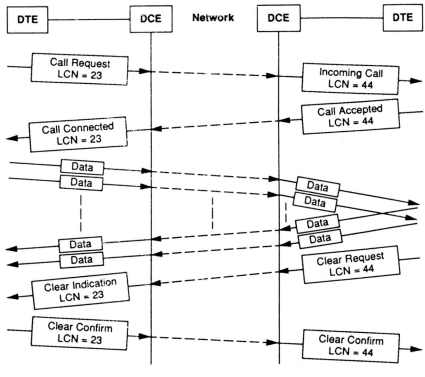

Note:
LCN   Logical channel number

Exhibit II-2-3.   A Switched Virtual Call

To terminate the session, either the DTE device or the DCE sends a clear-request packet. It is received as a clear-indication packet and confirmed with a clear-confirm packet. After the call is cleared, the logical channel numbers are made available for another session. To reestablish a connection, the two DTE devices must repeat the setup procedure.

Packet Format

The format of an X.25 call management packet is illustrated in Exhibit II-2-4. The fields and their functions are as follows:
- Logical channel group number and logical channel number—This field contains the logical channel numbers that identify the user's packet session to the network and the other user.
- General format identifier—This field is used to stipulate extended packet numbering (coded in bits 5 and 6 in the first octet of the

packet), to request end-to-end acknowledgment (coded with the D bit—bit 7 of the first octet of the packet—to 1), to designate a data packet as a qualified packet (coded with bit 8 of the first octet), and to request an extended ISDN-type address field for subsequent data packets (also coded with bit 8).

- Packet type identifier and control packet for data packets—This field identifies the type of packet (e.g., as a call request or a clear confirm). It is also used with data packets to code packet sequence numbers— P(S) for the sending sequence number, P(R) for the receiving sequence number, and the M bit to indicate that more data is coming.
- DTE addresses—This field contains codes of the called and calling DTE addresses.
- Facilities fields—These fields contain codes to identify the facilities to be used for the DTE-to-DTE session.
- Data field—This field contains user data or control information from an upper-layer protocol.

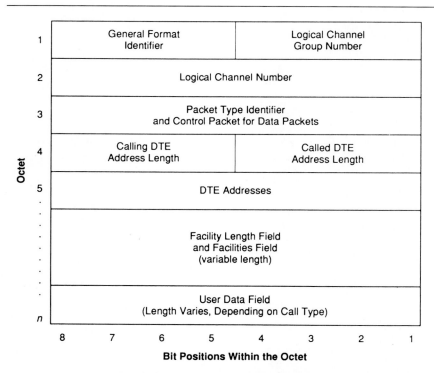

Exhibit II-2-4.   The X.25 Packet Format

INTRODUCTION TO THE X.25 FACILITIES

The X.25 facilities, or quality-of-service features, provide functions that range from the useful to the essential. They allow a user to tailor the way the network supports a session. The facilities are of four types:

- International facilities as specified in CCITT Recommendation X.2.
- CCITT-specified DTE facilities.
- Facilities offered by the originating public data network.
- Facilities offered by the destination public data network.

Some facilities are requested in the facilities field in the call-request packet. Users must identify which of these facilities they want to be able to use when they subscribe to the network.

CCITT Recommendation X.2 describes several facilities, classifies them as essential or additional (optional), and specifies whether they are applicable to switched virtual calls or to permanent virtual circuits. It also stipulates which facilities are to be used with a circuit-switched network and which with a packet-switched network.

Exhibit II-2-5 lists the facilities specified in X.2 and indicates their status. The heading User Class of Service in the exhibit refers to CCITT Recommendation X.1, which provides a standard for signaling rates on public data networks. The classes of service relevant to Exhibit II-2-5 are listed in Exhibit II-2-6.

The rest of this chapter provides a description of each of the facilities specified in X.25. For convenience, they have been grouped roughly according to the functions they perform, as follows:

- Call restriction facilities.
- Charging-related facilities.
- Quality-of-service facilities.
- Facilities for reliability, windows, and acknowledgments.
- Call destination management facilities.
- Miscellaneous facilities.

CALL RESTRICTION FACILITIES

Call restriction facilities include the incoming-calls-barred and outgoing-calls-barred facilities, the one-way logical channel outgoing and one-way logical channel incoming facilities, and the closed user group and bilateral closed user group facilities.

Incoming-Calls-Barred and Outgoing-Calls-Barred Facilities

The incoming-calls-barred facility prevents incoming calls from being presented to the DTE; the outgoing-calls-barred facility prevents the DCE from accepting outgoing calls from the DTE. Both facilities apply to all

Facility		User Class of Service			
		8–13		20–23	
		VC	PVC	VC	PVC
1.0	**Optional user facilities assigned for an agreed contractual period**				
1.1	Extended frame sequence numbering	A	A	—	—
1.2	Multilink procedure	A	A	—	—
1.3	Online facility registration	A	—	FS	—
1.4	Extended packet sequence numbering (modulo 128)	A	A	—	—
1.5	D-bit modification	A	A	FS	—
1.6	Packet retransmission	A	A	—	—
1.7	Incoming calls barred	E	—	A	—
1.8	Outgoing calls barred	E	—	A	—
1.9	One-way logical channel outgoing	E	—	—	—
1.10	One-way logical channel incoming	A	—	—	—
1.11	Nonstandard default packet sizes 16, 32, 64, 256, 512, 1,024, 2,048, 4,096	A	A	FS	FS
1.12	Nonstandard default window sizes	A	A	—	—
1.13	Default throughput classes assignment	A	A	FS	FS
1.14	Flow control parameter negotiation	E	—	FS	—
1.15	Throughput class negotiation	E	—	FS	—
1.16	Closed user group	E	—	E	—
1.17	Closed user group with outgoing access	A	—	A	—
1.18	Closed user group with incoming access	A	—	A	—
1.19	Incoming calls barred within a closed user group	A	—	A	—
1.20	Outgoing calls barred within a closed user group	A	—	A	—
1.21	Bilateral closed user group	A	—	A	—
1.22	Bilateral closed user group with outgoing access	A	—	A	—
1.23	Fast select acceptance	E	—	FS	—
1.24	Reverse charging acceptance	A	—	A	—
1.25	Local charging prevention	A	—	FS	—
1.26	Network user identification subscription	A	—	A	—
1.27	NUI override	A	—	—	—
1.28	Charging information	A	—	A	—
1.29	RPOA selection	A	—	A	—
1.30	Hunt group	A	—	A	—
1.31	Call redirection	A	—	FS	—
1.33	Long address acceptance	A	A	FS	—
1.34	Direct call	FS	—	A	—
2.0	**Optional user facilities on a per-call basis**				
2.1	Flow control parameter negotiation	E	—	—	—
2.2	Throughput class negotiation	E	—	—	—
2.3	Closed user group selection	E	—	E	—
2.4	Closed user group with outgoing access selection	A	—	FS	—

Exhibit II-2-5. Packet-Switched Data Transmission Facilities as Defined in X.2

		User Class of Service			
		8–13		20–23	
Facility		VC	PVC	VC	PVC
2.5	Bilateral closed user group selection	A	—	FS	—
2.6	Reverse charging	A	—	A	—
2.7	Fast select	E	—	FS	—
2.8	Abbreviated address calling	FS	—	A	—
2.9	Network user identification selection	A	—	A	—
2.10	Charging information	A	—	A	—
2.11	RPOA selection	A	—	A	—
2.12	Call deflection selection	A	—	—	—
2.13	Call redirection or deflection notification	A	—	FS	—
2.14	Called line address modified notification	A	—	FS	—
2.15	Transit delay selection and indication	E	—	—	—
2.16	Abbreviated address calling	FS	—	A	—

Notes:
A   Additional
E   Essential
FS   For further study
PVC   Permanent virtual circuit
VC   Virtual call

Exhibit II-2-5. (*Cont*)

---

logical channels at the DTE-DCE interface and cannot be changed on a per-call basis. A DTE device subscribing to incoming calls barred can initiate but cannot accept calls. A DTE device subscribing to outgoing calls barred can receive calls but cannot initiate them. Once a call is established, the session operates at full duplex.

Some network administrations use the calls-barred facility to restrict DTE access to the network on the basis of the permitted protocol-to-protocol agreement between DTE devices. This technique, called protocol screening, is enforced by the DCE or the packet assembler/disassembler (PAD). If the permitted protocols for one user do not include those of the other, the call is cleared. Other network administrations allow a call to go through only if the address of the called DTE device is the same as the address of the calling DTE device.

One-Way Logical Channel Outgoing and One-Way Logical Channel Incoming Facilities

The one-way logical channel outgoing facility restricts a logical channel to outgoing calls only; the one-way logical channel incoming facility restricts a logical channel to incoming calls only. These facilities are set when the user subscribes to the network, and they cannot be changed on a per-call basis. They provide more specific control than the calls-barred facility because they operate on a specific channel or channels. For example, an organization might allocate a certain number of channels to a time-sharing computer as incoming only to ensure access to customers who call in. In much the same way, a telephone-based private branch exchange can restrict some lines to incoming or outgoing calls only. Having the incoming-calls-barred facility is equivalent to having all virtual calls from a location one-way outgoing, and having the outgoing-calls-barred facility is equivalent to having all virtual calls one-way incoming.

Closed User Group Facilities

The closed user group facilities allow users to restrict access to and from a group of DTE devices, providing some security and privacy in an open network. These facilities allow a group of users to create, in effect, a virtual private network. A DTE device can belong to more than one closed user group, to a limit determined by the network. In addition, a DTE device that has subscribed to the closed user group facilities can call other closed user groups on a per-call basis. X.25 defines seven of these facilities.

Closed User Group Facility. This facility allows a DTE device to belong to one or more closed user groups, each established for a specific period of time. A DTE device that belongs to more than one closed user group must specify a preferential group.

User Class of Service	Data Signaling Rate
Synchronous Mode *(X.25 Interface)*	
8	2,400 bps
9	4,800 bps
10	9,600 bps
11	48,000 bps
12	1,200 bps
13	64,000 bps
Asynchronous Mode *(X.28 Interface)*	
20	50–300 bps
	10 or 11 units per character
21	75–1,200 bps
	10 units per character
22	1,200 bps
	10 units per character

Exhibit II-2-6. User Classes of Service

Incoming Calls Barred Within a Closed User Group Facility. This facility allows a DTE device to initiate calls to other members of a closed user group but not to receive calls from them. Using it is equivalent to having all logical channels one-way outgoing (originate only) within the group.

Outgoing Calls Barred Within a Closed User Group Facility. This facility allows a DTE device to receive calls from other members of the closed user group but not to initiate calls to them. Using this facility is equivalent to having all logical channels one-way incoming (terminate only) within the group.

Closed User Group with Incoming Access Facility. This facility allows a DTE device to receive calls from DTE devices that belong to the open part of the network (i.e., not to the closed user group) and from devices that belong to other closed user groups and have outgoing access.

Closed User Group with Outgoing Access Facility. This facility allows a DTE device to initiate calls to all devices that belong to the open part of the network and to devices that belong to other closed user groups and have incoming access. If the DTE device has a preferential closed user group, it specifies other groups with the closed user group selection facility.

Closed User Group Selection Facility. The DTE device uses this facility in a call request to specify a closed user group for a call. The DTE can use this facility only if the user has subscribed to the closed user group facility, the closed user group with outgoing access facility, or the closed user group with incoming access facility.

Closed User Group with Outgoing Access Selection Facility. Like the closed user group selection facility, this facility allows a DTE device to specify a closed user group for a call. It also indicates to the called DTE device that the calling device wants outgoing access. The called device receives the incoming call packet with the identification of the closed user group of the calling DTE device.

Bilateral Closed User Group Facilities

These facilities are similar to the closed user group facilities, but they allow users to restrict access to and from pairs of DTE devices. They support bilateral relationships between two devices and exclude access to or from others. There are three bilateral closed user group facilities.

Bilateral Closed User Group Facility. This facility allows a DTE device to belong to one or more bilateral closed user groups.

Bilateral Closed User Group with Outgoing Access Facility. This facility allows a DTE device to belong to one or more bilateral closed user groups and to initiate calls to the open part of the network.

Bilateral Closed User Group Selection Facility. The DTE device uses this facility in a call request to specify a bilateral closed user group for a virtual call.

CHARGING-RELATED FACILITIES

The charging-related facilities include the reverse charging and reverse charging acceptance facilities, the local charging prevention facility, the network user identification facilities, and the charging information facility.

Reverse Charging and Reverse Charging Acceptance Facilities

These facilities allow packet network charges to be billed to the called DTE; using them is like making a collect telephone call. They can be used with virtual calls and fast selects. Although the two facilities are closely related, they need not be used together. The reverse charging facility stipulates that the remote DTE device is to pay for the call and is requested by the calling DTE device on a per-call basis. The reverse charging acceptance facility authorizes the remote DCE to pass the calls on to the remote DTE device. Without it, the remote DCE does not pass the call on and the originating DTE device receives a clear-indication packet.

Some networks keep records of unsuccessful attempts to use the user data field to establish a reverse charge call. They then charge the calling DTE device for the attempts in order to discourage what would otherwise be free one-way transmission of data.

Local Charging Prevention Facility

This facility authorizes the DCE to prevent the establishment of calls for which the subscribing DTE device would ordinarily pay. For example, a sending DTE device might be prevented by this facility from sending a reverse charge to a DTE device that has subscribed to the reverse charging acceptance facility. The DCE could prevent calls requesting reverse charging by not passing them on to the DTE device, or (responding to the network user identification facility discussed next) it could charge a third party for the calls.

The local charging prevention facility is often used to force all calls from public telephone dial-up ports to use reverse charging because it is difficult to identify the originator of such calls. X.25 requires reverse charging if the call-request packet does not identify the charged party. In such cases, the remote DCE inserts the reverse charge facility code into the incoming call packet.

Network User Identification Facilities

The network user identification facilities let the calling DTE provide billing, security, and management information to the DCE on a per-call basis. They can also be used to invoke other subscribed facilities and to invoke a different set of facilities with each call, allowing users to tailor their use of the X.25 facilities.

The network user identification facilities work with network user identifiers. Each identifier is associated with a specific set of facilities. The DTE device transmits the identifier to the DCE, but the network never transmits the identifier to the remote DTE device. Furthermore, the calling address in the packet is independent of the user identifier, and the value of the identifier cannot be inferred from the address.

Networks use these facilities in many ways. The facilities make it possible, for example, to identify a network user regardless of the source of the call and can permit the user to make a call from a dial-up port without reversing the charges. Some vendors make products that use network user identification to prevent calls into specific ports unless the calling DTE device provides a valid network user identifier. Some networks use the identifiers to determine the billing for a call. The network user identification group contains three facilities.

Network User Identification Subscription Facility. This facility, established with the network for a specific period, allows a DTE device to use the network user identification selection facility to furnish information (in the call-request packet or the call-accepted packet) to the network on billing, security, and management. It may be used whether or not the DTE device is affected by the local charging prevention facility.

Network User Identification Override Facility. This facility is established with the network for a specific period and used on a per-call basis with the network user identification selection facility to override other facilities to which the user has subscribed. X.25 permits the overriding of only certain facilities in this way. These are:

- Nonstandard default packet sizes.
- Nonstandard default window sizes.
- Default throughput classes assignment.
- Flow control parameter negotiation (subscription time).
- Throughput class negotiation (subscription time).
- Closed user group.
- Bilateral closed user group.

- Bilateral closed user group with outgoing access.
- Charging information (subscription time).
- Recognized private operating agencies.

Network User Identification Selection Facility. A DTE device uses this facility to specify the network user identifier for a call that involves one of the other two network user identification facilities.

Charging Information Facility

With this facility, a DTE device obtains charging information from the DCE about any call charges it is responsible for. The DTE device can invoke the facility on a per-call basis or can subscribe to it for a specific period. If the DTE device has subscribed to the facility, the DCE will supply the information automatically; the user need not request it in the call request or call accepted packet.

QUALITY-OF-SERVICE FACILITIES

The quality-of-service facilities include the nonstandard default packet sizes facility, the nonstandard default window sizes facility, the flow control parameter negotiation facility, the default throughput classes assignment facility, the throughput class negotiation facility, and the transit delay selection and indication facility.

Nonstandard Default Packet Sizes Facility

This facility lets the user select nonstandard default packet sizes for an agreed period. The standard default size is 128 octets of user data. The size can be different for each direction of flow. Some networks that allow calls to be assigned priority levels also allow each level to use a different packet size.

Some vendors have different packet size options for the DTE-DCE interface, the DTE–packet assembler/disassembler interface, and the signaling terminal exchange–signaling terminal exchange (X.75 gateway) interface. Users should check the default sizes for each of these interfaces against the charges for a transmitted packet.

The larger the packet size, the larger the buffers must be in all the machines that process the packet. Larger packet sizes also reduce the opportunity for other users to share the channel, and, in the event of an error, the larger the packet size, the greater the amount of data that must be retransmitted. On the positive side, a larger packet reduces the ratio of overhead fields to user data. Most networks provide a default size of 128 octets and an optional size of 256 octets.

Another factor to consider is the effect of using different packet sizes at each end of a DTE-DTE connection. In Exhibit II-2-7, the DTE-DCE interface at A supports a packet size of 256 octets, and the DTE-DCE interface at B supports a packet size of 128 octets. The 256-octet packet is transported from interface A through the network to interface B. Here, the packet must be segmented into smaller, 128-octet packets.

The sequence of each packet is indicated in the P(S) field: the larger packet at interface A with P(S) = 3 is mapped into P(S) = 3 and P(S) = 4 for the two smaller packets at interface B. The local DCE and remote DTE have little problem with this procedure. They simply return the corresponding P(R) values of 4 and 5 to acknowledge the two packets.

If the D bit is set to 1, as in Exhibit II-2-8, however, the network (or the logic in the DCE nodes) must remember that packets at the remote interface numbered 3 and 4 in the P(S) fields must be mapped back to the local interface with a sequence number of 4 in order to keep the sequence correct at the local interface. These tasks can be written into bookkeeping software, but, because of the overhead involved, most networks restrict the use of different packet sizes to a single direction of transmission. If this service is needed, the transport layer or an intermediate gateway between networks can provide it.

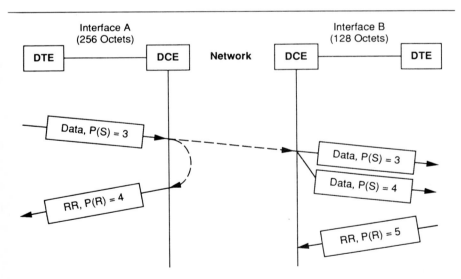

Note:
RR (receive ready) is used to acknowledge packets in the absence of data.

Exhibit II-2-7. A Call with a Different Packet Size at Each End, Acknowledgment with D=0

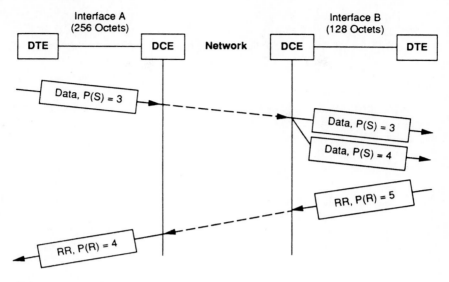

Note:
RR (receive ready) is used to acknowledge packets in the absence of data.

**Exhibit II-2-8.   A Call with a Different Packet Size at Each End,
Acknowledgment with D = 1**

Nonstandard Default Window Sizes Facility

This facility allows users to increase the window size for calls on the interface above the default size of 2 for an agreed period. Window sizes can be different at each end of the connection, but some networks require them to be the same for each direction of transmission across the DTE-DCE interface.

The DCE can change the DTE-DCE packet window size on the basis of the negotiated packet size and throughput class (discussed later). This action lets the network control its buffer requirements.

Flow Control Parameter Negotiation Facility

This facility allows window and packet sizes to be negotiated on a per-call basis for each direction of transmission. In many X.25 networks, the DTE device suggests packet sizes and window sizes during the call setup. The called device (if it subscribes to these facilities) may reply with a counter-proposal. If not, it is assumed the call-setup parameters are acceptable.

The network DCE may control the window and packet sizes. It might modify the parameters received from the sending DTE and send different values to the receiving DTE. Some networks require that the negotiated flow control parameters be the same for each direction of transmission.

X.25 stipulates the following rules for the flow control parameter negotiation facility:
- In the absence of this facility, the default values are used.
- The values in the call request packet can be different from the values in the incoming call packet (because, for example, the network might change them).
- The incoming call packet at the remote DTE device contains the values from which the negotiation starts.

Exhibit II-2-9 shows one example of the facility in operation. DTE device A has requested a window size of 6 and a packet size of 256 octets. The packet is transported to the remote DCE, and an incoming call packet is sent to the remote DTE with the same values. This DTE device attempts to negotiate the window size to 3 and the packet size to 128 octets. It places these values in the call-accepted packet, which is relayed through the network to the originating DTE device. The originating DCE then sends a call-connected packet to the originating DTE device.

In this example, the network did not alter the values that were negotiated between the two DTE devices; however, it does have this option. In addi-

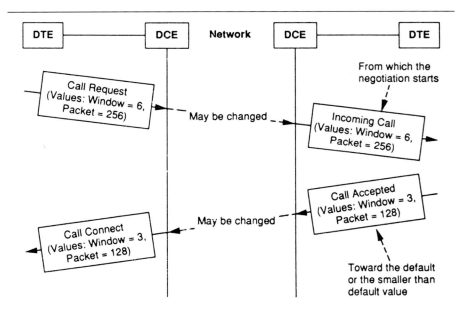

Exhibit II-2-9. Flow Control Parameter Negotiation

tion, the rules of X.25 require that the negotiation move toward the standard default values of 2 for window size and 128 for packet size.

To summarize, the negotiated window and packet sizes are determined by the following actions:

- The initial default values.
- The values in the call-request packet.
- The values in the call-accepted packet.
- Any changes made by the network.

Some networks and the products of some X.25 vendors accept a variety of packet sizes from the DTE device. If the packets are larger than permitted or preferred by the product (or a network), they are split into smaller packets and sent to the destination DTE device. At the DCE of the destination DTE they are recombined and sent to the destination DTE. Some products, however, do not perform packet splitting and recombination. For these, the network will clear a call that requests an unsupported packet size.

Default Throughput Classes Assignment Facility

This facility provides for the selection of one of the following throughput rates (in bits per second): 75; 150; 300; 600; 1,200; 4,800; 9,600; 19,200; and 48,000. X.25 stipulates these rates, but some networks may support others. The choice of the throughput class indicates the desired speed between the DTE device and the DCE.

The throughput rate defines the maximum amount of data that can be sent through the network when it is operating at full capacity. It can be affected by several factors:

- The number of active sessions in the network.
- The characteristics of the local and remote DTE-DCE interfaces, such as their line speeds and window sizes.
- The characteristics of the network.
- The use of flow control packets.
- The use of data packets that are less than the maximum length.
- The use of the D bit.

Throughput Class Negotiation Facility

This facility allows throughput rates to be negotiated on a per-call basis. The allowable negotiable rates (in bits per second) are 75; 150; 300; 600; 1,200; 2,400; 4,800; 9,600; 19,200; and 48,000. Many vendors, however, support higher speeds. Negotiation occurs during call setup through the call-request, incoming-call, call-accepted, and call-connected packets.

X.25 does not allow a negotiated rate to be higher than the subscribed default rate. Should a negotiated throughput rate be rejected (e.g., because it is higher than the default value or because the remote DTE device will not accept it), the call will not necessarily be cleared or blocked; the DCE can lower it to the default value.

The use of the facility is subject to the following:

- Throughput classes can be independent for each direction of flow or the same in both directions.
- The calling DTE device begins negotiations by placing a value in its call-request packet—The receiving DTE device can negotiate for the same or a lower rate; it cannot negotiate for a higher rate.
- The throughput class negotiation facility and the flow control parameter negotiation facility can both be used on the same call.
- Choosing small window and packet sizes may affect throughput.
- Use of the D bit may affect throughput.

Transmit Delay Selection and Indication Facility

This facility, established on a per-call basis, permits a DTE device to select a transmit delay time through the packet network. It can give an end user some control over response time in the network. The network must inform both the calling DTE device (in the call-connected packet) and the called DTE device (in the incoming-call packet) of the transit delay that will apply to the connection. This delay may be greater than, equal to, or smaller than the value specified in the call-request packet.

FACILITIES FOR RELIABILITY, WINDOWS, AND ACKNOWLEDGMENT

Facilities for reliability, windows, and acknowledgment include the D-bit modification facility, the packet retransmission facility, and the extended packet numbering facility.

D-Bit Modification Facility

During the 1970s, some networks implemented a procedure that provided a packet acknowledgment to the originating DCE from the receiving DCE that was then sent to the originating DTE device. This procedure is called internal network acknowledgment because the network assumes responsibility for providing acknowledgment within the network (i.e., between the originating and receiving DCE nodes).

The 1980 release of X.25 included a new feature called the D-bit (delivery bit) service. When the DTE sets the D bit to 1, it instructs the DCE (the network) to send the packet to the remote destination DTE device and not to acknowledge it. When the remote DTE device receives the packet with the D-bit value of 1, it responds with an acknowledgment, which is relayed through the network to the originating device. With this feature, X.25 provides end-to-end (DTE-to-DTE) acknowledgment, but it can be redundant on networks that provide internal network acknowledgment.

The D-bit modification facility allows DTE devices operating on the network's internal acknowledgment procedures to obtain this service. It applies to all virtual calls and permanent virtual circuits at the DTE-DCE interface.

The facility changes bit 7 from 0 to 1 in the general format identifier field in call-request, call-accepted, and data packets sent from the DTE device and from 1 to 0 for incoming-call, call-connected, and data packets sent to the DTE device.

Using the D Bit During Call Establishment (Exclusive of the D-Bit Modification Facility). The D-bit modification facility does not preclude the use of the D bit to obtain end-to-end acknowledgment between DTE devices. X.25 permits the calling device to set the D bit to 1 in a call-request packet to determine whether the called device will accept a D-bit procedure. The network does not act on this bit but passes it transparently to the called DTE device. The called DTE device can return a call-accepted packet with D = 1 to indicate that it is able to handle the D-bit confirmation. It returns D = 0 if it cannot handle the procedure.

Using the D Bit During Data Transfer (Exclusive of the D-Bit Modification Facility). When the D bit is set to 0 in a data packet, the local DCE returns a P(R) value in a receive-ready, receive-not-ready, reject, or data packet to the sending DTE device. In other words, the packet window of the DTE device is updated locally. This local acknowledgment does not imply any remote acknowledgment.

When the D bit is set to 1 in a data packet, the returned P(R) value implies that the remote DTE device has received a packet and that all data bits in a packet in which the D bit had originally been set to 1 have been received. Thus, even if the network segments a packet after it leaves the originating DTE device, the D bit confirms receipt of the original contents of the packet.

At first glance, it might seem that the D-bit procedure would be desirable for all data transmissions. After all, it would guarantee that the data had arrived safely at the destination. Using the procedure can create problems, however. For example, when D = 1, the local DTE device must rely on the remote DTE device to keep its transmit window open. Because X.25 defaults to a packet window of 2, the remote DTE device must be programmed to react quickly to D = 1. Otherwise, the local device might have to stop sending packets. Likewise, the local and remote DCE nodes should be programmed to handle D = 1 packets with minimum delay. This problem can be alleviated somewhat if the DTE device negotiates a larger packet window size (assuming the network and the remote DTE device accept larger windows).

Using the D bit may be unnecessary if an upper-layer protocol provides end-to-end acknowledgment. For example, the CCITT transport layer (class 4) provides such a service, as does the widely used transmission control procedure.

Exhibit II-2-10 illustrates how the D-bit procedure operates. As the exhibit shows, if a DTE device issues packets with $D = 1$ and then issues packets with $D = 0$, its local window may not be updated by the local DCE until the acknowledgment for the $D = 1$ packet is received from the remote DCE.

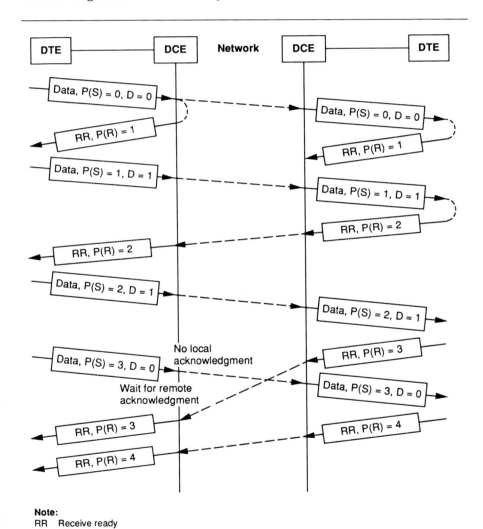

Note:
RR   Receive ready

Exhibit II-2-10.   D-Bit Procedure (Exclusive of the D-Bit Modification Facility)

This procedure might thus reduce throughput, but it could also be used to perform a periodic check with the remote DTE device. In other words, the DTE device could periodically issue a packet with $D = 1$ to determine whether all is well at the other side of the cloud.

Packet Retransmission Facility

This facility (also called the reject facility) applies to all logical channels at the DTE-DCE interface. It allows a DTE device (but not the DCE) to request transmission of one or more data packets. The DTE device specifies the logical channel number and a value for $P(R)$ in a reject packet. The DCE must then retransmit all packets from $P(R)$ to the next new packet. This facility works similarly to the go-back-n technique used by the line protocols at the data link level (the link access protocol, or LAP, reject frame), except that it functions at the packet network level. The packet-level reject facility is not implemented by many networks because the link level is responsible for error detection and the transmission of data. Moreover, X.25 has other conventions for rejecting packets, such as resets, restarts, clears, and the use of the $P(R)$ value in other packets.

Extended Packet Numbering Facility

This facility provides packet sequence numbering using modulo 128 (sequence numbers 1 through 127) for all channels at the DTE-DCE interface. In its absence, sequencing is done with modulo 8 (sequence numbers 1 through 7). Bits 5 and 6 in the general format identifier field are used to request this service (see Exhibit II-2-4).

This facility was added in 1984 to contend with the long propagation time of signals on satellite channels and on other media that have a very high bit transfer rate, such as optical fiber. With these media, the transmitting station can exhaust sequence numbers 1 through 7 before the receiving station has an opportunity to acknowledge the packets. X.25, however, does not allow a sequence number to be reused until the packet in which it was used before has been acknowledged. For example, a DTE device cannot send a packet with $P(S) = 3$ if the preceding packet that used this value has not been acknowledged. Thus, having only seven sequence numbers might require the DTE device or the DCE to stop sending packets, idling the virtual circuit. The extended packet numbering facility simply extends the range of sequence numbers available, allowing fuller use of the channel.

CALL DESTINATION MANAGEMENT FACILITIES

The call destination management facilities include the hunt group facility, the call redirection and call deflection facilities, and the called line address modified notification facility.

Hunt Group Facility

This facility distributes incoming calls across a designated grouping of DTE-DCE interfaces. This 1984 addition to X.25 gives users the ability to allocate multiple ports on a front-end processor or computer or on more than one front-end processor or computer for X.25 traffic. The DCE manages these multiple ports and is responsible for distributing calls across them. The way in which they are distributed is not within the purview of X.25.

The hunt group facility is valuable for organizations with large computing facilities that need flexibility in directing jobs to resources. It is similar in concept to the familiar port selector found in most installations.

X.25 requires selection of an incoming call if at least one logical channel is available, excluding one-way outgoing logical channels. A DTE device on a DTE-DCE interface belonging to a hunt group originates calls normally. The calling DTE device address in the incoming-call packet sent to the remote DTE device contains the hunt group address unless other provisions are made, such as the assignment of a specific DTE-DCE address. Permanent virtual circuits may be on the DTE-DCE interface that belongs to a hunt group, but they operate independently of the hunt group.

Exhibit II-2-11 shows one example of a hunt group in operation. DTE device 2 and DTE device 3 transmit packets to DTE device 1. Instead of using addresses A, B, or C in the packet address field, the sending devices

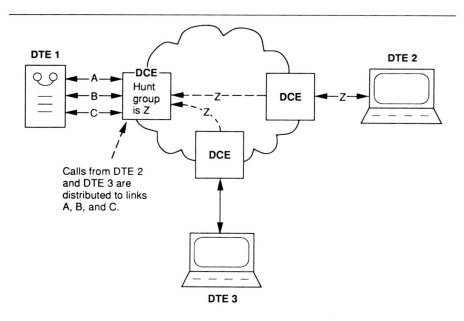

Exhibit II-2-11.   Calling a Hunt Group

use the hunt group address Z. The DCE that is servicing DTE device 1 receives the packets and determines that address Z is actually a hunt group address for ports A, B, and C. It then passes the packets to DTE device 1 across one of these links.

Networks administer hunt groups in several ways. Some place geographic restrictions on their use. Others require naming conventions for hunt group addresses. Subscribing to the hunt group facility may require subscribing to other facilities as well.

Call Redirection and Call Deflection Facilities

These facilities allow users to redirect or deflect calls from the called DTE device to another if the called device is out of order or busy or if it has requested a call redirection. Applications include rerouting calls to a backup DTE device to shield end users from problems and redirecting calls to different parts of a country to avoid time zone problems.

In the vocabulary of call redirection and call deflection, the destination DTE device is the originally called DTE device, and the DTE device receiving the call is the alternative DTE device. As Exhibit II-2-12a indicates, with call redirection, the originally called device never receives an incoming-call packet; with call deflection, the originally called device receives an incoming-call packet and then deflects it. The facilities allow a call to be redirected or deflected only once, but some networks allow further redirection or deflection to alternative devices. X.25 requires that the time required for call setup, including deflections and redirections, fall within established limits.

Call-Redirection Facility. This facility redirects calls when the DTE is busy or out of order. Some networks also allow for systematic call redirection for other criteria. In redirecting a call, the network might try one DTE device at a time from a list of alternative DTE nodes, or it might chain the alternative devices in a fixed order. In this second case, a call might go first to DTE device B, then to DTE device C, and then to DTE device D.

Call-Deflection Subscription Facility. This facility lets the DTE device request that an incoming call be deflected to an alternative DTE device. The network may require that the DTE respond with the deflection request within a certain time limit beyond which it will clear the call.

Call-Deflection Selection Facility. This facility, which is used on a per-call basis if the DTE device subscribes to the call deflection subscription facility, allows the originally called device to specify the alternative device. The following rules apply to call deflection:
- The originally called DTE device must return a clear-request packet to its DCE when it receives an incoming-call packet.

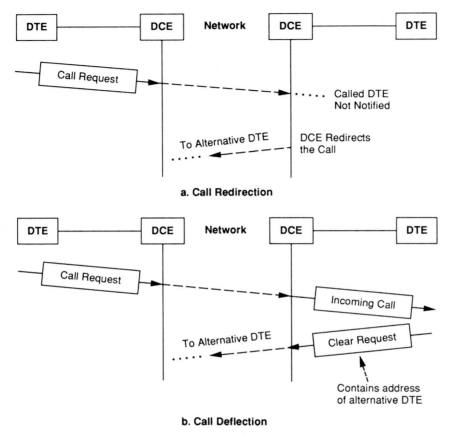

a. Call Redirection

b. Call Deflection

Exhibit II-2-12. Call Redirection and Call Deflection

- The clear-request packet must contain the address of the alternative DTE device and any data and facilities that came with the incoming call—It may also contain the call-redirection or call-deflection notification facility.
- The clear-request packet is not relayed to the calling DTE device.
- The incoming-call packet sent to the alternative DTE device must contain any data and facilities that were in the call-request packet from the calling DTE device.

Exhibit II-2-12b shows how call deflection works. DTE device A sends a call request to its DCE, which is relayed through the network to DTE device B as an incoming call. DTE device B deflects the call with a clear-request packet that contains the address of the alternative DTE device (DTE device

C). The network relays the signal to the alternative device's DCE, which sends an incoming-call packet to the device.

Call-Redirection or Call-Deflection Notification Facility. A DCE uses this facility to inform its DTE device that an incoming-call packet had been redirected or deflected from another DTE device. The packet also gives the reason for the redirection or deflection. Possible reasons for redirection are that the originally called DTE device is out of order, that the originally called DTE device is busy, or that the originally called DTE device had made a prior request that calls be redirected. The reason for deflection would be that the originally called DTE device had requested it.

Called Line Address Modified Notification

This facility allows the DCE to tell the calling DTE device why the called address in a call-connected or clear-indication packet is different from the address originally given in the DTE device's call-request packet.

The called DTE device can also use this facility when more than one address applies to a DTE-DCE interface (as is the case with hunt groups, for example). If the called DTE device address in the incoming-call packet applied to more than one valid address at the called DTE-DCE interface, the responding DTE device could use the called line address modified notification facility in the call-accepted or clear-request packet to indicate that the called DTE device address is different from the address in the incoming-call packet. If the address is not applicable to the interface, the DCE must clear the call.

The DCE at the originating DTE-DCE interface can provide the following reasons for the called line address modification to the calling DTE:
- Call distribution within a hunt group.
- Originally called DTE is out of order.
- Originally called DTE is busy.
- Prior request from the originally called DTE and the network.
- Called DTE originated.
- A deflection by the originally called DTE.

MISCELLANEOUS FACILITIES

Miscellaneous facilities include the online facility registration facility, the fast select and fast select acceptance facilities, the recognized private operating agencies facilities, and the long address acceptance facility.

Online Facility Registration Facility

This facility allows a DTE device, through a registration-request packet, to request facilities or to obtain the current parameters of the facilities. The

DCE responds with a registration-confirmation packet that contains the current values of all the facilities applicable to the DTE-DCE interface. Exhibit II-2-13 illustrates this process.

Some networks do not offer all the X.25 facilities, and some offer their own proprietary facilities. To avoid requesting facilities that are not available or are not allowed, the DTE device can transmit a registration-request packet to the DCE with no facilities values. The DCE will then return a registration-confirmation packet indicating facilities that can be negotiated. The DTE device can then modify these values in a subsequent registration-request packet. When the DCE returns the registration-confirmation packet, the facilities are in effect at the agreed values for subsequent virtual calls.

If the DTE device requests values for a facility that are out of the acceptable range, the DCE lists the acceptable values and a cause code in the registration-confirmation packet. If the DCE cannot accept a DTE request,

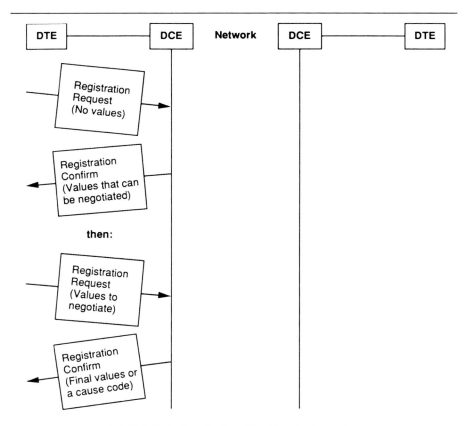

Exhibit II-2-13. Online Facility Registration

it leaves the facilities in question at their prior values. Reasons for not accommodating a request might be that it was in conflict with other facilities at the DTE-DCE interface or that the request packet was issued when a circuit was active (which would confuse the ongoing DTE-DCE dialogue).

Fast Select and Fast Select Acceptance Facilities

These facilities let users eliminate the overhead and delay involved in session establishment and disestablishment for applications that require few transactions or short sessions. The 1984 release of X.25 made fast select an essential facility, requiring vendors to implement it to be certified X.25 network suppliers. Most vendors have complied.

Among the applications that benefit from the fast select facility are inquiry and response applications that involve only one or two transactions. Some examples are point-of-sale updates, credit checks, and money transfers. These applications cannot use switched virtual calls effectively because of the overhead and delay involved in setting up and breaking down such calls. Because their use is occasional, they also cannot justify the cost of a permanent virtual circuit.

Fast select has two options: fast select call and fast select with immediate clear. A fast select call is similar to a virtual call (see Exhibit II-2-3) except that the call-setup packets can contain more user data. The DTE device can request fast select call on a per-call basis to the DCE with an appropriate request in the header of the call-request packet. The facility allows the packet to contain up to 128 bytes (octets) of user data. The called DTE device can respond with a clear-request or a call-accepted packet. If it responds with a call-accepted packet, the packet can also contain 128 bytes of user data. The session then continues with the usual data transfer and clearing procedures of a switched virtual call.

In fast select with immediate clear, the remote DTE device must respond with a clear-request packet, but the clear-request packet can contain 128 bytes of user data. Exhibit II-2-14 illustrates this option. As with the fast select option, the originating DTE device sends a call-request packet with user data. After it accepts the packet, the called device transmits a clear-request packet that also contains user data. The originating device receives it as a clear-indication packet and returns a clear-confirmation packet (which cannot contain user data). Thus the forward packet sets up the connection, and the reverse packet brings it down.

Recognized Private Operating Agencies Facilities

These facilities allow a calling DTE device to specify one or more recognized private operating agencies to handle a packet session. The recognized private operating agency is a packet network (value-added carrier) and acts as a transit network within a country or between countries.

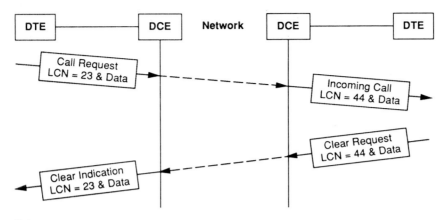

Note:
The logical channel number (LCN) is depicted in this exhibit
to show how it is used on each side of the network.

Exhibit II-2-14.   Fast Select Facility with Immediate Clear

All virtual calls involving more than one recognized private operating agency and one or more gateways require the recognized private operating agencies subscription facility. The recognized private operating agencies selection facility is used to designate a sequence of operating agencies (in the call-request packet) to handle an individual call. The selection facility overrides the subscription facility. Using the selection facility does not require subscribing to the subscription facility.

Long Address Acceptance Facility

This facility, added in the 1988 Blue Book release of X.25, permits the use of extended addresses in the call-setup and clear-request packets compatible with the E.164 ISDN addressing scheme. When this facility is in use, the DCE must always use the long address, whereas the DTE device has the option of using it or not.

CCITT-SPECIFIED FACILITIES

X.25 has seven facilities that support end-to-end signaling as defined by the open systems interconnection (OSI) network service standard. These are:
- Calling address extension.
- Called address extension.
- Minimum throughput rates.

Two-Way Logical Channels	Outgoing Calls Barred	Inconsistent
	Incoming Calls Barred	Inconsistent
Closed User Group	Closed User Group with Incoming Access	Inconsistent
	Closed User Group with Outgoing Access	Inconsistent
One-Way Outgoing Logical Channel	Outgoing Calls Barred	Inconsistent
	Incoming Calls Barred	Redundant
Incoming Calls Barred	One-Way Incoming Logical Channel	Inconsistent
	Reverse Charging Acceptance	Inconsistent
	Closed User Group with Incoming Access	Inconsistent
	Hunt Group	Inconsistent
	Fast Select Acceptance	Inconsistent
	Outgoing Calls Barred	Inconsistent
Outgoing Calls Barred	One-Way Incoming Logical Channel	Redundant
	Closed User Group with Outgoing Access	Inconsistent
Reverse Charging Acceptance	Local Charging Prevention	Inconsistent

Exhibit II-2-15. Inconsistent and Redundant Pairings of X.25 Facilities

---

- End-to-end transit delay.
- Priority.
- Protection.
- Expedited data negotiation.

The OSI 8208 standard contains more information about these facilities.

INCOMPATIBLE PAIRS OF FACILITIES

Some pairs of facilities are incompatible because they are redundant or logically inconsistent with each other. Exhibit II-2-15 identifies these combinations.

SUMMARY

As X.25 evolves, it continues to add facilities. Some vendors object to this constant change because it makes it difficult for them to keep their products current. Other vendors welcome it because it lets them create more marketable products. Without question, however, users benefit from the addition of facilities because it increases their options for interacting with the network and with each other.

Chapter II-2-2
Recommendation X.25: Changes for the 1988–1992 Cycle

Floyd Wilder

EVERY FOUR YEARS SINCE 1972, Study Group VII of the International Telephone and Telegraph Consultative Committee (CCITT) submits the results of its work to the plenary assembly for approval as recommended standards. Although many meetings take place during the four-year cycle, one last meeting occurs before the standards are put into final form. This meeting is attended by members of all the other study groups to ensure uniform treatment of all material presented. The most recent meeting took place in Melbourne, Australia in 1988. The purpose of this chapter is to identify and explain the changes contained in the 1988 CCITT Recommendation X.25, hereafter referred to as the X.25 Blue Book or the Blue Book (the color chosen for this four-year cycle). This chapter is not intended as a substitute for the Blue Book but can be used temporarily to augment the 1984 CCITT Recommendation X.25, hereafter referred to as the X.25 Red Book or the Red Book.

CCITT VOLUME VIII

The X.25 Blue Book is one of the X Series recommendations contained in Volume VIII, Fascicle VIII.2. The first change is that the recommendation is no longer contained in Volume VIII, Fascicle VIII.3, as it was in 1984, but is now in Fascicle VIII.2. The organization of the 1984 Volume VIII versus that of the 1988 volume is illustrated in Exhibit II-2-16.

Exhibit II-2-16 appears to indicate that Recommendations X.1 through X.32 are contained in the 1988 Fascicle VIII.2. Actually, Recommendation X.15 was deleted from this series and its contents (i.e., terms and definitions) moved to Fascicle I.3.

Fascicle	1984 Volume	1988 Volume
Fascicle VIII.1	Series V Recommendations	Series V Recommendations
Fascicle VIII.2	Recommendations X.1–X.15	Recommendations X.1–X.32
Fascicle VIII.3	Recommendations X.20–X.32	Recommendations X.40–X.181
Fascicle VIII.4	Recommendations X.40–X.181	Recommendations X.200–X.219
Fascicle VIII.5	Recommendations X.200–X.250	Recommendations X.220–X.290
Fascicle VIII.6	Recommendations X.300–X.353	Recommendations X.300–X.370
Fascicle VIII.7	Recommendations X.400–X.430	Recommendations X.400–X.420
Fascicle VIII.8	(Not Present)	Recommendations X.500–X.521

Exhibit II-2-16. Organization of CCITT Volume VIII: 1984 Versus 1988

Organization of the Recommendation

For the convenience of the reader, a copy of CCITT Recommendation A.20 from Volume I has been placed in the beginning of Volume VIII, Fascicle VIII.2.

The new organization may seem familiar because it is the same as the organization of the 1980 CCITT X.25 Yellow Book. For example, Recommendation A.20 was contained in the 1980 Yellow Book, Volume VIII, Fascicle VIII.2. This recommendation delineates the responsibilities of the CCITT, the International Standards Organization (ISO), and the International Electrotechnical Commission (IEC). Recommendation A.20 states that the CCITT must establish standards (called recommendations) in the field of data communications and that the ISO and the IEC must develop standards for data processing and office equipment. In addition, it defines areas in which collaboration between the organizations is required.

Many of the changes made in the X.25 Blue Book were in support of the ISO open systems interconnection (OSI) reference model. As a conciliatory gesture, the CCITT has replaced its term *level* with the ISO term *layer*. Therefore, it is now appropriate to refer to the seven layers instead of the three levels and four layers.

The organization of the X.25 Red Book versus that of the X.25 Blue Book is illustrated in Exhibit II-2-17. As can be seen in the exhibit, minor structural changes were made in the new book. The first change was the addition of paragraph 1.4, X.31 Interface. Paragraph 4.6, effects of the physical layer and the data link layer on the packet layer, was expanded and now includes the material from paragraph 3.5 in the Red Book. Therefore, paragraph 3.5 is not present in the X.25 Blue Book. Paragraph 6.25 was expanded to include the new call deflection facility. It also now contains the description of call redirection notification, which was in paragraph 6.27 in the X.25 Red Book. Paragraph 6.28 in the Red Book is now paragraph 6.27 in the X.25 Blue Book. Paragraph 6.28 of the Blue Book is new.

Section	X.25 Red Book	X.25 Blue Book
1	X.21, X.21 bis, and V Series	X.21, X.21 bis, and V Series X.31 Interface
2	Link Access Procedures: Paragraphs 2.1–2.7	Link Access Procedures: Paragraphs 2.1–2.7
3	Packet-Level Interface: Paragraphs 3.1–3.4 Effects of the Physical and Frame Levels on the Packet Level: Paragraph 3.5	Packet-Level Interface: Paragraphs 3.1–3.4 (Moved to Paragraph 4.6)
4	Procedures for Virtual Circuit: Paragraphs 4.1–4.6	Procedures for Virtual Circuit: ►Paragraphs 4.1–4.6
5	Packet Formats: Paragraphs 5.1–5.7	Packet Formats: Paragraphs 5.1–5.7
6	Optional User Facilities: Paragraphs 6.1–6.26 Call Redirection: Paragraph 6.27 Transit Delay: Paragraph 6.28	Optional User Facilities: ►Paragraphs 6.1–6.26 ►Transit Delay: Paragraph 6.27 TOA/NPI Address: Paragraph 6.28
7	Format of Facility/Registration: Paragraphs 7.1–7.3	Format of Facility/Registration: Paragraphs 7.1–7.3
Annex A	Range of LCNs	Range of LCNs
Annex B	Interface State Diagrams	Interface State Diagrams
Annex C	Action State Diagrams	Action State Diagrams
Annex D	DCE Time-Outs	DCE Time-Outs
Annex E	Diagnostic Fields	Diagnostic Fields
Annex F	Online Registration Applicability	Online Registration Applicability
Annex G	CCITT Facilities to Support OSI	CCITT Facilities to Support OSI
Annex H	(Not Present)	User Facilities Associated with NUI
Appendix I	Examples of Bit Patterns	Examples of Bit Patterns
Appendix II	(Not Present)	Explanation of N1 Derivation
Appendix III	(Not Present)	Examples of Multilink Procedure
Appendix IV	(Not Present)	Information on Addresses

Notes:

DCE	Data communications equipment
LCN	Logical channel number
NI	Network interface
NUI	Network user ID
TOA/NPI	Type of address–numbering plan identification

Exhibit II-2-17. Organization of the X.25 Red Book Versus That of the X.25 Blue Book

---

Annex H and Appendixes II, III, and IV have been added to the end of the volume.

CHANGES

The following sections describe the changes, deletions, and additions made for the 1988 X.25 Blue Book. There are many detailed examples and explanations of these changes; some familiarity with X.25 is necessary to understand the descriptions.

Although the appendixes do not constitute an integral part of the recommendation, changes to them are described in the same way as are changes to an annex. The changes are listed according to paragraph numbers in the Blue Book.

The X.25 Blue Book corrected many of the clerical errors found in the X.25 Red Book. In addition, there are fewer clerical errors in the Blue Book than in the Red Book. The only errors corrected in this chapter are those that might mislead the reader.

DTE-DCE Interface Characteristics (Physical Layer): Paragraph 1

As expected, the excitement and interest surrounding the integrated services digital network (ISDN) played a big part in the making of the X.25 Blue Book. Specifically, under the first section (Physical Interfaces), Sub-paragraph 1.4, X.31 Interface, was added. X.31 was mentioned in the Red Book but is not a physical interface specification like Recommendations X.21, X.21 bis, and V series interfaces. X.31 defines how X.21 and X.21 bis are handled by the ISDN. It was enhanced significantly in the Blue Book with the focus on ISDN. If the objective were simply to interface the ISDN with a packet-mode data terminal equipment (DTE) device, it would be appropriate to simply add a paragraph describing how to interface X.25 to Systems Network Architecture (SNA). Fortunately, the objective is for a packet-mode DTE device to be carried on an ISDN with X.25 packets to a packet-switched public data network (PSPDN).

Even ISDN promoters must concede that X.25 will not disappear tomorrow; X.31 states that data terminal equipment conforming to X.25 will be used, at least during the evolution of ISDN and possibly thereafter. This is not as bold as the 1980 treatment of X.21 bis, which was then given a much more limited life expectancy. Although Telenet, Transpac, Tymnet, and other networks will not be converting to ISDN anytime soon, Recommendations X.30, X.31, and X.32 and the appropriate I and Q Series recommendations should be read. The basic elements of interfacing a packet-mode device to an ISDN are illustrated in Exhibit II-2-18.

The following paragraphs define the basic elements of the ISDN interface presented in Exhibit II-2-18. Recommendations I.411, I.430, and X.325 contain more precise definitions.

Terminal Equipment. An ISDN has two types of terminal equipment, denoted as TE1 and TE2. TE1 is an ISDN terminal, and TE2 is a non-ISDN terminal. Non-ISDN terminals must be handled by a terminal adapter, whereas ISDN terminals are connected directly to an ISDN network termination.

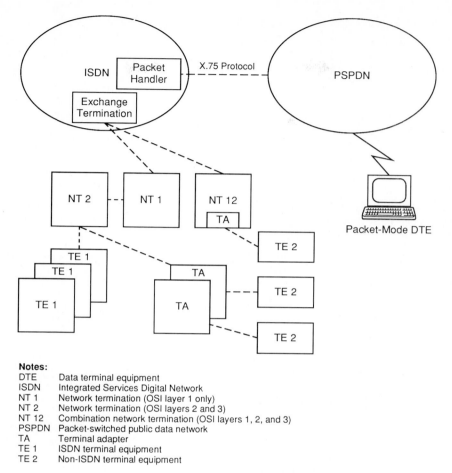

Notes:
DTE   Data terminal equipment
ISDN   Integrated Services Digital Network
NT 1   Network termination (OSI layer 1 only)
NT 2   Network termination (OSI layers 2 and 3)
NT 12   Combination network termination (OSI layers 1, 2, and 3)
PSPDN   Packet-switched public data network
TA   Terminal adapter
TE 1   ISDN terminal equipment
TE 2   Non-ISDN terminal equipment

Exhibit II-2-18.   X.31 ISDN Interface

Terminal Adapter.   A terminal adapter functions like a packet assembler/disassembler in an X.25 network (i.e., it provides a gateway between a dumb terminal and the PSPDN). The terminal adapter can reside in a unique piece of hardware on the user premises, at a network center, or in an ISDN network termination. As with a packet assembler/disassembler, terminal adapters provide speed conversion (rare adaptation), mapping to the Q.931 interface from X.21 (or from X.21 bis, X.20 bis, or V Series), and terminal control and data alignment. Terminal adapters can be designed to handle multiple classes of service (as defined in X.1) or only a single class of service—multiple terminal adapters may be required in this case.

Network Termination. There are two types of network termination devices. Type 1 interfaces only at layer 1, whereas type 2 functions at layers 2 and 3. A combination device has been defined to handle all three layers. The network termination device can contain one or more terminal adapters, depending on whether the administration opts for a multiple terminal adapter (i.e., one handling all classes of service) or provides a unique adapter for each class of service.

Packet Handler. The packet handler logically belongs to the ISDN but may be physically located in a node of the PSPDN. It is still the ISDN virtual circuit service. The protocol between the PSPDN and the ISDN is X.75. The door has been left ajar here by stating that X.75 (the internal network protocol), which was further defined as a functionally equivalent protocol, could be used when the PSPDN and the ISDN are handled by the same network provider (or by bilateral agreement of more than one provider).

Maximum Number of Bits in an I-Frame N1:
Paragraphs 2.4.8.5

No changes were made to the stated values for N1, but a helpful explanation of how the values were derived has been provided in Appendix II (which is new). The determining factor for the maximum N1 value of the data terminal equipment is the standard default packet size of a data packet (not a call-setup packet). As noted in Appendix II, Table II-1X.25, the DTE should support an N1 value not lower than 135 octets. Exhibit II-2-19 represents a summary of Table II-1/X.25 and shows how the N1 value is derived for the DTE.

Because a supported data packet size may be less than the standard default value of 128, the determining factor in establishing the data communications equipment (DCE) N1 value is the clear-request packet rather than the data packet. As noted in Appendix II, Table II-2/X.25, the value varies between 263 and 266 depending on the support of layer 2 modulo

Field Name	Field Length
Packet Header (Layer 3)	3
User Data (Layer 3)	128
Address (Layer 2)	1
Control (Layer 2)	1
Frame Check Sequence (Layer 2)	2
N1 Value	135

Exhibit II-2-19.   N1 Value for Data Terminal Equipment

Field Name	Field Length (octets)
Layer 3	
Header	3
Clearing Cause	1
Diagnostic Code	1
DTE Address Length	1
DTE Addresses	15
Facility Length	1
Facilities	109
Clear User Data	<u>128</u>
Layer 3 Total	259
Layer 2	
Address	1
Control	1 (or 2)*
Multilink Procedure	0 (or 2)†
Frame Check Sequence	<u>2</u>
Layer 2 Total	4
N1 Value	263

Notes:
*With level 2 module 128
† With multilink procedures
DTE   Data terminal equipment

Exhibit II-2-20.   N1 Value for Data Communications Equipment

---

128 and multilink procedures. Exhibit II-2-20 summarizes Table II-2/X.25 and shows how the N1 values are derived for the DCE.

Multilink Procedure: Paragraph 2.5

An overview paragraph and a diagram were added to Recommendation X.25 to explain the multilink architecture. Basically, the overview states that the multilink procedure (MLP) exists as an added upper sublayer of the data link layer, operating between the packet layer and multiple single-link procedure interfaces in the data link layer. Exhibit II-2-21 depicts this procedure. (The Blue Book diagram was slightly modified for clarification.)

Field of Application: Paragraph 2.5.1

Descriptions of the general features provided by the multilink architecture were added to the Blue Book. These include:

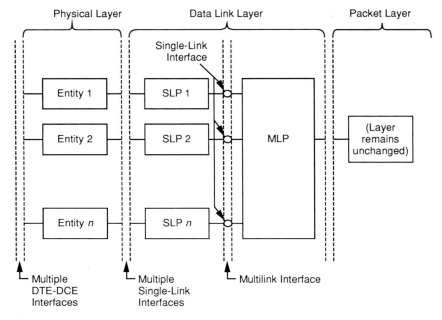

Exhibit II-2-21. The Multilink Procedure

- Achieving economy and reliability of service by providing multiple single-link procedures between DCE and DTE devices.
- Permitting addition and deletion of single-link procedures without interrupting the service provided by the multiple single-link procedures.
- Optimizing bandwidth use of a group of single-link procedures through load sharing.
- Achieving graceful degradation of service when a single-link procedure fails.
- Providing each multiple single-link procedure group with a single logical data link layer appearance to the packet layer.
- Providing resequencing of the received packets before delivering them to the packet layer.

Multilink Resetting Procedures: Paragraph 2.5.4.2

Although the multilink resetting procedure was not changed, an appendix was added to provide examples of DTE or DCE devices initiating the resetting procedures and of both multilink procedures initiating the resetting procedure simultaneously. Because it is unlikely that both the DCE and the DTE will simultaneously initiate a resetting procedure, only Figure III-1 from Appendix III is illustrated in this article (see Exhibit II-2-22). The arrangement is symmetric and represents the fastest possible multilink resetting procedure.

Transmitting Multilink Frames: Paragraph 2.5.4.3

A general topic was added (Paragraph 2.5.4.3.1) describing the responsibility of the transmitting DCE or DTE multilink procedure to control the flow

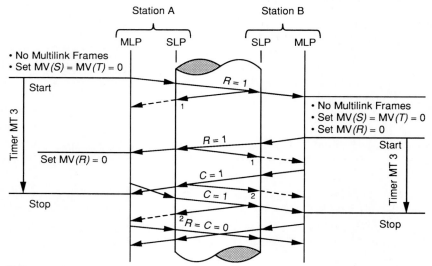

Notes:
[1]The SLP frame that acknowledges delivery of the multilink frame with R = 1
[2]The SLP frame that acknowledges delivery of the multilink frame with C = 1
MLP   Multilink procedure
MV(R) Multilink receive state variable
MV(S) Multilink sequence variable
MV(T) Multilink transmit variable
SLP   Single-link procedure

Exhibit II-2-22.   Multilink Resetting Procedure, Initiated by Station A While Station B Is Inactive

of packets from the packet layer, into multilink frames, and then to the single-link procedures. The functions of the transmitting DCE or DTE MLPs are to:

- Accept packets from the packet layer.
- Allocate multilink control fields containing the appropriate multilink sequence number, MN(S), to the packets.
- Ensure that the MN(S) is not assigned outside the multilink window (MW).
- Pass the resultant multilink frames to the single-link procedures for transmission.
- Accept acknowledgments of successful transmission from the single-link procedures.
- Monitor and recover from transmission failures or difficulties that occur at the single-link procedure sublayer.
- Accept flow control indications from the single-link procedures and take appropriate actions.

Receiving Multilink Frames: Paragraph 2.5.4.4

Receipt of an MLP frame with its MN(S) equal to the next expected multilink receive state variable MV(R) is typical and results in passing the packet from the MLP frame to the packet layer—in addition to one or more other packets received sequentially, but with the lowest MN(S) greater than the value of the MV(R). If a newly received MN(S) is equal to the MN(S) of a frame that has already been received, the new frame is discarded. If the received frame's MN(S) is not equal to the MV(R) but is less than MV(R) + MW + MX (MX is the receive MLP window guard region), the received frame is kept waiting for the arrival of a frame with an MN(S) that is equal to the MV(R). When a received frame's MN(S) is not equal to any of these variables, the frame is discarded. The paragraph has been amended to provide that in this event, further studies will be conducted on the recovery from a desynchronization that is greater than MX between the local and remote MLPs—that is, when the value of MN(S) that has been reassigned to new multilink frames at the remote multilink procedure is higher than MV(R) + MW + MX at the local multilink procedure. Exhibit II-2-23, which is a copy of Figure 3/X.25 in the Blue Book, depicts the detection of lost multilink frames.

Taking a Single-Link Procedure: Paragraph 2.5.4.5

When a single-link procedure is taken out of service, the recovery mechanism is based on initiating the multilink resetting procedures. Other recovery procedures will be studied further.

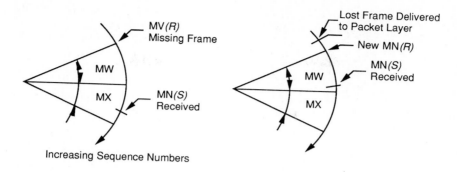

Notes:
MN*(S)*   Multilink sequence number
MV*(R)*   Multilink receive state variable
MW   Multilink window
MX   The receive multilink procedure window guard region

Exhibit II-2-23.   Detecting Lost Multilink Frames

Description of the Packet Layer DTE-DCE Interface: Paragraph 3.0

In the fourth paragraph, a sentence was deleted; it stated that as designated in Recommendation X.2, virtual call and permanent virtual circuit services must be provided by all networks. This sentence was deemed inappropriate in this particular section of the book.

Effects of the Physical and Data Link Layers on the Packet Layer: Paragraph 4.6

All of paragraph 3.5 in the Red Book was moved to paragraph 4.6. In addition, the existing description of this paragraph was enhanced and reorganized. The paragraphs under 4.6 are summarized in the following sections.

General Principles: Paragraph 4.6.1.   Only when an error recovery might result in a loss of data or a duplication of data are the higher layers notified.

Definition of an Out-of-Order Condition: Paragraph 4.6.2.   An out-of-order condition exists when the DCE cannot transmit or receive any frame because of a line fault or when the DTE responds with either a disconnect command or a disconnect mode. In the multilink procedure, when a single-link procedure results in a multilink procedure failure, it is considered an out-of-order condition.

Actions on the Packet Layer when an Out-of-Order Condition Is Detected: Paragraph 4.6.3. The DCE transmits a reset to the remote end with the out-of-order cause code for each permanent virtual circuit and a clear for each existing virtual call.

Actions on the Packet Layer During an Out-of-Order Condition: Paragraph 4.6.4. The DCE clears any incoming virtual calls and sends a reset on the permanent virtual circuit that was active.

Actions on the Packet Layer After Recovery: Paragraph 4.6.5. The DCE sends a restart indication packet with a "network operational" cause code to the local DTE and a reset with the remote DTE operational cause code to the remote end of permanent virtual circuits.

General Format Identifier: Paragraph 5.1.1

The general format identifier was modified to facilitate ISDN addressing. Namely, bit 8 of the identifier now has an alternative meaning for call-setup

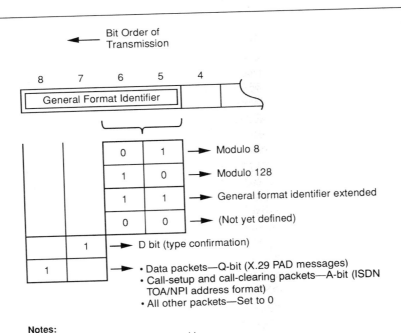

Notes:
PAD   Packet assembler/disassembler
TOA/NPI Type of address–numbering plan identification

Exhibit II-2-24.   General Format Identifier Field

packets. Previously, this bit was used as the qualifier bit in data packets and was set to 0 in all other packets. Now, bit 8 is called the address bit in call-setup and call-clearing packets. Exhibit II-2-24 illustrates the format of the general format identifier field.

Call-Setup and Clearing Packets: Paragraph 5.2

As described in the preceding section on general format identifier changes, a new ISDN address format has been introduced in X.25 for use with call-setup and clearing packets. In this section of the Blue Book, the description of the X.121 format has been enhanced and a better description of the placement of address digits (i.e., semioctets) in the packet has been added.

The term *address block* is used to describe the address lengths and the addresses. There is an address block format for the old address format (i.e., the X.121 format) and a new one for the ISDN address format. Although the formats are similar, their contents are defined by different recommendations. The ISDN address format is referred to as the type of address–numbering plan identification (TOA/NPI) subfield and is defined in Appendix IV of this recommendation and in Recommendations X.301, X.213, and E.164. The format of the TOA/NPI address block is illustrated in Exhibit II-2-25.

Through 1996, packet-mode DTE devices that operate according to Case B of Recommendation X.31 will be addressed by a maximum 12-digit address from the E.164 numbering plan. After 1996, they may have 15-digit E.164 addresses and TOA/NPI address procedures will be required (see Recommendations E.165 and E.166 in the Blue Book).

The maximum address-field length is 17 (decimal) in Exhibit II-2-25. Because this is an 8-bit field, it may now be expanded when needed. Subtracting two from this maximum for the TOA/NPI subfield leaves the current maximum address length of 15 digits. Many of the PSPDNs are already using 15-digit addresses. The first digit, either 0 or 1 depending on the network, signals the PSPDN that the address is not for this network (i.e., that it is for a different recognized private operating agency, or RPOA).

The called and calling address fields within an address block are examined next. Exhibit II-2-26 illustrates the format of a called address field, with a type of address, a numbering plan identification, and nine address digits (this number is arbitrary for illustration purposes). The calling address immediately follows. Because the called address contains an odd number of semioctets (9), the numbering plan identification subfield of the calling address will occupy bits 4, 3, 2, and 1 of the +5 octet, and the type of address subfield will occupy bits 8, 7, 6, and 5 of the +6 octet.

Many codes in the type of address and numbering plan identification fields are undefined; others are reserved for ISDN Q.931 to ensure cor-

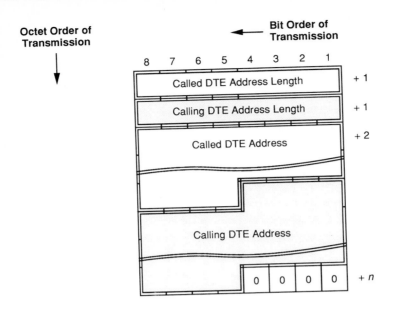

Octet Order of Transmission

Bit Order of Transmission

Notes:
The maximum value of the address lengths is 17 (decimal).
Each digit of an address is coded in a semioctet in binary-coded decimal with bit 5 or 1 being the low-order bit.
The high-order digit of the called address is coded in bits 8, 7, 6, and 5.
When present, the calling DTE address begins on the first semioctel following the end of the called DTE address.
When the total number of digits in the address fields is odd, a semioctel of value zero will be inserted after the calling address field to maintain octet alignment (as illustrated).
DTE   Data terminal equipment
TOA/NPI Type of address–numbering plan identification

Exhibit II-2-25.   TOA/NPI Address Block Format

respondence to the numbering plan in Recommendations F.69 and E.164. The type of address and numbering plan identification codes defined are illustrated in Exhibit II-2-27.

When the new TOA/NPI format is used in a call-setup or call-clearing packet, the A-bit (bit 8) of the general format identifier is set to 1. Because this is a subscription time option, existing packet-mode DTE devices can continue to address packets in the same way. When the address format used by one DTE device is different from the address format used by a remote DTE device, the network (if it supports the TOA/NPI address format) converts from one address format to the other. If the address format is not supported, however, it may either clear the call or send the incoming call request without a called address.

Facility Field: Paragraphs 5.2.2.4 and 5.2.4.2.3

These paragraphs were amended to state the possibility of adding a new field. Further study will dictate whether another value must be defined relative to the total number of octets in the packet.

Online Facility Registration: Paragraph 6.1

A remark was added regarding network user ID override. It stated that the registration confirmation of whether the network user ID override facility is supported by the network will undergo further study.

Closed User Group Selection: Paragraph 6.14.6

When the value of the index is 99 or less, some networks may permit the DTE to use either the basic or the extended format of the closed user group selection facility. A note to this effect was added.

Closed User Group with Outgoing Access Selection: Paragraph 6.14.7

A note identical to the one added in paragraph 6.14.6 was also added here.

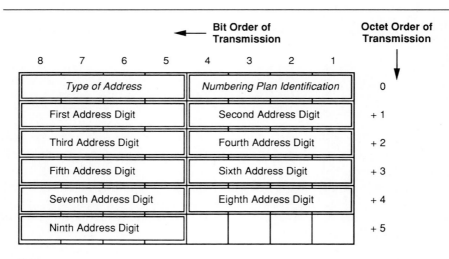

Note:
TOA/NPI   Type of address–numbering plan identification

Exhibit II-2-26.   TOA/NPI Called Address Format

Field	Primary Value	Meaning
TOA	0000	Network-dependent number. The address digits are organized according to the network numbering plan (e.g., prefix or escape code might be present).
TOA	0001	International number. Prefix or escape code not permitted in address digits.
TOA	0010	National number. Prefix or escape code not permitted in address digits.
NPI	0011	X.121 escape codes apply; when used, the TOA will be 0. This method is used because a similar function is not available with TOA/NPI. When defined, it will probably be an optional user facility.

Note:
NPI   Numbering plan identification
TOA   Type of address

Exhibit II-2-27.   Type of Address and Numbering Plan Identification Code Fields

---

Network User ID–Related Facilities: Paragraph 6.21

Network user IDs were first introduced in the X.25 Red Book with a skeleton description of the user facility. The X.25 Blue Book has enhanced the original description and added the network user ID override facility. The Blue Book has three subparagraphs that relate to subscription, override, and selection.

Subscription: Paragraph 6.21.1.   This is a temporary optional user facility for virtual calls. It enables the DTE to provide information to the network regarding such items as billing and security. It is initiated by the network user ID selection on a per-call basis.

Override: Paragraph 6.21.2.   This is also a temporary optional user facility for virtual calls. It enables the DTE to define network user IDs, each with a set of subscription-time optional user facilities that will override the existing selected options for the user. The override is initiated by the network user ID selection on a per-call basis. Annex H has been added to identify which optional user facilities may be associated with the network user override facility.

Selection: Paragraph 6.21.3.   This optional user facility (code C3 hexadecimal) is used to initiate (on a per-call basis) either a network user subscription (e.g., supplying billing and security) or an override of the existing subscription-time optional user facilities, provided the DTE is subscribed to that facility. The facility is better defined now, but the parameter field for the Class D code is described only as a format defined by the network administrator. Therefore, it is not yet a defined, usable facility.

RPOA-Related Facilities: Paragraph 6.23

This section was previously titled RPOA Selection. It is now better organized, in a manner that separates the optional RPOA subscription user facility, which is temporary, from the capability of RPOA selection by the DTE for a given virtual call only. There is no change in capability.

Call Redirection and Call Deflection–Related Facilities: Paragraph 6.25

This section previously addressed only call redirection. Call deflection, a new feature, is an optional user facility that permits the receiver of an incoming call to specify, in the clear request, the alternative DTE address to which the call is to be deflected. The originally called DTE device is responsible for setting the appropriate CCITT-specified facilities and call user data in the clear request.

The limitations on the amount of user data vary depending on whether fast select is present. After a call is deflected, the DCE device does not respond to the calling DTE device when the clear request is received from the called DTE. The calling DTE is notified of the deflection by use of the called-line address modification notification facility. (New addressing limitations of some networks are stated in Appendix IV, Paragraph IV.4.) For an interim period, some networks may not allow a deflected incoming-call packet's contents to be modified, in which case a deflecting DTE device is not permitted to use any user data or CCITT-defined facilities in the clear request.

Transit Delay Selection and Indication: Paragraph 6.27

This optional user facility was not implemented by PSPDNs in a timely fashion. Therefore, a waiver has been placed in the Blue Book stating that when the optional user facility is not supported by all networks, the transit delay indication applicable to the virtual call is not included in the incoming-call packet that is transmitted to the called DTE.

TOA/NPI Address Subscription: Paragraph 6.28

This new paragraph describes the optional user facility TOA/NPI address. When the TOA/NPI is subscribed to, the DCE device and the DTE device ensure that all address blocks in call requests and incoming calls are formatted according to the TOA/NPI format. When the DCE needs to transmit an incoming call with a TOA/NPI address block and the DTE has not subscribed to this facility, the DCE will not include the address; however, some public data network administrations may offer a subscription-time option that allows the call to be cleared using the incompatible destination cause.

General (Formats for Facility Registration Fields): Paragraph 7

CCITT-specified DTE facilities are described briefly in this paragraph with a reference to Annex G, which contains the actual description of these facilities. The entire section was added in 1984; two new facility codes were added in 1988. A summary of Table G-1/X.25, which defines the facility codes, is illustrated in Exhibit II-2-28.

All facility codes may be used in a clear request, but only when the call deflection selection facility is used. Because the changes to the address extension codes are identical for both the called and calling addresses, only the calling address extension code, illustrated in Exhibit II-2-29, is described.

Facility Function	Bits 8 7 6 5 4 3 2 1	Packet Type When Used
Calling address extension	1 1 0 0 1 0 1 1	Call request, incoming call, and clear request
Called address extension	1 1 0 0 1 0 0 1	All
Quality of service negotiation: • Minimum throughput class	0 0 0 0 1 0 1 0	Call request, incoming call, and clear request
• End-to-end transit delay	1 1 0 0 1 0 1 0	Call request, incoming call, call accepted, call connected, and clear request
• Priority*	1 1 0 1 0 0 1 0	Call request, incoming call, call accepted, call connected, and clear request
• Protection*	1 1 0 1 0 0 1 1	Call request, incoming call, call accepted, call connected, and clear request
Expedited data negotiation	0 0 0 0 1 0 1 1	All except clear indication

Note:
*Added in 1988.

Exhibit II-2-28. CCITT-Specified Data Terminal Equipment Facilities

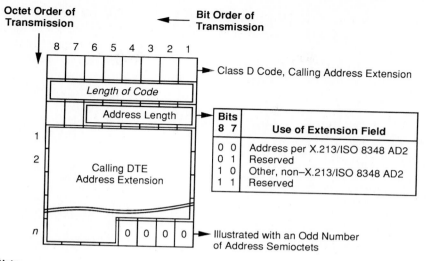

Note:
DTE   Data terminal equipment

Exhibit II-2-29.   Calling Address Extension Code

If the extension field is coded so that bit 8 is set to 1 and bit 7 is set to 0 (noted as other in Exhibit II-2-29), coding of the address extension field is not changed from that of the X.25 Red Book description. If both bits are set to 0, the octets of the address extension field are encoded using preferred binary encoding, described in Recommendation X.213.

One concern regards backward compatibility—specifically, whether these codes should be reversed so that current users will not have to make a change (i.e., coding bits 8 and 7 as 0 for non-X.213). Because the mechanism was not used before, the assignment of codes can be arbitrary without further impact to existing networks.

Receipt of an address whose length would set any of the high-order bits could crash a loosely coded program without a maximum field check. The CCITT previously placed a maximum value of 32 (decimal) on the address length field; it has been increased to 40 (decimal). Therefore, some networks may require a defensive patch to prevent receiving an incoming call with this length of an address field. This does not matter, however, because a defensive code is needed to prevent receiving an X.213/ISO-8208–formatted address extension. In addition, the ISO is perhaps the only user of this facility, which was specified this way in the 1984 version of ISO 8208 (not X.213); by 1992, the resulting problems will be corrected.

The priority facility (G.3.3.3) of CCITT-specified facilities is a new feature in the X.25 Blue Book. It is also new to ISO 8208 and X.213. This facility permits a calling DTE device to specify parameters concerning the priority on gaining a connection, data on connection, and maintaining a connection. Because the code is passed transparently by the DCE, the called DTE may act appropriately on the facility. The format of the priority facility is illustrated in Exhibit II-2-30.

A requested value is referred to as *target* in the call-request packet, *available* in the incoming-call packet, and *selected* in the call-accept and call-connect packets. The range of each of the six values (see Exhibit II-2-30) is from 0 to 255. Zero represents the lowest priority, 254 is the highest priority, and 255 is reserved for packets with unspecified priority.

The protection facility (G.3.3.4) is a new feature in the X.25 and X.213 Blue Books and in ISO 8208. Exhibit II-2-31 illustrates the format of the protection facility. Note that the acceptable levels (coded in octets 1 through *m*) are optional and not present on a call-accept or call-connect packet.

Coding of Facility Code Fields: Paragraph 7.2.1

New facility codes were added to Table 29/X.25 (previously Table 27/X.25) with notes. Exhibit II-2-32 is a summary of the table. Network user ID selection (Code C6 hexadecimal) under call accept has been amended to note that the facility code and associated facility parameter may be

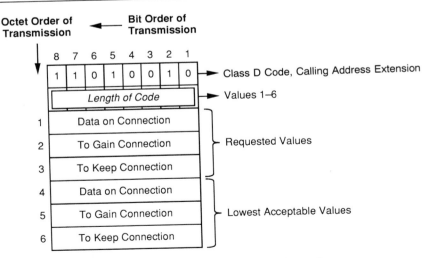

Exhibit II-2-30. CCITT-Specified Priority Facility

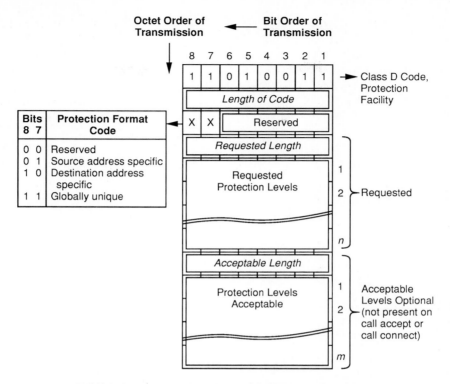

Exhibit II-2-31.   CCITT-Specified Protection Facility

present in a call-accept packet only in conjunction with the network user subscription facility. The call deflection code (Code D1 hexadecimal) was amended under clear request to note that the DTE is not allowed to use both call deflection selection and called-line address modification facilities in the same clear request, because the call is not yet at the destination.

Call Deflection Selection Facility: Paragraph 7.2.2.10

The call deflection selection facility (Class D) is illustrated in Exhibit II-2-33. The length-of-code octet gives the number of octets in the parameter field and is equal to $n + 2$, where n is the number of octets used to contain the alternative DTE address to which the call is being deflected. The next octet gives the reason for the deflection; the deflector may assign any value to the bits noted as X. These bits are passed transparently by the data communications equipment. If bits 7 and 8 are not set to 1, the data communications equipment forces them to 1.

Facility Function	Bits 8 7 6 5 4 3 2 1	Packet Type When Used
Flow Control Negotiation:		Call request, incoming
• Packet Size	0 1 0 0 0 0 1 0	call, call accept, and call
• Window Size	0 1 0 0 0 0 1 1	connect
Throughput Class Negotiation	0 0 0 0 0 0 1 0	Call request, incoming call, and call accept
Closed User Group Selection:		Call request and incoming
• Basic Format	0 0 0 0 0 0 1 1	call
• Extended Format	0 1 0 0 0 1 1 1	
Closed User Group with Out-going Access Selection:		Call request and incoming
• Basic Format	0 0 0 0 1 0 0 1	call
• Extended Format	0 1 0 0 1 0 0 0	
Bilateral Closed User Group	0 1 0 0 0 0 0 1	Call request and incoming call
Reverse Charging	0 0 0 0 0 0 0 1	Call request and incoming call
Fast Select	0 0 0 0 0 0 0 1	Call request and incoming call
Network User ID	1 1 0 0 0 1 1 0	Call request and call accept
Charging Information:		
• Requesting Service	0 0 0 0 0 1 0 0	Call request and call accept
• Monetary Unit	1 1 0 0 0 1 0 1	Clear indication and DCE clear confirmation
• Segment Count	1 1 0 0 0 0 1 0	Clear indication and DCE clear confirmation
• Call Duration	1 1 0 0 0 0 0 1	Clear indication and DCE clear confirmation
RPOA Selection:		Call request
• Basic Format	0 1 0 0 0 1 0 0	
• Extended Format	1 1 0 0 0 1 0 0	
Call Deflection Selection	1 1 0 1 0 0 0 1	Clear request
Called-Line Address Modified Notification	0 0 0 0 1 0 0 0	Call accept, call connect, clear request, and clear indication
Call Redirection or Deflection Notification	1 1 0 0 0 0 1 1	Incoming call
Transit Delay Selection and Indication	0 1 0 0 1 0 0 1	Call request, incoming call, and call connect
Marker	0 0 0 0 0 0 0 0	All except DCE clear confirmation

Notes:
DCE   Data communications equipment
RPOA Recognized private operating agency

Exhibit II-2-32.   Coding of Facility Code Fields

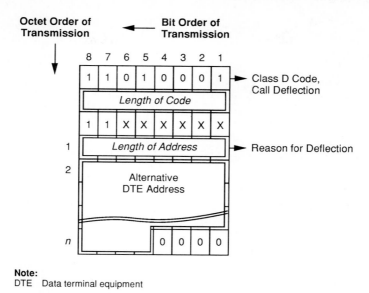

Exhibit II-2-33.   Call Deflection Selection Facility

The next octet gives the length of the new receiver's address, in semi-octets. If the number is odd, bits 4, 3, 2, and 1 of the last octet of the address are set to 0. The remaining octets contain the new receiver's address. If the calling DTE device and the new receiver have subscribed to the TOA/NPI facility, the maximum number of semioctets is 17 (decimal); otherwise, the maximum is 15 (decimal). If one has subscribed to the TOA/NPI format and the other has not, the network performs the conversion.

Call Redirection and Call Deflection Notification Facility: Paragraph 7.2.2.11

Although redirection and deflection provide different functions, the notification facility of each is the same. Therefore, the same facility code (Code C3 hexadecimal) is used for both call redirection notification and call deflection notification. The format of this Class D code is illustrated in Exhibit II-2-34.

The length-of-code octet gives the number of octets in the parameter field and is equal to $n + 2$, where n is the number of octets used to contain what was originally called the DTE address. The next octet gives the reason for the call deflection or the call redirection, which is followed by the length of the original DTE address, in semioctets. If the number is odd, bits 4, 3, 2, and

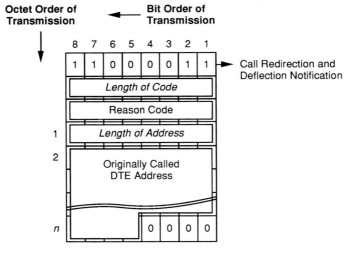

Note:
DTE   Data terminal equipment

Exhibit II-2-34.   Call Deflection and Call Redirection Notification Facility

---

1 of the last octet of the address are set to 0. The remaining octets contain the original DTE address.

If the calling DTE device and the new receiver have subscribed to the TOA/NPI facility, the maximum number of semioctets is 17 (decimal); otherwise, the maximum is 15 (decimal). If one has subscribed to the TOA/NPI format and the other has not, the network performs the conversion. Reason codes for call redirection or call deflection are illustrated in Exhibit II-2-35.

The bits denoted by Xs are set by the DTE in the call deflection selection facility. If bits 7 and 8 were not set to 1, the DCE sets their values to 1. Code value 7 (call distribution within a hunt group) may be used by some networks for network-dependent reasons that are not described in X.25 (i.e., they do not comply with Recommendation X.25).

CHANGES

The majority of modifications made in the 1988 X.25 Blue Book were in support of either ISDN or OSI. The following is a list of the most significant changes made to the 1988 X.25 Blue Book:
- Volume VIII was reorganized to accommodate the expanded OSI X.200 Series recommendations—Fascicle VIII.2, Recommendation X.25, had minor organization changes.

Reason	Bits 8 7 6 5 4 3 2 1
Originally Called DTE Busy	0 0 0 0 0 0 0 1
Call Distribution Within a Hunt Group	0 0 0 0 0 1 1 1
Originally Called DTE Out of Order	0 0 0 0 1 0 0 1
Systematic Call Redirection	0 0 0 0 1 1 1 1
Call Deflection by the Originally Called DTE	1 1 X X X X X X

Note:
DTE   Data terminal equipment

Exhibit II-2-35.   Reason Codes for Call Deflection or Call Redirection

---

- The X.31 Interface was placed in X.25 interfaces, and TOA/NPI addressing was added in support of ISDN.
- The CCITT-specified DTE priority and protection facilities were added, and the called and calling address extension facilities were modified, all in support of OSI.
- Annex H was added to itemize subscription-time user facilities that may be associated with the network user ID override facility.
- Appendix II was added to provide an example of how the N1 values were derived for DTE and DCE.
- Appendix III was added to clarify and give examples of multilink resetting procedure.
- Appendix IV was added to give examples of the X.121 and TOA/NPI address blocks and to define, by packet type, the addresses used.

PROBLEMS

Of all the changes contained in the X.25 Blue Book, only two have the potential to cause problems. Namely, they are adding the A-bit to the general format identifier and adding the 2-bit field in the address-length field of the calling and called address extension facilities.

Adding the A-bit in the general format identifier to signal a TOA/NPI address block should not affect current users because the DTE does not subscribe to this type of addressing until the feature is implemented. However, if the DCE needs to send a TOA/NPI address and the called DTE has not subscribed to this addressing, the DCE may either clear the call or send the incoming call without an address—with or without the A-bit set (not defined by CCITT). Therefore, users should be prepared to receive an incoming call with the A-bit set even when not subscribing to the TOA/NPI format.

In addition, certification tests will be modified to send an incoming call with the A-bit set (with and without a TOA/NPI address block) to see if it is handled properly. The proper response should be to send a clear request with a clearing cause code of incompatible destination (Code 21 hexadecimal). Without changing the existing network software, this could result in a clear request because of an invalid general format identifier (if defensively coded), a clear request because the address could not be found, or in the worst case, a call accept because an address match was made with an X.121 format address. If the emulation for certification is of the DTE, receipt of an invalid A-bit by the DCE should cause the call request to be discarded and a diagnostic packet returned with the diagnostic Code 67 (invalid called DTE address).

Adding the 2-bit use-of-extension field to the already defined address-length field for calling and called address extension facilities may cause problems if coded according to the X.25 Red Book. The potential problems range from a garbled address to disaster, depending on how defensively it was coded. Certification tests for the major public data networks will be modified to send an address extension facility with an address length greater than 128.

ACTION PLAN

To solve the A-bit problems, some administrations will offer a subscription-time option allowing the user to indicate that the DCE will clear the call when the called DTE device has not subscribed to the TOA/NPI addressing format and a call request specifies this addressing. Until the A-bit feature (or at least the mechanics of it) can be implemented, the subscription-time option to clear the call should be selected.

If the address extension facility feature is coded as stipulated in the X.25 Red Book, with the proper range check, the range check should be changed to allow a maximum of 40 (decimal) because the maximum value has been increased from 32 (decimal) to 40 (decimal). Until the range check is reset, truncated or garbled addresses could result. If the range check is not present, a defensive patch is needed until the 2-bit use-of-extension field is implemented. Without the range check, this is a potential time bomb and could result in disaster, depending on how it is coded.

Chapter II-2-3
T1 and Beyond

Walter J. Gill

T1 IS A WIDEBAND DIGITAL CARRIER FACILITY used for the transmission of digitized voice, digital data, and digitized image traffic. It operates at a 1.544M-bps signaling rate and uses industry-accepted signal and encoding standards. T1 is also referred to as T-carrier and DS-1.

TERMINOLOGY

T1 terminology was originally defined by AT&T. *T-carrier* refers to a multiplexing method used to carry multiple channels in one composite wideband digital signal. T-carrier uses time-division multiplexing. *T1* refers to twenty-four 64K-bps pulse-code modulation voice channels carried in a 1.544M-bps wideband signal. As originally used, a T-carrier system included pulse-code modulation encoding and multiplexing equipment as well as a T1 line (the transmission wire or cable and regenerative line repeaters). The D-type channel bank provides pulse-code modulation encoding and multiplexing. The D1 was the first pulse-code modulation channel bank. Encoding improvements resulted in the D1D and the D2 channel banks. The channel banks currently used are D3, D4, and D5.

DS-1 and T1

A T1 line carries a DS-1 signal (i.e., digital signal, level 1), which is the 1.544M-bps signal defined by AT&T. The specification for a DS-1 signal includes voltage, impedance, frequency, and pulse shape, but it does not include encoding or frame format. The frame format is determined by the equipment used (e.g., the D3 or D4 channel bank). The voltage and pulse shape for a DS-1 signal are specified at the DS-1 cross-connect point, a location known as DSX-1. DS-1 and DSX-1 specifications are included in various AT&T and Bell Communications Research (Bellcore) publications.

DS-0 Channels

DS-0 (i.e., digital signal, level zero) refers to the 64K-bps channels generated in a digital channel bank. A D-type channel bank encodes each analog voice channel into a 64K-bps DS-0 channel and multiplexes 24 such DS-0 channels along with framing bits to form a DS-1 signal. A DS-0 channel can also carry digital data. For example, 56K-bps data signals are commonly carried in DS-0 channels, with the remaining 8K bps of the 64K-bps DS-0 channel used for control. A DS-1 signal does not have to be composed of DS-0 channels.

T-CARRIER HIERARCHY

As wider-bandwidth digital facilities evolved, a hierarchy of T-carrier signals was developed, as shown in Exhibit II-2-36. T1C and T2 were developed to increase the carrying capacity of wire pairs beyond the capacity of T1 and are used primarily by telephone companies. T3 (or more precisely DS-3), which operates at 44.736M bps and carries 28 DS-1 signals, is becoming widely used by AT&T, the regional Bell operating companies (RBOCs), other common carriers, and private users on new microwave and light-wave facilities. Digital microwave radios typically carry two or three DS-3 signals. Optical-fiber light-wave systems carry from one to nine DS-3 signals.

In common usage, the distinctions between *T-carrier, T1*, and *DS-1* have become blurred to the extent that the terms are becoming de facto synonyms. This is also true of *T3* and *DS-3*. *T1* and *T-carrier* are popular and often misapplied terms. It is more correct to say that a wire, cable, microwave, light-wave, or satellite transmission facility supports or carries DS-1 or DS-3 signals than to say that it carries T1 or T3.

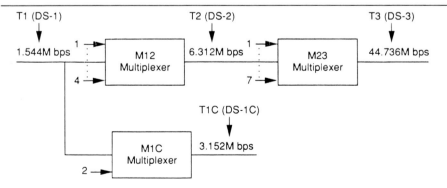

Exhibit II-2-36.   The T-Carrier Hierarchy

EQUIPMENT

Many organizations supply T1 equipment for private as well as telephone company use. The following sections include descriptions of T1 equipment; the lists are not intended to be inclusive.

Channel Banks

Exhibit II-2-37 shows a simple voice system with 24 voice circuits on a single T1 line. The channel bank converts the 24 analog voice circuits and their associated signaling to 64K-bps pulse-code modulation channels and multiplexes them into the 1.544M-bps DS-1 signal carried on the T1 line.

If the T1 line is obtained as a leased, tariffed line from AT&T or a regional Bell operating company, a channel service unit (CSU) must terminate the line at the customer's premises. Channel service units are not required on private T1 facilities. These units protect the public-network T1 facility and the customer-premise equipment, normalize transmit and receive voltage levels, and control maintenance and access. They must be certified under *FCC Rules and Regulations Part 68.*

Data Multiplexers

Exhibit II-2-38 shows a simple example of data multiplexing on a T1 line. A time-division multiplexer is used to combine twelve 56K-bps and sixty 9.6K-bps data circuits on the T1 line. Multiplexers that combine a variety of data speeds between 300 bps and 1.344M bps are available.

Currently available channel banks and time-division multiplexers enable voice and data to be combined. Data-port cards can be used in channel banks to permit one or more data circuits to be inserted in a 64K-bps slot typically used for pulse-code modulation-encoded voice. Voice cards can be used in time-division multiplexers to provide one or more voice circuits combined with the data. Typically, 32K-bps continuously variable slope delta modulation (CVSD) is used to encode the voice signals.

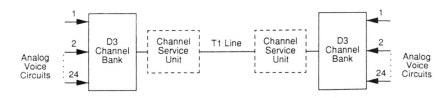

Exhibit II-2-37. Voice Channels on a T1 Line

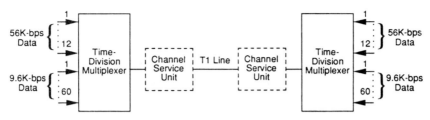

Exhibit II-2-38.   Data Multiplexing on a T1 Line

PBXs

Most digital PBXs now provide a direct DS-1 interface on their tie trunk sides. These DS-1 signals use the pulse-code modulation encoding formats used in D3 and D4 channel banks.

Voice Compression

Voice compression reduces the transmission rate for digitized voice signals. More than 24 voice channels can be put on a T1 line by means of compression. The most common method of voice compression is speech encoding at rates of less than 64K bps. Typical encoding schemes are continuously variable slope delta modulation and adaptive differential pulse-code modulation operating at a 32K-bps clock rate. These schemes result in a 2-to-1 compression relative to 64K-bps pulse-code modulation. Continuously variable slope delta modulation can be implemented by means of a low-cost, single-channel codec but provides lower-quality speech reproduction than adaptive differential pulse-code modulation.

Adaptive differential pulse-code modulation at 32K bps is the International Telephone and Telegraph Consultative Committee (CCITT)–recommended standard for voice compression. It is implemented by means of transcoders or codecs built into multiplexers. Transcoders convert the 64K-bps pulse-code modulation voice channels on two DS-1 signals to either forty-four or forty-eight 32K-bps adaptive differential pulse-code modulation voice channels on one DS-1 signal. Forty-four channels result when four separate 32K-bps channels are used for common-channel signaling. If each channel carries its own signaling (i.e., robbed-bit signaling), 48 channels can be obtained.

Adaptive differential pulse-code modulation encoding is also an integral part of multiplexers that integrate voice and data. These multiplexers operate on 64K-bps pulse-code modulation channels and convert them to 32K-bps adaptive differential pulse-code modulation channels. Individual voice channels can be assigned adaptive differential pulse-code modulation encoding or allowed to remain as pulse-code modulation signals, depending on how the channels are used in the network. This feature is important when voiceband data is involved. Adaptive differential pulse-code modulation encoding does not accurately carry voiceband data signals when they originate in data modems operating at rates higher than 4,800 bps. Therefore, if a voice channel carries a 9,600-bps data modem circuit, the channel should remain as 64K-bps pulse-code modulation; adaptive differential pulse-code modulation would not be selected.

Digital Access and Cross-Connect System

AT&T's digital access and cross-connect system (DACS) is a digital, time-slot interchange matrix switch that switches the 64K-bps (DS-0 level) channels among DS-1 signals. It is programmed in a static fashion either locally or remotely by an operator interface terminal. Equipment similar to AT&T's digital access and cross-connect system is available from independent manufacturers.

The digital access and cross-connect system was originally developed to provide test access to voice circuits contained on T1 lines terminating on 4ESS digital switches. Previously, back-to-back channel banks were used between the T1 lines and the digital switch. With this scheme, individual voice circuits could be accessed at the analog voice-frequency level for testing or cross-connection. The digital access and cross-connect system eliminated the channel banks and provided direct digital access for improved, automated maintenance. The typical voice toll traffic of AT&T and the regional Bell operating companies, usually referred to as message service, involves this application of the digital access and cross-connect system.

AT&T offers DACS-based service for private T1 lines, usually referred to as special services. The special services based on the digital access and cross-connect system are Accunet T1.5 with customer-controlled reconfiguration (CCR), Accunet Reserved 1.5, and Skynet. Customer-controlled reconfiguration permits customers to reconfigure the routing of DS-0 channels on their private T1 lines in a manual, static fashion. Accunet T1.5 does not require the use of a digital access and cross-connect system for premise-to-premise DS-1 links when DS-0 switching is not needed. Passage of an Accunet T1.5 line through a digital access and cross-connect system involves additional tariff charges.

Digital access and cross-connect system frames are always installed at AT&T or regional Bell central offices. A private T1 network must therefore include one or more digital access and cross-connect system hubs if the network configuration is controlled by or through AT&T. Access to an AT&T digital access and cross-connect system is provided through AT&T Accunet

T1.5 lines or through regional Bell operating company or other common carrier intra–local access and transport area (LATA) T1 lines. A combination of both is usually used.

Transmission Resource Managers

Private networks, which are often implemented because of the control and flexibility they provide, are much more complex than the simple voice and data applications shown in Exhibits II-2-37 and II-2-38. A fully interconnected T1 backbone network resembles a mesh topology rather than a point-to-point topology. Mesh topologies, such as those that involve transmission resource managers, are more flexible and reliable than point-to-point topologies because they permit the sharing of T1 facilities and provide alternative routes in case of T1 failure or bandwidth saturation.

Transmission resource managers multiplex voice, data, and video traffic and manage the T1 bandwidth in a complex network. They provide dynamic bandwidth allocation, automatic alternate routing, and flexible multiplexing of voice, data, and image. In some cases, transmission resource managers also provide voice compression. They can manage from 4 to 96 T1 lines and provide software control and diagnostic functions that can be locally or remotely controlled. Exhibit II-2-39 shows a backbone T1 network that uses transmission resource managers.

PROVIDERS OF T1 FACILITIES

T1 lines are available from many providers. Although AT&T is the largest organization that offers private-line T1 facilities, it does not necessarily provide the most available or economical facilities. Various types and sources of T1 should be examined before a private T1 network is implemented. Fortunately, the DS-1 signal and its DSX-1 cross-connect point are standardized, so a mix of T1 types and sources is a workable network solution. Types of T1 facilities include terrestrial or satellite, leased or user-owned, and cable, microwave, or light-wave.

AT&T

AT&T provides leased facilities under tariffs approved by the FCC. The T1 services offered by AT&T are Accunet T1.5, Accunet Reserved 1.5, and Skynet. Accunet T1.5 is a terrestrial (the T stands for terrestrial, not for T1) 1.544M-bps private-line service that may or may not pass through an AT&T digital access and cross-connect system, depending on the customer's preference. Accunet Reserved 1.5 is a 1.544M-bps service that a customer uses only when required—for example, as a backup for other T1 facilities and for infrequent, scheduled videoconferences. Accunet Reserved T1

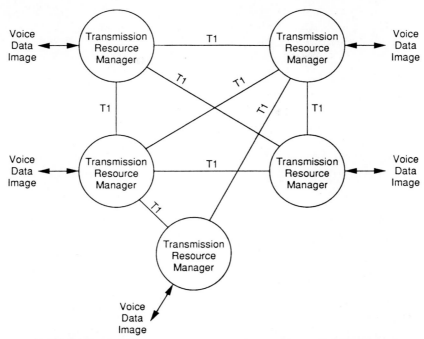

Exhibit II-2-39.   Backbone T1 Network Using Transmission Resource Managers in a Mesh Topology

lines always go through DACS frames for configuration control. Skynet provides T1 circuits over AT&T communications satellites and can provide dedicated, full-time T1 lines or can be used in conjunction with Accunet Reserved 1.5. AT&T terrestrial T1 facilities usually use cable or microwave transmission media. AT&T also has extensive light-wave facilities under construction. AT&T's rates for Accunet services are contained in tariffs filed with the FCC. Although Accunet T1.5 rates have been declining, the communications manager should obtain current rates from recent tariff filings.

Regional Bell Operating Companies

The regional Bell operating companies provide leased T1 facilities under tariffs that are usually approved by state public utility commissions or the FCC. Most of these services are similar to AT&T's Accunet T1.5. In many cases, the regional Bell T1 facilities are used to connect the customer's premises to the AT&T point of presence for Accunet T1.5 or Accunet Reserved 1.5 access. Regional Bell T1 facilities are almost always terrestrial and may use any combination of wire, cable, microwave, or light-wave transmission media. Customers usually do not know and should not be particularly concerned about which type of transmission medium is used. Users can expect excellent quality from any T1 transmission medium when it is properly installed and maintained.

Other Common Carriers

Other common carriers also provide various T1 services. Most of the facilities of these companies are optical-fiber light-wave systems that may be combined with cable and digital microwave. Several of these companies are offering T3 as well as T1 leased facilities.

Private Microwave Equipment Manufacturers

T1 links for private use were first provided by microwave equipment manufacturers. Private microwave transmission equipment carrying T1 was used primarily to connect corporate facilities within a metropolitan area or to connect locations separated by one or two line-of-sight microwave hops (i.e., 20 to 50 miles). Private networks now use digital microwave for long-haul as well as short-haul links. Equipment is available that can carry from one DS-1 signal to two DS-3 signals (i.e., 56 DS-1 signals). Microwave operates in frequency bands assigned and controlled by the FCC; microwave users must obtain station licenses from the FCC.

NEW FEATURES AND CAPABILITIES

AT&T has devised several features that improve the use and maintenance of T1 lines. These features include the extended superframe format, zero suppression, and clear-channel capability.

Extended Superframe Format

Most T1 systems currently use the standard frame format established for D3 and D4 channel banks. This format consists of a superframe made up of twelve 193-bit frames. Each 193-bit frame contains 192 information bits and 1 frame bit (F bit). The F bit (the 193rd bit) carries a 12-bit pattern over the superframe that identifies the frame locations in the superframe. The 192 information bits are subdivided into twenty-four 8-bit bytes that, in the case of D3 or D4 channel banks, represent the twenty-four 64K-bps pulse-code modulation voice channels.

The least significant bit in the 8-bit pulse-code modulation word is used in every sixth frame for signaling. These bits are the A/B bits, and this method of inserting the signaling bits in the encoded voice word is called robbed-bit signaling. The location of the robbed bit is determined by the F-bit pattern in the superframe structure.

The newer extended superframe (ESF, formerly known as Fe) T1 format uses twenty-four rather than twelve 193-bit frames. The 193rd bit is used for frame synchronization and is also multiplexed over the 24 frames to provide a cyclic redundancy check (CRC-6) performance-measuring channel and a 4K-bps data link. T1 providers that incorporate the extended superframe format use the CRC-6 channel to monitor line performance without interrupting customer traffic. The 4K-bps data link is used by these providers for maintenance supervision and control. The extended superframe format is now offered on some equipment and can be expected to be broadly supported by equipment suppliers and T1 providers in the near future.

Zero Suppression

T1 lines do not have a separate clock line for sending end-to-end timing as do RS-232 data circuits. Rather, the timing information is inherent in the information pulses carried on the line. On a T1 line, 1 bits are represented by pulses and 0 bits are represented by the absence of pulses. Transmission of an excessive number of 0 bits results in insufficient timing information and can result in signal errors or signal loss.

AT&T has established a specification for DS-1 signals that limits the number of consecutive zeros permitted to 15 and that requires an average pulse density of at least 12.5%. In voice applications of T1, the encoding scheme used in D3 and D4 channel banks guarantees compliance with AT&T's zero-suppression specification. When a variety of data and image applications are carried on a T1 line, however, it is not always possible to ensure that the composite DS-1 bit stream meets the zero-suppression standard.

A zero-suppression code is designed into the line format of all T-carrier lines (including T1C, T2, and T3) except T1. AT&T has devised the binary eight zero substitution (B8ZS) method, which modifies the T1 line format to permit arbitrary source encoding without violating the zero-suppression specification. For B8ZS to be incorporated, all existing T1 multiplexing and transmission equipment must be upgraded. For this reason, B8ZS has not been widely implemented by AT&T or the regional Bell operating companies. Independent providers of T1 are implementing B8ZS in their new facilities, and AT&T and the regional Bell companies are expected to provide B8ZS on new facilities in the near future. To accommodate non-B8ZS T1 lines, most equipment manufacturers build some form of zero suppression into their equipment.

Clear-Channel Capability

Clear-channel capability, as defined by AT&T and the regional Bell companies, refers to the ability of T1 equipment to transmit 64K-bps DS-0 channels within DS-1 signals without regard to bit pattern; clear-channel capability, in other words, permits the transmission of the all-zero, 8-bit word. On

T1 lines without B8ZS or some other form of zero suppression, the transmission of two consecutive all-zero words violates the 15-zero restriction and the transmission of too many all-zero words violates the 12.5% pulse-density constraint. Clear-channel capability cannot be attained without B8ZS or some other form of zero suppression. However, when such zero-suppression methods as B8ZS are applied to the T1 lines, clear-channel capability is not limited to the 64K-bps DS-0 channel but can be generalized to include any channel data rate.

The Synchronous Optical Network

The synchronous optical network (SONET) transmission standard has been proposed by Bellcore to define a hierarchy of signals carried by broadband optical-fiber transmission facilities. Two sets of signals, designated STS-N (STS stands for synchronous transport signal) and OC-N (OC stands for optical carrier), are defined, where N is an integer from 1 to 255. Currently, only the values of $N=1, 3, 9, 12, 18, 24, 36$, and 48 are recognized. STS-N, the Nth level synchronous transport signal, is constructed by byte interleaving N STS-1 signals. The rate of an STS-1 signal is 51.840M bps. OC-N is the Nth level optical carrier; it is the signal transmitted over the optical-fiber facility. An OC-N is the optical signal that results from an electrical to optical conversion of an STS-N signal.

The transport of standard rate signals at less than the STS-1 rate is accommodated by virtual tributaries (VT). VT1.5 (1.728M bps) carries synchronous and asynchronous DS-1 (1.544M bps) signals. VT2 (2.304M bps) carries synchronous and asynchronous Conference of European Postal and Telecommunications Administrations (CEPT) 2.048M-bps signals. Similarly, VT3 (3.456M bps) and VT6 (6.912M bps) carry DS-1C (3.152M bps) and DS-2 (6.314M bps) signals, respectively. Provision is also made in the SONET STS-1 format for carrying DS-3 (44.736M bps) signals.

STANDARDS

In most cases, AT&T T-carrier standards have become de facto industry standards. However, the postdivestiture environment and the internationalization of communications standards are causing a move toward consensus in standards setting. AT&T and Bellcore continue to be major participants in standards activities.

The CCITT

The two sets of CCITT recommendations that relate to digital transmission are the North American standard and the CEPT standard. The North American standard is based on the 1.544M-bps T1 rate and is used in the US,

Canada, and Japan. The CEPT standard is based on a primary rate of 2.048M bps and is used in Europe, Africa, and parts of Asia. Recommendation G.703 defines the North American T1 standard, which is almost identical to the AT&T DS-1 and DSX-1 specifications. Recommendation G.703 also defines the CEPT 2.048M-bps signal.

The ANSI/ECSA T1 Committee

The American National Standards Institute (ANSI) and the Exchange Carriers Standards Association (ECSA) have established the T1 Committee, which has the charter to set standards for digital transmission. (*T1* derives from ANSI committee designation methods and only coincidentally matches the AT&T T1 nomenclature.) The ANSI/ECSA T1 Committee comprises several working subcommittees, designated T1E1, T1M1, T1Q1, T1S1, T1X1, and T1Y1. Each subcommittee meets regularly and receives broad-based participation from carriers and manufacturers.

THE IMPACT OF ISDN

Although deregulation and the Bell divestiture have reduced the drive toward the integrated services digital network (ISDN) in the US, the ISDN concept is still receiving attention and has a definite impact on T1 and private networks.

The ISDN Primary Rate

The ISDN primary-access or primary-rate interface refers to transmission between network nodes or between customer-premise equipment (e.g., a PBX) and a network switching node. In North America, the ISDN primary rate is the 1.544M-bps DS-1 signal. The format of the primary-rate signal, however, differs substantially from the format used in the D3 and D4 channel banks.

The ISDN primary-rate signaling uses twenty-three 64K-bps voice (or data) channels and one 64K-bps signaling channel. This format is referred to as 23B+D. The B channels are the voice and data channels; the D channel contains the signaling and supervision associated with the 23 B channels. This common-channel signaling method provides more extensive network signaling than the A/B bit method and sets the stage for clear-64K-bps-channel capability. In private network applications, the 23B+ D format will be used on T1 lines to interconnect PBXs and to connect PBXs to telephone company central office switches (ISDN nodes). The ISDN D-channel signaling is based on CCITT Recommendations Q.921 and Q.931.

In addition to the D and B channels, the ISDN includes other higher-speed channels for use with primary-rate signals; these are the H0 384K-bps channel and the H11 1.536M-bps channel. The H12 1.920M-bps channel is for operation with the 2.048M-bps CEPT primary-rate signal.

Broadband ISDN

The definition of standards for broadband ISDN is being carried on within CCITT Study Group XVIII. Progress in broadband ISDN is driven by the need for applications that exceed the usual voice and data traffic accommodated by ISDN. These applications include video conferencing, high-speed fac-simile, ultrafast data transmission, television, and, eventually, high-definition television.

Broadband ISDN will most likely use the SONET signal structure for trans-port and switching. A new high-speed channel, H4, operating at 135.168M bps, has been defined for broadband ISDN. The H4 payload signal will be carried at a 139.246M-bps gross rate.

Switching technology should make rates to STS-12 (622.080M bps) pos-sible. The switching structure for broadband ISDN is expected to be based on a combination of circuit switching and broadband packet switching. Time-sharing multiplexing will combine the low-rate STS-N signals that make up the STS-12 signal, and the low-rate signals will be directed through circuit switching within the transport network. The control of user traffic within STS-1 and STS-3 signals will involve either time-division multiplexing and circuit switching or asynchronous time-division multiplexing and broadband packet switching. Broadband packet switching will be used to multiplex and switch a variety of low-speed and high-speed applications. The advantage of broadband packet switching is that it provides flexibility, an efficient mix of bursty traffic, and rate independence. One particular type of broadband packet switching is a technique known as asynchronous transfer mode, which uses fixed packet lengths and is gaining widespread international support.

The terminology for these multiplexing and switching techniques is still fluid.

Digital Multiplexed Interface

The digital multiplexed interface (DMI) is a 23B+D DS-1 transmission method devised by AT&T, Hewlett Packard, and other vendors as an inter-im step toward the eventual ISDN primary-rate signal. The intended applica-tion of the digital multiplexed interface is a DS-1 connection between host computers and PBXs that provides the flexibility to accommodate products from multiple vendors. As currently specified, the DMI D channel uses bit-oriented signaling (BOS) that essentially puts the D3-D4 A/B signaling bits in a common channel. This is referred to as DMI-BOS. A second D-channel definition, based on message-oriented signaling (MOS), uses

CCITT Recommendations Q.921 and Q.931. This definition, referred to as DMI-MOS, will provide a migration path to the ISDN 23B+D primary rate when the rate is fully defined.

T1 APPLICATIONS

Most communications managers maintain their voice traffic on a combination of such AT&T and regional Bell operating company services as Centrex, WATS, and Electronic Tandem Network and on private lines supplied by AT&T, the regional Bell companies, and the other common carriers. Data traffic is typically spread over voiceband modems on leased 3002-series private lines as well as on AT&T's Dataphone Digital Service and equivalent data circuits provided by the regional Bell companies at 2.4K, 4.8K, 9.6K, and 56K bps. Multiplexers are used to combine lower-speed data channels on the Dataphone Digital Service circuits. Thus, Dataphone Digital Service, especially at 56K bps, provides a form of bulk-transmission bandwidth that can be used for customer-configured data networks.

The solution for a growing number of organizations is a corporate communications network based on a T1 backbone. Such networks use equipment and transmission facilities from several sources to fully integrate voice, data, and video traffic. For example, many metropolitan-area private networks that primarily use T1 on digital microwave to integrate voice and data traffic among corporate locations have been installed. Such networks have economic payback periods of a few months to two years relative to telephone company–provided services.

Comparisons can also be made among the various tariffed services, including private-line voice (3002 series), data (on Dataphone Digital Service), and T1. Exhibit II-2-40 compares the costs of these services when all are chosen from AT&T tariff offerings. It lists the number of private voice lines or the number of 56K-bps data lines whose cost is approximately equal to the cost of one T1 line over the same distance. Over medium distances (e.g., 500 miles), twenty-four 3002 voice lines cost about the same as one T1 line carrying 24 voice channels. At shorter distances, the T1 line is more economical; it carries 24 voice channels for the same cost as sixteen 3002 private lines. At long distances, adaptive differential pulse-code modulation voice compression must be used to put 44 voice channels on a T1 line before the cost is equal to that of forty-four 3002 lines. The values listed for the numbers of lines are approximations; actual comparisons depend on the costs of customer-premise equipment and local access lines as well as long-haul tariffs. In general, however, the trade-offs shown in Exhibit II-2-40 typically result from such comparisons.

The comparison for data circuits is more dramatic. A T1 line can carry twenty-four 56K-bps data circuits for the same cost as eight individual

	Distance in Miles		
	100	500	2,000
Number of T1 lines	1	1	1
Number of 3002 private voice lines	16	24	44*
Number of 56K-bps private Dataphone Digital Service data lines	8	9	12

Note:
*With compression

Exhibit II-2-40.   Economic Comparison of Individual Voice and Data Lines with T1 on an Equal Cost Basis

56K-bps Dataphone Digital Service lines for short and medium distances and 12 individual 56K-bps Dataphone Digital Service lines at longer distances.

SUMMARY

The divestiture of the Bell System and the growing importance of communications are causing many organizations to implement fully integrated private communications networks. Such networks are based on digital technology: digital voice, digital data, digitized image, and digital transmission (T1 and T1-based derivatives).

T1-related issues transcend conventional data communications problems and solutions. The communications manager must therefore learn to think and function in an expanded role, with total responsibility for business communications. In this role, the communications manager should address the following issues:

- What level of integration is appropriate?
- How must corporate charters and controls be modified to support an integrated network?
- What standards should be complied with, and when should innovation be encouraged?
- Does the planned network have adequate provision for gateways to public network services?
- What trends should be watched relative to common-carrier virtual-network service offerings compared with private network capabilities?

Chapter II-2-4
An Update on ISDN
William Heflin

THE UNVEILING OF THE INTEGRATED SERVICES DIGITAL NETWORK (ISDN) is proceeding slowly but surely—the local exchange carriers are exchanging mechanical and analog central offices in favor of digital exchanges with the appropriate software. The three major long-distance carriers are deploying Signaling System 7 throughout their national networks and are gradually extending this service internationally.

US federal and local government agencies are either installing or studying ISDN for incorporation into their procurement documents. Additional standards to extend bandwidth have been agreed on by the International Telephone and Telegraph Consultative Committee (CCITT). European progress on ISDN implementation continues, with Germany leading the Conference of European Postal and Telecommunications Administrations in terms of deployment.

ISDN OVERVIEW

ISDN provides end-to-end connectivity supporting a wide range of services, including voice and nonvoice, that users can access with a limited set of standard multipurpose user network interfaces.

The basic rate interface allows the user to have two 64K-bps clear voice-data B channels and a 16K-bps signaling-data D channel. The term *clear* implies that the derived channels will not have signaling or other control tones within the 64K-bps digital channel.

The primary rate interface accommodates users requiring many connections to the communications system. The North American version consists of twenty-three 64K-bps B voice-data channels and one 64K-bps D signaling-data channel. The European version contains thirty 64K-bps voice-data channels with one 64K-bps D channel for signaling, operation, administration, and maintenance and another 64K-bps channel for framing. In the North American version, the D channel must be the 24th channel.

All of today's communications services—voice, packet-switched, and straight data transfer—will be available on ISDN. Packet-switched messages may be exchanged using the X.25 protocol on the D channel or by using the link access protocol B and X.25 on the B channel. The B channel can also be submultiplexed to allow data transfer at 2.4K, 4.8K, 8K, 9.6K, 16K, 19.6K, and 32K bps. Clear 64K bps can be transferred by using a full B channel, and channels can be grouped to provide 384K-, 1,526K-, and 1,920K-bps data transfer rates. Further CCITT standards work is establishing a broadband version of ISDN for higher data transfer rates and full-motion television.

The objective of worldwide ISDN connectivity depends directly on the accuracy and compatibility of international signaling protocols and systems. The CCITT has specified that the seven-layer, open systems interconnection (OSI) protocol model be used along with its standard common channel—Signaling System 7.

ISDN INVESTMENT REQUIREMENTS

ISDN development is divided into the following four main phases:
- Digital central offices with the appropriate software to allow switching of the basic rate interface, primary rate interface, and packet-switched data services.
- Interexchange digital circuits (T1 carriers with extended frame format) and Signaling System 7 for circuits between exchanges.
- Digital circuits over digital microwave or fiber optics, or analog-to-digital converted long-distance circuits controlled by Signaling System 7.
- Customer premises equipment that is compatible with or converted to ISDN requirements (i.e., voice telephones, data sets, personal computers, or computer terminals and video screens).

Common carriers must have several expensive products or programs in place to offer basic or primary rate services. The local exchange carriers must have digital central offices with the appropriate software programming to provide these services. Remote digital central offices that can handle a limited number of customers have been engineered by major manufacturers.

Generally, digital trunking (T-type carriers) and a digital central office must be present in the local area before ISDN services may be offered. The local central offices must be interconnected by Signaling System 7 and the backbone long-distance network must be similarly equipped. The three major players in long-distance network—AT&T, MCI, and US Sprint—have all incorporated Signaling System 7, and each carrier is extending ISDN service to international destinations.

All of the Bell operating companies and most of the larger independent telephone companies have digital central offices in their major metro-politan areas. Northern Telecom and AT&T have the largest installed digital central office bases that have been programmed for ISDN services. Siemens, Ericsson, Alcatel, and Stromberg-Carlson are also in the digital central office business. Japanese communications manufacturers have not been very successful selling their digital central office technology in the US. Among local exchange carriers, the Bell operating companies and larger independents have performed ISDN field trials working with one or more of the digital central office manufacturers and have gained valuable equipment and service information in the process.

Most local exchange carriers plan to convert their electromechanical central offices to electronic central offices with lower maintenance costs. The decision to incorporate digital central offices equipped with ISDN software, however, is a several-hundred-million-dollar decision for all but the smallest local exchange carrier.

Local exchange carriers have, in many instances, converted their electromechanical offices to analog electronic offices. These analog offices provide ordinary telephone service with a low maintenance cost. Bellcore, the research and engineering arm of the Bell operating companies, has been working on an adaptive system that allows analog offices to provide ISDN services. Local exchange carriers, faced with the enormous investment decision of digitizing their operations, have decided to proceed slowly and cautiously. Because their rates are controlled by state regulatory commissions, the local exchange carriers have been steadily informing the commissions about ISDN services and costs and, in most metropolitan areas, have filed tariffs to cover the services.

ISDN FIELD TRIALS

After ISDN was developed, real-life testing was required to determine whether it could perform as anticipated. A variety of field trials took place involving sophisticated customers with business requirements that might be met with ISDN services.

McDonald's Corp

One of the most comprehensive field trials involved the Oak Brook IL headquarters of McDonald's Corp, its local exchange carrier, Illinois Bell, and AT&T Network Systems. This field test was set up for central office–based ISDN, or Centrex ISDN. The central office used was the AT&T 5ESS switch with the 5E4(2) software generic.

The initial network had three primary locations: the corporate home office, the Lodge, and Hamburger University. A new home office was also planned and was integrated into the network at a later date. McDonald's wanted to consolidate at least 15 separate voice and data networks into a more efficient and manageable single network. AT&T installed a terminal that permitted the corporation to monitor its ISDN network on premises. With this terminal, McDonald's could test the ISDN or analog lines, change translations on either line, and report trouble to the central office.

All McDonald's employees in the home office were equipped with ISDN voice-data terminals. The typical configuration was for each user to have one B channel for voice, one B channel for circuit-switched data, and one D channel for signaling and packet-switched data. Each user's voice channel was set up for call forwarding, hold, drop, conference, transfer, electronic directory, speed calling, message retrieval, call pickup, and time and date.

The ISDN terminals were equipped with the CCITT standardized eight-wire ISDN interface. In this interface, four wires are used for sending and receiving, two are used for power, and two remain unassigned. This mechanical arrangement greatly simplifies the problem of moves and changes. When a move is called for, the terminal is physically moved, plugged into a new location using twisted-pair copper wiring, and the new location number translation is performed on McDonald's terminal monitor.

McDonald's decided to convert their office automation system to an ISDN-compatible system. For this purpose, McDonald's selected an AT&T system using 3B2/600 processors that run remote file sharing under UNIX System V, Release 3.1. Included in the new system were standard facilities (e.g., word processing, data base management, graphics, spreadsheets, electronic mail, and time management). The office automation system is accessed by users through the D-channel X.25 packet-switched or B-channel circuit-switched data connections, depending on the basic rate interface configuration.

To complete the system, rapid access to hard-copy output is necessary. Individual laser printers are distributed throughout the complex so that no employee is more than 25 ft from a printer. Each printer is connected to the system by an ISDN virtual circuit or a B-channel circuit-switched connection.

The McDonald's office automation system also allows for easy implementation of distributed processing. A microcomputer version of the system allows documents to be generated and modified on personal computers. These documents may be uploaded from a personal computer to the system for printing and distribution or, alternatively, files in the office automation system can be downloaded to the personal computer for further changes or offline storage. IBM, IBM-compatible, or Apple Macintosh microcomputers can be used in this particular system.

If microcomputers are to be used in the ISDN, they must be compatible with ISDN transmission and signaling standards. Those computers not designed for ISDN require a terminal adapter interface. Terminal adapters for the most popular personal computers are currently available or will be soon.

With ISDN, microcomputer-to-microcomputer communications is much more efficient. File transfer is accomplished using 9.6K-bps and 19.2K-bps rates. Connections are made on a demand basis by dialing a four-digit directory code, and microcomputer-to-host connections are performed in the same manner. McDonald's uses facsimile transmission to communicate with its regional and international offices, in addition to its customers and service providers. The corporation employs a centralized facsimile network based on a CCITT Group III standard. Distribution from the central location to the user is performed electronically wherever possible using Group IV 56K bps or 64K bps. When facsimile information is available in an electronic format (e.g., a document on the office automation system), it may be transferred by using the file transfer format.

McDonald's has implemented an electronic publishing system based on an Ethernet Transmission Control Protocol and Internet Protocol (TCP/IP) network and X.25 gateways. Terminals, microcomputers, and remote file servers can access the electronic publishing system by circuit-switched or packet-switched connections. In addition, users from different McDonald's locations in the Oak Brook area can all have access, by ISDN, to the electronic publishing system.

Although the initial focus has been on the McDonald's home office, the local ISDN will function with the external communications networks, including the existing analog network, through modems and access to the public packet-switched network. McDonald's will use both of these networks to contact their accounting center, regional offices, and company-owned stores. The company-owned stores have installed AT&T 6386 computers for the collection of sales and inventory information. This information will be transferred on a routine basis to the regional offices and the home office by X.25.

When ISDN becomes available on a regional basis, McDonald's intends to incorporate ISDN in each region using the Centrex approach or by using digital PBXs that support ISDN standards and interfaces. Eventually, all regions will be linked to the home office by ISDN using the long-distance carriers.

Chevron Corp

One of the largest companies on the West Coast—Chevron Corp—also has operations all over the world. This field trial differs from the McDonald's trial because Chevron installed ISDN using a PBX in their subsidiary, Chevron Information Technology Corp, located in San Ramon CA. The field trial involved Chevron Information Technology, Pacific Bell, and Northern Telecom. It brought 50 basic rate interface lines into four buildings and provided links among a Chevron host computer, personal computers, and home workstations.

Chevron Information Technology has approximately 15,000 systems network architecture terminals, and expects to save about $500 each year with each terminal. Before ISDN was implemented, each terminal was moved an average of once every two years at a cost of $1,000 per move. Using ISDN, the moving procedure is simplified to plugging the terminal into its new location and informing the PBX of the move. Chevron has ordered 60 SL-1 PBXs from Northern Telecom and 60 Meridian Mail application packages, which will be installed throughout the US at various Chevron operation sites. Eventually, the 60 PBXs will be equipped with primary rate interface capabilities and interconnected into a national network.

AT&T

American Transtech, the wholly owned telemarketing subsidiary of AT&T, was selected by its parent corporation to act as a test bed for a variety of ISDN services including automatic incoming number identification. One of the telemarketing services provided by American Transtech is supplying dealer locations to prospective customers calling a toll-free number. Before ISDN was implemented, a pool of agents had to ask customers for either their telephone number or zip code, which would then be keyed into the computer data base to retrieve dealer address data.

With the automatic incoming number identification, the data base correlation was performed by the PBX computer and the result was displayed on a screen at the operator's position. Over the three-month trial period, both productivity and accuracy improved. Call transaction time dropped by 12 seconds, resulting in a saving of $2\frac{1}{2}$ hours per agent per week, or 3.6 cents per call. Eliminating the need for agents to enter calling party identification data reduced the error rate to zero from 10% to 17% in terms of menu selection and populating number fields. The results pleased everyone involved. The agents liked the system because it eliminated the repetitive and error-prone keypunching formerly needed to obtain dealer location data. Management liked the improved accuracy and cost savings, and the customers approved because of the speed of response.

FIELD TRIAL ANALYSIS

In all three preceding ISDN field trials, both customers and carriers gained the knowledge they sought with respect to ISDN services and costs. In addition, when the appropriate tariffs were filed, the trials turned into communications realities.

These examples represent just a few of the many field trials performed. In cases in which the customer was sophisticated and the trials were extensive with respect to the services offered, the majority became regularly tariffed extensions of the customer's communications services.

All Bell operating companies have ISDN offerings available in their service areas. With AT&T, MCI, and US Sprint completing the installation of Signaling System 7 in their networks and extending the service to ISDN service areas abroad, a worldwide digital network is inevitable.

GOVERNMENT USE OF ISDN

Led by the National Institute of Standards and Technology in Gaithersburg MD several federal government agencies are planning a national ISDN called ISDN Net. Although the planning is open to any interested government communicator, the air force bases at Griffiss (Rome NY), Mather (Rancho Cordova CA) and Barksdale (Shreveport LA) as well as Pensacola Naval Air Station (FL), and the Goddard Space Flight Center (Greenbelt MD) are leading the effort. Both basic rate and primary rate interface services are to be included in the ISDN Net, in addition to a variety of ISDN switches and transmission equipment.

INTERNATIONAL IMPLEMENTATION OF ISDN

As a result of the joint efforts of the Bundespost and the major German communications manufacturers (Siemens and SEL Alcatel), Germany has made the most progress in establishing national ISDN. Successful ISDN trials were held in Stuttgart and Mannheim in 1988 and early 1989. Later in 1989, local and long-distance exchanges were added in Frankfurt, Berlin, Hannover, Dusseldorf, Munich, and Nuremburg.

The Bundespost has 6,200 local central offices in its network and 472 long-distance switching exchanges. Over the years, it has been converting its central offices and exchanges to a digital format.

France has held communications modernization as a matter of high national priority. France Telecom opened the world's first ISDN in 1987 in Brittany. ISDN is being marketed in France under the name of NUMERIS. It is available in Paris and its western suburbs, Lyons, Marseilles, and Lille. ISDN market forecasts are for 150,000 access lines by 1992 and between 500,000 and 700,000 by 1995.

The Netherlands, long one of the chief trading nations of the world, intends to have one of the most modern communications networks in

Europe. A pilot project for ISDN was put into service in the Rotterdam area with voice, visual data, computer data, and facsimile services offered. PTT Telecom BV plans to replace two-thirds of its cables with fiber-optic cables during the early 1990s.

The United Kingdom, Belgium, Italy, and Spain are all working on ISDN projects. In the Far East, Japan, Hong Kong, and Singapore are planning for ISDN. The industrialized countries will deploy ISDN first, but the Third World countries are very aware of the technological communications trends and will install digital networks when they can afford them.

Section II-3
Technology

TODAY'S TECHNOLOGY IS THE BASIS for tomorrow's products. Equipment and software now being developed and tested in the laboratory, as well as early products first venturing into the marketplace, will be common, and even taken for granted, in 5 or 10 years. This section examines changing technology in existing products (e.g., modems and protocols), current technology as it is being applied in relatively new products (e.g., electronic mail, voice and data applications of PBXs), and emerging, still maturing technology (e.g., optical fiber options and digital compression algorithms).

Chapter II-3-1, "Advances in Modem Modulation Techniques," summarizes the current state of modem modulation techniques. Despite opinions that the modem is fast becoming a dinosaur, demand for these ubiquitous devices remains high. The demand, to some extent, is fueled by the ability of the vendors to continually extract more bits per second from the available bandwidth and to continue to add useful features, especially ones that facilitate diagnostic and management functions. Chapter II-3-2, "The Practical Side of Voice and Data Integration," addresses an increasingly important issue. Advances in technology probably make inevitable the integration of the two major information services—human speech and machine-generated data. Organizational changes and integration techniques are discussed in this useful chapter.

Communications protocols, like modems, will not disappear. They keep getting better, meaning more efficient in terms of cost and performance. Chapter II-3-3, "Communications Protocols," reviews their history, characteristics, features, advantages, and shortcomings. The chapter also offers valuable selection criteria.

Electronic mail is a technology whose time has come. The emergence of standards, such as X.400, and products built to those standards will accelerate growth. Chapter II-3-4, "Electronic Mail Systems," is an excellent primer on this technology. It reviews types of electronic mail and examines benefits, needs, issues, and costs. The chapter concludes with an evaluation checklist.

Chapter II-3-5 of this section, "Trends in Fiber-Optic Technology," updates the communications system manager on one of today's dynamic technologies. Fiber is rapidly being deployed in the long-haul telephone transmission infrastructure, including the transoceanic cables. Fiber is supplanting coaxial cable as the medium of choice in local area, backbone, and metropolitan networks. Chapter II-3-5 explores the status of fiber technology and discusses the cost-effective application of fiber in communications systems.

The transmission of images is now as critical an element as voice and data in the communications manager's plans. Unfortunately, image transmission consumes large amounts of bandwidth. To reduce this expensive characteristic, digital signals representing images are usually compressed before transmission. Chapter II-3-6, "An Overview of Digital Image Bandwidth Compression," is an easy-to-read tutorial on the subject. The chapter discusses image compression methods and identifies and explains compression issues as they affect the human visual system.

PBX technology has made, and continues to make, rapid strides from being an all-voice technology to one that combines voice, data, and image-capable technology. The PBX can be an important element of the contemporary communications system. The PBX can certainly handle voice-switching requirements. It can also play a role as an alternative to a local network. It occupies a unique position as a gateway between the local and long-haul distribution systems. The final chapter of this section, Chapter II-3-7, "The PBX in Perspective," provides a very useful insight into the current and future possibilities of PBX networking technology.

Chapter II-3-1
Advances in Modem Modulation Techniques

Gilbert Held

THE BASIC ELEMENTS OF A MODEM INCLUDE a transmitter, a receiver, and a power supply that provides the voltage to operate the modem's circuitry. The transmitter contains modulation, amplification, filtering, and wave-shape circuitry that converts digital pulses into an analog signal for transmission over a telephone line. The receiver contains a demodulator and its associated circuitry, which convert the received analog signal into a series of digital pulses for computer and terminal processing.

THE TELEPHONE NETWORK

Data can be transmitted over the public switched telephone network (PSTN) or over leased lines. The PSTN is a two-wire system that provides random call routing and access to any location connected to the network for the duration of the call. The PSTN usually provides only half-duplex transmission. In comparison, a leased line is usually a four-wire system installed between two locations that provides fixed routing and full-duplex transmission. Because the routing of leased lines is fixed, the telephone company can condition these lines to optimize their performance. The temporary circuits established for the duration of a call routed over the PSTN cannot be conditioned.

COMMUNICATIONS CONCEPTS

The following sections discuss several communications concepts, providing a foundation for the subsequent examination of modem modulation and demodulation techniques.

Comparing Bits per Second with Baud

Bits per second (bps) is a measurement of data throughput. Baud is a measurement of signal change. If each bit entering a modem results in one signal change, then one bps equals one baud. If the modem is designed so that a pair of bits is represented by one signal change, then two bps equals one baud. The baud rate is then one-half the data expressed in bps.

The data rate of a modem is expressed in bps and provides a measurement of the capacity of the modem to transmit information per unit time. The baud rate and the method of encoding many bits into a baud provides information concerning modem compatibility.

The Nyquist Limit

During the 1920s, Nyquist discovered a relationship between the bandwidth of a noiseless channel and its capacity in bits per second. This relationship is expressed in the equation:

$$C = 2^*B$$

where:
* = convolution
B = the bandwidth of the channel in hertz
C = the capacity of the channel in bps

Both leased lines and PSTN circuits have a bandwidth of 3,000 Hz. According to the Nyquist equation, the data transfer capacity of these lines would be limited to 6,000 bps before one signal change would interfere with another. Because an oscillating modulation technique immediately halves the signaling rate, most modems operate at one-half the Nyquist limit or less; the actual data transfer rate of a telephone line is limited to 3,000 bps. Because the Nyquist limit only applies to the signaling (baud) rate, an increased amount of information carried by each signal (i.e., more bits packed into each signal change) increases the number of transmitted bits per second. A technique that encoded two bits into one baud could permit the data transfer capacity of the line to each 6,000 bps without violating the Nyquist limit, and a technique that encoded three bits per signal change could permit the transmission rate to reach 9,000 bps.

THE MODULATION PROCESS

In the modulation process, an analog signal known as the carrier is modified to impress the original digital information onto the signal. The carrier is usually a sine wave, represented by the equation:

$$a = A \sin (2\pi ft + \theta)$$

where:
a = instantaneous value of voltage at time

A = maximum amplitude
f = frequency
t = time
θ = phase

Characteristics of the carrier that can be altered include the amplitude in amplitude modulation (AM), the frequency in frequency modulation (FM), and the phase angle in phase modulation (PM). Exhibit II-3-1 illustrates the effect of each modulation technique on the carrier.

Amplitude Modulation.   The amplitude of the signal is altered to correspond to binary ones and zeros. One amplitude represents a binary 1 and a second amplitude represents a binary 0.

Frequency Modulation.   The frequency of the carrier is altered to correspond to the binary ones and zeros. One frequency is used to represent a binary 1, and a second frequency is used to represent a binary 0.

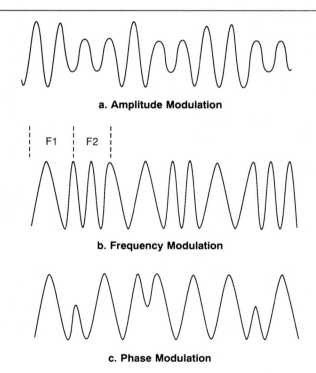

a. Amplitude Modulation

F1   F2

b. Frequency Modulation

c. Phase Modulation

Exhibit II-3-1.   Amplitude, Frequency, and Phase Modulation

Phase Modulation.   The phase of the carrier is varied based on the content of the bits received by the modem. The simplest method of phase modulation is varying the phase between two phase values, with one phase representing a binary 1, and the second representing a binary 0. In this method of phase modulation, one bps equals one baud.

Because it is relatively easy to change the phase angle of the carrier, phase modulation is typically used by modem designers to pack many bits into one signal change. When two bits are used to represent one baud, the modem must be capable of generating four phase angle changes, with each phase angle change used to represent each of the dibit pairs: 00, 01, 10, and 11. In this modulation technique, the baud rate is one-half the data rate, which is expressed as bits per second, and the method of encoding is known as dibit encoding. The process of using three bits to represent one baud is known as tribit encoding. In this case, the baud rate is one-third the bit rate. Because eight tribits ranging from binary 000 to binary 111 can occur, tribit encoding requires eight phase angles—one to represent each of the eight possible tribit combinations.

Both dibit and tribit encoding are also known as multilevel coding. Because the phase angle is shifted to represent each dibit or tribit, this technique is also known as phase shift keying (PSK). Exhibit II-3-2 lists the common phase angle shifts used in modems employing multilevel PSK. Because a low signaling rate permits a higher data rate, phase modulation is typically used in high-speed modems. The differences in possible phase angle changes decrease as the number of bits encoded per phase change increases. For dibit encoding, the phase angle differences are 90° apart. The optimum distance between angle changes is halved to 45° in tribit encoding (see Exhibit II-3-2). Because each increase in the number of bits in a baud

Bits Transmitted	Possible Phase Angle Values (Degrees)		
00	0	45	90
01	90	135	0
10	180	225	270
11	270	315	180
000	0	22.5	45
001	45	67.5	0
010	90	112.5	90
011	135	157.5	135
100	180	202.5	180
101	225	247.5	225
110	270	292.5	270
111	315	337.5	315

Exhibit II-3-2.   Common Phase Angle Values: Multilevel Phase Shift Keying

doubles the number of possible phase angles while halving the distance between phase angle changes, the circuitry required to discriminate the phase changes becomes more complex and expensive. Therefore, modem designers have focused their attention on the use of combined modulation methods that require less complex circuitry.

COMBINED MODULATION METHODS

Techniques that incorporate a large number of bits into each signal change provide high data transfer rates at a reasonable cost. One such technique is quadrature amplitude modulation (QAM), which combines phase and amplitude modulation. Exhibit II-3-3 illustrates the original version of QAM, in which a combination of 12 phase angles and three amplitudes was used to produce 16 signal states. Because of its ability to support high-speed data transmission, QAM is employed in most synchronous high-speed modems.

Plotting the signal points illustrated in Exhibit II-3-3 results in the representation of all data samples possible in the original version of QAM. The points show the signal structure of the modulation technique. These points

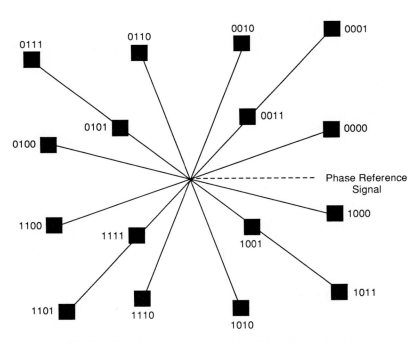

Exhibit II-3-3.   Quadrature Amplitude Modulation

are often described as the constellation pattern. By examining the constellation pattern of a modem, it is possible to predetermine the modem's susceptibility to certain transmission impairments. For example, phase jitter, which causes certain signal points to rotate about the origin, can result in one signal being misinterpreted for another, which would cause four bits to be received in error.

NEW TECHNIQUES

Modems based on trellis-coded modulation (TCM) employ an encoder that adds a redundant code bit to each symbol interval, resulting in only certain sequences of signal points being valid. If an impairment occurs that causes a signal point to be shifted, the modem's receiver will compare the observed point with all valid points and select the valid signal point closest to the observed signal. As a result, a TCM modem is only half as susceptible to as much noise power as a conventional QAM modem. TCM modems operating at 14,400 bps employ a modified QAM process, in which data bits are collected into a 6-bit symbol 2,400 times per second. The incorporation of the redundant code bit into the modulation process causes a 128-point signal constellation pattern to be formed. Other TCM modems offer data rates to 16,800 bps over leased lines and 9,600 bps over the PSTN.

MODEM OPERATIONS AND STANDARDS

In the US, most low- to medium-speed modems are compatible with the de facto Bell System modem standards. In most European countries, the postal, telegraph, and telephone administration regulates the use of modems connected to both the PSTN and leased lines. Because the recommendations of the International Telephone and Telegraph Consultative Committee (CCITT) are followed in Europe, most European modems are not compatible with modems manufactured in the US. Certain CCITT recommendations, however, have been followed for modems designed for use within the US.

300-bps Modems

Most asynchronous 300-bps modems use a frequency shift keying (FSK) modulation technique, in which one frequency of the carrier represents a mark, or binary 1, and the other frequency represents a space, or binary 0. Two pairs of frequencies are employed to allow the bandwidth of the communications channel to be split in two, enabling full-duplex transmission over the two-wire PSTN. Exhibit II-3-4 illustrates FSK for 300-bps Bell System–compatible modems. The lower pair of frequencies is the origination channel, the higher pair of frequencies the answer channel. Because terminals typically originate calls, a modem connected to a terminal would

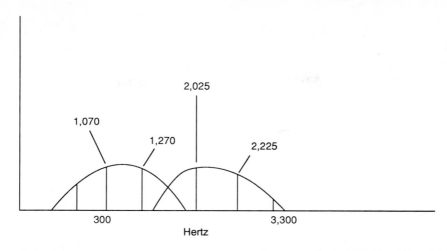

Exhibit II-3-4.   FSK Operation for 300-bps Bell System–Compatible Modems

---

transmit a mark at 1,270 Hz and a space at 1,070 Hz. The receiver would be tuned to receive a mark at 2,225 Hz and a space at 2,025 Hz. A modem connected to a mainframe that usually receives calls would transmit a mark at 2,225 Hz and a space at 2,025 Hz, and the receiver would be tuned to receive a mark at 1,270 Hz and a space at 1,070 Hz. Many 300-bps modems have an originate/answer switch that allows the modem to be used with a terminal or mainframe, and intelligent 300-bps modems can have a mode set under program control.

The FSK operation for the 300-bps modem (see Exhibit II-3-4) corresponds to Bell System 103/113-type modems. In comparison, most European 300-bps modems operate in conformance with the CCITT V.21 recommendation. Although the V.21 recommendation is also based on FSK, the V.21 frequency assignments differ from those of Bell System 103/113-type modems. Exhibit II-3-5 lists frequency assignments for Bell System 103/113 and CCITT V.21 modems. Most 300-bps modems built into microcomputers sold for operation in the US are compatible with Bell

---

Modem Type	Originate	Answer
Bell System 103/113	Mark 1,270 Space 1,070	Mark 2,225 Space 2,025
CCITT V.21	Mark 980 Space 1,180	Mark 1,650 Space 1,850

Exhibit II-3-5.   300-bps Frequency Assignments (Hz)

445

System 103/113, and most European value-added carriers use V.21 modems to accept data calls at 300 bps. For these reasons, US-built 300-bps modems are usually unable to access US mainframes through a European value-added carrier.

300- to 1,800-bps Modems

Several Bell System and CCITT V modems operate in the range of 300 to 1,800 bps. Some of these modems (e.g., the Bell System 212A and the CCITT V.22) are capable of operating at one of two data rates. Others (e.g., the Bell System 202 and the CCITT V.23) operate at only one data rate.

Bell System 212A and CCITT V.22 Modems. The 212A modem can operate either asynchronously or synchronously over the PSTN. For asynchronous transmission at 300 bps, the 212A modem uses FSK modulation and frequency assignments compatible with Bell System 103/113-type modems. At 1,200 bps, dibit PSK modulation enables the modem to operate either asynchronously or synchronously. Exhibit II-3-6 shows the phase shift encoding used by 212A-type modems.

CCITT V.22 modems operate at 1,200 bps over both the PSTN and leased lines and have a fallback data rate of 600 bps. V.22 modems use four-phase PSK modulation at 1,200 bps and two-phase PSK modulation at 600 bps. Both the operating speed and modulation technique differ for Bell System 212A-type and V.22 modems. At 1,200 bps, V.22 modems can operate in one of three modes: the mode of operation is governed by whether data is transmitted asynchronously or synchronously or whether an alternate phase change set is used for asynchronous transmission. In two of these

	Dibit Values	Phase Shift	
Modem	00	90	
212A-Type	01	0	
	10	180	
	11	270	
Bell System 201B/C	00	225	
	01	315	
	11	45	
	10	135	
V.26		**Pattern A**	**Pattern B**
	00	0	45
	01	90	135
	11	180	225
	10	270	315

Exhibit II-3-6.   Modem Phase Shift Encoding

modes, the modulation technique is exactly the same as that used by a Bell System 212A-type modem. The frequency of the carrier tones used by 212A and V.22 modems, however, is slightly different.

When a 212A modem answers a call, it transmits a tone of 2,225 Hz, which the originating modem should recognize. When a V.22 modem answers a call, it first sends a tone of 2,100 Hz, after which it transmits a 2,400 Hz tone that is also incompatible with a 212A tone. However, the 2,400 Hz tone is followed by a burst of data with a primary frequency of about 2,250 Hz, which is close enough to the 212A frequency of 2,225 Hz to allow most 212A modems to respond. Therefore, most 212A-type modems can usually communicate with V.22 modems at 1,200 bps.

Bell System 202 Modems.   Modems that conform to the Bell System 202 operating characteristics operate at 1,200 bps in a half-duplex mode over the PSTN and at data rates to 1,800 bps over leased lines in either a half- or full-duplex mode. Modems in the 202 series use FSK modulation, in which a mark is represented by 1,200 Hz and a space by 2,200 Hz. Modems in the 202 series have an optional 5-bps reverse channel for switched network use that employs amplitude modulation for data transmission. The reverse channel can be used to transmit short acknowledgments in the reverse direction without changing the direction of transmission. Use of the reverse channel eliminates the overhead associated with line turnarounds.

V.23 Modems.   Modems that conform to the CCITT V.23 recommendation are similar to, but not compatible with, 202 modems. V.23 modems can be used to transmit data over the PSTN at 600 or 1,200 bps. Both asynchronous and synchronous transmission is supported by FSK; however, the frequency assignments for marks and spaces differ from frequency assignments for 202-type modems. In addition, the V.23 optional reverse channel operates at 75 bps, whereas a 202 reverse channel operates at 5 bps.

2,400-bps Modems

Modems that operate at 2,400 bps include the Bell System 201 series, the CCITT V.26 series, and the V.22bis modem (a secondary variation of the V.22 recommendation). The Bell System 201 and CCITT V.26 series modems are designed for synchronous transmission at 2,400 bps. The V.22bis recommendation defines 2,400-bps asynchronous transmission.

Bell System 201B/C Modems.   The 201 series includes the 201B and 201C modems. Both types use dibit PSK modulation: the phase shifts are based on the dibit values listed in Exhibit II-3-6.

The 201B modem is designed for half- or full-duplex synchronous transmission at 2,400 bps over leased lines. In comparison, the 201C is designed for half-duplex transmission over PSTN. A more modern version of the 201C is marketed by AT&T as the 2024A, which is compatible with the 201C.

V.26 Modems.   The V.26 recommendation defines a 2,400-bps synchronous modem for use on a four-wire leased line. Modems operating according to the V.26 recommendation employ dibit PSK modulation and use one of two recommended coding patterns. Exhibit II-3-6 lists the phase shift based on the dibit values for each of the V.26 coding patterns. A comparison of the phase patterns that result from the dibit values listed in Exhibit II-3-6 shows that 201 and V.26 modems are incompatible.

Two similar CCITT recommendations to V.26 that warrant attention are the V.26bis and the V.26ter. V.26bis defines a dual-speed 2,400/1,200-bps modem for use on the PSTN. At 2,400 bps, the modulation and coding method corresponds to the V.26 recommendation for pattern B (see Exhibit II-3-6). At the reduced data rate of 1,200 bps, a two-phase PSK modulation technique is used. A binary 0 is represented by a 90° phase shift and a binary 1 is represented by a 270° phase shift. Modems conforming to the V.26ter recommendation, a third variation of the V.26 recommendation, use the same phase shift scheme as the V.26 modem but also incorporate an echo-canceling technique that allows transmitted and received signals to occupy the same bandwidth. The V.26ter modem is capable of full-duplex operations at 2,400 bps over the PSTN.

V.22bis Modems.   The CCITT V.22bis recommendation defines modems designed for asynchronous data transmission at 2,400 bps over the PSTN, with a fallback rate of 1,200 bps. The use of the V.22bis modem is rapidly increasing in both the US and Europe. The answering tone incompatibility between Bell System 212A-type modems and CCITT V.22 modems extends to V.22bis modems manufactured in the US and abroad. V.22bis modems built for operation in the US follow the telephone network specifications that require the modem's answer tone to be sent at 2,225 Hz. In most European countries that follow CCITT recommendations, the modem's answer tone is 2,100 Hz. Therefore, a V.22bis modem manufactured in the US is CCITT V.22bis compatible but not identical. As a result, communication between US and European V.22bis modems is not ensured. In addition, some US V.22bis modem manufacturers include a third data rate offering, which follows the Bell System 103/113 modulation technique for 300-bps operations. Because the CCITT V.22bis recommendation does not include a 300-bps operating rate, US-built and European V.22bis modems are incompatible at 300 bps.

4,800-bps Modems

The Bell System 208 series and CCITT V.27 modems represent the most common types of modems designed for synchronous data transmission at

4,800 bps. The 208A is designed for either full- or half-duplex operations over leased lines, and the 208B is designed for half-duplex operations over the PSTN. V.27 modems are designed for operation over a four-wire leased line.

Both 208-type and CCITT V.27 modems pack data three bits at a time and encode the resulting tribit as one of eight possible phase angles. Because 208-type and V.27 modems use different phase angles to represent a tribit value, however, they are incompatible.

9,600-bps Modems

Three commonly used modems that operate at 9,600 bps are the Bell System 209, the CCITT V.29, and the CCITT V.32.

The Bell System 209 and CCITT V.29 modems are designed to operate in a full-duplex, synchronous mode over leased lines. The Bell System 209 modem uses a QAM technique and includes a built-in synchronous multiplexer, which can multiplex four data streams into one aggregate 9,600-bps data stream. Vendors offering 209-compatible modems usually offer the multiplexer capability as a separate option.

With the exception of Bell System 209-type modems, most 9,600-bps modems adhere to the CCITT V.29 recommendation. This recommendation defines data transmission at 9,600 bps for full- or half-duplex operations over leased lines, with fallback data rates of 7,200 and 4,800 bps. At 9,600 bps, the serial data stream is divided into groups of four bits. The first bit in the group is used to determine the amplitude of the signal, and the remaining three bits are encoded as the phase change. Because of its popularity, the V.29 is the only modem available worldwide that affords users international compatibility.

The CCITT V.32 recommendation defines full-duplex 9,600-bps transmission over the PSTN. The V.32 recommendation specifies a modified QAM technique and is based on an echo-canceling technique that permits transmitted and received signals to occupy the same bandwidth. Intelligence is incorporated into the modem's receiver to permit it to cancel out the effects of its own transmitted signal. V.32 and V.32 bis modems afford users international compatibility over the PSTN.

Nonstandard Modems

Computer users require faster transmission throughput to support file transfer and interactive full-screen display operations. To satisfy this need, several vendors have designed proprietary operating modems to achieve data rates that were unheard of several years ago. The key to achieving high-speed data transmission is the incorporation of data compression algorithms as well as the use of proprietary modulation techniques. The result is an increase in possible operating rates over the switched telephone network.

Compression-Performing Modems

Data compression and decompression algorithms permit data to be compressed before transmission and then expanded to its original form at the receiving modem. Because compression decreases the amount of data requiring transmission, the modem can accept more data than it is capable of transmitting. Therefore, a V.29 modem incorporating data compression with a 2:1 compression ratio is theoretically capable of transmitting data at 19.2K bps, even though the modem operates at 9.6K bps. Because the compression efficiency depends on the susceptibility of the data to the compression algorithms, the modem actually operates at a variable data rate. When compression is not possible, the modem operates at 9.6K bps. The actual throughput of the device increases as the data input becomes more susceptible to compression.

Packetized-Ensemble-Protocol Modems

A second type of nonstandard modem became commercially available in 1986. The packetized-ensemble-protocol modem combines a high-speed microprocessor with approximately 70,000 lines of instructions built into read-only memory chips on the modem board.

With the packetized ensemble protocol, the originating modem simultaneously transmits 512 tones onto the line. The receiving modem evaluates the tones and the effect of noise on the entire voice bandwidth and reports unusable frequencies to the originating device. The originating modem then selects a transmission format suitable to the useful tones, employing 2-bit, 4-bit, or 6-bit QAM, and packetizes the data before transmission. Exhibit II-3-7 illustrates the carrier use of a packetized-ensemble-protocol modem.

As an example of the efficiency of this type of modem, if 400 tones are available for a 6-bit QAM scheme, the packet size is 400 × 6, or 2,400 bits. If each of the 400 tones is varied four times per second, a data rate of approximately 10K bps is obtained. The modem automatically generates a 16-bit cyclic redundancy check to detect errors, which is added to each transmitted packet. At the receiving modem, a similar cyclic redundancy check is performed. If the transmitted and locally generated cyclic redundancy check characters do not match, the receiving modem requests the transmitting modem to retransmit the packet, resulting in error correction by retransmission.

Key advantages of a packetized-ensemble-protocol modem include the ability to automatically adjust to usable frequencies, increasing the use of

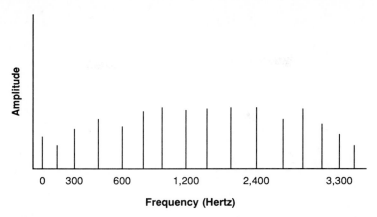

Exhibit II-3-7.   Carrier Tones for the Packetized-Ensemble-Protocol Modem

the line bandwidth, and the ability to lower the fallback rate in small increments. Exhibit II-3-8 illustrates how this type of modem loses the ability to transmit on one or more tones as the noise level on a circuit increases. This results in a slight decrease in the data rate of the modem. In comparison, a conventional 9.6K-bps modem is designed to fall back to a predefined fraction of its main data rate, which is typically 7,200 bps or 4,800 bps.

The original packetized-ensemble-protocol modem was designed by the Telebit Corp and is marketed as the Trailblazer. Several other vendors market similar modems under license, using different trade names.

In addition to compatibility with other packetized-ensemble-protocol modems, these devices are compatible with V.22bis, V.22, 212A, and 103 modems. This permits a terminal or microcomputer to use the device for high-speed file transfer operations when connected to another packetized-ensemble-protocol modem. The user can also access information utilities, other microcomputers, and mainframes that are connected to industry-standard modems.

Asymmetrical Modems

Borrowing an older design concept, several vendors have introduced new modems based on the use of asymmetrical technology. These modems contain two channels, which in the early days of model development were known as the primary and secondary channels.

Originally, modems with a secondary channel were used for remote batch transmission. The primary high-speed channel was used to transmit data to a mainframe while a lower-speed secondary channel was used to

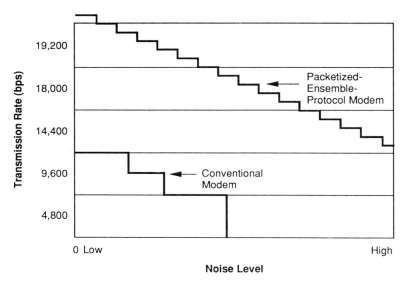

Exhibit II-3-8. Transmission Rate Versus Noise Level

acknowledge each transmitted block. Because the acknowledgments were much shorter than the transmitted data blocks, it was possible to obtain efficient full-duplex transmission, even though the secondary channel might have one-tenth the bandwidth of the primary channel.

During the late 1980s, several modem vendors realized that high-speed transmission was necessary to transfer files and refresh microcomputer screens. Transmission in the opposite direction, however, is typically limited by the user's typing speed. Vendors, therefore, developed a new category of devices that use wide and narrow channels to transmit in two directions simultaneously, as illustrated in Exhibit II-3-9. The wide bandwidth channel permits a data rate of 9.6K bps; the narrow bandwidth channel supports a data rate of 300 bps.

Asymmetrical modems differ from older modems with secondary channels because they incorporate logic to monitor the output of attached devices and to reverse the channels. They permit an attached terminal device to access the higher-speed (wider-bandwidth) channel when necessary. Although standards do not exist for asymmetrical modems, several manufacturers are attempting to formulate common frequency assignments for channels. Because a 9.6K bps asymmetrical modem costs approximately half the price of a V.32 modem, it appears that consumer use of this type of modem may result in the development of standards.

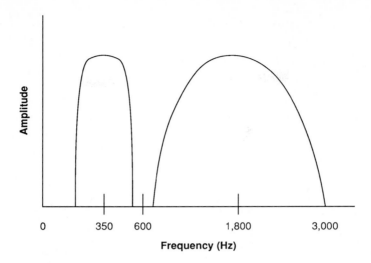

Exhibit II-3-9. Asymmetrical Modem Channel Assignment

An asymmetrical modem encodes data onto two carrier signals. High-speed data is encoded onto an 1,800-Hz carrier; low-speed data is encoded onto a 350-Hz carrier.

MODEM SELECTION

Compatibility is the most important factor in the modem selection process. When selecting a modem for international communications, the data communications manager should examine differences in modems designed to operate over leased lines and those designed to operate over the PSTN, as well as the regulations of telephone companies in other countries. In some countries, the postal, telephone, and telegraph organization provides a list of approved modems; in other countries, it supplies the modem. The postal, telephone, and telegraph administrations that supply modems usually offer CCITT V.29 modems, which are available in the US and which have evolved into a worldwide standard for 9,600-bps transmission. For transmission at lower data rates, postal, telephone, and telegraph organizations usually supply one or more of the CCITT V series modems. These modems are not commonly available in the US; however, several US vendors offer other members of the V series that can communicate overseas on leased lines.

Options for international communications over the PSTN include attempting a call from the US with a 212 or V.22bis modem operating at 1,200 or 2,400 bps and using the appropriate V series modem.

Communications with a US-built modem over the PSTN in other countries may be difficult because of the differences in regulations and the varying structures of telephone systems in other countries. In most European countries, low-speed Bell System 103/113-type modems are illegal because their operating frequencies are not compatible with the PSTN.

The acquisition of nonstandard modems employing proprietary technology should be conducted with caution. Some vendors (e.g., Telebit) license their technology to other vendors, allowing several modem manufacturers to market compatible nonstandard modems. Unfortunately, other vendors use proprietary technology that is restricted to their product offering only.

One of the key areas that the data communications manager should keep in mind when examining such proprietary modems is downward compatibility. Downward compatibility refers to operating rates (other than the modem's highest data rate) at which the modem can function as a Bell System- or CCITT-compatible device. As previously discussed, the Telebit modem is compatible with V.22bis, V.22, 212A, and 103 modems and therefore can be used with a large base of non-Telebit modems.

SUMMARY

The communications systems manager should focus careful attention on modem compatibility to ensure that communications requirements can be met. Factors that the communications manager should consider when selecting a modem include data transmission rates, line facilities, call origination and destination, telephone company structures in the countries to be serviced, and postal, telephone, and telegraph organization regulations. With proper planning and coordination with the appropriate regulatory administrations, most communications barriers can be overcome.

Chapter II-3-2
The Practical Side of Voice and Data Integration
Nathan J. Muller

INTEGRATING VOICE AND DATA over high-capacity backbone networks sounds like a good idea. Being able to fulfill the bandwidth requirements of the MIS and telecommunications departments over the same digital facilities makes for efficient channel fills, which can meet the organization's bottom-line objective for economical long-haul transport.

Despite all the hype surrounding the concept, however, voice and data integration is still an elusive goal for many organizations. Even when integration is accomplished with some degree of success, managing voice and data separately proves to be the final hurdle that frequently cannot be cleared. This is because voice and data have fundamentally different transmission requirements.

Whenever round-trip delay in a voice network exceeds 32 ms, echo cancelers are needed. The degree to which echo impedes communication is determined by the combination of amplitude and delay. Although telephone users can tolerate a relatively high degree of delay—as much as 40 ms of round-trip delay if the echo is not too loud—the same amount of delay can cause supervisory problems for PBX tie lines, creating a condition known as glare. Glare occurs when both ends of a trunk are seized at the same time by different users, blocking the call. Wide variations in delay at multiple locations can make glare a frequent and particularly annoying problem.

MULTIPLEXERS

Despite the onslaught of digital PBXs in recent years, T1 multiplexing still represents the single most important technology for the simultaneous trans-

455

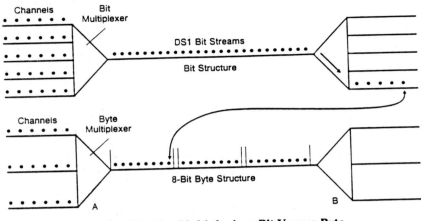

Exhibit II-3-10. Multiplexing: Bit Versus Byte

mission of voice and data. Even within this category, however, there are significant equipment differences. Multiplexers are based on bit or byte structures. The choice of a multiplexer can be critical in controlling the amount of delay on the network (See Exhibit II-3-10).

Multiplexers that conform to the byte-framing structure offer maximum flexibility in implementing public network interconnectivity, restoration, and alternative routing at the network transport level. As shown in Exhibit II-3-10, however, the process of accumulating bits in the multiplexer's buffer (point A) for bundling into bytes is a cause of delay, which can affect the quality of voice and data. The benefits of using bit technology include minimal nodal delays, high aggregate efficiency, and broader selection of channel data types and rates. Specially equipped multiplexers now offer the best of bit and byte technologies, permitting optimal access to both public and private facilities.

The Impact of Delay on Voice

Exhibit II-3-11 illustrates a 1,400-mile, 4-node multiplexer network operating at a channel rate of 32K bps. The formula for deriving nodal delay is as follows:

$$\text{Delay} = \frac{\text{Number of bits buffered}}{\text{Channel rate}}$$

Assuming the network has a typical byte-oriented multiplexer with 7 bytes (56 bits) of buffer delay per channel at each node and assuming voice is compressed to 32K bps, the formula yields the following result:

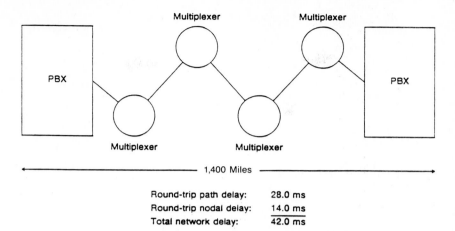

Round-trip path delay: 28.0 ms
Round-trip nodal delay: 14.0 ms
Total network delay: 42.0 ms

Exhibit II-3-11. Voice Communications Through a Typical Byte-Oriented Multiplexer at 32K bps

---

$$\text{Delay} = \frac{56 \text{ bits}}{32,000 \text{ bps}} = 1.75 \text{ ms}$$

A nodal delay of 1.75 ms per node in one direction equals 7 ms of delay each way, or 14 ms round trip.

Path delay usually amounts to 1 ms per 100 miles. In a 1,400-mile network, the total path delay one way would be 14 ms, or 28 ms round trip. Adding round-trip path delay to round-trip nodal delay gives 42 ms of network delay, which greatly exceeds the 32-ms threshold at which echo cancelers are required.

Exhibit II-3-12 illustrates the performance of a bit-oriented multiplexer operating at a channel rate of 32K bps on the same type of voice network. The round-trip path delay remains the same as in the previous example at 28 ms. Nodal delay, however, is significantly lower—only 0.25 ms per node, which equals 2 ms of round-trip nodal delay. Adding round-trip path delay and round-trip nodal delay results in 30 ms of network delay, which means that echo cancelers are not required.

The choice of multiplexer determines whether the cost of echo cancelers must be factored into the overall network cost.

The Impact of Delay on Data

Delay also affects the performance of data networks. Exhibit II-3-13 illustrates a 200-mile, seven-node statewide network operating at a channel speed of 2,400 bps with a multipoint tail circuit. In this case, the typical

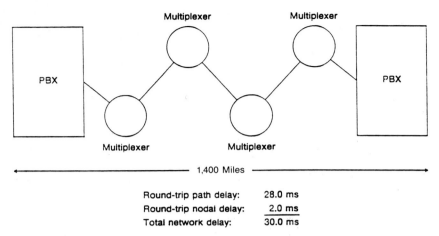

Round-trip path delay:	28.0 ms
Round-trip nodal delay:	2.0 ms
Total network delay:	30.0 ms

Exhibit II-3-12. Voice Communications Through a Bit-Oriented Multiplexer at 32K bps

byte-oriented multiplexer is responsible for 23.3 ms (56 bits/2,400 bps) of delay per node. This translates into one-way nodal delay of 163 ms and a round-trip delay of 326 ms. Adding 4 ms of round-trip path delay on the 200-mile route yields 330 ms of network delay. Assuming, for the sake of simplicity, that there are only 10 drop locations with a total network delay of 330 ms, the byte-oriented multiplexer adds 3.3 seconds to round-robin polling, which may make host-terminal response time objectives difficult to achieve.

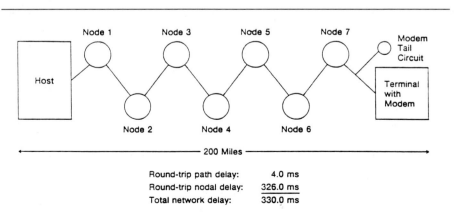

Round-trip path delay:	4.0 ms
Round-trip nodal delay:	326.0 ms
Total network delay:	330.0 ms

Exhibit II-3-13. Data Communications Through a Typical Byte-Oriented Multiplexer at 2,400 bps

Exhibit II-3-14 illustrates the same network using a bit-oriented multi-plexer. The round-trip path delay remains the same at 4 ms for the 200-mile route. The nodal delay, however, is only 3.3 ms, which produces a round-trip nodal delay of only 46.6 ms. When the 4 ms of round-trip path delay is added in, network delay is only 50.6 ms. This results in the addition of only a half-second to round-robin polling, putting host-terminal response time objectives within reach.

Delay can affect the performance of networks in other ways. Consider the issue of rerouting to circumvent a congested node. Another route cannot be randomly selected simply because it happens to be available. All reason-able routes should be checked for sources of delay, and the route chosen must deliver the data to its destination on time. Otherwise, the performance of the network degrades after rerouting. For example, byte-oriented multi-plexers pose no problems when used in point-to-point configurations. When these products are used for large multinode networks, however, users experience problems with throughput, which is largely a function of delay.

Delay is such an important parameter for both voice and data that con-cerns about design tolerance and monitoring are equally valid for both. Unless delays are monitored, however, it is impossible to attribute the cause of changes in performance to the carrier or to users. The ability to measure internodal delay through the network management system allows the network specialist to request better routes from the carrier and to monitor the result of that change. If the source of delay is internal, the network traffic can be rerouted around points of congestion or restrictions can be placed on access.

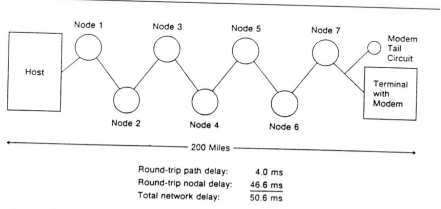

Round-trip path delay:	4.0 ms
Round-trip nodal delay:	46.6 ms
Total network delay:	50.6 ms

Exhibit II-3-14. Data Communications Through a Bit-Oriented Multiplexer at 2,400 bps

The Impact of Errors on Voice and Data

As noted, delays in voice transmission are more acceptable than delays in data transmission. Voice transmission can also tolerate more error. Should line impairments introduce errors in voice transmission over ordinary phone lines, the listener can compensate and still hear intelligible speech. Even when voice is digitized for transmission over high-grade leased lines, it is highly resistant to errors, largely because the listener can make the necessary adjustments. When the error rate exceeds a tolerable threshold, either party on the line may hang up and call back. It is likely that a better-quality circuit will be selected to handle the second call.

With data transmission, however, the only business machines that can compensate for errors are those equipped with special programs that are designed to check for errors and make the necessary corrections. But there is a price to pay for this intelligence. Some transmission protocols initiate retransmissions on error detection. When this happens often enough, throughput suffers, making response time objectives difficult to achieve. Even when retransmissions are not required, as in forward error-correction schemes, throughput suffers because extra time is required by the receiving device to perform the necessary processing and extra data is required for the forward error-correcting scheme to work. This, too, may make host-terminal response time objectives difficult to achieve.

Implications for Network Design

The differing transmission requirements of voice and data must be considered in any effort to build, expand, or upgrade the corporate communications network. Although it is technically feasible to combine voice and data over the same network, integration may be desirable only if they can be managed separately. The transport management system of a networking T1 multiplexer should be able to reroute data when the error rate exceeds a predetermined threshold and leave voice traffic where it is. In this way, the total load can be more evenly distributed over available facilities.

The capability to assign priorities to voice and data traffic is also important. For example, on a high-capacity network, two levels of service for data and a third for voice may be designated. Critical data would have the highest priority in terms of response time and error thresholds and would take precedence over other classes of traffic (e.g., voice and lower-priority data) for recovery. Because routine data can tolerate a longer response time and higher error rate, the point at which restoration is implemented can be prolonged. Because voice is more tolerant than data with regard to errors and delay, restoration may not be necessary at all.

The ability to assign priorities to traffic and reroute only when necessary ensures maximum channel fills, which improves the efficiency of the entire network and reduces the cost of operation. All these factors must be considered in the selection of a multiplexer.

BIT VERSUS BYTE ARCHITECTURES

If voice traffic must be moved through the public network, a byte-oriented T1 multiplexer is required. If data must be moved over private facilities, a bit-oriented T1 multiplexer can do the job faster and more efficiently.

Byte-oriented multiplexers take advantage of current and future carrier services that provide access to the public network through such services as the digital access and cross-connect system. The byte structure is also compatible with AT&T's extended superframe format, which is offered on its Accunet T1.5 circuits. Although it may be desirable to choose a byte-oriented multiplexer to take advantage of carrier service offerings or to position the network to do so at some time in the future, byte-only multiplexers are not without drawbacks.

The byte-oriented multiplexer can consume as much as 5% of the T1 bandwidth for overhead functions. This may not appear to be very significant until translated into dollars. For example, a 712-mile Accunet T1.5 circuit between Chicago and New York could cost $16,000 a month, including local loop charges at both ends. With a 5% overhead, the user could pay $800 per month for bandwidth that cannot be used. Put another way, the user is paying for 24 channels but is only able to use 23. These losses are compounded on a multinode T1 network and compounded further by any increases in distance.

For example, an Accunet T1.5 circuit between New York and San Francisco could cost $38,700 a month, including local loop charges at both ends. With 5% of the bandwidth consumed in overhead, the user pays $1,935 per month—and $23,220 per year—more for that circuit than necessary.

Take the case of several 56K-bps circuits on a network in which the bandwidth is managed through a byte-oriented multiplexer. Assuming that the multiplexer abides by AT&T's specifications for subchannel derivation, there are five 9.6K-bps data channels in the 56K-bps circuit. Notice that this subrate scheme immediately deprives the user of 14% of the bandwidth (8K bps), which is consumed in overhead. When this is added to the 5% of lost bandwidth that the multiplexer uses, the result is 19% of unusable bandwidth on a 56K-bps circuit. With multiple long-haul 56K-bps links going through multiple byte-oriented multiplexers, the annual dollar losses are nothing less than astounding.

A bit-oriented multiplexer, on the other hand, does not present such problems, because it uses all of the available bandwidth for production data. Using voice-compression techniques (e.g., adaptive differential pulse-code modulation [ADPCM]) to double the amount of voice channels on the circuit can derive a full 48 channels, instead of only 44 under the byte orientation. This is because when an ADPCM device is used for point-to-point applications, the robbed-bit mode can be invoked to yield the maximum throughput of 48 voice channels. In the robbed-bit mode, the eighth bit of each of the twenty-four 8-bit time slots is used to convey such voice-related signaling information as on-hook and off-hook status.

When the channels are switched in the bundle mode through a carrier's digital access and cross-connect system, however, only 44 voice channels are available over the T1 pipe, rather than 48 with the robbed-bit mode. This is because the remaining four channels are reserved for supervisory signaling and network management functions.

POSSIBLE SOLUTIONS

Bit and byte technologies involve trade-offs that affect the cost and flexibility of today's networks. Although bit technology provides maximum bandwidth efficiency and cost savings, byte technology permits access to public networks through the digital access and cross-connect system and the most advantageous use of carrier offerings. The ideal solution would be to have one multiplexer that takes advantage of both bit and byte architectures, allowing users to reap the benefits of private and public networks. Several vendors now provide ways for their multiplexers to do exactly that. These approaches, however, differ widely in efficiency and economy.

For example, vendor A might claim that its multiplexer operates in both proprietary and network-compatible environments. Closer inspection, however, could reveal a very clumsy implementation strategy. The vendor offers only a single-aggregate multiplexer intended for point-to-point configurations. When paired with another of its systems, the vendor can deliver on its promise of public network compatibility, but without an integral multiplex function. Interconnection of the system with a D4 channel bank does not provide channelization, but the resulting equipment configuration is severely limited by the channel bank's lack of overall network management. Although the system can be used to establish a backbone network with public network compatibility, the user cannot have channel terminations without investing further in the single-aggregate multiplexer. The purchase of two complementary systems is required for multinode solutions that include both public network and channel terminations.

Vendor B, on the other hand, might require the addition of only one card to its networking T1 multiplexer to deliver hybrid networking capability, rather than the addition of an entire box as required with vendor A. The difference is one of providing a systems-level solution as opposed to providing mere interconnectivity through any means possible, regardless of cost.

ACTION PLAN

Integrating voice and data over the high-capacity backbone network does more than just sound like a good idea, it is a good idea. Successful implementation, however, hinges on the capabilities of the multiplexer to monitor for errors and delay as well as to set priorities for and control voice and data separately. This way, the needs of MIS and telecommunications can be properly addressed without compromising the bottom-line objectives of the organization. The multiplexer must provide interconnectivity and management control through the public network so that users can benefit from new and existing carrier offerings, while extending the reach of private facilities economically.

Such capabilities are not offered by the latest generation of digital PBXs, which is why many users have not fared well with their voice and data integration strategies. To ensure success, a careful needs assessment must be performed to determine the current and foreseeable needs of MIS and telecommunications. A consultant should be used to bridge the communication gap if the two groups are not speaking the same language.

A detailed appraisal of vendor offerings is also required. Vendor assurances of voice and data integration are not enough; an evaluation of the multiplexer's transport management system is necessary. Specifically, users should look for:

- The ability to make qualitative measurements to help minimize the harmful effects of error and delay.
- The ability to manage voice and data separately and to set priorities for both types of traffic.
- A multiplexer that is compatible on both the channel and aggregate sides with the digital access and cross-connect system—This way, the organization can take full advantage of public network offerings while maintaining the benefits of private network management.

Chapter II-3-3
Communications Protocols

Jerry Gitomer

A COMMUNICATIONS PROTOCOL is a set of rules governing the transmission of information across communications lines. These rules control transmission speed, the number of data bits that each character comprises, what parity is to be used, which control characters will be used, and when the various control characters will be used. To communicate, both the sending and receiving devices must observe the same protocol, either directly through compatible systems software and hardware or indirectly through auxiliary protocol converters.

Whenever two devices must pass information back and forth, they must observe a common protocol. When a computer processor and a peripheral controller communicate, for example, both observe a common channel protocol; all devices attached to a local area network (LAN) or wide area network (WAN) would observe the LAN or WAN protocol; and devices that are connected by long-haul communications lines observe communications protocols. (Long-haul communications include all local and long-distance dial-up telephone and telegraph services.)

Although all devices require protocols to communicate with each other and the system processor may be forced to deal with several different types of protocols at any one time, each protocol is unique and accommodates a unique environment.

TYPES OF COMMUNICATIONS PROTOCOLS

The three basic types of communications protocols are asynchronous, synchronous, and bit synchronous. All other protocols are only proprietary variations or versions of the three basic types. With an asynchronous communications protocol, characters are sent one at a time and the two communicating devices need not be synchronized with each other. With a synchronous communications protocol, the two communicating devices are synchronized with each other and one or more characters are grouped into a message or packet, which is transmitted as a single entity. Bit-synchronous protocols are an updated form of synchronous protocols designed to fit the operating characteristics of today's higher-speed communications facilities. Exhibit II-3-15 lists the characteristics of each type of protocol.

Asynchronous Protocols

Asynchronous protocols were originally developed for sending telegraph messages by way of communications devices with alphanumeric keyboards that generated and transmitted encoded characters. Previously, messages were sent by Morse code, with each character's code pattern keyed in by the operator. The keyboard-based devices reduced error rates and enabled telegraph companies to hire less-skilled individuals at lower salaries.

The properties of asynchronous protocols as listed in Exhibit II-3-15 are attributable to the nature of the communications devices available before the computer. These devices lacked data buffers and processing logic. When the operator pressed a key, the associated bit pattern was immedi-

---

Asynchronous
Characters are transmitted one at a time.
Each character is prefixed with a start bit.
Bits within a character are transmitted at fixed time intervals.
Each character is suffixed with one or two stop bits.
There is no fixed time between the transmission of successive characters.
No error checking is specified in the protocol.

Synchronous
Two or more synchronizing characters precede each transmission, or message.
Characters are transmitted at a constant rate.
The receiving device synchronizes itself with the transmission rate of the transmitting device.
Each message has a control code that indicates the end of a message to the receiving device.
The receiving device must acknowledge every message.
Error detection is specified as part of the protocol.

Bit-Synchronous
Messages are enveloped in packets.
Each packet includes a header that describes the packet.
Messages can be broken up into several packets by the protocol handlers.
Packets may arrive at the receiving device in a different order than they are sent by the transmitting device.
Communication is synchronized.
Sophisticated error detection and correction techniques are specified as part of the protocol.

Exhibit II-3-15.   Protocol Characteristics

ately transmitted across the communications lines. When the communications device received a bit pattern, it determined what character the bit pattern represented and then triggered the print mechanism. Because the operator could not type at a constant speed and the communications devices could not store the characters keyed by the operator, a protocol was required to negotiate the variable time interval between characters. The lack of processing logic required that each character be preceded by a signal (i.e., the start bit) that was used to turn on the receiving communications device.

The lack of error checking in asynchronous communications protocols does not present a problem if the terminal device operates in full-duplex mode. In full-duplex mode, every character sent to the computer is echoed back to the terminal screen, where it is visually verified by the operator.

In addition to being used for long-haul communications, asynchronous protocols are used instead of channels and buses to connect terminals to micro- and minicomputers.

The most common asynchronous communications protocols are the basic TTY protocols as originally specified by the Teletype Corp more than 50 years ago, Digital Equipment Corp's VT+100 protocol, which is an outgrowth of the TTY protocol, and the American National Standards Institute (ANSI) X3.64 protocol. As so often happens in the data processing world, the legal standard, ANSI X3.64 in this case, has lost out to a de facto standard, Digital Equipment Corp's VT+100. The VT+100 protocol, by convention, encompasses the VT+100's screen controls as well as the actual communications protocol.

The X3.64 protocol, described and defined by the X3.64 ANSI subcommittee, specifies more than 90 functions. Vendors offering terminals that do not support an entire set of functions may still claim that their terminals are ANSI X3.64 terminals. If the subcommittee eventually specifies subset levels, as was done with the COBOL standard, purchasers will know that X3.64 terminals provided by different vendors will have the same characteristics and functional capabilities. In the meantime, terminals supporting the VT+100 protocol will continue to dominate the market because they can be purchased from several vendors and mixed or matched without difficulty.

Synchronous Protocols

Synchronous communications protocols were developed during the 1960s to provide more efficient and error-free communications for transaction processing systems. With these protocols, both the transmitting and receiving devices are assumed to have limited data buffers and built-in processing logic. An implicit design assumption is that most transmitted messages consist of several characters.

In addition to the major characteristics listed in Exhibit II-3-15, synchronous protocols have additional rules pertaining to the recognition and correction of errors. For example, every message that is sent using a synchronous protocol includes additional encoding, generally some form of cyclic redundancy check (CRC), that enables the receiving device to verify the correctness of the transmission. Each message is acknowledged (ACKed) if received correctly or negatively acknowledged (NACKed) if received incorrectly. Most synchronous protocols also define several special embedded control codes, which allow a computer program to exercise full control over the terminal devices.

Bit-Synchronous Protocols

Bit-synchronous communications protocols were developed during the 1970s to improve line use for data transmission in switched networks, to take full advantage of the capabilities of intelligent communications networks, and to eliminate propagation delay problems with satellite communications. (Propagation delay is the noticeable time increase that results from the additional distance that signals travel when satellite links are used.)

In a typical scheme, a message may be broken up into as many as seven packets. Because the packets may arrive at the receiving device in a different order than the one in which they were sent, each packet contains a packet identification number in the header, a flag indicating whether or not the packet is the last one in the message, and a message identification number. As each packet arrives, the receiving device applies the error-detection algorithm and attempts to correct any errors. If an error cannot be corrected, the packet containing the error is transmitted. The receiving device stores the error-free packets in a buffer as they are received. (All but the last packet in a message have one standard length.) When the last

Bit-synchronous protocols are designed to compensate for modern communications practices implemented by the various long-haul service providers (e.g., telephone, cable, telegraph, and satellite communications vendors). Modern switched communications networks may deliver messages in a different order than the one in which the service vendor receives them. This is because messages are broken up into packets and each packet is transmitted over whatever facilities are available at that time. Thus, some packets in a message may travel over satellite communications, others over cable and wires, and still others microwave. As a result, the transmission time for packets in the same message is not uniform. Bit-synchronous protocols and the communications devices that support them compensate automatically for this difference in sequence and in a manner undetected by users and other software components of the system.

packet is received, the device acknowledges receipt and makes the message available if all of the packets have been received; if they have not, the device requests retransmission. Finally, if a packet from a different message is received before all of the packets of the current message, the receiving device requests retransmission. This scheme requires significant processing capability and data storage memory at both ends of the communications line.

The most common bit-synchronous protocols are IBM's synchronous data link control (SDLC) and the US government's high-level data link control (HDLC). The differences between the various bit-synchronous protocols are so minor that many mainframe vendors support the IBM SDLC either in addition to or instead of their own proprietary protocols.

ADVANTAGES AND DISADVANTAGES

Asynchronous protocols are the simplest of the three basic types, because the sending and receiving devices need not be synchronized, characters are transmitted independently of one another, and the devices require virtually no intelligence or data storage. However, because they do not confirm the arrival or correctness of the information sent, asynchronous protocols are generally limited to low-speed (9,600 bps or lower) communications and the direct connection of terminals to micro- and minicomputers. Much of the micro- and minicomputer software has been designed under the assumption that every keystroke will be transmitted independently to the computer where the program can access it and take immediate action. Such programs will run inefficiently—and perhaps incorrectly—when the micro- or minicomputer supports remote terminals through synchronous or bit-synchronous protocols.

A worst-case example is a UNIX implementation on a large-scale mainframe with several remote terminals running the Vi screen editor program. Vi analyzes each keystroke it receives and makes an immediate response if appropriate. This is necessary because the terminal user can move anywhere in a file and on the current screen with terse commands. In response to a two- or three-keystroke sequence, Vi may have to reposition itself in a file and transmit a new screen image to the terminal. For example, the two-stroke sequence 1G causes Vi to move to the beginning of the file being edited and display the first screen of data from the file. The sequence $G causes Vi to move to the end of the file and display the last screen of data. No end-of-command, carriage return, transmit, or enter key is required. In this example, even the most efficient large-scale mainframe front-end communications programs (required to handle synchronous and bit-synchronous communications as well as asynchronous communications) execute more than 50,000 instructions per message. Because each individual keystroke must be treated as a separate message in a Vi program, it doesn't take very many Vi users to overload even a supercomputer.

Synchronous protocols (though significantly more efficient than asynchronous protocols when transmitting several characters at a time) are less efficient when small numbers of characters are transmitted at a time. Synchronous protocols, however, offer positive confirmation of the arrival and correctness of information. Synchronous protocols were designed with the expectation that dedicated leased lines would be used. Maximum message lengths (usually less than 128 bytes) are based on the high cost of data storage at the time the protocols were first defined. The requirement for the receiving device to positively or negatively acknowledge receipt of every message—even a 1-byte message—reflects such thinking and is one reason synchronous communications protocols tend to be inefficient unless the application is designed to optimize the protocol characteristics.

The following is a worst-case example illustrating what happens when a program is not designed around the peculiarities of a synchronous communications protocol. A major vendor offered an order entry program on a four-user small business computer. The processing of each line item on an order required four separate pairs of messages between the terminal and the computer. During the process, 180 bytes were transmitted; of the 180 bytes, 36 were data and 144 (80%) were control characters. Minor program modifications reduced the communications traffic from four pairs of messages to one pair of messages, decreased the number of bytes transmitted from 180 to 72, and more than doubled the number of terminals the computer could handle with adequate response time. Because most synchronous protocols include extensive rules for managing terminal devices and data transmission, the synchronous protocols of mainframe vendors—and their terminal devices—are generally unique.

Bit-synchronous communications protocols are designed to facilitate efficient transmission of large messages over switched networks. If that transmission is successful, an acknowledgment is not required until seven 128-byte packets have been received. The built-in compensation for packets arriving in a different order from the one in which they were sent is another means of achieving efficiency. In addition, the error-detection and correction facilities of these protocols are superior to those of the synchronous protocols. The primary disadvantage of bit-synchronous communications protocols is their significant amount of overhead per message, which makes the transmission of short messages inefficient when compared to other basic protocols.

CHOICES AVAILABLE

Unless an organization's interest in communications is limited to expanding its current network, it generally can choose any of the three basic protocols.

Some circumstances may warrant the use of more than one type of protocol. In general, most mainframe vendor terminals require either synchronous or bit-synchronous communications protocols; the native mode of most minicomputer terminals and personal computers is an asynchronous communications protocol.

Whether two devices are directly connected to each other or use a switched network, they must either support a common communications protocol or communicate through a protocol translator. Protocol translators can either be attached to the terminal devices that require them or to a communications processor through which all traffic is routed. If there are only a few incompatible devices or if they are all in one location, the least expensive method is to attach the protocol converter to each device or to a multiplexer serving the location.

If several locations require different protocols, a front-end communications processor that can perform the required protocol translations may prove the least expensive approach. For example, one installation uses Unisys front-end communications processors to connect several hundred asynchronous terminals, IBM synchronous terminals, Unisys synchronous terminals, and a high-speed broadband LAN to several Digital Equipment Corp, IBM, and Unisys computers. The front-end processors permit any terminal in the network to be connected to any computer system that it is entitled to access.

Although this installation is larger than most, the principle applies to most general-purpose computers and to any computer that can be used as a front-end processor.

In the case of a public data network such as Telenet or Tymnet, only those protocols supported by the network that are being used are applicable. These networks are usually intended to support low-speed asynchronous terminals, though their long-haul communications networks use bit-synchronous protocols. Public data networks can support these protocols by using communications concentrators, which accept asynchronous communications from their customers, packetize the information, transmit it using a bit-synchronous protocol, depacketize it at the receiving end, and then distribute it to the destination terminals using asynchronous protocols.

Some public data networks provide similar services for selected synchronous protocols. In some cases, they support bit-synchronous communications protocols directly from one customer site to another—provided that the customer is using the same bit-synchronous protocol that the service is using.

SELECTION CRITERIA

The best communications protocol is the one that performs all of the necessary functions at the lowest total cost. However, determining the lowest total cost is not easy. Cost, in this case, includes the communications devices, any communications processors, transmission, and undetected errors.

At any functioning level, terminals using asynchronous communications protocols are the least expensive. Typically, when vendors offer comparable asynchronous, synchronous, or bit-synchronous terminals, the asynchronous terminals sell for about half the price of the others. The cost difference is, for the most part, because the asynchronous terminal consists of nothing but off-the-shelf components while the synchronous and bit-synchronous terminals have custom keyboards and unique electronics.

Although it is not generally recognized, the nature of the terminal often has a significant impact on operator training costs. For example, the typical synchronous or bit-synchronous terminal, which requires the operator's understanding of several special-purpose keys that are meant to make the terminal more efficient, actually results in operator training costs that can easily exceed the cost of the terminal. For example, the airlines spend a full week training travel agents to look up flight schedules and to book seats.

Transmission costs depend on the number of bytes—including overhead and error retransmission bytes—that are transmitted. From a practical viewpoint, transmission errors usually depend on the quality of the service provided by the long-haul carrier. In most regions of the US, this equates to error-free 2,400-bps transmission over dial-up telephone circuits. For directly connected terminals, an error-free 19,200-bps rate is easily attained, provided the terminals and computer can support it.

The overhead for an asynchronous communications protocol is two additional bits (one start bit and one stop bit) per character transmitted. Thus, 10 bits are transmitted for each 8-bit character sent and the overhead rate, if there are no transmission errors, is fixed at 20%.

At best, a synchronous protocol requires the transmission of two synchronizing bytes, one CRC byte, one end-of-message signal byte, two additional synchronizing bytes, and one ACK or NACK byte for each message transmitted. Thus, 7 bytes are dedicated to overhead regardless of whether the message is 1 byte or 128 bytes long. In addition, synchronous communications protocols have a higher overhead than asynchronous communications protocols for messages of less than 36 characters.

Bit-synchronous communications protocols have an even higher overhead than synchronous communications protocols but are much more efficient and effective in dealing with errors.

Undetected errors are a problem with any communications protocol. Communications devices have excessive error rates compared with other devices attached to computer systems, and some types of errors, though rare, are undetectable by either the cyclic redundancy check of the synchronous protocols or the more sophisticated error correction codes of the bit-synchronous protocols. The real concern, however, is the asynchronous communications protocol operated in half-duplex mode, because it provides no error detection or correction and has a high probability of undetected errors. The question is whether such errors are significant and, if so, how to resolve the problem.

Errors that affect a particular application tend to be the most destructive. For example, if accounts receivable, inventory, or financial data is being transmitted, even one undetected error can prove costly. The most economical solution under these circumstances is to operate in full-duplex mode so that the computer repeats every keystroke it receives from the terminal. Managers should note that this effectively halves transmission speed.

Data communications managers should account for the cost of front-end communications processors as well as transmission overhead. For applications in which the messages are usually short (i.e., an average length of less than 64 bytes), asynchronous communications protocols are the least expensive. Conversely, if messages are generally long (i.e., more than 64 bytes) bit-synchronous communications protocols tend to be the least expensive. The industry is beginning to favor bit-synchronous communications protocols over synchronous communications protocols; however, progress is slow, because most of the mainframe terminals in service today are synchronous terminals that cannot be economically upgraded to bit-synchronous. In many cases, the investment in new terminals will exceed the savings that can be realized through the use of bit-synchronous communications protocols.

SUMMARY

If an organization uses only one computer vendor's equipment, the preceding selection criteria can aid in choosing from among the protocols that vendor supports. However, most organizations purchase equipment from more than one vendor and thus require a common communications protocol so that the various kinds of equipment can communicate with each other. Under these circumstances, the common protocol, even if it is not the best technical solution, should be the protocol of choice. The long-term benefits of using only one communications protocol more than compensate for any temporary operating inefficiencies. If no common protocol exists, it is advisable to standardize by using one vendor's protocol and either front-end communications processors or protocol translators. The objective is to permit any terminal in the organization to communicate with any of the organization's computers it is entitled to communicate with.

Chapter II-3-4
Electronic Mail Systems

Robert L. Perry

ELECTRONIC MAIL USED TO REFER to relatively expensive dedicated systems of little more than terminals for point-to-point delivery of short electronic memos—glorified telex machines, requiring terminal rental and message and use fees. During the past several years, however, electronic mail has advanced technologically, and it can now be broadly defined as any system that includes the electronic transfer of text, images, information, or voice from a sender to a receiver's electronic mailbox (i.e., a user or terminal location). For example, Texas Instruments links its 50,000 employees worldwide with electronic mail and uses its own satellite communications systems to handle message traffic. Employees at Software Publishing use 10 different electronic mail services to communicate with other employees, and sales and marketing personnel use electronic mail to communicate with outside salespeople and firms that do business with the software house.

TYPES OF ELECTRONIC MAIL

Eight types of electronic mail are available:
- National and international electronic mail services (e.g., Western Union, Tymshare's OnTyme, GTE Telemail, and British Telecom's Dialcom).
- Proprietary, in-house systems consisting of dumb terminals and a central computer dedicated to electronic mail and message management.
- Off-the-shelf electronic mail software for in-house mainframe or mini-computer networks, which allows users with access to terminals to communicate with other users on the same network—More than 30 software vendors offer such packages for their mainframe and mini-computer hosts.
- ASCII-based information networks (e.g., CompuServe, Dow Jones, and The Source) that employ integrated electronic messaging, electronic conferencing, and closed user groups.

- Combined electronic mail and traditional post office or hand delivery of documents—A user can create an electronic document and transmit it to a receiver electronically, by courier, or by postal delivery. All the major electronic mail services provide combined delivery.
- Integrated electronic mail and facsimile transmission—This method combines electronic messaging with electronic delivery, facsimile transmission, hand delivery, or a combination of the three. This service is available nationally and internationally.
- Electronic mail software for microcomputers—Each leading electronic mail service vendor offers its own package, and dozens of other packages have been introduced by independent software houses and computer manufacturers. However, almost all of these packages simply turn microcomputers into dumb terminal emulators that work with one or more electronic mail mainframe systems.
- An emerging second generation of electronic mail software for microcomputers—The new generation will be highly intelligent, working with multimedia messages (i.e., text, graphics, images, voice, data), will allow messages to be created and stored offline and will handle much larger messages than currently can be sent.

The X.400 MESSAGE HANDLING STANDARD

The International Telephone and Telegraph Consultative Committee (CCITT) X.400 Message Handling Standard established the first working model for interconnecting all electronic mail systems. The goal is to create interconnections among electronic mail systems that are as transparent as telephone calls.

Dozens of leading vendors have been working to introduce hardware and software that implements the X.400 models and recommendations, and X.400-based systems have begun to enter the market; many more are expected. As still-unresolved X.400 issues are decided, the volume of communications between different companies and between distant branches of the same company will increase dramatically. In fact, the long-range impact on business communications should be as great as that of the telephone and the communications satellite.

ELECTRONIC MAIL BENEFITS AND DISADVANTAGES

Electronic mail offers the following benefits:
- Accessibility—With the right terminal, users can access their mailboxes or be reached from anywhere in the world. As X.400 becomes a reality, electronic mail accessibility will increase.
- Reliability—Electronic mail service is quick, dependable, and perhaps more effective than the telephone network.

- Accurate retrieval—Electronic mail messages tend to be stored more accurately and for longer periods than telephone or personal messages.
- Broadcasting—Used with hand document delivery, electronic mail systems can reach almost anyone in hours or, at most, a few days. For example, electronic mail broadcasting can be used to arrange meeting schedules without dozens of unproductive telephone calls.
- Reduced need for meetings—Electronic mail can save time and reduce the demand for in-person business meetings.
- Telephone productivity—Electronic mail sharply reduces telephone tag and leaves a permanent record of exchanges.
- More productive travel time—Teleconferencing and message broadcasting can eliminate some travel and allow traveling managers to use their time more productively.
- Freedom from time constraints—Unlike telephone calls, electronic mail can be sent and retrieved at the convenience of both the sender and receiver, an especially important feature for organizations with offices in different time zones.
- Speed—Electronic mail sends information faster than any printed document or message delivery method and almost as fast as a telephone call.

A properly implemented in-house electronic mail network can save managers 8% of their time (i.e., approximately three hours per week) and 6% of a clerical employee's time (about 2.5 hours per week). In total salaries, this means managers can make better use of about $10,000 per year; support staff, about $2,000 per year. It is difficult to know whether employees would translate these savings into greater profits for the organization, but an electronic mail network makes these profits more likely.

Electronic mail has its disadvantages, however. In many organizations, it replaces more informal communication, including casual telephone conversations, personal meetings, and the office rumor mill. In these cases, employees merely use a newer (and more expensive) method to exchange relatively unimportant information.

Furthermore, employees often do not check their mailboxes for days or weeks, missing timely messages. Most electronic mail systems do not notify users of waiting messages; some do, particularly those operating through an office automation or mainframe system. Some electronic mail network providers offer telephone alert services that call recipients when an important message is waiting. And some users send messages electronically and hand-deliver them. In addition, MCI sells a device called a PostmatiQ that checks an organization's mailboxes and either downloads its information for storage or prints it for mail delivery. In the future, microcomputer electronic mail software will probably flash a light on the microcomputer screen or use a similar technique to remind users to check their electronic mail-boxes.

Another electronic mail shortcoming is its lack of standards. Although X.400 will provide a common model for access and message structure, currently each network has a different procedure and each software package has its own sets of commands. It is often impossible or very difficult to send messages to different systems; consequently, employees who use electronic mail often subscribe to many systems. Some microcomputer software (e.g., PFS Access from Software Publishing) performs all the necessary format conversions and translations to allow users to send one message to receivers on different networks.

Furthermore, most existing networks are text or character based and cannot transmit such binary information as spreadsheets, images, graphics, or text from many word processors. Many vendors are competing to solve this problem, but it will be several years before a standard becomes widely accepted. Although new technology will iron out many of the problems that have plagued electronic mail, an organization must still manage an electronic mail system carefully and consistently to achieve the best results.

ASSESSING THE NEED FOR AN ELECTRONIC MAIL SYSTEM

The first and most important step is to determine whether the organization needs electronic mail. This evaluation requires the cooperation of MIS, communications, office automation, word processing, the mailroom, and any other department responsible for business communications. The manager should question these departments regarding the number of locations, frequency and methods of corporate communications, effectiveness of such communications, and their cost. Organizations with many locations spending more than 25% of their time and telephone and mail costs on communicating with one another are prime candidates for electronic mail (e.g., insurance companies with many field representatives and pharmaceutical firms with large direct sales forces).

Next, specific departments and functions that could benefit from electronic mail should be identified. Some popular ones are discussed in the following sections.

Managerial- and Executive-Level Communications. Electronic mail could connect the organization's executives, key executive staff managers, and operations managers at all organization locations. A pilot program based on this application might link the CEO to the managers and executives who report directly to the CEO.

Sales Management and the Sales and Marketing Department. Combinations of electronic mail and hand-delivery systems can enable sales departments to communicate effectively with even their remotest sales representatives.

Supplier-Manufacturer or Vendor-Customer Communications. In organizations using such new inventory control methods as just-in-time (JIT) manufacturing and automated warehouses, electronic mail immediately transmits important information (e.g., orders, invoices, and information on shipments, deliveries, and delays). Similarly, electronic data interchange (EDI) can send purchase orders, invoices, bills of lading, and related information from terminal to terminal. In fact, General Motors and Ford are encouraging their suppliers to install and use EDI terminals instead of telephones.

Interdepartmental and Cross-Functional Project Team Communications. Organizations with project teams spread across many buildings or locations can use electronic mail to exchange ideas quickly and easily.

Vendor-to-Dealer Communications. Apple Computer uses this application to disburse product and company news through General Electric's Quik-Comm network.

Company–to–Trade Association Communications. Many trade associations have established electronic mail networks for their members to communicate with others in the same industry. Members of professional groups can use electronic mail to discuss their shared responsibilities.

KEY MANAGEMENT ISSUES

In many organizations, the implementation of an electronic mail network requires the close cooperation of the MIS, communications, and office automation departments. To avoid conflict, the various departments must decide which executive will have final authority over the network, to which budgets electronic mail expenditures will be charged, who will supervise and direct any new employees hired to manage the electronic mail network, who will supervise network maintenance, and who will act as liaison between the organization and public electronic mail network providers.

Many organizations try to keep management of electronic mail simple. For example, an organization with a mainframe-based electronic mail system might make electronic mail an adjunct to MIS; in this case, electronic mail is simply another software application to maintain. Organizations that use public networks may keep authority at the departmental level and charge electronic mail costs to departmental telephone or communications budgets.

An electronic mail network must have complete support of executive management and preferably the board of directors, because an extensive electronic mail network can change an organization's operations as dramatically as the telephone and microcomputer have. MIS, communications, and executive management must also consider the added demand of an electronic mail system on an organization's information processing and communications resources. An extensive in-house network can substantially increase demand on mainframe ports and storage capacity; electronic mail packages for microcomputers can increase their memory requirement. It may be necessary to buy many plug-in memory boards, modems, communications, or electronic mail software packages.

In addition, the demand for and costs of electronic mail system management, maintenance, and security must be considered. With critical organization marketing and sales information being sent across networks, the communications or MIS manager may be held responsible for the security of this information.

In the future it may be necessary to integrate in-house electronic mail with public networks. Currently, these connections are difficult to make, but X.400 standards will make this task much easier during the next several years; the communications manager should be sure to track these improvements.

SELECTING A NETWORK

The organization should next analyze its installed base of mainframe hosts, minicomputers, terminals, microcomputers, and workstations with communications capabilities. Organizations with many 3270 terminals but few microcomputers may lean toward using a mainframe-based electronic mail package. An organization with an extensive single-vendor office automation system may want to add that vendor's electronic mail utility. Clearly, most organizations retrofit electronic mail systems onto existing hardware.

The organization's communications capabilities should also be determined. Many organizations already have gateways and bridges between their mainframes, minicomputers, and microcomputers that can be used for an electronic mail system, but compatibility problems can still occur with the electronic mail software packages running on the various systems. Using public electronic mail networks avoids these compatibility problems but adds cost. Exhibit II-3-16 provides a checklist of issues to examine when selecting an electronic mail system.

Organizations planning to establish in-house systems can choose to run their systems either as time-sharing service bureaus on a dedicated machine or as utilities for users already connected to existing machines. Dedicated machines can support a maximum of about 30 users per mainframe communications port. An electronic mail utility package with a system of hardwired users, however, can support as many users as the number of terminals connected to the system. This is an important cost consideration.

When evaluating an electronic mail service for their organizations, MIS managers should consider the following factors.

Current Environment
- Computer systems:
  —Number and type of mainframes, terminals, and communications links.
  —Number and type of minicomputers, terminals, and links.
  —Number, types, and communications capabilities of microcomputers.
  —Interconnections among mainframes, minicomputers, and microcomputers.
- Communications patterns:
  —Number of facilities, offices, and factories.
  —Employees, terminals, and microcomputers at each location.
  —Average daily messages (including telephone, mail, and electronic messages) among all organization locations.
  —Annual cost of interoffice telephone, mail, and document delivery.
  —Annual cost of outside communications with suppliers and customers.

Vendor Electronic Mail Systems
- Is new hardware required? What kind, and how much will it cost?
- Is new software required? How many users will it accommodate?
- What are the maintenance, service, installation, and training requirements?
- Does the system provide interconnections to public networks? What are the requirements and cost?
- Is it expandable and easy to use, and will the vendor provide future software updates?
- What are its security procedures?
- Are communications protocols supported?
- Does it provide text editing functions?
- Are there automatic functions for sending and receiving messages?
- Is there a maximum or minimum number of system users or a point at which the system is no longer cost-effective?

Public Electronic Mail Networks
- What are the fee or charge methods (e.g., online time or connect cost; character storage, transmission, or receipt; monthly subscription fee; and message delivery)?
- What is the cost per unit and the number of characters per unit?
- What are the security procedures?
- What are the editing functions?
- Does the system provide interconnections with microcomputer software?
- Does it offer automatic forwarding, batch transmission, or file conversion?
- Can it provide multiple receiver transmissions?
- Does it offer automatic message notification?
- Does it have a telex interface?
- Can it provide access to public data bases? At what cost?
- Can it provide access to wire services? At what cost?
- Does it allow hand delivery of documents? At what cost?
- Does it provide a facsimile delivery interface? At what cost?
- Are other states and countries served?
- Does it have a forms generation capability?
- Can it transmit binary data files?

Exhibit II-3-16.   Electronic Mail Service Evaluation Checklist

Mainframe electronic mail software can cost $15,000 or more, but if an organization has hundreds or thousands of terminals, the cost per user will be very low. Furthermore, software requires relatively little mainframe communications and mass storage resources, and the same cost per user applies to any host-, local area network–, or minicomputer-based network.

ESTIMATING NETWORK COSTS

Estimating network costs is not easy because many costs are difficult to understand, especially the fees charged by public network providers. Public network costs can include time spent online (also called connection charges); charges per character, block, or unit of characters transmitted; and message delivery, message storage (per message or character), general, and administrative charges. Because many services define their charges differently, it is very difficult to compare services directly.

In addition, because services are often aimed at different markets, an organization may need to use a particular service, even though it is not the least expensive. For example, some services have a strong international network, whereas others act more as a connection to Telex and other services. When estimating electronic mail network costs, the communications manager should consider the costs outlined in Exhibit II-3-17.

In-House Network
Mainframe electronic mail software
Minicomputer software
Necessary gateways or bridges
Workstation software
Workstation hardware boards and modems
Additional host storage
Additional host processors
Additional host communications or front-end processors
Additional printers
Personnel training
Outside consulting and planning fees
Additional electronic mail network management staff
MIS staff and management time and involvement
Communications staff and management time and involvement
Executive time and involvement
Telephone line and local or long-distance call charges

Public Electronic Mail Services
Use charges
Time charges
Message or character transmission charges
Character or message storage
Long-distance or local telephone charges
Administrative, general, or start-up charges
Combined electronic and hand delivery of documents costs
Special services (e.g., teleconferencing or closed user groups)
Monthly subscription fees
Dedicated electronic mail terminal lease, fee, or rental costs

Exhibit II-3-17.   Network Costs

IMPLEMENTING AN ELECTRONIC MAIL NETWORK

The next step is to develop a pilot program to demonstrate the benefits that electronic mail can bring the organization. The key to such a program is to choose a department, function, or activity in which an immediate, visible gain can be realized. Pilot program expenses should be minimized by renting, leasing, or borrowing software and by using existing resources (e.g., terminals, microcomputers, modems, and communications software) whenever possible. During the program, representatives from other departments should be brought in to review how the system works, and proposals should be solicited from each department, suggesting how its staff could benefit from an electronic mail system. During this step, support must be generated among top executives, and employees should become familiar with the system and shown how it can help them with their work.

A pilot program should last from six months to a year. If successful, the program should be gradually expanded from one department to another throughout the organization. It is most effective and cost-efficient to expand the program through the installed base of terminals and microcomputers, because trends indicate that workstations will eventually replace most mainframe terminals. However, an organization's timetable for replacing terminals with microcomputers should determine this matter.

Organizations planning significant changes in MIS resources during the next several years may find it simple and inexpensive to make electronic mail a small part of the complete changeover. Electronic mail functions should be designed into such major changes as the installation of a large-scale office automation system or a microcomputer network. IBM's document exchange standard, LU (Logical Unit) 6.2, and its Document Interchange and Document Content Architectures (DIA and DCA) will gradually transform electronic mail functions, permitting the transmission of not only messages but also programs, digital data, images, and graphics.

Organizations that communicate heavily with suppliers and customers should also investigate how to integrate their in-house systems with outside services. For example, microcomputers can be used to save on outside service charges by creating and storing messages offline and then transmitting them in a rapid burst at the most advantageous and least expensive time. In the past, a head office transmitting a daily sales report to 50 branch offices had to both compose the message and enter instructions to send it online, at high expense. Currently, such organizations can save significant portions of these time and storage charge costs.

An organization with electronic mail in only some of its branches can save time by combining electronic and hand-delivery methods. The same message can be sent to all the branches because the public service distinguishes which versions should be delivered electronically, by regular courier, or by facsimile.

THE FUTURE OF ELECTRONIC MAIL

The future of electronic mail is bright. Microcomputers and workstations will become the dominant electronic mail terminals. The X.400 standard will eventually make all networks so transparent that users will not know or care whether the microcomputers they are using are connected to an in-house host, a public network, or a facsimile network. As electronic mail networks using the X.400 standard proliferate worldwide, directories of electronic mail users will become as prevalent as telephone directories. Public vendors, which already offer numerous electronic mail services, will continue to add services as the fierce competition continues.

Electronic mail workstations will perform an increasing number of network functions. Second-generation electronic mail software will handle all forms of digital information, provide automatic dialing and answering, define all network parameters, and determine the most cost-effective route for each message to follow—much like least-cost routing in telephone PABX systems. Electronic mail software will also handle such program and data conversions as hard and soft carriage returns, which will also save end-user time and effort.

Eventually, however, electronic mail will be absorbed into the growing drive toward transparent data interchanges among different and previously incompatible systems. IBM's strong push toward document-exchange standards and the rapid acceptance of this trend by key competitors herald the day when electronic mail will become just another file transmitted along with programs, facsimile, graphics, voice, and video among many different hosts all over the world. This will not be realized overnight, and organizations that can discern clear benefits from implementing electronic mail now should do so, but they should purchase systems that can grow and change with the rapidly evolving electronic mail technology.

ACTION PLAN

The abundance of electronic mail services and technology can be confusing. Managers evaluating the many electronic mail opportunities should do so carefully, ensuring that the systems they choose can accommodate technological change and easily integrate microcomputers and workstations. They should select an electronic mail system to serve a defined business need that will bring immediate and continuing productivity improvements and a provable return on the organization's investment. They should also plan the electronic mail system to coordinate with the increasingly interconnected worldwide networks of electronic mail services.

Chapter II-3-5
Trends in Fiber-Optic Technology

Otto I. Szentesi

DURING THE PAST 10 YEARS, fiber-optic transmission has become the medium of choice in telephony trunking applications, including long-haul and interoffice trunks, underwater systems linking continents, feeders connecting digital remotes to central offices, and metropolitan area networks. More recently, fiber-optic transport systems have become a key component of premises networks. In the context of this chapter, premises includes campus, interbuilding, and intrabuilding applications.

The successful premises communications designer must pay careful attention to the strategic benefits of incorporating fiber optics into a system's plan. It is both economically sound and practical to use fiber optics as the backbone of many types of premises systems. Estimates show that in the US, approximately 80% of buildings under construction that measure more than 100,000 square feet will have fiber-optic backbone networks.

Fiber optics is a communications technique that transmits light through a hair-thin glass fiber rather than transmitting electrical currents through conventional copper wire. In this process, electrical energy from a computer interface is converted to light energy by a light source, usually light-emitting diodes (LEDs) or semiconductor injection lasers.

Once emitted from the light source, light is coupled into the optical fiber, which guides the lightwave. The center element of the optical fiber is called the core and the outer structure is the cladding (see Exhibit II-3-18). The light beam is transmitted through the fiber because the refractive index of the core exceeds that of the cladding.

There are three types of fiber: step index multimode, graded index multimode, and single mode (see Exhibit II-3-19). In step index multimode fiber, the refractive index changes abruptly at the interface between the core and the cladding. Light is guided through the fiber in multiple paths, or modes.

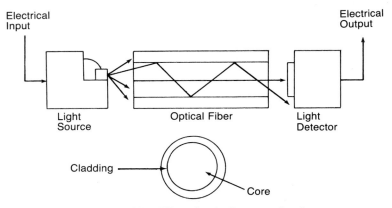

Exhibit II-3-18.   Fiber-Optic Communications

As the light beam approaches the cladding, the cladding acts as a mirror and reflects the light back into the core. Light rays traveling at different angles to the fiber axis travel different path lengths in traversing a finite length of fiber. This causes the signal to spread out along the fiber and limits the bandwidth in step index multimode fibers to approximately 25 MHz-km. (MHz-km is the product of bandwidth multiplied by distance.)

In graded index multimode fibers, the refractive index of the core decreases toward the core-to-cladding interface. Through tailoring of the index profile, all modes can have virtually the same net velocity along the fiber. The bandwidths of graded index fibers can exceed 1 GHz-km.

In single-mode fibers, the core size is small, as is the difference between the core and cladding, and only one mode is allowed to propagate. The bandwidths of such fibers can exceed 10 GHz-km.

Light exiting the fiber is projected onto a light-detecting device, which produces an electrical current. Light-detecting devices are either P insulated N-channel (PIN) diodes or avalanche photodiodes (APDs). The PIN diode produces electrical current in proportion to the amount of light energy projected onto it. The APD is a more complex device that amplifies light energy while converting it to electrical energy. This amplification process is important in a system that spans long distances between amplifying stages.

BENEFITS OF FIBER OPTICS

As with other technologies, a fiber-optic system must be carefully examined for its cost-effectiveness in typical installations. Nevertheless, fiber-optic cables provide unique performance capabilities in many applications. Fi-

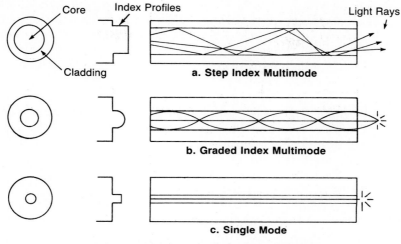

Exhibit II-3-19.   Common Types of Fiber

ber-optic cables offer features superior to those of twisted-pair wires or coaxial cables. Several of those features and their advantages are:

- Wide bandwidth—Increased data carrying capability, future expansion capability, and low cost per channel.
- Low attenuation—Longer transmission distances.
- Immunity to electrically generated noise—Reliability and a lower bit-error rate; no crosstalk.
- Resistance to taps—System security.
- Total electrical isolation—Elimination of ground loops.
- Compactness—Space and weight savings and lower installation costs.
- Environmental stability—High resistance to environmental factors, such as fire or corrosion.

These advantages are discussed in the following sections.

Wide Bandwidth

The greatest advantage of fiber-optic cable is its wide bandwidth. The ultimate information-carrying capacity of any transmission system is limited by the carrier frequency. Therefore, the theoretical upper limit for lightwave transmission is on the order of 10,000G bps. Practical systems operate far from the theoretical limit, but commercially available systems can transmit data rates in excess of 1.7G bps. Such high rates allow both high-speed data transfer between processors and the multiplexing of several low-speed channels for subsequent transmission over the fiber-optic medium.

Whereas the limited bandwidths of twisted-pair wiring and coaxial cable force computer networks into parallel transmission to obtain a desirable throughput rate, the larger bandwidth of fiber-optic cable enables them to transfer data serially by multiplexing a 16-bit data bus onto one cable. This reduces connector hardware and wiring requirements. The bandwidth capabilities of fiber-optic cable far outstrip those of twisted-pair wiring and exceed those of even the most expensive coaxial cable. The most publicized applications using this bandwidth capability are telephone company installations, which use fiber-optic techniques that allow the multiplexing of 24,000 voice-grade channels on a single optical fiber. In the laboratory, using advanced multiplexing techniques, the equivalent of 400,000 simultaneous telephone calls (approximately 26G bps) have been transmitted on a single piece of fiber-optic cable.

As shown in Exhibit II-3-20, signal attenuation in a fiber-optic cable is relatively independent of frequency when compared with signal attenuation in a copper-wire system. Signal attenuation in coaxial cable, even for the highest grade, increases rapidly with frequency. By contrast, attenuation in the optical waveguide is flat over a range exceeding 1 GHz in some cases. Fiber-optic cables containing graded index fibers are currently being manufactured and used with bandwidths in excess of 1 GHz-km, and single-mode cables are available with bandwidths greater than 10 GHz-km. It must be emphasized, however, that the useful bandwidth of optical fibers is

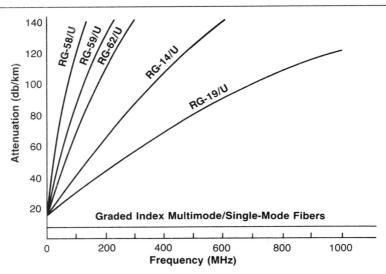

Exhibit II-3-20. Comparison of the Bandwidths and Attenuation of Coaxial Cable and Optical Waveguide

strongly influenced by the characteristics of the light source, primarily its wavelength and spectral width.

Low Attenuation

Optical fibers do not transmit all wavelengths of light with the same efficiency. The attenuation of light signals is much higher for visible light (i.e., wavelengths from 400 nm to 700 nm) than for light in the near-infrared region (i.e., wavelengths from 700 nm to 1,600 nm). Even within this near-infrared region, there are wavelength bands of decreased transmission efficiency, which leave only several wavelengths at which fibers can operate with very low attenuation. These wavelength regions that are most suitable for optical communications are called windows. The most commonly used windows are found near 850 nm, 1,300 nm, and 1,550 nm.

Multimode fibers currently sold are designed for operation at 850 nm (i.e., first window), 1,300 nm (i.e., second window), or both (i.e., dual window). Characteristically, the attenuation of a dual-window fiber is lower at 1,300 nm than at 850 nm. The bandwidth can be equal in both windows or larger in the second window.

Single-mode fibers have lower attenuation and higher bandwidth than multimode fibers. They are usually designed for operation at a wavelength of 1,300 nm. However, the wavelength of 1,550 nm is gaining acceptance in long-haul telephony applications because the attenuation is lower.

The size of optical fibers is typically designated in micrometers by the core diameter followed by the outside diameter of the cladding. For example, 8.7/125 indicates a single-mode fiber with a core diameter of 8.7 μm (0.0003 in) and an outside diameter of 125 μm. Common multimode fiber sizes are 50/125, 62.5/125, and 100/140 μm. Usually, the smaller the core size, the higher the available bandwidth; the larger the core size, the more easily light passes into a fiber. These fibers are then coated with plastic and cabled. The typical optical performance ranges for fibers are summarized in Exhibit II-3-21.

NOISE IMMUNITY

Fiber-optic cables are essentially immune to electrically generated noise; radio interference from electric motors, relays, power cables, or other inductive fields; and radio or radar transmission sources. Such immunities are significant because they reduce bit-error rates. A typical system using fiber-optic cables has a bit-error rate exceeding 10^{-9} as compared with the 10^{-6} bit-error rate usually found in metallic connections. Electromagnetic noise does not affect the fiber-optic cable; therefore, its bit-error rate depends only on the signal-to-noise ratio within the fiber-optic system. In

Fiber Size (μm)	Wavelength (nm)	Attenuation Range (db/km)	Bandwidth Range (MHz-km)
8.7/125	1,300	0.35–1.0	20,000
	1,550	0.25–1.0	
50/125	850	2.7–4.0	400–1,000
	1,300	0.8–3.0	400–1,500
62.5/125	850	3.3–5.0	100–200
	1,300	0.9–3.0	200–800
100/140	850	4.0–6.0	100–300
	1,300	2.0–5.0	100–500

Exhibit II-3-21.   Optical Performance Ranges

addition, because of the higher noise immunity, a system designed with fiber-optic cables does not require as many error checks as a wire system. Not only are overall data transfer rates increased by dramatically reducing error-checking overhead, but overall system performance is improved by reducing retransmission.

Fiber-optic cables do not aggravate the electromagnetic environment in which they are installed—a factor that can help simplify system design. The nonradiation features of fiber-optic cable enhance system reliability by virtually eliminating crosstalk and thus reducing the bit-error rate. Inter-channel isolation of 90 decibels or more is readily attainable in a multifiber optic cable. This feature becomes cost-effective for applications used, for example, in train depots where the electronic monitoring of trains is momentarily interrupted by electrical or diesel electrical generations and in manufacturing plants that have equipment driven by electrical motors or arcing equipment in the processing plant.

Security

Security is an important consideration. Copper wire can be tapped by directly connecting to the wire or wrapping a coil around the wire. An optical link, however, can be tapped only by directly accessing the fiber, and this is difficult to accomplish. If a readable amount of light is removed from a fiber, a power loss in the fiber-optic system occurs and can be easily detected.

In addition, because the fiber-optic cable is nonconductive, signals and electromagnetic noise do not radiate from it, and the cable resists conventional tapping techniques, reducing its potential as a security problem. These security features become cost-effective in government, military, banking, and commercial applications that require secure data transmissions.

Isolation

Fiber-optic cables isolate transmitters and receivers, eliminating the need for a common ground. The nonconductive qualities of fiber-optic cables benefit data communications systems because the equipment is totally isolated from one point to another. Because the cable is made of glass or plastic, electrical isolation problems, such as ground loops in an installation, are eliminated and the amount of noise in the electronic system is decreased.

Moreover, because the nonconductive qualities of fiber-optic cables prevent such dangers as a short circuit or spark, these cables are suitable for systems installed in dangerous gas atmospheres, such as petroleum refineries or chemical processing plants. In contrast, a copper-wire connection between remote pieces of equipment (e.g., between an electronic-based test station or recording instrument and a central processor) increases the danger of a spark causing an explosion or fire.

Space and Weight

The small diameters of fiber-optic cables offer substantial size and weight advantages over metallic cables of equivalent bandwidths. For example, a 10-fiber cable has less than a 3/8-in diameter.

The tensile strengths of available single-fiber cable, an RG-58/U coaxial cable, and a twisted-pair cable are 90, 60, and 40 pounds, respectively. The small size and light weight of the fiber-optic cable facilitate shipping, handling, and storage and solve the problem of pulling cable through crowded ducts. Weight and size criteria are important in aircraft applications, for example, in which lighter weight and smaller size can increase the overall operating range or payload of the aircraft. Fiber-optic cable usually can be installed with little or no difficulty in applications whose crowded conduits prevent installation of bulkier metallic cables.

Environmental Factors

The use of fiber-optic cable is generally less restricted by harsh environments than that of its metallic counterparts, and it is not as fragile or brittle as might be expected. Fiber-optic cables usually can be protected by proper jacketing of the signal-carrying glass or plastic member. For example, high tensile strength can be provided by the strength of members surrounding the fiber and within an outer protective jacket. Fiber-optic cable can be manufactured to solve most cabling problems, including those in aerial, burial, underwater, duct, and in-building applications.

Fiber-optic cable is more corrosion resistant than copper wire because it contains no metallic conductors. Hydrofluoric acid is the only chemical that affects optical fiber. Corrosion resistance is a particular concern at splicing locations, where complete protection from the environment is difficult. In addition, optical fiber can withstand greater temperatures than copper wire can. Even when fire melts the outside jacket of the surrounding fiber, a fiber-optic system usually can still operate.

FIBER-OPTIC STANDARDS

Significant progress in standardization is accelerating the use of fiber optics in premises applications. The Electronic Industries Association's (EIA) Working Group on Commercial and Industrial Building Wiring Standard (i.e., TR41.8.1) has defined two media—62.5/125 μm fiber and copper. Fiber is seen as the medium of choice for backbone systems.

A summary of current local area network (LAN) systems is given in Exhibit II-3-22. Institute of Electrical and Electronics Engineers (IEEE) Standard 802.3 describes a baseband LAN with a bus topology. It is essentially a formalization of Ethernet. The IEEE 802.3 Committee has established the 10BASE-F Task Force to standardize fiber-optic Ethernet networks.

IEEE Standard 802.5 describes a baseband network with a ring topology and a token-passing accessing technique. It is a formalization of the proposed IBM LAN and calls for a 4M bps LAN running on twisted-pair, coaxial cable, or fiber optics. IBM has released a 16M bps version of its Token Ring, and the 802.5 Committee has accepted proposals from the 802.5J Task Force for fiber-optic Token Ring standards.

CURRENT FIBER-OPTIC APPLICATIONS

Fiber-optic applications in the data communications environment can be classified into two broad categories: point-to-point and LANs. These categories are discussed in the following sections.

Point-to-Point Applications

Currently, most of the fiber-optic communications systems are point-to-point applications. These consist of two nodes, or communications devices, that communicate directly and exclusively with each other. To operate, most point-to-point applications require a fiber pair (one to transmit and one to receive), except in cases such as one-directional video applications.

Point-to-point applications for fiber optics exploit the fiber's bandwidth and low attenuation, allowing a signal to be sent longer distances at a faster speed than is possible on coaxial cable. (Coaxial cable is the copper equivalent to which fiber is most frequently compared.)

Other typical point-to-point applications benefit from fiber's intrinsic advantages, such as immunity from electromagnetic interference, lightning

LAN Type	Status of Standard	Access Method	Topology	Signaling Scheme (Data Rates)	Transmission Media	Application	Commercial Example
802.3	Published 7/85	CSMA/CD	Bus	Baseband (50M bps) Broadband (5M bps, 10M bps)	75 Ω Coax 50 Ω Coax (Thincoax) Broadband Coax Twisted-Wire Pair* Optical Fiber*	Front-End LANs	Xerox Ethernet
802.4	Published 7/85	Token Passing	Bus	Broadband (5M bps, 10M bps)	Coax Optical Fiber*	Manufacturing Control Systems	MAP
802.5	Published 7/85	Token Passing	Ring	Baseband (4M bps, 16M bps)	Twisted-Wire Pair Optical Fiber*	Front-End LANs	IBM Token Passing Ring
802.6	Under Development	Slotted	Ring	Baseband (12M bps, 50M bps)	Optical Fiber	Metropolitan Area Networks Back-End LANs	None Available
FDDI	Under Development	Token Passing	Ring	Baseband (100M bps)	Optical Fiber	Back-End LANs Backbone Transmission	None Available
FDDI-II	Under Development	Token Passing plus Dedicated TDM	Ring	Baseband (100M bps)	Optical Fiber	Wideband Voice/Data Systems	None Available

Notes:
CSMA/CD   Carrier-sense multiple access/collision detection
FDDI   Fiber distributed data interface
TDM   Time-division multiplexed
*Under consideration

Exhibit II-3-22. Summary of Current LAN Systems

effects, and ground loops, and its superior security characteristics. Even at speeds acceptable for coaxial and twisted-pair wire, fiber optics is used when the small cable size allows significant cost savings in cable installation. Some point-to-point applications include the following:

- Channel extender—A fiber-optic channel extender allows a computer mainframe controller to communicate directly with another controller or other peripherals at much greater distances than copper. It also allows for such things as a remotely located, high-speed, channel-attached printer, which was not possible before fiber optics.
- Voice/data multiplexer—A fiber-optic multiplexer combines several communications channels into a single bit stream for transmission over the fiber; channels are then separated at the other end. Common applications include T1 communications (i.e., 24 multiplexed digital voice channels), large PBX internodal links, and RS-232 and 3270 transmission.
- Modem—A fiber-optic modem converts electric signals to fiber-optic signals. Usually, fiber modems are used to extend distances between computers and peripheral devices (e.g., a microcomputer printer sharing over RS-232). Fiber optics can eliminate the antenna effect of comparable copper runs whereby electric heaters, power surges, and fluorescent lights can sometimes cause errors.
- Computer-aided design and manufacturing (CAD/CAM)—Fiber-optic links are often found in these systems, in which three-dimensional high-resolution graphics require high data rates.
- Video—Fiber-optic links are often used for video links. Broadcast-quality video transmission over fiber is not susceptible to ground loop and hum problems and does not require the use of equalizers.
- Security video—Fiber-optic point-to-point links are often used in outdoor environments to minimize lightning damage and interference.
- LAN bridging—Communication between LANs is often implemented through the use of a fiber-optic point-to-point bridge between two or more copper LANs. Fiber extends the permissible distance between bridges used for this purpose.

LAN Applications

Although less common than other applications, the installation and use of fiber-optic LANs is growing rapidly. In a fiber-optic LAN, more than two nodes communicate with one another over the fiber-optic cable. The nodes are logically arranged in a ring, star, or bus configuration. Some examples of fiber-optic LANs include:

- Ethernet.
- Token ring.

- Time-division multiplexed (TDM) ring.
- Fiber distributed data interface (FDDI).

PERFORMANCE ADVANTAGES

A useful formula for calculating the performance capabilities of twisted-pair wiring and coaxial and fiber-optic cables is the product of the bandwidth multiplied by the distance. The standard bandwidth/distance parameters are 1 MHz-km for common twisted-pair wiring, 20 MHz-km for coaxial cable, and 400 MHz-km for fiber-optic cable. A general cost/performance factor can be computed by taking the average cable cost (10¢ per ft for twisted-pair wiring, $3 per ft for coaxial cable, and 75¢ per ft for fiber-optic cable) and dividing the bandwidth/distance parameter into the cost per kilometer. The following relationship between the cost per Megahertz and the bandwidth for a kilometer shows that the fiber-optic cable has a definite cost/performance advantage over its metallic counterparts:

- Twisted-pair wiring—$300 per Megahertz per km.
- Coaxial cable—$450 per Megahertz per km.
- Fiber-optic cable—$5 per Megahertz per km.

The cost of a typical installation of line drivers or limited-distance modems (LDMs) can also be compared with that of a fiber-optic installation. For applications requiring distances greater than that supported by the RS-232, 50-ft-limit line drivers are usually used. Line drivers are required to overcome the signal distortion that limits the distance and speed attainable with cable. Extended-distance transmission produces pulse-rounding—a condition in which the edges of a square wave pulse are distorted because the high-frequency element is lost. In addition, transmission over extended distances increases signal attenuation, resulting in marginal reception and lost or erroneous bits. These adverse effects may be partially overcome by upgrading from twisted-pair wiring to coaxial cable. Cost/performance trade-offs can result in the selection of line drivers for each end of the interface to improve system performance. Line drivers usually operate at speeds up to 19.2K bps. Special units that operate at data rates up to 1.5M bps are also available; distance may range from 100 ft to several mi.

If line drivers cannot be used to solve a communications problem, LDMs are a less expensive and simpler alternative to a conventional telephone modem. The cost advantage of LDMs increases with the data rate because the three major functions performed by conventional modems can be relaxed. At low data rates, the cost difference is not great; however, at high speeds, the cost difference between an LDM and a telephone modem can be thousands of dollars. LDMs operating in a full-duplex mode require four-wire systems. Most LDMs use twisted-pair wiring; however, some transmitting at higher data rates (from 19.2K to 1M bps) use coaxial cables to eliminate crosstalk and electromagnetic interference.

When customer-owned cable is used within a link interfacing an LDM, either twisted-pair wiring or coaxial cable can be used. Selecting the cabling technique to use with LDMs is a cost/performance trade-off.

In addition to direct connection, line drivers, and LDMs, new interfacing techniques that use fiber-optic cables are available. Some companies are producing RS-232–compatible standard links for interfacing MIS equipment to data terminals. The fiber-optic data link converts an RS-232 standard signal to optical-encoded information, which is then transmitted over a fiber-optic cable and reconverted from optical data back to the RS-232 standard data format. The RS-232 fiber-optic links available are plug-compatible with the RS-232 standard interfaces.

Cost Analysis

Fiber-optic data links are less expensive than drivers and LDMs and outweigh them in performance. When a remote terminal is required to operate at higher data rates (e.g., 19.2K bps), a fiber-optic data link system has the expansion capability to meet this requirement. For the line driver and LDM, however, end modules must be replaced with a higher-performance device. In addition, the twisted-pair wiring cannot usually provide the performance capability necessary to transmit at these higher data rates. Therefore, along with replacing the end links, the user must upgrade from twisted-pair wiring to coaxial cable.

PLANNING FOR A FIBER-OPTIC SYSTEM

With the increasing use of data communications and accelerating communications speeds, fiber-optics technology can be cost-effective in many applications. Although a fiber-optic system may be slightly more expensive to install than a wire-based system, its expansion capability justifies its application. When planning for a fiber-optic system, MIS managers must consider the variables described in the following sections.

Fiber Type

Many manufacturers of optical transmission equipment have already determined the maximum distance over which their particular systems can operate and can therefore recommend a specific core size and performance of fiber to be used for given lengths and data rates. There are, however, two typical situations where this decision is not as clear-cut. In many cases of premises network design, fiber selection is made before the active components are selected. In other situations, systems may have to be designed for potential upgrades for which the active elements are not yet available.

In general, multimode fiber is best suited for premises applications where links are short (i.e., less than 5 km) and have many connectors. Of the three sizes of multimode graded index fiber currently used in communications, 62.5/125 μm has become the de facto standard for LAN and premises applications. It offers the best combination of available attenuation and bandwidth to meet present and future requirements and is endorsed by major OEM suppliers. For upgradability, dual-window fiber should be specified. For example, a fiber installed to meet Ethernet (10M bps) in the first window can also be selected to support a future upgrade to an FDDI system of 100M bps in the second window.

Single-mode fibers are usually selected for long distances (i.e., greater than 5 km) or very high bit rates or both. Single-mode fibers are actually less expensive than multimode fibers. The only reason why they are not used for all applications is that the cost of the associated terminal equipment (e.g., transmitters and receivers) is higher than their multimode equivalents.

Cable Construction

Cable construction and environment—whether the cable will be installed in a duct or in an aerial, buried, or underwater installation—must be determined. Any type of cable construction using the copper-wire method can be applied to optical fibers. Cables have been used for aerial, buried, and underwater applications as well as in direct plow-in applications in which rodent-proofing was required. Fiber-optic cables can also have tensile load strengths of more than 6,000 lb as well as a fire-retardant capability.

For indoor applications, cables that comply with the stringent requirements of the 1987 National Electrical Code (NEC) are available. Recent changes to the NEC affect designations and ratings of both fiber and communications cables. The NEC specifies the requirements for general-purpose, riser, and plenum applications.

Number of Fibers

During cable selection, the number of fibers required must be determined. Multifiber cables containing 2, 6, and 10 fibers are common; cables with as many as 420 fibers have been constructed. The first consideration should be the future expansion capability and the number of channels of information that will be used with fiber optics. The number of fibers to be installed in the cable can then be specified.

Repeaters

Although the fiber-optic cabling methods have less data loss than copper-wire techniques, repeater or amplification stages are required to span long distances. Most fiber-optic communications equipment can easily communicate up to 1 km (3,000 ft) or more. Many long-distance communications (i.e., from 30 to 80 km) are possible without any repeater or amplification stages, but special interfacing equipment is needed to transmit over these long distances. In the premises environment, it is usually more cost-effective to use higher performance terminal equipment than to use repeaters.

Splicing

In the premises environment, the designer or installer can avoid the requirement of fiber-to-fiber field splicing by installing a continuous length of cable, which is usually the most economical and convenient solution. However, because of the cable plant layout, length, raceway congestion, or requirements to transition between nonlisted and UL-listed cable types at the building entrance point, splices cannot always be avoided.

Field splicing methods for optical fibers can be grouped into three major categories:

- Fusion—The two fibers to be joined are typically fused with an electric arc.
- Single fiber mechanical—The fibers to be joined are aligned with a mechanical alignment device and usually held in position with adhesive.
- Mechanical multifiber array—Several fibers are held in a linear array and are joined to a similar set using a mechanical holder and adhesive.

Each method has both advantages and disadvantages. The choice depends on the installer's equipment, preference, training, application, and volume of fiber-optic splicing. Each of these methods is field proven and has excellent long-term reliability.

Connectors

The final step in the installation of a fiber-optic cable is to terminate it, which can be done in various ways. A variety of equipment, hardware, and connectors is also available.

Currently, optical fibers can be terminated three basic ways: pigtail splicing, field connectorizing, and installing preconnectorized cable assemblies. Field connectorization has often been avoided because of the complexity of the procedure and labor time requirements. As technology has evolved, however, new products and techniques have reduced this complexity. For example, new glass-in-ceramic connectors, available in ST- and FC-compatible versions, can now be field-installed in less than five minutes, making this termination method highly attractive. In the premises environment, the ST-type connector has become the connector of choice for multimode applications.

Communications Equipment

There are many suppliers of fiber-optic equipment. Their products include:
- Fiber-optic extenders:
  —RS-232 modems and multiplexers.
  —IBM 3270 modems and multiplexers (some extend up to 6.2 mi).
  —T1 modems and multiplexers.
- Fiber-optic LAN extenders:
  —Ethernet repeater-bridge links.
  —Token ring extenders-repeaters.
- Fiber-optic LANs:
  —Ethernet.
  —Token ring.
  —Manufacturing automation protocol (MAP).
  —FDDI.
  —TDM.

In many cases, fiber-optic implementation is the most economical way of meeting communications requirements. In other cases, there may be a premium. For example, a user may have to pay 10% to 20% more for a fiber-optic backbone in place of the traditional coaxial backbone in an Ethernet network. This is a relatively small penalty, however, for upgradability to FDDI.

SUMMARY

Fiber optics is no longer a laboratory curiosity but a proven communications medium. The use of fiber-optic cabling techniques can increase performance by offering increased bandwidth, enhanced security, and freedom from electrically generated noise. Equally important is the improved cost/performance ratio that fiber optics will bring to future applications.

Chapter II-3-6

An Overview of Digital Image Bandwidth Compression

Heidi A. Peterson
Edward J. Delp

MANY APPLICATIONS REQUIRE the transmission or storage of images. Examples of image-transmission applications include video teleconferencing, facsimile transmission of printed matter, military communications, and deaf communication. Satellites transmit image data for remote sensing, the study of weather patterns, and military reconnaissance. Satellite links are also used to transmit television programs around the world.

The data rates proposed for the planned high-speed integrated services digital network (ISDN) range from approximately 144K bps to 1.984M bps. The data rate required for uncompressed image transmission, however, often exceeds even this capacity. For example, an uncompressed 512-by-512 pixel, 256-gray-level monochrome image, which corresponds to the approximate resolution of a single frame of National Television Systems Committee (NTSC) monochrome video, requires more than 14 sec to transmit over a 144K-bps channel. An NTSC color image of the same resolution requires approximately twice that amount of time. A 16-color, 512-by-512 pixel graphics image requires approximately 7 sec.

The bandwidth requirements for video (i.e., time-varying imagery) transmission are even higher. Transmitting monochrome NTSC video (30 frames per sec) in digital format requires approximately 60M bps; digital color NTSC requires at least 100M bps. A compact disc that stores one hour of stereo music at 16 bits per channel sampled at 40 kHz could store only 45 sec of digital video.

Medium-resolution monochrome teleconferencing (e.g., 256 by 256 pixels, with six bits per pixel) at 10 to 15 frames per sec requires approximately 5M bps for uncompressed transmission. High-definition television (HDTV) requires approximately 400M bps for uncompressed transmission.

Transmitting real-time video or large volumes of high-quality still images in a reasonable amount of time, even with ISDN, therefore requires image-data compression. The only other alternative is to dramatically increase the channel data rate. Higher data-rate channels, however, would undoubtedly be more expensive to use than their lower data-rate counterparts, because the ISDN user will be charged according to the capacity used rather than connect time. As a result, image compression should be economically desirable for almost any image-transmission application on ISDN.

Applications that require the storage of digitized images include the computer archiving of medical X-rays, fingerprints, architectural drawings, weather maps, satellite images, and geological survey maps. Even given the capacity of computer storage available today, however, the storage requirements of such applications make storing uncompressed images impractical.

The goal of image compression is to represent an image with as few bits as possible without noticeable degradation in the decoded (or uncompressed) image. Image-data compression can be thought of as a two-step process, as shown in Exhibit II-3-23. In the first step, a digitized image is represented by a sequence of messages that can be chosen in a wide variety of ways; however, they must be chosen so that a reasonable approximation of the original image can be reconstructed from the sequence. In the second step, the message sequence is binary coded to reduce redundancy. The overall goal is to generate a coded version of the image that contains all important image information with absolutely no redundancy.

IMAGE COMPRESSION METHODS

Any image compression method can be broadly classified as either statistically based (i.e., algebraic) or symbolically based (i.e., structural). Statistically based image compression is based on the principles of information

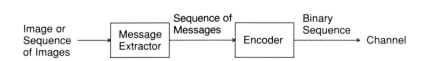

Exhibit II-3-23.   General Image Coder

theory; the techniques used usually involve very localized, pixel-oriented features of the image. Limitations of the statistical approach prevent it from achieving low bit rates (i.e., rates of about 0.1 bits per pixel). These limitations led to the development of symbolically based image-compression (also called second-generation image-coding) methods, which employ computer-vision and image-understanding techniques and properties of the human visual system to achieve very low data rates. Symbolically based methods emphasize the geometric structure of the image scene, whereas statistically based methods emphasize the algebraic structure of the pixels in the image.

An image-compression method can be further classified on the basis of the techniques it employs, the type of image to which it is applied, and the distortion it introduces in the image.

Techniques. A method can be adaptive or nonadaptive, depending on whether it employs adaptive or nonadaptive compression techniques. Adaptive techniques compensate for spatial variations in the characteristics of an image (or time variations in the case of images that vary over time) by changing some parameters of the coder depending on the location in space and time within the image. Nonadaptive techniques do not.

Many image-compression methods are implemented on a block basis, in which the image is partitioned into blocks and each of these blocks is coded separately. Block coding derives from rate-distortion theory, discussed in a later section of this chapter. Dividing the image into blocks helps the image-compression algorithm adapt to local image statistics and allows all blocks to be coded in parallel. Block coding is especially attractive when used with a computationally intensive coding algorithm. A disadvantage, however, is that the borders of the blocks are often visible in the decoded image.

Type of Image. Any image-compression method can be applied to digital video (i.e., time-varying imagery) by applying the chosen compression method to each of the frames of the image sequence, thereby coding the digital video signal as a sequence of single-frame images. Exploiting the temporal (i.e., frame-to-frame) redundancy in the image sequence often makes substantial reductions in the data rate possible. A block-coding method, for example, can use three-dimensional blocks (two dimensions in space and one in time) for time-varying imagery. Other techniques that exploit the temporal redundancy in digital video can be quite sophisticated. One such technique is motion-compensated coding, in which only those portions of the image that have changed from one frame to the next in the image sequence are coded.

An approach to compressing color imagery is to decompose the image into three component images (e.g., luminance, hue, and saturation) and then code these three images individually. A lower data rate can often be achieved by exploiting the spectral and temporal redundancy among the component color signals. Some compression techniques encode the composite NTSC color baseband video signal directly.

Distortion. Some compression methods are distortionless, some are nondistortionless. With a distortionless compression method, the decoded image is a perfect version of the original image. Nearly all distortionless methods are based on information-theoretic principles and usually attain data rates in the neighborhood of two to four bits per pixel. Nondistortionless compression methods allow differences between the decoded image and the original image, but the distortion introduced in the decoded image must be kept as unobtrusive as possible. Nondistortionless compression techniques usually attain much lower data rates than distortionless techniques. An important issue in image compression is accurately measuring the severity of the distortion caused in the image by the coding and decoding process.

GENERAL ISSUES IN IMAGE COMPRESSION

Three of the most important issues in image coder design are the image model, the image-quality measure (i.e., the distortion measure), and the impact of the application and the human visual system on coder design.

The Image Model

To design a compression algorithm that performs well for a class of images, some characteristics of the images—an image model—must be assumed. If the model of the image is not accurate, the compression method based on the model will most likely not work well. The problem of finding an adequate model for a natural scene is not simple, and it is even more difficult for time-varying scenes (i.e., digital video).

Many researchers have modeled images as random fields, which involves modeling the pixel statistics of the image. This approach has proved difficult, however, largely because of the highly nonstationary nature of images. Image-pixel statistics can change dramatically as the temporal and spatial position within the image changes. Furthermore, some information in an image may not be readily represented with pixel statistics. For example, expressing the fact that a particular scene is composed entirely of triangles of different sizes and orientations is difficult to do with pixel statistics. An additional complication with the random-field approach is that different types of images have very different pixel statistics.

Because of these complications, the success of the random-field approach has been limited despite the considerable effort that has been spent on it. Better statistical models might result from approaches that consider an image to be the output of many sources, each with its own type of statistics. Another promising approach is to model the statistics of a more global feature of the image, such as the edges in the image, rather than the pixel statistics. An example of this approach is an image model generated by random tessellations of the image plane.

The Image Distortion Measure

As stated earlier, every image-compression method can be classified as either distortionless or nondistortionless. A distortionless compression method, which results in a decoded image that is identical to the original image, can be evaluated solely on the basis of the merits of the compression algorithm (e.g., its data rate, the number and complexity of computations it requires, and its susceptibility to channel errors). Evaluating nondistortionless compression methods fairly requires measuring the quality (or distortion) of the decoded image, which requires a method for measuring the severity of the degradation to the image caused by the coding and decoding process.

A measurement of distortion is necessarily a function of the original image and the decoded image, but the specific method used to measure distortion can vary greatly, depending on the application. In some applications, the edges in the image may be very important, for example, and it is vital that they be unaffected by the coding and decoding process. An image-distortion measure used to evaluate compression schemes for these applications, therefore, should heavily weigh the accuracy of the edges in the decoded image. In general, the characteristics of the decoded image that are important for a given application should be reflected in the image-distortion measure used to evaluate compression schemes for that application. Because humans are the most common viewers of decoded images, any measure of image distortion should model human judgment of image distortion. The development of such a measure is not straightforward, however, because it is difficult to derive an analytical expression that quantifies the degradation of image characteristics that are important to humans.

This difficulty has led to the use of such traditional mathematical measures of image distortion as mean-square error. Often, these traditional distortion measures are modified in simple ways that take into account the application—they may be modified to reflect an assumed human visual-system model, for example. A major shortcoming of traditional mathematical measures of image quality, however, is that they are applied to the individual pixels of the image, and few pixelwise mathematical image-quality measures have a consistently high correlation with human judgment of image quality. Measures that correlate well with human judgment of image quality need to be more globally based.

Another approach to the measurement of distortion is to use the subjective evaluations of human viewers. There are two basic types of subjective evaluation methods: rating-scale and comparison. In the rating-scale method, the subject views a sequence of images and assigns each a qualitative score such as *good, fair,* or *poor.* Rating may also be based on an impairment scale, in which viewers assign scores such as *imperceptible, slightly annoying,* or *very annoying.*

In the comparison method, the subject compares the distorted image to a reference image and judges the distorted image using a scale that ranges from *much better* to *same* to *worse.* The comparison method can also be implemented by allowing the subject to add standard impairments to a reference image until the reference image is judged to be of the same quality as the distorted image.

Other variables having nothing to do with actual image quality can also affect subjective image evaluation. These include ambient illumination, viewing distance, and viewer fatigue. These variables have been investigated and standards are being defined for the subjective assessment of image quality.

The Impact of the Application and the Human Visual System

A basic understanding of the applications for which imagery will be used is necessary not only for specifying an accurate image-distortion measure but also for the whole process of designing an image coder. Understanding the application helps designers determine how to hide the inaccuracies introduced into the image by the coding and decoding process. Fewer bits can be used to code those parts of the image that are unimportant to the observer and more bits can be used to code the parts that are important. To tailor a compression scheme in such a way, however, requires an accurate model of the image observer for the application under consideration.

Assuming for simplicity an application for which the observer is a typical human, the image coder should ideally use very few bits to encode information to which the human visual system is least sensitive and more bits to encode information to which the human visual system is most sensitive. The more that is known about the characteristics of the human visual system, the better should be the design of the coder.

The human visual system is very complex, however, and the visibility of distortion in an image is a function of many factors: the nature of the distortion itself, the intensity and busyness of the image in the vicinity (in both space and time) of the distortion, and the lighting in the room where

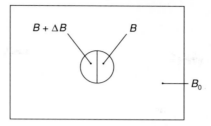

SOURCE: W.F. Schreiber, "Picture Coding," *Proceedings of the IEEE* 55 (March 1967), pp 320–330.

Exhibit II-3-24.   Test Pattern for Measuring the Contrast Sensitivity of the Human Visual System

---

the image is viewed. If more than one distortion is introduced into an image, as is usually the case, the interplay of these multiple distortions can become very complicated.

Capturing the complexity of the human visual system in image-compression algorithms is difficult, but doing so is critical to attaining algorithms that result in very low bit-rates with little image distortion.

The characteristics of human vision that are most important in the development of image-compression techniques are those that relate to the sensitivity of the human viewer to noise and distortion in images. One such characteristic that has been studied extensively is the contrast sensitivity of the eye, which is measured by showing a subject a test pattern and varying the intensity of neighboring regions in the test pattern until the difference in intensity is just noticeable. The configuration shown in Exhibit II-3-24 can be used to provide a simple measurement of contrast sensitivity. The just-noticeable difference, ΔB, is measured as a function of B and B_0. The results of a study of contrast sensitivity are shown in Exhibit II-3-25. The fraction $B/\Delta B$, known as the Weber fraction, is obtained as a function of B_0 and B. The graph in Exhibit II-3-25 shows that the human visual system has greatly reduced contrast sensitivity in very bright or very dark intensity regions of an image, and that the eye is most sensitive to contrast in a range of about 2.2 log units, centered about the background brightness. The eye is less sensitive to contrast as B_0 moves away from B. Knowledge of the variations in contrast sensitivity of the eye can be useful for such things as nonuniform quantization of images and the design of image-distortion measures.

Another important characteristic of human vision is the spatio-temporal frequency response of the human visual system. This response, often referred to as the modulation-transfer function (MTF), is measured by presenting a test subject with a periodic wave, usually a sine wave or a square wave, and varying the modulation of this wave until the threshold of visi-

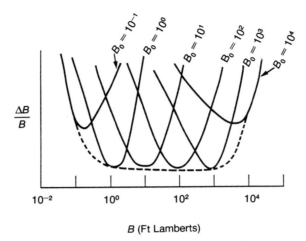

B (Ft Lamberts)

SOURCE: W.F. Schreiber, "Picture Coding," *Proceedings of the IEEE* 55 (March 1967), pp 320–330.

Exhibit II-3-25.   Contrast Sensitivity of the Human Visual System for the Test Pattern of Exhibit II-3-24

---

bility is determined. (The modulation of a periodic wave is the ratio of the wave's amplitude to its average value.) The modulation can be spatial (i.e., a series of light and dark bands across the image plane—to measure the spatial modulation-transfer function) or temporal (i.e., an image plane that varies in time from light to dark—to measure the temporal modulation-transfer function). The value of the modulation-transfer function at a particular frequency is the threshold modulation at which a stimulus of that frequency is just visible.

Recent studies have shown that the spatial and temporal responses of the human visual system are closely interrelated. The spatio-temporal modulation-transfer function is generally measured using either a flickering grating or a grating moving across the field of view. When a sine wave moving across the field of view is used as the stimulus, it results in the spatio-temporal modulation-transfer function shown in Exhibit II-3-26. The nonuniform frequency response of the human visual system, as demonstrated by the spatio-temporal modulation-transfer function, affects many aspects of the human perception of images. One consequence is that the eye is less sensitive to distortion in those parts of a scene that are moving. Research has also shown that the human viewer takes a substantial fraction of a second to recover spatial acuity after a scene change; reducing spatial resolution for as long as 0.75 sec after a scene change is not noticeable to a human observer.

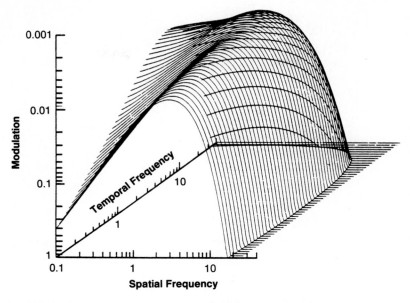

SOURCE: W.H. Kelly, "Motion and Vision II: Stabilized Spatio-Temporal Threshold Surface," *Journal of the Optical Society of America* 69 (October 1979), pp 1,340–1,349.

Exhibit II-3-26. The Spatio-Temporal Modulation-Transfer Function of the Human Visual System

---

If an absolutely complete description of the spatio-temporal response of the human visual system were available, the visibility of any type of degradation in an image could be determined and there would be no need for subjective observer tests of image quality. The complexity of human vision has so far precluded a complete description. Nonetheless, some general statements can be made about the response of the human visual system to noise or distortion in images:

- Distortion is most visible in those portions of the image that are constant in intensity—The more complicated a part of the image, the less visible noise will be there. Spatial busyness in an image, therefore, has a masking effect on distortion. Temporal busyness in an image also affects the visibility of distortion, but in a more complicated way.
- The sensitivity of the human visual system to distortion varies depending on the way the distortion is correlated with the image—Quantization noise in an image, for example, is more annoying than a similar quantity of random noise. This fact complicates the image coder design process, because many types of distortion introduced by the

coding and decoding process are correlated with the image in ways to which the human visual system is highly sensitive.

- The human visual system is more sensitive to distortion that is structured in some way than to distortion that occurs randomly in the image plane—The distortion that occurs along the grid of block boundaries that results from compression, for example, is more annoying to a human viewer than the same quantity of distortion distributed randomly in the image plane.
- The sensitivity of the human visual system to noise is affected by the frequency spectrum of the noise.
- The presence of any noise in an image reduces the contrast and sharpness of the image and degrades its quality significantly.

STATISTICALLY BASED IMAGE-COMPRESSION TECHNIQUES

Much of the first 25 years of work in image compression, beginning in 1960, fits into the statistically based category. Statistically based image-coding techniques address the image-compression problem from the point of view of information theory, with a focus on the elimination of statistical redundancy in the pixels in the image. A block diagram of a general statistical image-compression system is shown in Exhibit II-3-27. The ideal preprocessor in the exhibit maps the pixels into independent data, as prescribed by rate-distortion theory. In practice, however, the best a preprocessor can produce is uncorrelated data, as, for example, by taking the discrete Fourier transform of the image pixels.

Rate-distortion theory defines the optimum coder to be the one that attains the best possible data rate for a given image distortion. Research has shown that for any data source, lower data rates can be achieved by coding blocks of data, rather than individual data points. The optimal coder is one for which $N \to \infty$, where N is the length of the block of data being coded. (Block coders based on this reasoning are now more popularly known as vector quantizers.) A coder with an infinite block length is obviously impossible, and even a coder with a reasonably long block length is difficult to design and implement. It can be shown, however, that if the data samples are statistically independent, N block-length-one coders are nearly as good (within about 0.25 bits per sample) as one block-length-N coders for the squared error-distortion measure. If the data samples can be transformed

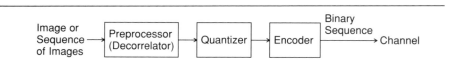

Exhibit II-3-27.   Statistically Based Image Coder

to be statistically independent, then nearly optimum coder performance can be achieved with a block-length-one coder (i.e., a simple quantizer).

These facts are the basis for all statistically based image-compression techniques and underlie the reasoning behind the development of the discrete Karhunen-Loeve transform (KLT). For Gaussian distributed pixels, this transforms the data so that the samples are independent. These transformed pixels can then be coded nearly optimally using a simple quantizer. Another example of this same type of reasoning is predictive coding—if the image pixels can be modeled as a Markov random process, then the differences between consecutive pixels are independent. These differences can also be coded nearly optimally using a simple quantizer.

Unfortunately, the application of rate-distortion theory to image compression has problems. Two factors are important in the design of statistically based coders: a valid random-field model of the image and a valid distortion measure. As mentioned earlier, however, a simple statistical model of an image does not exist, and an accurate distortion measure is not presently known for images. Despite these problems, there have been many successful statistically based image-compression techniques proposed. These techniques generally fall into three categories: predictive coding, transform coding, and interpolative and extrapolative coding.

Predictive Image-Compression Techniques

The philosophy behind predictive image-compression techniques—also known as differential pulse-code modulation (DPCM)—is that if the pixels can be modeled as a Markov process, the differences between consecutive pixels are statistically independent and a simple quantizer will be nearly optimum. For predictive coding techniques, the value of a pixel is predicted based on the values of a neighboring group of pixels. The group of pixels on which the prediction is based can be spatially distributed or, for digital video, temporally distributed. The error in the prediction is quantized and coded for transmission. (Delta modulation is a variation of this technique.)

Predictive coding results in data rates of from one to two bits per pixel. Predictive coding methods can be made adaptive by varying the prediction algorithm or the difference quantizer. Adaptive predictive coding achieves bit rates 10% to 20% lower than nonadaptive predictive coding.

Transform Image-Compression Techniques

The motivation behind transform-image compression, which involves applying a transformation to an image before coding, is to convert the statistically dependent image pixels into independent transform coefficients. However, because it is usually impossible to obtain independent transform coefficients, a transform is used that results in nearly uncorrelated transform coefficients. After the transformation is performed on the image pixels, the next step is to quantize the transform coefficients; the quantized values and locations of the coefficients are then encoded for transmission. Some examples of transforms used for image compression include Karhunen-Loeve, Fourier, Hadamard, and cosine; cosine transform techniques are the most popular.

Bit rates of slightly less than one bit per pixel can be achieved with transform image-compression methods. Transform coding can be made adaptive by varying the way the coefficients are quantized or by varying the transformation used. These adaptive algorithms can improve the data rate by about 25%.

A disadvantage of transform coding is that the number of floating-point computations required to perform the image-pixel transformation can be quite large. Fast transform algorithms have therefore been developed and are often used for transform-image coding. Transform-coding algorithms are nearly always implemented on a blockwise basis.

Interpolative and Extrapolative Image-Compression Techniques

Interpolative and extrapolative image-compression techniques obtain a subset of the pixels in an image by subsampling the image. This subset is then transmitted, and the decoder interpolates or extrapolates to fill in the missing pixels. The subsampling of the image can be performed in either of the spatial dimensions, in the temporal dimension, or in any combination of these. The interpolation function can use straight lines or higher-order polynomials, but with higher-order polynomials it may be necessary to transmit polynomial coefficients in addition to the subset of image pixels. This class of compression techniques can be made adaptive by varying the degree to which the image is subsampled, the direction of the subsampling, or the function used to do the interpolation or extrapolation. Interpolative compression techniques achieve bit rates in the neighborhood of two bits per pixel.

SYMBOLICALLY BASED IMAGE-COMPRESSION TECHNIQUES

The bit rates that can be achieved with statistically based compression methods seem to have reached a low recently of about 0.5 bits per pixel. For many applications, however, data rates as low as 0.1 to 0.01 bits per pixel are desirable. Symbolically based image compression was designed to attain these very low bit rates. A block diagram of a general symbolic image-compression system is shown in Exhibit II-3-28.

Symbolically based image-compression methods examine global rather than local pixel-oriented features of the image. Features such as the size, shape, or orientation of objects in the image scene and the relationships

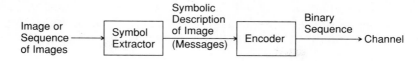

Exhibit II-3-28.   Symbolically Based Image Coder

---

among these objects are used to symbolically describe the image. Techniques from image analysis, computer vision, and artificial intelligence, combined with findings about the properties of the human visual system, aid in the development of symbolically based image-compression methods.

The ultimate goal envisioned for the symbol extractor (Exhibit II-3-28) is for it to be able to obtain a complete high-level description of the image scene. Such a description might take the form of a list of scene attributes along the lines of, There is a chair in the upper left corner of the scene, or, A man in a red shirt is running from left to right in the scene while turning his head and looking at the camera. These very high-level descriptions deal with scene content rather than actual image-pixel values. When the symbol extractor has completed its descriptions, the encoder efficiently encodes them.

The current state of the art in symbolic image compression is not yet capable of such complicated scene descriptions. Issues such as the optimal symbolic description of an image, the lowest achievable data rate for a given image, and the way distortion manifests itself in the decoded image remain the subject of research.

Three symbolically based image-compression methods that have been developed are the synthetic-high system, thought to be one of the earliest symbolically based image-compression techniques; the region growing–based method; and a human visual system–based segmentation image coder.

Synthetic-High Image Compression

The synthetic-high method of image compression, originally applied to an analog image signal, is designed to decompose an image into a high-frequency component (containing edge information) and a low-frequency component (containing general area brightness information). The two components of the image are coded separately, using two different methods. The low-pass (low-frequency) image component is sampled according to the two-dimensional sampling theorem at a very low rate, and the samples are then coded.

For the high-pass (high-frequency) component, an edge detector locates edges in the original image, and this edge information is then used to establish a threshold and determine which edge points are important. The locations and magnitudes of important edges are then coded.

At the decoder, the image is reconstructed first by a filter that synthesizes the high-pass part of the image from the edge information, and then by the addition of the low-pass component. This method leads to data rates that are slightly less than one bit per pixel. Since it was first proposed in 1959, many other compression methods have been proposed that make use of the same basic principle.

Region Growing–Based Image Compression

The region growing–based method of generating a symbolic representation of an image separates image pixels into two classes: contour and texture. This classification is accomplished in two stages: preprocessing and segmentation.

Preprocessing, which eliminates granularity in the image, must be done carefully to avoid altering the edges in the image. After preprocessing, a region growing is performed as the first step toward generating a segmented image.

The next step is the elimination of artifacts in the image that result from region growing. At this point, the image consists entirely of regions separated by contours (i.e., it is a segmented image). The image is then processed to eliminate regions that are so small or so weakly contrasted with their neighbors as to be insignificant. The end result of image segmentation is an image with contours separating segments with textures; these segments serve as the messages that represent the image.

The final step in the process is to code the contours and textures of the image segments. The textures of the regions are described by two-dimensional polynomials with order determined by the allowable approximation error and the cost of coding the polynomial coefficients. The information describing the straight lines, circle segments, and the polynomials is encoded to represent the image. When the image is decoded and reconstructed, granularity, in the form of pseudorandom noise, is added to the image to give it a more natural look. This region-growing method of image compression yields a data rate of less than half a bit per pixel.

Segmentation-Based Image Compression Using Human Visual System Properties

Segmentation-based image compression using human visual-system properties is similar to the region-growing method in that the image is partitioned into regions and information describing these regions is then coded.

It differs, however, in that it incorporates human visual-system properties and uses a new technique to represent the image segments for coding.

Image segmentation is performed in two steps, both based on human visual-system properties. A variation of centroid-linkage region growing is used to segment the image initially. This region-growing process incorporates a threshold based on the contrast sensitivity of the human visual system. In the second step, a nonlinear MTF-based filter is applied to the segmented image to eliminate all regions that contrast insignificantly with neighboring regions.

Once the image segmentation is complete, the next step in the algorithm is to represent the image segments for coding. The shape of each image segment is conveyed by a morphological skeleton of the segment, which is a thin line caricature that summarizes shape and gives information about size, orientation, and connectivity. The morphological skeleton of each image segment is obtained using mathematical morphology. These skeletons, along with the average intensities of the image segments, form a compact symbolic representation of the image. The last step in the compression process is to binary encode the set of pairs (i.e., skeleton, average intensity) using a source-coding technique such as Huffman coding. Data rates of 0.12 bits per pixel are achievable with this compression technique.

SUMMARY

This chapter discusses some of the important issues in image compression and provides an overview of past approaches to the image-compression problem. It also examines a new approach, symbolically based image compression, that can lead to lower data rates than have been achieved with more traditional methods. Even with the introduction of high-speed channels such as those provided by ISDN, conservation of bandwidth remains an important economic consideration for image transmission, and even with the capacity of current storage technologies, it remains economically advantageous to store digital images with as few bits as possible. For these reasons, image compression will remain important for the storage and transmission of both conventional images and those produced by the newly emerging technologies based on high-quality digital video.

Chapter II-3-7
The PBX in Perspective

Nathan J. Muller

IN THE SIMPLEST TERMS, the PBX is a circuit switch that uses control signaling to perform three basic functions:

- In response to a call request, it establishes end-to-end connectivity among its subscribers (on-net) and from its own subscribers to remote subscribers (off-net) through intermediate nodes that can consist of other PBXs or central office switches on the public telephone network—The connected path is dedicated to the user for the duration of the call.
- It supervises the circuit to detect call request, answer, signaling, busy, and disconnect (hang-up) signals.
- It clears the path upon call termination (disconnect) so that another user can use the resources available over that circuit.

These functions closely parallel those of the central office switch. In fact, the PBX evolved from the operator-controlled switchboards that were used on the public telephone network. The first of these simple devices was installed in 1878 by the Bell Telephone Co to serve 21 subscribers in New Haven. The operator had full responsibility for answering call requests, setting up the appropriate connections, supervising the call to confirm answer and disconnect, and clearing the path on call completion. Interconnectivity among subscribers was accomplished through cable connections at a patch panel.

Today's PBXs are much more complicated, of course, but they provide the same basic capabilities as the first generation of circuit-switching devices. The process of receiving call requests, setting up the appropriate connections, and clearing the paths upon call completion, however, is now entirely automated. Because the intelligence necessary to perform these tasks resides on the user's premises, the PBX allows organizations to exercise more control over internal operations and to incorporate communications planning into their long-range business strategies. Exhibit II-3-29 illustrates the role a PBX plays in a corporate voice network.

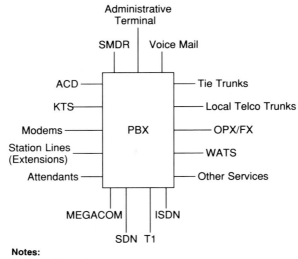

Notes:

ACD	Automatic call distributor
KTS	Key telephone system
SDN	Software defined network
ISDN	Integrated services digital network
WATS	Wide area telecommunications service
OPX	Off-premises extension
FX	Foreign exchange
SMDR	Station message-detail recording

Exhibit II-3-29.   The PBX in the Corporate Voice Network

In a little more than 100 years, PBXs have evolved from simple patch panels to sophisticated systems capable of integrating voice and data. The history of PBXs can be divided into four generations, as follows:

- First-generation—Operator-controlled patch panels.
- Second generation—Systems capable of automatic dialing and space-division switching. These evolved from such electromechanical central office switches as step-by-step (Strowger) and cross-bar.
- Third generation—Systems capable of supporting distributed architectures as well as other features. These systems used electronic components under stored-program control instead of electromechanical switches.
- Fourth generation—Computer-based systems capable of such features as automatic call distribution and voice mail. These systems use time-division switching, which permits the integration of voice and data for T1 trunking.

Capabilities such as stored-program control, advanced processing power, large-scale integration, and high-capacity memory not only have reduced the cost of PBX ownership but have also endowed PBXs with an array of sophisticated features. In fact, among the dozens of products available, hundreds of features are offered that can be implemented directly from the telephone keypad. Examples are listed in Exhibit II-3-30. Other capabilities, which promote the efficiency and administration of the PBX, are typically transparent to most users. Some representative examples of these appear in Exhibit II-3-31. Add-on capabilities, which require the purchase of adjunct computers and other specialized hardware and software, include voice mail, a message center, automated attendant-station equipment, local area networking, and communications management systems. Today, standalone PBXs from a single vendor may not be adequate for an organization's communications needs. Companies with multiple locations require many PBXs linked over an efficient, reliable network.

PBX COMPONENTS

In addition to line and trunk interfaces, three major elements make up the typical PBX: processor, memory, and switching matrix. Together these elements allow users to place calls anywhere on the public or private network without human intervention.

Add-on conference—Allows the user to establish another connection while continuing a call already in progress.

Call forwarding—Allows a station to forward incoming calls to another station, even when the first station is busy or unattended.

Call holding—Allows the user to put the first party on hold to answer an incoming call.

Call waiting—Lets the user know that an incoming call is waiting. While a call is in progress, the user will hear a special tone that indicates another call has come through.

Camp-on—Allows the user to wait for a busy line to become idle, at which time a ring signal notifies both parties that the connection has been made.

Last-number redial—Allows the user to press one or two buttons on the keypad to activate dialing of the previously dialed number.

Message waiting—Allows the user to signal an unattended station that a call has been placed. An indicator tells the user of the called station that a message is waiting.

Speed dialing—Allows the user to place calls with an abbreviated number. This feature also allows users to enter a specified number of speed-dial numbers into the main data base. These numbers may be private or shared among all users. Entering and storing additional speed-dial numbers is accomplished through the telephone keypad.

Exhibit II-3-30. Common PBX Features Under Direct User Control

Automated attendant—Allows the system to answer incoming calls and prompt the caller to dial an extension or leave a voice message without going through the operator.

Automatic call distribution (ACD)—Allows incoming calls to be shared among a number of stations so that the calls are answered in the order of their arrival. This is usually an optional capability, but it may be integral to the PBX or purchased separately as a standalone device from numerous non-PBX vendors.

Automatic least-cost routing—Ensures that calls are completed over the most economic route available. This feature may be programmed to restrict some employees to the least expensive carrier, while permitting executives to choose whichever carrier they want.

Call-detail recording (CDR)—Allows the PBX to record information about selected types of calls, including the collection of detailed traffic statistics. Additional hardware and software are usually required to arrange this data into meaningful management reports.

Call pickup—Allows incoming calls made to an unattended station to be picked up by any other station in the same trunk group.

Class of service restrictions—Control access to certain services or shared resources. Access to long distance services, for example, can be restricted by area code or exchange. Access to the modem pool for transmission over analog lines may be similarly controlled. A modem pool is a shelf of card-mounted modems that are shared among many users who have infrequent transmission needs.

Data base redundancy—Allows the instructions stored on one circuit card to be copied to another card as a protection against loss.

Direct inward dialing (DID)—Allows incoming calls to bypass the attendant and ring directly on a specific station.

Direct outward dialing (DOD)—Allows outgoing calls to bypass the attendant for completion anywhere over the public telephone network.

Hunting—Routes calls automatically to an alternative station when the called station is busy.

Music on hold—Provides assurance to callers that the connection is still being maintained. During peak hours, incoming calls might have to wait in queue until handed off by the automatic call distributor to the next available station operator.

Power-fail transfer—Permits the continuation of communications paths to the external network during a power failure. This capability works in conjunction with an uninterruptible power supply (UPS) that starts up within a few milliseconds of detecting a power outage.

System redundancy—Allows the sharing of the switching load, so that in the event of failure, another processor can take over all system functions.

Exhibit II-3-31. Common PBX Capabilities Transparent to the User

---

Processor

The processor is responsible for controlling the various operations of the PBX, including monitoring all lines and trunks that provide connections, establishing line-to-line and line-to-trunk paths through the switching matrix, and clearing connections on call completion. The processor also controls such optional capabilities as voice mail and the recording of billing and traffic information. Because the processor is programmable, features and services can be added or changed at an administrative terminal.

Many PBX systems have additional processors as backups, which increases the reliability of the system. Programs and configuration information are automatically downloaded from the main processor to the standby processor to ensure uninterrupted service in the event of a failure.

Memory

The processor works with memory to implement the functions of the PBX, which are defined in software rather than hardware. The operating instructions and system-configuration information are contained in nonvolatile memory; volatile memory (RAM) is used for the temporary storage of frequently used programs and for work space. In most systems, if a system failure destroys the contents of nonvolatile memory, a reserve program—also stored in nonvolatile memory—is put into operation automatically. Some systems, however, require that a spare program be loaded manually after a catastrophic failure. When the system is restarted, information stored in nonvolatile memory is automatically dumped into RAM for access by the redundant processor.

Switching Matrix

The switching matrix, which is controlled by the processor, interconnects lines and trunks using either space-division switching or time-division switching. (In proper usage, the terms *line* and *trunk* have different meanings: a line refers to the link between each station—telephone set—and the switch, whereas a trunk refers to the link between switches. The term *tie line*, therefore, should really be *tie trunk*.)

Space-division switching originated in the analog environment. As its name implies, a space-division switch sets up signal paths that are physically separate from one another (i.e., divided by space). Each connection establishes a physical point-to-point circuit through the switch that is dedicated entirely to the transfer of signals between the two endpoints. The basic building block of the space-division switch is a metallic crosspoint or semiconductor gate that can be enabled and disabled by the processor or control unit. Interconnection is therefore achieved between any two lines by enabling the appropriate crosspoint.

The single-stage space-division switch, illustrated in Exhibit II-3-32, because it always has a path available to connect input to output, is virtually nonblocking. (Blocking refers to the inaccessibility of a switch due to the unavailability of the crosspoints that establish the connections between various endpoints. In theory, all switches can experience blocking no matter how they are designed; in practice, however, some switch designs are less prone to blocking than others.) This type of switch, however, is accept-

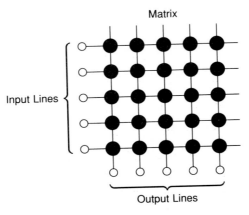

Matrix

Input Lines

Output Lines

Exhibit II-3-32.   Single-Stage Space-Division Switch

able for small configurations only; as the number of crosspoints increases (to accommodate growth), the matrix becomes increasingly costly and increasingly less efficient.

An alternative space-division switch design is the multistage crosspoint matrix. Although this type of switch requires a more complex control scheme than does a single-stage space-division switch, its use of multiple stages reduces the number of crosspoints and increases their efficiency. Because it provides more than one path through the network to connect two endpoints, this type of switch also increases reliability. Its drawback is that it cannot eliminate the possibility of blocking entirely. Exhibit II-3-33 illustrates a three-stage space-division switch.

Time-division switching—a digital technology—is the preferred switching technique employed in the latest generation of PBXs. With digital technology, voice signals are sampled at a rate of 8,000 times per second and encoded for transmission using such techniques as pulse-code modulation (PCM). Analog signals can therefore be digitized, and several digitized analog signals, each corresponding to a voice conversation, can be interleaved (multiplexed) for simultaneous transmission through the matrix. The benefits of digital time-division extend beyond the matrix to the link itself. Not only is a digitized voice signal easier to switch, it is easier to store, recognize, synthesize, multiplex, concentrate, and integrate with other digital streams. Furthermore, using pulse-code modulation to encode an analog signal into digital form makes the signal compatible with the D4 multiplexing format used by telephone companies for transmissions through the public network.

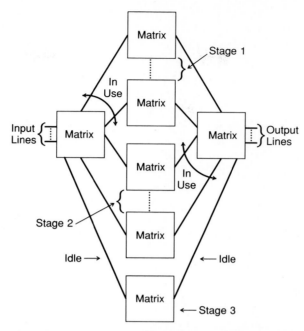

Exhibit II-3-33. Three-Stage Space-Division Switch

In the future, vendors are expected to provide digital PBXs with the ability to implement—through add-in cards—various voice-compression schemes that increase the efficiency of voice communications. One such scheme, the International Telephone and Telegraph Consultative Committee (CCITT) voice compression standard, adaptive differential pulse-code modulation (ADPCM), reduces the 8-bit words of pulse-code modulation to 4-bit words, doubling the number of voice channels available on a digital line.

ANALOG VERSUS DIGITAL

One factor behind the transition from analog to digital PBXs is the growing presence of microcomputers in the office environment. Such industry analysts as the Gartner Group of Stamford CT estimate that there is an installed base of 26 million microcomputers, many of which are networked over the local area networks (LANs). Some PBX vendors suggest that microcomputers and peripheral devices can be linked economically through a digital PBX. For example, AT&T's low-end System 25 is compatible with microcomputers, asynchronous ASCII terminal, printers, and fax machines.

In addition to marketing it as an electronic tandem network endpoint, AT&T is also positioning the System 25 as a gateway to StarLAN, AT&T's LAN.

Another factor in the migration from analog to digital switching is the anticipation of the integrated services digital network (ISDN), which promises both end-to-end digital connectivity and a greatly simplified means of moving voice, data, and video through the network. Several PBX vendors claim that their products are ISDN-ready—that is, they are able to accommodate the special interfaces required to provide primary-rate ISDN services when they become available.

The accelerated pace of ISDN implementation by AT&T and the regional telephone companies is a compelling reason to choose digital over analog PBXs. This does not mean, however, that there are no sound reasons for retaining or even purchasing analog systems. For an organization that lacks a commitment to ISDN or wishes to postpone making a decision about ISDN until potential applications become better defined, choosing an analog PBX that can be upgraded to digital, thus preserving much of the original hardware investment, may be the most prudent course of action. Furthermore, analog PBXs have several advantages over digital systems—in particular, they are substantially lower in price. They also have the benefit of many years of software development behind them, ensuring reliability in a variety of configurations and operating environments. In contrast, digital systems are still grappling with software bugs. Many analog systems use English-language, menu-driven procedures for administrative functions, whereas digital systems use high-level programming languages that require special abbreviations, mnemonic codes, and arcane procedures.

Certain organizations have an immediate or foreseeable need for a digital PBX, whereas others may not need digital equipment for some time. Although digital systems offer considerably faster call setup, they offer no appreciable improvement in voice quality over analog PBXs. An analog PBX also provides the same basic features as a digital PBX and can efficiently handle the occasional need for data communications through a modem pool. Thus, in the final analysis, the choice between digital and analog PBXs hinges on a thorough assessment of the complexity of the organization's current and future networking needs.

CENTRALIZED VERSUS DISTRIBUTED ARCHITECTURE

The most common PBX network involves a centralized switching system or hub from which all stations are connected, as shown in Exhibit II-3-34. Telephones, terminals, and other PBXs are connected to the central hub in a star arrangement. Having a central point of control simplifies the implementation of the numbering plan as well as the allocation and use of

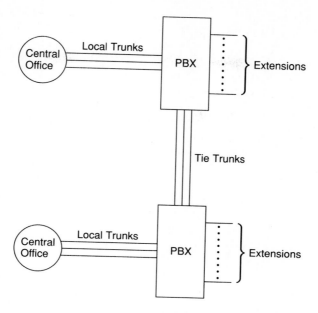

Exhibit II-3-34. Centralized PBX Architecture

network resources. Overall system administration is also simplified under a centralized system. However, if the hub fails in such a system, many users can be left without service.

Under the distributed approach, a number of smaller sites are integrated into a larger network. In the architecture shown in Exhibit II-3-35, each PBX is equipped with its own processor, memory matrix, and line and trunk interfaces. These nodes may be interconnected with each other over analog tie trunks or digital T1 lines that carry multiplexed tie trunks. The key aspect of this architecture is its distributed control; there are no master or slave relationships. Each node functions independently and contains routing information for all other nodes. Special link protocols send information about the caller through the network to the receiving end, which lets each node on the network handle the call appropriately, on the basis of such information as type of originating equipment and class of service.

One disadvantage of the distributed PBX is that the software needed for its operational control is complex and almost always requires some customization and debugging. This can become a problem for management, if only because software problems generally are much harder to fix than hardware problems.

Proponents of distributed architecture claim that it is easier to isolate and repair hardware problems in a distributed system than it is with a centralized architecture. Diagnostic tools now available with most PBXs, however, facilitate fault isolation regardless of architecture. Furthermore, hardware repairs are often made more quickly in centralized PBXs simply because a technician is usually available on site, which is not the case for decentralized PBXs.

The argument that a distributed architecture provides a high degree of protection against catastrophic failure is strong. PBXs with a centralized architecture, however, can be adequately protected against the possibility

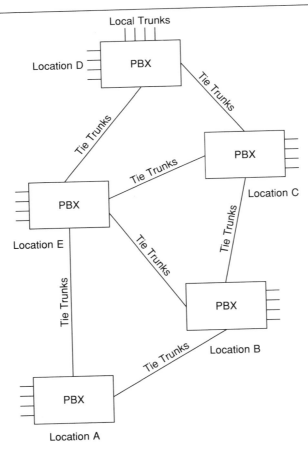

Exhibit II-3-35. Distributed PBX Architecture

of catastrophic failure. Northern Telecom's SL-1RT, for example, offers redundant common equipment, CPU, and memory for medical, utility, and military applications. In addition, the system continually runs self-diagnostics of its CPU, memory, and power units, making it essentially a fault-tolerant PBX. The SL-1RT sells for $650 to $800 per line; distributed systems sell for $750 to $1,000 per line.

A distributed architecture under centrally weighted control offers a compromise between centralized and distributed architectures. In this scheme, low-level functions (e.g., call setup) are handled by PBXs at each node. Administrative functions are also handled locally. The central processor handles the management and supervision of the entire network. It is used to collect billing and traffic information from subordinate nodes through a polling arrangement, to add or reconfigure lines and trunks, and to provide access to other networks.

MIGRATION STRATEGIES

AT&T

AT&T's System 75 is a PBX designed for small to medium-sized organizations; its System 85 is a PBX designed for large organizations. These two systems had been criticized for their incompatibility. Recognizing the need to protect the investments of customers who had bought a System 75 and needed to move up to a System 85 or who needed to connect a System 75 PBX to a System 85 PBX, the company introduced the Definity 75/85 PBX in early 1989.

The Definity 75/85's Generic 1 software and hardware package doubles a System 75's port and traffic-handling capacity and makes it compatible with a System 85. Options provided by the package include redundant common control, ISDN primary-rate interface (PRI), and a more comprehensive networking package than that provided by the System 75. It allows users to upgrade while protecting 90% of their original investment in the System 75.

The Definity 75/85's Generic 2 package provides an upgrade for either a System 85 or for a System 75 that has been upgraded with the Generic 1 package. It provides space-saving cabinets that support microprocessor-based high-density circuit-card packs. When combined with a cache-memory card, these packs improve system call processing by 20% and increase system reloading time twentyfold. The package uses a microcomputer or minicomputer to make systems management more efficient and user friendly. It also provides an ISDN primary-rate interface option; the Definity 75/85, however, does not provide dedicated DS1 connections and transparent signaling, which severely limits its ability to support data.

Other Vendors

Other vendors have not yet matched the flexibility of migration that AT&T has provided with the Definity 75/85. Migration from Northern Telecom's SL-1ST or SL-RT to its larger SL-1NT/ST models, for example, requires a total change of equipment cabinets; no migration path exists at all from the SL-1 to the larger SL-100 system. AT&T's offering will force other vendors to assess the feasibility of providing only discrete products or models. Migrational flexibility will become increasingly important as users develop hybrid networks that consist of both public and private elements. The type of modular approach that AT&T has adopted will also help users take advantage of ISDN on an incremental basis as it becomes available in their service areas.

COMPUTER-CONTROLLED PBXS

Some manufacturers are integrating add-on equipment that was once optional into their PBXs. As a result, the functions and features of PBXs, computers, and voice-processing applications are merging into a single system. Proprietary interfaces allow control signals to pass between the PBX and the computer, making possible, for example, a telemarketing application in which a computer selects likely contacts from a data base for a sales rep to call. The computer would not only select the next contact but would pass the phone number to the PBX together with instructions for setting up the call and selecting a carrier to route it.

Other add-on options being considered for integration with the PBX are voice mail and station message-detail recording (SMDR). Putting these systems into PBX shelves and tying them directly to the switch's bus would eliminate the cost of purchasing separate processors and reduce the demand for limited floor space in equipment rooms. It would also eliminate potential points of failure on the network.

DATA APPLICATIONS

Traditionally, the PBX has been used to handle voice communications over organizational networks and the circuit-switched telephone network. In this context, the basic function of the PBX is to implement the supervision required to set up and clear the connections over which calls are placed. The PBX sets up an end-to-end physical connection for each call before communication begins, and the path remains dedicated to the users until it is terminated by an on-hook condition. During the call, signaling does not continue. Consequently, a PBX provides only transparent service with no mechanism for error control when sending data. It also provides no mechanism for flow control, so the users at each end may have to negotiate a common speed at which to set their data communications equipment if the network does not do it for them.

The nature of circuit switching limits the value of PBXs for data transmission in other ways:

- Several seconds may elapse on a PBX before a communications path is established—This amount of delay may be unacceptable for certain applications, especially transaction processing.
- Bandwidth for PBXs is limited to 64K bps for synchronous data transmission and 19.2K bps for asynchronous transmission—This prevents the PBX from being used for many types of applications, including batch, high-speed file transfer, host-to-host communications, and the transmission of high-resolution graphics or facsimile.
- Even the most advanced voice-data PBXs provide only minimal support for protocol conversion and terminal emulation—Although data-only PBXs continue to expand these areas of support, because they are circuit-switched devices they have many of the same deficiencies as voice-only PBXs, including a long call-setup time and limited bandwidth.
- Bursty traffic, characteristic of transaction processing, is more efficiently handled by packet switches than PBXs.
- The PBX is ill-suited for certain applications, such as broadcast distribution or resource sharing, that typically require simultaneous communications among multiple devices.

In spite of these limitations, PBX vendors have been trying to convince customers that the PBX makes an effective engine for LANs and is a satisfactory alternative for integrating voice and data over wide area networks (WANs). One reason for this may be that the PBX market is slow, forcing vendors to derive revenue from the sale of add-on features to installed systems. One company, for example, has claimed that it can derive six dollars from add-on sales for every dollar of PBX sales.

WIDE AREA PBX NETWORKS

Wide area voice networks are of basically four types. Three of these are PBX-based; of these, one is based on access codes, another on tandem hubs, and the third on AT&T's electronic tandem network. The fourth type—the virtual network—uses the intelligence built into public networks run by AT&T, MCI, and US Sprint to provide the benefits of a private network.

Access Code Networks

This simplest form of voice network uses PBXs connected by tie trunks. To call a remote location, the user picks up the phone, receives a dial tone from the local PBX, and dials the access code assigned to the remote PBX, such as 8. On receiving a dial tone from the remote PBX, the user then dials the desired extension number. A user in New York City, for example, could dial 8-4032 to call an extension in Boston. If the call failed to go through on the first attempt, the user could either try again later or reroute the call through another PBX. To use an access code network, the user must know the required codes and understand how to reroute calls when the primary route is busy.

As a call is set up along the network, a trunk is seized at each intervening PBX, blocking access to the trunk by other users. If the call is blocked at any node, the user must hang up and start all over again, usually selecting another route. Many users can therefore tie up many trunks without ever completing their calls. If this occurs frequently, users will become frustrated and begin to use the public network, thereby defeating the purpose of installing the PBX network.

Tandem Hub Networks

The tandem hub network involves designating a central hub to which all other PBXs are connected in a star arrangement. This type of network offers a simpler dialing plan than an access code network; the user typically dials 8 to seize a trunk at the hub, a three-digit code corresponding to the desired location on the network, and the four-digit extension of the called party. Despite this unified dialing plan, however, tandem hub networks have several drawbacks. If the hub suffers a catastrophic failure, communications among the other nodes is cut off because a star configuration cannot provide alternative routing, except through the public network. Equipping the hub with enough redundancy to protect against catastrophic failure is expensive and cannot prevent the failure of other nodes.

Electronic Tandem Networks

A new concept in wide area voice networking was introduced by AT&T during the mid-1970s—the electronic tandem network. This design uses software to define various aspects of the network, thereby overcoming many of the limitations of previous designs. It offers such benefits as:

- Unified numbering plan—The same 7- or 10-digit number can be used to reach a destination from anywhere on the network, regardless of the origin of the call in the network.
- Automatic alternative routing—A predefined alternative path is selected automatically when the most direct path is not available.
- Overflow routing—Calls are routed over the facilities of the public network if they cannot be routed over the corporate network.
- Traveling class mark—The class of service assigned to each station is passed from the originating PBX to the remote PBX to ensure that

specified calling privileges are maintained, regardless of where the call terminates on the network.

Virtual Networks

Virtual networks, which have emerged recently, allow organizations to obtain the advantages of a private voice network through the public network. Because the software required to implement virtual networks resides at the carrier's serving office, users have the flexibility of choosing PBXs from more than one manufacturer.

Exhibit II-3-36 illustrates AT&Ts software defined network, which offers a variety of speciai access arrangements for specific levels of traffic. Access possibilities include a single voice frequency channel, 24 voice channels through a DS1 link, 44 voice channels through a T1 link equipped with

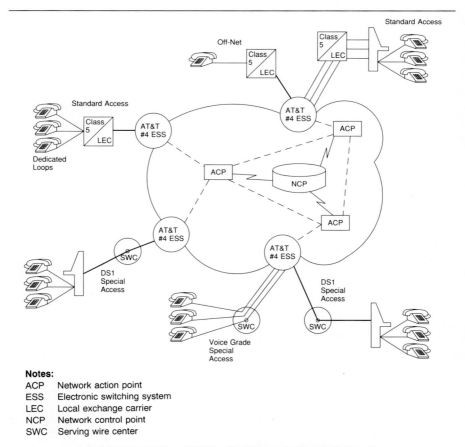

Notes:

ACP Network action point
ESS Electronic switching system
LEC Local exchange carrier
NCP Network control point
SWC Serving wire center

Exhibit II-3-36. AT&T's Software Defined Network

bit-compression multiplexers, and connection to off-net services through a capability that allows a DS1 link to be split into its component DS0 channels at the serving office. A software defined network consists of a network action point connected to the PBX through dedicated switched lines. The action points connect to the carrier's network control point, where the subscribing organization's seven-digit on-net number is converted to the appropriate code for routing through the network.

The PBXs in AT&T's software defined network are connected to an AT&T point of presence through various local access arrangements. The private network exists as a virtual entity on the carrier's backbone network. The carrier assumes the responsibility for translating digits from a customer-specific numbering plan to the carrier's own numbering plan and vice versa, making all routing and failures transparent to the user.

The software defined network also provides reports on configuration, use, and equipment status. For example, AT&T's expanded service management system lets communications managers control various aspects of the network service on an on-premises terminal and a 9.6K-bps private-line connection to AT&T's software defined network's centralized network information data base. Through the data base, a manager can change and delete authorization codes; authorize the use of such capabilities as international dialing by caller, work group, or department; and activate the software defined network's flexible routing feature, which allows the user to direct traffic from one site to another. This last feature, for example, would allow calls to an East Coast sales office to be answered by the West Coast sales office after the East Coast office had closed for the day.

AT&T's network information and management feature allows users to obtain software defined network reports containing call-detail and net-work-use summaries. These reports, similar to those available through PBX and Centrex data collection and reporting systems, help users track costs, bill departments, identify network traffic trends, and review network performance. In addition, users may use the network information and management feature to request access-line status and schedule transmission tests.

VOICE AND DATA INTEGRATION

Integrating voice and data at the PBX is difficult because the traffic patterns for voice are typically very different from those for data. If a PBX configuration has been optimized to handle voice calls efficiently and economically, it cannot readily be applied to data calls.

Local Area PBX Networking

As noted earlier, some vendors of integrated voice-data PBXs are attempting to position their products as LAN alternatives. The use of PBXs for local area data transport, however, is only practical or efficient in certain narrowly defined circumstances, such as when:

- Only a small number of dumb terminals require access to a local host or server.
- The use of terminals is intermittent and only low to moderate speed is required.
- The sustained throughput between the terminal and the host is 64K bps or less.
- The applications require complete delay and protocol transparency, achievable only with a direct connection.

Choosing between a LAN and a PBX also depends on the delay tolerances of the various applications. During peak hours, the PBX is susceptible to blocking, which may manifest itself in one of two ways: the connection attempt fails entirely or the connection setup process is extended, both of which result in longer delays. Once the connection is established, however, no amount of additional traffic can impair the performance of connections in progress.

Almost the opposite is true of LAN switching. A LAN is designed to accommodate the establishment of multiple connections over the same facility. If the initial configuration is properly optimized, there should be no appreciable connection-setup delay. As devices are added, however, all users begin to experience noticeable delays during periods of heavy traffic, regardless of when their connections were established. Also, with more traffic on a LAN, the chance of collisions becomes greater, thereby increasing the number of retransmissions and degrading throughput even more. When the amount of delay exceeds predetermined response-time thresholds, however, some LANs are capable of initiating congestion-control procedures to block access attempts. PBXs do not have comparable capabilities.

During times of heavy traffic, the PBX will penalize only a few users by increasing their call-setup time or by blocking connection attempts. Given the same circumstances, LANs penalize all users by increasing transport delay. Consequently, the choice between a PBX or LAN for local area networking should be based on the type of applications the device is intended to support. Exhibit II-3-37 lists the advantages and disadvantages of a PBX for local area networking; Exhibit II-3-38 lists the advantages and disadvantages of LAN switching.

Many users remain skeptical about the use of their PBXs for local area networking. Many of AT&T's customers, for example, apparently do not accept the company's plans to make the PBX the centerpiece of the future LAN, mostly because they do not want to use a PBX as a LAN server or data concentrator for microcomputers, preferring to purchase these devices separately.

Advantages	Disadvantages
Long circuit holding time; ideal for voice and batch transactions	Inefficient for short, bursty traffic from a large number of terminals
Penalizes only a few users during periods of heavy traffic by limiting access attempts	Error rate increases with distance
	No store-and-forward capability; no error control
	No flow control; prone to congestion
	Time required for call setup prevents rerouting of data during transmission

Exhibit II-3-37.   Advantages and Disadvantages of a PBX for Local Area Networking

Hybrid Switches

Some manufacturers have developed PBXs that use separate connection subsystems (i.e., time-division multiplexing buses) for voice and data. These hybrid switches contain a voice bus—a traditional voice-only time-circuit switch—and a second bus that supports packet switching of data traffic using bandwidth only when required. Switches based on this hybrid architecture appear to overcome the separate limitations of PBXs and LANs in the following ways:

- They provide a high capacity for bursty traffic and provide error protection for all data traffic.

Advantages	Disadvantages
Supports many users on the same facility; ideal for short, bursty traffic from many terminals	Penalizes many users during periods of heavy traffic, because access cannot be controlled
Error rate is low because of the relatively short distances involved	* Does not support voice traffic
No barriers to access; no queuing delay	
Flow control limits congestion; supports proper control of end-to-end device connections and protocol matching	

Exhibit II-3-38.   Advantages and Disadvantages of LAN Switching (Ethernet or Token-Ring) for Local Area Networking

534

- They have no adverse performance impact on voice calls, regardless of data-traffic level.
- They reduce port consumption by eliminating the need for a stand-alone packet-switching converter or packet assembler/disassembler for the X.25 interface and for protocol emulation.

In the hybrid approach, the data circuit switch is connected to the LAN and configured as another server with full interconnection to other resources on the LAN.

PBXS VERSUS MULTIPLEXERS

T1 multiplexing is the single most important technology for the simultaneous transmission of voice and data because of its transport-management capabilities. Today's PBXs offer T1 interfaces that can provide connectivity to systems network architecture (SNA) and X.25 networks, but from a management perspective, private networks built around a PBX are architecturally deficient. Although data and voice channels can be consolidated onto DS1 facilities through PBX T1 interfaces, the PBX cannot route the channels through the central office digital access and cross-connect system (DACS) and still permit users to exercise autonomous management and control. Although PBX vendors provide T1 interfaces, their principal concern has been to maintain compatibility with public network standards for voice transmission (i.e., 64K-bps pulse-code modulation and D4 framing). They have not been concerned with bandwidth management and have ignored transport management altogether. Carriers have also paid little attention to bandwidth efficiency because they have not considered data transport a major part of their business until recently.

Intelligent T1 networking multiplexers with sophisticated transport-management capabilities have grown in popularity among private network users because they address these issues. PBXs are still less efficient and more costly than LANs when used for data transmission and are totally inadequate in managing voice and data together at the transport level. For these reasons, they will not soon replace LANs and multiplexers except in certain narrowly defined circumstances.

THE PBX AND ISDN

In an ISDN environment, the PBX has some potential as a special-purpose local area network server for desktop microcomputers and for voice and data integration. Primary-rate (23B + D) and basic-rate (2B + D) ISDN are currently offered through the Centrex exchanges of telephone carriers. The bandwidth of primary rate ISDN corresponds to the bandwidth of T1, making it suitable for wide area networks. An ISDN-equipped PBX can provide a gateway to remote resources on a wide area network. Basic-rate ISDN

535

offers considerably less bandwidth than primary-rate ISDN, making it applicable to local area networks and creating the potential for integrating voice, data, and control signals over a single pair of wires.

Each ISDN B channel can be used for voice or data at any given moment. In an ISDN-equipped PBX, changes in the use of a particular channel can be implemented by software in the PBX on a call-by-call basis. As an alternative, channel changes can be requested on an ad hoc basis by terminal equipment on the D channel, making it possible for connections between cluster controllers and terminals also to be managed by the PBX.

Voice and data integration on an ISDN-equipped PBX can also simplify terminal moves and changes. Data terminals and telephones can simply be plugged into new outlets and the changes programmed at the PBX on an administrative terminal; cables need not be rerouted. Eventually, it will be possible for ISDN equipment to communicate all network changes automatically on the D channel, thereby enforcing a user's service class no matter where the user's equipment has been relocated.

AT&T implements primary-rate ISDN on its System 85 and Definity 75/85 PBXs. In this implementation, illustrated in Exhibit II-3-39, digital lines from the customer terminate at an AT&T service node consisting of a 4ESS (electronic switching system) central office switch and a digital access and cross-connect system. ISDN access to switched services (e.g., MEGACOM, MEGACOM 800, and ACCUNET switched digital services) is provided through the 4ESS switch. Another digital link transports voice to the digital access and cross-connect system, where the D channel is routed to the 4ESS switch. The digital access and cross-connect system routes individual DS0 channels to the 4ESS switch, to AT&T's ACCUNET packet service, or to another digital access and cross-connect system that provides access to private-line services. This implementation of ISDN has some serious limitations:

- A full 64K-bps channel is needed for each data application, whether or not that much bandwidth is actually used.
- Unlike a digital access and cross-connect system, a 4ESS switch won't be able to offer DS0 bundling until tariffs are available for the ISDN H channels—Therefore, certain applications requiring substantial bandwidth (e.g., the 384K bps of H0 and the 1.536M bps of H11) cannot be supported by a 4ESS switch.
- No end-to-end network management paths are available, so that the level of user control is less than that available with private networks.
- No loopback capability is available, and consequently, no provision for internal diagnostics.

Overcoming these limitations requires the use of a T1 networking multiplexer for ISDN instead of a PBX.

Before primary-rate ISDN becomes feasible as a backbone for linking PBXs in a wide area network, the current islands of ISDN must be con-

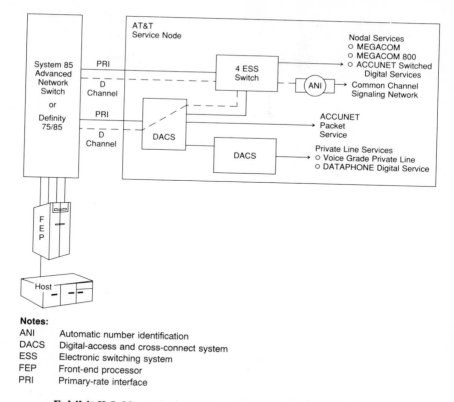

Notes:

ANI	Automatic number identification
DACS	Digital-access and cross-connect system
ESS	Electronic switching system
FEP	Front-end processor
PRI	Primary-rate interface

Exhibit II-3-39. AT&T's Primary Rate ISDN Implementation

nected through signaling system #7. When that happens, carriers could offer more network management, control, and flexibility than they can with current virtual private network services. Some private network users, however, do not want to wait and have begun limited implementations of ISDN on their own PBX networks. If a multitude of differing implementations of ISDN emerge, however, seamless interconnectivity of all devices on a network—the goal of ISDN—may become increasingly difficult to achieve.

SUMMARY

For the foreseeable future, the PBX will remain the cornerstone of the corporate voice network. LANs, however, appear to be more efficient and economical for data transfer than PBXs. The traffic patterns of voice and data are different, and experience has demonstrated that the PBX is inadequate for managing voice and data on the same network. This task is better accomplished at the transport level by high-end T1 multiplexers or through emerging services offered through the public network.

Part III
Local and Metropolitan Networks

LOCAL AREA NETWORKS (LANs) have become causes of concern with many communications systems managers. This is because the communications manager has sole and complete responsibility for the operational effectiveness of the LAN—there is no outside common carrier to blame when things go wrong or to lean on for technical support, as there is with a wide area network (WAN).

There has been a veritable explosion in the deployment of microcomputers in offices and of microcomputer-based workstations in factories and development laboratories. As a result, a huge and growing market has been created for local networking systems.

In the architectural sense, local area networks are really not networks. They are actually multipeer data links that permit connections to be established using a shared physical transmission medium. The medium is usually coaxial cable or twisted-pair wire or, increasingly, fiber-optic cable. An end user's computer is attached to the transmission medium through an interface device that may be either a standalone unit or integrated into the user's computer. This actual transmission equipment is then supplemented with various application-oriented software, servers associated with file, print, or other functions, and network management processes. Internetwork communications and connections off the networks are established through various bridges, routers, and gateways, depending on topology and functions required.

Most local area networks are truly local in nature; they span departments, multifloor organizations, and campus environments. To satisfy these requirements, a variety of topologies, access methods, and protocols have been developed. The most popular of these are Ethernet (supported in particular by DEC and Xerox) and the increasingly popular token-ring approach supported heavily by IBM. Local networks are well documented by both national and international standards. These have been based, primarily, on the work of the Institute of Electrical and Electronics Engineers (IEEE) 802 Committee.

A new type of local network, the metropolitan area network (MAN), is undergoing development. As the name implies, a metropolitan network is bigger than a local network but smaller than a wide area network. In reality, the boundaries are very blurred. Local networks can be interconnected seamlessly to span continents. Wide area networks can be deployed in campus environments and metropolitan networks can have the characteristics of both local and wide area networks. The terminology has come to have more to do with technology than with geography.

Although the LAN market has been dominated by the IBM, Xerox, and DEC approaches, many innovations have come and continue to come from the smaller players such as 3COM, Novell, and others. With the access methodology, transmission medium characteristics, and protocols in place, there is new emphasis on application development issues unique to the local networking environment (e.g., performance, shared resource queuing, and security).

There is no reason to expect that the interest in local or metropolitan area networking will decrease in the future. Most projections forecast continued growth and unabated investment in associated research and development and software. Speeds, however, are likely to continue to increase, especially with the deployment of more fiber-based networks.

A local network trend that will undoubtedly increase in importance is the interconnection of local area networks by means of higher-speed wide area networks. This is one of the major sources of pressure for the development of metropolitan networks. Common carriers, in particular, see metropolitan networks as a business opportunity, allowing them to continue providing local services by connecting buildings across urban and suburban areas.

Part III looks at two aspects of the local networking scene. Section 1 contains a series of chapters on LAN techniques and practices. Section 2 looks specifically at metropolitan area networking, including the fiber distributed data interface, or FDDI.

Section III-1
Local Network
Methodologies

IT HAS BEEN SAID that one of the nice things about local area networks (LANs) is that there are so many of them. This refers not to shipments but to the bewildering number of choices that are available to the prospective user. Chapter III-1-1, "The Manager's Guide to Selecting a Local Area Network," helps to sort out the many choices by providing guides to the communications manager for evaluating and selecting an appropriate solution. The chapter begins with a discussion of the rationale for a local network. It then suggests a procedure for selecting and maintaining a solution to local networking needs. The chapter has a cookbook approach: it presents a six-step procedure that defines the selection process from initiation through planning, analysis, recommendation, implementation, and monitoring. At each step, the author provides useful hints and checklists that will prove beneficial to anyone planning a local network procurement.

The section next examines some practical aspects of the local environment. "Expanding Token-Ring Networks" is the subject of Chapter III-1-2. The realm of local networking seems never to be limited to the local area. As soon as users become comfortable with their new LAN, they begin to press for the capability to interconnect with resources and applications that are located beyond the boundaries of their immediate LAN. This means that the communications system manager needs to be prepared to provide some local-to-local or local-to-wide area connectivity solution. In this chapter, the author explores the expansion requirements and alternative expansion solutions for token-ring networks. After a brief review of token-ring technology, repeaters, bridges, and routers are explored. Each is discussed in terms of functions, differences, and impact on performance.

Chapter III-1-3 concludes this section with a discussion of "Failure Analysis in Token-Ring LANs." This chapter addresses an issue of importance to every communications network administrator: what to do when the network fails. It begins with a brief overview of token-ring operation. It then looks at two categories of failure and how they manifest themselves. The two categories of failures are hard failures and soft failures. The discussion of soft failures is particularly helpful because these are the intermittent, insidious problems that plague the network but defy analysis. Following the discussion of failure modes, the chapter examines tools and methodologies for pinpointing the cause of failure.

Chapter III-1-1
The Manager's Guide to Selecting a Local Area Network

Karen L. Rancourt

A LOCAL AREA NETWORK (LAN) CONSISTS OF the connecting medium (e.g., cable) between computers, printers and storage devices, the interfaces between the devices, and the software that manages the movement of data throughout the network. As the term *local area* implies, LAN systems are commonly used by work groups or departments among several offices in a single building or among several buildings housed in a complex or within a few miles of each other. One indication that the growth and acceptance of networking is assured is that as new commercial buildings are being constructed, LAN conduits are being included in the plans along with heating ducts and electrical and telephone wiring.

The key to selecting a successful LAN is for everything on the network—computers, printers, storage devices, device interfaces, and network software—to be compatible; that is, all parts of the LAN must be able to communicate with each other. Because final versions of industry standards have not in all cases been established, care must be taken to ensure that vendors' claims of compatibility are accurate for specific situations.

A SIX-STEP PROCEDURE FOR SELECTING A LOCAL AREA NETWORK

The LAN selection process is similar to many projects that require the cooperative companywide participation of people with different needs and technical backgrounds. The six steps that should be followed for selecting a LAN are:
1. Initiation.
2. Planning.

3. Analysis.
4. Recommendations and approval.
5. Implementation.
6. Evaluation and monitoring.

STEP 1: INITIATION

Defining the Client Group or Sponsor

During this phase of the network project, the client group or sponsor is identified as the party interested in selecting a LAN. The client may be:

- A functional group in the company (e.g., the finance and administration department).
- A work group (e.g., the finance and administration payroll group).
- A cross-functional group with something in common (e.g., marketing and sales groups that use the same application or data base).
- A senior manager (the more senior the better).

It is advantageous to have a client group or sponsor that is not only enthusiastic and respected but influential throughout the company. The decisions that result from a networking selection project could be ones that an organization will have to live with for many years.

Identifying the LAN Project Coordinator

After a project sponsor is designated, the next task is to identify the project coordinator. For example, will this project be coordinated by the communications system manager, the MIS manager, or will it be a shared responsibility? Because of the unique circumstances and factors in each company, there are no rules of thumb for answering this question. Nevertheless, it is certainly important for all involved managers to be supportive of and involved with networking projects in the company. The communications department will be called on to provide some of the technical expertise required, the information center staff will probably be involved because of its access to and experience in working with nontechnical end users, and the communications staff will provide expertise in the local network technology and standards.

Adding Technical Experts to the Project

Once a project coordinator and client group or sponsor have been identified, other users who have or will have an interest in the outcome of the project should be included. For example, if the client is the finance and administration payroll group that will be using a personnel multiuser data base, users from the human resources department who also use this data base should be represented in the project.

Technical experts are also an essential part of the networking project team, especially during the analysis and remaining phases of the project. These experts are most likely to come from the data communications and MIS organizations, and they should be involved throughout the project—not just during those sessions that require direct input. It is important for the technical staff to have firsthand knowledge of the context of the users' needs and expectations; otherwise, the technical staff could configure great solutions for the wrong situation.

Finally, those who are responsible for collecting, analyzing, and preparing the project data need to be brought on board. It is advisable to select some of these people from the client group. Although they may need guidance, their sense of ownership in the project will increase if they are actively involved.

Forming an Ad Hoc Committee

Overall, the networking project team is likely to be an ad hoc group (i.e., it has a mission with a specific ending point). In this case, ad hoc members are responsible for bringing the project through the evaluation and monitoring phase. Subsequently, certain project members will remain active (e.g., staff responsible for networking maintenance), but the project group as a whole serves its purpose and disperses. It is also helpful for this group to have a name, such as the finance and administration networking project. Total ad hoc committee membership should never exceed 10. If, for whatever reason, more people desire involvement, the project coordinator can be creative, by involving them as alternative representatives or, if they are senior people, by having them contribute in an advisory capacity.

STEP 2: PLANNING

Interviewing Users to Determine Need

To begin this step, it is important to have users focus on both their immediate and future needs. Users should not be constrained by current technology. For example, the networking team should prohibit such statements as, "Well, I won't even mention this because the technology doesn't exist." At this point, wish lists are important and can be a valuable source of information as long as users understand they are stating ideal requirements that may or may not be met. Another reason to collect wish-list information is that networking technology is evolving and changing rapidly. What is currently technically infeasible could become reality in the near future.

For these reasons, it is important to structure open-ended interviews with users, encouraging them to be both realistic and imaginative. For example, users should be asked the following types of questions:

- Would the sharing of information, tools, and resources make you more productive? In what ways do you think they would help you be more productive?
- If you could design a hypothetical, ideal network, what would it be like? Discuss all the people, computers, services, and internal and external data bases you would want access to. How would this access increase your effectiveness and productivity?

Defining the Users' Major Work Tasks

In addition to collecting data on users' perceptions of how networking might increase their individual productivity, it is important to gather such specific, detailed information as what they do, how they do it, when they do it, information they need to do it, sources of this information, and information bottlenecks. Later, during the analysis phase, their input will prove invaluable in defining basic networking needs at both the individual and work group levels.

Define Networking Restrictions or Constraints

A networking project can be affected by geographical, physical, financial, or time-related restrictions. Therefore, all restrictions or constraints need to be identified up front. For example, if a specific type of cable has already been installed, management might decide to use only this type of cable throughout a facility. If future plans call for linking several buildings by means of networks, configuration decisions will have to take this into account. If certain purchases are to come out of the capital expenditure budget, planners need to know about any fiscal constraints.

Learning from Others' Experiences

In networking projects, an often underused source of information is network users from other companies. No one knows the potential pitfalls of making networking decisions better than those who have been through a similar process. In addition, network users in other installations can offer advice from a hands-on perspective. Their experience and advice can be invaluable. Locating these people can be accomplished in various ways, including contacting friends in other companies, asking vendors for customer contacts, attending networking seminars or trade shows, and tracking down contacts mentioned in professional publications.

Drafting the Ideal Configuration

Once the information has been gathered from users, the networking team can define the ideal configuration from a functional perspective. This means all functionality desired by users is included in this configuration regardless of whether the technology actually exists. Later, detailed analysis may indicate that the technology does in fact exist or is forthcoming. At this point, the configuration should reflect what is needed and preferred, not what is technologically feasible. It is important that project members and users understand that this ideal configuration is a model, not the end product.

STEP 3: ANALYSIS

Selecting and Evaluating Potential Vendors

During this step of the networking selection process, the emphasis is on matching user requirements with existing technology and products. The project constraints and restrictions identified earlier should now be examined in detail. For example, a company may have a policy that US vendors are given top priority in purchases; therefore, only certain vendors need be considered initially.

The networking team must also pay close attention to user input in developing a vendor checklist, carefully detailing the requirements for each of the major categories that compose the network. Major categories include:

- User interfaces.
- File-sharing devices (e.g., network server).
- Printer services.
- Communications requirements.
- Data transmission requirements.
- LAN-to-LAN connections.
- Minicomputer and mainframe connections.
- Personal computer or workstation compatibility requirements.
- Security.
- Network administration.
- Documentation.
- Support, service, and warranty.
- Application compatibility.

Within each category, a complete set of requirements should be listed. For example, in the user interfaces category the requirements might include:

- Easy log-on and log-off procedures.
- Menu-based access to network services.
- Availability of help screens.
- Availability of status information on each workstation.
- User-defined default options.

- Explicit error messages.
- Capability for advanced users to shortcut certain steps.

Next to each requirement there should be space for a checklist in which the networking team can indicate *yes, no,* or *future* (i.e., the date that the vendor claims the requirement can be met). For each vendor under consideration, a checklist of each major category is completed. At this point, certain vendors can be eliminated because they do not meet basic requirements.

Rating Vendors

Once the first cut has been completed, two procedures are then followed to rate the vendors still under consideration. First, each vendor is rated on the basis of its overall ability to meet the individual critical requirements. For example, one convenient method of evaluation would be to rate the vendors in the following manner:

5 = Fully meets requirement
3 = Partially fulfills requirement
1 = Unsatisfactorily fulfills requirement

Second, a standard configuration is developed for each vendor, and cost estimates are given for each alternative.

Narrowing Down the Vendor List

Using the results of the ratings and cost estimates, the networking team can once again condense the list of potential vendors. At this point, it is advisable to learn as much as possible about each vendor, including:

- History of the vendor company (e.g., annual revenues, independent financial analysts' ratings).
- Profile of the company (e.g., size, growth, management team).
- Company's reputation (e.g., how other customers, especially those with installations similar to those being considered, rate the vendor).
- Support, service, and warranty provided—This information can be obtained by meeting and talking with vendor staff and management.

In addition, the networking team should borrow as many products as possible from each vendor to conduct an in-house, hands-on analysis. Some vendors are willing to loan equipment readily, others are not. A vendor's response to such a request can in itself prove enlightening.

At this point, proposed agreements are negotiated with each vendor. These agreements should address specific purchases, associated costs, payment schedules, delivery schedules, support, service, and warranty as well as any unique circumstances. In some companies, the legal department will be required to review the proposed terms and conditions, especially if licenses are involved.

STEP 4: RECOMMENDATIONS AND APPROVAL

Preparing Project Recommendations

During this stage, a report summarizing the networking project recommendations is developed. It should include the following items:
- A brief description of project objectives.
- A list of participants (i.e., groups or individuals) and an explanation of their involvement.
- A brief description of the steps taken in the project.
- Any recommendations made, including configuration definition, technology requirements, vendors, cost estimates, and staff requirements.
- The benefits to be derived from implementing recommendations.
- A discussion of any risks associated with the project and of any plans to address them.
- The proposed implementation plan, including scheduling, training, documentation, materials, and changes in administrative policies and procedures.

Obtaining Necessary Approval

The approval cycle varies among companies. In some cases, the report of project recommendations may be submitted to one or more boards or committees for approval; in other cases, one manager's signature may be all that is required. Regardless of individual policies and procedures, the networking team must clearly define the steps of the approval cycle at the outset and keep all committees or individuals with signatory power apprised of project progress. This may be accomplished through written progress reports, informal discussions, or regularly scheduled updates presented at meetings. When ongoing communication has been handled adequately, project approval is usually a straightforward process. Minor changes may have to be made, but major revisions or analyses should not be required.

STEP 5: IMPLEMENTATION

The major components of the implementation phase are threefold:
- Defining and scheduling the mechanics of the networking project (e.g., installation of hardware, software, devices, and wiring) and the resources required for and allocated to this purpose.
- Defining and addressing conversion and transition activities (e.g., production downtime, parallel systems during changeover, administrative policies and procedures affected).
- Defining and scheduling required training for users and network personnel.

The key to successful implementation depends on providing detailed information on who needs to do what, when they need to do it, and what resources are needed to do it.

Establishing a Contingency Plan

The network team should spend some time discussing those factors that could have a negative effect on project implementation and should establish a contingency plan. For example, certain product deliveries are critical (e.g., the network cable), whereas others are less serious (e.g., only three of the six personal computers that were ordered arrived on time). Contingency planning should not be an exercise that leaves everyone thoroughly depressed; rather, it should be an opportunity to emphasize that schedule deviations are inevitable.

By identifying critical activities, the network team can ensure that energy is focused and not wasted. For example, if cabling is a critical activity, the team should talk with the cabling installers regularly—every day if need be. The point is to recognize that implementation plans cannot be carved in stone; there are too many factors involved that are beyond the control of the best planners. Users and management need to be aware of possible schedule delays.

STEP 6: EVALUATION AND MONITORING

This phase of the network project is ongoing. The two major areas of concern are postimplementation network monitoring and network project evaluation.

Monitoring Network Activities

Hardware, software, and network cables and devices must be monitored regularly. Trained personnel should be held responsible for tracking network use and activities and making adjustments for system overloads. In addition, administrative policies and procedures need to be in place to address system problems and failures. Users must know what to do when they encounter problems (e.g., what diagnostic procedures should be used at the workstation, when and how to contact the system administrator). System administrators and other network personnel have to be trained in system troubleshooting and problem resolution as well as in procedures for recording, escalating to a higher management level, and closing out problems.

Evaluating the Network Project

To determine whether networking goals and objectives are being met and costs and anticipated benefits are being attained as planned, representatives of the network administrator and the client group who have close contact with both users and system administrators should be assigned this task.

Generally, the more successful the initial networking project, the more likely it is that the demand for additional networks will increase. As networking needs increase, the more important a consolidated and comprehensive networking strategy is to the company.

FUTURE NETWORKING PROJECTS

At this point, it is advisable to bring the original networking project to a close. The ad hoc members' efforts and contributions should be recognized. An informal party or get-together might be planned for them, and senior management should be included. A note of thanks along with a description of their work in a company newsletter or the MIS bulletin is a possibility. Regardless of how it is handled, this is a perfect opportunity to highlight the effectiveness of the team effort and educate others in the organization about the benefits of networking. The design and implementation of a successful network project in a company is cause for celebration.

Subsequently, a permanent, core networking coordinator or team should be established to control the strategy, growth, and implementation of future networking projects and to oversee current ones. For those networking needs that require in-depth planning and analysis, new ad hoc committees can be formed, serve their purpose, and disband. In other cases, networking needs can be easily met by the core networking individual or team.

In all cases, selecting and maintaining LANs requires the cooperative and participative efforts of users, network administrators, data communications staff, and the MIS department. By following the guidelines described in this chapter, the manager and other networking personnel can help ensure that successful partnerships result and the benefits of LANs are obtained.

ACTION PLAN

In selecting and maintaining a LAN, the communications systems manager can help make the process successful by implementing the following steps:
- Step 1—Initiation:
  —Define the client group or sponsor.
  —Identify the LAN project coordinator.
  —Add technical experts to the project.
  —Form an ad hoc committee.
- Step 2—Planning:
  —Interview users to determine needs.
  —Define the users' major work tasks.
  —Define networking restrictions or constraints.

-Poll other network users and learn from their experiences.

-Draft the ideal configuration.

- Step 3—Analysis:

  -Select and evaluate potential vendors.

  -Rate vendors.

  -Narrow down the vendor list.

- Step 4—Recommendations and approval:

  -Prepare project recommendations.

  -Obtain necessary approval.

- Step 5—Implementation:

  -Define and schedule installation of hardware, software, devices, and wiring.

  -Define and address system conversion and transition activities.

  -Define and schedule training for users and network personnel.

  -Establish a contingency plan.

- Step 6—Evaluation and monitoring:

  -Monitor network activities.

  -Complete a network evaluation.

  -Provide for future networking projects.

Chapter III-1-2
Expanding Token-Ring Networks

Anand Rao

THIS CHAPTER EXAMINES METHODS that can be used to expand the range of IBM Token-Ring local area networks (LANs). It first reviews the operating basics of LANs of this type, then examines the devices that are available to extend them.

TOKEN-RING BASICS

IBM's Token-Ring LAN is a baseband network that can use any of several cabling media, including unshielded twisted-pair cable. IBM's product and the standardized Institute of Electrical and Electronics Engineers (IEEE) 802.5 version are usually configured as a star-wired ring, using two independent paths—one for regular (i.e., primary) communications and the other used as a backup.

A unidirectional, point-to-point signal transmission methodology (with backup capabilities built in) is used. Wiring concentrators positioned along the network provide connections to the user areas. The use of the concentrators makes it possible to isolate failed links or components and use the network management system to configure backup options. Additional links can be added or existing links can be removed at any time without significantly altering the remainder of the LAN configuration.

Although IBM has made several enhancements for greater speed and flexibility, Token Ring has essentially remained a product for installation in a single building, with a maximum of 260 users per LAN. To connect one LAN to another, to increase the number of users per LAN, or to connect users in different buildings requires the addition of connectivity devices (e.g., repeaters, bridges, and routers). Before discussing these devices, the operation of the Token-Ring network in its basic star-wire topology is discussed.

Data Transmission Techniques

Data transmitted on a Token-Ring LAN is encoded into frames. Each frame contains segments for data link control, the data, and the physical trailer. The data link control segment contains a physical header of variable length that includes control fields, the destination address, and the source address. The data link control segment is followed by the information segment (also of variable length), which is followed by a 6-byte physical trailer segment containing control fields that signify the end of the frame.

The IEEE 802.5 standard provides automatic acknowledgment that a frame has been copied by the addressee. When the destination node copies a frame, it generates a copied bit. The source node deletes the frame from the ring when it detects the copied bit during its frame-scanning procedures.

Data movement is accomplished through the Token-Ring access control procedure. A token, a standard signal containing the access control field, is passed from node to node. The node that has data to transmit captures the token, changes its status to that of a frame, and transmits it to the next node. The change in status of the token is brought about by the addition of the destination address and routing information to the access control field in the token. This transforms the access control field to a physical header. The data to be transmitted is added immediately after the physical header and is followed by the physical trailer.

As the frame moves along the ring, its physical header is examined by every node to determine its address. If the address in the physical header matches the node's address, the node copies the frame from the ring and changes the physical trailer to copied status. Forwarding of the frame continues (with no more copying, because the address does not match the address of any of the following nodes and the physical trailer contains the copied bit) until it reaches the node that sent it. The sender removes the information added to the physical header and the data, changes the status of the frame to a token, and transmits it along the network.

Each node has a unique address and can copy only the data that is addressed to it. It is possible, however, to issue frames with a broadcast address indicating that all the nodes along the way must copy the data.

The nodes are primarily involved in processing data; token handling (i.e., frame recognition, token generation, error checking, address decoding, frame buffering, fault detection in the ring, and frame conversion) is performed by a ring interface adapter that is connected to each node.

EXPANDING THE TOKEN RING

As stated earlier, the Token-Ring LAN is essentially a product for a single building. Even within a building, additional devices are needed to extend

554

the LAN over several floors. To connect one Token-Ring LAN to another requires additional connectivity products. Some of the products that extend and interconnect Token-Ring LANs are repeaters, bridges, and routers.

Repeaters

A repeater is used to distribute signals beyond the distance limits of the ring interface adapters. (Most adapters can transmit the frame or token to a distance of approximately 770 m.) Repeaters can use either copper wire or optical fiber technology. Copper wire repeaters have a range of an additional 770 m. Optical repeaters can drive the signals as far as 2 km.

Bridges

If nodes are widely dispersed or their number exceeds the maximum allowed in one ring, one solution is to create and interconnect separate LANs. Bridges, which connect the separate LANs, examine the data encoded in the physical header and determine the ring in which the addressed node is located. By examining all the frames that pass through it, a bridge can construct an address table. It then becomes easy to determine whether an incoming frame is addressed to a local node or one on an adjoining ring. By copying and buffering the frames that are destined for the adjoining ring, the bridge facilitates smooth transmission of frames.

A bridge operates at the media access control sublayer of the open systems interconnection (OSI) protocol stack and works independently of the upper layers. Although the primary function of bridges is to logically connect two adjacent LANs, they can nevertheless be employed to build extended Token-Ring networks. Bridges can be divided into two main types: static bridges and learning bridges.

Static Bridges.   The connection functions of static bridges are restricted by address tables that must be created by programmers or operators. Static bridges perform no independent connectivity functions and strictly adhere to the information contained in the address tables. Although static bridges are fast and error free, they are not flexible. The address table must be reconfigured when any node is moved, deleted, or added. In addition, the number of nodes in a ring is restricted by the size of the address table.

Learning Bridges.   A learning bridge has the capacity to modify the address table and learn the location of the nodes relative to the bridge. The bridge can address all nodes, even when a node's address is unknown. It can configure itself and does not require previous knowledge of the location of an addressed node. Learning bridges are more tolerant than static bridges of network faults and do not need to be manually reconfigured when nodes are added, deleted, or moved.

Measuring Bridge Performance

A bridge should be capable of processing both the source and destination addresses of the frames. The source address table is constantly scanned to determine the active nodes in the ring; some bridges can delete addresses that are not in use. Because the bridge constantly compares the source and destination addresses and generates statistics on the flow of traffic, it becomes an important data base of the traffic on the network.

Both static and learning bridges have certain limits, and for them to operate efficiently requires judicious network management. Network designers and managers should take into account the data-handling capacities of bridges when making plans to allow inter-ring traffic. It is possible for the number of data frames arriving at a bridge to exceed its buffering capacity; the results are frame collisions and network congestion.

Quality of service is measured by a bridge's ability to minimize frame loss, reduce traffic delays, delete old frames, perform error checking efficiently, filter unsuitably large frames, determine the priority of a frame, and generate meaningful data on the traffic.

Routers

Routers operate at OSI layer 3, the network layer. Routers' functions are similar to bridges (i.e., frame routing and forwarding) but are performed at the network level, and they also have address filtering functions that are beyond the capabilities of bridges. In addition, they perform their functions differently than bridges. In a Token-Ring network, bridges do not have to be addressed; their presence is transparent to the transmitting and receiving nodes. A router needs to be addressed by a node when the node requests a specific type of service. A router can select the most desirable path between LANs depending on the parameters set by the operators, such as transit delay, congestion avoidance, shortest hop (between rings), and transmit count.

Routers, because they make decisions, can introduce certain delays of their own, leading to the formation of a frame queue. The selection of a suitable router depends on its frame processing rate, which should be appropriate to the projected inter-ring traffic.

Categories of Routing

Directing network data through bridges and routers requires the use of routing algorithms; these algorithms are classified as source routing algorithms, adaptive or learning algorithms, or spanning tree algorithms.

The Source Routing Algorithm. The source routing algorithm routes frames through a multiple-ring local network. Operation of a source routing algorithm requires the following information:

- Source and destination addresses—Typically, these are the designated identification numbers of the nodes.
- Ring and bridge numbers—The numbers of the destination and intermediate rings and bridges.
- Path exploration information—A test (i.e., XID) command to the local ring to determine on-ring information and a test command to all rings to determine off-ring information.
- Best route found—All potential routes that meet the set parameters are reported back to the source.

Use of a source routing algorithm with a bridge improves bridge performance, because the bridge does not have to perform table maintenance. In addition, it is possible to eliminate transmission loops by using the source routing algorithm. Longer sessions are possible at the workstation level because the test command establishes a firm connection. At the network level, source routing introduces some delays because of the station identification traffic inherent in the frames.

The Adaptive or Learning Algorithm. This algorithm works through routing information that is determined by the bridge and stored in the table. An adaptive algorithm has the following unique features:

- The frame header does not include ring or bridge numbers—The routing mechanism treats the bridged networks as one network.
- The frame header does not contain a routing field—The routing path is decided by the bridges.

The path (i.e., routing) exploration process includes the following steps:

- A learning period, to determine the best possible route.
- An unlearning period, to discard unusable routes.
- Ring determination through response to the XID, to determine whether a given ring is functioning.
- Off-ring pool or broadcast determination, to determine the addresses of nonfunctioning rings through flooding (i.e., broadcasting) frames to determine the addresses of unknown rings.

The adaptive routing algorithm makes the bridges less transparent, because they are addressed in each frame. This implies that the bridge tables used with the adaptive algorithm are more complex to build. Adaptive routing has no special effect on the workstations as compared with source routing. The performance level of the workstation is slightly reduced because there is a heavier processing load, and network performance is improved because there is lighter XID traffic. The network configuration is strictly nonloop because the system does not tolerate redundancy.

The Spanning Tree Algorithm. The spanning tree algorithm is generally used in addition to adaptive routing in multiring environments. The algorithm ensures data integrity but at the cost of speed, because it requires probing messages to sense the connectivity of bridges throughout the extended network. Potential loops are eliminated by turning off some bridges. The spanning tree algorithm also dictates that a workstation be designated as a standby node, for use as needed.

The spanning tree algorithm has the same overall effect as adaptive routing, though it is more dependable in multiring environments. Network designers can use either bridges or routers to provide connections between rings. It is generally accepted that bridges offer better performance than routers, though routers provide better control and more options for ring management. On large networks, both bridges and routers may be used.

SUMMARY

The traffic data generated by bridges and routers is helpful in monitoring the segment and overall performance of the Token-Ring LAN. The network manager can then make choices between repeaters, bridges, and routers and their operating algorithms so as to attain the best combination of communications priority, network speed, and accuracy.

Chapter III-1-3
Failure Analysis in Token-Ring LANs

L. Michael Lumpkin

LOCATING AND ELIMINATING FAILURES in token-ring LANs is a complex task; it requires an understanding of the general principles of network technology and of the token-ring protocol, specifically. This chapter describes the types of failures that can occur on token rings and the conditions under which they can occur. It also describes the token-ring protocol's mechanisms and other analysis systems that isolate and allow recovery from network failures.

TOKEN-RING OPERATION

A token-ring LAN is basically a set of point-to-point communications links that form a closed electrical circuit. Unlike Ethernet, which requires stations to contend for access, token-ring nodes cooperate to send and receive each other's data. The upstream and downstream neighbors of a node function as signal repeaters and therefore directly affect communications and management of the entire LAN.

The term *ring* is used for descriptive purposes and to differentiate a token ring from the more linear Ethernet design. The actual physical shape of a token-ring network is relatively unimportant. The network designer can lay out the network in any shape that fulfills users' functional needs, as long as the points of origin and termination form a logical and physical loop.

A continuous stream of serial data is transmitted through the ring, moving from one communications link (i.e., station) to another. This condition exists as long as the ring is active and working properly. A station need not be talking to another station or sending its own instructions to a device (e.g., a printer) to keep data moving on the LAN. It simply receives and transmits data as it comes along.

The data can take the form of three types of signals: a token, a data frame, and a fill pattern. When a station needs access to the ring, it waits for and takes possession of the token and then transmits data through the network to a destination address. Frames consist of formatted data that is communicated among nodes on the LAN. Fill patterns are signals transmitted by a station when it is not receiving or transmitting data. These signals simply fill the intervals between data and tokens in order to maintain physical synchronization.

DISCOVERING AND ELIMINATING THE CAUSES OF FAILURE

There are two categories of token-ring failures that can affect the whole network, at least to the point of diminishing its performance: hard failures and soft failures. These are discussed in the following sections.

Hard Failures

Hard failures are caused by the failure of one or more hardware components; they account for the majority of network downtime. Hard failures disrupt communication among all stations on the network, interrupting the movement of data between stations. When this occurs, the station that is immediately downstream from the fault (when it detects a signal loss) attempts to restore communication following procedures defined by the token-ring protocol.

For example, if the ring consists of three stations, A, B, and C, all connected by a single multistation access unit, the data stream's direction is counterclockwise from A to B, B to C, and C to A. If a cable fault occurs between the multistation access unit and station A, the station detects a signal loss and attempts to claim the token and reinitialize the ring, as shown in Exhibit III-1-1. If it is unsuccessful, station A begins to transmit a periodic distress signal, called a beacon. The beacon frame identifies the failure domain, which includes station C's transmitter, station A's receiver, the multistation access unit, and all interconnection cables. Station A transmits the beacon until station A's receiver picks it up; this can occur only if the cable fault is corrected.

Communication between stations A and B and between B and C continue during this process. When stations B and C receive beacon frames from station A, they enter the beacon repeat state. Once all stations on the ring are in the beacon repeat state, no additional stations are allowed into the ring. This ensures that the configuration of the network and the fault domain remain constant, thereby simplifying fault isolation and error recovery.

At this point, station C detects that it is identified by the beacon as the last known upstream neighbor of station A. Station C removes itself from the

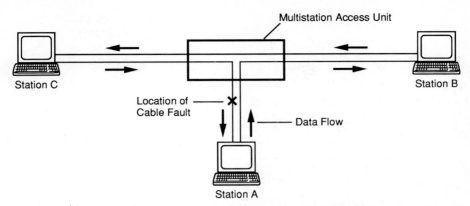

Exhibit III-1-1. Cable Fault Between a Multistation Access Unit and Station A

ring and performs a self-test to determine the condition of its adapter card and the cables between the card and the multistation access unit (see Exhibit III-1-2). Because the faulty cable lies between the multistation access unit and station A, station C passes the self-test and reinserts into the ring.

Station A repeats the process; when it fails the self-test, it exits the ring (see Exhibit III-1-3). Either station B or station C claims the token and restores communication.

If the fault is in the multistation access unit or in the backbone or cable plant, the token-ring protocol cannot isolate or resolve the fault. The network administrator would not receive an error message, and manual in-

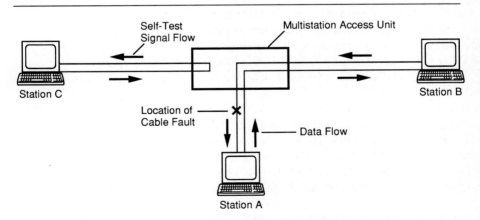

Exhibit III-1-2. Station C Exits Ring and Performs Self-Test

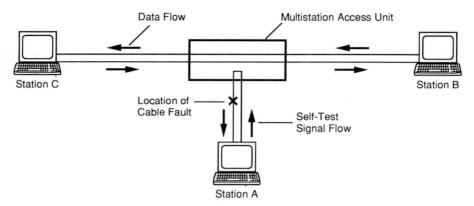

Exhibit III-1-3.   Station A Fails Self-Test and Exits the Ring

---

tervention would be needed (see Exhibit III-1-4). By a process of elimination, stations would be physically unplugged until the faulty cable is isolated.

Soft Failures

Soft failures are intermittent events that, as they accumulate, cause overall network performance to degrade. There are two types of soft failures: normal and abnormal. Normal failures are expected on token-ring LANs because they are directly related to stations inserting or exiting from the ring. Abnormal failures are those that cannot be correlated directly to a station insertion or exit.

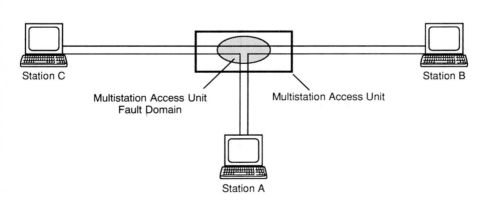

Exhibit III-1-4.   Fault Inherent to the Multistation Access Unit

The ability to differentiate between abnormal and normal failures is a key issue in token-ring analysis. One of the unique characteristics of token rings is that normal soft failures are always occurring on them. The vast majority of these are insignificant because they can be correlated to stations entering and exiting the ring. Abnormal failures can cause errors that are identical to those caused by normal failures. The network analysis system must therefore be able to note normal failures and differentiate these from abnormal failures, so that the network administrator can determine the appropriate course of action.

Normal Soft Failures.   When a station enters the ring, it applies a voltage to the multistation access unit; this opens a relay, and the station becomes part of the ring. The insertion of the station disrupts communication on the ring because the station does not have a way to physically synchronize itself with ring activity. When a station inserts into the ring, it literally breaks the ring for as much as 5 ms, which disrupts communications on the entire network. Stations exiting the ring cause the same disruption. In both cases, the brief disruption can result in a soft error, such as a burst error or fragmented frame. Other soft errors include the following:
- Token errors—These occur when a station insertion or exit corrupts the token. Token errors can also be caused by network noise. It is typical for one or two token errors to occur per station insertion or exit.
- Lost frame errors—These occur when a frame is corrupted by a station insertion or exit. The frequency of lost frames is very low. Not more than one frame should be corrupted by station insertion or exit.
- Line errors—These are reported by the first station to repeat or copy a corrupt token or frame.

Abnormal Soft Failures.   The occurrence of abnormal failures during operation of a token ring cannot be anticipated. Generally, they are more severe than normal failures, affecting overall network performance. They also can indicate marginal hardware operating conditions. Typical abnormal soft errors include the following:
- Burst errors—These are the result of network noise and can cause frames to be lost or fragmented. As many as five burst errors are typical per station insertion and exit.
- Station insert and exit errors—Too high an incidence of these is considered an abnormal soft failure. The cause may be a user frequently rebooting or turning off a station, or a hardware failure in a station.
- High error counts—These normally indicate such marginal hardware conditions as a deteriorating multistation access unit relay, a high-resistance adapter card connection, or a station with a marginal power supply.

- Soft errors not correlated to station insertion or exit—These may be associated with ring performance conditions or file server performance.

TOOLS FOR EFFECTIVE TOKEN-RING ANALYSIS

Token-ring operation is relatively simple once the ring is configured and running. Initialization and error recovery are the complex aspects of token-ring operation. The token-ring protocol is very susceptible to faults in the open systems interconnection (OSI) physical and data link layers; in fact, 60% of all network failures occur at these lower levels. Network administrators therefore need an analysis system that concentrates on the lower OSI layers. The system should provide comprehensive, rapid, and detailed analysis of every aspect of network operation.

A token-ring analysis system must also be able to monitor and analyze activity at the network's maximum bandwidth. An idiosyncrasy of token rings is the constant occurrence of events that the administrator must be made aware of. Having a full bandwidth monitoring capability ensures that all events—normal or abnormal—are captured, time stamped, analyzed, and reported. This allows rapid error recovery, and a system with this type of monitoring capability is also a useful tool for planning network alterations.

Capturing and filtering every event on a token-ring network requires an analyzer with extremely fine time-stamp granularity. Token-ring events occur in rapid succession and for very brief periods of time. The analyzer must be able to report the events in the precise sequence in which they happened. The correlation of events to specific failures is crucial to error recovery, and this must be provided by the analyzer. Otherwise, a series of several events could appear to be the cause of one failure.

Other analysis system capabilities are of greatest value when the ring experiences faults that require rapid isolation. The failure mode description is one such feature, because when the ring begins beaconing, new stations cannot enter the ring until the failure domain has been isolated and the fault resolved. For time-critical analysis, it is necessary to have an analysis system that can enter the ring during beaconing.

Such a system should not depend entirely on the use of a standard token-ring adapter card, which would be recognized and restricted as if it were another station. The system should be a specialized implementation of some upper-layer functions of the token-ring protocol, but with a concentration of analytical capabilities at the physical layer and media access control sublayer.

Some adapter card–based systems can enter the ring during beaconing, but in doing so, they alter the conditions under which the error occurs. Because token rings pass signals along communications paths, any change to the path will affect the integrity of the signal. The entry of a card-based system alters not only the physical configuration of the ring but also the logical configuration. When the card-based analyzer inserts into the ring, it changes the last known upstream neighbor of the beaconing station. As a result, the network failure may appear to be resolved when in fact it is temporarily disguised. The faulty condition will reappear when the analyzer exits the ring. The solution is an analyzer that does not depend entirely on standard token-ring adapter cards.

Once inserted into the ring, the optimal analysis system will capture the massive amounts of data typically found on token-ring LANs. It must correlate specific soft errors to the insertion or exit of stations and identify errors caused by an abnormal condition. The system thereby enables the administrator to determine whether the error is hardware related or caused by heavy user traffic or interference (i.e., noise).

Being able to simultaneously capture and generate heavy network traffic provides a secondary benefit to the communications systems manager. If the analysis system can generate traffic that approaches the full network bandwidth, massive amounts of data can be downloaded to connectivity devices (i.e., bridges) to test the actual throughput of the device. Such tests can prevent configuration problems before they occur.

SUMMARY

An administrator of a token-ring network needs specialized diagnostic tools to detect, analyze, and remedy network disorders. Many of the network analysis solutions promoted for use on token rings are mediocre adaptations of technology developed for use with Ethernet LANs. Both types of networks have failure modes, but the symptoms and causes of network disorders are essentially different from each other. Because token rings are very susceptible to errors at the physical layer and media access control sublayer, it is crucial for a token-ring analysis system to have capabilities that focus on these layers.

The continuous stream of data that is characteristic of token rings requires special capabilities as well. The system must be able to constantly capture and filter data and provide the administrator with a means of predicting errors. An ideal system would perform some upper-layer analysis but also would have the following specialized functions:

- Capture of 100% of the token ring's events.
- Fine time-stamp granularity for accurate reporting.
- Correlation of network events to normal and abnormal conditions.
- The ability to enter a beaconing ring.
- The ability to enter the ring without altering the ring configuration.
- Simultaneous capture of data and generation of full-bandwidth traffic to test throughput of bridges, routers, and other connectivity devices.

Section III-2
Metropolitan Area Standards and Technology

METROPOLITAN AREA NETWORKS (MANs) are networks with many of the characteristics of local networks but with much larger geographic coverage and, usually, higher data rates. They are designed to span campuses, urban cores, and, of course, metropolitan areas. The four chapters in this section address this subject.

Chapter III-2-1, "The IEEE 802.6 Metropolitan Area Network Distributed-Queue, Dual-Bus Protocol," introduces the DQDB approach to metropolitan network design. The characteristics of the network are explained, as are the elements and components that constitute an implementation of this network.

Chapter III-2-2, "Metropolitan Area Network Standards from IEEE 802.6," continues our discussion of the DQDB MAN technique. This is a higher-level, less technical view of the 802.6 network. It discusses the need for such networks, applications, architectural considerations, standards activities, and timetables.

Chapter III-2-3, "Performance Evaluation of Metropolitan Area Networks," discusses the characteristics of three types of metropolitan area networks: the Manhattan Street Network, the HR4-NET, and the Queued Packet and Synchronous Switch. The emphasis is on the comparative performance of these MAN approaches.

Finally, Chapter III-2-4, "An Overview of FDDI," introduces the second generation of local area networks. FDDI is a 100M-bps fiber-based network suitable for deployment over distances to 200 Km. This chapter serves as a primer on the subject by examining topology, protocols, physical interfaces, and other aspects of this approach.

Chapter III-2-1
The IEEE 802.6 Metropolitan Area Network Distributed-Queue, Dual-Bus Protocol

THE PROLIFERATION OF LOCAL COMPUTER NETWORKS has underscored the need for interconnecting these networks reliably and efficiently. The method of interconnection must satisfy requirements of distance, high speed, and standardized protocols and services. Interconnections must cover a large geographic area. They must be able to operate at greater than 20M bps to serve as backbones for the installed local area networks (LANs), which are invariably heterogeneous. Therefore, a metropolitan area network (MAN) based on an international networking standard is a highly desirable method of providing the interworking element for multivendor LANs. In addition, because of the changing nature of network traffic from data only to a mixture of data, voice, and video traffic, these MANs must support multimedia services. Furthermore, as MANs are deployed in the public domain, support of voice services assists the integration of data and voice over a common set of equipment, thereby reducing maintenance and administrative costs.

Considerable work is under way in the standards groups to define the features of a broadband integrated services digital network (ISDN) that provides a universal and seamless connectivity for multimedia services like the one that the public telephone network currently provides for voice services. There is, however, an immediate need to interconnect LANs and support other high-speed communications services across a metropolitan area, which demands an interim solution. This interim network must act as a migration path to a broadband ISDN and may, following large-scale deployment of broadband ISDN, act as an access method to these networks.

The IEEE project 802.6 committee is working on a MAN that attempts to address these issues. This chapter reviews the work that has been done by the committee and presents an overview of the technical aspects of the proposed standard.

THE DISTRIBUTED-QUEUE, DUAL-BUS PROTOCOL

A distributed-queue, dual-bus protocol is the access protocol specified in the 802.6 MAN standard. The protocol can support such traffic types as data, voice, and video. The distributed-queue, dual-bus subnetwork is used as a public network controlled by the operating companies or as a private backbone network within the customer premises. It also serves as a LAN.

The 802.6 MAN operates on a shared medium with two unidirectional buses that flow in opposite directions. Depending on the type of traffic, a node on the distributed-queue, dual-bus subnetwork can queue to gain access to the medium by using a distributed-queue, arbitrated access method or by requesting a fixed amount of bandwidth through a prearbitrated access method. Data is transmitted on the medium in fixed-size units called slots, which are 53 bytes long (a 52-byte data slot plus a 1-byte access control field).

Distributed-queue, dual-bus subnetworks are connected with a dual-port or multiport bridge. (The multiport bridge, which could be based on the fast packet-switching technology being developed for broadband ISDN, will be considered by the 802.6 committee in the future.) Distributed-queue, dual-bus private networks are connected to the public network by point-to-point links, which are distributed-queue, dual-bus subnetworks with two nodes (see Exhibit III-2-1).

To support services across a metropolitan area, a single distributed-queue, dual-bus subnetwork may range from a few kilometers to more than 50 kilometers in diameter. The subnetwork can be implemented to operate at a variety of speeds greater than 1M bps (a typical implementation of a distributed-queue, dual-bus subnetwork operates from 34M bps to 150M bps).

Network Topology

Nodes in a distributed-queue, dual-bus subnetwork are connected to a pair of buses flowing in opposite directions. The subnetwork operates in

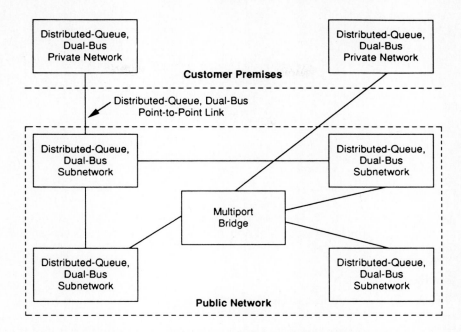

Exhibit III-2-1. Distributed-Queue, Dual-Bus Subnetworks

one of two topologies: open bus or looped bus. The node, or head, at the beginning of each bus generates empty slots to be used on the bus (see Exhibit III-2-2). In the looped-bus topology, the two heads of buses are located at the same node (see Exhibit III-2-3). The two topologies provide the same service, but the looped bus can reconfigure to provide full connectivity if a fault occurs (see Exhibits III-2-4 and III-2-5).

Data Types

All data on a distributed-queue, dual-bus subnetwork is carried in fixed-size data units called slots. The slot size and format align with the broadband ISDN cell size proposed by the International Telephone and Telegraph Consultative Committee (CCITT) SGXVIII committee. This was done to facilitate eventual migration to the broadband ISDN. The payload of each 52-byte slot is called a segment.

The two types of slots generated in a distributed-queue, dual-bus subnetwork are queued arbitrated and prearbitrated. Nodes gain access to the queued arbitrated slots by using the distributed-queue access protocol. Nodes gain access to the prearbitrated slots by requesting and receiving bandwidth from a call/connection control entity (the functions of which

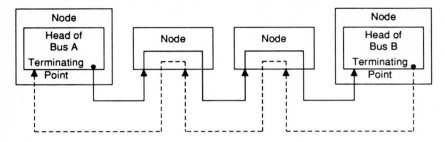

Exhibit III-2-2. Open-Bus Topology

have yet to be defined). The slot generator, a function of the head of bus, marks the prearbitrated slots, and only nodes that were previously assigned these slots can use them.

Services

The distributed-queue, dual-bus layer is intended to provide a range of services, including connectionless data transfer, connection-oriented data transfer, and isochronous data transfer. Convergence functions adapt the underlying medium access services to provide a specific service to a user. The standard specifies the convergence function, which provides connectionless media access control data service to the logical link control sublayer and offers guidelines for the provision of an isochronous service. The

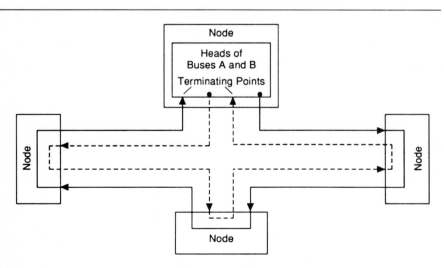

Exhibit III-2-3. Looped-Bus Topology

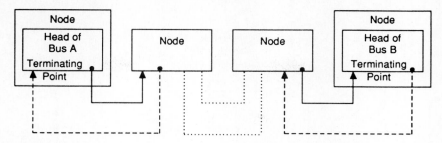

Notes:
The link between the two nodes in the open-bus configuration is broken. The subnetwork is partitioned into two disjointed subnetworks.

Exhibit III-2-4.   Open-Bus Reconfiguration

---

connection-oriented data service is under study. Additional convergence functions may be defined as future services are developed or defined.

Connectionless Media Access Control Service.   The logical link control sublayer operating over the distributed-queue, dual-bus layer provides the service of the open systems interconnection (OSI) data link layer, which supports link-level data communications between two open systems. The connectionless media access control service supports the transport of frames as long as 9,188 bytes. This frame size can encapsulate all types of 802 LAN packets except for the 18K-byte 802.5. The service is provided to the logical link control sublayer (802.2 LLC). It is the media access control service because it is compatible with the access control service provided by other 802 LAN standards. The function providing this service is called the convergence function. Connectionless media access control data is transmitted and received in queued arbitrated slots using the distributed queued arbitrated access protocol.

Connection-Oriented Service.   The connection-oriented service supports the transport of 52-byte segments between two or more nodes sharing a virtual channel connection with no guarantee of a constant interarrival time. This service is similar to asynchronous transfer, the proposed transport mechanism for broadband ISDN. Connection-oriented data is transmitted and received in queued arbitrated slots using the distributed queued arbitrated access protocol.

Isochronous Service.   The isochronous interface is used by isochronous service user entities that require a constant interarrival time over an isochronous (e.g., voice) connection. Isochronous data is transmitted and received by bytes in prearbitrated segments.

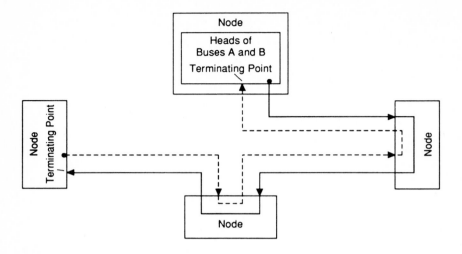

Note:
The link between the two nodes in the looped-bus configuration is broken. The subnetwork is reconfigured to an open-bus configuration. Connections along all nodes are maintained.

Exhibit III-2-5.   Looped-Bus Reconfiguration

Exhibit III-2-6 shows the distributed-queue, dual-bus layer partitioned into the access arbitration block (queued arbitrated, prearbitrated, and common functions) and convergence functions to various service users.

PHYSICAL ATTRIBUTES

The distributed-queue, dual-bus access layer is independent of the physical layer. Therefore, a variety of distributed-queue, dual-bus networks can be built using the same access layer yet operating at different data rates, depending on the transmission systems selected.

Three transmission systems have been considered to date:

- The American National Standards Institute (ANSI) DS3—This system transmits data at 44.736M bps over 75-ohm coaxial cable or fiber.
- The ANSI SONET STS-3c—This system transmits data at 155.52M bps over single-mode fiber.
- The CCITT G.703—This system transmits data at 34.368M bps and 139.264M bps over a metallic medium.

The physical layer incorporates a convergence protocol that provides a consistent service to the distributed-queue, dual-bus layer, regardless of the transmission system used (see Exhibit III-2-7). This protocol provides a mapping function, operation and maintenance parameters, and information to the distributed-queue, dual-bus layer. Each transmission system has a different physical layer convergence protocol.

Network Components

Several functions are associated with operating a distributed-queue, dual-bus subnetwork. A single node can perform one or more of these functions. Two major functions are described in this section: slot generation and timing generation.

Slot Generation. Every 125 µs, a slot generator transmits multiple slots to the shared medium (the number of slots generated depends on the physical transmission rate). Nodes read and copy data from the slots; they also gain access to the subnetwork by writing to the slots.

In the open-bus topology, there are two slot generators that serve as the head of bus A and head of bus B. In the looped-bus topology, only one default slot generator serves as the head of both bus A and bus B; any node on the subnetwork can become the slot generator if the designated slot generator fails.

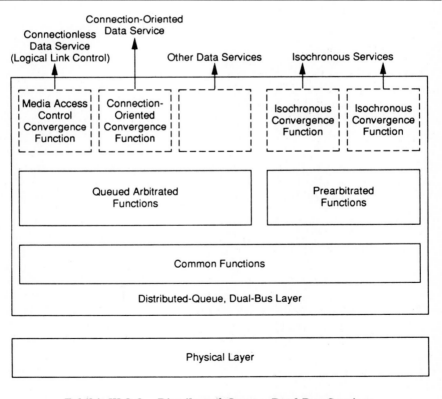

Exhibit III-2-6. Distributed-Queue, Dual-Bus Services

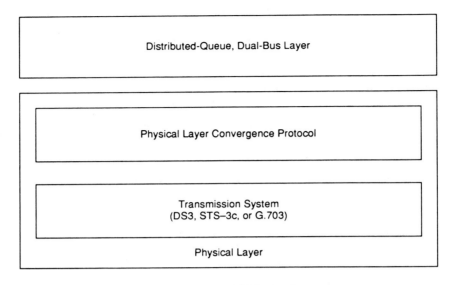

Exhibit III-2-7. Physical Layer Structure

Timing Generation. The distributed-queue, dual-bus subnetwork operation is based on a 125–µs clock to allow for isochronous services (the 125µs interval reflects the 8-kHz public networking frequency required by voice services). The clock is either generated internally within the subnetwork or extracted from an external timing source (e.g., from the public network).

The timing generator function is performed at the physical layer convergence protocol level and is activated on one node in each subnetwork. Other nodes can be used as standby if the designated timing generator fails.

Node Configuration

A node is connected to the distributed-queue, dual-bus subnetwork as either a single node or a node in a cluster. A single node has its own dedicated pair of transmission links (see Exhibit III-2-8), whereas a cluster consists of multiple nodes connected serially in the same box. Nodes in a cluster share optical electronic, power supply, and configuration control components and two transmission links (see Exhibit III-2-9).

DISTRIBUTED-QUEUE, DUAL-BUS LAYER ACCESS
CONTROL PROCEDURES

The distributed-queue, dual-bus layer provides two general methods to access the medium. For traffic that requires guaranteed bandwidth (e.g.,

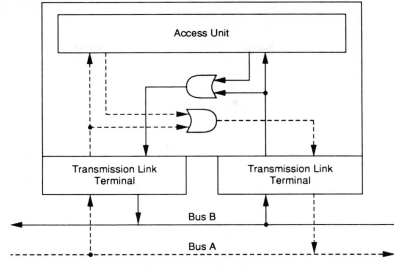

Exhibit III-2-8.   Single Node

traffic with a constant flow of data), the subnetwork provides a preallocated (prearbitrated) bandwidth. For bursty data that has less stringent requirements on delay, the users contend for bandwidth using the distributed-queued arbitrated access protocol.

The Queued Arbitrated Access Protocol

A node gains access to transmit by putting itself in a queue. There are two queues, one for each bus. They are monitored and controlled by each node on the subnetwork. In each node, the distributed queue on bus A is maintained by counting requests passing on bus B and counting free slots passing on bus A. The difference between the two sums is the number of outstanding requests for slots on bus A. This value also indicates the position of the node in the queue to transmit on bus A. The distributed queue on bus B operates symmetrically.

The distributed-queue, dual-bus algorithm for bus access provides a mechanism under which a distributed queue is maintained in a subnetwork. Nodes with data to transmit in the queued arbitrated slots must join this queue.

The mechanism that passes information among nodes to maintain the distributed queue is the request field of the access control field of the slot (see Exhibit III-2-10). The request field allows a node to convey its intent to transmit data to upstream nodes. This is done by setting the request bit of

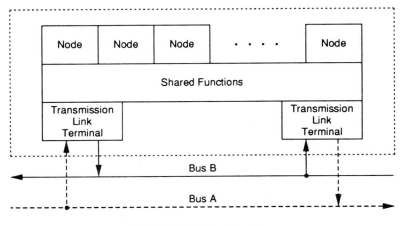

Exhibit III-2-9. Cluster

a slot on the bus opposite from the one on which data is to be sent. Each node keeps a tally of downstream requests for each bus by monitoring the request field of slots that pass by on the other bus.

The behavior of a single node is addressed here first. As shown in Exhibit III-2-10, the request field is 4 bits wide, but for simplicity of description, a 1-bit request field is assumed. In addition, it is assumed that data transfer is on only one bus (bus A). The terms *upstream* and *downstream* refer to the direction of this bus. When a node is idling (i.e., has no data to send), a count is made of the number of requests (i.e., slots with the request bit set) that pass by on bus B. This count is kept in a request counter, which is decremented by one for every empty slot (i.e., a slot with the busy bit not set) that passes by on bus A. The request counter keeps a tally of the number of downstream queued requests and balances it with the number of empty slots that pass by and satisfy the requests (see Exhibit III-2-11). The request counter is never allowed to become negative. If more empty slots pass by on one bus than request bits on the opposite bus, the counter remains at 0.

When the node wishes to transmit data, it joins the distributed queue by first putting its request on bus B (to inform upstream nodes) and waits for its turn in the queue. The node's position in the queue is determined by the value of the request counter. This value is transferred to the countdown counter, and the request counter is reset to 0. The countdown counter decrements for every empty slot that passes by on bus A; the node transmits its own data when this counter reaches 0. The request counter, on the other hand, no longer decrements for empty slots on bus A but simply counts the number of requests on bus B (see Exhibit III-2-12). The dual bus is symmetric; a node with data to transmit on bus B goes through a process equivalent to the one described with a separate set of request and countdown counters. To send another slot of data, the node must repeat the process.

As shown in Exhibit III-2-10, the request field is 4 bits wide, allowing for 4 access priority levels; 0 is the lowest priority, and 3 is the highest. Each node on the subnetwork has a request counter and a countdown counter for each priority level. When a node is idling, the request counter increments for requests of priorities equal to or greater than its own priority level. When a node is queuing, each request counter increments for the requests of its own priority level; the countdown counter increments for requests of a higher priority level. For example, when a node is idling, the request counter of priority level 1 counts the requests of priority levels 1, 2, and 3.

In summary, to gain access on a bus, a node:

- Sends a request to tell the upstream nodes that it wishes to transmit data.
- Puts itself in the distributed queue by using the countdown counter.
- Maintains the distributed queue with the request counter.
- Allows enough empty slots through to satisfy the needs of nodes that are ahead in the distributed queue.
- Sends its own data when its turn comes.

Queued Arbitrated Access Operation. The following example demonstrates the operation of a distributed queue of a five-node dual-bus network. Again, for simplicity, a 1-bit request field is assumed. It is also as-

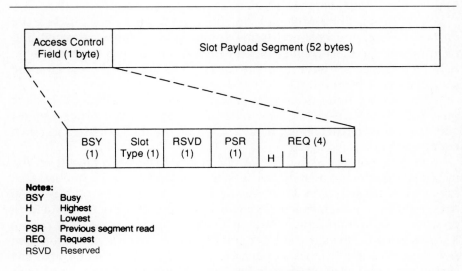

Notes:
BSY   Busy
H   Highest
L   Lowest
PSR   Previous segment read
REQ   Request
RSVD Reserved

Exhibit III-2-10.   Access Control Field

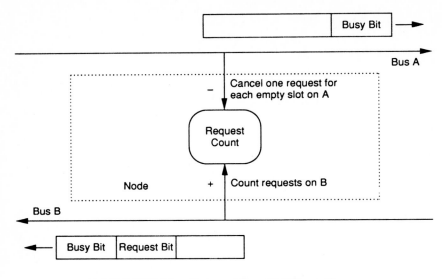

Exhibit III-2-11.   Request Counter Operation

sumed that all request counters start with the value 0 and that all slots are busy (see Exhibit III-2-13).

Node 5 wishes to transmit data on bus A. It sets the request bit in a slot on bus B. Every node that detects this request bit increments its request counter by one. Node 5 also transfers its request counter (with a value of 0) to its countdown counter. It clears its request counter to 0.

Node 2 wishes to transmit data on bus A. It sets the request bit in a slot on bus B. Only the upstream node (node 1) sees the request bit and increments its request counter. Node 1's request counter is incremented to 2. Node 2 copies its request counter value to its countdown counter and sets its request counter to 0.

Node 3 wishes to transmit data on bus A. It sets the request bit in a slot on bus B. Nodes 1 and 2 see this request bit and increment their request counters. The request counter of node 2 is incremented, but its countdown counter value is unchanged.

In the situation in Exhibit III-2-14, the nodes have been placed on the distributed queue but no data has been transmitted. Node 5 is queued first, followed by nodes 2 and 3. The next empty slot passes by on bus A. All nodes decrement their request counters and countdown counters appropriately. Nodes 2 and 3 decrement their countdown counters but do not transmit because the value of their countdown counters is not 0 when the slot is received. Node 5 has a countdown counter value of 0 and uses this slot to transmit data. The countdown counters of nodes 2 and 3 are also 0. In such a case, the position of the nodes on this bus determines the or-

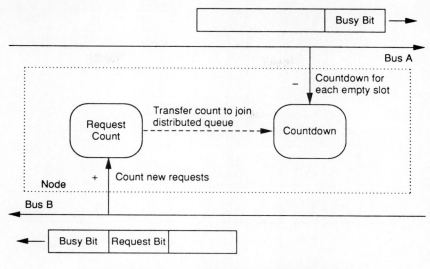

Exhibit III-2-12. Countdown Counter Operation

der in which they are served (the most upstream node is served first).

The next empty slot causes the request counters and countdown counters to be decremented. Node 2 gains access to the bus. Node 3 is able to send its data in the next empty slot.

The Prearbitrated Access Protocol

Access to the prearbitrated slot is based on a request-and-assign procedure. When a node needs to establish an isochronous connection with another node, it requests a connection from a call/connection control entity. These procedures are beyond the scope of the current 802.6 standard; they may be based on such a signaling scheme as Q.931.

The slot generator (or default slot generator) of the subnetwork is notified by the call/connection control entity to generate the appropriate prearbitrated slots. Nodes are assigned to write to 1 or more bytes in a prearbitrated slot. They are also assigned to read from 1 or more bytes in another prearbitrated slot.

THE INTERNAL STRUCTURE OF THE DISTRIBUTED-QUEUE, DUAL-BUS LAYER

The distributed-queue, dual-bus layer is divided into three functional blocks: convergence, media access control, and layer management entity (see Exhibit III-2-15).

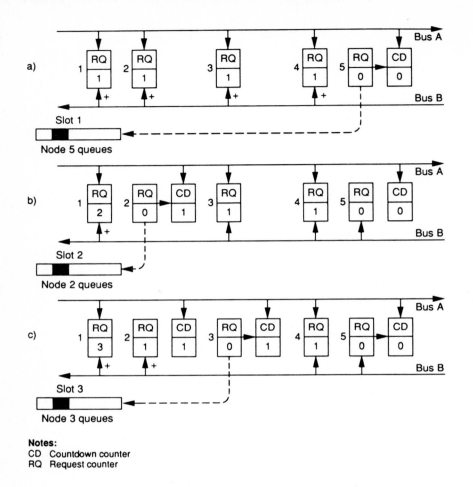

Notes:
CD Countdown counter
RQ Request counter

Exhibit III-2-13. Nodes Queuing for Slots

The convergence block. The convergence block provides the necessary translation between the service requirements of the users and the segment-based transport. Therefore, the convergence functions adapt data from the users to the formats required by the distributed-queue, dual-bus access layer. Conversely, they also extract data from the distributed-queue, dual-bus segments and present the information to the users.

The standard specifies the media access control convergence function, which provides the connnectionless datagram service to the logical link control sublayer. It also provides guidelines for an isochronous convergence function. The connection-oriented convergence function is under study.

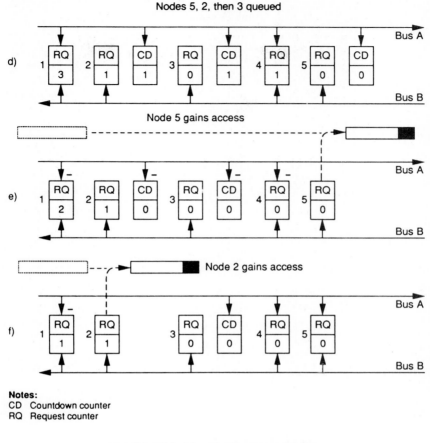

Notes:
CD   Countdown counter
RQ   Request counter

Exhibit III-2-14.   Nodes Transmitting

The Media Access Control Block.   The media access control block is responsible for transmitting and receiving slots to and from the buses using the prearbitrated and queued arbitrated access procedures. The media access control block is divided into three subentities: common functions, queued arbitrated functions, and prearbitrated functions.

The Layer Management Entity Block.   The layer management entity block is composed of two elements: a distributed-queue, dual-bus layer management and a layer management interface to network management interface.

The distributed-queue, dual-bus layer management entity handles the functions that are required to provide distributed-queue, dual-bus layer

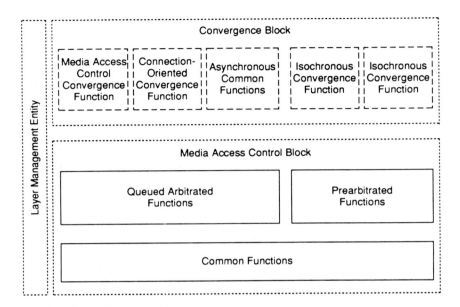

Exhibit III-2-15.   Distributed-Queue, Dual-Bus Layer Internal Structure

management when services to higher layers cannot be guaranteed to be fully operable (e.g., when the subnetwork is being initialized). Work in this area is currently in progress.

The interface protocol for layer and network management is reserved for future study while its primitives are being defined in the standard.

The Convergence Block

The Media Access Control Convergence Function.   The media access control convergence function is responsible for segmenting the logical link control protocol data unit and appending the appropriate headers and trailers before transmitting data (data is transmitted in multiple segments if its overall size is greater than one segment length). The media access control convergence function also reassembles the segments provided by the distributed-queue, dual-bus layer to provide the standard media access control service required by the logical link control sublayer.

Connection-Oriented Convergence Functions.   A connection-oriented convergence function is responsible for adapting the slot-based service provided by the distributed-queue, dual-bus layer (payload of 48 bytes) to the connection-oriented data services required by the asynchronous transfer mode (payload size is to be determined).

Isochronous Convergence Functions. An isochronous convergence function is responsible for providing the buffering needed to transmit and receive bytes (in prearbitrated slots) between the distributed-queue, dual-bus layer and the isochronous service user. Because the service provided by the distributed-queue, dual-bus layer to the isochronous convergence function does not guarantee a constant interarrival time, the isochronous convergence function must provide the necessary smoothing.

The Media Access Control Block

Queued Arbitrated Functions. The queued arbitrated functions provide service for the media access control convergence function and the connection-oriented convergence with the queued arbitrated segments.

The queued arbitrated functions provide a data transfer service of 48-byte payloads. The queued arbitrated functions accept the payload from a media access control or connection-oriented convergence function and add the appropriate header to the payload to create a queued arbitrated segment. Queued arbitrated segments received by queued arbitrated functions destined for the node are stripped of the header, and the payload is passed to the appropriate convergence function.

Prearbitrated Functions. The prearbitrated functions provide service for the isochronous convergence functions with the prearbitrated segments. The use of the prearbitrated segments by isochronous convergence functions requires the establishment of a connection. The prearbitrated functions can support the transfer of 1 or more bytes to provide variable service rates and the sharing of one segment by multiple isochronous users. The prearbitrated functions accept a byte from an isochronous convergence function and write it into the preallocated positions within the payload of a prearbitrated segment. To receive a byte, the prearbitrated functions copy the byte from the preallocated position within the payload of a prearbitrated segment and pass it to the isochronous convergence function.

Common Functions. After receiving a slot from the bus, the common functions process the slot header and relay the slot payload to queued arbitrated or prearbitrated functions, depending on the slot type. For transmission, the common functions include data from queued arbitrated and prearbitrated functions in the slot after modifying the slot header according to the access protocol.

The common functions also perform other functions to maintain the operation of the subnetwork. The functions that have been specified thus far are configuration control and reservation of message identifiers (used for the segmentation and reassembly process, which is described later in this chapter).

THE MEDIA ACCESS CONTROL SERVICE DATA UNIT

Structure

Project 802.6 is viewed as the mechanism to bridge lower-speed 802 LANs to the public network. Because of this, the subnetwork is required to provide necessary media access control service to the IEEE 802.2 logical link control sublayer.

As shown in Exhibit III-2-16, the logical link control protocol data unit is enveloped in a header and trailer to form an initial media access control protocol data unit. The initial media access control protocol data unit is divided into 48-byte fragments to fit in the payload of a slot (although the payload of a slot is 52 bytes, header fields reduce the effective payload size to 48 bytes). If the initial media access control protocol data unit is larger than 20 bytes, it is divided into multiple fragments. Each fragment is enveloped in a 2-byte header and 2-byte trailer to form a derived media access control protocol data unit. The derived media access control protocol data unit is the fixed-size segment payload that is transmitted on the bus. The media access control convergence function is required to perform the segmentation process to divide one logical link control protocol data unit into one or more derived media access control protocol data units as described. The convergence function is also required to reassemble the received segments before presenting data to the logical link control sublayer.

Each derived media access control protocol data unit is enveloped into a segment by the queued arbitrated functions. Each segment is in turn enveloped into a slot in the common functions.

Segment Type. The segment type indicates the order in which the segments are divided and transmitted to provide a reliable scheme for reassembling the segments at the media access control convergence function of the receiving end.

There are four segment types:
- Beginning of message.
- Continuation of message.
- End of message.
- Single-segment message.

If the initial media access control protocol data unit has 20 bytes of data or less in the information field, the entire protocol data unit can be transmitted in 1 segment.

Destination and Source Addresses. Connectionless datagrams must contain addressing information, and in 802.6, the address field is fixed at 64 bits wide to accommodate 16-, 48-, and 60-bit addresses. The 60-bit address specified in 802.6 is based on the ISDN E.164 public addressing scheme with 15 binary coded decimal (BCD) digits. Each node on a

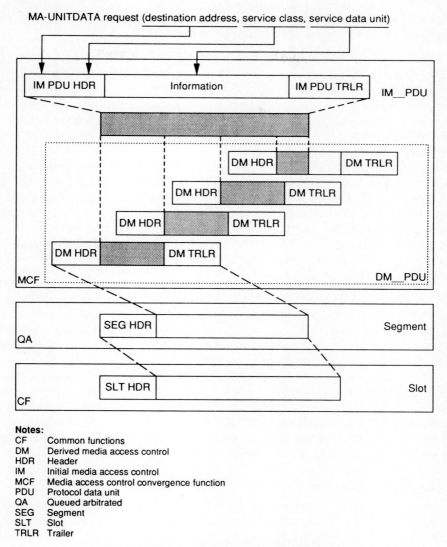

MA-UNITDATA request (destination address, service class, service data unit)

Notes:

CF	Common functions
DM	Derived media access control
HDR	Header
IM	Initial media access control
MCF	Media access control convergence function
PDU	Protocol data unit
QA	Queued arbitrated
SEG	Segment
SLT	Slot
TRLR	Trailer

Exhibit III-2-16. Protocol Data Unit Structure

---

distributed-queue, dual-bus subnetwork must have a 48-bit address be-cause it is the required 802 address. Addresses of 16 and 60 bits are op-tional.

Nodes on a distributed-queue, dual-bus use the 48-bit addresses as do other 802 LANs. Nodes on a distributed-queue, dual-bus MAN are assigned 60-bit addresses within its subnetwork and are managed by the telephone operating companies. The destination and source addresses are parts of the initial media access control protocol data unit.

The Message Identifier. If the initial media access control protocol data unit is too long to fit into one slot and must be segmented, the source and destination addresses are transmitted in the beginning-of-message segment only. The message identifier field provides the labeling mechanism for the segments following the beginning-of-message segment. At the receiving end, the message identifier allows the node to reassemble the datagram from the incoming segments.

The message identifiers must be unique within the subnetwork to prevent segments from different sources with the same destination from being mixed up. A distributed page-allocation algorithm is defined to allow nodes to obtain pages of message identifiers for use. Each page contains four message identifiers—one for each priority level of the distributed-queue, dual-bus access protocol. The message identifier is part of the derived media access control protocol data unit header.

The Virtual Channel Identifier. The virtual channel identifier is used along with the message identifier to identify the convergence function to direct segment payload. Because there is no destination address or source address associated with an isochronous connection, the virtual channel identifier and the byte offset are used to identify the byte in a prearbitrated slot that has been preassigned to a node.

There is at least one virtual channel identifier for each convergence function. For the connectionless service, the media access control convergence function, the default virtual channel identifier that requires support is the hexadecimal value FFFFF. For the isochronous services, at least two virtual channel identifiers are associated with each connection. The virtual channel identifier is a part of the segment header.

Detailed Header Structure

The relationship between the different structures and fields that constitute a media access control protocol data unit is shown in Exhibit III-2-17. The numbers in roman font denote the field size in bytes; those in italic denote the field size in bits.

Media Access Control Service Data Unit Transmission Flow

The slot generator generates multiple empty queued arbitrated slots within each 125-μs interval. A node gains access to transmit the derived media access control protocol data units of a service data unit to the subnetwork in these queued arbitrated slots on the basis of the distributed-queue protocol. The node must queue up to transmit each derived media

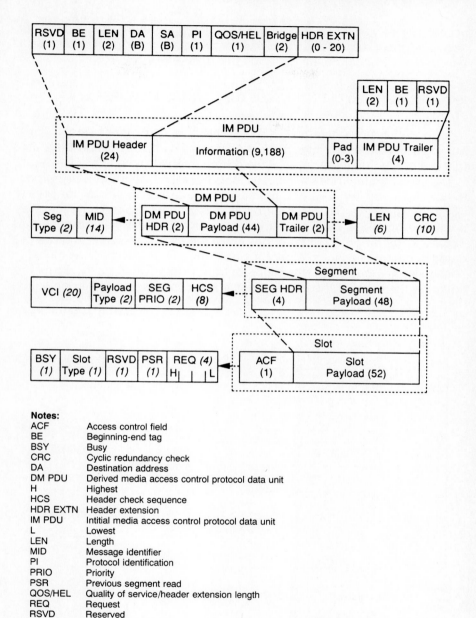

Notes:

ACF	Access control field
BE	Beginning-end tag
BSY	Busy
CRC	Cyclic redundancy check
DA	Destination address
DM PDU	Derived media access control protocol data unit
H	Highest
HCS	Header check sequence
HDR EXTN	Header extension
IM PDU	Intitial media access control protocol data unit
L	Lowest
LEN	Length
MID	Message identifier
PI	Protocol identification
PRIO	Priority
PSR	Previous segment read
QOS/HEL	Quality of service/header extension length
REQ	Request
RSVD	Reserved
SA	Source address
SEG	Segment
VCI	Virtual channel identifier

Exhibit III-2-17.   Media Access Control Protocol Data Unit Structure

access control protocol data unit individually; therefore, derived media access control protocol data units from a media access control service data unit may not appear contiguously on the medium.

ISOCHRONOUS DATA UNIT STRUCTURE AND TRANSMISSION FLOW

Each isochronous data unit is represented as 1 or more bytes of data. They are transferred between the isochronous service user and the distributed-queue, dual-bus layer at fixed time intervals (see Exhibit III-2-18).

The prearbitrated functions receive data from the isochronous users over the isochronous convergence function and transmit them on the preassigned positions in passing prearbitrated segments. Conversely, the prearbitrated functions read bytes from preassigned positions and relay the data to the appropriate isochronous convergence functions. Each prearbitrated segment is enveloped into a slot in the common functions block.

The slot generator generates prearbitrated slots within each 125-µs interval on the basis of the isochronous connection requests. A node gains

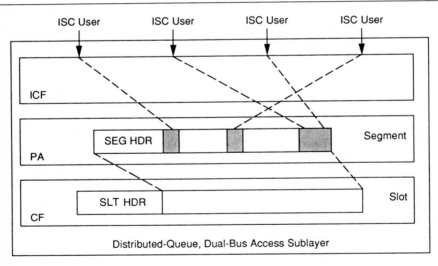

Notes:

CF	Common functions
HDR	Header
ICF	Isochronous convergence function
ISC	Isochronous
PA	Prearbitrated
SEG	Segment
SLT	Slot

Exhibit III-2-18.   Isochronous Data Unit Structure

access to transmit isochronous service data in the preassigned positions in these prearbitrated slots according to the isochronous connection protocol.

DATA FLOW IN THE DISTRIBUTED-QUEUE, DUAL-BUS LAYER

In this section, two examples are considered: transmitting a packet from the logical link control sublayer and transmitting a byte from an isochronous user.

Transmitting a Packet from the Logical Link Control Sublayer

A packet sent from the logical link control sublayer is processed in the following manner. In the media access control convergence function block, the packet is enveloped with the initial media access control protocol data unit header and trailer to form an initial media access control protocol data unit (see Exhibit III-2-19). The initial media access control protocol data unit

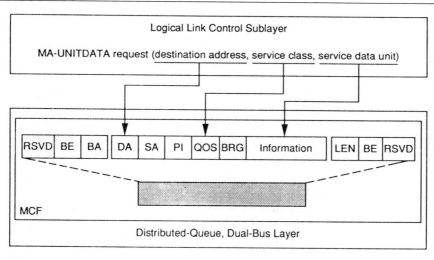

Notes:

BA	Buffer allocation size
BE	Beginning-end tag
BRG	Bridging
DA	Destination address
IM PDU	Initial media access control protocol data unit
LEN	Length
MCF	Media access control convergence function
PI	Protocol identification
QOS	Quality of service
RSVD	Reserved
SA	Source address

Exhibit III-2-19.   Processing a Packet in MCF: Creating an IM PDU

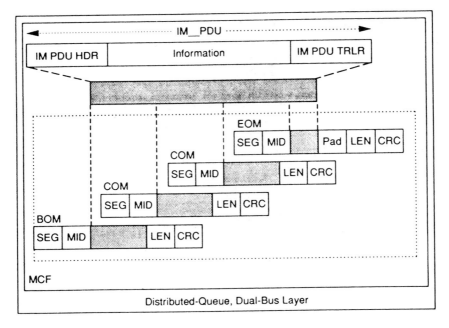

Notes:

BOM	Beginning of message
COM	Continuation of message
CRC	Cyclic redundancy check
EOM	End of message
HDR	Header
IM PDU	Initial media access control protocol data unit
LEN	Length
MCF	Media access control convergence function
MID	Message identifier
SEG	Segment
TRLR	Trailer

Exhibit III-2-20. Processing a Packet in MCF: Creating Derived Media Access Control Protocol Data Units

is then segmented into multiple units to fit the fixed-size slots. Each unit forms the payload of a derived media access control protocol data unit. Headers and trailers are then added (see Exhibit III-2-20).

In the queued arbitrated functions block, the derived media access control protocol data unit is enveloped into a segment including a header field. For the current standard, these fields are set to predefined values, as shown in Exhibit III-2-21.

In the common functions block, the queued arbitrated segment is enveloped into a slot. The busy bit of the slot header (the access control field)

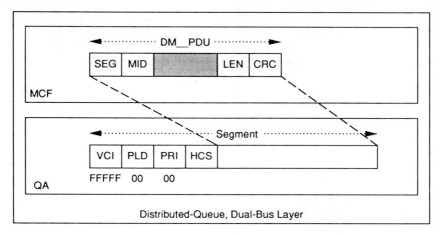

Notes:

CRC	Cyclic redundancy check
DM PDU	Derived media access control protocol data unit
HCS	Header check sequence
LEN	Length
MCF	Media access control convergence function
MID	Message identifier
PLD	Payload type
PRI	Priority
QA	Queued arbitrated
SEG	Segment
VCI	Virtual channel identifier

Exhibit III-2-21. Processing a DM PDU in the Queued Arbitrated Functions Block: Creating A Segment

---

is set to 1 to mark the slot as used, and the slot type is set to 0 to denote a queued arbitrated slot. The previous-segment-read bit is set at 0 and the request field is left unchanged (see Exhibit III-2-22). The slot is then transmitted to the shared medium according to the signals from the physical layer convergence protocol entity.

Transmitting Data from the Isochronous User

A byte or series of bytes sent from an isochronous user is processed as follows. Bytes are written on the basis of the information table set up in the isochronous convergence functions after the isochronous connection has been established between the calling and called isochronous users. The transmitted bytes are buffered by the prearbitrated functions block and written to the appropriate positions in the preassigned segments as they arrive. In each prearbitrated segment, a particular byte is written to and from the segment on the basis of the offset value specified (Exhibit III-2-23

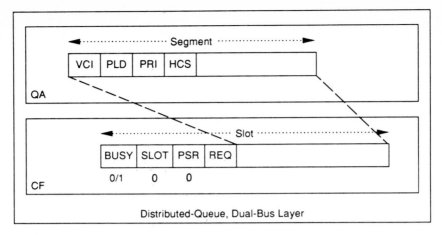

Exhibit III-2-22.   Processing a Segment in the Common Functions Block: Creating a Slot

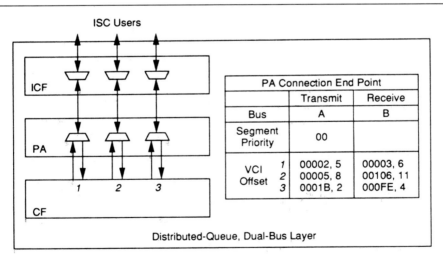

PA Connection End Point		
	Transmit	Receive
Bus	A	B
Segment Priority	00	
VCI Offset 1	00002, 5	00003, 6
VCI Offset 2	00005, 8	00106, 11
VCI Offset 3	0001B, 2	000FE, 4

Exhibit III-2-23.   Transmitting Isochronous Data

594

shows data transmitted in the fifth bytes of the prearbitrated segment with the virtual channel identifier of 00002 on bus A).

In the common functions block, the prearbitrated segments are enveloped in slots. The busy bit of the slot header (i.e., the access control field) is set to 1 to mark the slot as used, and the slot type (i.e., SLOT) is set to 1 to denote a queued arbitrated slot. The slot is then transmitted to the shared medium according to the signals from the physical layer convergence protocol entity.

SUMMARY

This chapter summarizes the features of the emerging MAN currently being defined by the IEEE 802.6 committee. The need for a fast, efficient, and transparent interconnection for information interchange across the public network is clear. The survival of most companies and corporations increasingly depends on maintaining up-to-date information and on transferring information in the most timely and efficient manner. Although broadband ISDN must be developed further before it becomes a feasible solution to this problem, 802.6 is an attempt to bridge the gap and offers an evolutionary route to a universally networked environment.

Chapter III-2-2
Metropolitan Area Network Standards from IEEE 802.6

James F. Mollenauer

LOCAL AREA NETWORKS (LANs) have been part of the computing environment for more than 10 years, but metropolitan area networks (MANs) are a relatively new concept. This chapter explains what MANs are, evaluates their benefits, and describes what the standards committee IEEE 802.6 is doing in the area of standards for these networks.

Conventional computer networks are digital data adaptations of existing telephone networks. Telephone wiring's ability to carry data was exploited in the very earliest days of computers: data transmission over telephone lines was demonstrated by Bell Telephone Laboratories before World War II. The expansion of digital technology to encompass voice as well as data has blurred the distinction between voice and data, but the underlying analog telephone technology is still there, with point-to-point connections, multidrop circuits, and multiplexing over high-speed trunks. Full implementation of digital systems like integrated services digital network (ISDN) is still far away.

Local area networks were a departure from telephone technology. Their development was based more on the internal bus used by most computers than on the telephone system. A computer's bus is typically allocated to different devices on a request-grant basis, with both the CPU and I/O devices able to use it to pass data in and out of memory. The LAN simply serializes the bus and establishes a distributed bus allocation method to handle bursty requests from many devices on the same line. It is very well adapted to computer traffic, which is inherently bursty and delay tolerant, as opposed to voice and video, which require dedicated bandwidth and cannot tolerate delay.

Metropolitan area networks extend LAN concepts to accommodate a city and its suburbs; the specific bus-allocation protocols generally used in LANs cannot accommodate networks of this size without substantial loss of efficiency. MAN protocols are tailored to larger networks, without attempting to be efficient on a continental or satellite-link scale as well.

Metropolitan area networks must also be faster than LANs. It is assumed that they will tie together a complex set of LANs, large computers, PBXs, and other high-volume data sources. Speeds in the range of 5M to 10M bps are insufficient. Optical fiber (and also the older coaxial cable technology) can handle bit rates of hundreds of megabits per second without undue expense.

Finally, metropolitan area networks must be tailored to handle such fixed-bandwidth services as voice and video. Within a building, there would be little expense saved in routing voice calls over a LAN. Over metropolitan distances, however, the situation is different. Many organizations are located in buildings scattered over a city and its suburbs; in these cases, most of the bits passed between the buildings are digitized voice, not data. Cost-effective use of a MAN dictates that the PBXs in the buildings be connected to the network, eliminating the expense of leased voice-grade and T1 circuits between the sites.

SUPPORT OF METROPOLITAN AREA NETWORKS

Another distinction between local and metropolitan networks is that the LAN is strictly a private network. It is installed by one organization on its premises and is usually maintained by that organization or by an outside firm under contract. Even though standardization has benefits, many proprietary designs have succeeded by providing standard interfaces (e.g., RS-232). Although the internal LAN protocols are proprietary, the standard interfaces make it possible to attach existing terminals and computers to the network. The speed capability of the RS-232 interface, however, is very limited, and equipment attached in this way loses the benefit of the high burst speeds attainable over LANs.

Metropolitan networks have a geographic extent that requires them to cross public rights of way, and they can connect equipment belonging to a diverse set of organizations. Except for networks installed on campuses (university or industrial), permission must be sought to cross streets and to run on existing poles. Even for strictly private networks, this is a considerable burden, one that is vastly lightened if the network is installed by an organization that already has the needed rights of way.

There are only two organizations in most areas that have the rights of way and information-carrying experience: the cable television operators and the telephone company. Interest on the part of the cable television industry has been minimal, leaving the telephone industry to provide MAN services. This industry is now strongly in favor of metropolitan networks, after a period of regarding them as potential competitors. The MAN standards committee (IEEE 802.6) now enjoys the participation of most of the Bell operating companies, AT&T and Bell Communications Research, Telecom Australia, Bell Canada, and British Telecom, as well as a host of industry suppliers. The technology is emerging, but the deployment of networks by the telephone companies will depend on the existence of standards through which they can specify equipment ordered from manufacturers.

USE OF MANs

The uses of MANs that can be foreseen now are probably a very small subset of the functions they will actually be called on to perform in the future. Radically different applications may call for different systems, but right now it appears that the capacity of fiber-based systems will be difficult to exhaust for some time to come, probably not before the end of the century.

The anticipated uses of MANs include the following:
- Interconnection of LANs.
- Connection of large computers.
- Interconnection of PBXs.
- Gateway to wide area networks.
- Transmission of computer-aided design files, high-resolution graphics, medical imaging services.
- Compressed video distribution.

It is very possible that networks will fall into two large categories analogous to private and switched telephone lines. A private MAN will exist as a dedicated circuit stopping off at a corporation's buildings scattered around a metropolitan area. It will be able to carry all interbuilding voice traffic and drop the off-network calls at a telephone office. Similarly, data will be carried from building to building, usually in packets but optionally on fixed-bandwidth channels. The network will be able to carry digital video (for conferencing), using compression because uncompressed video requires 90M bps in North America and more in Europe.

A public MAN, by comparison, will interconnect premises of many organizations. Beyond installation and maintenance, it needs an operating organization to provide for billing as well as to enforce responsible behavior on the part of users (i.e., to provide security and privacy mechanisms). Billing will preferably be on a use-sensitive basis, requiring monitoring of packet source addresses and fixed channel allocation requests.

INITIAL MAN DEPLOYMENT: BELLCORE'S SWITCHED MULTIMEGABIT DATA SERVICE

To meet urgent demands for high-speed data transmission, Bellcore, on behalf of the Bell operating companies, has developed a service known as switched multimegabit data service (SMDS). SMDS makes use of the IEEE 802.6 protocols but implements only the connectionless packet service. Voice is assumed to be a less urgent issue, given that there are existing facilities to provide for it. Through a credit mechanism, additions to the IEEE 802.6 specification are also made in terms of guaranteed bandwidth availability. Data fed into SMDS in excess of the bandwidth specified in the contract may be discarded if there is congestion on the network.

To provide privacy and take advantage of current fiber layout patterns, separate fibers will run from central offices to each customer. The distributed-queue, dual bus (DQDB) protocol will be used on the fibers, permitting multiple attachments within the customer's premises (in the same building or in the vicinity). Initially, the speed will be 45M bps to take advantage of existing DS3 transmission electronics. This layout is shown in Exhibit III-2-24. It is expected that synchronous optical network SONET transmission at 155M bps will be available later.

ARCHITECTURE

Although metropolitan networks are in some respects wider-area versions of local networks, they have additional requirements that affect their architecture. The first requirement has an impact on the design of the media access control (MAC) protocols; the second affects segmentation of the network.

The need to carry voice on the MAN imposes significant constraints on the choice of protocols. In contrast to the generally bursty requirements of computer data, voice generates a constant stream of digits. Delay requirements for voice are stringent because of echo problems; in the worst case, the end-to-end delay over the MAN may be restricted to 2 milliseconds maximum. This is impossible to guarantee with existing IEEE 802 protocols.

Privacy issues affect MAN architectural considerations. A private MAN, interconnecting only the offices of one company, is as secure as current telephone service. A competitor wishing to eavesdrop would need to intercept the cable physically on a pole or underground. But if the MAN cable were public, carrying data for many users, every user's data would pass through every node on the network. This is clearly unacceptable for most customers. The solution is to separate the MAN into a transport network and an access network. The transport network carries every user's data but

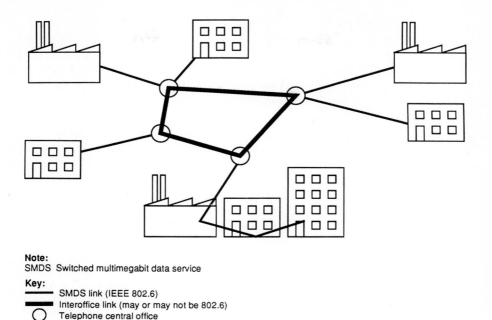

Note:
SMDS Switched multimegabit data service

Key:
─── SMDS link (IEEE 802.6)
▬▬ Interoffice link (may or may not be 802.6)
○   Telephone central office

Exhibit III-2-24. SMDS Metropolitan Area Network

stays on public property. Users are connected to it by means of access networks that are bridged from the customers' premises to the transport segment. At least initially, this bridging will be done at the telephone offices, and the access network carries only one user's data. This is shown in Exhibit III-2-25.

Because the bridging is done at the media access control (MAC) level, the connection between the access and transport networks is invisible to the user. The use of MAC–level bridging to enlarge a LAN is a relatively new IEEE 802 development. Although it does not conform perfectly to the OSI reference model of data communications, it has performance advantages relative to network-layer bridging and also requires less knowledge of network details on the part of each node.

The transport network can use IEEE 802.6 technology, interconnecting telephone company buildings. The bridge operates at the MAC level, where it is invisible to the layers of communications protocol above the MAN or LAN. Data circulating on the MAN toward other destinations is ignored by the bridge, which copies to the access network only the data addressed to that subnetwork. Outbound, the bridge forwards only data addressed to the MAN and points beyond; it ignores data intended for other users within the premises.

The access network need not be an IEEE 802.6 network: it could be an extension of the user's internal LAN, wired to the connection point in a nearby vault. If it is an IEEE 802.6 network, the bridge can be simpler, but other IEEE 802 networks or proprietary protocols could be used depending on the user's needs. The distance covered by the access network can range from close to zero to several miles. For example, with SMDS, the shared portion of the MAN runs only between telephone company central offices, with dedicated access networks carrying the DQDB protocol to the customer.

Additional structure can also be expected to evolve in the MAN in terms of backbone networks. Although a single network might satisfy the communications needs of a metropolitan area initially, as traffic grows, multiple networks will be necessary. Additional MAC-level bridging might be possible but only to a point. A backbone hierarchy similar to that found in telephone systems would serve better. In fact, the technology for the higher elements in the hierarchy might be quite different. AT&T has made one such proposal, for a multiport switch. It is described later in this chapter.

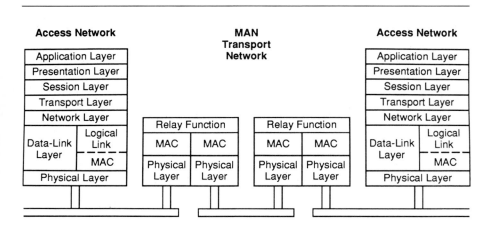

Notes:
MAC Media access control
MAN Metropolitan Area Network

Exhibit III-2-25. MAN Subdivided into Transport and Access Subnetworks, Connected by MAC-Level Bridges

THE HISTORY OF IEEE 802.6

The MAN group was started in 1982, about two years after Project 802 was chartered. The original backing for the metropolitan network came from the communications satellite industry, which wanted inexpensive high-speed links between ground stations and customers in the general vicinity. The satellite business never blossomed as much as its early backers predicted, but support for the idea of a MAN continued with a focus on the cable television industry and data sources that were not necessarily satellite dishes.

The MAN committee was in fact the only working group of IEEE 802 not tied to a particular protocol. Its charter encompassed choosing the MAC protocol as well as defining the medium. Several different protocols have been studied thoroughly, and media ranging from coaxial cable to radio to optical fiber have been proposed in particular cases.

As it turned out, interest on the part of the cable television industry was minimal. Some system operators were very interested in employing their cables for data, but most were unwilling to go beyond entertainment services.

Work on cable television–oriented standards was dropped in favor of optical fiber late in 1984, when several vendors, led by Burroughs Corp, proposed a fiber-based ring. The protocol was based on fixed time slots that would accommodate fixed-bandwidth applications (e.g., voice or compressed video). The ring speed was adjusted to match the telephone hierarchy at the DS3 speed of 45M bps. To accommodate such speeds, the charter of the MAN committee was changed by the IEEE, removing the 20M-bps upper limit that had existed in the scope of project 802. This limitation had been useful in the LAN area, but a metropolitan backbone clearly needed to employ higher speeds if they were technically available. Voice and video were accommodated in the ring by the use of time slotting (i.e., the division of the traffic into fixed time periods synchronized to the voice digitizing rate of 8,000 samples per second). Fixed bandwidth allocation implied permanent allocation of a slot for the duration of the connection, while bursty packet service could acquire slots on a demand basis.

However, economic issues took control just when both the standards document and the silicon implementation were nearing completion. Burroughs merged with Univac Corp, and in the process MAN support lost its funding. With the loss of its most visible backer, the slotted ring scheme was delayed, despite interest on the part of Nippon Telephone and Telegraph in Japan.

Two new proposals took the place of the slotted ring: one from Telecom Australia known as the distributed-queue, dual bus (DQDB), and one from Integrated Networks called multiplexed slot and token (MST). MST was based on use of the packet media access control protocol of the fiber distributed data interface, a 100M-bps fiber-optic LAN standardized outside the IEEE. This was overlaid with fixed slots for voice and video; it also added speed flexibility: rather than a fixed 100M bps, it could run at any of several telephone-hierarchy speeds.

In late 1987, the committee achieved a consensus and the distributed queue dual bus was selected as the proposal on which to base the MAN standard.

THE DUAL BUS PROPOSAL

The dual bus proposal was in fact an evolution of a coaxial cable–based system known as Fasnet, which was presented to the MAN committee by Bell Laboratories in 1983. In the Australian form, it runs on optical fiber at 155M bps in each direction, although segments can run on coaxial cable or even microwave radio if the links have sufficient bandwidth. The dual bus is actually laid out as a ring, with the provision that it can be star wired somewhat like the IEEE 802.5 token-passing ring. The cable can be routed so as to return periodically to central offices for ease of maintenance; in between these offices the nodes can be clustered or attached individually in string-of-pearls fashion. An alternative is used by SMDS: individual customers are served by separate DQDB subnetworks connecting them with central offices, with shared facilities running only between the central offices. This solves the privacy issue but has high optical fiber costs; as a result, the links out to the customer are open rather than looped buses.

The departure from earlier ring systems is that the network, although cabled as a ring, is logically a bus: one node on the net does not repeat the incoming data. As a result, packets on the MAN do not have to be explicitly removed—they just fall off at the end of the network. The end node (node 0) generates the slot framing for each fiber, performing the function of the master node in a ring system (see Exhibit III-2-26).

The great advantage of this bus topology is reliability. If any node or line segment should fail, the opening in the ring is moved to the location of the failure. In Exhibit III-2-27 a break between nodes 2 and 3 causes the opening in the ring to move to the site of the break. The nodes on either side of the break (nodes 2 and 3) take up the bus-end function of slot generation. Operation continues at full speed with no increase in latency. In any case, such failures should be very rare, because bridging relays can protect against the failure of a node or cluster.

This fail-safe operation contrasts to most ring systems, which maintain a second fiber or wire solely for backup and, as a result, cut the potential capacity of the network in half.

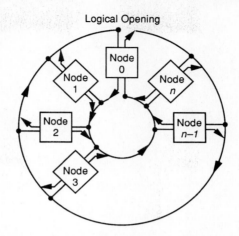

Exhibit III-2-26.   Directional Connection of Nodes to the Dual Bus

A second major advantage of the DQDB network is in scheduling, as shown schematically in Exhibit III-2-28. Requests coming from the right, and empty slots traveling to the right, are counted in the up-down counter. When the station wishes to transmit to the right, it takes a sample of the up-down counter and counts down to zero on empty slots. At zero, it is enabled to transmit. The system takes advantage of the bidirectional nature of the bus to maintain a distributed queue of nodes waiting for access. Under light load, the system provides quick access to the network, as in Ethernet, but under heavy load, the efficiency approaches 100% with bounded delay times. In this respect, it is similar to a token system.

This is similar to waiting for a bus at rush hour. No seats are vacant on the buses going by, but to go eastbound, it is possible to send a message on a westbound bus. The message tells the first eastbound bus with an available seat that space should be held.

In the case of the network, the dual-bus controller maintains an up-down counter for each direction of transmission. The counter counts up for each slot request arriving from the direction of the destination, and down for each vacant slot going toward the destination on the other fiber. If the counter is already zero when a station wishes to transmit, it can use the next vacant slot. If there are requests pending, the counter will have a nonzero value. The controller takes a sample of the counter (which continues to count up and down for benefit of future requests) and separately counts down as vacant slots come by. These vacant slots satisfy the requests that

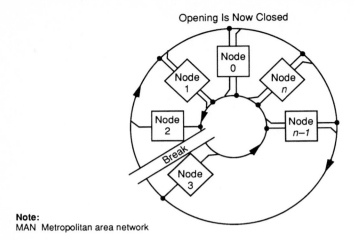

Note:
MAN Metropolitan area network

Exhibit III-2-27.   Dual-Bus MAN: Logical Opening Moved to Site of Break

were already pending when the station decided to transmit. When the sample has counted down to zero, the next vacant slot is available for use in transmitting.

The performance of this distributed reservation method has been shown to be equivalent to a centralized first-come, first-served queue, with only small differences that are attributable to the time for the reservation to propagate on the cable. Monopolizing the medium is prevented by not permitting a station to have more than one reservation outstanding at any time. Once a packet is transmitted, another reservation can be made.

Because of the opening in the ring, there is a slight asymmetry in the access times, depending on whether a node is near the end or the center of the bus. Under regular operating conditions (50% to 80% of capacity), this asymmetry is not significant, on the order of 5 to 10 microseconds difference between the center and the end. Under overload conditions, the difference becomes greater, but four priority levels are available for the distributed queue, and these can be assigned in a way that restores reasonable fairness.

WORLDWIDE INTERCONNECTION: SYNERGY WITH ISDN

While the IEEE 802.6 standards have been under development, the worldwide telephone industry has been gearing up to provide the integrated services digital network (ISDN). ISDN is intended to provide data and digital voice service using the same facilities, with compatibility on an inter-

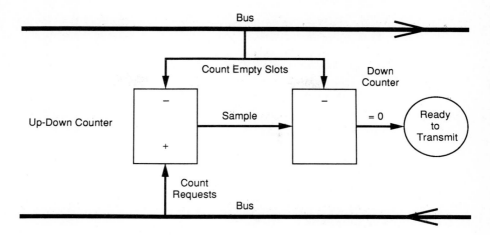

Bus

Count Empty Slots

Down Counter

Up-Down Counter

−

Sample

−

= 0

Ready to Transmit

+

Count Requests

Bus

Exhibit III-2-28.   Access with Distributed Queue

national scale. Initial implementations will provide speed capabilities comparable to existing facilities (64K bps and 1.5M bps), but a future generation known as broadband ISDN will go much higher. It will run at fiber-optic speeds—155M bps.

With the exception of digital video, very few applications supply or absorb data at full line speed. Multiplexing must be used to assemble enough load to make broadband ISDN useful. The IEEE 802.6 MAN is ideal for this purpose, providing a very flexible method for allocating the bandwidth either in packets on demand or in fixed allocation isochronous channels.

Coordination of the work of IEEE 802.6 with broadband ISDN is now under way. A formal liaison has been created with American Standards Committee X1S1.1 (formerly X1T1.1), which formulates US input to the broadband ISDN standards process. The results of this process have been excellent: the same segment and header fields have been adopted by both IEEE 802.6 and ASC T1S1. The IEEE 802.6 standard has been designated as the preferred multiplexing arrangement for broadband ISDN. Given the concurrence of the full International Telephone and Telegraph Consultative Committee (CCITT), the international agency standardizing ISDN, this means that segments can be copied directly from the metropolitan area network to broadband ISDN with essentially no change. The boundary between the MAN and WAN can be made invisible to the user.

IEEE 802.6 has made a 60-bit address field optional, in addition to the mandatory 48-bit address. (According to IEEE 802 practice, 16-bit addresses also exist as an option.) The 60-bit field accommodates the ISDN address as defined in CCITT recommendation E.164. Within the MAN, any address is interpreted as a flat address space without structure; each node either receives the passing data or ignores it, depending on whether the address matches or not. But once the segment is moved to broadband ISDN, the E.164 address can be interpreted hierarchically, with its embedded country code, area code, and local number. In effect, ISDN can provide a global bridge between MANs that is independent of the type of higher-layer protocol, whether standard or proprietary. SMDS uses the 60-bit address exclusively.

Header formats for DQDB have been modified to accommodate broadband ISDN in a number of ways. The most significant is the inclusion of a virtual circuit identifier field of 20 bits. The DQDB protocol does not require this field, but it will be used once the packet goes beyond the dual bus system into broadband ISDN.

The resulting format is a segment of 48 bytes payload and 5 bytes header. In the connectionless service, provisions must also be made for segmentation, because the original service data unit received from higher layers may exceed 48 bytes in size. (In fact, sizes to 9,144 bytes are permitted.) Therefore, the first 2 bytes of the 48 carry a 10-bit label to guide reassembly and a 4-bit sequence number, along with a field that indicates whether the segment represents a single-segment message or whether it is a beginning, continuation, or end of a multisegment message. The last two bytes carry redundancy check and padding-length information. This format is shown in Exhibit III-2-29.

AT&T AND THE MULTIPORT BRIDGE

In addition to serving as a multiplexing facility for broadband ISDN, the metropolitan area network should find synergy with advanced work in the telephone industry in the switching area. A great deal of work is going on in the industry in the area of asynchronous transfer mode or fast packet switches.

These switches are being designed on the assumption that the input arrives at fiber-optic speeds. In general, they are based on switching the data (or a packetized voice signal) in the packet rather than at a bit or byte level. They read and interpret the packet header while the packet is in motion, using it to route the packet through the switching fabric to its destination. The architecture of these switches is highly repetitive, making them relatively inexpensive and easy to construct as integrated circuits.

The segment (or packet) size on the dual bus can be tailored for compatibility with the switch architecture. The 5-and-64 agreement between 802.6 and T1S1 paves the way for a central asynchronous transfer mode switch to act as a multiport media access control–level bridge to interconnect a number of MAN loops.

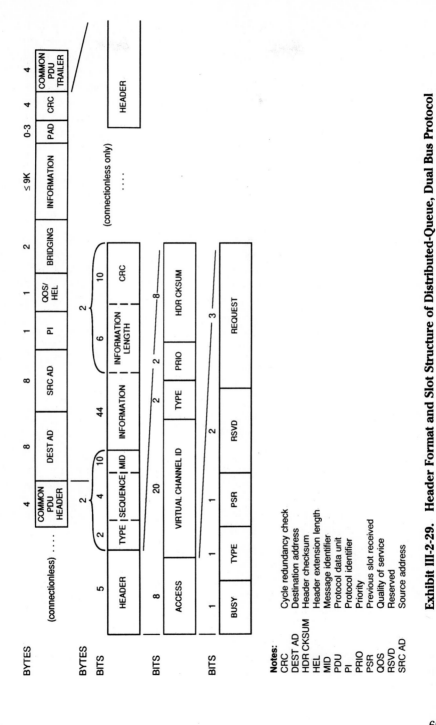

Exhibit III-2-29. Header Format and Slot Structure of Distributed-Queue, Dual Bus Protocol

Notes:
CRC   Cycle redundancy check
DEST AD   Destination address
HDR CKSUM   Header checksum
HEL   Header extension length
MID   Message identifier
PDU   Protocol data unit
PI   Protocol identifier
PRIO   Priority
PSR   Previous slot received
QOS   Quality of service
RSVD   Reserved
SRC AD   Source address

AT&T has made such a proposal to the committee. The proposal has been accepted as a work item and a task force has been chartered to pursue the issues. The multiport bridge proposal represents a technique to interconnect a set of dual bus loops in a hierarchical way. An alternative hierarchy would be to use a second-level dual bus as a backbone. With sufficient localization of traffic, most data would not need to travel over the backbone, and a loop of the same speed would suffice, at least temporarily. The technology of the multiport bridge, however, is very open ended with respect to capacity, and it provides a growth path that should be adequate well into the next century.

SUMMARY

Just as the local area network standards were in use before they reached the full international standard level, IEEE 802.6 MANs will be implemented in the near future. Pilot networks are in operation in Perth, Australia and in Philadelphia; these serve both as full-speed technical prototypes and as market tests for services to be provided over the network. These networks and their production versions will provide citywide systems with the same advantages now available with LANs and voice PBXs.

Chapter III-2-3
Performance Evaluation of Metropolitan Area Networks

Johnny S. K. Wong

ENGINEERING WORKSTATIONS use three types of networks: the wide area network (WAN), metropolitan area network (MAN), and local area network (LAN). The main difference among the three is the way in which they are geographically distributed.

WANs are characterized by their large geographic span. They provide customers with the ability to access a wide range of hardware and software resources. Because the nodes in a communications network are physically distributed over a large geographical area, the communicating links are relatively slow and unreliable. WANs are also known as store-and-forward networks because each node is responsible for storing and forwarding packets to the neighboring nodes in the network. This requires routing and flow control mechanisms. The irregular mesh structure of the WAN system can create problems that are much more complex than those of a LAN.

LANs connect a variety of devices within a very small geographical area (e.g., a single building or several adjacent buildings). Because all the devices in these systems are close to each other, the communications links have higher speeds and lower error rates.

The devices in a LAN share a single transmission medium, minimizing routing and flow control problems. Access control to the communications

	LAN	MAN	WAN
Geographic Span	Building or campus	Metropolitan area	Nationwide
Topology	Common bus	Regular mesh common bus	Irregular mesh
Speed	Very high	High	Low
Error Rate	Low	Medium	High
Flow Control	Simple	Medium	Complex
Routing Algorithm	Simple	Medium	Complex
Media Access	Random scheduled	Scheduled	None
Ownership	Private	Private and public	Public

Exhibit III-2-30. Comparison of LANs, MANs, and WANs

medium, however, is an important design issue. Media access control algorithms are protocols designed to allow only one device to attempt transmission at a time. Three media access control standards have been developed by the Institute of Electrical and Electronics Engineers (IEEE) 802 Committee: 802.3 for carrier-sense multiple access with collision detection (CSMA/CD), 802.4 for token bus, and 802.5 for token ring configurations.

MANs are intended to connect high-speed LANs over a metropolitan-sized region. MANs, like WANs and LANs, are based on two basic types of structures: passive bus transmission technology or point-to-point connection with a simple switching function. The systems based on a passive bus afford simple access protocols and grant reliable transmission during station failures; but their extension is somewhat limited. Point-to-point systems do not have this limitation but suffer from processing overhead at intermediate nodes and reconfigurability problems when stations fail. Exhibit III-2-30 provides a comparison of the three network types. Three examples of MANs are discussed in the following sections. The Manhattan Street Network and the HR^4-NET are point-to-point mesh systems, and the Queued Packet and Synchronous Switch is a dual-bus network.

MANHATTAN STREET NETWORK

The Manhattan Street Network (MSN), developed by N. F. Maxemchuk, is a rectangular mesh with alternate directed rows and columns. Opposite nodes on the left and right boundaries are connected to each other, as are opposite nodes on the upper and lower boundaries, as shown in Exhibit III-2-31.

The number of rows and columns is even, with two links entering and two links leaving each node. This node structure and the access strategy are similar to those in a bidirectional slotted loop system. The principal difference between the MSN and a loop network is that in the MSN a routing decision must be made for each packet transmitted at each node.

The regular structure of an MSN has several special features that allow it to achieve higher throughput and to support more nodes than conventional loop and bus networks. These include:

- Multiple paths between any two nodes—This provides higher reliability and flexibility.
- Traffic locality—Mesh networks, unlike the conventional LAN, increase the throughput by decreasing the fraction of the network capacity needed to transmit information between the source and destination. Sources that communicate frequently can be clustered into communities of interest that do not interfere with one another.
- A short distance between any two nodes—On the average, fewer links in the network are used to connect source and destination because of the wraparound connections between boundaries and the grid structure of the network.

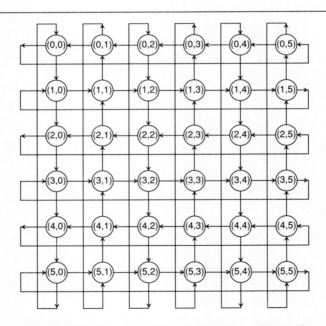

Exhibit III-2-31. 6 × 6 Manhattan Street Network

System Operation and Access Strategy

The packet size in any slotted system is fixed. For each output line, a transit packet that originated from an input line or a packet that originated from the source node is periodically transmitted. At each node, the packets from the input lines are delayed so that they arrive at the switch at the same time that the node transmits a packet. The node switches each of the incoming packets not destined for the node to one of the output links. If two packets are destined for the same link, one is forced to take the other output link. This strategy guarantees that packets are never lost, even if there is no buffer; however, the buffers at the nodes can help reduce packet misdirections.

Packets from the source are transmitted only when there is an empty slot on an output link. The node controls the source so that packets do not arrive faster than they are transmitted, and the rate available to the source decreases when the network is busy. A packet that is misdirected takes a longer path to its destination, preventing new packets from entering the network. Therefore, the access strategy behaves like a flow control mechanism.

In the MSN, each time a packet is misdirected, the length of the patch to the destination is increased by no more than four links. In addition, there are many nodes for whichever outgoing link provides the same path length to the destination—when a packet may take either link, the probability that any packet will need to be misdirected decreases.

Routing

The MSN routing algorithm can be divided into two categories: deterministic and random. Deterministic routing requires that nodes calculate the relative address and determine the preferred path.

Once the calculation is completed, the packet selects the preferred path from a node. If there are two preferred paths from a node, it selects either path. If there are no preferred paths, it selects the best available path. This guarantees the shortest path to the destination—if there is no conflict. When two packets have the same preferred path, one of them must be deflected. The decision could be based on random selection between two packets or on additional routing information (e.g., hop count). In the second case, the packet with the larger hop count should be considered first because it has consumed more network resources.

When random routing is selected, a packet chooses links with equal probability, and each node checks to see if the packet is at the destination. There are two advantages to using random routing rather than deterministic routing: it is extremely easy to implement, because it is not necessary to calculate the relative address and the preferred path; and it is extremely tolerant of network irregularities. Topology change due to link failure, node failure, or adding nodes has no impact on the routing algorithm. Random routing, however, is inefficient; a packet has no knowledge that the destination is just one link away and might choose the wrong link. This results in low throughput.

HR⁴-NET

HR⁴-NET, proposed by F. Borgonovo and E. Cadorin, is a rectangular mesh-structured network like the MSN, except that each node is linked to each of its four neighbors through full-duplex connection lines. An illustration of this network is provided in Exhibit III-2-32.

In addition to all the features contained in the MSN, full-duplex communications links, all operating at the same speed, guarantee that the capacity of the outgoing links at every node does not exceed the capacity of the incoming links should any link fail. This prevents packet loss caused by lack of storage; when a link fails, each node still has the same incoming and outgoing capacity. Full-duplex links also provide greater connectivity and allow an even shorter path length between two nodes, unlike the MSN's unidirectional links, which sometimes take two more hops to reach the destination.

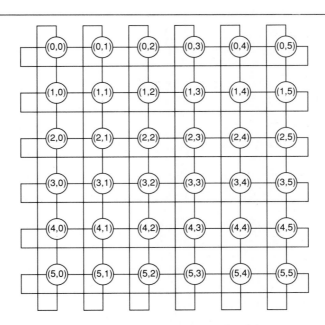

Exhibit III-2-32. 6 × 6 HR⁴-NET

Routing

Deterministic routing calculates the relative address and finds the shortest path to the destination. This may be effective in maximizing the throughput of the network. This routing technique is complex, however, and may be too time-consuming to be applicable in MANs. In addition, deterministic routing requires knowledge of node locations, and this knowledge must be updated if nodes are added or dropped from the network.

These drawbacks are completely eliminated by a totally random routing, which does not require knowledge about the location of nodes. Instead, it selects the output link of every packet at random. Unfortunately, the network throughput allowed by the random routing is very low.

Borgonovo and Cadorin proposed a routing procedure capable of maintaining the advantages of the random routing while achieving a network throughput that favorably compares with the shortest-path algorithm. The network is organized into low-level (L) rings (i.e., streets) and high-level (H) rings (i.e., avenues). At each node, a packet is considered an H-packet if its destination ring address does not match the ring address of the node. In that case, the packet must be relayed over the H-ring. Otherwise, it is considered an L-packet and it must be relayed over the L-ring to reach its destination.

A packet is satisfied by a routing procedure if it is routed to the output link (e.g., an H-packet is forwarded to an H-link and an L-packet to an L-link), provided that the output link is not the packet incoming link. If there are several equally good assignments of packets to outgoing links, the HR^4-NET routing strategy chooses randomly among the possible alternatives, trying to locally satisfy the maximum number of packets.

QUEUED PACKET AND SYNCHRONOUS SWITCH

The Queued Packet and Synchronous Switch (QPSX) is a distributed network based on two contradirectional buses, as shown in Exhibit III-2-33. The buses originate and terminate at the network controllers but have no through connections. A station is connected to the buses by access units that perform the distributed queuing protocol. This is a reservation-based access scheme that forms a single distributed queue, controlling the access of packets to the network in essentially the same order as they are generated. If no packets are waiting, access is immediate. Under conditions of light load, it behaves like CSMA/CD. Under heavy load, the bus use approaches 100%, and it is similar to the token-based protocol, but without the overhead of a token traveling around the network. The single queue property also eliminates unnecessary delay, maximizing bus use.

Advantages of this passive bus structure include:
- The dual-bus structure allows full-duplex communications between each pair of nodes.

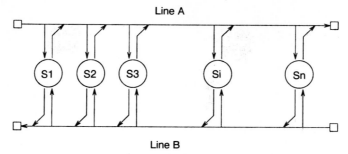

Line A

Line B

Exhibit III-2-33. QPSX Dual Bus Architecture

- The bus structure of the QPSX is inherently reliable—The operability of the network is independent of the operation of the individual nodes.
- The routing decision is made only once—Therefore, the processing overhead is much less than for a point-to-point network.
- A shared-medium network does not require a large expense—This facilitates network expansion, with the initial investment mainly in cabling.
- With an appropriate access scheme like the distributed queuing protocol, the bandwidth and delay requirement of the voice packet can be easily satisfied.

The two significant performance measures are throughput and delay. Whereas the conventional media access control schemes make it sensitive to network size, the distributed queuing protocol employed by the QPSX is much less sensitive to network size.

Distributed Queuing

The communications channel is a time-slotted packet system. Controllers at the bus ends generate all timing signals to ensure synchronization. Two bits are used to implement the protocol. The BUSY bit indicates that the slot is being used. The REQ bit indicates a request for transmission.

The protocol, shown in Exhibit III-2-34, can be described as follows (S_i accessing bus A):

- IDLE state (when there is no packet for transmission):
  —The REQ bits received on bus B are tracked.
  —The REQ counter is incremented for each REQ bit received. This indicates the number of packets queued for transmission on bus A downstream from S_i.
  —The REQ counter is decremented for each empty packet that passes S_i on bus A.

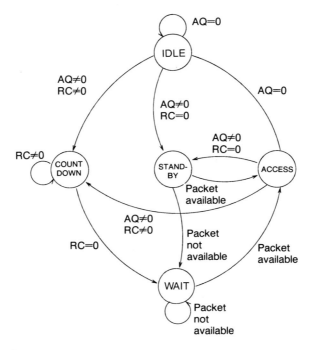

Notes:
AQ   Number of packets queued
RC   REQ bit count

Exhibit III-2-34.   State Transmission Diagram for Packet Access

- COUNTDOWN state (when S_i has a packet for transmission and the REQ count is nonzero):
  - —The REQ counter is transferred to the CD counter, and the REQ counter is cleared.
  - —The CD counter is decremented by one for each empty packet that passes on bus A.
  - —The REQ counter is incremented for each received REQ bit.
  - —Transmission of the REQ bit is initiated to reserve a free packet (it will wait until there is no contention for the REQ bit).
- WAIT and ACCESS state (when the CD counter goes to zero):
  - —The next free packet on bus A transmits its queued packet.
  - —If available, the BUSY bit is set and the packet is transmitted.
  - —If there are no more packets, an IDLE state is initiated.
  - —If there are more packets for transmission and the REQ counter is not zero, a COUNTDOWN state is initiated.

- STANDBY state (when a new packet moves into the first queue position and the REQ counter is zero):
  —Access is attempted in the next packet.
  —If the packet is free, it is transmitted without a REQ bit.
  —If the packet is busy, the sending of a REQ bit is initiated, followed by the WAIT state.

Compared with existing packet access protocols, which have no current knowledge about the state of the system, distributed queuing represents a significant advance in performance because it ensures that all stations have sufficient information about the state of the network at all times. Packets know exactly when to access without waiting for control information. Insensitivity to network size also makes it more suitable in the larger and higher-speed networks of the future.

PERFORMANCE EVALUATION AND COMPARISON

A performance study by simulation model and queuing analysis for the QPSX and the MSN is described in this section. HR⁴-NET is not included because of its similarity to the MSN.

Simulation Model

The number of nodes in the simulation model is 36. Each node generates packets with probability p at each slot. Those not accepted by the network in the same slot are discarded. In operating conditions they are necessarily queued. The QPSX normalized propagation delay a is set to 1, so that the packet length equals the link length in bits. The deterministic routing algorithm is assumed for the MSN.

The performance measures are absolute throughput, delay, and network use, where absolute throughput is defined as the mean number of packets received by the destination nodes per slot time and delay is defined as the sum of queuing delay and transit delay. Network use, or normalized throughput, is the average busy ratio of the links.

The basic simulation model for the MSN is shown in Exhibit III-2-35. Initially, the program first reads SLAM control statements, which specify the characteristics of the model (e.g., the number of runs, the time limit of each run, the seeds of the random number generator, and several monitoring options). After initializing the general variables, the program then calls the events according to the event time associated with them. One event deals with the packet generation for each node; the other is the packet arrival event. For each arriving packet, the program first calculates the relative address of the destination. It then calls the scheduling subroutine to decide whether the packet has reached the destination, should be dropped, or routed to the next node. It repeats the same procedures

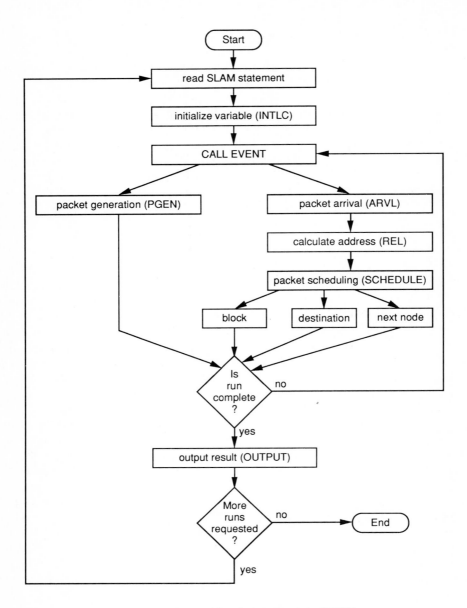

Exhibit III-2-35. Simulation Model of MSN

until a time limit has been reached. The simulation is printed out, and it enters the next run.

The model for the QPSX is shown in Exhibit III-2-36. The reading of SLAM statements, initialization of variables, calling of packet generation events, and output printing are the same procedures as for the MSN. The only difference is the QPSX protocol. Based on the node information and bus status, each node can be in IDLE state if it has no packet for transmission, in COUNTDOWN state if it is waiting for all previously queued downstream packets that have been transmitted, in WAIT state when it is waiting for an empty slot, or in ACCESS state when it finally gets a free slot to transmit.

Exhibits III-2-37 and III-2-38 show absolute throughput (i.e, packets per second) and delay as a function of node input rate p. When input rate increases, the QPSX soon reaches its maximum throughput and the queuing delay increases accordingly based on the single queue property of the QPSX. The MSN delay is also an increasing function of the input rate. That is due to the deflection effect—when the preferred route is already occupied, the packet will be deflected to the other link. This results in longer transit delay. In order to minimize this deflection effect, higher priority is given to a single preference packet. The use of a buffer can also reduce the number of deflections and increase the throughput.

The results of the simulation show that the MSN has better performance than the QPSX for both throughput and delay. This is because:

- The point-to-point structure of the MSN allows two nodes to transmit packets at the same time without interfering with each other, whereas the QPSX allows only one packet on the bus at any time.
- The multiple-path property of the MSN gives more flexibility and reliability, whereas each packet in the QPSX has only one outgoing choice.

Another important factor is the length of slot time. In the simulation model, the same slot time for the MSN and the QPSX is used, although they can be different in real situations. Each intermediate node in the MSN needs to perform the routing decision, whereas in the QPSX, the routing decision is made only once at the source node. This computation overhead may result in larger slot time and reduce the throughput difference. For example, if the slot time of the MSN is twice that of the QPSX, the throughput is the MSN must be divided by two in order to compare to the throughput of the QPSX.

The normalized throughput or network use, which is defined as the ratio of absolute throughput to maximum throughput that can be achieved by the network, should be considered. The maximum throughput for the QPSX with $a = 1$ is 2 packets per slot or dual bus. For the MSN, the maximum throughput is $2n^2/d$, where n^2 is the network size and d is the average number of hops between two nodes. The value $2n^2$ represents the total

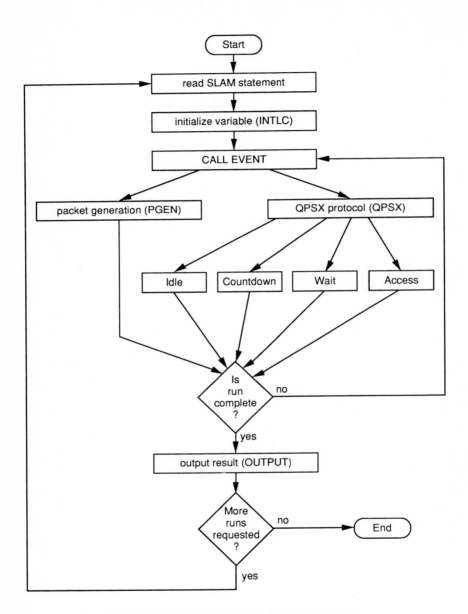

Exhibit III-2-36.   Simulation Model of QPSX

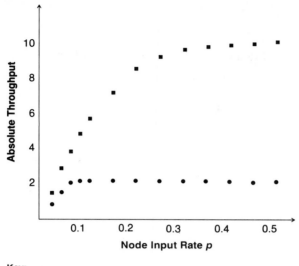

Key:
■ Manhattan Street Network
● Queued Packet and Synchronous Switch

Exhibit III-2-37.   Absolute Throughput Versus Node Input Rate *p*

number of links (i.e., the maximum number of packets that can exist on the route at the same time). On the average, however, each packet needs *d* hops to reach its destination. Therefore $2n^2/d$ is reached. Exhibit III-2-39 shows that the QPSX has much better bus use than the MSN. The results also confirm the arguments for a QPSX. That is, when input load is very light, the packet can gain access to the bus almost immediately. When input load becomes higher, it behaves like an ordered queue, which has 100% bus use.

The effect of the access level flow control is displayed in Exhibit III-2-40. Because a QPSX behaves like a single queue, when the input rate gets higher, the entire network reaches the saturation point without a dropping curve. For an MSN, because the deflection effect makes the network behave like a flow control mechanism, the high input rate will not cause a dropping curve.

Queuing Analysis of QPSX

Because the distributed queuing protocol ensures that there are no wasted slots on the network whenever there is a packet queued, it is possible to model the network average delay on a single-server queuing

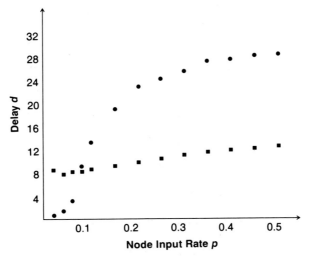

Key:
● Queued Packet and Synchronous Switch
■ Manhattan Street Network

Exhibit III-2-38. Delay *d* Versus Node Input Rate *p*

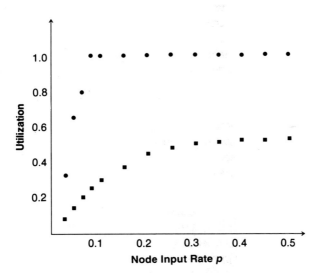

Key:
● Queued Packet and Synchronous Switch
■ Manhattan Street Network

Exhibit III-2-39. Use Versus Node Input Rate *p*

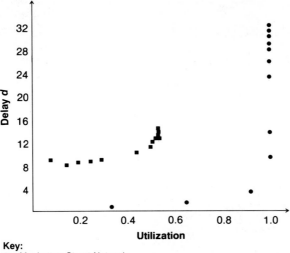

Key:
■ Manhattan Street Network
● Queued Packet and Synchronous Switch

Exhibit III-2-40.   Delay *d* Versus Use

---

system. The server is a deterministic slotted server with the service time assumed to be one unit. The packet generation process is assumed to be Poisson and all buffers are infinite.

W is the average waiting time and λ is the average arrival rate; $\rho = \lambda/1$ is the network use (or load). This leads to the following equation:

$$\overline{W} = \frac{1}{2} + \rho\overline{W}$$

The first term on the right-hand side of the equation is the average time between the arrival of a packet and the start of the next slot. The second term is, by Little's Law, the average number of packets awaiting transmission at the moment the packet arrives. The equation then becomes:

$$\overline{W} = \frac{1}{2(1 - \rho)}$$

This result differs from that corresponding to the standard *m/d/*1 queuing system by the additional term ½. The added term results from the synchronous nature of the server and represents the average time an arriving packet must wait before the start of the next time slot.

Exhibit III-2-41 compares the delay results from the queuing model of the second equation and simulation result. In this equation, *r* is the node input rate, while ρ is the total network input rate. The relationship between these two parameters is $p = \rho/36$ when the network size is 36.

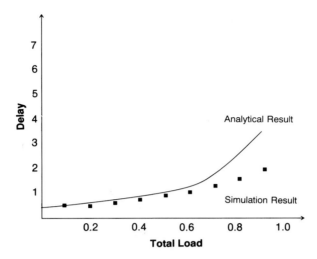

Exhibit III-2-41. **Delay Versus Total Load p Using Queuing Analysis of QPSX**

SUMMARY

MSNs and HR⁴-NETs are regular point-to-point mesh toroidal networks with uni- and bidirectional links, whereas the QPSX is a dual-bus multiaccess network. This topology difference results in higher hardware costs but better traffic locality for MSNs and HR⁴-NETs. All three networks have high reliability and reconfigurability and are insensitive to network expansion. Flow control problems are simplified because of their special structures. For example, they are deadlock-free. The access control for the QPSX is the distributed queuing protocol, which treats the entire network as a single queue and regulates the order of transmission so that no slot is wasted. MSNs and HR⁴-NETs need a more complex routing algorithm. The outgoing link selections are 4, 2, and 1 for the HR⁴-NET, MSN, and QPSX respectively. Because the distributed queuing protocol makes an entire network behave like a single queue, bus use could be very high.

Chapter III-2-4
An Overview of FDDI

John F. Mazzaferro

THE FIBER DISTRIBUTED DATA INTERFACE (FDDI) network is based on dual rings of optical fiber that connect stations over distances of 100 kilometers. The token-ring network physically consists of a series of point-to-point links, each less than two kilometers long, between neighboring stations. When a station becomes active on the ring, it must accurately repeat all frames it receives to ensure the integrity of the ring. The primary ring is used for data transmissions; the secondary ring is used as a backup in the event of a link or station failure (see Exhibit III-2-42).

Access to the network is controlled by a rotating token. The token is a control signal, composed of a unique symbol sequence, that circulates on the network after each information transmission. After detecting the token, any station may capture the token (and the right to transmit) by removing it from the ring. The station may then transmit information until its token-hold timer expires or it has no more information to transmit. At the completion of its transmission sequence, the station issues a new token, providing other stations with the opportunity to gain access to the ring.

If a failure occurs, recovery is supported by two mechanisms. The first, called wrapback, involves using the secondary counterrotating ring to bypass the failing link or station. This reconfiguration occurs automatically and is an integral part of the FDDI specification. If a failure occurs between stations 2 and 3 as shown in Exhibit III-2-43, the ring detects an inoperable condition caused by expiring token-rotation timers. The ring is forced into error recovery, using both the primary and secondary fiber-optic cables to bypass the break and maintain continuity on the network (see Exhibit III-2-43).

The second recovery mechanism is an optical bypass switch, which is used to completely bypass nodes on the primary and secondary rings (see Exhibit III-2-44). This is most effective when power is lost at a station or a station is anticipated to be inactive for an extended period.

FDDI consists of three primary entities that conform to the open systems interconnection (OSI) model, as illustrated in Exhibit III-2-45: the physical

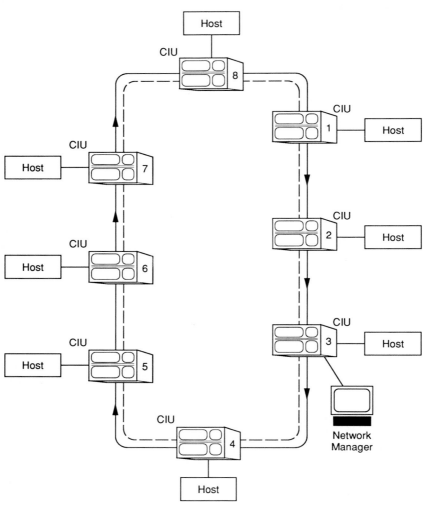

Note:
CIU   Channel interface unit

Key:
———————— Primary ring
— — — — Secondary ring

Exhibit III-2-42.   FDDI Dual Counterrotating Fiber-Optic Ring

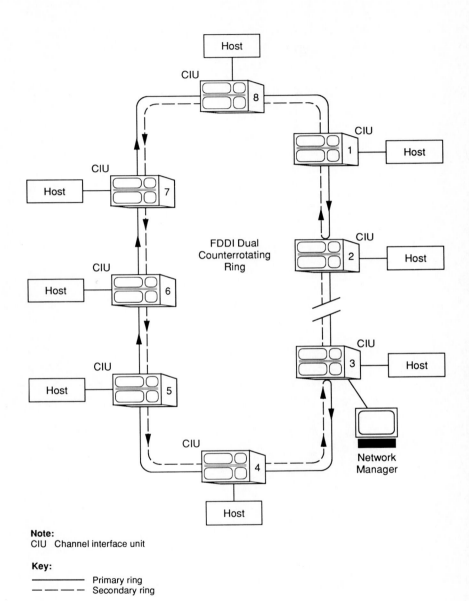

Note:
CIU   Channel interface unit

Key:
———————   Primary ring
— — — — —   Secondary ring

Exhibit III-2-43.   FDDI Ring Using Secondary Ring

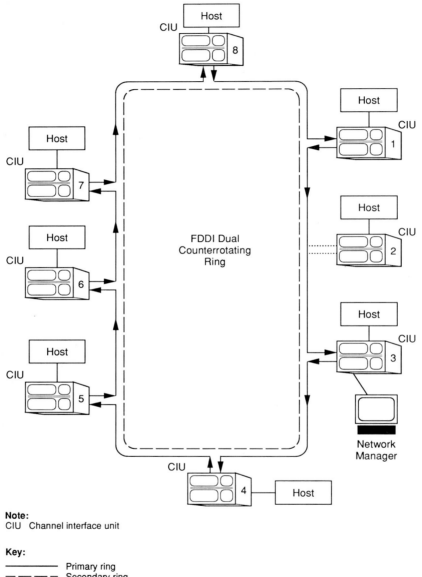

Note:
CIU   Channel interface unit

Key:

——————— Primary ring
— — — — Secondary ring
:::::::::::::: Optical bypass
          connection

Exhibit III-2-44.   FDDI Ring with Optical Bypass Switch

OSI	FDDI	
Application		
Presentation		
Session		
Transport		
Network		
Data Link	802.2 Logical Link Control Media Access Control	Station Management
Physical	Physical Layer Protocol Physical Medium Dependent	

Exhibit III-2-45. Layers of the OSI Model with FDDI

layer, the data link layer, and station management. The physical layer is subdivided into:
- A physical medium dependent layer, which provides the point-to-point communications between stations in the network—The physical medium dependent layer defines and characterizes the fiber-optic drivers and receivers, cables, connectors, and physical hardware.
- A physical layer protocol layer, which provides a connection between the physical medium dependent layer and the data link layer—Information communicated on the interface medium is encoded by the physical layer protocol into transmission codes. The physical layer protocol provides the synchronization of incoming and outgoing data and their clocks.

The data link layer is subdivided into:
- A media access control standard, which provides fair and deterministic access to the medium, address recognition and generation, and verification of frame-check sequences—Its main function is the delivery of frames, including frame insertion, repetition, and removal.
- A logical link control standard, which provides a common protocol to provide the required data assurance services between media access control and the network layer.

A station management standard provides the control necessary at the station level to manage the processes under way in the various FDDI layers so a station can work cooperatively on a ring. This standard provides such services as control of station initialization, configuration management, fault isolation and recovery, and scheduling procedures.

OPTICAL FIBER OVERVIEW

The optical interface contains the building blocks of a network—optical fiber, the transceiver, and connectors. Fiber-optic technology is a significant step in the evolution of electronic communications. Copper cable is less able to meet the requirements of increasing speeds and transmission distances. In the areas of signal integrity and information-carrying capacity, fiber optics offers many advantages over copper cable. The primary benefits of fiber optics include:

- Wide bandwidth—Optical-fiber bandwidth, the information-carrying capability of the fiber, is more than 100M bps.
- Low loss—Optical fibers offer low signal attenuation.
- Electromagnetic immunity—The fiber is a dielectric and therefore not affected by electromagnetic energy. Benefits of such immunity include the elimination of crosstalk, ground loops, and signal distortion in hostile environments.
- Security—Fiber-optic cable is almost impossible to tap without detection. Other eavesdropping techniques are also difficult because the fiber does not radiate electromagnetic energy.

Fiber-Optic Cable

An optical fiber is a thin, flexible glass or plastic waveguide through which light is transmitted. As illustrated in Exhibit III-2-46, an optical fiber has three main components:

- A core—The central region of an optical fiber, through which the light is transmitted.
- Cladding—The dielectric material surrounding the core of an optical waveguide. The cladding must have a lower index of refraction than the core (i.e., it must be easier for light to travel in the core than in the cladding).
- A jacket—A protective material extruded directly on the fiber coating to protect it from the environment and physical damage.

The FDDI physical medium dependent layer specification recommends an optical fiber with a core diameter of 62.5 microns and a cladding diameter of 125 microns.

There are two primary types of optical-fiber cable—single-mode and multimode. As illustrated in Exhibit III-2-45, the core size varies greatly between single-mode and multimode optical cable. Modes are discrete

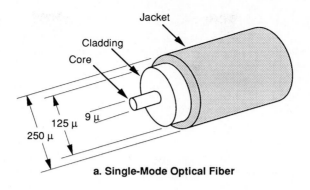

a. Single-Mode Optical Fiber

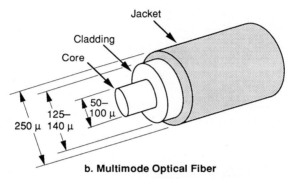

b. Multimode Optical Fiber

Exhibit III-2-46.   Fiber-Optic Cable Mode Types

optical waves, or signals, that propagate in the fiber. In a single-mode fiber, only the fundamental mode can propagate. Using a laser light source, single-mode fiber is particularly suitable for very high speed transmission (e.g., 1.0 GHz) over long distances. As the cost of a laser light source decreases, single-mode solutions are becoming more attractive.

In multimode cable, a larger number of light modes are coupled into the cable, making it suitable for the less costly light-emitting diode (LED) light source. Multimode graded-index optical cable bends the light so that all modes have a similar effective speed, providing the ability to achieve a high bandwidth (e.g., 500 MHz) and distances of several kilometers. These characteristics meet the FDDI requirements. Exhibit III-2-47 depicts graded-index cable propagating a signal through the optical waveguide.

Another important consideration is the light-gathering capability of the fiber, or its ability to accept enough light to maintain the required signal integrity. This light-gathering capability is called the numeric aperture. Although a fiber with a high numeric aperture gathers more light than a fiber with a low numeric aperture, it also has greater dispersion. Matching the numeric aperture of system components (sources, fibers, and receivers) minimizes losses in a fiber-optic system (see Exhibit III-2-48). FDDI specifies a nominal numeric aperture of 0.275.

The optical waveguide specifications build a set of guidelines to support the two primary parameters of optical transmission: attenuation loss and bandwidth. Attenuation loss is the decrease in magnitude of the power of a signal between two points. Attenuation is usually measured in decibels per kilometer (dB/km) at a specific wavelength. The lower the attenuation number, the better the fiber. Typically, FDDI-grade optical cable has a loss of less than 2.5 dB/km at 1,300 nanometers (nm).

The other primary parameter of optical transmission is bandwidth, the measure of information-carrying capacity, which is usually stated in terms of megahertz per kilometer (MHz/km) at a particular wavelength. FDDI specifies a minimum bandwidth of 500 MHz/km at 1,300 nm for optical fiber used in the ring.

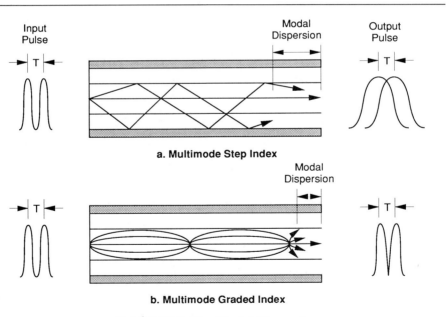

a. Multimode Step Index

b. Multimode Graded Index

Exhibit III-2-47.   Modal Dispersion

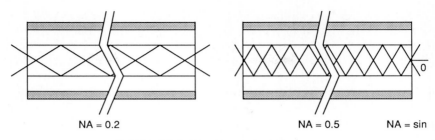

NA = 0.2   NA = 0.5   NA = sin

Exhibit III-2-48.　Numeric Aperture

Optical Signals

All optical links consist of an optical transmitter, the optical fiber, and an optical receiver. The transmitter is an electronic device that receives a modulated electrical signal and converts it to a modulated light signal, which is then launched into an optical fiber. The receivers convert the incoming optical signal back to an electrical signal for processing by the network station.

Transmitters and receivers are often physically packaged together to form a transceiver or transducer. The minimum specifications for FDDI transceivers are:

- The output level must be at least −16 dBm—This is the minimum peak-source power supplied by the transmitter.
- Receiver sensitivity must be at least −27 dBm—This is the minimum power required by a receiver for a specified bit error rate.
- Attenuation must be no more than 11 dBm—This is the decrease in optical power between two points. It is the difference between the minimum power output of the transmitter and the sensitivity of the receiver, or the power budget.

The optical transmitter is usually one of two devices: an LED or a laser diode. LEDs were chosen as the transmitters by the FDDI standards committee because of their availability, reliability, bandwidth, distance, and cost criteria. It is important to match the transmitter's characteristics with those of the optical fiber. Characteristics include center wavelength, spectral width, average power, and modulation frequency.

Optical fibers do not transmit all wavelengths of light with the same efficiency. The attenuation of light signals is much higher for visible light (wavelengths of 400 to 700 nm) than for light in the near-infrared region (wavelengths from 700 to 1,600 nm). Within the near-infrared region, there are wavelength bands at which fibers can operate with a very low loss. Low-loss wavelength areas used for optical communications are near 850,

1,300, and 1,550 nm. FDDI has specified the 1,300-nm window as the operating wavelength to be used. The actual center wavelength frequency of the 1,300-nm window is 1,330 nm.

The total power emitted by a transmitter, however, is not confined to just the center wavelength. It is distributed over a range of wavelengths around the center wavelength. This range is quantified as the spectral width, which is measured in nanometers. Spectral widths vary from narrow to wide, depending on the type of source used. FDDI specifies a typical spectral width of 140 nm.

Electrical Interface Transmission Specifications

FDDI electrical interface transmission specifications are as follows:

- 100M bps—Some products can reach data rates of 100M bps across the optical links.
- 125M baud—This is the rate at which data is physically being clocked across the network. The data is nonreturn-to-zero-inverted encoded for speed and efficiency.
- 4B/5B encoding—The four-to-five-bits encoding scheme is used to ensure that unique and valid transmission occurs. Four data bits are expanded to five bits and named a symbol. This four-to-five expansion explains the difference between the 100M-bps transfer rate and the 125M-baud transmission rate.

Optical Connectors

An FDDI station is attached to the fiber-optic medium by a media interface connector. The media connection between adjacent stations consists of a duplex fiber-optic cable assembly attached to the respective media interface connectors at the stations. To ensure interconnectability between conforming FDDI stations, a connector mating standard is defined. It consists of the following:

- A media interface connector—This duplex fiber-optic connector receptacle provides keying appropriate to the station type and intended service. When the media interface connector plug is viewed from the front with the keying on top, the left ferrule is the station optical input port, and the right ferrule is the station optical output port (see Exhibit III-2-49).
- Connector types—The media interface connector receptacles of a station are keyed to prevent improper plug attachment. Four keys are defined for media interface connector receptacles. Exhibit III-2-50 illustrates the receptacle keys as viewed from the front of the receptacle. The media interface connector A cable with a red key contains

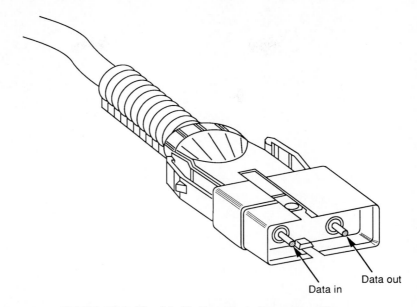

Data out

Data in

Exhibit III-2-49.   Media Interface Connector Plug

the primary-in/secondary-out optical fibers, and the media interface connector B cable with a blue key contains the secondary-in/primary-out fibers.

- A duplex-to-duplex coupling unit—This unit allows direct mating of primary and secondary cables for manually bypassing nodes (see Exhibit III-2-51). It can be used when sufficient optical power is available because no active components are included. High-loss connection occurs with this unit.

OPTICAL BYPASS

The optical bypass switch is used to bypass stations on the primary and secondary rings (see Exhibit III-2-52). The switch is automatically activated when the network station is powered down. It can also be manually activated to bypass a nonfunctional station or to perform diagnostic tests on a station.

The optical bypass switch integrates two fully reversing, electrically operated switches, creating a self-test path that permits a station's transmitter to connect to its receiver. The optical bypass switch is provided completely terminated with two fixed-shroud duplex connectors on cable pigtails.

Station Class	Attachment Type	Key Type	Front View
Dual	MIC A MIC B	A B	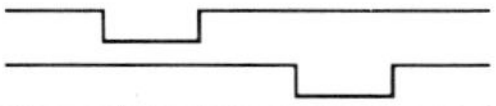
Single	MIC S	S	
Concentrator	MIC A MIC B MIC M	A B M	

Note:
MIC   Media interface connector

Exhibit III-2-50.   Media Interface Connector Types

TOKEN-RING OPERATION

The FDDI token-ring network consists of a set of stations serially connected by optical fiber to form a closed loop, or ring. A station is a device that is attached to the ring, possibly with a channel interface that communicates with other devices on that ring. Each station receives information from its nearest upstream neighbor and retransmits that information to its downstream neighbor. If a station is not transmitting its own information, it sends the received information, exactly as received, to its downstream neighbor. Therefore, a station sends information to any other station by sending it through the stations that are between them on the ring.

Information to be transmitted on the ring is placed in a frame. The frame contains the originating station's address, the destination address, and some control information. When a station recognizes the destination address as its own, it copies the information as it is received into the buffer and forwards the information on the ring. Only the originating station can remove the information, or frame, from the ring when the station recognizes its own address as the source address. If the originating station is alerted that the information was received with errors, it retransmits the information from its buffer.

The token is used to signify the right to transmit data. It is a control signal, a unique symbol sequence, that circulates on the ring after each information transmission. Any station detecting the token may capture the token by removing it from the ring and may then transmit one or more frames of information. When transmission is complete, the station issues a new token. A token-holding timer limits the length of time that a station may use the medium to transmit information before a new token must be issued to the ring.

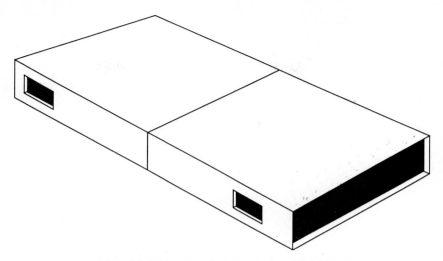

Exhibit III-2-51. Duplex-to-Duplex Coupling Unit

Multiple priority levels are available for independent and dynamic assignment, depending on the class of service. The classes of service may be synchronous (used for such applications as real-time voice), asynchronous (used for interactive applications), or immediate (used for such extraordinary applications as ring recovery). Synchronous transmission guarantees each station the right to transmit data to its maximum allotment. Unused synchronous allotments are transferred onto the ring for use as asynchronous transmission. A station may transmit asynchronous data until the unused bandwidth is used up or until the token-holding timer expires. The allocation of ring bandwidth for synchronous data occurs by mutual agreement among users of the ring.

Various mechanisms are provided to restore the ring to regular operation in the event of failure. Failure can occur as a result of transmission errors or a break on the physical ring itself (i.e., improper addition or removal of a station). Error detection and recovery is a function performed by each station attached to the ring. The network stations identify and isolate the problem (either a family station or a faulty fiber-optic link), bypass the failure with the use of both the primary and secondary cables, and restore the ring to an operational state.

Protocol Data Units

Two protocol data unit (PDU) formats are used by the media access control: tokens and frames. Tokens are the means by which the right to

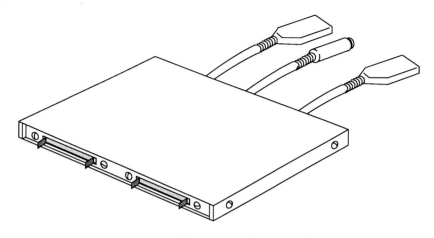

Exhibit III-2-52. Optical Bypass Switch

transmit is passed from one station to another. They are composed of four parts: the preamble, the starting delimiter, frame control, and the ending delimiter (see Exhibit III-2-53). The preamble contains 16 or more idle symbols that provide a pattern to establish and maintain clock synchronization among stations on the ring (idle = 11111). The starting delimiter consists of a specific pattern. The J symbol (J = 11000) is the first symbol of the starting delimiter sequence; the K symbol (K = 10001) is the second. No frame or token is considered valid unless it starts with this explicit sequence. The frame control indicates the token (as opposed to a frame containing information addressed to a station). The ending delimiter consists of two consecutive T-symbols to indicate the end of the token frame (T = 01101).

The frame is the vehicle used to transmit packets of information from one station to another. Like a token, it contains a preamble, a starting

PA	SD	FC	ED

Notes:
ED Ending delimiter (2 symbols)
FC Frame control (2 symbols)
PA Preamble (16 or more symbols)
SD Starting delimiter (2 symbols)

Exhibit III-2-53. Token Format

PA	SD	FC	DA	SA	INFO	FCS	ED	FS

Notes:
DA   Destination address (4 or 12 symbols)
ED   Ending Delimiter (1 symbol)
FC   Frame control (2 symbols)
FCS   Frame-check sequence (8 symbols)
FS   Frame status (3 or more symbols)
INFO   Information (0 or more symbol pairs)
PA   Preamble (16 or more symbols)
SA   Source address (4 or 12 symbols)
SD   Starting delimiter (2 symbols)

Exhibit III-2-54. Frame Format

---

delimiter, frame control, and an ending delimiter as well as source and destination addresses, an information field, and a frame status field (see Exhibit III-2-54). The preamble and starting delimiter are similar to those of tokens.

The frame control defines the type of frame and its associated control functions. The two symbols are separated into a class bit, a media access control address length bit, and format and control bits that specifically define the frame (see Exhibit III-2-55). The class (C) bit indicates whether the frame contains synchronous or asynchronous data. The frame address length (L) bit indicates the length of the media access control address for both the destination and the source address fields as being 16 or 48 bits long. The frame format (FF) and frame control (NNNN) bits are used to indicate the type of frame and the reason for transmission. They consist of the following:

- The media access control beacon frame—This frame indicates the need for corrective action and is sent as a result of serious ring failure (e.g., loss of signal or a jabbering station). It is used to identify the location of the fault.

C	L	F	F	N	N	N	N

Notes:
C   Class bit
FF   Format bits
L   Address length bit
NNNN Control bits

Exhibit III-2-55. Frame Control Bit Definitions

- The media access control claim frame—This frame is transmitted during start-up or error recovery to determine the station that creates the new token and initializes or reinitializes the ring. The station inspects the source address of all claim frames received; if the source address matches its own address, its claim was successful and it generates a new token to send out on the ring.
- The logical link control frame—This ensures that messages are delivered and guarantees that all frames are received and information arrives at the destination in the order sent. Logical link control frames are frequently used for communications between host computers and applications.
- The station management frame—The station management function monitors activity and controls such overall station activity as initialization, activation, performance monitoring, and error control. To control network operation, the station management frame communicates with other station management entities on the ring. Station management frames can include information regarding the allocation of network bandwidth, administration of addressing, and network configuration.
- The reserved-for-implementor frame—This frame permits the transmission of specialized frames across the network and is defined by the network implementor.

The destination address is a 48-bit field that identifies the station for which the frame is intended (see Exhibit III-2-56). The field's first two bits are for control purposes; the other 46 bits are the address. The first bit identifies the address as a specific station or a group of stations, and the second bit identifies the address as either universally or locally administered. An individual address identifies a unique station on the ring; a group address defines multiple destinations for a frame. A group address to all stations on a ring is known as a broadcast address. Administering 48-bit addresses can be done locally or through a universal authority. With uni-

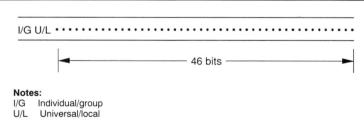

Notes:
I/G Individual/group
U/L Universal/local

Exhibit III-2-56. Destination Address

versal administration, each station address is globally distinct from all other addresses. The assignment of these addresses is performed by the Institute of Electrical and Electronics Engineers (IEEE) and the equipment manufacturer. Local administration requires the individual stations on the ring to have different addresses.

The source address is composed of two control bits that define the type of address and the 48-bit address. The source address is the media access control address used to identify the station originating the frame and is the same format and length as the destination address of that frame.

The information field contains the data packet. Interpretation of the information is made by the destination (i.e., media access control, logical link control, or station management). The length of the information field is variable and can be as long as approximately 4,500 bytes (this limitation is imposed by FDDI's physical layer).

The frame-check sequence determines the transmission integrity of the received message. The transmitter generates a polynomial based on the data contained in the frame control, destination address, source address, information, and frame-check sequence fields. The receiver also calculates the polynomial and compares it with the transmitted frame-check sequence to detect any transmission errors that have occurred.

The ending delimiter consists of a single T-symbol and indicates the end of the frame.

The frame status consists of three control indicators used for error detection, address recognition, and frame copied. The three indicators are:

- The error-detected (E) indicator—The E indicator is set by a repeating station when an error is detected. All stations on a ring inspect repeated frames for errors.
- The address-recognized (A) indicator—The A indicator is set when a station recognizes the destination address as its own individual or group address.
- The frame-copied (C) indicator—The C indicator is set when a station recognizes the destination address of a frame as its own and copies the frame into its frame buffer.

Frame Transmission

Access to the ring is controlled by passing a token around the ring. The token gives the downstream station (the next receiving station) the opportunity to transmit a frame or sequence of frames. When data needs to be transmitted, a packet is prepared and queued while waiting for the receipt of a token. Upon receiving the token, the transmitting station strips the token from the ring and begins transmitting its queued frames. A station may transmit frames until it has no more frames to transmit or until the token-holding timer has expired. After transmission of the frame is complete, the station immediately transmits a new token, allowing a downstream station to begin its transmissions.

Stations that are not transmitting must repeat the incoming symbol stream. While repeating the incoming symbol stream, each station determines whether the information is intended for that station. This is done by matching the destination address to its own address or a relevant group address. If it matches, symbols received subsequent to the frame-check sequence are processed by the media access protocol, and the frame-copied bit is set. When the originating station recognizes its address as the source address, it checks to confirm that the frame was received without errors and that it was copied by the destination station. If the frame contains errors when it is received by the destination station, the originating station must wait until it can capture the token before retransmitting the frame.

Frame Stripping

Each transmitting station is responsible for removing frames that it originated from the ring. When a station receives a frame whose source address matches its own, it replaces all further data with idle symbols. This process is called frame stripping. The process of stripping leaves remnants of frames consisting of the preamble, starting delimiter, destination address, and source address fields, followed by idle symbols. The remnants do not have any effect on the ring because certain criteria must be met to indicate a valid frame, including recognition of an ending delimiter. Remnants are removed from the ring after encountering a transmitting station.

Ring Scheduling

To guarantee that a station is able to capture the token and transmit information, two classes of services are supported: synchronous and asynchronous transmissions.

The media access control transmitter maintains a timer to control ring scheduling within each station. An acceptable target token-rotation time is negotiated during the claim token-bidding process. The target token-rotation time of the station requiring the most frequent service wins the bid. When ring initialization occurs, the target token-rotation time (which equals half the interval between each synchronous station's guaranteed access to the token) is loaded into each station's token-rotation timer. If the token arrives at a station before the token-rotation timer has expired, either synchronous or asynchronous data may be transmitted. If the token-rotation timer has expired, the station can transmit only synchronous traffic.

Synchronous Transmission. The synchronous class of service is for applications with bandwidth and response time limits that are predictable (e.g., voice or video). Allocation of synchronous bandwidth is established by station management. Initially, each station has a zero allocation (percentage of target token-rotation time). It must request an increase in its allocation from station management if synchronous traffic is to be transmitted. The sum of all stations' current allocations must not exceed the maximum usable synchronous bandwidth of the ring. Support for synchronous transmission is optional and is not required for interoperability.

Asynchronous Transmission. Asynchronous transmission, usually employed for host-to-host communications, derives its bandwidth from the pool of unused synchronous bandwidth. The asynchronous class of service is for those applications with unpredictable bandwidth requirements for transmitting large amounts of data.

Whenever a token is captured, the value of the token-rotation timer is saved in the token-holding timer. This is the amount of time allowed for the station to transmit asynchronous data on the captured token. Synchronous transmissions get top priority in FDDI, but asynchronous transmissions can be grouped into eight priority levels. Each level is associated with a time value related to its importance. This evenly allocates the asynchronous bandwidth to each level as needed. Setting priority levels for asynchronous transmission is optional.

Ring Monitoring

Each station continuously monitors the ring for invalid conditions. Ring initialization is required as a consequence of inactivity or incorrect activity on the ring. Ring inactivity is typically detected by expiration of the valid-transmission timer in the media access control receiver. Incorrect ring activity is usually detected by counting successive expirations of the token-rotation timer either with the late counter in the media access control transmitter or by station management processes.

Claim Token Process. Any station detecting a requirement for ring initialization can initiate the claim token process. This process determines which station is responsible for issuing the token and the ring timing. A claim frame contains the station's bid for the target token-rotation time. When a station receives an incoming claim frame, it compares the bid for time with its own. If the receiving station's bid for time is quicker, it sends out a claim frame containing its own time. If a station receives a claim frame containing a higher bid for ring time, it repeats that claim frame onto the ring. The claim token process is completed when all other stations have yielded and one station receives its own claim frames returning from their circuit around the ring. At this point, the winning station initializes the ring by issuing the initial token.

Each station times the claim token process by setting a timer to a value capable of permitting stable ring recovery. If the timer expires while a station is in the claim token process, the claim token process has failed to recover the ring. This is usually the result of a break or failure in the ring. At this point, external intervention is required, and the station initiates the beacon process.

Beacon Process.   The beacon process is initiated when a station detects that the claim token process has failed or on request from station management. In this case, the ring has probably been physically interrupted and may have been globally reconfigured (e.g., one logical ring may have been partitioned into two, or several logical rings may have been joined into one). Intervention external to the media access control must be invoked to restore the logical ring. The purpose of the beacon process is to signal to all remaining stations that a significant logical break has occurred and to provide diagnostic or other assistance to the restoration process (through station management). When the location of the break (or failure) has been identified, the stations on the network reconfigure the ring in the wrapback mode, using both the primary and secondary cables.

On entering the beacon process, a station continuously transmits beacon frames. A station continues to send beacon frames until it receives a beacon frame from an upstream station. If a station receives another station's beacon, it simply repeats the frame. Ultimately, the only station left sending beacon frames on the ring is the station located downstream from the fault. At this point, the ring is in a wrapback mode. When a station in the beacon process receives its own beacon frames, it assumes that the logical ring has been restored and initiates the claim token process to quickly recover the ring.

Initialization Process.   Ring initialization begins when one station successfully completes the claim token process. That station proceeds to initialize the ring by setting the target token-rotation time because it was the station that won the bidding during the claim frame process. Then it resets its token-rotation timer and issues an initial token.

The purpose of the initial rotation of the token is to align both the target token-rotation time values and the token-rotation timer in all stations on the ring. No station captures the initial token or transmits frames. In each station, on receipt (and repetition) of the initial token, the token-rotation timer is reset. This allows stations to transmit synchronous data on the second token rotation while inhibiting asynchronous transmissions. Beginning with the second token rotation, each station accumulates current synchronous bandwidth use (as opposed to allocated limits) in the token-rotation timer. Transmission of asynchronous data is possible on the third and subsequent token rotations, subject to the availability of unused synchronous bandwidth.

Structure

The media access control consists of two cooperating asynchronous processes within each station—the receiver and the transmitter. The need for separate receiver and transmitter processes arises from the requirement that certain functions (recognizing media access control frames, detecting a station's own address, and capturing the frame status) must be performed concurrently with and asynchronously to the states of the transmitter. The error-detection functions are specified in the receiver, and the recovery functions are specified in the transmitter. This allows a station to perform self-monitoring for loop-back testing and to avoid unnecessarily compromising ring integrity.

The physical layer protocol provides the physical connection between the media access control and the ring. The physical layer protocol receives data from the media access control and converts that data into the symbol code to be transmitted over the network. Information that is received from the network is converted into a format that is readable by the media access control.

Part IV
Communications Systems Management

AS USED HERE, SYSTEMS MANAGEMENT refers to managing the conversion of a plan for a communications system into a functioning, effective communications system. This plan incorporates users, common carriers, product and service vendors, the work of standards bodies, and not insignificantly, the requirements of regulators and the manager's own organization.

Users are, of course, the raison d'être for the entire network. They are also, directly or indirectly, the source of the money that pays for the network. As a customer, each user deserves consideration. Users frequently want, however, a communications system that offers infinite capacity and zero response time. Meeting this requirement can make life difficult for communications systems managers.

Common carriers are an important element in any but the most local of networks. Simply tracking and understanding the service offerings of a common-carrier network facility provider is a daunting task for the manager.

Vendors of products and services are eager to assist the systems manager with the procurement effort. Many vendors, however, make their promises in very small print or hedge their offerings by using interesting descriptions, such as *OSI conformant* or *ISDN ready*. The manager must read the small print and be informed enough to ask the questions that will lead to accurate performance information.

Standards bodies are specifically directed at making life easier and products cheaper for the communications systems manager. They are also, however, responsive to their members, a large percentage of which are vendors, not users. This can lead to a surprising variety of so-called standard communications solutions.

Regulators also can introduce complexities to communications management. They do this by issuing rules defining what kind of services can be made available where, by whom, and under what conditions.

COMMUNICATIONS SYSTEMS MANAGEMENT

A manager's own organization can also increase the difficulty of planning and implementing a communications system. In a very hierarchical, top-down organization, it is not unusual for one segment of the organization to be planning an expansion of a communications system while another segment is planning a reduction of personnel or budgets.

The communications systems manager is and will continue to be a critical element in the communications-based environment of the 1990s. To succeed, the manager needs the organizational structure and the processes that can respond to the rapidly changing and complex environment. The manager must therefore work within the organization to formulate strategies that will support effective network deployment. The manager must also work outside the organization to bring together the diverse elements so as to produce a cost-effective communications network for the organization.

The first section in this part of the handbook supports the selection and procurement process. Section IV-2 addresses guidelines for implementation, and the final section discusses several topics related to the ongoing operation of the system.

Section IV-1
Selection and Procurement Considerations

THE SELECTION AND PROCUREMENT of the elements of a communications system spans a wide variety of activities. Some, such as choosing a modem, can seem mundane and simple. Others, such as choosing a system integrator, may seem complex and exciting. Both, however, can be equally important and critical to the success of the system. Successful networks are the result of a proper combination of the simple and the complex, of the mundane and the exciting. The adage "For want of a nail, the shoe was lost" is certainly the quintessence of the procurement and selection process.

Although digital services are beginning to represent a significant percentage of transmission facilities, analog transmission services still predominate. Therefore, there continues to be a need for modems that condition digital business machine signals for transmission over analog facilities. There are many different manufacturers and models of modems on the market. Chapter IV-1-1, "Modem Selection Factors," offers 15 criteria for the successful procurement of modems. These range from user application requirements to the technical aspects of modem selection to the recurring and nonrecurring cost elements.

Advances in modem technology now routinely permit higher transmission speeds on analog lines. In addition, digital lines, which operate at higher transmission speeds than analog facilities, are becoming more available. As a result, the available transmission capacity continues to increase. The need to use this capacity more efficiently is a new procurement requirement, which can be fulfilled by the statistical multiplexer. Chapter IV-1-2, "Selecting and Evaluating Statistical Multiplexers," examines the capabilities, limitations, and operational features of these devices. The chapter includes a list of features and related parameters that need to be considered when an organization is evaluating vendors.

Cost control is an important element of the communications manager's task. It is also one that is easily demonstrated and measured because costs are a highlight of every management presentation or review. The checklists provided in Chapter IV-1-3, "Reducing the Costs of Equipment and Outside Services," guide communications systems managers through a detailed review of potential cost-saving items and functions. Broad categories include use of equipment and acquisition of services, supplies, and physical facilities.

The last chapter of this section addresses another major procurement issue: how the organization should locate and select the expertise necessary to design and install a complex network. Chapter IV-1-4, "How to Choose a Systems Integrator," discusses the various types of systems integrators (e.g., consultants, service providers, and hardware vendors). It then presents a graphic model for evaluating and choosing the systems integration services that are available.

Chapter IV-1-1
Modem Selection Factors

John E. Gudgel

MOST DATA COMMUNICATIONS ANALYSTS predict the eventual development of a worldwide digital communications network. Yet, despite the boasting by long-distance carriers of their digital fiber-optic networks and the publicity surrounding the integrated services digital network (ISDN), the truth is that, for the immediate future, most data traffic will continue to be transmitted over analog communications facilities for the following reasons:

- ISDN is still in its infancy—Telephone companies have only recently begun to run ISDN trials, and it will be several years before ISDN becomes widely available.
- Digital facilities generally remain expensive—Digital circuits can cost 50% more than comparable analog lines.
- Digital services are not available in all areas—Carriers must have special equipment in their central offices to provide digital services. This equipment is not yet installed in all locations.
- Digital communications still requires the use of a modemlike device— A data service unit (DSU) is needed to shape and amplify a data signal so that it can be transmitted reliably over a digital facility. Such devices are not as widely available as modems, nor do they typically offer as many functions.

For the next few years, communications transmissions will continue to be analog, requiring the continued use of modems. This chapter discusses factors that must be addressed in evaluating modems and provides some suggestions to help the communications manager make sensible and cost-effective decisions when purchasing a modem.

THE USER APPLICATION

The first and perhaps most important factor for the communications manager to consider before choosing a modem is the users' applications and their communications requirements. Identifying the specific users provides useful information about the applications, the people in charge of the project, and their level of knowledge. Different types of equipment may be appropriate for data center personnel experienced in dealing with data communications equipment and for office workers who know virtually nothing about modems and care only about logging on to a remote computer.

Detailed knowledge about specific applications is also critical. Some pertinent facts include:

- The application type—The communications manager should categorize the application, assess the magnitude of the job, and determine how much equipment, money, and effort are required to satisfy the users' needs. Important points to consider are whether the application requires point-to-point or multipoint communications and whether an interactive or store-and-forward system is needed.
- Traffic volume and intensity—The communications manager must determine how much data will be transmitted and how often transmission will occur. It is important to distinguish between a modem that is used occasionally to access a remote data base and one that is used around the clock, seven days a week, to transfer critical files or support network operations.
- User response time expectations—Acceptable response time can be a subjective measure of user satisfaction. For example, a response time delay of 30 seconds may be adequate for such a store-and-forward application as electronic mail but totally unsatisfactory to an interactive remote job entry (RJE) user. Managers should examine the requirements of the application and educate the users regarding the costs of and limitations to any proposed modem solution.

THE COMMUNICATIONS FACILITIES

Communications facilities can be divided into three categories: local (e.g., in-house), public switched (e.g., dial-up), and public nonswitched (e.g., leased). Correspondingly, three categories of modems are used to transmit data over these facilities: limited-distance modems, dial-up modems, and leased-line modems. The type of modem required depends on the type of communications facility.

Limited-Distance Modems. These devices, which are generally used to transmit data through in-house facilities, are typically less expensive than dial-up or leased-line modems because they do not usually incorporate the line-equalization techniques needed by the more sophisticated devices to successfully operate over long distances on the public network. Instead, limited-distance modems take advantage of the electrical characteristics of copper wire to transmit both synchronous and asynchronous data at speeds of 19K bps or higher to distances of 20 miles.

Dial-Up Modems. These are used to transmit data over the public-switched telephone network. They are characteristically equipped with special features that allow them to operate efficiently and cost-effectively in this environment. For example, most dial-up modems are designed to answer and set up a connection automatically for incoming calls, dropping the line when the call is completed. Most dial-up modems have automatic redial, stored number recall, and other features commonly associated with modern voice phone service.

Because of the limitations and variability of public-switched lines, dial-up data transmission generally takes place at speeds of 9.6K bps or lower. Modern technology continues to advance public-switched line capacity, however, and dial-up rates greater than 9.6K bps may be available in the near future.

Leased-Line Modems. These are used on nonswitched facilities— dedicated private circuits that are leased from the telephone company. Unlike dial-up facilities, whose costs increase with use, the cost of a leased line is fixed. Dedicated lines are commonly used by companies with large volumes of traffic between two or more locations.

Leased-line modems differ from dial-up modems in several ways. Because leased-line prices are fixed, a call can be maintained continuously at no additional cost. Leased-line modems do not require such dial-up modem features as automatic answer and disconnect. In addition, because the transmission path is dedicated to a single customer, leased lines are uniformly cleaner than dial-up facilities and characteristically are capable of supporting faster data rates. Leased-line modems are available that support speeds of 19K bps, approximately twice the current rate for the fastest dial-up modem.

THE TRANSMISSION MEDIUM

Most electronic data communications continues to be accomplished using copper wire; two- or four-wire copper is still used by all telephone companies for intracity and intercity circuits as well as for many local area networks. Several other transmission media are being increasingly used for data transmission, however, including optical fiber, coaxial cable, and microwave, satellite, and FM radio broadcasts.

The transmission medium used by a telephone company to provide an intercity connection is basically transparent to the user. Although microwave transmission quality can be affected by weather conditions and satellite transmissions have an inherent delay, these factors do not usually influence a user's modem selection. The transmission medium is more important to local data settings in which users provide their own communications facilities—many local area networks require modems especially designed to operate over optical fiber, coaxial cable, or radio. The communications manager should identify the transmission technologies being used before making a local data network modem purchase.

DATA TRANSMISSION DISTANCE

Most dial-up and dedicated modems are designed to work over all public facilities, regardless of the distances involved. Limited-distance modems, however, are more restricted. Those transmitting at higher speeds over longer distances generally cost more than units working at lower data rates over shorter distances. For example, a limited-distance modem that transmits data at 56K bps for a distance of more than 10 mi can cost three times as much as a unit operating at 9.6K bps over distances of less than 5 mi. The speed and range of these units depend somewhat on the gauge of wire used, but distance has a decisive impact on the cost of the device.

DATA RATE AND THROUGHPUT REQUIREMENTS

The data rate, or speed, of the modem used is one of the major factors affecting data communications response time. The technical specifications and sales brochures for most modems list a maximum data rate at which the unit is designed to operate; this is the fastest speed at which the unit can transmit data over a circuit when conditions are optimal. Low-speed modems operate from 300 to 1,200 bps; medium-speed modems are 2,400-, 4,800-, and 9,600-bps units; and high-speed modems can operate as fast as 14.4K bps and 19.2K bps.

Not surprisingly, conditions are rarely optimal, and the actual data rate, or throughput, of a modem is almost always less than the optimum rate advertised. The optimum data rate can still be an accurate indication of whether a modem can meet a user's response time requirements. For example, the solid line in Exhibit IV-1-1 shows that increasing the modem data rate decreases the time needed to transmit a 1M-byte file. The theoretical optimum data rate can be used to estimate the actual transmission time, depending on the efficiency of the modem arrangement. The dotted line in Exhibit IV-1-1 shows the actual file transmission time at various data rates, assuming the unit operates at 80% efficiency.

The throughput efficiency can be improved in several ways. The quality of the circuit can be improved through the use of line conditioning. Im-

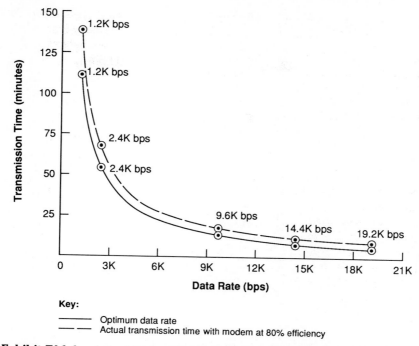

Exhibit IV-1-1. Actual Transmission Time of 1M-byte File Compared with Optimum Data Rate

proved quality means fewer errors, resulting in fewer transmissions and higher average throughput. Conditioning, which must be specified when the circuit is ordered, is a way of engineering a circuit so that it experiences less error-causing interference. Exhibit IV-1-2 lists the common types of conditioning and their principal uses. Another way of improving data rate efficiency and overall throughput is through the use of error-correcting codes. These codes can reduce or eliminate the need to retransmit a message when an error occurs, but they also add overhead to meaningful data, which can actually degrade overall throughput, especially on a relatively error-free line.

SYNCHRONIZATION

Synchronization, which involves the predictability and continuity of data transmission, is generally a function of the types of terminals and processing equipment used and of the nature of the applications. The two most common types of transmission are asynchronous and synchronous. Asyn-

Conditioning Type	Principal Uses
B1	• 3002-type voicegrade circuit of less than five miles. • Data rate of 2,400 bps to 4,800 bps. Bit error rate not to exceed 1:100,000. • Limits attenuation and envelope-delay distortions. • Used with early types of Bell 201, 202, and 208 modems.
B2	• 3002-type voicegrade circuit of less than five miles. • Data rate to 9,600 bps. Bit error rate not to exceed 1:100,000. • Limits attenuation and envelope-delay distortions.
C1	• Used on point-to-point, multipoint, and switched facilities of unlimited distance. • Data rate of 1,200 bps. Bit error rate not to exceed 1:100,000. Permits 2,400-bps operation for modems with fixed equalizers. • Further limits attenuation and envelope-delay distortions. • Used with early types of Bell 202-series modems.
C2	• Used on point-to-point, multipoint, and switched facilities of unlimited distance. • Permits 4,800-bps operation for modems with manual or adjustable equalizers. Bit error rate not to exceed 1:10,000. • Further limits attenuation and envelope-delay distortions. • Used with some types of Bell 201, 202, and 203 modems.
C4	• Used on two-, three-, and four-point private facilities of unlimited distance. • Permits 2,400- to 9,600-bps operation for modems with adaptive equalizers. • Further limits attenuation and envelope-delay distortions.
D1	• Used on point-to-point facilities of unlimited distance. • Limits ambient noise and harmonic distortion. • Originally created to allow 9,600-bps operation with Bell 209-series modems. • Currently specified by many modem vendors for high-speed modem transmission to 19.2K bps.

Exhibit IV-1-2.   Common Line-Conditioning Types

chronous transmission occurs in random spurts and is characterized by start bits, which indicate the transmission of a character, and stop bits, which verify that the transmission has been completed. Synchronous transmission, on the other hand, is continuous and predictable. Sending and receiving ends are in phase and know when data is sent and when it is received.

Asynchronous transmission is typically used in terminal-to-host communications when speed is not a critical factor. Asynchronous modems are slower than synchronous modems—the upper limit of their operation is about 2,400 bps—but they are also less expensive. Synchronous modems can operate at much higher speeds—to 560K bps—and are generally used in networking and host-to-host communications. Synchronous modems require costly clocking circuitry, however, to keep each end in phase and are therefore more expensive.

DIRECTIONAL MODE

Some modems can only transmit or only receive data; others can do both simultaneously. Which directional mode is required depends on what host or terminal equipment is at either end of the line.

A simplex modem transmits data in one direction only, which may be adequate for a field office that needs to send information to but not receive it from its home office. Most applications, however, need two-way or duplex communications. One option is a half-duplex modem that allows data to be transmitted in two directions, but only one way at a time. The alternative is a full-duplex modem, which can transmit data in both directions simultaneously. The advantage of full-duplex modems over half-duplex modems is that response time is usually faster because there is no line-turnaround delay. Half-duplex modems, however, are usually less expensive than full-duplex ones and can use the entire bandwidth for a one-way transmission. Full-duplex modems must share the bandwidth for simultaneous two-way transmissions. Because speed is in part a function of bandwidth, half-duplex systems may be faster for some applications.

NETWORKING

Modems are commonly used in data communications networks, which can consist of such point-to-point circuits as an X.25 packet-switching configuration or such multipoint lines as those found in a typical Systems Network Architecture (SNA) or bisynchronous arrangement.

An X.25 packet-switching network, as illustrated in Exhibit IV-1-3, can consist of hundreds of point-to-point circuits connecting locations around the world. In a typical configuration, local low- to medium-speed tail circuits connect terminals or hosts to strategically located concentrators (e.g., packet assemblers/disassemblers or packet switches). These concentrators are then interconnected by high-speed backbone trunks. In such an environment, modem reliability is a critical issue because modem failure can result in hundreds of users losing network access. As a precaution, many of these networks are equipped with alternate routing paths, network management systems to monitor modem status, hot modem spares, and dial backup units.

With a multipoint network, as shown in Exhibit IV-1-4, modem reliability is also a concern. Once again, the failure of a key modem can affect hundreds of users. Because most of these networks are built around a

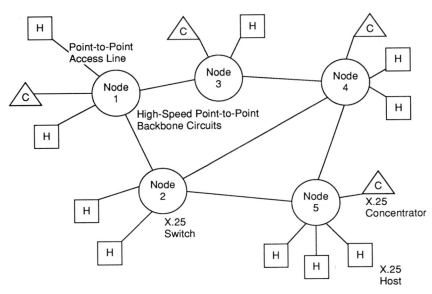

Exhibit IV-1-3. Typical X.25 Packet-Switching Network

centralized processor, however, the impact of a modem failure really depends on where in the circuit the problem occurs. Multipoint configurations
(e.g., IBM SNA networks) differ from X.25 networks in their dependence on
modem polling, which always introduces an element of delay that is not
present in point-to-point networks. In a properly configured multipoint network, this delay is relatively transparent to the user; but if there are too
many modems attached to a multipoint circuit or if the polling sequence is
poorly conceived, users may experience unacceptable response time. Two
important factors that must be considered in deploying modems in a multipoint network are the unit's distribution and its polling capabilities.

PHYSICAL INTERFACES

A modem interface is an electrical connection that allows a terminal or
host computer to be attached to the modem. The type of modem interface
required depends on several factors, including the distance between the
terminal or host equipment and the modem, the user's data rate requirements, and the particular interfaces supported on the user's equipment.
This last factor is especially important because the interface presented by
the modem must be compatible with that available on the user's equipment.

Manufacturers have avoided compatibility problems despite the number of different makes and models of both computers and modems by adopting industry or international interface standards. The most familiar of these standards in North America is the Electronic Industries Association (EIA) RS-232 connector. This connector has 25 pins, each of which is assigned a particular circuit-controlling function. The EIA specification for RS-232 allows for both synchronous and asynchronous data transmission at speeds of 19.2K bps over distances of 50 feet or less. In reality, an RS-232 interface can be used to transmit data over several hundred feet if specially shielded cable is used.

The International Telegraph and Telephone Consultative Committee (CCITT) has developed interface standards known as V.24 and V.28, which are functionally compatible with EIA RS-232. Adherence to these standards allows data communications equipment to be plug-compatible worldwide. Exhibit IV-1-5 lists some of the other interface standards being used in the US and internationally.

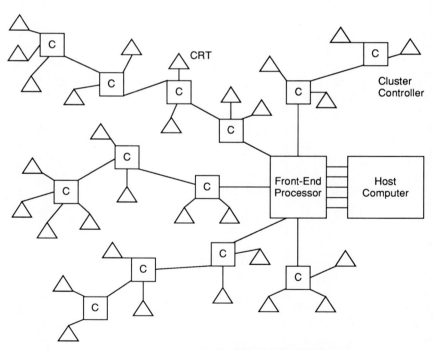

Exhibit IV-1-4.   Typical SNA Multipoint Network

Standard	Issuing Organization	Description
RS-232	EIA (North America)	Interface specification for connecting data terminal equipment (DTE) to data communications equipment (DCE) at data rates to 19.2K bps over distances to 50 feet. Characterized by a 25-pin connector.
RS-449, -422, and -423	EIA	Interface specification for connecting DTE to DCE at data rates to 2M bps over distances to 4,000 feet. Characterized by a 37-pin connector. Intended to eventually replace RS-232 but have not become generally accepted.
RS-530	EIA	New EIA interface standard intended to replace RS-449. Designed to allow data rates as high as 2M bps at distances to 200 feet. Characterized by a D-shaped 25-pin connector.
V.24 and V.28	CCITT (International)	Functionally similar to RS-232.
V.35	CCITT	International interface for high-speed transmission to 4.8K bps. Characterized by a rectangular 34-pin connector.
X.21	CCITT	Interface specification for connecting DTE to DCE for synchronous operation on public data networks. Characterized by a 15-pin connector.
MIL-STD-188	Department of Defense (US)	Provides technical interface parameters for military communications systems.

Exhibit IV-1-5. Physical Interface Standards

MODEM COMPATIBILITY AND STANDARDS

Not only must terminal or host equipment be able to interface with a modem, but the modems at either end of a circuit must be compatible with one another. Industry and international standards have been created to enable manufacturers to design and market compatible modem products.

The most influential industry-based producer of modem standards is AT&T—known before divestiture as the Bell System. Over the years, the Bell

System developed many modem products that later became the basis of global standards, including the Bell 103, 201, 202, 208, and 212 series. Exhibit IV-1-6 presents a list of some of the Bell System standard modems and their specifications.

Another industry player with a large impact on the development of modem standards is Hayes Microcomputer Corp. Hayes's major success was the development of the AT command set, a dialing protocol that governs the way a computer sends commands to the modem. Many modem manufacturers have adopted variations of the AT command set, and with the increase in microcomputers and dial-up network modem access, it has become a de facto standard within the user community.

For users with global communications requirements, it is important that the equipment chosen conform to international standards. In some instances, an international standard is fully compatible with a US industry standard; for example, the CCITT V.26 standard for 2,400-bps dial-up operation is functionally compatible with most Bell 201B modems. On the other hand, the CCITT has also defined several modem standards that do not

Modem Type	Maximum Data Rate (bps)	Operating Mode	Facilities Type	Synchronization Type
103 Series	300	Half duplex or full duplex	Dial-up or private line	Asynchronous
108	300	Half duplex or full duplex	Dial-up or private line	Asynchronous
113A and 113B	300	Half duplex or full duplex	Dial-up	Asynchronous
202C, 202D, and 202R	1,200 (1,800 with C2 conditioning)	Half duplex or full duplex	Dial-up or private line	Asynchronous
202S	1,200	Half duplex or simplex	Dial-up	Asynchronous
202T	1,200 (1,800 with C2 conditioning)	Half duplex or full duplex	Private line	Asynchronous or synchronous
212 and 212A	1,200	Full duplex	Dial-up	Asynchronous or synchronous
201B	2,400 with C2 conditioning	Half duplex or full duplex	Private line	Synchronous
201C	2,400	Half duplex or full duplex	Private line	Synchronous
208A	4,800	Half duplex or full duplex	Private line	Synchronous
208B	4,800	Half duplex	Dial-up	Synchronous
209	9,600 with D1 conditioning	Half duplex or full duplex	Private line	Synchronous

Exhibit IV-1-6. Bell System Modem Specifications

CCITT Standard	Description	Approximate Bell Equivalent
V.21	Asynchronous, low-speed (0 to 300 bps), full-duplex transmission over switched facilities	Bell 103
V.22	Asynchronous, low-speed (1,200 bps), full-duplex transmission over either switched or leased facilities	Bell 212
V.23	Asynchronous, low-speed (600 to 1,200 bps), half-duplex transmission over switched facilities	Bell 202
V.26	Synchronous, medium-speed (2,400 bps), half- or full-duplex transmission over leased facilities	Bell 201B
V.26 bis	Synchronous, low- to medium-speed (1,200 to 2,400 bps), half-duplex transmission over switched facilities	Bell 201C
V.27	Synchronous, medium-speed (4,800 bps), full-duplex transmission over leased facilities	Bell 208A
V.29	Synchronous, medium-speed (9,600 bps), full-duplex transmission over leased facilities	Bell 209

Exhibit IV-1-7. CCITT Modem Standards

have any US industry–accepted equivalent. These standards were proposed and designed to offer a great deal of flexibility to accommodate future technological innovation. They are so flexible, however, that manufacturers have released CCITT-compatible modems that cannot communicate with other manufacturers' CCITT-compatible devices. Therefore, it is important to examine the interconnectivity of modems marketed by different vendors. Exhibit IV-1-7 describes some of the more common CCITT modem standards.

MODEM FEATURES

Modems come equipped with features that manufacturers add to their product to make it more attractive. Some of these features are rather extravagant; others are practically essential. A few of these features are discussed in this section.

Dial-Up Modem Features. Dial-up modem manufacturers have developed a variety of special features that support the unique needs of dial-up access users. Many users find it convenient to have a modem that stores frequently called phone numbers and automatically dials the number at the touch of a button. Users who access numerous data bases may need to adjust the speed of their modems (from 300 to 1,200 to 2,400 bps) to conform with the operator's dial-up speed restrictions. Some dial-up modems that are used only to receive calls can automatically answer a call, make the connection to the host, and automatically disconnect from the computer when the call terminates.

Multiplexing.   Some private-line modems support multiple ports that allow several devices to share the bandwidth of a single circuit. This built-in multiplexing capability can be an expensive feature, but the cost can be offset by eliminating the need for multiple private circuits.

Security Features.   Because modems are often used to transmit sensitive information, security features are important. Some modems encrypt data that is transmitted over a circuit so that it cannot be read by anyone not authorized to do so. Because modems are often the gateways to a computer system, communications managers use them as a first line of defense against hackers. Some modems require users to enter passwords and IDs; others can be programmed to automatically call back a user at a pre-programmed phone number once access is requested and the user is identified. Finally, some modems offer features to enhance the physical security of the modem itself; these include tamperproof front panels and soft-strappable modems that make it more difficult for an intruder to change the unit's settings.

Dial Backup.   Because leased circuits can become noisy or fail completely, making it impossible to transmit data correctly, it is recommended that a dial-backup mechanism be installed to use the public-switched telephone network for temporarily reestablishing data communications. A dial-backup component is often built into the modem and is automatically activated when a line failure occurs.

Hot Spare.   Modems occasionally fail without warning. If a modem connection is particularly vital to company operations, a hot spare that can automatically be placed in operation when required should be purchased.

Equalization and Error Correction.   Users and providers of data communications are always concerned with the integrity of transmitted data. Line hits and phase shifts can cause errors resulting in delay-causing re-transmissions. To compensate for these changes in line quality, many modems are equipped with equalizers that automatically adjust for frequency response and envelope delays to preserve the transmission rate. Other modems use code that can recognize and correct errors, eliminating the need for retransmissions.

Asynchronous-to-Synchronous Conversion.   Some communications environments use both synchronous and asynchronous devices to communicate. An asynchronous-to-synchronous converter allows the modem to support either type of device, eliminating the need for separate asynchronous and synchronous modems.

DIAGNOSTIC CAPABILITIES

In the current communications environment, multiple equipment and circuit providers may be involved in a single data communications connection. Many managers must therefore be able to test and control their own modems and lines. Diagnostic modem capabilities can help to quickly isolate problems, even in a multivendor environment. Diagnostic modems come with a variety of features, some of which are discussed here.

Local and Remote Self-Test.   This test allows a user to verify that both the local and remote modems are working and able to transmit and receive data or that a communications problem is being caused by a modem.

Local and Remote Digital Loopback.   This procedure checks the integrity of the cable or circuit that connects the modem to the data terminal or host. A failed test indicates that the equipment at one end is malfunctioning or that a cable has failed or come loose.

Local and Remote Analog Loopback.   Another potential point of failure is the analog telephone line or the inside wiring that connects the modems to one another. A failed test indicates that there is a break somewhere in the circuit.

End-to-End Bit Error–Rate Test.   Data transmission problems can be caused by degradation of the circuit quality. These problems typically cannot be detected by any of the previously mentioned tests because neither the equipment nor the circuit has actually failed. A bit error–rate test can reveal that errors occurring during transmission are most likely being caused by interference or some other line-quality problem.

A communications manager should examine not only what types of diagnostic tests are available but how those tests are run. Some manufacturers run tests over the same channel used for data transmission, disrupting data communications during the test. Other vendors have designed their units to employ an unused portion of the channel to run the test— these diagnostics do not disrupt communications. The choice of diagnostic capabilities ultimately depends on the users' needs, the importance of the circuit, and budget restrictions.

MODEM MANAGEMENT

Some managers require total control over their modem assets. Several vendors offer sophisticated modem-management systems that make it possible to monitor and control all the modems in a network from a centralized network operations center. With such a system, the network operations center can proactively monitor network components, run diagnostic tests, detect and isolate problems, report and track trouble areas, collect network statistics, update inventories and network data bases, and generally perform all of the tasks needed to ensure that the network operates efficiently. These systems, however, can be expensive.

SUPPORT ISSUES

The communications manager should consider two types of support issues when selecting a modem. The first is site support, provided by users, which is necessary to operate the modem on the premises. The second is vendor support, which should be available during and after installation to maintain and service the unit and ensure that it performs according to specifications.

Site Support. Issues involving site support include power availability, environmental control, space allocation, and security. Electrical power is usually not an issue with a modem installation because most modems plug into standard three-pronged outlets. Because a modem occasionally requires a nonstandard power outlet, however, the communications manager should confirm that the correct power is available before scheduling a modem installation.

Most modems are designed to work in a fairly broad range of temperature and humidity, and most modems are installed in climate-controlled computer labs or business offices. If the modem is to be installed in a location where temperature or humidity extremes are anticipated, however, the unit selected should be designed to operate under these conditions.

Space is another important site-support issue. This is usually not a consideration when only a few modems are involved, because single standalone modems can be placed out of the way, on or underneath a desk. Space does become a problem, however, when several modems are installed at a site. The communications manager may want to consider rack-mounted modems that can be neatly installed into a cabinet, an arrangement that is not only neater and more efficient but can also save money. Rack-mounted modems are often less expensive than their standalone counterparts because they can share power supplies and do not require special housings.

Organizations should be concerned about who has access to the modems at a particular site if the modems are used to transmit confidential or sensitive data. Access security is another site-support issue that must be addressed.

Vendor Support. Vendor-support responsibilities can include equipment installation, user training, warranty guarantees, and postwarranty maintenance and repair.

Some modems are easy to operate and can be installed without vendor assistance; others are more complicated and may require installation by a company-trained technician. Before buying a modem, a communications manager should first determine whether installation assistance is available and, if so, what services are supplied at what cost. For example, for a dial-up modem installation, a technician may be needed to work with the user to make sure the modem strap settings are set to operate with the application. For a leased-line modem, technicians may be required at both ends of the circuit and may need to work with the telephone company to resolve problems.

User training can also be a significant vendor-support issue. Some modems have numerous strap settings, options, and diagnostic tests that can be manipulated only by a properly trained technician. Before purchasing a modem, the manager should determine the difficulty of use and, if training is required, the extent of training needed and whether its cost is included in the purchase price.

Most manufacturers warrant their equipment for a specified period of time, typically from 90 days to one year from date of purchase. During this warranty period, a vendor usually repairs or replaces defective equipment free of charge, particularly in clear-cut cases in which equipment obviously fails. Disputes can arise, however, particularly when the equipment does not actually fail but does not fully meet the users' performance expectations. A formal request for proposal or specification that clearly defines the performance expectations can help alleviate these problems. Vendors should be required to respond to the proposal before receiving a contract, and the performance specifications should then be clearly reiterated in the purchase agreement.

Once the warranty has expired, most users need an established method for servicing their modems when problems arise. A maintenance agreement is the most common way to achieve this. With a maintenance agreement, the vendor repairs any modem that fails for a set monthly fee. The advantages of a maintenance agreement are that the costs are fixed and known—there are no variable costs for parts and labor—and that most maintenance contracts include guarantees that service technicians will be on site within a specified time after the trouble is reported (typically four hours or less).

An alternative to a maintenance contract is to contract for service on a time-and-materials basis. In this arrangement, the user pays the parts and labor charges for any repairs. The advantage of this arrangement is that if the unit does not break, the user does not pay anything. The disadvantage is that it can be difficult to control repair costs because modem failure is unpredictable. A few repair calls on a time-and-materials basis can cost more than an annual maintenance agreement. In addition, because vendors often give priority to maintenance agreement customers, time-and-materials clients may wait longer for service technicians to arrive at their site.

Some users who use their modems only occasionally may not require on-site maintenance; they can mail broken modems to a repair center for servicing. This is usually the least expensive way to repair a modem, but it is also the slowest. It can take months before a unit is fixed and returned to service.

RECURRING AND NONRECURRING OPERATING COSTS

Once all the previous factors have been evaluated and several modems have been identified that can satisfy the users' needs, the communications manager should objectively compare costs. Relevant costs include not only the modem purchase or lease price but also directly related expenses (e.g., installation and maintenance) and auxiliary costs (e.g., telephone lines), which can be dramatically reduced by choosing the right modem. For example, the right choice might be a modem that can operate at sufficient speed without a conditioned line, or a modem with a built-in multiplexer, eliminating the need for multiple circuits.

The total operating costs for the candidate modems should be compared through the use of the present-value economic analysis available in many spreadsheet packages, taking into account such factors as recurring and nonrecurring equipment and line expenses, equipment depreciation, tax savings, and the cost of funds. The total present-value operating costs of each modem can then be weighed along with all other considerations to help the communications manager make a practical and cost-effective modem selection.

ACTION PLAN

When evaluating modems and other data communications equipment for purchase, the communications manager should:
- Talk to the users to understand their applications—The best modem selection is one that cost-effectively satisfies users' needs.
- Do preliminary research by reading product brochures and by talking with vendors and their customers.
- Write a detailed specification outlining technical requirements and user expectations.
- Evaluate vendor proposals, paying close attention to technical capabilities and vendor support.
- Run a spreadsheet present-value analysis to compare the total operating costs of various units.

Following these suggestions results in the purchase of the most applicable and cost-efficient modem for the organization.

Chapter IV-1-2
Selecting and Evaluating Statistical Multiplexers

Gilbert Held

THE PRIMARY PURPOSE OF A TRADITIONAL time-division multiplexer (TDM) is to scan for the presence of data on each line input to the device. If a character is presented to the device, the multiplexer's logic places that character into a fixed position by time onto the high-speed line connecting that multiplexer with another multiplexer at the distant end of the transmission link. Each complete scan of the input lines connected to the multiplexer results in the transmission of a frame of information onto the high-speed line: each character input to the multiplexer is placed into a predefined position on the output frame.

If no activity occurs on an input line during the multiplexer's scan, it places a null character into the frame to enable the multiplexer at the remote location to demultiplex the data on the high-speed line. The remote multiplexer demultiplexes the data by examining each message frame and extracting the characters in the frame and routing them to a specific output line on the basis of their position in the frame. Null characters encountered in the frame are used for positioning to ensure that the data is demultiplexed correctly.

Exhibit IV-1-8 illustrates multiplexing and demultiplexing by a pair of TDMs used to concentrate data from two terminals onto one line. In the upper part of the exhibit (which illustrates the multiplexing process), the letter b denotes the absence of line activity from a terminal; the character ƀ denotes the null character inserted into a message frame. The three character times used illustrate the development of three composite multiplexer frames that are transmitted onto the high-speed line.

The lower part of the exhibit illustrates the demultiplexing process. The insertion of null characters into the composite message frame enables the

Multiplexing

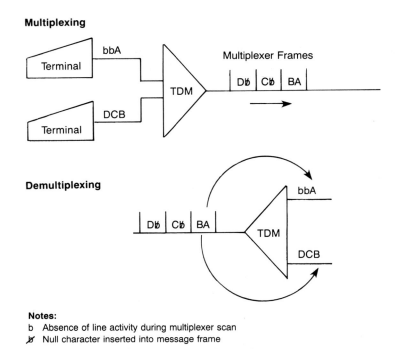

Demultiplexing

Notes:
b Absence of line activity during multiplexer scan
ƀ Null character inserted into message frame

Exhibit IV-1-8.   Multiplexing and Demultiplexing by TDMs

TDM to demultiplex the data on the high-speed line correctly, ensuring that the data entered in the first multiplexer is output onto the correct line of the second multiplexer.

Although the use of null characters ensures correct demultiplexing, this technique is basically inefficient because of the typical characteristics of terminal data traffic activity. First, terminals rarely operate continuously. Second, when a terminal is online, the terminal operator typically pauses during a session to consult a manual or examine data received in response to a query. Although no data flows to or from the terminal during these times, the TDM continues to insert null characters into the message frames so that the demultiplexing of the frames by character position is conducted correctly. Therefore, most of the characters in a multiplexer's frame are null, which results in inefficient use of the high-speed line.

STDM OPERATIONS

To use high-speed lines more effectively, statistical time-division multiplexers (STDMs) first buffer several characters before they are transmitted.

A line address and character count are then added to each character group to ensure that the statistical multiplexer receiving the multiplexed data can demultiplex it correctly. Exhibit IV-1-9 shows the statistical multiplexing process. Because each character group includes a line address and character count, the multiplexer need not insert null characters into the frame to enable demultiplexing by position.

PERFORMANCE COMPARISONS

Statistical multiplexers can support more data sources than traditional TDMs can because they transmit data only when data sources are active. Exhibit IV-1-10 compares the servicing of terminals by a TDM and an STDM. In the top part of the exhibit, eight 1,200-bps terminals serviced by a traditional TDM require a composite line speed of 8 × 1,200, or 9,600 bps, to operate.

In addition to taking advantage of the inactivity of data sources, a statistical multiplexer strips the start, stop, and parity bits from asynchronous data when it constructs the synchronous message frame. Some also compress the data before it is transmitted.

The service ratio of a statistical multiplexer denotes its capability; that is, its overall performance level in comparison to a TDM. The lower part of Exhibit IV-1-10 shows an STDM with a service ratio of 2 to 1. Over a period of time, this multiplexer has twice the capability of a TDM because it can support twice the number of data sources on the average. If connected to a 9,600-bps modem, the STDM typically can support sixteen 1,200-bps terminals because some will be inactive at any given time. However, if 10 terminal users simultaneously listed a long file, data would flow into the multiplexer at 12,000 bps (10 × 1,200) and leave the multiplexer at a lower data rate (if the STDM is connected to a 9,600-bps modem). This would

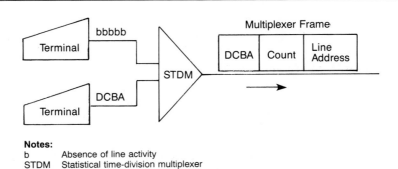

Notes:
b   Absence of line activity
STDM   Statistical time-division multiplexer

Exhibit IV-1-9.   Statistical Multiplexer Operation

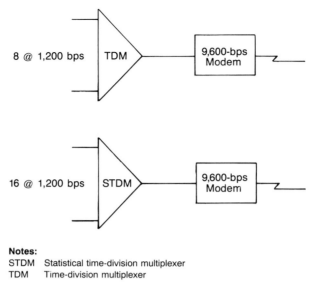

Notes:
STDM   Statistical time-division multiplexer
TDM   Time-division multiplexer

Exhibit IV-1-10. Multiplexer Performance Comparison

---

cause data to fill the STDM's buffer and, if not checked, would cause data to be lost when the buffer eventually overflows.

Flow Control

Multiplexer vendors have incorporated several flow control procedures into their products to prevent STDM buffers from overflowing. The most common procedures result in the STDM transmitting the XOFF character or dropping the clear-to-send (CTS) signal to inhibit data from flowing into the multiplexer. When the multiplexer's buffer is emptied to a predefined level, the STDM transmits an XON character or raises the CTS signal to enable transmission to the multiplexer to resume.

The transmission of the XOFF/XON sequence and the raising and lowering of the CTS signal apply only to asynchronous data sources. To control the flow of synchronous data, the STDM must lower and raise the clocking signal, a more difficult operation offered by only a few vendors.

Service Ratio Differences

Synchronous transmission denotes blocks of data with characters placed in sequence in each block; there are no gaps in this mode of transmission. In contrast, a terminal operator transmitting data asynchronously may pause between characters to think before pressing each key on the terminal. Therefore, the service ratio of STDMs for asynchronous data is higher than for synchronous data. STDM asynchronous service ratios typically range between 2:1 and 3.5:1; synchronous service ratios range between 1.25:1 and 2:1, depending on the STDM's efficiency and built-in features, including bit stripping and data compression.

Data Source Support

Although some statistical multiplexers support only asynchronous data, others support both asynchronous and synchronous data sources. When a statistical multiplexer supports synchronous data sources, it is extremely important to determine the method the vendor uses to implement such support.

Some vendors use a band pass channel to support synchronous data sources. In these cases, the synchronous data is not multiplexed statistically, and the data rate of the synchronous input limits the multiplexer's capability to support asynchronous transmission. Exhibit IV-1-11 illustrates the effect of multiplexing synchronous data by means of a band pass channel. When such a channel is used, a fixed portion of each message frame is reserved for the exclusive multiplexing of synchronous data. The portion reserved is proportional to the data rate of the synchronous input to the STDM. Only the remainder of the message frame is then available for multiplexing all other data sources.

As an example of the limitations of band pass multiplexing, an STDM that is connected to a 9,600-bps modem supports a synchronous terminal operating at 7,200 bps. If band pass multiplexing is used, only 2,400 bps is available in the multiplexer for multiplexing other data sources. In comparison, if an STDM statistically multiplexes synchronous data and has a service ratio of 1.5 to 1, a 7,200-bps synchronous input to the STDM would, on the average, take up 4,800 bps of the 9,600-bps operating line. When the

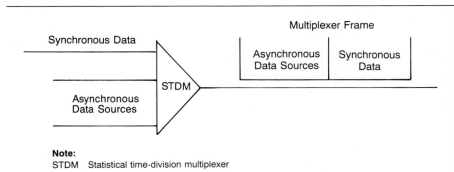

Note:
STDM   Statistical time-division multiplexer

Exhibit IV-1-11.   The Use of a Band Pass Channel to Multiplex Synchronous Data

synchronous data source is inactive, other data sources the STDM services flow through the system more efficiently. In contrast, the band pass channel always requires a predefined portion of the high-speed line to be reserved for synchronous data, regardless of the activity of the data source.

Problems in Multiplexer Use

Some of the problems associated with the use of statistical multiplexers include the random occurrence of increased transmission delays and, occasionally, protocol time-outs. To illustrate how these problems can occur, the statistical multiplexer in Exhibit IV-1-12 is servicing five data sources. The multiplexer is connected to two synchronous and three asynchronous data sources that have an aggregate transmission throughput of 38.4K bps. The multiplexer is shown connected to a remote site by a modem operating at 19.2K bps.

Although the aggregate transmission throughput of the individual data sources exceeds the line capacity, there are minimal throughput delays and only a small possibility of a protocol time-out. There are two reasons for this: First, statistical multiplexers have a large data-handling capacity; second, at any given time there is a high probability that one or more data sources are inactive. Even when active, if an asynchronous terminal operating at 9,600 bps is being used for data entry, its effective data rate is on the order of tens of bits per second even though it is transmitting at 4,800 or 9,600 bps to the multiplexer. Only when one or both synchronous devices start transmitting and one or more asynchronous devices are used to list a long file or similar data-intensive operation will the multiplexer's buffers begin to fill.

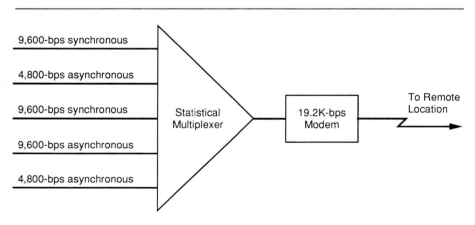

Exhibit IV-1-12.   Statistical Multiplexers

To prevent a buffer overflow, statistical multiplexers typically use several flow control procedures, including XON/XOFF, ENQ/ACK, or lowering and raising different RS-232 control signals. When transmissions from one data source are randomly inhibited, the throughput of other data sources is affected. In addition, depending on the activity of the data sources, delays through the statistical multiplexer can increase to the point at which they cause protocol time-outs.

To ascertain the potential delays and probability of protocol time-outs, one of the best tools is the round-trip delay test built into many statistical multiplexers. To use this feature, a remote channel is put into loop-back mode, then the round-trip delay test is conducted. The result is a display of information similar to the following:

Round-Trip Delay Channel XX = 1.3 Seconds

If a round-trip delay test is conducted when most of the statistical multiplexer channels are active, a reasonable estimate can be obtained for the amount of time it takes a terminal connected to the multiplexer to receive an acknowledgment of a transmitted data block. Exhibit IV-1-13 lists the results of a series of tests conducted every hour during a busy day.

If bisynchronous data is being transmitted on one of the synchronous channels of the statistical multiplexer, the 10 round-trip delay tests lead to the following observations. First, delays increase in the late morning (i.e., before lunch) and in the afternoon as people attempt to complete jobs before quitting time. Running some jobs in the early morning or during lunch would therefore increase the protocol efficiency, because each block has to be acknowledged. Under the circumstances, if the bisynchronous time-out value is equal to 1.75 sec, on the basis of the round-trip delay time tests, bisynchronous data transmitted at 4 PM would have experienced a time-out, resulting in a break in the communications session through the statistical multiplexer. To avoid time-outs, the time-out value should be raised to 2 sec.

Time	Delay (sec)	Time	Delay (sec)
8 AM	0.3	1 PM	0.8
9 AM	0.5	2 PM	1.2
10 AM	1.3	3 PM	1.6
11 AM	1.5	4 PM	1.9
Noon	0.6	5 PM	0.5

Note:
STDM   Statistical time-division multiplexer

Exhibit IV-1-13.   STDM Round-Trip Delay Test Results

SELECTION AND PROCUREMENT CONSIDERATIONS

Another problem associated with statistical multiplexers is a random increase in terminal response times, which appear as noticeable pauses in screen output. This situation is also associated with the aggregate transmission through the multiplexer building up to a level that adversely affects data flow. Fortunately, most multiplexers generate a variety of system reports that can be used to isolate and correct such problems.

Exhibit IV-1-14 illustrates a buffer utilization report, obtained by connecting a terminal device to the master console port on most multiplexers or as part of a system report. The effect on terminal response time of data buildup in the multiplexer can be computed with the information in the buffer utilization report, as follows.

The buffer utilization report generates a summary of the average buffer contents during a predefined time interval. In Exhibit IV-1-14 the HH:MM:SS indicates the hour, minute, and second at which the 10-min average time interval ends (in this case, the interval is exactly 10 min). The 9,600 HS LINK entry means that the high-speed communications path or link between multiplexers operates at 9,600 bps.

The OUTBOUND XMT BUFFER refers to the buffer in the local multiplexer at the computer site through which data transmitted from computer ports must flow to reach the high-speed link. The exhibit shows the average occupancy of the outbound buffer to be 1,420 characters. This means that to reach the link, the first character flowing from the computer port must wait until 1,420 characters are emptied from the buffer onto the line. It is possible to compute the response time delay contributed by the outbound buffer: with an average of 1,420 characters in the buffer and each character represented by 8 bits, 1,420 characters × 8 bits per character, or 11,360 bits, must be transmitted. The link operates at 9,600 bps, so the output buffer occupancy results in a delay of 1.18 sec (11,360 ÷ 9,600).

The INBOUND XMT BUFFER refers to the buffer on the remote multiplexer through which data from terminals flows to the computer. As might be expected, terminal users enter a relatively small amount of data (which will execute programs that generate computer reports), so the occupancy of the inbound transmit buffer will be small relative to the occupancy of the outbound buffer.

HH:MM:SS   10 MINUTE AVG 9,600 HS LINK

OUTBOUND XMT BUFFER   1,420 CHAR

INBOUND XMT BUFFER   87 CHAR

Exhibit IV-1-14.   Sample Buffer Utilization Report

Exhibit IV-1-14 shows the average occupancy of the inbound buffer as 87 characters. This means that terminal data must wait until 87 characters from the inbound buffer are transmitted until a character from the terminal can reach the line. Therefore, the inbound buffer delay is 87 × 8 ÷ 9,600, or 0.07 sec. At these rates of occupancy, the inbound and outbound buffers together contribute approximately 1.25 sec to the response time observed by a terminal user between entering a query and receiving the first character of the computer's response.

Although 1.25 sec may not be excessive for some applications, a response time of this length may be unacceptable for others. The preceding methodology indicates how the data provided by multiplexer reports or report elements can be used to ascertain the effect of potential buffer overload conditions. If the computed delay were much higher, the data transmission rate between multiplexers might be increased to lower the average occupancy of the transmit buffers. Alternatively, one or more users could be placed on a different communications system to lower the quantity of data transferred through multiplexers. A third option would be to identify terminal users that print large files and schedule their operations for times in which many other terminal users are inactive. By selecting one of these alternatives, both the buffer occupancy level of each multiplexer and the data flow delays through those devices can be reduced.

Switching and Port Contention

Switching and port contention usually are available with the more sophisticated statistical multiplexers. Switching capability is also referred to as alternative routing. It requires the multiplexer to support multiple high-speed lines, whose connection to the multiplexer is known as a node. Thus, switching capability typically refers to the multiplexer's ability to support multiple nodes. Exhibit IV-1-15 illustrates how alternative routing can be used to compensate for a circuit outage. If the line connecting locations 1 and 3 becomes inoperative, an alternate route through location 2 can be established if the STDMs support data switching.

Port contention is usually incorporated into large-capacity multinodal STDMs designed for installation at a central computer facility. Such STDMs may demultiplex data from hundreds of data channels. However, because many data channels are usually inactive at a given time, providing a port at the central site for each data channel on the remote multiplexers is a waste of resources. Therefore, the STDM at the central site contains a smaller number of ports than the number of channels of the distant multiplexers connected to it. The STDM at the central site then connects the data sources entered through remote multiplexer channels to the available ports on a demand basis. If no ports are available, the STDM may issue a NO PORTS

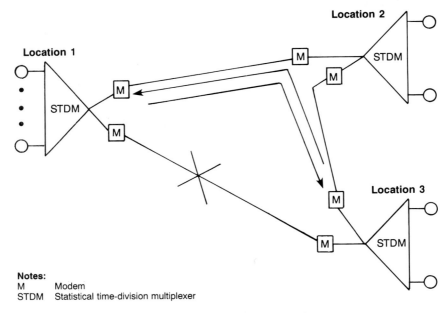

Notes:
M   Modem
STDM   Statistical time-division multiplexer

Exhibit IV-1-15.   Alternate Routing

AVAILABLE message and disconnect the user or put the user into a queue until a port becomes available.

Features to Consider

Exhibit IV-1-16 lists the features the communications manager should consider when evaluating statistical multiplexers. Many of these features have been discussed in earlier sections of this chapter. The following sections discuss features that primarily govern the type of terminal devices that the statistical multiplexer can support efficiently. These features include auto baud detect, flyback delay, and echoplexing.

Auto Baud Detect.   This refers to a multiplexer's ability to measure the pulse width of a data source. Because the data rate is proportional to the pulse width, auto baud detect enables the multiplexer to recognize and adjust to different-speed terminals accessing the device over the switched telephone network.

Flyback Delay.   On electromechanical printers, a delay time is required between sending a carriage return to the terminal and sending the first character of the next line to be printed. The delay time enables the ter-

	YES	NO	PAR	NA	ACT	NOTE
3. If more than one system is used, the level of use is continually reviewed to determine whether work load redistribution between the systems would reduce bottlenecks and personnel attention because of crunches						
4. Service availability schedules are evaluated frequently to determine whether services are actually needed during the entire time they are provided						
5. Automated job scheduling aids are used whenever possible to maximize job mix instead of leaving scheduling to the discretion of remote users or local computer operators						
6. Software aids (e.g., optimizers, simulators, and design aids) are used periodically to: a. Identify and analyze configuration bottlenecks						
b. Analyze configuration component performance						
c. Analyze the impact of new or changed hardware and software						
d. Identify intermittent malfunctions						
e. Analyze transmission peaking						
7. Computer console logs are frequently reviewed to determine whether unauthorized processing has occurred						
8. Controls exist in hardware and software architecture and in operating procedures to prohibit (or at least discourage) users and technical support staff from: a. Intentionally misusing or crashing the system						
b. Exploiting holes in the operating software						
c. Violating job classes and priorities						
d. Avoiding resource charges						
e. Using restricted resources						
f. Otherwise violating security						
9. Online capability to monitor system status is available only to authorized personnel						

	YES	NO	PAR	NA	ACT	NOTE
10. Preventive maintenance for equipment is done before heavy use periods to ensure reliability						
11. Preventive maintenance is not habitually postponed because of processing schedule pressures						
12. Realistic performance standards have been set						
13. Mandatory standards are in effect to reduce costs (e.g., for naming and retaining files, leveling work loads, classifying jobs)						
14. Data communications staff are periodically sent to vendor-provided refresher classes to help them maximize operating performance						

ACQUISITION

The wide range of equipment and outside services available today offers a multitude of cost-reduction possibilities. The size of the market, changing costs, and number of equipment vendors and leasing deals that are available can be overwhelming. Therefore, in making acquisitions, the manager must keep in mind that what appears to be cheapest in terms of acquisition cost may not be the least expensive in the long run, when operational performance, service, and versatility to accommodate future growth are factored in.

	YES	NO	PAR	NA	ACT	NOTE
1. A legal expert is consulted when contracts or agreements are complex, have long-term effects, support critical resources, or involve large expenses						
2. All contracts and agreements ensure that the organization will have the best financial position if unforeseen events occur (e.g., poor service performance, reduced use, or saturation)						
3. The communications manager is involved in negotiations on contract and agreement terms and conditions to ensure that they correspond to communications needs						
4. If applicable, contracts and agreements are negotiated on a national basis for greater leverage						
5. Leases are prepared to provide the option of upgrading equipment without incurring a penalty and to avoid being billed for redundant equipment during changeovers						

	YES	NO	PAR	NA	ACT	NOTE
6. Lease contracts and agreements provide the following: a. Provisions for early cancellation without penalties						
b. Upgrade provisions to allow equipment to be replaced with more up-to-date units						
c. Purchase options at fair market value						
d. A renewal discount from base-term rental rates						
e. A limitation on transportation-out costs at the end of the lease						
f. An option for the lessee to relocate or move equipment or to assign it to a third party or another company division						
g. A renewal notice period instead of an automatic renewal						
h. The right to add options from a different manufacturer or lessor						
i. The right to add features						
j. Reliability and performance standards that must be met						
k. Specifically defined maintenance coverage that can be renegotiated at any time						
l. Personal property taxes borne by the lessor						
m. Income tax credit (ITC) benefits applied to the lessee						
n. A patent indemnification provision to protect the lessee						
o. The possibility for the lessee to sell time or services to a third party						
p. The ability to add options at a later date that will be coterminous with the original lease						
q. The ability to terminate or renew a device or service without having to terminate or renew other devices and services at the same time						
r. A clear description of the device or service to prevent later misunderstanding						
s. Any limitations on use clearly spelled out						
t. Insurance protection						
u. Identification of training, manuals, supplies, materials, and parts to be provided initially and subsequently						

SELECTION AND PROCUREMENT CONSIDERATIONS

	YES	NO	PAR	NA	ACT	NOTE
7. All software is obtained first on a trial basis to ensure that it will perform as advertised						
8. Buyout is considered when evaluating leasing						
9. Arrangements are made with vendors to bring in additional equipment during peak processing times						
10. Items with long lead times are ordered far enough in advance						
11. There is a formal process for ordering equipment in advance from vendors						
12. The formal process includes a list of all equipment that: a. Shows device name and model, order confirmation deadline, delivery date, cancellation date (without penalty), cost, reason for being ordered (e.g., increased work load, cost improvement, decreased work load, obsolete technology, better performance)						
b. Is updated monthly and distributed to the center, the users (if pertinent), application development, and support management						
13. A monthly meeting is held with vendors, management, and responsible technicians to: a. Generate an updated equipment ordering list						
b. Identify actions to be taken on items for which the disposition status is not clear						
c. Initiate order or cancellation memos to vendors						
d. Relate anticipated equipment needs to service performance work-load forecasting and cost reduction						
14. Backlogs are reviewed periodically to determine whether delivery dates are still valid (i.e., moved up, pushed out, or canceled)						
15. There is a clear understanding of approval limits for each management level so that approvals are not unduly delayed because of misdirection						
16. Evaluations for purchasing and selling consider income tax credit (ITC) as well as book loss						
17. Capital funding requirements are identified and presented to senior management for approvals in advance to avoid last-minute bottlenecks						

	YES	NO	PAR	NA	ACT	NOTE
18. Technical and engineering support personnel must justify to senior management well in advance recommendations for changing the acquisition method or current equipment or services or for adding equipment or services						
19. Long-term leases or purchase arrangements for hardware and software consider technological obsolescence						
20. The resale value of existing equipment is continually monitored to determine whether it is advantageous to sell the equipment before a capacity upgrade must be made						
21. There is a corporate staff control function that publishes and maintains guidelines and requirements for: a. Obtaining approval for new hardware and software						
b. Developing new applications so that time delays are minimized when approvals are sought						
22. A tickler file is kept on all leased and rented software and facilities to trigger action in advance of termination and renewal deadlines						
23. There is a master contract and agreement file to which the communications manager has access						
24. There is an inventory of all outstanding contracts and agreements listing device or service, type of agreement, term, initiation date, termination date, rates, penalties, or any special conditions						
25. Although dealing with several vendors may reduce procurement costs, the potential costs of finger pointing between vendors and system interfacing are taken into consideration						
26. Bids are solicited from multiple vendors whenever possible to find the lowest cost source						
27. Evaluation of new hardware and software considers future support, upgradability, and compatibility						
28. It is a formal practice to evaluate all acquisition methods (e.g., short-term rental, lease, purchase) for new resources even when one appears to be superior to the others, to ensure that every possibility has been considered and to check the possibility of special arrangements with vendors						

689

OUTSIDE SERVICES AND SUPPLIES

Communications systems managers often renew service contracts year after year without considering alternatives. The manager must evaluate the quality of service before renewing services and must take appropriate action if performance has been unsatisfactory or if prices are too high.

Generally, however, cost reductions are pursued more for outside services and suppliers than for other categories because:

- They are usually itemized in budget plans and involve out-of-pocket costs that single them out more than other, more obscure, cost-reduction opportunities.
- Large organizations have purchasing departments that specialize in negotiating for the lowest-cost vendors.
- Vendor competition is usually keen.
- Purchase order and invoice activities continually emphasize the transactions that are occurring.
- Even moderately expense-conscious management maintains a vigilance over outside services and supplies.

Outside services and suppliers require continual control and monitoring. There may be areas in which increases in some outside expenses can bring decreases in other expense areas, resulting in a bottom-line expense reduction. Therefore, it is important to audit outside services periodically to ensure that the return justifies their cost.

	YES	NO	PAR	NA	ACT	NOTE
1. Meetings are held regularly to evaluate vendor maintenance performance and to determine whether maintenance coverage is excessive for the results obtained						
2. Special servicing arrangements are set up with hardware and communications vendors in anticipation of peak or highly critical processing periods to place them on standby						
3. Vendor maintenance invoices are verified and signed off by management to ensure accuracy						
4. Corrective maintenance is scheduled during regular working hours whenever possible to avoid special vendor charges						
5. Third-party maintenance has been explored						
6. User management is periodically given a list of the reports it receives so it can determine whether it actually needs them—or whether they are as useful as they should be						

	YES	NO	PAR	NA	ACT	NOTE
7. Supply inventories are periodically reviewed to ensure that they are as low as possible without incurring shortages						
8. Costs of services and equipment are directly charged to the users or at least frequently made known to the users to increase their cost consciousness						
9. Services, equipment, and supplies facilitate rather than restrict accomplishment of duties						
10. Employee referral awards and increased emphasis on direct hiring are used to reduce employment agency fees						
11. There is a list of approved vendors that helps ensure that only reliable vendors are used						
12. The company draws from its in-house expertise whenever possible rather than pay for outside consultants						
13. Outside services obtained by users are periodically reviewed to determine whether higher-quality or lower-cost services could be provided in-house						
14. The organization takes advantage of opportunities provided by vendors to reduce materials and supplies costs by volume ordering or bill payment within 30 days						
15. Small-capital expense items are marked with identification numbers, and a list is kept that identifies the user of each item						
16. Periodic meetings are held with vendors to: a. Correct poor performance						
b. Plan changes						
c. Improve interfacing						
17. Periodic audits ensure that hardware and software manuals are current and properly used and are not being stolen						
18. If utility costs are too high, areas where use might be reduced are investigated						
19. Non-common-carrier equipment and services are investigated periodically to determine whether they could be more cost-effective than common-carrier facilities						

SELECTION AND PROCUREMENT CONSIDERATIONS

	YES	NO	PAR	NA	ACT	NOTE
20. Data transmission line capacity is reviewed periodically to determine whether it can be reduced without affecting service adversely						
21. Attempts have been made to determine whether other local companies have telecommunications networks with transmission capabilities and geographic coverage that offer the potential for line sharing						
22. Audiovisual training is used whenever possible to reduce outside course fees and travel and living expenses						
23. The organization takes advantage of no-cost consulting and advisory services offered by hardware and software suppliers to improve use, avoid upgrades, and reduce services						
24. Outside consultants are placed on retainer periodically to review overall production and financial performance, technical support activities, and technological plans						
25. All specialized technical staff members are used sufficiently to justify retaining them as full-time employees; there is no need to obtain the same services from a contractor on a retainer or project-by-project basis						
26. There are no currently purchased services (e.g., equipment maintenance, security) that could be performed by current employees (or new employees) at lower cost						
27. Consultants are contracted on a fixed-results basis rather than by time and material						
28. The communications manager records and audits the time of consultants who are used on a retainer basis						

Chapter IV-1-4
How to Choose a Systems Integrator

Nathan J. Muller

ANNUAL INVESTMENT IN COMMUNICATIONS NETWORK TECHNOLOGY—hardware, software, and associated services and support—has grown to about $200 billion a year. According to conservative projections, it may reach $300 billion in the early 1990s. This rapidly growing investment reflects the conviction of most executives that communications networks are critical to corporate success. The corporate network is no longer seen only as an overhead expense; now it is considered to be a strategic resource—the key to cost control, improved customer service, new business opportunities, and increased market share.

Installing an effective network, however, is no small undertaking. Needs must be assessed, budgets hammered out, consultants hired, plans drawn, technologies reviewed, requests for proposals (RFPs) written, vendors evaluated, feasibility studies conducted, equipment installed, lines leased, and users trained. Diverse products from many vendors must be seamlessly interconnected. Finally, all this complexity must be transparent to the users.

Few organizations have sufficient expertise to accomplish this formidable task alone. Many are turning to a type of outside service that specializes in network installation: the systems integrator.

THE SYSTEMS INTEGRATOR

A systems integrator brings objectivity and expertise to the task of network implementation. It ensures the compatibility of the disparate hardware and software elements that make up a network—customizing interfaces when necessary—at a price the customer can afford. A systems integrator should have extensive experience in management information systems, voice and data communications, and project management; it should be familiar with a variety of customer needs and operating environments; and should be stable and financially secure.

VENDORS OF SYSTEMS INTEGRATOR SERVICES

Systems integrators usually fall into one of six categories:

- Communications service providers.
- Consultants.
- Accounting and management consulting firms.
- MIS shops.
- Traditional service firms.
- Hardware vendors.

Each category has specific strengths and weaknesses that a client should consider when choosing a systems integrator. The wrong selection can delay implementation of the network and inflate operating costs.

Communications Service Providers

Communications service providers (e.g., AT&T) and the regional holding companies (RHCs) have systems integration units. A strength of these companies is their experience in voice switching and transport—they have a long history of success, for example, in joining private networks to the public network—which they hope to translate into complete systems integration services.

Systems integration, however, calls for expertise in many areas. Successfully implementing a customer's business applications in the multivendor hardware environment typical of most organizations requires a strong MIS background. The RHCs traditionally have not offered MIS expertise to their customers and have been criticized for a lack of understanding of customer requirements beyond the PBX. Much of this weakness was due to the regulatory constraints under which these companies operated, and they have taken steps to counter it. These include engaging in team arrangements with computer makers, acquiring service firms with a national presence, and engaging in aggressive personnel recruiting efforts to fill gaps in their technical expertise.

Consultants

This category includes contract programmers and independent consulting firms. The consulting arms of the major accounting firms are discussed separately.

The activities of contract programmers include the design and writing of programs for applications software, operating systems, media conversions, and communications protocols between microcomputers and mainframes. Contract programmers play a valuable role in systems integration, but usually as subcontractors to a primary contractor, who typically assumes full financial responsibility and risk management for the entire project.

Independent consulting firms engaged in systems integration claim objectivity as their strength. This is because they are not associated with any vendor or service provider and can presumably keep their customers' best interests in mind when evaluating hardware, software, and services. However, although objectivity is indeed a valuable asset during the vendor selection process, systems integrators are often called in after the selection of hardware and software has been made. Therefore, a decision to choose an independent consulting firm for systems integration must rest on other criteria: financial resources, areas of technical expertise, organizational stability, and project management skills.

Consultants who talk about "vendor bashing" or "cutting carriers down to size" should be avoided. This approach reveals poor interpersonal skills, which can unnecessarily prolong a systems integration project and drive up costs. In contrast, a highly skilled communicator who is focused on the project at hand can be of great benefit. Effective performance in systems integration hinges on the ability to interact well with multiple vendors, carriers, consultants, and subcontractors, as well as a client's own staff.

Most independent consulting firms have very limited financial and organizational resources to draw on. This restricts the size and complexity of systems integration projects they can handle. It is up to the customer to discover, preferably not from experience, what that threshold is.

Consulting firms (and other providers of systems integration services, including communications carriers) may seek to broaden their qualifications by teaming up with another organization with complementary expertise. It is incumbent on the customer to ascertain the nature of such relationships. Complementary relationships should already be in place and not have been hastily thrown together just to obtain a specific contract. Both parties should have evaluated each other's strengths and weaknesses long before entering into any formal or informal arrangement. The team should be able to demonstrate the integrity of the relationship by providing a list of projects that they have completed together. A check of references may reveal flaws in the relationship, signaling the need for further investigation.

Accounting and Management Consulting Firms

In recent years, the major accounting firms have expanded the scope of their activities from accounting and management services to MIS and communications consulting. They bring a broad range of technical expertise and an intimate knowledge of client operations to systems integration. Not being aligned with either hardware vendors or service providers, they also claim objectivity.

Some major accounting firm clients, however, dispute this claim of objectivity. They argue that there is often a conflict of interest between the communications consulting branch of a firm and the other branches. The accounting side of the house, for example, may uncover problems that require the services of the communications consulting group and the communications consulting group may uncover problems that need the services of the accounting group. Although the extent of such practices differs from firm to firm, the customer really has no way of predicting it. A check of references may alert the customer to potential problems, but only if the firm's behavior was clearly aggressive.

MIS Shops

Sometimes a company's in-house MIS staff may become expert at systems integration or some aspect of it; the in-house staff might then form a separate profit center offering systems integration services.

Hiring such firms may be worthwhile if a systems integration project is confined to the MIS or data processing center. For a project that involves wide area networking, however, a firm whose experience is based in MIS may prove inadequate. The technologies that differentiate local area networks (LANs) from wide area networks (WANs) are so dissimilar that specialized expertise is required for each. Even providing the links between LANs and WANs through such devices as gateways, bridges, and routers requires more expertise than is usually found among MIS professionals.

Traditional Service Firms

Traditional service firms have branched into systems integration from their base as providers of MIS services. As such, they have the same weakness as in-house staff in dealing with projects that involve wide area networks. However, they have established long-term strategic relationships with other vendors to provide expertise in this and other areas in which they are weak.

Another problem with traditional service vendors is that they may use their role as systems integrator to sell their own products ranging from management services to software and processing services. Also, because they are so big and have expensive infrastructures, they are more interested in large projects than small ones. Consequently, they might not give a small or medium-sized project adequate attention.

Hardware Vendors

Both computer makers and makers of data communications equipment offer systems integration services. Computer makers have strong expertise and integration experience in the area of MIS but often lack depth in communications. Some computer manufacturers have attempted to address this imbalance by acquiring smaller companies with expertise in building and managing networks.

Makers of data communications equipment are in a position to bridge MIS and communications. As such, they occupy a strategic position in the industry, especially for users building networks with both private and public elements. An independent maker of data communications equipment that is not aligned either with a communications carrier or with a computer manufacturer might for this reason be a good candidate to provide systems integration services.

A major concern with hardware vendors—whether of computer equipment or of communications equipment—acting as systems integrators is the potential for conflict of interest this creates. Will the vendor offer objective advice, or use its position as systems integrator to encourage the purchase of its own products? Recognizing these concerns, vendors usually handle systems integration through a separate division or subsidiary whose charter separates it from the sales and marketing of the products of the parent company.

Potential conflicts of interest are also reduced by the fact that vendors recognize that users increasingly expect interconnectability. A vendor will be more successful if its products can link to those of other vendors, and a vendor acting as systems integrator must be able to reach outside its own product line. The participation of vendors in international standards organizations, their support for open systems interconnection (OSI) through membership in the industry consortium Corporation for Open Systems, and adherence to International Telephone and Telegraph Consultative Committee (CCITT) recommendations on the requirements for modem communications, packet switching, and integrated services digital network (ISDN) all reflect this recognition.

A MODEL FOR EVALUATING SYSTEMS INTEGRATION SERVICES

Exhibit IV-1-17 presents a model for choosing a systems integrator and evaluating its performance. The discussion that follows elaborates on the steps in the process, ending with the core of the model, the criteria for evaluating systems integrator candidates.

A systems integration project starts with the formulation of organizational objectives (A in Exhibit IV-1-17). A simple objective might be to serve customers faster and more efficiently. A more complicated objective might be to reorganize the company to permit a faster response to changing market conditions. Any systems integration project must support these objectives.

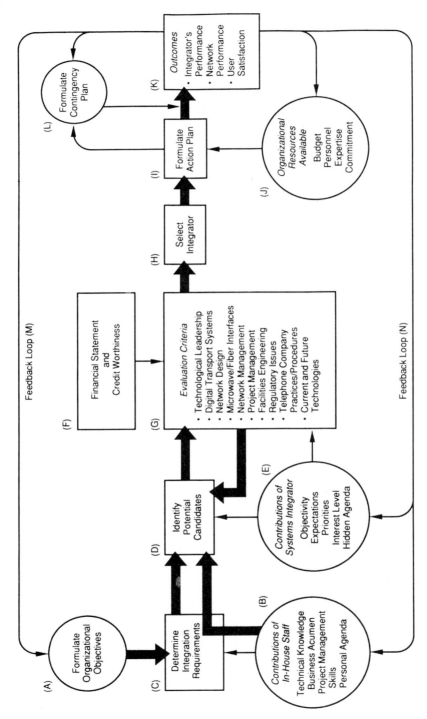

Exhibit IV-1-17. A Model for Evaluating Systems Integration Services

With the organization's objectives in mind, in-house MIS and communications staff draw on their knowledge of the organization's business, their technical knowledge, and their project management skills to determine the nature and scope of a proposed integration project (B in Exhibit IV-1-17). In considering the proposals of in-house staff, senior management should be aware of the personal agendas of those involved: their hopes for a promotion or raise for a job well done, or their expectations of obtaining additional staff or of increasing their visibility among fellow employees or peers in the industry.

After other departments affected by the project are consulted, an initial plan is developed with tentative hardware and software requirements and a schedule with all project milestones and target dates for completion (C in Exhibit IV-1-17). At this point, RFPs may be developed and sent to appropriate systems integration candidates (D in Exhibit IV-1-17) who respond with specific information on experience, expertise, methodology, pricing, and availability.

The integrator's response (E in Exhibit IV-1-17) reflects its objectivity, priorities, expectations, and interest level. For example, a vendor may respond with no bid because it cannot possibly meet the project deadline given its commitments to other customers. The integrator's expectations would include the price and terms it hopes to exact from the customer as well as intangible factors such as the publicity and market value that would come from a job well done. The integrator's decision to respond to the RFP may also be influenced by its willingness to be evaluated according to the hiring organization's criteria (F and G in Exhibit IV-1-17). Finally, the integrator may have a hidden agenda, such as using the project to sell its other products.

The final selection of the integrator (H in Exhibit IV-1-17) should be based on financial, technical, and managerial criteria (F and G in Exhibit IV-1-17) established when the RFP is developed. These criteria are discussed at greater length later in this chapter.

When the integrator has been selected, the evaluation team and integrator formulate an action plan (I in Exhibit IV-1-17) with delivery, installation, and cutover dates and a training schedule. At this point also, a contingency plan for missed deadlines or poor performance should be developed (L in Exhibit IV-1-17). The action plan should take into account the organization's budget for the project, the availability of personnel to work with the integrator, and the technical expertise available to support the plan (J in Exhibit IV-1-17).

On completion of the project, the outcome is evaluated in terms of the integrator's performance, the performance of the network, and the satisfaction of its users (K in Exhibit IV-1-17). If the outcome of the project is not satisfactory, a number of options are available, depending on what went wrong and how. If the integrator missed important deadlines, it might incur a financial penalty as specified in the contingency plan. Alternatively, achieving a satisfactory solution might require additional organizational resources and a reformulation of the action plan. Whether satisfactory or not, the outcome of the project may reveal the need for other integration projects, leading back to the beginning of the process (feedback loops M and N in Exhibit IV-1-17).

Evaluation Criteria

The best systems integrator is the one with the broadest experience in a variety of technical fields, customer applications, and operating environments. Contenders should be evaluated in terms of their technological leadership and their experience with digital transport systems, network design, microwave and fiber-optic interfaces, network management, project management, facilities engineering, regulatory issues, telephone company practices and procedures, and current and future technologies. The following discussion looks at each of these areas.

Technological Leadership. Clients should confirm vendor or carrier claims of technological leadership by delving into their products or services to see how much responsibility they really had for their development. Does the company hold any patents or software copyrights? Does it license proprietary technology to other manufacturers? Does it supply key subsystems for any product already on the market? Does it design and build key components based on large-scale or very-large-scale integration to give its products price and performance advantages? If the answer to these questions is no, the organization should clarify what the vendor means when it describes its products or services as advanced, state-of-the-art, leading edge, innovative, or unique. Many times these words and phrases are just marketing jargon designed to attract inquiries.

Clients should inquire about the performance record of products or service offerings already installed in networks similar to theirs. Validating the answers by checking with references will help determine product or service quality, levels of customer satisfaction, and responsiveness to changing customer requirements. All of these are key aspects of technological leadership.

Digital Transport Systems. Clients should look for demonstrated expertise in the design, engineering, and implementation of large digital transport systems for a variety of rigorous applications, including retail, financial services, manufacturing, and government. Evidence of such experience should include contract awards and successfully completed projects that involve installing and implementing multimode digital backbone networks.

Clients should ask for references from organizations that use hardware and software products similar to their own. When calling references, they should inquire about the firm's ability to meet work schedules, deal with multiple vendors, get along with client staff members, and stay within budget.

Clients should also look at the integrator's experience in designing and installing hybrid networks consisting of such diverse elements as time-division multiplexers, statistical time-division multiplexers, and X.25 switches and packet assembler-disassemblers. With these building blocks, users may combine several architectures, adding precision and control to network operations and positioning the network to accommodate such new communication services as ISDN. These building blocks can also allow network expansion to include satellite microwave and fiber-optic links. Even if the need for these services seems remote, it is preferable to have a relationship with a systems integrator who has the experience to provide them when they are needed.

An effective systems integrator has experience in integrating local area networks with wide area networks using such devices as gateways, bridges, and routers. Experience with these network elements helps to establish the breadth and depth of a firm's knowledge of communications protocols, management systems, and specialized interfaces. Experience integrating international networks is also valuable.

Network Design.   The integrator's network design tools should be capable of taking into consideration factors such as line topology, traffic load, facility costs, equipment types, communications protocols, hubbing arrangements, and the differing performance parameters of both voice and data. These factors must be considered together with such variables as switch performance, which includes queuing, blocking, and reliability. The design tool should also take into account the type of traffic that the network must support: voice, data, image, full motion video, or any combination of these. This is important because most of the current computer tools for network design make assumptions about certain aspects of a network that may not apply to particular situations.

Microwave and Fiber-Optic Interfaces.   Because of the increasing sophistication of networks, systems integrators should have expertise in connecting voice and data digital networks with microwave and fiber-optic systems. If the integrator is a hardware vendor, this expertise can be determined by finding out how its products are being deployed by its customers.

For example, if the vendor makes a diagnostic modem that has been successfully incorporated into a network that connects an oil company's offshore drilling rigs in the Gulf of Mexico with its control facility in Texas, it may be safe to conclude that the vendor knows something about connecting networks with microwave facilities. Likewise, if a vendor has the capability to equip its T1 products with fiber termination cards to provide customers with high-capacity fiber links, it may be safe to conclude it has experience connecting networks to fiber optic transmission systems. In all such cases, of course, references should be checked to be sure the clients are satisfied.

If a firm has no direct experience in fiber-optic or microwave connections it may have strategic alliances with other hardware vendors that do. These alliances should be checked.

Network Management.   Network management systems unify computer and communications resources and transform them into strategic assets with which to improve a company's competitive position. Selecting a systems integrator with experience in designing and implementing network management systems may be critical to the success of the integration plan.

Some data communications firms have demonstrated their expertise in network management by integrating host-based management systems (e.g., IBM's NetView) with proprietary modem/multiplexer management systems that also permit ties into AT&T's unified network management architecture, which unifies the various management systems of that company's transmission products. When OSI standards are also stabilized, these vendors intend to provide links into the network environments of other vendors. A vendor's expertise in developing such network management systems gives it a unique perspective from which to provide advice on all aspects of network operations.

Project Management.   The systems integrator's project management team should have extensive experience in all facets of data and voice communications, including system design, product development, integration, installation, and problem resolution. The team should also have experience in network change management, which may be an ongoing activity, depending on the size and complexity of the network in question.

A dedicated project manager should be assigned to oversee all aspects of an integration project, including its implementation. For the evaluation process, the systems integrator should provide the résumé of the project manager it plans to assign to a project along with an organizational chart showing lines of responsibility. The integrator should not change the project manager or the reporting structure without the client's knowledge and approval.

The project management team should be well versed in the use of computerized planning tools to coordinate and track systems implementation plans. PERT, Gantt, CPM, or other project management tools should be used to develop and review implementation progress. Once the equipment is installed and integrated into an entire system, the project management team should develop and oversee customized acceptance testing with the client's staff. The project management team should also prescribe appropriate levels of staff training, and be prepared to implement training if required.

The project management team should be able to assist in identifying specific inventory control and internal billing requirements, and then to recommend, develop, or customize software to meet those needs.

Facilities Engineering. To successfully integrate diverse products from many manufacturers, a systems integrator should be staffed with professionals who understand all aspects of facilities engineering, including planning, scheduling, and coordinating the preparation of a site as well as managing a total implementation plan. These abilities help ensure that a site meets all environmental, space, and power requirements before any equipment arrives.

The systems integrator should be able to configure, stage, integrate, and test entire systems before delivery to a site. This proactive project planning can eliminate potential problems early in the systems integration process when they are easier and less costly to correct.

Regulatory Issues. The systems integrator should track all regulatory issues that may affect its clients and its ability to meet client needs, including interLATA (local access and transport area) and intraLATA tariffs and carrier service offerings. By keeping track of tariffs, the systems integrator can show clients how to design or redesign their networks to take advantage of tariff anomalies, which can translate into substantial savings.

Telephone Company Practices and Procedures. Because today's networks typically span the serving areas of many telephone companies, the systems integrator must be completely familiar with their differing practices and procedures, including administrative and technical practices and procedures governing the installation of various devices to the local loop and in-house facilities. The systems integrator should track and study Bellcore technical advisories and technical requirements publications, including those governing new equipment building specifications, which cover heat dissipation, power consumption, relay rack mounting, and alarm arrangements in central offices. The systems integrator should also be familiar with the common language equipment identifier coding scheme used by telephone companies.

The project management team should include a former central office engineer or an experienced technician who knows the language of telephone companies and is familiar with their practices, procedures, and organizational structure.

Current and Future Technologies.   There is nothing more frustrating than dealing with professionals who do not keep abreast of new developments in technology. A systems integrator needs a formidable background on the transmission requirements that support key applications (e.g., video conferencing, electronic mail, electronic document interchange, and remote data base access). Its expertise should not be limited to standard terrestrial copper; it should include microwave, satellite, and fiber optics as well.

ACTION PLAN

The systems integrator should have an infrastructure organized for action, with methodologies in place, planning tools available, and the required expertise on staff. The candidate must also be able to show that it has successfully completed projects similar to those of the client. Proposals should reflect an understanding of the client's industry, competitive situation, corporate culture, and information systems environment. The client's objective is to engage a systems integrator that it can treat as a partner and trust to create the network that best meets its interests.

Finally, clients should not overlook the possibility of engaging more than one type of systems integrator. For example, if in-house MIS and communications staff have strong project management skills, they might oversee an integration project involving several outside providers, including consultants, hardware vendors, and contract programmers. Such an arrangement combines centralized control with the benefits of the specialized expertise of outside providers. It also helps avoid conflict-of-interest problems.

Section IV-2
Implementation Guidelines

THROUGH IMPLEMENTATION, the planning, definition, and design of a communications system are turned into reality. The implementation phase of a communications networking project is when all of the elements come together—and it is when the communications manager either succeeds or fails. This section explores several implementation issues with the objective of providing useful guidelines to the communications systems manager.

Chapter IV-2-1, "Data Network Design Fundamentals," introduces the issues that become important once the organization starts to move toward implementation of an actual network: design tradeoffs, determining network objectives, and generating an optimum network design. The goal is to create a network with suitable performance, maximum reliability, and minimum cost.

No implementation program is complete until the communications system has been acceptance tested. Acceptance testing determines a system's readiness for operational use; the test program's goal is to avoid going online with an inadequate or poorly implemented system that fails to meet the organization's objective. "System Acceptance Testing," Chapter IV-2-2, thoroughly delineates the benefits and objectives of testing and the consequences of not testing. It describes an approach to an acceptance test plan and provides helpful tips on conducting the test.

Chapter IV-2-3, "An Integrated Communications Cabling Network," is a case study of a cabling network designed to support voice, data, and image applications. (This is an increasingly common mix of information sources.) The chapter describes the plan, the requirements, alternatives, and elements of the network cabling, outlets, and wiring closets. The chapter concludes with discussions of the required physical transmission facilities and management issues.

The section turns, finally, to a very current issue: implementation of a functional open systems interconnection (OSI) architecture—in this case, the Manufacturing Automation Protocol (MAP). Chapter IV-2-4, "The Man-

ufacturing Automation Protocol and the Technical and Office Protocol,"
examines the MAP functional architecture as well as the associated Technical and Office Protocol (TOP). These profiles are an attempt to create a
set of specifications for manufacturing and office applications that will
comply with OSI and therefore result in less expensive implementations.
The chapter provides some background on MAP/TOP and discusses the
relationships with the US government's functional profile, called GOSIP.

Chapter IV-2-1
Data Network Design Fundamentals

Ellen Koskinen-Dodgson
Greg Koskinen-Dodgson

A SYSTEMATIC APPROACH TO NETWORK DESIGN is essential if the objective of the network is to obtain a desired level of performance and reliability at an acceptable cost. This chapter presents a detailed, practical method for designing a data network that will optimize performance and reliability while minimizing cost.

The network design issues discussed in this chapter include:

- The design trade-offs that must be kept in balance.
- The network options available.
- How to determine network objectives.
- The design information needed and the level of accuracy required.
- How to generate and optimize network design alternatives.
- What constitutes the best design and how to get final approval for it.
- What part the network designer should play in implementation and follow-up.

THE DESIGN TRADE-OFFS

A designer ideally should be able to provide users with an inexpensive, fast, and extremely reliable network. In the real world, however, designers are faced with classic trade-offs among cost, performance, and reliability. Data networks must be designed on the basis of the user's individual priorities.

Cost is often the most important factor. The designer should know how much the user is prepared to spend and should present the user with several network designs, ranging from those that are more reliable to those with higher throughput. When confronted with the cost of an initial network design, many users revise their reliability and performance objectives significantly. For example, in some cases, accepting slightly poorer response time can result in a 50% reduction in total network cost. It is usually during this process of assigning priorities to features that the user comprehends the relationship between network design trade-offs.

THE NETWORK OPTIONS

The selection of the network topology to be implemented is usually a compromise between existing hardware and software constraints, cost, performance, reliability, and ease of design and maintenance. Five different network topologies are commonly used.

The Star Topology. The star network is the simplest network to design and implement, but it is usually the most expensive. All terminals communicate through dedicated lines to a central site; this topology has the best performance and is the easiest to install and maintain.

The Ring Topology. The ring topology is generally not used for large networks because of the cost required to make it robust. It is an economical topology for such local area networks as IBM's Token-Ring Network. Ring topologies have difficulties dealing with breaks in the line; if the ring is broken because a site or one of the lines connecting the sites is down, the entire network can be disrupted.

Centralized Tree Topology. The relatively high cost of data communications lines has led vendors to implement polled-multidrop protocols, in which many devices share one access line that runs to the host, as depicted in Exhibit IV-2-1. The host or communications controller governs access to the line by polling each device sequentially or in another order. This topology is very common for interactive terminal networks. This centralized tree topology is harder to design, has poorer performance, and is less reliable than a star network, but it is far less expensive.

Centralized Hierarchical Topology. A centralized hierarchical network, as shown in Exhibit IV-2-2, is commonly implemented with concentrators or multiplexers forming the backbone of a terminal network. The multiplexer allows many devices to share a single long-distance link to a central host. The local access topology may be a star (point-to-point), a second layer of multiplexing, or a multidrop configuration, depending on hardware and software constraints as well as site requirements and designer choice.

Mesh or Distributed Hierarchical Topology. The distributed mesh network, illustrated in Exhibit IV-2-3, can be used well in installations with multiple hosts or computer sites because it usually supports continued service during periods of link and site failures. The cost of a distributed

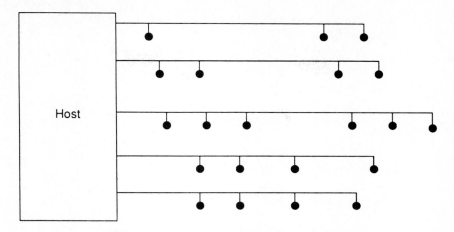

Exhibit IV-2-1.   A Centralized Tree Topology

mesh network is typically higher than that of other topologies, but its reliability is vastly superior because it requires hardware and software in each host or communications controller.

THE DESIGN PROCESS

The design process itself contains five major steps:
- Establishing network objectives.
- Gathering design information.

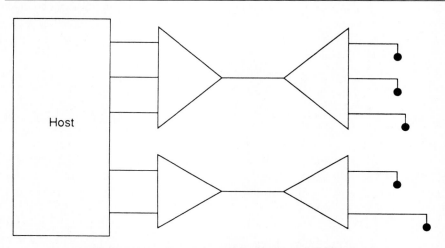

Exhibit IV-2-2.   A Centralized Hierarchical Topology

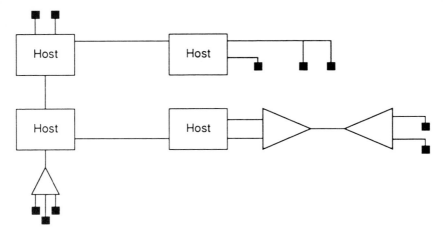

Exhibit IV-2-3.   A Mesh or Distributed Hierarchical Topology

---

- Generating and optimizing network configurations.
- Selecting the design.
- Obtaining approval for the design.

These issues are discussed in the following sections, as is the relationship between the design process and the implementation and operation of the network.

Establishing Network Objectives

The most important step in the design process is determining the overall network objectives. Users often do not have a clear picture of network objectives because they focus on finding a solution to their data communications problems, not on designing a network.

Several issues affect the identification of network objectives: the network users, the network applications, the system being replaced by the network, corporate philosophy, cost savings, life expectancy, and performance, cost, and reliability objectives. Understanding these issues allows the network designer to develop relevant network objectives.

The Network Users.   The designer must clearly understand who the network will serve and what purpose it will fulfill. For example, if the network is to provide online stock prices and transactions, reliability, not cost, is the prime consideration.

Another important consideration is whether the user department will be able to manage and troubleshoot the network or whether a third party will be needed. It may be appropriate to recommend a value-added network (VAN) to reduce the requirement for maintenance and administration by the user.

Network Applications.   Knowing the type of traffic to be carried on the network is critical to the network designer. The following questions should be researched:
- Are the applications CPU-intensive?
- What are the traffic patterns for each application?
- Is the application interactive, requiring a fast response?
- Is the application vital, with a need for redundancy?
- Can the response time for the application be improved by a faster network or would a faster CPU be a wiser investment?

The System Being Replaced by the Network.   The network being designed may be replacing a smaller network, a manual procedure, or nothing at all (i.e., it may introduce new functions). Knowing what the network replaces provides the designer with an understanding of what human constraints should be considered. For example, if the network is replacing an existing system, the new design should allow users to function in a similar way with the new system, if possible.

Corporate Philosophy.   Corporate objectives may have an impact on the design of the network. A conservative management team may choose to avoid leading-edge technology, whereas a high-profile company may want to be seen as a technological leader. It is important for a network designer to understand a company's corporate image.

Cost Savings.   Savings in the form of productivity improvement or reduction in costs for maintenance, monthly line charges, hardware, or other expenses should be identified. If there is no tangible or intangible saving, the designer should reevaluate the justification for implementing the network.

Life Expectancy.   Determining the life expectancy of the network should greatly influence the amount of time, effort, and money spent on the network design. For example, extra time spent in finding the best solution for a temporary problem may end up costing more in design time than the expected saving.

Performance, Cost, and Reliability Objectives.   Establishing user objectives for performance, reliability, and cost allows the designer to develop several suitable options. It is imperative that the designer and the user use the same terms of reference when discussing these objectives. A set of formal definitions allows the designer and the user to evaluate the performance, cost, and reliability of the existing network as well as that of the new network.

The first definition required is for actual response time. An appropriate definition of response time could be the time from when the user presses the last character of input until the user receives the first character of the response on the screen or the time between the user's pressing the send key and the receipt of the entire response and unlocking of the keyboard.

The second required definition is how to evaluate expected response time—as the average response time or as the 90th or 95th percentile of response time (e.g., 90% of all transactions must have a response time of less than five seconds).

Response time is a function of data network protocols and components as well as host processing time and can best be discussed using graphs. It is often useful to indicate on these graphs the minimum possible response time (i.e., when local devices attach directly to the CPU) to provide a realistic basis for comparing network response times.

The third definition needed is the method used to calculate the cost of the proposed network. To provide a just comparison between designs, the total cost should include all initial expenditures, including those for hardware, software, and installation, as well as any other one-time charges. The total should also include the cost to expand and maintain the network over a given time period (e.g., five years). These operating costs should include maintenance contracts, hardware, software, installation, and other one-time charges over the five-year period, with future costs expressed in present values.

A design with greater reliability, lower downtime, and a higher purchase cost may in fact be less expensive overall than a cheaper but less reliable design. The key to comparing designs is determining the cost of downtime: the higher the cost of downtime, the more attractive the higher-cost, higher-reliability design.

The final required definition is how to evaluate the availability of the network. This calculation must take the availability of all network components into consideration, including terminals, CPUs, lines, modems, multiplexers, and front-ends.

The availability of the network can be defined as:

$$\text{Network Availability} = \frac{\text{Number of available terminal hours}}{\text{Total number of terminal hours}}$$

This definition works well when a terminal is either available or unavailable. The definition is more difficult to establish if a terminal is available but performance is severely limited in a network that contains some redundancy.

Gathering Design Information

The accuracy of the final design depends on the accuracy of the input data, but obtaining accurate information can be difficult. In fact, gathering design information is the most difficult and time-consuming step in the design process because the necessary information often does not exist or exists in an unusable form.

Converting raw data into useful information requires that certain assumptions be made. For example, data may exist about past traffic patterns; the assumption must be made that this data does have a relationship to future traffic patterns. Raw data gathered for one hour of the day or one day of the week must be turned into assumptions about the rest of the week, month, and year.

The designer and the user must agree on what information should be gathered and, more important, on what assumptions they can make about the significance of this information. The design information required includes data regarding:

- Terminal and processor sites.
- Transaction volumes and types.
- Peak traffic.
- Growth forecasts.
- Network constraints.
- Vendors.

These are discussed in the following sections.

Terminal and Processor Sites.   Detailed information about existing and proposed network locations is required, including street address, telephone number (needed for line rating purposes), quantities and types of existing equipment and lines, current line capacity, current line volume, and anticipated growth rate.

Transaction Volumes and Types.   It is important to identify all interactive transactions and set up a message-size probability distribution, which can be used to generate such elementary statistics as average message size and standard deviation. These statistics are used in the design process. Although many networks have monitors that can generate transaction information automatically, the designer may need to gather this information manually or to use separate test equipment, because the use of network monitors can have a negative effect on network performance during peak hours—the time when statistics are required.

Peak Traffic.   When sizing a network, the designer needs information on the acceptable performance of the network during peak periods.

713

If the designer has no specific network information and must make estimates, the table in Exhibit IV-2-4 provides a reasonable approximation of traffic distribution.

Growth Forecasts. Predicting the future is an art rather than a science, and growth forecasts will not always be accurate. Sensitivity analyses should therefore be conducted to identify the effect of an incorrect forecast. For example, such an analysis might indicate that a 20% increase over the forecast would result in a requirement to reconfigure the network in two months rather than in 12 to 18 months. The designer may decide to initially configure the network in anticipation of that unplanned growth as insurance against such a change. As an alternative, the designer might monitor the network traffic growth very carefully, reacting only if unusual growth is occurring.

Several traditional sources of information for growth forecasts should be examined. User estimates, which are a prime source of data, are generally low, because users tend to believe that they will increase their network use very little over time.

Previous trends are another major source of data for growth forecasts, but they should be used cautiously—contributing factors can change dramatically. A drop in the price of technology can radically affect growth forecasts by altering the number of potential users wishing to access the network. Reorganization, changes in senior management, changes in data communications policies, and other events can reduce the usefulness of previous trends as factors that can be used to forecast growth.

User steering committees—a group of users meeting to discuss growth planning—are an underused source of valuable forecasting information. These committees can provide such information as a list of departments that have an increased data processing budget or that are developing new applications.

Length of Day (Hours)	Percent of Transactions in Peak Hour
4	34%
5	27%
6	23%
7	19%
8	17%
12	14%
24	12%

Exhibit IV-2-4. Average Network Traffic During Peak Periods

Network Constraints.   The task of designing a network is often further complicated by the fact that the design is an addition or modification to an existing network. Decisions made or equipment purchased previously must be accommodated. For example, one or more of the computers in the network may not be able to support multidrop polled lines.

Vendors.   It is essential to obtain all information about the operation of the proposed equipment or facilities and all cost quotations from the vendor in writing. The time limit during which the cost quotation is valid must be specified, and the quotation must include any one-time charges (e.g., installation or data base setup). The designer should keep in mind, when obtaining a quote for a tariffed item, that if the quote and tariff disagree, the tariff prevails. The designer should also understand the vendor's policy regarding trading in or upgrading the equipment and software.

The designer should thoroughly investigate the vendor's list of references. When contacting references, the designer should ask about the circumstances surrounding the installation to determine whether they are similar and therefore relevant.

Vendor service is an often-overlooked area. Designers should investigate the following items:

- The number of representatives qualified to service the piece of equipment under consideration.
- The number of local service representatives.
- The types of service contracts available.
- The speed of service guaranteed.
- The number of installations supported by the local representatives.
- A reference list of customers using their services.

Generating and Optimizing a Configuration

The designer must first choose a topology and then generate a network configuration that provides adequate performance. The cost and reliability of that configuration must be calculated and compared with the network objectives, and the configuration modified, if necessary, to achieve those objectives.

For example, for a multidrop configuration, the designer would first generate response-time curves for the system and then, using this information, generate a multidrop design layout to support a required level of performance. The design would also allow for a growth capacity of perhaps 18 months. The designer would then calculate the cost and reliability of that configuration.

Typically, the designer should produce at least three alternative designs, all of which should meet the design objectives for cost, performance, and reliability. Multiple designs provide some level of confidence that the designer did not simply accept the first feasible solution.

Selecting the Design

The best design has the right balance of performance, reliability, and cost for the organization. It is important to discuss the alternative designs with the user and to conform to the organization's design standards for performance, reliability, and cost. The best design also contains some safeguards for the future, ensuring that the design has the flexibility to adapt to future changes in network requirements.

Obtaining Approval for the Design

Designers should realize that those who make the final decisions about network design are not primarily technical people. Frequently, it is the best presentation rather than the best design that wins ultimate approval.

The most common mistake made in technical presentations is stressing features—fiber optics, digital transmission, data compression, and satellites. The emphasis instead should be on benefits—the time and money the new network will save, the reduced frustration levels, and the reduced error rates.

IMPLEMENTATION AND FOLLOW-UP

Implementation is not usually part of the design process, but the designer should be involved if possible because this results in better network designs. Involvement in the implementation process allows the designer to interact with the operations staff, which has a twofold benefit. It allows the designer to become aware of such real-world considerations as actual line error rates and equipment that does not meet its performance objectives. It also makes the designer more available to the operations staff for clarifications and even design changes.

After the network has been implemented, a crucial consideration is to test the performance of the network and compare it with the performance predicted by the designer's network model. Any difference immediately after installation means that either the traffic data was incorrect or the network model did not accurately reflect the network behavior. In either case, this information must be fed back into the design process. The designer should also maintain contact with operations personnel to track the reliability of the network components and the actual growth rate of traffic and devices on the network. Changes in growth rate must be noted as soon as possible, because an upgrade may be required earlier than expected.

ACTION PLAN

The effective design of a data network is a multistep process. Specifically, the designer should:

- Determine the network objectives in as much detail as time allows.
- Gather detailed design information on existing equipment types and locations, traffic patterns, and growth forecasts.
- Generate and optimize network configurations.
- Select the design on the basis of performance, cost, and reliability trade-offs.
- Obtain approval for the design through a brief, clear presentation that stresses benefits rather than features.
- Maintain contact with the operations staff during installation and beyond.
- Verify the performance of the network model used.
- Feed all information received back into the design process for future design work.

Chapter IV-2-2
System Acceptance Testing

INADEQUATE ACCEPTANCE TESTING is a major reason that communications systems often fail to meet business objectives. Because effective acceptance testing treats the new system as both a system and a business process, the test plan must address not only verification of all functions but also the adequacy of user procedures and training as well as verification of operational considerations.

Selecting the best individual to assume responsibility for acceptance testing may prove difficult, however, and could lead to inadequate testing or no testing at all. Because the developers of communications systems are concerned primarily with hardware and software, they may overlook non-technical considerations. If the user department is unable to use a system, that system will fail regardless of its technical sophistication. A system that cannot deliver timely results is likewise doomed.

Furthermore, communications systems engineers typically are under considerable pressure to complete a system. Estimates of the level of effort required to develop a system are generally too low. During the final stages of development, shortcuts are often taken and omissions made. Because the purpose of acceptance testing is to verify a system's completeness and readiness for implementation, development personnel obviously are not in the best position to conduct the acceptance test—and the user lacks the expertise required for the task. The communications manager, however, can often assume this responsibility.

The communications manager has experience with the systems technology required to plan and conduct an effective acceptance test. In addition, the manager can maintain the objectivity of a user: the communications department is, in fact, a system user and must assume primary responsibility for user support. The manager, therefore, has a vital interest in validating the system.

A low-quality communications system can destroy carefully created production schedules, degrade online response time, and prevent the communications department from attaining its service-level objectives. In addition, poor implementation can weaken morale when employees become frustrated working overtime to make the system work. When implementation problems are controlled, processing delays are minimized, which benefits the entire organization as well as the user department.

OBJECTIVES OF ACCEPTANCE TESTING

The primary objective of acceptance testing is to ensure that the communications system will enable the end user to attain business objectives. This does not necessarily mean that a system will be totally errorless. Although zero defects is desirable, it usually is not a cost-effective goal—that is, the cost of ensuring zero defects often exceeds the benefits. Minor deficiencies can be grouped and addressed in a subsequent release, or version, of the system. In fact, the communications department should consider planning a release specifically for such maintenance to follow implementation. The time frame should be a function of the complexity of the system—it could be four weeks, or three to six months.

Most communications departments have implemented a formal systems development methodology dictating several user-approval steps. These are primarily performed at the end of the system requirements definition and the system functional specifications design. A frequently overlooked task in these methodologies is user approval of acceptance test criteria. As noted, development personnel face an obvious conflict of interest in developing rigorous test criteria, and the end user frequently lacks the knowledge to do so. The manager, however, can bridge this gap.

One major element of the acceptance criteria is a classification scheme for action requests originating during the test and a criterion for accepting or rejecting the system based on the number of requests of each classification. Action requests may be product deficiencies caused by defective system specifications or by improper implementation of correct specifications. They can also be enhancements. A typical classification scheme is shown in Exhibit IV-2-5.

The manager has a vested interest in test criteria because these standards frequently help to define the service level a user expects. These expectations are the criteria by which the manager will ultimately be judged. The following paragraphs review acceptance-testing objectives.

Meeting User Requirements. Before development even begins, the manager should establish that the specifications fully meet user requirements. Creation and walkthrough of an acceptance test plan can frequently help clarify a user's needs and ensure that the initial specifications are valid.

Classification	Definition
1	Major system impact, cannot continue (e.g., system ABEND or wrong results)
2	Seriously degrades system, can continue testing with "work-arounds"
3	Minor system impact, operational or manual "work-arounds" available
4	Minor cosmetic defect (e.g., misspelling of literals on a screen)
5	Not in original specifications, can be deferred as enhancement for future version

Exhibit IV-2-5. Typical Test Action Request Classification Scheme

Defects removed in the design stage usually save the most money in the development effort.

Meeting Design Specifications. After the acceptance criteria have been established, the manager must determine whether the new system operates according to its design specifications, which in turn must meet the acceptance criteria. If design specifications do not meet those criteria, either the business needs have changed during development or the methodology used is deficient. If the methodology is at fault, the manager can revise the methodology to include acceptance testing.

Staff. A major objective of any acceptance test is to verify the ability of the technical staff to use the new system. Disruptions can result if the staff cannot run or support the new system, and erroneous output caused by faulty operations can create havoc among users. Included in these considerations is the ability to support microcomputer-based systems.

Real-Time Systems. In the case of real-time systems, the ability of the staff to run the system is critical because any error typically becomes immediately apparent to the end user. In addition, disruptions of the end user's business process can result in staggering costs. For example, an idle assembly line can cost thousands of dollars per minute, and an incorrect product run can cost hundreds of thousands of dollars in both lost revenues and wastage.

To achieve optimal performance in these systems, response time as well as proper classification of transaction classes must be verified. Equally important is verification that the new system will not overload the current computer configuration. If the system is just below the critical point on the response curve, even a minor increase in the work load can be too much. Response time considerations are equally important, if not more critical, with systems that run on department minicomputers.

Microcomputer Systems. Microcomputer-based systems are really a special subset of real-time systems. As such, all of the real-time considerations apply, but there are significant additional considerations. User training, both in the specific application and in other applications the user will run on the same microcomputer, is critical. System errors can easily be introduced by a user who is doing something wrong in another application or at the operating system level. Common errors are typically the result of deleting, renaming, or moving files or of editing files outside of the tested system.

A second major consideration for microcomputer-based systems is that by their nature they tend to be modified easily by users. Therefore, a validation technique is required to determine whether a system or one of its files has been modified. This helps significantly in the ongoing support effort. It can also serve as a basis for allowing the user to modify the system and then execute a regression test to validate the modification.

Preventing Scheduling Problems. The critical elements of the operating schedule must be compatible and represent realistic expectations. Detecting and eliminating scheduling problems before implementation can remove a major cause of friction between personnel in the various departments participating in system operation.

Verifying User Readiness. As mentioned, one of the most important objectives of the acceptance test is to verify the user's ability to use the new system. An improperly trained user staff can negate the benefits of a well-designed and implemented system. Problems arise when, for example, an improperly equipped or staffed user department must run the new system in extended parallel operation with the old system. Parallel operation often requires double work by the computer operations and user departments. Users often inaccurately estimate the effect of such operation on their departments. Planners often underestimate the additional effort required of the user department to learn to operate new terminals.

Backup, Recovery, and Degraded Operations. No system is perfect—eventually, any system will fail. Therefore, procedures for backup, recovery, and degraded performance should be tested. This should help users understand the operational implications of a system failure and minimize its impact on live operations. The procedures almost always involve manual work-around procedures implemented by the user community.

PLANNING THE ACCEPTANCE TEST

Planning the acceptance test is central to the eventual success of the system. A test plan is best constructed during the early phases of systems development, when sufficient time can be devoted to it. Its design can begin after the conceptual design has been completed. When the detail design phase has been finalized, the acceptance test plan should be near completion.

Before planning can begin, the individual responsible for creating the plan must be familiar with the proposed system. This planner should study all design documentation and interview the system designers and users.

Acceptance test planning comprises the following steps:

1. Identifying the tasks required to accomplish the acceptance test.
2. Estimating the employee hours required to complete each task.
3. Assigning an individual the responsibility for completing each task.
4. Scheduling start and completion dates for each task and distributing copies of the test schedule to those involved in the acceptance test.
5. Obtaining formal user approval of the acceptance criteria and test plan.

System planners can identify tasks by dividing the acceptance test into subsystems and determining the tasks for each subsystem. (Exhibit IV-2-6 provides a list of representative subsystem components.) The selection of subsystem components is accomplished by dividing the entire system into functionally independent sections. The breakdown used by the systems development staff can be helpful in this activity.

After the major components of the system have been determined, the tasks required to test each component can be selected. The following is a list of typical acceptance-testing tasks:

Subsystem	Component
Computer system	Master file conversion creation
	Edit/update
	Transaction processing
	Reporting
	Backup/restore
	Audit trail
	Computer resource requirements
Operations department	Operating procedures
	Operating schedule
	Balancing
	File retention
	Conversion/transition procedures
User department	Operating procedures
	Balancing
	Error handling
	Document retention
	Conversion/transition procedures
	Degraded system procedures

Exhibit IV-2-6.   Typical Acceptance Testing Subsystem Components

- Reviewing documentation.
- Reviewing procedures.
- Reviewing schedules.
- Reviewing training.
- Creating test data.
- Reviewing test data.
- Documenting test data.
- Determining test data results.
- Evaluating test run results.
- Presenting plan and criteria to users and developers.
- Interviewing user personnel.
- Analyzing user personnel readiness.
- Analyzing documentation.
- Analyzing procedures.
- Determining schedule feasibility.
- Estimating input volume.
- Estimating input preparation time.
- Estimating file size.
- Estimating run times.
- Obtaining formal approval of acceptance criteria/plan.

Each task should have a specific product that can be reviewed to measure the completeness of the task. The individual in charge of acceptance testing should regularly review task progress.

Estimating the time required to finish each task is one of the most difficult activities in the planning process. Testing tasks are particularly difficult to estimate because of the relatively infrequent use of the acceptance test in systems development cycles. However, many activities (e.g., preparing test data) are mechanical in nature and can be estimated reliably. The smaller the task, the easier it is to estimate. In general, the more detailed the task list, the better the overall estimate. Experience has demonstrated that estimates are consistently low. Interestingly, it is also true that as the individual task size diminishes, the total estimate rises. Based on both observations, it follows that as the task size approaches zero (in terms of employee-hour effort), the total project estimate approaches the true value.

To properly control the acceptance-testing project, responsibility for specific task completion should be assigned to individual employees. If tasks are assigned to departments rather than to individuals, the tasks are usually neglected. Individual employees often have more incentive to complete a task on time if their names appear on a planning document or if the completed plan and subsequent progress reports are sent to supervisors.

The level of effort required to schedule start and completion dates for each task relates to the size of the system being tested. A small acceptance test can be scheduled simply with the aid of a calendar and a vacation

Acceptance Test Plan							
Prepared by: _____			Date: _____				
Revised by: _____			Date: _____				
Activity	Assigned to	Estimated Hours	Actual Hours	Activity Dates			
				Start		Complete	
				Scheduled	Actual	Scheduled	Actual

Exhibit IV-2-7. Sample Acceptance Test Documentation Form

schedule. A large system may require the aid of a sophisticated PERT/CPM analysis. One frequent mistake is to base a plan on a 40-hour week. Such interruptions as routine tasks and illness reduce the true work week to about 30 to 32 hours. The project's priority also affects available staff time.

Acceptance test planning results should be documented and sent to all interested individuals. Exhibit IV-2-7 represents a sample document that could be used for developing an acceptance test plan as well as for monitoring and reporting the test's progress. Most microcomputer scheduling packages have report formats that show which personnel are responsible and work hours and dates; therefore, these packages can also be used for monitoring progress.

Like all plans, an acceptance test plan is subject to frequent change. For maximum value, it must be updated regularly. If an outdated plan is discarded rather than updated, the benefits of the acceptance test may be substantially reduced. Consequently, a major responsibility of the test leader is to keep the plan current and to announce schedule and activity changes.

PREPARING TEST DATA

Much of the effort required to conduct an acceptance test lies in the preparation of the test data, which is designed primarily to demonstrate the proper functioning of a computer system. The development of effective test data is a complex process. Several products available are designed specifically to generate test data. Accurate system test data should be used during the acceptance test and then preserved for use in testing future system modifications.

The communications system software program type, complexity, and unique requirements and constraints must all be considered in test data development. A system cannot be tested adequately with a single run. A series of runs must be designed to verify proper operation. The test data should also be designed to check the adequacy of system controls and system error and exception reports. The following sections present guidelines for preparing test data for interprogram communications and for several types of programs.

System Flow.   Interprogram communications are a typical source of defects. A range of transactions should be input to the initiating program to produce communication files and calls that test all cases appropriate for the receiving program's type.

Edit Programs.   A range of transactions should be prepared for each data element that can be processed by the edit function. The transactions should cover such situations as:
- Numeric fields containing all nines, all zeros, all blanks, and alphabetic data.
- Numeric values greater than valid upper limits and lower than valid lower limits.
- Numeric values equal to valid upper and lower limits.
- Numeric values with both positive and negative signs.
- Valid and erroneous combinations of values in related data fields.

Update Programs.   Various combinations of file maintenance transactions should be prepared to test the functioning of update programs in the system. Attempts to do the following should be tested:
- Create a master record with a zero-value key.
- Create a master record with a key containing all nines.
- Create a master record with a key containing all blanks.
- Create a master record with all possible data combinations.
- Create a master record with a key equal to a record that already exists on the file.
- Change data in a nonexistent record.
- Change data in the first and last record on the file.
- Add a record with a key lower than the lowest existing record and greater than the largest existing record.
- Create and change a record in the same run.
- Delete a nonexistent record.
- Create and delete a record in the same run.
- Change and delete a record in the same run.

Calculation Programs.   Although calculations may occur in different types of programs, each calculation function must be treated separately. For each system calculation, attempts should be made to:
- Cause a condition for multiplication and division by zero.
- Cause an arithmetic overflow to occur.
- Cause results with high, low, and average values to occur.
- Create out-of-balance conditions.

Report Programs.   Report programs are tested with data that creates various conditions. Report programs should be tested by attempts to:
- Create report values that are negative.
- Create report values that are all nines.
- Create report values that are all zeros.
- Create report values that exceed design specifications for width.
- Create an out-of-balance condition.

Sources of Test Data

A parallel operation can be used to augment or replace the preparation of acceptance test data. In general, although a parallel operation is a poor substitute for carefully designed test data, the use of a parallel operation to augment synthetic data can be a valuable addition to the acceptance test. The parallel operation permits evaluation of the system in a real-life contest (with allowances for extra efforts caused by parallel operation—such as double data entry).

Parallel operation should be carefully evaluated in the context of each situation before it is selected as part of the acceptance-testing procedure. Having insufficient user department personnel or equipment often makes it impossible to conduct a valid parallel operation.

During the preparation of acceptance test data, an acceptable result for each piece of data must be predicted. Although time-consuming, the task is critical to the successful completion of an acceptance test. In addition, a minimum level of performance should be established to determine whether the system has passed the acceptance test.

RUNNING THE ACCEPTANCE TEST

How the acceptance test is run can affect its outcome. The user and operations departments should run the system exactly as they would in a usual operational environment. This tests not only the system but also user and operator manuals and the operator run books.

Personnel associated with system operation should perform their usual functions. All input should be formatted exactly as in the user documentation or procedures. The development team should not be involved in running the test except under abnormal conditions (e.g., system failure or inadequate documentation).

The individual supervising the acceptance test must carefully monitor the test run. Because acceptance testing occurs late in the development cycle (when deadlines are most critical), personnel will often be urged to implement the system quickly. Despite pressure from others to curtail the acceptance test in favor of implementation, the acceptance test supervisor must refuse. It may be necessary at the eleventh hour to reemphasize the advantages of the acceptance test in order to complete the test successfully.

This is an area in which the manager should take personal interest. It is common for tempers to wear thin and for otherwise positive, ongoing relationships to suffer during a test. The manager should help maintain relationships during the test and ensure that they are fully reestablished following the test's completion.

EVALUATING TEST RESULTS

The communications system manager must plan to minimize the adverse impact of these pressures on the validity of the acceptance test. Realistic schedules, use of independent personnel (e.g., operations personnel rather than the original developers) to conduct the test, and integration of the review process into the actual testing help maintain the test's validity.

The value of developing acceptance criteria early on becomes evident when test results are evaluated. Without previous agreement on acceptance criteria, it would be difficult to convince management to abort an implementation for reasons other than catastrophic system failure during the acceptance test. The prevailing mood during implementation usually favors deemphasizing problems and proceeding with the system. To counteract false optimism, the manager must ensure that acceptance criteria are documented and used in evaluating acceptance test results.

ACTION PLAN

The following are important considerations for ensuring effective acceptance testing:

- The ultimate criterion for accepting a system into production is whether it accomplishes the end user's business objectives.
- An acceptance test must exercise more than just the programs and transactions—It must validate the documentation and business procedures that the end user will use.
- Acceptance testing should be conducted by the operations and end-user personnel who will actually use the production version and in the environment in which it will be used.

- A standard should be established so that the acceptance test plan will be part of the original external design specifications—This minimizes wasted programming effort and often crystallizes vague user requirements.
- Specific acceptance criteria must always be included in a test plan and must be approved before any testing is started.

Chapter IV-2-3
An Integrated Communications Cabling Network

David P. Levin

THE VOICE AND DATA COMMUNICATIONS REQUIREMENTS of an organization change over time, and new software and hardware are needed to accommodate these changes. This chapter presents a case study of the design, construction, and implementation of an integrated communications cabling network at a typical small company, Nutter, McClennen, and Fish, a Boston-based law firm (hereafter referred to as Nutter or the firm). This firm recently implemented in its new headquarters an integrated communications cabling network based on a hybrid Wangnet broadband network with extensive twisted-pair cabling. The firm's relocation involved four major steps:

- Relocating and upgrading an existing Northern Telecom SL-1 private branch exchange (PBX) for voice communications.
- Installing new Wang host processors in a newly designed data center and network control center.
- Installing an integrated communications cabling network for support of data communications, voice communications, facsimile transmissions, copier tracking device communications, text scanners, and private line voice and data communications.
- Installing separate cabling networks for devices not economically supported on the integrated network, including audiovisual communications, access security, fire detection, and paging.

THE CONCEPTUAL RELOCATION PLAN

Many users elect to install a new telephone PBX when relocating. Nutter had a substantial investment in its Northern Telecom SL-1 PBX, however, and decided instead to upgrade its existing system during the move. The telephone switch was located within one portion of the data center to take advantage of a single voice and data network control facility.

A small data center of approximately 1,500 sq ft was designed; it included a raised floor, controlled access, and preaction fire suppression sprinklers. Two 10-ton air conditioners provide cooling for all of the data center components, including the printer room and the supervisor's room.

The data center's electrical power is controlled through the use of a 50-kilovolt-ampere (kVA) uninterruptible power supply with 20 minutes of battery backup. Although the firm's management initially questioned the value of such a supply, doubts were removed when more than six power failures occurred during host testing.

A new Wang processor, the VS 300, was installed in the data center along with new disk and tape drives. This new processor had to be installed and fully tested in the new location before the move—a difficult task because of various construction-related delays.

The most difficult relocation task involved the design and implementation of an integrated communications cabling network. Working with a computer-communications engineering and management consulting company, Nutter began surveying its requirements. Nutter's extensive commitment to Wang for information processing indicated that a Wangnet broadband network would allow the most efficient integration. The firm also had a Digital Equipment Corp minicomputer that supported the majority of the information processing in the trusts, wills, and estates department. In addition, many MS-DOS–compatible microcomputers were appearing throughout the firm, which indicated that a local area network (LAN) might be necessary in the near future.

LOCAL AND REMOTE COMMUNICATIONS REQUIREMENTS

The firm's future communications requirements—focused primarily on local and remote data communications—are summarized in Exhibit IV-2-8. Local data communications requirements at the time of the move primarily included Wang-based terminal-to-host and printer-to-host transmissions. The cabling network was required to support a single channel of Wang microcomputer networking at speeds of 1M bps to 10M bps within two years and multiple channel networking within 10 years—well within the firm's 15-year lease. The same level of local area networking for IBM and other MS-DOS–compatible microcomputers was also required.

The Digital Equipment Corp (DEC) minicomputer remained temporarily in separate office space; the firm was planning to relocate its department to an adjacent floor within five years. Within two years of the initial move, Nutter required support for DEC-compatible asynchronous ASCII terminals

	Required at Time of Move	Supported Within 2 Years	Supported Within 10 Years
Local Data Communications Connection Requirements			
Wang terminal to local Wang host	X	X	X
Wang printer to local Wang host	X	X	X
Wang microcomputer to local Wang host	X	X	X
Wang microcomputer to microcomputer (single channel)		X	X
Wang microcomputer to microcomputer (multichannel)			X
IBM microcomputer to microcomputer (single channel)		X	X
IBM microcomputer to microcomputer (multichannel)			X
Asynchronous terminal to local Digital Equipment Corp host			X
Serial printer to local Digital Equipment Corp host			X
Asynchronous terminal to local multiplexer		X	X
Serial printer to local multiplexer		X	X
TWX and Telex terminals to local transmission line	X	X	X
Synchronous scanner to local host	X	X	X
Remote Data Communications Connection Requirements			
Local Wang host to remote Wang host	X	X	X
Local Digital Equipment Corp host to remote time-sharing host		X	X
Local microcomputer to remote host at 1K to 2K bps	X	X	X
Local asynchronous terminal to remote host	X	X	X
Local multiplexer to remote Digital Equipment Corp host		X	X
Local Video Communications Support Requirements			
Video training			X
Video conferencing			X
Security surveillance			X
Other Communications Support Requirements			
Facsimile transceiver access to local transmission line	X	X	X
Copier use tracking	X	X	X
Paging	X	X	X

Exhibit IV-2-8.   Projected Local and Remote Data Communications Requirements

---

and serial printers connecting to a local multiplexer. Communications support between these devices and a local DEC minicomputer was required within 10 years.

Additional local communications requirements at the time of the move included support for text scanners, TWX and Telex terminals, and facsimile transceivers connecting to local hosts and leased transmission data lines. Nutter uses a copier tracking system from Copitrak to control photocopying costs and reimburse clients. The 17 photocopy machines throughout the four floors each have a data collection and access device, which requires that client and user codes be entered. The collection devices are wired in a tree topology and polled by a microcomputer through the serial communications port. A single digital line driver is able to drive a network with several thousand feet of twisted-pair cable.

The firm uses a paging service to help locate key personnel—a capability required at the time of the move. Local video communications support—including video training, video conferencing, and video security surveillance—were not required at the time of the move or within two years, but support was required within 10 years.

Remote data communications requirements included support for local Wang hosts communicating with remote Wang hosts in other offices at the time of the move and local microcomputers and asynchronous terminals communicating with a remote host at 1,200 to 2,400 bps. Support for local Digital hosts communicating with remote time-sharing hosts and for local multiplexers communicating with remote Digital hosts was required within two years but not at the time of the move.

STATION CABLING ALTERNATIVES

The cabling support alternatives for Nutter's various local communications requirements are summarized in Exhibit IV-2-9. Twisted-pair cable was required for voice transmissions between the Northern Telecom SL-1 PBX and its station instruments. The problem remaining was how to design an effective hybrid cabling scheme for all the firm's requirements. The twisted-pair cabling network could be supplemented by a baseband coaxial cabling network, a broadband coaxial cabling network, or a combination of the two, or the twisted-pair network could be used unsupplemented.

Twisted-Pair Station Cabling

Twisted-pair station cabling supports zero-slot LANs (i.e., LANs that require no interface card within the microcomputers and that operate through the serial or parallel port). When used with zero-slot or PBX-based LANs, twisted-pair station cabling typically operates at maximum backbone speeds from 20K bps to 300K bps, making data base information sharing difficult from a performance viewpoint.

Twisted-pair station cabling supports media-sharing LANs. These networks support backbone speeds ranging from 400K bps to 10M bps and effectively maintain the sharing of information stored in a microcomputer file server. When evaluating Nutter's communications requirements and how twisted-pair station cabling could meet these needs, the best of the twisted-pair local area networking capabilities were used during the comparison.

In Exhibit IV-2-9, the twisted-pair station cabling capabilities for each local communications requirement are rated as providing full function, limited function, or no support. Twisted-pair station cabling provides full-function support for the firm's local voice communications requirements and a few data communications requirements, including connection of asynchronous terminals and serial printers to a local DEC host to a local

	Cabling Combinations			
	A	**B**	**C**	**D**
Cable Components				
Twisted-pair cabling	Yes	Yes	Yes	Yes
Baseband coaxial cabling	No	No	Yes	Yes
Broadband coaxial cabling	No	Yes	No	Yes
Local Data Communications Connection Requirements				
Wang terminal to local Wang host	Y-L	Y-F	Y-L	Y-F
Wang printer to local Wang host	Y-L	Y-F	Y-L	Y-F
Wang microcomputer to local Wang host	Y-L	Y-F	Y-L	Y-F
Wang microcomputer to microcomputer (single channel)	Y-L	Y-F	Y-L	Y-F
Wang microcomputer to microcomputer (multichannel)	N	Y-F	N	Y-F
IBM microcomputer to microcomputer (single channel)	Y-F	Y-F	Y-F	Y-F
IBM microcomputer to microcomputer (multichannel)	N	Y-F	N	Y-F
Asynchronous terminal to local Digital Equipment Corp host	Y-F	Y-F	Y-F	Y-F
Serial printer to local Digital Equipment Corp host	Y-F	Y-F	Y-F	Y-F
Asynchronous terminal to local multiplexer	Y-F	Y-F	Y-F	Y-F
Serial printer to local multiplexer	Y-F	Y-F	Y-F	Y-F
TWX and Telex terminals to local transmission line	Y-F	Y-F	Y-F	Y-F
Synchronous scanner to local host	Y-F	Y-F	Y-F	Y-F
Local Voice Communications Connection Requirements				
PBX station to PBX switch	Y-F	Y-F	Y-F	Y-F
Auxiliary station to telephone company demarcation	Y-F	Y-F	Y-F	Y-F
Local Video Communications Support Requirements				
Video training	N	Y-F	N	Y-F
Video conferencing	Y-L	Y-F	Y-L	Y-F
Security surveillance	N	Y-F	N	Y-F
Other Communications Support Requirements				
Facsimile transceiver access to local transmission line	Y-F	Y-F	Y-F	Y-F
Copier use tracking	Y-F	Y-F	Y-F	Y-F
Paging	Y-F	Y-F	Y-F	Y-F

Notes:
A   Twisted-pair cabling only
B   Broadband coaxial cabling with twisted-pair cabling
C   Baseband coaxial cabling with twisted-pair cabling
D   Broadband and baseband coaxial cabling with twisted-pair cabling
Y-F Yes, with full-function support
Y-L Yes, with limited-function support
N   No support

Exhibit IV-2-9. Cabling Support for Local Communications Requirements

---

multiplexer, TWX and Telex terminals, and single-channel microcomputer LANs.

Twisted-pair station cabling provides limited support for Wang terminals, printers, and microcomputers communicating with a local Wang host. This twisted-pair cabling support is provided using baluns and home-run cabling that connects each station cable to the computer room through feeders.

Twisted-pair cabling requires each device to be specifically connected to a Wang host. With multiple Wang hosts, terminal connectivity to the other hosts requires processor resources rather than network resources to accomplish this type of transmission.

Twisted-pair station cabling supports single-channel LANs of Wang microcomputers using Banyan's Vines, a connection device comarketed by Wang. Twisted-pair station cabling does not support multichannel microcomputer LANs because the underlying baseband technology permits only one data transmission signal on the cable at a time. Baseband networks use time-division technology to service multiple users, whereas a multichannel operation also requires frequency-division multiplexing.

Twisted-pair station cabling supports full-motion video conferencing on a limited basis if an expensive coder-decoder (e.g., in the $25,000 to $50,000 range) is used to convert the video and audio signals into a digital data stream between 56K bps and 1.544M bps. Twisted-pair station cabling does not support video training or security surveillance communications, the cost of which typically cannot justify coder-decoders.

Twisted-pair station cabling offers full-function support for facsimile transceivers accessing a local transmission line. All transmission lines are terminated in the computer room's main demarcation panel and cross-connected through the station distribution cabling to the appropriate outlet. Twisted-pair station cabling supports the Copitrak copier tracking devices using two pairs of cable. Copier tracking devices are connected to each other in a satellite closet and connected to other floors using two pairs of feeder cabling.

Although paging is supported by twisted-pair cabling, the attenuation of audio signals is considerably greater than that of digital computer signals. Cable with greater conductor sizes, typically 18 AWG (American Wire Gauge), is used instead of telephone cable, which is typically 22, 24, or 26 AWG. The serial cabling of paging speakers is difficult to adapt to the star topology of twisted-pair communications station cabling without introducing unacceptable signal loss. Paging cannot be economically included in the design of an integrated communications station cabling network because it requires separate cabling.

The specific design of the twisted-pair cabling network—including the wire size, the number of conductors, the number and type of connectors, the type of cross-connect panels, and the type of shielding—depended on whether other coaxial or fiber-optic cable would be used. Fiber-optic cable was considered unacceptable. Because Nutter did not have a specific application for fiber-optic station cabling at the time of the move, it was highly probable that when the firm attempted to use the fiber, the electronics used to drive the fiber would be incompatible. In addition, the cost of the electronics needed to take advantage of the tremendous fiber bandwidth is not justified at the workstation level, based on Nutter's current and future requirements.

Supplementing with Baseband Coaxial Cabling

An unsupplemented twisted-pair cabling network could not meet all of Nutter's two-year requirements. The alternatives for supplementing twisted-pair cable were baseband coaxial cable, broadband coaxial cable, or a combination of the two. Exhibit IV-2-9 shows that the station cabling alternative that includes twisted-pair, baseband, and broadband coaxial cabling provides full-function support for all of Nutter's local communications requirements. The cost of using all three cabling types is excessive, yet this approach is still taken by many large organizations that cannot decide which of the other two alternatives to use.

A system that supplements twisted-pair cabling with baseband coaxial cabling does not support multichannel microcomputer LANs because of the limitations inherent in baseband technology. The twisted-pair and baseband coaxial cabling hybrid does not support the firm's local video communications requirements, and when used with Wang terminals and printers communicating with a local Wang host, the firm's baseband LANs do not support the 4M-bps transmission speeds and the proprietary protocol required by local Wang workstations.

The hybrid offered only limited functional support of the firm's primary local data communications requirements. At the design stage, Wang did not comarket Banyan products, which include a baseband coaxial LAN with a gateway into a Wang VS minicomputer. This solution is still not feasible for connecting Nutter's installed base of Wang workstations and printers. The hybrid seemed to offer few functional advantages over unsupplemented twisted-pair cabling in the context of Nutter's requirements, and it required additional expense. Therefore, it was not an economical choice.

Supplementing with Broadband Coaxial Cabling

Wang's product for local workstations communicating with one or more local Wang hosts is Wangnet, a dual broadband coaxial cable–based network engineered to use cable television modulation technology. Exhibit IV-2-10 depicts how Wang workstations and printers are connected to Wang host processors using the Wangnet broadband LAN. Up to eight Wang workstations are connected to network multiplexers using dual baseband coaxial cable connections. The network multiplexer is a radio frequency terminal server, similar to units made by Digital Equipment Corp, and is connected to Wangnet taps using dual broadband coaxial drop cables. Although Wang offers radio frequency workstations that do not

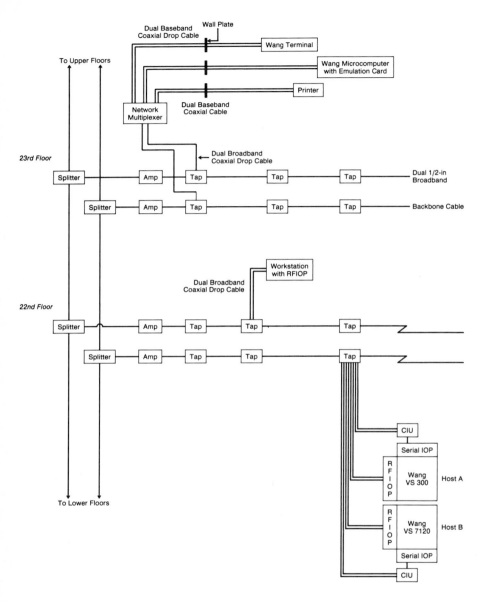

Notes:
CIU   Channel interface unit
IOP   Input/output processor
RFIOP   Radio frequency input/output processor

Exhibit IV-2-10.   Connecting Wang Workstations to Hosts Using Wangnet

require connection to a network multiplexer, these devices are costly and have been available only for the last four years.

A Wang host processor is connected to Wangnet using a radio frequency input/output processor that allows any terminal attached to a network multiplexer to communicate with any Wang host. Wang host processors may also be connected to the Wangband channel (216 to 243 MHz) of Wangnet to facilitate communications among the processors. Terminal-to-host communication occurs at 12M bps on the peripheral band (100 to 150 MHz). Wangnet uses one backbone cable for transmission signals and the second backbone cable for receive signals.

Wangnet has a total potential bandwidth of 440 MHz and does not require the allocation of a guard band, as does a single backbone broadband cable television LAN. Wangnet provides the ability to locate host processors and workstations throughout a building or campus of buildings. Remote Wangnet allows remote workstations to communicate with multiple local hosts using a single leased line and Wang's proprietary X.25 protocol.

Combining the capabilities of Wangnet broadband networking and twisted-pair cabling provided the most cost-effective solution to Nutter's station cabling requirements. The most significant drawback of Wangnet concerns the network's use of baseband services instead of broadband services.

Station cabling is typically installed in a star configuration that terminates in a few distribution closets on each floor. If network multiplexers are located within a limited number of communications closets, baseband coaxial station cables can be used to connect each user outlet to the network multiplexer in the distribution closet. Manual coaxial patch fields are used to terminate the dual baseband coaxial station cables. Coaxial jumper cables connect specific user outlets to a network multiplexer.

With this cabling scheme, only baseband Wangnet services are extended to the user outlet. A second coaxial cable must be connected from a backbone tap to each user outlet for Wangnet broadband services.

Station Cabling User Outlet

The details of Nutter's detailed station cabling user outlet are depicted in Exhibit IV-2-11. Each user outlet consists of a stainless steel wall plate with four connectors. Two nonkeyed RJ-11 telephone jacks are used to terminate the twisted-pair station cabling. One six-pair shielded cable was selected, with three pairs of cable terminating on each RJ-11 jack. Siemons M66 punch-down blocks with 89B mounting brackets are used for all twisted-pair station cabling and feeder cabling terminations.

TNC and BNC coaxial connectors are employed at each user outlet, as opposed to left-hand and right-hand threaded F-59 screw connectors used

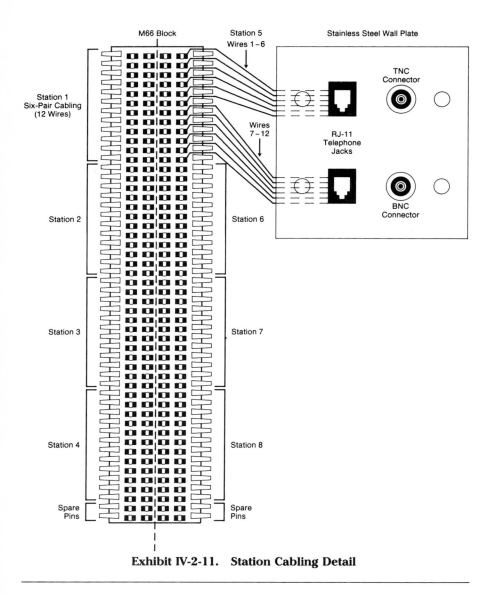

Exhibit IV-2-11.   Station Cabling Detail

for Wangnet's radio frequency–based broadband connections. Because the majority of Nutter's terminals and printers require connection to network multiplexers, baseband coaxial cable was run between the distribution closets and each user outlet. BNC and TNC connectors were chosen for the existing terminal equipment to allow the use of existing device-to–wall plate cables.

THE COMMUNICATIONS DISTRIBUTION CLOSET

The use of three communications distribution closets—one main closet and two satellites—on each floor was central to the success of Nutter's integrated communications cabling network. Exhibit IV-2-12 illustrates a typical satellite communications distribution closet layout. Each closet occupies approximately 30 sq ft of space and requires air-conditioning 24 hours a day, 7 days a week. Wall-mounted components include the Wangnet component board, the coaxial patch panel, M66 termination blocks, and electrical plug molds to power auxiliary transformers for telephone instrument speakerphones and displays.

Each communications distribution closet is designed to contain two electrical equipment cabinets with movable casters. Each cabinet has electrical plug molds to power components (e.g., network multiplexers and microcomputer LAN servers). Exhibit IV-2-13 details the wall-mounted components in a typical communications distribution closet at Nutter.

The connection between Wangnet broadband taps and multiple user outlets that function as signal splitters is detailed in Exhibit IV-2-14. Network multiplexers located in the equipment cabinet are connected to the multi-

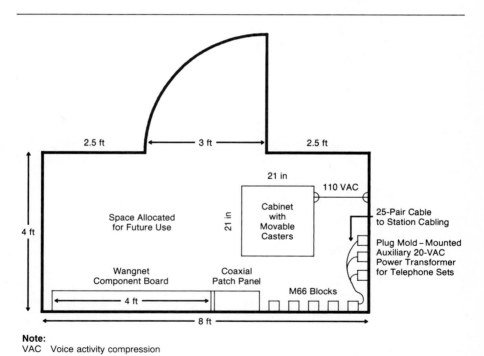

Note:
VAC   Voice activity compression

Exhibit IV-2-12.   Satellite Communications Closet Layout

ple user outlets with broadband coaxial drop cable. Baseband coaxial jumper cables connect each network multiplexer port to specific user outlet appearances on the coaxial patch panel. Transformers are plugged into wall-mounted plug mold strips, with the 20-volt output wired to the M66 termination blocks using a short piece of 25-pair feeder cable. Cross-connection jumper cable attaches the transformer termination block to the appropriate pair of station cables.

To take full advantage of the broadband network capabilities, radio frequency services as well as baseband services must be brought to each user outlet. Nutter could not justify the additional expense, however, of bringing

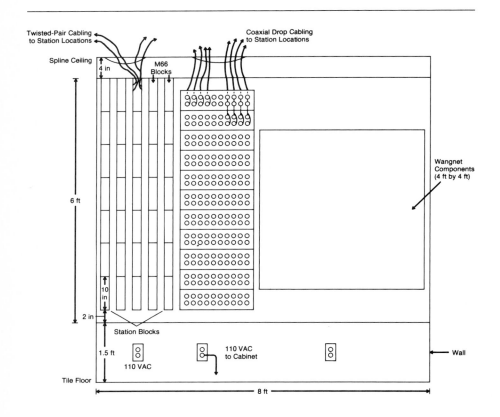

Note:
VAC   Voice activity compression

Exhibit IV-2-13.   Wall-Mounted Components in a Satellite Communications Closet: Front View

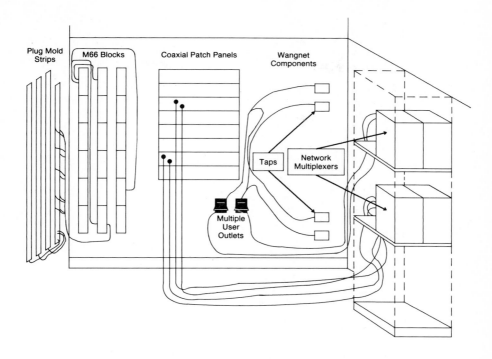

**Exhibit IV-2-14.   Satellite Communications Closet Network Multiplexer
Interconnection**

a second set of broadband coaxial drop cables to each of the approximately 600 user outlets. The solution, therefore, was to expand the use of the manual patch panel to allow the connection of coaxial cables to additional Wangnet broadband taps located within each distribution closet. One coaxial drop cable can now deliver either baseband or broadband services to the user outlet, depending on its coaxial patch panel connection. Wang was reluctant to accept these modifications to the standard Wangnet design; because non-Wang broadband devices are not part of Wangnet maintenance, however, the design was eventually accepted.

TWISTED-PAIR FEEDER CABLING PLAN

The interfloor and intrafloor twisted-pair feeder cabling is detailed in Exhibit IV-2-15. Each communications distribution closet in the circle portion of the floor serves as the main distribution closet, because these were the base building telephone closets. The exception to this rule was the 17th floor, where the computer room serves as the main distribution closet for that floor and as the main distribution frame for Nutter's entire twisted-pair cabling network.

Each satellite communications closet is connected to the floor's main closet by 100 pairs of twisted-pair cable, consisting of four separately sheathed and shielded 25-pair cables. In addition, each satellite closet connects to the floor's main closet with 24 pairs of feeder cable consisting of four separately sheathed and shielded 6-pair cables.

The 100-pair feeder is used for voice communications, and the 24-pair feeder is used for data communications applications. Voice and data signals do not share the same feeder cables. Each main distribution closet connects to the main distribution frame in the computer room with eight pairs of 25-pair voice feeders (200 pairs total) and four pairs of 6-pair data feeders (24 pairs total).

Exhibit IV-2-16 shows the twisted-pair feeder cabling cross-connection panel layout for the 15th floor main communications distribution closet.

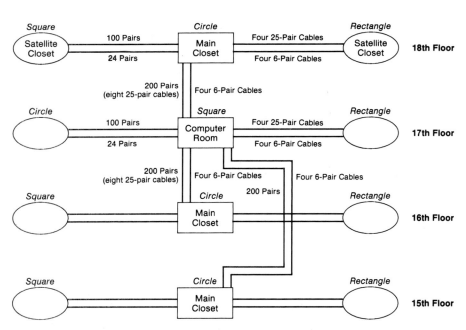

Exhibit IV-2-15.   Interfloor and Intrafloor Twisted-Pair Feeder Cabling

The left column of blocks and the left side of the second column of blocks are used to terminate 72 6-pair (12-conductor) station cables from user outlets in the immediate vicinity of the closet. The right side of these blocks is used for two sets of 24 pairs (four 6-pair cables) of data feeder cables from the two satellite closets and one set of 24 pairs to the main distribution frame. Spare block capacity has been allocated for terminating 75 pairs of data feeder cables.

The two columns of termination blocks on the right side of the exhibit are

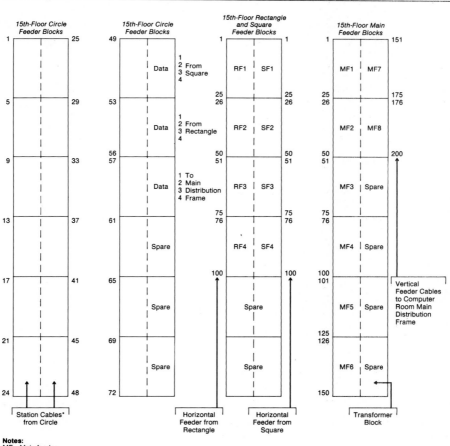

Notes:
MF   Main feeder
RF   Rectangle feeder
SF   Square feeder
*There are 72 station cables (24 six-pair cables) for each station.

Exhibit IV-2-16.   Twisted-Pair Feeder Cabling Cross-Connection Panel Layout

used for the feeders to the 15th floor's rectangle distribution closet and square distribution closet and to the main distribution closet and the auxiliary power transformers for the telephone stations in the immediate vicinity of the closet. The 100 pairs of horizontal feeder cable from the rectangle and square distribution closets consume a total of four M66 blocks. Spare capacity has been provided to terminate an additional 100 pairs of horizontal feeder cable.

The entire right column of termination blocks is devoted to terminating the vertical feeder cables to the main distribution frame in the computer room. The six M66 blocks terminate the existing 200 pairs with space for an additional 50 pairs. Space has also been provided within these blocks for a maximum of 50 auxiliary power transformers (most closets have between 15 and 28 transformers in operation).

ADDRESSING CABLE NETWORK MANAGEMENT REQUIREMENTS

The optimal time to begin the management and documentation of a cabling network is at the inception of the detailed design stage. Initial network design specifications produce the following documents, which form the basis of the as-built network documentation:

- Blueprints of all floors, indicating numbered outlet locations, distribution closet locations, and backbone cable runs with taps.
- A written description of the cabling network plan, including a general description, an explanation of the twisted-pair network, an explanation of the coaxial cable-based network, cable types used, and explanations of each type of station termination used.
- A conceptual representation of the cabling network (sometimes called a grand display).
- An intrafloor and interfloor feeder cabling diagram.
- Communications distribution closet layouts, including closet component interconnection charts and a detail of wall-mounted components.
- Broadband network signal level calculations, including a listing of design criteria, outlet signal levels, an amplifier setup summary, and a display of each network branch.
- Cable network outlet interconnection documentation explaining how each pair of station cabling wires is interconnected in its local communications distribution closet.
- A separate set of blueprints for each specially dedicated cabling network, including the paging network, the video surveillance card-access security network, and the fire detection and suppression network.

Exhibit IV-2-17 depicts Nutter's cable network interconnection documentation, listing the manner in which each pair of twisted-pair station cable and each dual coaxial station cable is connected in the local communications distribution closet. Because each closet is fully documented with all station and feeder cabling terminations, a technician can easily follow a connection from the user outlet and through the satellite distribution closet, the floor's main distribution closet, and the main distribution frame in the computer room to its final interconnection with a data and voice device or transmission line.

When shopping for a cable network management product, the buyer can easily be misled about the need for a customized software product, with every conceivable bell and whistle, running on a mainframe or minicomputer. Many of these software products sell for $20,000 to $250,000 and require a host that costs $100,000 to $1 million. Users find that a considerable amount of work is required to maintain the currency of this type of comprehensive cable network management facility. If the work required to maintain the data base becomes burdensome, the network management tool loses its effectiveness.

The management approach used by Nutter can be recommended for others who want a comprehensive yet cost-effective tool. The primary tasks in cable network management use word processing, basic data base manipulation, and image processing with output on 36-by-48-in paper (i.e., E-size blueprint paper). The ideal hardware platform for a cable network management facility is a microcomputer with an industry standard microcomputer operating system.

For word processing and basic data base manipulation, Nutter uses Lotus Corp's Symphony, which is used to document all the cable network interconnections, including twisted-pair feeder cabling, twisted-pair station cabling, and coaxial station cabling.

Image processing is done with Autodesk's AutoCAD software. It costs approximately $2,500 and requires a microcomputer with an Intel 80286 microprocessor, an 80287 coprocessor, a digitizing pad with a mouse, and a high-resolution video display (e.g., one using IBM's video graphics adapter). The approximate cost of the appropriately outfitted microcomputer is $5,000. Blueprints are produced by a Houston Instruments DMP-62 plotter, which costs approximately $5,000. AutoCAD may be used to produce digitized floor plans and various other diagrams required of a cable network management facility.

THE PHYSICAL CONSTRUCTION OF THE DATA CENTER

Nutter's information processing resources, including all computer and communications components, are located in the firm's data center. Exhibit

ID	FLR	AREA	OFFICE	PERSON/ROOM	STATION CABLE	PAIR	HOZ PR	VER PR	EXT	PHONE TYPE	SWITCH PORT	TERMINAL TYPE	S/N	WP POS	WANG PATCH POS	NETMUX POS	ID	PORT	CPU	LOGICAL ADDRESS
15001	15	Clr.	102	Smith, Norma	5C001J	Brown		5MF001	2386	2112	04070	0W6+								
15002	15	Clr.	101	Brown, Suzie	5C002J	Blue		5MF002	2489	2112	17077	0W6+								
15003	15	Clr.	100	Petersen, Kathryn	5C003J	Blue		5MF003	2415	2112	04060	0W6+								
15004	15	Clr.	99	Hanlon, Maria	5C004J	Blue		5MF004	2399	2112	05065									
15005	15	Clr.	----	Martin, Jack	5C005J	Blue		5MF005	2614	2112	18076									
15006	15	Clr.	----	OPERATIONS AREA	5C006J	Blue / Orange			Balun											
15007	15	Clr.	----	Martin, Jack	5C007J	Blue		5MF007	PVT-											
15008	15	Clr.	95B	Ryan, Lisa	5C008J	Blue		5MF008	2446	2018	00066	0W2+								
15009	15	Clr.	----		5C009J	Blue / Orange		5MF009 / 5MF010		COPY TRACK / COPY TRACK										
15010	15	Clr.	----	SUPPLIES	5C010J	Blue		5MF011	2434	2009	02066	0W2+								
15011	15	Clr.	95A	Farmer, Doreen	5C011J	Blue		5MF012	2688	2018	03052									
15012	15	Clr.	----	Frey, Al	5C012J	Blue		5MF013	2624	2009	18063									
15013	15	Clr.	----	SUPPLIES	5C013W	Blue		5MF014	2465	U1	02096									
15014	15	Clr.	----	SICK ROOM	5C014J	Blue		5MF015	2404	U1	17094									
15015	15	Clr.	----	MESSAGE CENTER	5C015J	Blue		5MF016	2576	SL-1	05045									
15016	15	Clr.	----	Lori, Anne	5C016J	Orange		5MF017 / 5MF019	2229	U3	03090									
15017	15	Clr.	----	Dunn, Anna	5C017J	Blue		5MF020	2286	U3	02090									
15018	15	Clr.	----	Stewart, Darnley	5C018J	Blue		5MF021	2280	2009	16067									
15019	15	Clr.	----	AUX ROOM	5C019J	Blue		5MF022	2327	U1	16084									
15020	15	Clr.	54	McGowan, Janet	5C020J	Blue		5MF023	2397	2112	01065									
15021	15	Clr.	53	Cetrone, Michael	5C021J	Blue		5MF024	2243	2112	01075									
15022	15	Clr.	----	COURTESY PHONE	5C022W	Blue		5MF025	2668	U1										
15023	15	Clr.	----	CONFERENCE ROOM	5C023J	Blue		5MF026	2575	2112	18067									
15024	15	Clr.	51	Poppel, Matthew	5C024J	Blue		5MF027	2262	2112	18086									
15025	15	Clr.	48	Josephson, Anne	5C025J	Blue		5MF028	2258	2112	19033									
15026	15	Clr.	----	CONFERENCE ROOM	5C026J	Blue		5MF029	2649	U1										
15027	15	Clr.	----	CONFERENCE ROOM	5C027J	Blue		5MF030	2455	2112	03066									
15028	15	Clr.	----	Hurwitz and Pucker	5C028J	Blue		5MF031	2629	2112										
15029	15	Clr.	45	Blute, Joseph	5C029J	Blue		5MF032	2591	2112	19026									
15030	15	Clr.	----	CONFERENCE ROOM	5C030J	Blue		5MF033	2650	U1										

Notes:
HOZ PR Horizontal pair
NETMUX Network multiplexer
VER PR Vertical pair

Exhibit IV-2-17. Cable Interconnection Documentation

748

IV-2-18 shows a layout for Nutter's data center and word processing area, which was built entirely on a raised floor to facilitate data center expansion. The data center's design includes a separate printer room, which helps prevent printer-generated debris from contaminating disk drives. The supervisor's office is somewhat small, but it provides a convenient work space otherwise not available in the immediate area. The true value of the supervisor's office became clear during the move, when it served as an emergency technician's workbench.

Air-conditioning for Nutter's data center is provided by two 10-ton Liebert air-conditioning units; this is enough capacity to allow either unit to individually cool the entire center.

Fire detection is provided by smoke and temperature sensors in the ceiling and by panic buttons at each door. Fire suppression is handled by preaction fire cycle sprinklers, which sound a 60-second audible warning before the sprinklers are loaded with water. Another audible warning is sounded just before the water is released into the data center. Nutter elected not to use a dry fire suppressant (e.g., Halon) because of its expense: Boston building codes require sprinklers to be installed as well as a dry suppressant because it is possible for a fire to reignite after the dry suppressant has been expelled.

Exhibit IV-2-19 shows the wall-mounted communications components (on painted, fire-retardant, plywood backboards) inside Nutter's data center, which include the cross-connection panels for the floor's main distribution closet, local station cabling terminations, and main demarcation panel feeder terminations from each of the other floors. Additional components include a battery control panel for the telephone PBX, New England Telephone Co trunk termination blocks, Northern Telecom's PBX interconnection blocks, Wangnet power converters, paging equipment interconnections, and temperature alarms.

CENTRALIZING DATA COMMUNICATIONS COMPONENTS

The data communications components in the data center that cannot be wall mounted are installed in three electronic equipment cabinets (see Exhibit IV-2-20). The cabinet on the left houses the Wangnet broadband head-end components, which consist of amplifiers, taps, a test-point access panel, and a power converter. The center cabinet houses a multiplexer for asynchronous terminals connected to the remote Digital Equipment Corp minicomputer.

A Gandalf modem rack houses various short- and long-haul modems operating at 1,200, 2,400, 4,800, and 9,600 bps. The cabinet on the right houses several Gandalf and Codex standalone modems and network multi-

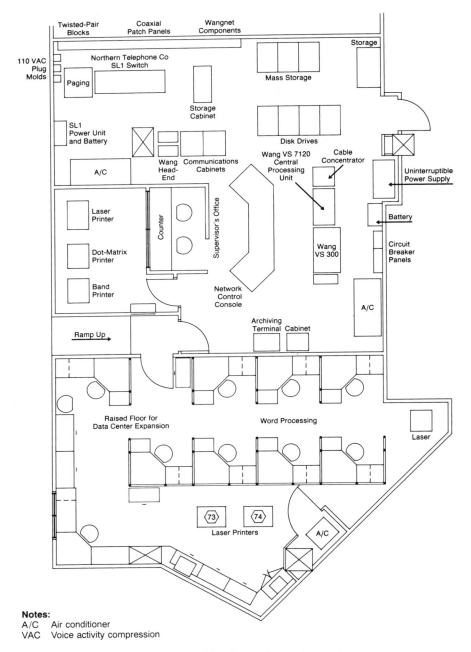

Exhibit IV-2-18.   Data Center Layout

Notes:
A/C   Air conditioner
VAC   Voice activity compression

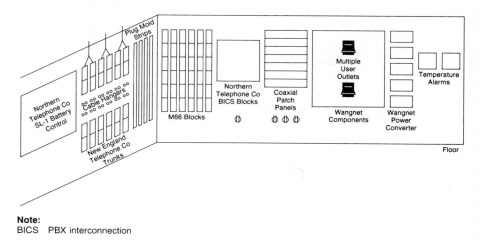

Note:
BICS   PBX interconnection

Exhibit IV-2-19.   Wall-Mounted Data Center Components

plexers for local Wang workstations. Wang 6554 communications controllers are located at the bottom of the data communications cabinets.

BUILDING THE NETWORK CONTROL CONSOLE

The focal point for managing Nutter's information processing resources is the network control console (see Exhibit IV-2-21). The console is centrally located in the data center to allow visual surveillance of the entire space, including the printer room and supervisor's office.

A turret in the center of the console houses a digital interface patch panel and digital interface monitor (also called a data scope). All switched and leased data lines are wired through the digital patch panel to facilitate network troubleshooting.

The console supports one Wang terminal and one Wang microcomputer that act as data control consoles to the Wang VS 300 and VS 7120 hosts. Two telephones are included, one on each side of the turret. The countertop overhangs the front of the console, providing ample room for paperwork.

THE NEED FOR EFFECTIVE PROJECT MANAGEMENT

Effective project management is fundamental to the success of building a comprehensive cabling network. Nutter's project team used Harvard Total Project Manager II, an MS-DOS software package that costs less than $500, to assist with project management tasks.

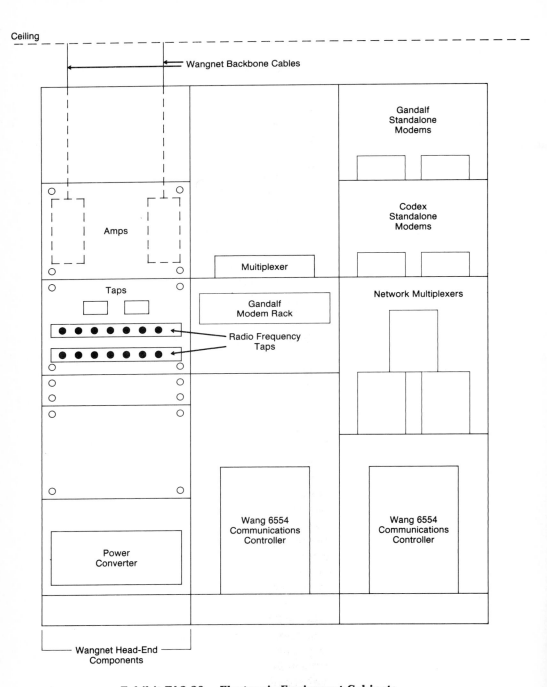

Exhibit IV-2-20.   Electronic Equipment Cabinets

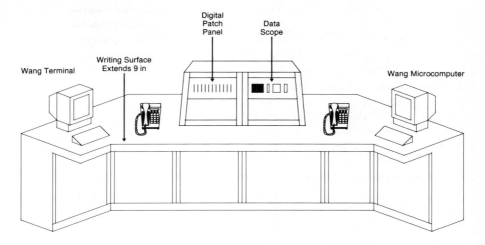

Exhibit IV-2-21.   Network Control Console

There is often needless worry about starting the local network cabling installation as early as possible. The cabling schedule, however, is determined by the general construction schedule. A competent cable installer is prepared to provide several crews of workers during certain weeks and no workers during other weeks.

The detailed cabling network design should begin approximately four months before the start of general construction. Delays in the construction schedule are typically caused by end users' allowing an insufficient amount of time to accomplish the detailed design and procurement tasks.

Exhibit IV-2-22 illustrates Nutter's project time line as created by the Harvard software package. Although the Wangnet cable design specifications took only two weeks to create, design approval consumed an additional three weeks. During the fifth week, requests for proposals were created, including one for the unionized labor needed to install the cabling network. All equipment and construction materials were procured on Nutter's behalf by their communications consulting engineering firm. Vendors were requested to submit bid proposals within three weeks of receiving the bid proposal package.

Vendor proposals were reviewed during a two-week period, and a short vendor list was subsequently created. Two weeks were devoted to getting the final bids from vendors on the short list, a practice many end users fail to put to their advantage. The project bid award occurred during the 13th week, and the following week was then devoted to reviewing the general

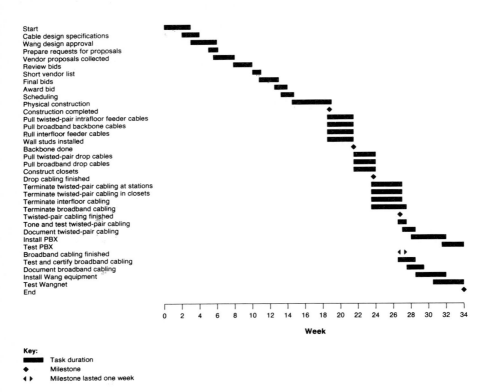

Start
Cable design specifications
Wang design approval
Prepare requests for proposals
Vendor proposals collected
Review bids
Short vendor list
Final bids
Award bid
Scheduling
Physical construction
Construction completed
Pull twisted-pair intrafloor feeder cables
Pull broadband backbone cables
Pull interfloor feeder cables
Wall studs installed
Backbone done
Pull twisted-pair drop cables
Pull broadband drop cables
Construct closets
Drop cabling finished
Terminate twisted-pair cabling at stations
Terminate twisted-pair cabling in closets
Terminate interfloor cabling
Terminate broadband cabling
Twisted-pair cabling finished
Tone and test twisted-pair cabling
Document twisted-pair cabling
Install PBX
Test PBX
Broadband cabling finished
Test and certify broadband cabling
Document broadband cabling
Install Wang equipment
Test Wangnet
End

0   2   4   6   8   10 12 14 16 18 20 22 24 26 28 30 32 34

Week

Key:
■■■■■   Task duration
◆   Milestone
◀▶   Milestone lasted one week

Exhibit IV-2-22.   Cabling Network Project Time Line

construction schedule and developing a cabling installation schedule with the chosen vendor.

Nutter's construction schedule allowed installation of the broadband backbone to begin after all base building construction was completed and after the majority of the tenant work—including ceiling plumbing and sprinkler piping and sheet metal for heating, ventilation, and air-conditioning—was finished. In building renovations, broadband backbone is typically run while the ceiling is still under active construction; however, this increases the probability of damage to the backbone cable and the need for subsequent rework.

General construction ended during week 19, as indicated by the diamond-shaped milestone symbol. Installation of the Wangnet broadband backbone, intrafloor twisted-pair feeder cabling, and interfloor feeder cabling began simultaneously during week 19.

Installation was completed before the general contractor finished installing wall studs. When the wall studs were in place, the cable installer pulled the coaxial and twisted-pair drop cables. Closet construction, including wall-mounted communications components, was finished between the 21st and 24th weeks. All drop cabling was completely installed by week 24, before the hung ceilings were closed.

Between weeks 24 and 27, the twisted-pair cabling was terminated at the station locations and in the communications closets. The interfloor twisted-pair feeder cabling and Wangnet broadband drop cables were also terminated during this time. Two weeks were devoted to the testing of the twisted-pair and broadband networks and to the preparation of the associated network documentation.

The installation of Wang computer and communications components began during week 28. The installation and testing of telephone components began during week 29. Cutover was successfully accomplished during the move, the weekend of week 34.

SUPPORTING DIGITAL EQUIPMENT CORP CONNECTIVITY

As soon as Nutter moved into the new office, it became necessary to support Digital Equipment Corp terminals and printers communicating with the DEC PDP 11/84 minicomputer used by the firm's trusts, wills, and estates department. Nutter was considering moving this department to another floor of the building and needed to know whether the network could support this new requirement.

Using Wangnet Broadband Modems

The cabling network provided several alternatives for supporting the required Digital Equipment connectivity. Exhibit IV-2-23 depicts how microcomputers and Digital Equipment Corp terminals can be connected to a Digital host processor using Wangnet and Wang's SIM-2 asynchronous radio frequency broadband Wangnet modems. The Digital terminals are connected to the Wang modems using an RS-232 cable. The Wang modem is attached to the Wangnet by a dual coaxial drop cable. In Nutter's cabling network, the dual coaxial drop cable is connected to a radio frequency broadband tap through the closet coaxial cable patch panel.

When attaching a microcomputer to the Digital Equipment Corp host, the connection to the Wang network multiplexer in the closet is provided by the second twisted-pair station connection and baluns. The Wang asynchronous broadband modem connects to the microcomputer's serial communications port (COM1 or COM2) using an RS-232 cable. The modem is connected to Wangnet's radio frequency services by a dual coaxial drop cable and appropriate coaxial patch panel closet interconnections.

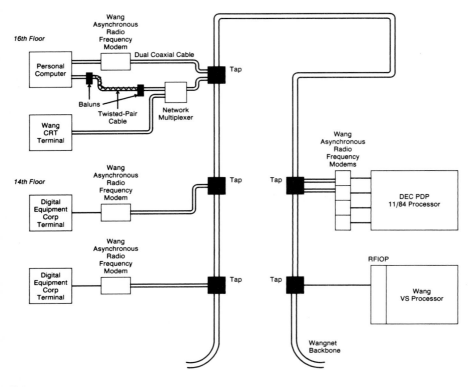

Note:
RFIOP   Radio frequency input/output processor

Exhibit IV-2-23.   Digital Equipment Corp Connectivity Using Wangnet Broadband Modems

On the host side, each Digital Equipment Corp minicomputer terminal port requires a port on the Wang SIM-8 multiport asynchronous broadband modem. Each broadband terminal port connects to each host port using RS-232 cable. Each broadband modem requires one dual coaxial drop cable and appropriate coaxial patch panel interconnections. In this setup, Wangnet provides contention-based access to the Digital Equipment minicomputer's terminal ports. More terminals may be connected to the network than host ports; the terminals contend for host ports on a first-come first-served basis.

Terminals can be located anywhere in Nutter's office. Moving a terminal simply involves moving the modem with the terminal, reconnecting the modem in the new office, and making the appropriate coaxial patch panel closet interconnections.

The advantage of this approach is Wangnet's ability to provide Digital Equipment host port service on a contention basis. The disadvantage is its high cost—typically $1,600 per broadband modem at the terminal end.

Using Short-Haul Modems

Digital Equipment connectivity can be provided at Nutter using short-haul modems and the second twisted-pair station connector of the network (see Exhibit IV-2-24). The Digital Equipment terminal is connected to the short-haul modem using an RS-232 cable, and the short-haul modem is attached to the second connector (i.e., the bottom connector) of the station wall plate using a telephone company RJ-11 station cable.

The appropriate closet interconnections route the two-pair signal to the communications closet closest to the Digital Equipment minicomputer. Each newly installed 25-pair feeder cable provides the connections for a group of 12 terminals from the closet to rack-mounted short-haul modems located near the Digital minicomputer. RS-232 cables connect each rack-mounted short-haul modem to the Digital minicomputer terminal ports.

The disadvantage of this approach is the inability to provide host port service on a contention basis. With short-haul modems, each terminal requires its own host port. The advantage is its low cost—typically $80 to $300 per short-haul modem.

Using Short-Haul Modems and a Port Contention Switch

Exhibit IV-2-25 shows how Digital Equipment connectivity can be provided at Nutter using short-haul modems and a port contention switch. In this arrangement, the Digital terminal is connected to the short-haul modem and the modem is attached to the second connector (i.e., the bottom connector) of the station wall plate.

Each 25-pair feeder cable provides the connections for 12 terminals' short-haul modems from the closet to rack-mounted short-haul modems located near the Digital Equipment minicomputer. RS-232 cables connect each rack-mounted short-haul modem to a terminal port on the port contention switch and each host port on the port contention switch to a Digital minicomputer terminal port.

The advantage of this arrangement is the ability to provide host port service on a contention basis using short-haul modems and an asynchronous port contention data switch. The disadvantage is the cumbersome amount of cabling between the short-haul modems, the port contention data switch, and the host.

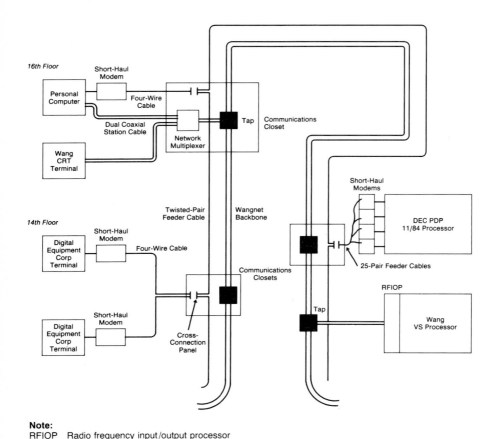

Note:
RFIOP   Radio frequency input/output processor

Exhibit IV-2-24.   Digital Equipment Corp Connectivity Using Short-Haul Modems

Using Only a Port Contention Switch

Several vendors offer port contention data switches that integrate the host-side short-haul modem into the terminal interface board within the port contention switch. Other vendors offer terminal interface boards for their port contention data switches that eliminate the need for short-haul modems. In either case, the result is a network configuration similar to that shown in Exhibit IV-2-26.

The advantage of this approach is the use of short-haul modem and port contention networking with the elimination of the host-side cabling problem. The disadvantage of the extended-distance terminal interface board is the need for two more twisted-pair cables per terminal if the control leads

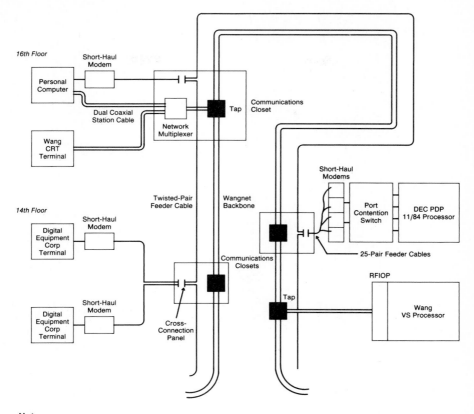

Note:
RFIOP   Radio frequency input/output processor

Exhibit IV-2-25.   Digital Equipment Corp Connectivity Using Short-Haul Modems and a Port Contention Switch

of the Electronic Industries Association (EIA) interface (e.g., data terminal ready, data set ready, clear to send, request to send, and carrier detect) are required.

COMPARING ALTERNATIVE COSTS

On the basis of Nutter's requirements, the two most feasible alternatives for supporting Digital Equipment Corp connectivity were to use broadband Wangnet modems or to use a port contention switch with short-haul modems. Exhibit IV-2-27 details the estimated costs for supporting 20 terminals and associated host ports with each alternative.

The Wangnet alternative uses 20 Wang SIM-2 asynchronous radio fre-

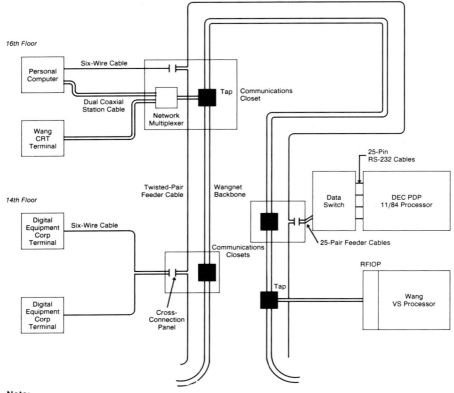

Note:
RFIOP   Radio frequency input/output processor

Exhibit IV-2-26.   Digital Equipment Corp Connectivity Using Only a Port Contention Switch

---

quency broadband modems ($1,600 each); three Wang SIM-8 multiport asynchronous broadband modems ($3,600 each) are required to support up to 24 host ports. A Wang translator is also needed, and a Wang administrative terminal is required for network management. The total cost of the Wangnet alternative, including installation, is around $66,300.

The port contention switch and short-haul modem alternative uses 20 pairs of short-haul modems; the terminal-side modems cost $100 each, and the central-site modems (which have loop-back capabilities for network management) cost $300 each. The port contention switch with port cards is estimated to cost $12,400. The total cost of the port contention and short-haul modem alternative is around $25,100, which is considerably less than the Wangnet alternative.

Materials and Services	Amount ($)
20 Wang SIM-2 asynchronous radio frequency broadband modems for terminals	32,000
3 Wang SIM-8 multiport asynchronous broadband modems for Digital Equipment Corp ports	10,800
1 Wang SIM-ADMIN administrative terminal	11,000
1 Wang SIM-XLTR channel translator	3,500
20 terminal-to-modem cables	500
20 modem-to–Digital Equipment Corp port cables	500
20 modem-to–wall plate dual coaxial cables	600
20 tap-to-modem dual coaxial cables	600
Total materials cost	59,500
Installation and system check:	
Wang installation	1,800
Non-Wang installation	1,400
Sales tax and shipping	3,600
Total cost	$66,300

a. Wangnet Alternative

Materials and Services	Amount ($)
20 pairs of short-haul modems	6,000
20 terminal-to-modem cables	500
20 modem-to–Digital Equipment Corp port cables	500
20 modem-to–wall plate telephone company cables	400
20 feeder-to-modem rack cables	400
1 port contention switch	6,000
4 contention switch port cards	6,400
20 switch-to–Digital Equipment Corp port cables	600
Total materials cost	20,800
Installation and system check:	
Wang installation	0
Non-Wang installation	2,800
Sales tax and shipping	1,500
Total cost	$25,100

b. Port Contention Switch and Short-Haul Modem Alternative

Exhibit IV-2-27.   Digital Equipment Corp Connectivity Cost Alternatives

---

The primary reason for the cost variance is the difference in cost between a vendor's proprietary broadband modem with relatively complex circuitry and a commodity-type asynchronous short-haul modem with relatively simple circuitry. The asynchronous port contention switch is also a commodity-type data communications component, produced by many highly competitive vendors.

THE FINAL CONFIGURATION

In the end, the Digital Equipment Corp minicomputer and part of Nutter's trusts, wills, and estates department were moved into available space in a different building. The remaining part of the trusts department moved to another floor of Nutter's main office and required access to the now remote Digital Equipment minicomputer.

The four alternatives for connecting Digital Equipment Corp terminals

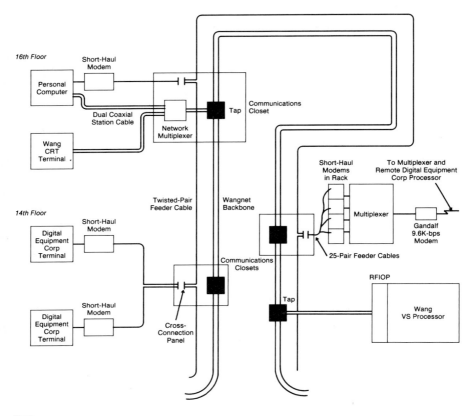

Note:
RFIOP   Radio frequency input/output processor

Exhibit IV-2-28.   The Final Digital Equipment Corp Connectivity Configuration

with a local Digital host apply to connections with a remote Digital host. The difference is that a statistical time-division multiplexer replaces the local Digital processor. A second statistical multiplexer is located next to the Digital minicomputer, connected by a leased line and pair of modems.

The resulting data communications configuration uses short-haul modems to connect Digital terminals and microcomputers to a statistical multiplexer (see Exhibit IV-2-28). The multiplexer was connected to a Gandalf 9.6K-bps modem and leased data line to the other Nutter building, where the Digital minicomputer is currently located.

Chapter IV-2-4

The Manufacturing Automation Protocol and the Technical and Office Protocol

Vincent C. Jones

THE MANUFACTURING AUTOMATION PROTOCOL (MAP) and Technical and Office Protocol (TOP) represent an attempt to create a complete set of specifications for factory and office networks that conforms to the open systems interconnection (OSI) model proposed by the International Standards Organization (ISO). This chapter discusses the history of the MAP/TOP movement, describes and compares early versions of MAP and TOP with MAP 3.0 and TOP 3.0, and discusses the relationship between MAP/TOP and the government open systems interconnection profile (GOSIP).

GOALS

The goal of the MAP/TOP movement was captured in the specification for MAP version 2.1:

> The driving force behind the General Motors MAP effort is the need for compatibility of communications to integrate the many factory floor devices. These devices are now provided by many different vendors, and it is our continued intention to use many vendors in the future. It is GM's goal to provide an environment for multiple vendors to participate on a standard communications network.

General Motors (GM) had realized earlier that the problem of incompatible devices threatened its plans for automation. By 1984, however, the number of programmable devices in use in GM plants had grown dramatically and the problem was getting out of control.

The company could not meet its plans for computer-integrated manufacturing with a proprietary network because no single vendor could meet all its needs. Relying on multiple vendors was impossible because of communications incompatibilities, and developing customized solutions would be so expensive that it would allow for only a few functions to be addressed.

GM was not alone in this quandary; virtually every large manufacturer has had to confront it. Nor was the problem limited to manufacturers; organizations trying to implement office automation also faced it—and some still do. Vendors' offerings for such basic services as electronic mail, for example, are often incompatible with each other. Users often cannot retrieve the information they need to perform a task because it is located on a system that is incompatible with theirs. The MAP/TOP movement's goal is to take control of distributed computing away from computer vendors and give it to user organizations.

The state of computer automation today is comparable to the tire and automobile industry 50 years ago. Then, the owner of a Stutz Bearcat that needed new tires could get them only from Stutz and could choose only from the options Stutz offered. The same was true of other manufacturers. This situation raised costs in two ways. First, the engineers designing a car had to design wheels and tires also; they could not rely on existing standards because none existed. Second, owners with special needs (e.g., for extra-wide balloon tires for soft ground) had to have them custom made because the demand was insufficient to justify production of special tires for each brand and model of car.

In contrast, current standards, although they have reduced the number of tire sizes available, have dramatically increased the variety of types available. Almost any type of tire is readily available from many tire manufacturers for just about any vehicle. Car owners can select cars suitable for their transportation needs and tires for specific driving conditions.

The goal of the MAP/TOP effort is to create similar conditions for computer automation users, freeing them from dependence on a single vendor and allowing them to select computers and other automation hardware that fit their needs.

EARLY MAP/TOP

The first commercially available versions of MAP/TOP were MAP versions 2.1 and 2.2 and TOP version 1.0. All three are based on the protocol suite shown in Exhibit IV-2-29. MAP 2.1 established the baseline capability for MAP/TOP. It provides three application-level services: file transfer, manu-

Layers	MAP 2.1 Protocols	TOP 1.0 Protocols
Layer 7: Application	ISO FTAM (DP 8571) File Transfer Protocol, Manufacturing Messaging Format Standard, and Common Application Service Elements	ISO FTAM (DP 8571) File Transfer Protocol
Layer 6: Presentation	ASCII and Binary Encoding Only	
Layer 5: Session	ISO Session (IS 8327), Kernel and Full-Duplex Functional Units Only	
Layer 4: Transport	ISO Transport (IS 8073), Class 4 Only	
Layer 3: Network	ISO Connectionless Internet (DIS 8473), with Optional Use of CCITT X.25 as a Subnetwork Under ISO 8473	
Layer 2: Data Link	ISO Logical Link Control (DIS 8802/2) (IEEE 802.2, Type 1, Class 1); Optional Use of CCITT X.25 Requires Appropriate Link and Physical Layers for X.25	
Layer 1: Physical	IEEE 802.4 Token-Passing Bus, for 10M-bps Broadband Media	ISO CSMA/CD (DIS 8802/3) CSMA/CD Media Access Control, for 10M-bps Baseband Media

Exhibit IV-2-29.   MAP 2.1 and TOP 1.0 Protocol Suites

facturing, and interprocess communications. These services are provided at the top of a protocol stack built on international standards under development by the ISO. The layer 1 specification for physical media uses the Institute of Electrical and Electronics Engineers (IEEE) 802.4 standard, which has become the international standard for broadband, token-passing, local area networks (LANs). The layer 3 specification for communications among geographically dispersed sites uses the International Telephone and Telegraph Consultative Committee (CCITT) X.25 standard for packet-switched networks.

As shown in Exhibit IV-2-30, MAP 2.2 provided the same application-level services as version 2.1 but added carrier-band media as an option at the

Layers	Standard MAP 2.1 Stack	MAP 2.2: Enhanced Performance Architecture, and MiniMAP
Layer 7: Application	ISO FTAM (DP 8571) File Transfer Protocol, Manufacturing Messaging Format Standard, and Common Application Service Elements	Equivalent to MAP 2.1
Layer 6: Presentation	ASCII and Binary Encoding Only	Bypassed
Layer 5: Session	ISO Session (IS 8327), Kernel and Full-Duplex Functional Units	Bypassed
Layer 4: Transport	ISO Transport (IS 8073), Class 4 Only	Bypassed
Layer 3: Network	ISO Connectionless Internet (DIS 8473), with Optional Use of X.25 Under ISO 8473	Bypassed
Layer 2: Data Link	ISO Logical Link Control (DIS 8802/2) (IEEE 802.2, Type 1, Class 1), with Appropriate Link and Physical Layers for X.25	IEEE Single-Frame Confirmed Service (IEEE 802.2, Type 3, Class 3)
Layer 1: Physical	IEEE 802.4 Token-Passing Bus, for 10M-bps Broadband Media or 5M-bps Carrier-Band Media	

Exhibit IV-2-30.   MAP 2.2 Protocol Suite

physical layer and introduced the enhanced performance architecture and miniMAP options. The enhanced performance architecture permits the optional bypassing of several layers of the OSI reference model to promote faster response time for certain real-time applications. MiniMAP similarly provides for limited-function devices that require access only to layers 1, 2, and 7.

TOP 1.0 is an adaptation of MAP 2.1 for the office environment. It provides for file transfer on an IEEE 802.3 (Ethernet) LAN. It is primarily used to provide access to MAP 2.X networks from an Ethernet-based office. With the exception of miniMAP, the only differences among MAP 2.1, MAP 2.2, and TOP 1.0 are in the application services each provides and in the LAN media and media-access methods each supports. Even at the application and physical layers, the three share many specifications. All three have the same file transfer, access, and management (FTAM) standard and the same implementation of IEEE 802.4. The specifications of each for layers 2 through 6 of the OSI reference model are identical.

Because the prototype MAP 2.1 was issued before stable international standards were available for several layers, MAP 2.X and TOP 1.0 have major limitations. Specifically, the protocols for file transfer and interprocess communications are based on draft proposals that have since been revised. In 1985, work had not even begun on an international manufacturing protocol, so MAP 2.X includes a unique manufacturing protocol developed by GM.

MAP's designers were aware of these limitations but felt it was better to proceed than to wait for the international committees. As a result, the goal of vendor-independent networking was achieved years sooner than would otherwise have been possible. The implementation of the early versions of MAP also helped the standards committees identify and correct deficiencies in their draft proposals that might otherwise have been overlooked.

The problem with implementing draft standards, however, is that the standards have since changed. As a result, MAP/TOP 3.0, based on the stable versions of the international standards, is incompatible with the earlier versions of MAP/TOP. In anticipation of this problem, MAP 2.1 was deliberately restricted to minimize the risk to vendors of investing in the development of products that would become obsolete. In particular, the file transfer service, based on the ISO FTAM protocol, supports only transparent binary bit stream and simple text transfer, and the interprocess communications facility, based on the ISO common applications service elements, provides no data translation services—the user application must determine how data being transferred is to be represented in binary octets.

For similar reasons, MAP 2.1's network management and directory services are just sufficient to allow a reasonably sized network to operate under stable conditions. Proprietary extensions to the specification are required for management tools to provide such capabilities as directory updates, failure reports, and diagnostics.

The manufacturing messaging format standard (MMFS), the manufacturing messaging service of MAP 2.1, suffers from the opposite problem in that it is too powerful. Developed by GM before OSI protocols for most layers were available, the MMFS assumes a network with only minimal transport service and includes many of the functions that OSI networks handle in presentation-, session-, and transport-layer protocols. In MAP/TOP 3.0, the MMFS has been replaced by the ISO manufacturing messaging specification (MMS), which assumes a network with a full OSI stack and does not provide the lower layer functions available in the MMFS.

MAP 3.0

The final versions of MAP 3.0 and TOP 3.0 were published in 1988. All the services in MAP 2.1 are available in MAP 3.0 but have been considerably enhanced. Among the enhancements are the following:

- The addition of remote file access, improved file management, and the ability to handle structured files and multiple data bases to network file transfer services.
- The extension of the interface to network interprocess communications to permit access to more underlying capabilities.
- The replacement of the GM-developed MMFS with the ISO standard MMS.

All core protocols in MAP 3.0 are based on international standards. All additional protocols, with the exception of network management, are based on either international standards or draft international standards. The standards for the three user application services—manufacturing messaging, file transfer, and association control—are:

- The EIA RS-511 MMS—This provides for communications tuned to the needs of factory floor devices (e.g., for programmable controllers, numerical controllers, and robots).
- ISO FTAM—This provides the ability to transfer files from one machine to another, to retrieve files on remote devices as if they were local, and to manage such file attributes as names and protection schemes.
- ISO association control service elements (ACSEs)—These allow interprocess communications to support distributed applications that are not explicitly supported by other protocols.

The physical layer standard for MAP 3.0 remains the same as that for MAP 2.1: the IEEE 802.4 broadband cable, token-passing bus. This continuity protects the investment in cable of MAP 2.X system users.

The network management and directory services in MAP 3.0, although they have been substantially upgraded from the 2.X services, are still incomplete because accepted international standards for these services are not yet in place.

Exhibit IV-2-31 illustrates the protocols that make up MAP 3.0 and their interrelationships.

Because of substantial incompatibilities between MAP 2.X and MAP 3.0, the upgrade was traumatic for users and vendors. To avoid additional trauma, the MAP users group has issued a stability statement that promises a six-year moratorium on incompatible extensions.

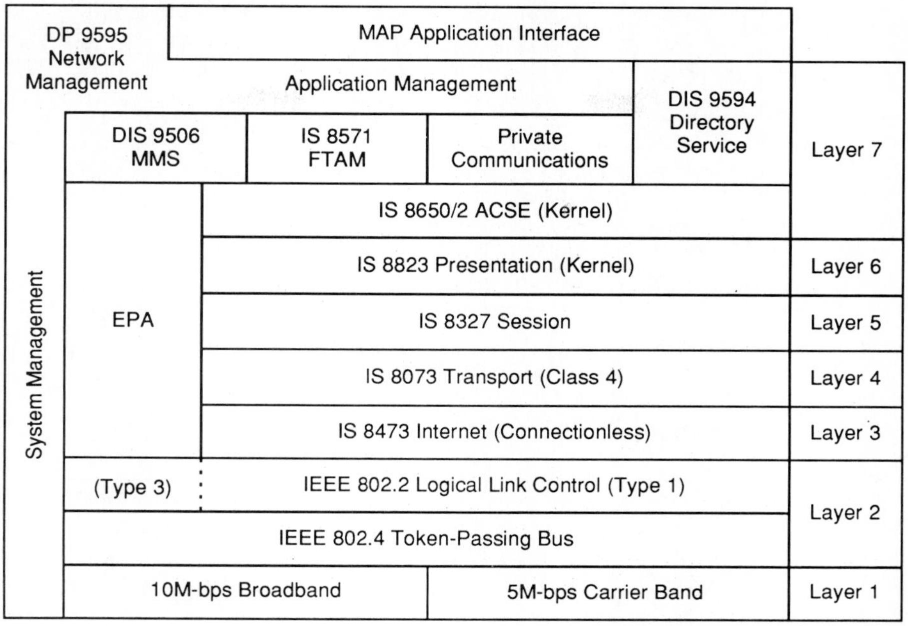

System Management	DP 9595 Network Management	MAP Application Interface			
		Application Management		DIS 9594 Directory Service	Layer 7
		DIS 9506 MMS	IS 8571 FTAM	Private Communications	
	EPA	IS 8650/2 ACSE (Kernel)			
		IS 8823 Presentation (Kernel)			Layer 6
		IS 8327 Session			Layer 5
		IS 8073 Transport (Class 4)			Layer 4
		IS 8473 Internet (Connectionless)			Layer 3
	(Type 3)	IEEE 802.2 Logical Link Control (Type 1)			Layer 2
	IEEE 802.4 Token-Passing Bus				
	10M-bps Broadband		5M-bps Carrier Band		Layer 1

Notes:
DIS   Draft International Standard
IS   International Standard

Exhibit IV-2-31.   MAP 3.0 Protocol Suites

TOP 3.0

Unlike TOP 1.0, which was a subset of MAP 2.1, TOP 3.0 is a superset of MAP 3.0. All the capabilities of MAP 3.0 except manufacturing messaging, mini-MAP, and the enhanced performance architecture are also available in TOP 3.0. A comparison of Exhibit IV-2-32, which shows the TOP 3.0 protocol stack, with Exhibit IV-2-31, which shows the MAP 3.0 protocol stack, reveals the TOP 3.0 uses the same protocol specifications as MAP 3.0 at every layer for all functions they share.

An underlying assumption of the specifications for both MAP 3.0 and TOP 3.0 is that MAP and TOP networks should be interoperable. The differences between them in layers 2 through 6 involve compatible extensions that permit the support of additional capabilities in the application layer and additional cable plants in the physical layer. All political and technical

IS 8613 ODA/ODIF	DIS 8632 CGM	IS 7942 GKS-2D	ANSI IGES v3.0

DP 9595 Network Management	MAP/TOP Application Interface				
	Application Management			DIS 9594 Directory Service	Layer 7
	CCITT X.400 MHS (1984)	IS 8571 FTAM	IS 9041 Virtual Terminal		
	IS 8650/2 ACSE (Kernel)				
	IS 8823 Presentation (Kernel)				Layer 6
	IS 8327 Session				Layer 5
	IS 8073 Transport (Class 4)				Layer 4
	IS 8473 Internet (Connectionless)				Layer 3
	IEEE 802.2 Logical Link Control (Type 1)			CCITT X.25	Layer 2
	IEEE 802.3 CSMA/CD	IEEE 802.4 Token Bus	IEEE 802.5 Token Ring		
	10BASE5	10BROAD36	10M-bps Broadband	5M-bps Carrier Band	4M-bps Shielded Twisted Pair	Layer 1

(left vertical label: System Management)

Note:
IS   International Standard

Exhibit IV-2-32.   TOP 3.0 Protocol Suites

reasons for the continued independence of the MAP and TOP disappeared years ago. The only reason they have not completely merged is that they have incompatible document-generation systems; the costs involved in overcoming this incompatibility have so far outweighed the benefits.

The features of TOP rival those of any proprietary network available today for office and engineering use. In addition to all the capabilities of MAP 3.0, TOP 3.0 includes a wide range of protocols that support engineering and office applications. The application services TOP 3.0 provides are:

- The ISO basic class virtual terminal protocol—This permits a terminal on any machine to log on to any other machine as a simple character-mode terminal.

- ISO FTAM (as defined in the discussion of MAP 3.0).
- The CCITT X.400 message-handling service—This permits electronic mail transfers of all kinds of documents within facilities, between facilities, and internationally.

The information exchange protocols TOP 3.0 provides are:

- ISO office document architecture and office document interchange format (ODA/ODIF)—These protocols permit formatted documents like those produced by word processing and desktop publishing programs to be transmitted independently of the package used to create them. They include support for text, raster graphics, and line graphics data.
- The American National Standards Institute (ANSI) initial graphics exchange standard (IGES)—This protocol permits the exchange of product design information among CAD/CAM workstations independently of the CAD/CAM package used to create the information.
- The ISO computer graphics metafile (CGM)—This permits the exchange of geometric graphics independently of the devices used to create them.

TOP 3.0 provides many more physical media options than TOP 1.0. These include the IEEE 802.3 (Ethernet) protocol on 10BROAD36 as well as 10BASE5 media, the IEEE 802.4 token-bus protocol on 10M-bps broadband and 5M-bps carrier-band media, and, for compatibility with IBM networks, the IEEE 802.5 token-ring protocol on 4M-bps twisted-pair media. The preferred protocol is IEEE 802.5 on 10BASE5 media. The other protocols and media are recommended only when 802.3 10base5 is not practical. The specification requires that any alternative implementations be able to interoperate through bridges with an 802.3 10BASE5 network.

TOP committees are currently studying many extensions to version 3.0. The fiber distributed data interface (FDDI) standard will be incorporated at the physical layer when it is complete. Likewise, fiber optic-based additions to the IEEE 802 series are being considered as they are defined. At the network layer, a work group is studying better use of connection-oriented services. At the application layer, the critical services of transaction processing and remote data base access, for which international standards are being defined, could be available in the early 1990s. Committees are also evaluating electronic data interchange (EDI) and the standard generalized markup language for inclusion in TOP and are considering upgrading from version 3 to version 4 of IGES.

MAP, TOP, AND GOSIP

Recognizing the importance of the open systems movement for government computer users, the US government has developed the government open systems interconnection profile (GOSIP). As Exhibit IV-2-33 indicates,

CCITT X.400 (Red Book, 1984)		IS 8571 FTAM		Layer 7
Public Data Network ⋮ Private Network		IS 8650/2 ACSE (Kernel)		
		IS 8823 Presentation (Kernel)		Layer 6
IS 8327 Session				Layer 5
(Class 0)⋮ (Class 4)	IS 8073 Transport			Layer 4
CCITT X.25	IS 8473 Internet (Connectionless)			Layer 3
			X.25 PLP	
	IEEE 802.2 Logical Link Control (Type 1)		HDLC LAP-B	Layer 2
	IEEE 802.3 CSMA/CD	IEEE 802.4 Token Bus	IEEE 802.5 Token Ring	MIL-STD 188-114-A	EIA RS-232D	Layer 1

Exhibit IV-2-33.   GOSIP Protocol Suites

---

the GOSIP protocol stack is closely related to that of TOP 3.0. This similarity is no coincidence; it reflects close cooperation between the MAP/TOP community and the National Institute of Standards and Technology (NIST).

GOSIP is a proper subset of TOP 3.0, with the exception of extensions for addressing assignments and security specified by NIST. GOSIP includes those aspects of TOP 3.0 that were stable in 1987; other services, including the virtual terminal service and directory service, will be added as they mature.

Other governments have made similar commitments to the open systems movement. The UK has developed its own OSI profile, which is also called GOSIP, and other countries have equivalent specifications.

SUMMARY

The goal of the MAP/TOP movement has been to create specifications for manufacturing, office, and technical products based entirely on accepted international standards. The shift from MAP 2.X and TOP 1.0, which are based largely on draft proposal standards, to MAP 3.0 and TOP 3.0, which use international standard protocols at all layers, represents a significant step in the achievement of that goal. Users and vendors can expect a series of 3.X releases as the standards for additional capabilities (e.g., security and concurrency control) are developed, refined, and added to the specifications.

Section IV-3
System Operations

FOLLOWING THE IMPLEMENTATION of the communications system, the manager has the ongoing task of operating, monitoring, and controlling the network. This section discusses the operations, management, and administration processes of a communications system. These functions are sometimes collectively referred to as OMA, OMAP, or OMP. The key to successful operation of an effective system probably lies with network management.

Chapter IV-3-1, "A Manager's Perspective of Network Management," covers not only technical issues but the network's impact on the organizational structure and environment. The communications manager is also likely to be concerned with product availability and future directions of management protocols and tools; each of these is covered in this chapter. The chapter reviews the environment, including objects to be managed and methods to be used, describes the elements of network management, and concludes with the current status of standardization and future trends.

Often, a case study of a problem's solution is useful to the manager. Although the study may not be directly applicable, if the concepts and principles that are applied tend to be the same, the case study may provide hints and approaches that can be applied in the manager's specific case. Chapter IV-3-2, "Managing the Building of a Network Control Center," offers a detailed description of the task of planning, procuring, and implementing a network control center for a large multihost data center.

The management of a communications system must include processes that can measure the effectiveness of network control mechanisms. This monitoring or auditing process becomes even more important as the distribution of network resources increases the risks. "Evaluating the Effectiveness of Data Communications Network Controls," Chapter IV-3-3, explores five areas of communications system control requirements: data integrity, data security, disaster recovery, hardware, and administration. For each requirement, the chapter explores the risks, analyzes the controls, and enumerates solutions. A risk analysis checklist is included.

The monitoring and operations functions would be virtually nonexistent if nothing ever went wrong. The majority of the management and monitoring functions are aimed at detecting and recovering from abnormal conditions that threaten the integrity of the network. Detection of an abnormality usually requires some type of maintenance, which can be time-consuming and expensive. One alternative that should be considered by the communications systems manager is third-party maintenance. Third-party maintenance may be the answer in multivendor installations, when the vendor does not have extensive support capabilities, or when the vendor's support is geographically remote. Chapter IV-3-4 discusses "The Third-Party Maintenance Alternative." The author begins with an explanation of the need, explores the characteristics of the industry, defines the advantages and disadvantages of third-party maintenance, and offers useful guidelines for contracts and checklists.

Chapter IV-3-1
A Manager's Perspective of Network Management

Celia A. Joseph
Kurudi H. Muralidhar

MOST DISCUSSIONS OF NETWORK MANAGEMENT center on technical issues. Managers, however, have a different perspective. They need to be aware of major technical issues, but they must also determine how to make network management work in their organizational environment. In this chapter, the issues that are examined include: how network management fits into an organizational environment, what products are available now, and where network management is going in the future.

As used in this chapter, the term *network management* is limited to the management of data communications resources that use standard interfaces—that is, to open systems. Proprietary network management schemes are not considered. Although network management functions apply to all networks, some functions should be tailored to take into account the special needs of the applications running on the network. This chapter is oriented to the needs of local area networks (LANs) and uses the Manufacturing Automation Protocol (MAP) and Technical and Office Protocol (TOP) version 3.0 network management specifications.

What is network management and why is it important? Network management can be defined as the coordination, monitoring, and control of the distributed resources in a network. Network management is important because it can provide a wide range of information about a network and a powerful set of tools for running it. The types of information that network management can provide include network configuration and status, network performance and trends, and current and historical data on network operations. Using network management tools, a network administrator can modify the network configuration, change its status, adjust parameters to tune its performance, and determine the location of faults and the best way to correct them. As organizations rely more heavily on networks and networks become increasingly complex, management information and tools become critical to the organization's operation.

The chapter begins with background information on network management, including the elements needed to put it into place, and the basic functions it provides. Next, it gives an example of a network manager application. It then examines the status of network management today, reviewing what current products can do and what capabilities should be forthcoming. This is followed by a discussion of the long-term direction of network management. The chapter ends with a short list of issues that a manager should consider before implementing a network management system.

BACKGROUND

As just defined, network management is the coordination, monitoring, and control of the resources that allow communications to take place over a local area network. The term *resources* usually refers to network components, such as workstations and file servers, although it may also refer to the elements within a component, such as protocol layers or any configurable parameters that the component may have.

More specifically, network management involves ensuring the correct operation of the network, monitoring the use of network resources, maintaining network components in good working order, planning for changes to the network, and producing a variety of information on network operations, such as periodic or ad hoc reports.

The Network Management Environment

Network management is frequently viewed as only a technical problem. However, the ultimate responsibility for management resides with people, not machines. Many people within an organization are involved with network management, including users of the network who may need access to current LAN status information, managers throughout the organization who may be concerned about the effect the LAN's performance may have on the performance of the parts of the organization they are responsible for, and the actual communications system administrator in charge of the day-to-day operation of the LAN. Thus, the integrated management environment within which network management resides is a combination of human, social, organizational, and technological resources.

Exhibit IV-3-1 shows the main components of this integrated management environment. These include:
- Users of the network—Those interested in the operation and use of the network.
- Network and system resources—End systems (e.g., workstations, controllers, file servers), relay systems (e.g., bridges, routers, and gate-

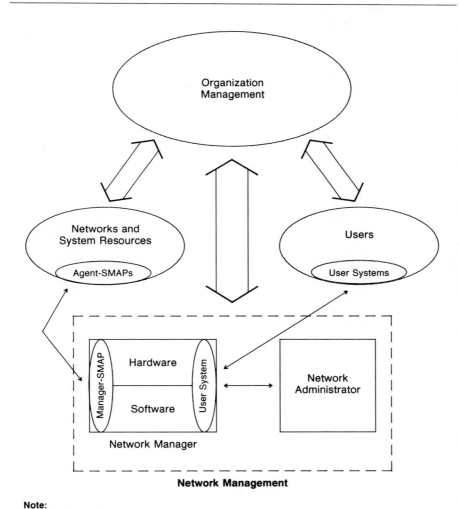

Network Management

Note:
SMAP   System Management application process

Exhibit IV-3-1.   The Network Management Environment

ways), and network components—in other words, the collection of objects that require managing.

- The organization's management—Directly affects network management by setting policies for the organization that have an impact on network structure and activities. Management, for example, would decide whether or not to implement open systems interconnection (OSI) or whether or not to adopt a centralized or distributed management structure.
- Network management—A combination of human, software, and hardware elements. The human element consists of the network administrators who make decisions on network management. The software and hardware elements are the automated network management tools that provide management capabilities for the network.

Network Management Elements

Exhibit IV-3-2 shows an example of a manufacturing LAN with network management elements. Each device has a LAN component that permits it to attach to the LAN. In this example, all the LAN components use the standard, layered interface protocols specified by MAP and TOP. Some devices may share a LAN component. As defined by the standards groups, network management fits into and around the layered protocols in the LAN interfaces. However, more is needed to put network management in place than an element in a LAN component's interface. The full set of elements needed to implement network management consists of the following:

- The communications system administrator—The person who uses the network manager to perform network management functions.
- The network manager application—An automated tool with a special user system that the communications system administrator uses to monitor and control LAN activities. The LAN may have more than one network manager application.
- The agent–system management application process (agent-SMAP) and the manager–system management application process (manager-SMAP)—Agent-SMAP is a program that resides in each LAN component, manages the resources within the LAN component, and communicates with the manager-SMAP. The manager-SMAP is an analogous program that resides in the network manager applications LAN component.
- The network management protocol—A set of rules that defines how a manager-SMAP communicates with agent-SMAPs. This protocol is sometimes called the manager-agent protocol.
- The management information base (MIB)—The data base of information that each network device maintains on its own resources. In

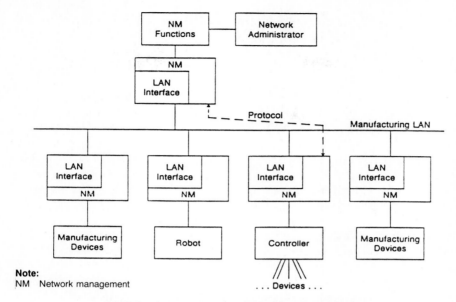

Exhibit IV-3-2.   Network Management Elements

Exhibit IV-3-2, the MIB is included in the network management (NM) element of each device. In addition, the network manager maintains an information base for the domain for which it is responsible, which is included in the box marked NM functions.

• The management domain—The set of all agent-SMAPs that report to the same manager-SMAP, or in other words, the set of devices that a network manager application will manage. If the LAN has a single network manager application, then all devices will be in that manager's domain. However, a LAN may have multiple manager applications, in which case a domain must be defined for each manager and an agent may be in more than one manager's domain.

Network Management Functions

The exact functions that network management should provide are still being defined by the standards groups. So far, they have agreed on a set of basic functions that include configuration and name management, fault management, performance management, accounting management, and security management.

Configuration and name management are mechanisms to determine and control the characteristics and state of a LAN and to associate names with managed resources. Some of the services that configuration manage-

779

ment provides include setting LAN parameters, initializing and terminating LAN resources, collecting data on LAN status for reports, and changing LAN configuration. Some of the services that name management provides include naming the resources to be managed and managing name assignments.

Fault management includes the mechanisms to detect, isolate, and recover from or bypass faults in the LAN. The way fault management is performed depends on the LAN's application. For example, in some manufacturing applications, LAN downtime is intolerable. In these cases, fault management should be proactive, forecasting probable faults and emphasizing preventive maintenance. For LANs in which downtime is not so catastrophic, fault management could be reactive, acting in response to faults only as they occur and emphasizing accurate diagnosis and rapid repair.

Performance management includes mechanisms to monitor and tune the LAN's performance as defined by user-set criteria. Some environments may need special performance metrics. The factory environment, for example, is typically hostile to communications equipment because it is noisy, dirty, and has wide variations in temperature. Thus, instead of classic performance metrics that assume error-free operations, these LANs should use preformability metrics that provide a means of measuring performance and reliability in a unified manner.

Accounting management includes mechanisms for controlling and monitoring charges for the use of communications resources. Security management includes mechanisms for ensuring that network management operates properly and for protecting LAN resources. Although accounting management and security management mechanisms have been defined, they have not yet been included in the MAP/TOP specifications and are not discussed further in this chapter.

AN EXAMPLE NETWORK MANAGER

The standards committees have defined the basic functions that network management should provide, but they have not defined how these functions should be applied to the management of a particular type of network or network application. The standards provide a set of basic services, but they do not specify how to implement or use those services.

The example of a network manager application that follows is intended to suggest how these services could be used. The application is an ideal in that some of the services described are not yet implementable.

The example system supports the three network management functions specified in MAP/TOP 3.0: configuration management, fault management, and performance management. The system provides the following configuration management services:

- Defining the LAN topology—The system provides tools to assist in setting up the network's design and in initially configuring the LAN components. For each type of component, the system suggests how the component's configurable parameters should be set.
- Displaying the LAN topology—Given the initial LAN design, the system generates a display that shows the location and the operational status of each device in the LAN.
- Reading current values—The system lets the network administrator request the current value of any communications resource within the LAN components. For example, it might tell the administrator the number of messages that have been sent or received by a specific network device.
- Setting values—The system lets the network administrator modify the value of the configurable parameters in the LAN components. The administrator could, for example, change the status of a device from active to inactive, or change the retransmission parameters of the transport layer.
- Adding or deleting devices—The system supports dynamic changes to the LAN's topology, permitting components to be addressed and removed.

The system provides the following fault management services:

- Detecting and giving notices of faults—The system gives the network administrator prompt notice that a fault has occurred somewhere in the LAN. It has an option that allows it to change from reactive to proactive fault management. In proactive mode, the system alerts the administrator of its fault predictions.
- Isolating faults—The system helps the network administrator determine where a fault has occurred in the LAN topology, which device failed, and which portion of the device failed. It does this in one of two modes: an automated mode in which the system isolates the fault without intervention from the network administrator, or an assistant mode in which the system provides suggestions to guide the network administrator in locating the fault.
- Correcting or bypassing faults—In automatic mode, the system corrects the fault or implements a bypass without intervention from the network administrator. In assistant mode, the system gives advice on how to correct or bypass the fault, but the network administrator makes the changes.

The system provides the following performance management services:

- Collecting statistics—The system maintains a set of current and historical statistics on each device in the network.
- Evaluating performance—The system uses the statistics to calculate performance metrics and evaluates these metrics against predefined user criteria for performance.

- Reporting—The system generates text reports and graphic displays of the statistics and performance evaluation results. These reports can be periodic or on demand.
- Tuning performance—The system can tune the network in either of two modes: automatic or assistant. In automatic mode, the system dynamically monitors performance levels and adjusts parameters when they move out of range. In assistant mode, the system provides advice to the network administrator on how to tune performance, but the network administrator makes any changes manually.

THE CURRENT STATUS OF LAN NETWORK MANAGEMENT

The MAP/TOP Users' Group has led the OSI community in defining the mechanisms needed to provide the functions in the example network management just described. The MAP/TOP 3.0 specification includes an application layer network manager, a set of basic network management services, and a network management protocol. Other standards organizations, most notably the International Standards Organization (ISO), are progressing slowly in fully defining all of the functions of network management.

Products currently available for network management include cable monitors, modem monitors, protocol monitors and analyzers, and configuration support tools.

- Cable monitors monitor the status of low-level devices on the cable plant, such as amplifiers and power supplies—They may also monitor the radio frequency levels on cables.
- Modem monitors keep track of the status of the modems used in the LAN.
- Protocol monitors and analyzers passively collect information on the protocol transactions of one or more network layers—They may calculate statistics, such as the average level of specific types of network traffic, and generate reports. Analyzers may also perform such additional functions as identifying patterns in network traffic or capturing a specific type of traffic for closer examination.
- Configuration support tools provide offline assistance in configuring the operating characteristics of LAN devices—Some of these products can also download software to LAN devices.

Several companies are developing network management systems for open systems that are oriented toward managing telecommunications networks. These include AT&T's unified network management architecture (UNMA), IBM's NetView, and Codex's 9800 network management system.

AT&Ts unified network management architecture is a virtual network management system that provides network management information at a customer site by linking the local site to a remote network control center through a protocol based on an OSI profile. Only a few vendors supply products conforming to this architecture. IBM's NetView is an integrated network management product. NetView's main component is its command facility, which includes data collection, monitoring, and control functions. A number of vendors supply NetView-based products. The Codex 9800 network management system is based on the evolving OSI standards for network management and is currently limited to managing Codex products only.

Emerging Capabilities

Network management capabilities that should be appearing in products soon include expanded network management functions, full ISO protocol support, nonstandard interface support, and multiple network manager support.

Expanded Network Management Functions. Current LAN network management products focus on managing the lower protocol layers. To expand network management functions to cover all protocol layers, network management must be implemented in two areas: the network devices being managed and the network manager application. The network devices that implement MAP/TOP 3.0 include network management functions. For network management applications, forthcoming systems should provide tools that interact dynamically with the LAN to manage its components. These tools will provide flexible network management functions that can take the special needs of a LAN application into account.

Protocol Suites. Network management products should support all the ISO protocols. These may be tailored to a specific platform (e.g., MAP or TOP).

Nonstandard Interfaces. Not all LAN interfaces use standard protocols. The devices using these interfaces, however, must be managed. The MAP/TOP committees have acknowledged this problem, although they have not yet solved it. As a result, creative vendor solutions may drive the standards efforts in this area.

Multiple Managers. Organizations with a distributed management structure or geographically separated LANs may wish to use more than one network manager. The standards committees will be addressing this issue in the future.

THE OUTLOOK FOR LAN NETWORK MANAGEMENT

The work of standards organizations on network management standards is progressing slowly. The ISO will probably not complete work in this area until the mid-1990s. The MAP/TOP Users' Group has defined some facilities sooner and may spur the standards organizations to faster progress.

Once the standards have been defined, building workable network management systems will still be a difficult job. The network management systems will require technologies from several areas, including artificial intelligence. Some of the applicable technologies are discrete event modeling, statistical pattern recognition, sensor fusion, control theory, distributed artificial intelligence, diagnostic reasoning, and game theory. The relation of each of these technologies to network management is as follows:

- Discrete event modeling permits detailed dynamic experiments with complex systems that can help answer what-if questions—Stochastic activity networks, for example, are particularly useful for modeling networks.
- Statistical pattern recognition permits the monitoring of the quality of processes and products and makes it possible to correct problems before the system drifts out of acceptable bounds.
- Sensor fusion synthesizes the output of several sensors to derive parameters that are not available from a single sensor.
- Control theory includes a variety of techniques for distributed control.
- Distributed artificial intelligence includes techniques for distributed problem solving.
- Diagnostic reasoning includes techniques for determining problem causes and solutions.
- Game theory includes techniques for reaching optimal solutions for games with multiple players, which may be particularly useful for multiple manager systems.

Some of the areas of network management to which these technologies are relevant are configuration management, performance management, and fault management. Applications of configuration management include helping configure a wide range of LAN devices, dynamically adding and deleting devices to and from the LAN, and dynamically setting or modifying LAN device characteristics.

Performance management applications include dynamically evaluating the LAN's performance and identifying problem areas, suggesting key parameters to watch for evaluating performance, suggesting parameter ranges when changes are needed, and dynamically tuning the LAN's performance.

Fault management applications include detecting and isolating a specific type of fault and suggesting corrections or detours for specific types of faults. Fault management systems should learn how to respond to increasingly complex faults from experience.

SUMMARY

Network management can provide a wide range of services. Determining which are best for a particular organization and LAN application, however, requires careful consideration of many organizational and technical issues. Among the organizational issues that the communications manager should consider when planning network management services are the following:

- Training—Who in the organization needs training in network management? What types of training should these people receive? What types of training should network administrators receive?
- Organizational commitment to OSI—How committed is the organization to the concept of open systems? How many devices with non-standard interfaces will have to be managed? What is the schedule, if any, for phasing out these devices?
- Budget priorities—What are the organization's investment goals? Is the emphasis on the long-range or the short-range goals? Network management and open systems are relatively high short-term expenses with long-term payoffs.
- Management information needs—Who will be permitted access to network management information? What types of information are needed? In what form? How often?
- Security—What level of security should be used to protect network management information and facilities?
- Management architecture—How should the network management structure relate to the organization's management structure? Is a centralized or distributed structure more appropriate?
- Management mode—Should the network manager work in a proactive or reactive mode?

Among the technical issues that the communications manager should consider when planning network management services are the following:

- Network management architecture—Will the network management architecture be centralized or distributed? If multiple managers are needed, is the organization willing to do research on how to define the functions that have not yet been addressed by the standards groups?
- Missing pieces—Two key areas have not yet been addressed by the standards groups: how to deal with multiple managers and how to manage devices with nonstandard interfaces.
- Application characteristics—What characteristics of the organization's LAN application have special network management requirements? LANs with real-time traffic, for example, need performance maintained within strict limits.
- Expert systems—Which expert system tools are applicable to the organization's network management requirements? How critical are these requirements? Should the organization consider pushing technology development in these areas by funding research or conducting its own research?

Chapter IV-3-2
Managing the Building of a Network Control Center

David Levin

BEFORE CONSTRUCTION OF ITS NEW NETWORK CENTER, the network management resources of the Metropolitan Transit Authority (MTA), a New York State Public Authority that operates New York City's commuter railroads, subways, buses, and bridges, were limited. Diagnostic devices consisted of an antiquated D-502 interactive digital interface monitor and a Spectron T-51 tape mass storage unit. The MTA did have several IBM software diagnostic products installed—including network communications control facility, network problem determination aid, and network logical data manager. Only the basic capabilities of these tools were used, and network management functions, such as they were, were distributed among different sections of the MTA's data center.

THE PROJECT PLAN

The project to construct a new network control center was divided into three major phases with the following work tasks identified within each phase:
- Phase one—Define network control needs and requirements:
  —Task 1. Specify the recommended hardware components.
  —Task 2. Create performance specifications of major components.
- Phase two—Component selection and procurement:
  —Task 3. Make sole-source and competitive bid procurements.
- Phase three—Implementation:
  —Task 4. Development of detailed implementation plan.
  —Task 5. Design of network control console and cable runs.

—Task 6. General construction and implementation.

—Task 7. Documentation and operator training.

The critical path method of project management was used to monitor and control tasks within the project. Development of the management plan involved a consulting company, MTA personnel from various departments, and personnel within the MTA's associated agencies (e.g., the New York City Transit Authority). A detailed work plan was developed to define specific project goals, personnel responsibilities, time schedules, and specific delivery requirements for each task. Status reports were prepared and disseminated throughout the duration of the project, keeping all team members—including appropriate MTA management personnel—continually informed of progress.

Although usually a difficult task in the private sector, building the MTA's network control center posed even more difficulties because of public sector guidelines calling for procurement by competitive bidding whenever possible.

Task Descriptions

The primary task (task 1) involved specifying, in generic form, the hardware for the new network control center. This task was composed of two parts: the documentation of the current status of communications hardware and data lines making up the network, and the recommendation of equipment to expand network management capabilities. Recommendations covered digital switches, analog access switches, digital and analog diagnostic instruments, equipment cabinets, control consoles, terminals, and printers. Less expensive, but equally critical, components included cables, specified by type and quantity, equipment shelves, and cabling supplies.

Task 2 consisted of creating detailed functional specifications, again in generic form, for the various network control center components. Most of the hardware was procured through a competitive bidding process based on issuance of requests for proposals with detailed requirements specifications for each component. The components were divided into six groups:

- Digital switching equipment.
- Analog switching equipment.
- Analog diagnostic instruments.
- Digital diagnostic instruments.
- Network control console cabinetry (including equipment racks).
- Cables and cabling supplies.

Task 3 consisted of the selection and procurement of network control center equipment. Some (e.g., front-end processors, local terminal controllers, control terminals, printers, data service units–channel service units (DSU-CSUs), and selected long- and short-haul modems) were obtained on a sole-source basis. The installation of new data lines was performed by New York Telephone, and IBM supplied front-end processors, control terminals, and various required printers. MTA management decided to standardize on IBM host and front-end processor products before building the control center. AT&T provided modems and DSU-CSUs.

Task 3 activities included the design of new telephone company data line terminations and chambers (including New York Telephone 820 arrangements), cable connections from 829 terminators to the demarcation points, and space for AT&T DSUs, CSUs, and modems. In addition, the configuration of front-end processors, terminal control consoles, and network control printers was reviewed and recommended.

Task 3 also included evaluating the vendor bids according to the MTA's requirements and modifying those bids to facilitate comparisons. Comparisons and equivalent price estimates were compiled for each vendor's bid, and some were eliminated. The features and capabilities of the acceptable bids were then reviewed and cost comparisons were created. A written report summarizing bid comparisons was prepared for MTA management after negotiating best and final bids from top vendors. During vendor selection, acceptable product delivery schedules were a prerequisite.

A detailed implementation plan—the key to successful installation of the control center—was developed in task 4. Activities included developing plans for data line installation and cutover, final central-site equipment layout, and dismantling of the old facilities. The integration of network control components required final detailed design. Additional work included planning the configuration and installation of the digital switch, the analog access switch, coaxial patch panels, and terminal controllers. Final planning to refine the floor plan and equipment layout was also necessary.

Task 5 included the design of the network control console (the focal point of the network control center) and cable interconnections. The network operators control, monitor, and troubleshoot the network from this point. In addition, the design of cable runs and interconnections should allow for effective troubleshooting and the orderly addition of new cabling. Planning for the installation and wiring of analog and digital diagnostic instruments was included as part of the overall design for cable runs and interconnections. A formal scheme for cable labeling and component naming was also developed.

Task 6 consisted of managing the general construction and network control center implementation. Activities included supervising the New York Telephone installation, testing, and cutover of the data lines. Equipment cabinets and the control center console required physical assembly. Installation and testing of front-end processors, data service units, analog modems, the matrix switch, and the bus and tag cabling for front-end processors were provided by the vendors. MTA personnel installed the analog access switch, analog and digital diagnostic devices, and cabling.

Task 7 involved creating as-built documentation and providing operator training on the control center components. The following equipment was documented: central-site network components, data lines, signal cabling, and signal cabling interconnections. Operating procedures were written for analog and digital diagnostic devices, the digital matrix switch, and the analog access switch. Training included basic network control concepts, diagnostic instrument use, operation of the maintenance and operator subsystem with IBM 3725 front-end processors, diagnostic troubleshooting, change management control, and initiating recovery procedures using IBM software.

Configuring the Front-End Processors

Significant effort was devoted to optimally configuring the front-end processors. Four options were formulated, each affecting the network's overall functional abilities and operational characteristics. In addition, the cost differential among the alternatives exceeded $300,000. It was decided to also use two IBM 3710 controllers—reducing the cost of asynchronous port support by approximately $50,000. The four options were:

- Three 3705 and three 3725 front-end processors with embedded back-up ports.
- Four 3725 front-end processors with embedded backup ports.
- Four active 3725 front-end processors with one cold standby 3725.
- Three active 3725 front-end processors and one cold standby 3725.

Configurations that used only 3725 front-end processors were preferred to mixing 3705s and 3725s. The 3725 front-end processor supported later releases of IBM network software and provided an easier migration to the new network control center. Configurations that employed the cold standby backup processor were preferred to those that used the embedded backup port strategy because they provided more efficient sparing in the event of an entire front-end processor failure.

The third option, four active 3725s and one on cold standby, was recommended because it satisfied the MTA's projected front-end port requirements for several years and eliminated the need for short-term upgrades. Because of the planned relocation of the data center during 1989, however, MTA management decided to support only the immediate (i.e., 1987) requirements, and the fourth option was chosen. Functional characteristics for options 3 and 4 were similar.

Projecting Port and Processor Requirements

Exhibit IV-3-3 summarizes the communications controller port requirements from 1985 through 1989, divided between active (in use) and backup ports and according to protocol and speed. The total controller port require-

Line Control	Speed (bps)	Existing		1986		1987		1988		1989	
		In Use	Backup	In Use	Backup	In Use	Backup	In Use	Backup	In Use	Backup
Asynchronous	134	9	3	9	3	9	3	9	3	9	3
	300	17	2	17	2	17	2	17	2	17	2
	600	1	0	1	0	1	0	1	0	1	0
	1,200	6	0	6	0	11	0	16	10	16	10
	2,400	0	0	20	0	24	0	26	0	26	0
	4,800	0	0	5	0	15	0	19	0	24	0
Subtotal		33	5	58	5	77	5	88	15	93	15
Remote Job Entry	2,400	1	1	1	1	1	1	1	1	1	1
	4,800	16	11	16	11	16	11	19	14	19	14
	9,600	24	4	25	5	26	6	26	6	27	7
	56,000	1	1	3	2	4	2	5	2	6	2
Subtotal		42	17	45	19	47	20	51	23	53	24
Bisynchronous	1,200	0	0	0	0	0	0	0	0	0	0
	2,400	15	0	15	0	15	0	15	0	15	0
	4,800	68	14	78	14	79	9	78	9	77	9
	9,600	46	7	60	7	62	7	62	7	63	7
	56,000	0	0	0	0	0	0	0	0	0	0
Subtotal		129	21	153	21	156	16	155	16	155	16
Synchronous Data Link Control	1,200	0	0	0	0	0	0	0	0	0	0
	2,400	0	0	0	0	0	0	0	0	0	0
	4,800	7	0	7	0	7	0	12	0	22	0
	9,600	39	17	89	44	179	64	191	66	202	66
	56,000	2	0	9	7	9	7	11	9	16	14
Subtotal		48	17	105	51	195	71	214	75	240	80
Total		252	60	361	96	475	112	508	129	541	135
Grand Total (in use plus backup)		312		457		587		637		676	

Exhibit IV-3-3. Front-End Processor Port Requirements for All Agencies

ment was projected to grow from 312 to 676 ports by the end of 1989. The majority of data traffic flows through the MTA data center at 9,600 bps; 4,800 bps is the next most common speed. The trend is toward increased use of 9,600-bps and 56K-bps data. The total number of 56K-bps ports in 1989, however, was projected to be less than 5% of the total number of ports.

The 3725 front-end processor supports 1.544M-bps (T1) links, but this support consumes most of the processing power. The T1 data rate is used by equipment in the network control center, but the MTA did not project using it as a direct link to the front end.

The network control center had to support five types of protocols; asynchronous protocols; remote job entry protocols, including IBM 2780 and 3780; bisynchronous protocols, including 3271 BSC; IBM's synchronous data link control protocol; and other high-level data link control protocols.

All protocols (except the high-level data link control) were supported by the IBM front-end processors. The use of asynchronous protocols was expected to grow approximately threefold in the number of ports by 1989, primarily as a result of the increased number of personal computers requiring mainframe access. Remote job entry and bisynchronous protocol use was expected to remain relatively constant, with little future growth predicted. Synchronous data link control protocol use was expected to grow rapidly, increasing in port requirements fivefold by 1989, and making it the predominant MTA data communications network.

High-level data link control protocol use, particularly in the form of X.25 links, is not shown in Exhibit IV-3-3. Although not used by the MTA in the front-end processor, this protocol was supported by the 3725 processor. The protocols were also used by other data communications devices (e.g., multiplexers).

Communications Component Sparing Requirements

The three categories of component sparing required by the MTA data center—analog-digital line sparing, modem-codec sparing, and communications processor-multiplexer sparing—are depicted in Exhibit IV-3-4. Analog-digital line sparing places a switching device between two circuits—a primary and a backup—and one data communications equipment device (e.g., a modem). When the primary circuit fails, the backup circuit is switched to the same modem, which eliminates the duplication of data communications equipment when backup lines are used. The data center supports several critical user sites using two circuits with only one active at any time.

The second component sparing requirement is modem-codec sparing. Codecs are digital coders and decoders or channel service units–data service units—a requirement for leased digital service. This sparing technique places switching devices on both the telephone company–line side and the digital side of each data communications equipment device, and spare

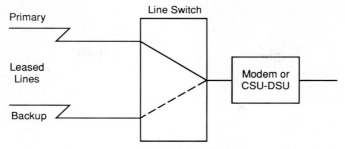

a. Analog-Digital Line Sparing

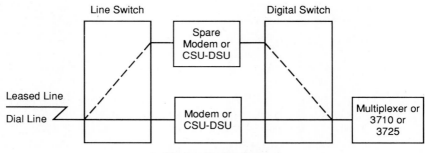

b. Modem-Codec Sparing

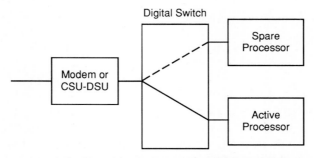

c. Communications Processor–Multiplexer Sparing

Notes:
CSU   Channel service unit
DSU   Data service unit

Exhibit IV-3-4.   Component Sparing Requirements

---

devices may be connected in a similar fashion. When a device fails, a spare device can be switched into place, allowing for the pooling of spares.

The third component sparing requirement involves communications processors—sparing an entire communications processor in addition to the individual ports within the processor. To do this, a switching device is needed between the digital side of a data communications equipment device and the ports on the communications processors. In the event of a communications processor failure, the digital data streams are switched to the spare processor. Data center management chose a cold standby strategy for backup of front-end processing resources. Multiplexers that are part of this generic sparing category include statistical, time division, and wideband.

Data center switching requirements can be divided into three categories: bus and tag channel switching; digital switching including RS-232 and V.35 interfaces; and telephone company four-wire switching including voice-grade analog, low-speed leased digital, and wideband leased digital transmission channels. Bus and tag channel switching at the data center is handled by IBM 2814 and 3814 channel switches and was not moved into the new network control center.

The MTA data center's generic digital data communications and telephone company data line switching requirements are summarized in Exhibit IV-3-5. The basic requirements for each interface type included both diagnostic and reconfiguration activities. Digital diagnostics require access to the digital interface, enabling attachment of the appropriate diagnostic device. The extensive digital diagnostic activities routinely performed at the data center include bit error-rate testing, monitoring data line traffic, emulating data devices and protocols, monitoring RS-232 and V.35 interfaces, and performance monitoring. Exhibit IV-3-5 also describes the functional requirements—ranging from regular access to a small number of interfaces for short periods of time to constant monitoring of many interfaces at all times—of interface access for each diagnostic activity.

The digital interface reconfiguration capabilities are also quite extensive. The five primary activities include front-end processor sparing, front-end processor port sparing, central-site multiplexer switching, central-site protocol converter switching, and disaster recovery.

The data center needed to regularly reconfigure its front-end processors for backup, maintenance, and network testing purposes. The network control center had to be able to easily reconfigure the connection of front-end ports and modems. In addition, regular reconfiguration of the connections between multiplexers and associated front-end ports had to be supported. Connections among protocol converters and other types of communications processors and the modems and codecs of data lines required digital interface flexibility on a regular basis. Implementation of disaster recovery capabilities required the infrequent reconfiguration of all data communications lines.

The requirements for telephone company line interface access and reconfiguration were considerably less complex than for the digital interfaces. Interface access was required on a regular basis for a short period of time to facilitate diagnostic and facilities benchmark activities.

Activity	Functional Requirements
Digital Diagnostics	
Bit error-rate testing	Regular access to small number of lines for 5–10 minutes per test
Monitor data traffic	Regular access to small number of lines for 5–30 minutes per test
Emulate data traffic	Regular access to small number of lines for 5–30 minutes per test
Monitor RS-232 and V.35 interfaces	Regular monitoring of many lines at all times
Performance monitoring	Constant monitoring of many lines at all times
Digital Interface Reconfiguration	
Front-end processor sparing	Regular reconfiguration of front ends for backup and testing
Front-end processor port sparing	Regular reconnection of lines to front-end processor ports plus modem substitution
Central-site multiplexer switching	Regular reconfiguration of selected ports for multiplexer sparing
Central-site protocol converter switching	Regular reconfiguration of selected front-end processor ports to protocol converters
Disaster recovery	Infrequent reconfiguration of all lines
Line Diagnostics and Interface Reconfiguration	
Line diagnostics	Regular access to selected lines for testing
Line reconfiguration	Regular reconnection for modem and codec sparing
	Regular reconfiguration of selected lines for backup

Exhibit IV-3-5.   Summary of Generic Data Communications Switching Requirements

The generic switching requirements for telephone company line interface reconfiguration included regular reconnection for modem and codec sparing purposes. Regular reconfiguration of selected data lines for backup was also an important requirement.

Telephone company line interface access requirements included extensive access to the data communications equipment side of the data line in a terminating test connection. The majority of digital interface testing is accomplished by passively monitoring the interface, whereas telephone line testing is achieved by terminating the line into the diagnostic device. Use of a telephone line interface access switch was recommended as the generic technique to meet these requirements in a cost-effective manner.

Digital Switching Requirements

The MTA's digital interface access and reconfiguration requirements include manual patching, electromechanical bulk switching, electromechanical switching with control of individual ports, and electronic matrix switching. The data center's requirement for front-end processor sparing, either with embedded ports or cold standby, meant that manual patching would be possible only if it were supplemented with another digital switch facility. The remaining digital interface access and reconfiguration alternatives were electromechanical bulk switching with manual patching, individual port switching with manual patching, electronic matrix switching with manual patching, and electronic matrix switching without manual patching.

Patching.   Patching allows simultaneous access to any number of interfaces. Disadvantages relate to the architecture's inability to perform a timely, large-scale reconfiguration. Digital patching is difficult to incorporate into the control console because of its large physical size and the tendency of the patch field to grow. Considering its limited capabilities, patching is a costly option.

Electromechanical Bulk Switching.   Electromechanical bulk switching limitations include the inability to switch selected port groups, particularly when the port section is constantly changing. The number of switch modules required to effect one-to-three sparing was considered difficult to manage and very costly for the level of operation.

Electronic Matrix Switch.   The electronic matrix switch was the alternative recommended to meet the MTA's digital interface access diagnostic and reconfiguration requirements. The most significant advantage over electromechanical switching, with or without supplemental patching, is the ability to rapidly reconfigure large-scale port connections in front-end processor sparing. This capability was also required for sparing multiplexers, protocol converters, and other communications processors. Matrix switch technology can meet the requirements for an unlimited number of diagnostic ports installed on a modular basis including monitor, data circuit termination, and data terminal equipment test ports. Different port configurations are stored in the digital matrix switch's memory and are quickly invoked under operator command.

Electronic matrix switch technology combines digital interface access and reconfiguration functions into a single piece of hardware—making this switching device a critical link in the flow of network traffic. As such, a high level of component redundancy is required, regardless of the chosen product.

The MTA needed many of the capabilities offered by the latest generation of matrix switch technology, including a large (2,000 to 4,000) port capacity, a nonblocked architecture that allows any-to-any port connectivity, a modular architecture that allows port capacity to be added without interrupting service for recabling, a passive and active port monitoring capability, a signal alarming of ports, a user-friendly operator interface, a comprehensive security capability to prevent unauthorized access to switch functions, a switch event logging feature, and a management reporting capability.

Recommended Digital Matrix Switch Configuration

The recommended digital matrix switch configuration block diagram is illustrated in Exhibit IV-3-6. The switch is connected to all front-end processor ports and to the digital side of all data communications equipment devices. It is also connected to both the input and output ports of two IBM 3710 network controllers, all low- and medium-speed multiplexers, and the RS-232 and V.35 ports on the T1 multiplexers. Immediate connection of the high-speed 1.544M-bps trunk ports to the digital matrix switch was not recommended, but the switch would have to support T1 transmission rates in the future.

Requirements for the baseline configuration (front-end option 4) and the 1989 growth configuration (front-end option 3) are summarized in Exhibit IV-3-7. Port counts are divided by device or facility type and by interface type. The minimum data center functional matrix switch requirements included initial support for 1,200 to 1,500 ports. Initial requirements also included 1,100 to 1,400 RS-232 ports and approximately 100 V.35 ports.

None of the digital matrix switch vendors initially satisfied the MTA's warranty requirements, and therefore all vendors were asked to submit clarifications to their bids. The three best submissions were T-Bar's Distributed Switch Matrix Galaxy Plus, Dynatech's CTM-2000, and Bytex's Autoswitch 4000. The Bytex Autoswitch 4000 was the most inexpensive product that fully met matrix switch requirement specifications.

Telephone Company Line Switching Requirements

The variety of solutions to the MTA's telephone company interface access and reconfiguration requirements included manual patching, electromechanical bulk switching, electromechanical switching with control of individual ports, and electronic matrix switching. Data center requirements for data line sparing and modem substitution were very limited compared with their digital counterparts. As a result, manual patching was possible but limited. The important issue was whether or not the MTA required a supplementary line switching facility. The optimum telephone company

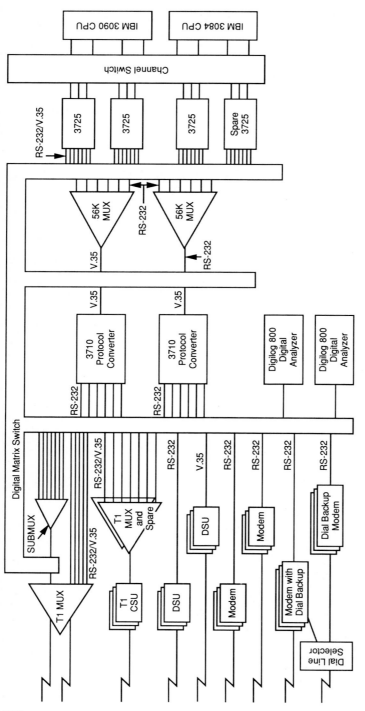

Notes:

CPU	Central processing unit
CSU	Channel service unit
DSU	Data service unit
MUX	Multiplexer
SUBMUX	Submultiplexer

Exhibit IV-3-6.   Recommended Digital Matrix Switch Connections

line interface access and reconfiguration strategy involved a decision among these choices:
- Manual patching alone.
- Electromechanical access switch with manual patching.
- Electronic matrix switch with manual patching.
- Electronic matrix switch alone.

Patching offers the advantage of simultaneous access to any or all lines. The disadvantages of patching during large-scale reconfigurations were not significant because bulk line interface reconfigurations were not required. Because of its small physical size, analog line patching is easier to centralize on the control console than digital patching. A typical analog patch panel for 25 four-wire leased lines occupies 1.75 in of vertical rack space. Line patching is very cost-effective when the capabilities are viewed as insurance at a modest one-time cost of about $25 per line. Line patching uses inexpensive telephone technology, benefiting from the economies of scale that accompany this mature industry.

Electronic matrix switch technology was not recommended for line interface access and reconfiguration because of cost and functional reasons. For analog data line signals, electronic matrix switch technology uses analog-to-digital converters to digitize the analog waveform and move it through the switch. This transformation is reversed with a digital-to-analog conversion as the signal leaves the matrix switch and is presented to the modem. Converting the analog data communications signal was not recommended, however, particularly considering that the conversion was performed twice.

Current electronic matrix switches are unable to support the four-wire digital transmission line signal because of problems with network clocking. The timing of digital service channels is controlled by a master clock in the AT&T communications network and the matrix switch is currently unable to accept this external clocking.

Manual patching and electromechanical access switching (the latter using electromechanical switching relays) provide a copper-to-copper connection between the input and output port. Patches and electromechanical switches do not convert the telephone company signal in any way.

Manual patching, supplemented by an electromechanical access switch, was recommended to alleviate unnecessary wear caused by overuse of the patch fields. A dual-bus electromechanical access switch was recommended to allow the simultaneous testing or reconfiguration of two analog interfaces. The line access switch was the approved method of connecting diagnostic instruments and temporarily connecting substitute central-site modems or data service units. The cost of a dual-bus line access switch was approximately $125 per line.

Baseline Case	Port Type	Quantity	Ports per Device		Total Ports		Total Ports
			RS-232	V.35	RS-232	V.35	
Device							
Active 3725 front-end processors	DTE	3	128	8	384	24	408
Modems and multiplexer ports	DCE				384	22	406
3710 network controllers	DTE	2		1	0	2	2
Cold standby 3725 front-end processor	DTE	1	128	8	128	8	136
3710 network controllers	DTE	2	40	1	80	2	82
Modems (dial and leased line)	DCE				80		80
56K-bps multiplexers	DTE	10		1	0	10	10
Modems and multiplexer ports	DCE				0	10	10
Spare 56K-bps multiplexers	DTE	1	10	1	10	1	11
Spare modems and DSUs	DCE	15	1		15	0	15
Spare modems and DSUs	DCE	5		1	0	5	5
Dial backup modems and DSUs	DCE	10	1		10	0	10
Dial backup modems and DSUs	DCE	5		1	0	5	5
DTE Total					602	47	649
DCE Total					489	42	531
Grand Total					1091	89	1180

Exhibit IV-3-7.   Digital Matrix Switch Port Requirements

Growth Case	Port Type	Quantity	Ports per Device		Total Ports		Total Ports
			RS-232	V.35	RS-232	V.35	
Device							
Active 3725 front-end processors	DTE	4	128	8	512	32	544
Modems and multiplexer ports	DCE				512	30	542
3710 network controllers	DTE	2		1	0	2	2
Cold standby 3725 front-end processor	DTE	1	128	8	128	8	136
3710 network controllers	DTE	2	48	1	96	2	98
Modems (dial and leased line)	DCE				96		96
56K-bps multiplexers	DTE	15		1	0	15	15
Modems and multiplexer ports	DCE				0	15	15
Spare 56K-bps multiplexers	DTE	2	10	1	20	2	22
Spare modems and DSUs	DCE	25	1		25	0	25
Spare modems and DSUs	DCE	10		1	0	10	10
Dial backup modems and DSUs	DCE	10	1		10	0	10
Dial backup modems and DSUs	DCE	5		1	0	5	5
DTE Total					756	61	817
DCE Total					643	60	703
Grand Total					1399	121	1520

Exhibit IV-3-7. *(Cont)*

Notes:
DCE   Data communications equipment
DSU   Data service unit
DTE   Data terminal equipment

Three vendors offered products consistent with the MTA's requirements: Dynatech, T-Bar, and Ameritec. Performance specifications were developed for the four-wire line access switch and the four-wire line patch fields. Ameritec's Modem AM-6 Test Access Switch was the most inexpensive product fully meeting the requirement specifications.

OVERVIEW OF NETWORK MANAGEMENT ALTERNATIVES

Network management uses either sidestream or mainstream diagnostic techniques. The primary focus of most sidestream-based solutions is the physical interface, whereas mainstream methods typically use both physical and logical network views.

Sidestream. Sidestream network management techniques rely on sideband communication. Sidestream test modules sample and analyze data from the RS-232 interface between each front-end processor port and modem. The test module gathers diagnostic information about modems, data service units, channel service units, multiplexers, packet switches, and telephone company circuits. This information is transmitted to the central site through a secondary transmission channel that is frequency-division multiplexed with the primary data channel.

Secondary-channel modem tests fall into two categories: nondisruptive tests that use the sideband channel and disruptive tests that use the primary data channel. Nondisruptive tests include measurement of analog parameters, testing of remote RS-232 modem-to-terminal interfaces, and disabling a remote tributary modem's transmitter on a multidrop line. The disruptive tests include analog and digital loopback tests, data transmission tests, modem self-tests, and multipoint polling tests.

Sidestream network management products require separate computer processing capabilities, usually in the form of a minicomputer or high-performance microcomputer. The central processor performs the monitoring tasks and provides data base storage for the collection of alarm statistics and performance measurements.

Mainstream. Mainstream network management products consist of incremental hardware and software functions implemented among existing network components. These software programs and optional hardware components feed problem-determination and performance measurement data to a set of network management application programs residing within the mainframe central processing unit or a dedicated network management host.

Network management messages are transmitted within the network— intermixed with the user's data traffic. Information contained within the headers of these alert messages is used by various software components to route diagnostic messages to the appropriate network management application program. Problems reported by a mainstream alerting mechanism include terminal failures, terminal-to-controller in-house wiring failures, remote and local controller hardware and microcode failures, remote and local modem failures, telephone company line problems caused by poor line quality and line failures, and host access method–application software failures.

Mainstream network management is provided through IBM equipment incorporated into the MTA's network: front-end processors, network controllers, terminal control units, terminal subsystem control units, and modems and data service units.

IBM software that provides network management includes Network Communications Control Facility, Network Problem Determination Application, Network Logical Data Manager, Link Problem Determination Application, and Network Performance Monitor.

RECOMMENDED NETWORK MANAGEMENT STRATEGY

The network management strategy recommended was the continuing use of the mainstream method provided by IBM hardware and software. Justification was based on more comprehensive problem detection. Because network management facilities are an integral part of the hardware and software that constitute the network, mainstream network management operation can identify more problems in the network component. Sidestream network management reports are limited to the transmission components, telephone lines, modems, and controller interfaces.

Problem isolation with mainstream network management techniques is also more extensive for the same reason. If the reported problem cannot be traced to a line or modem, the sidestream method provides no information regarding other data communications components. From the user's viewpoint, if part of the isolation process requires testing, mainstream techniques are less disruptive than sidestream techniques.

Response time measurements obtained through sidestream monitoring are not as precise as those obtained through mainstream monitoring. Sidestream monitoring views only the middle of the communications path between user and host and can accurately determine only the host turnaround time. Sidestream network management is only able to estimate the network delay portion of response time by monitoring the time interval between successive terminal polls. One half of the poll-to-poll delay is generally assumed to be the average time that elapses between the user pressing enter and the arrival of a poll message at the terminal. Response time is then approximated by adding the host delay and half the poll-to-poll time. Mainstream monitoring is performed at the user terminal and measures the exact interval between pressing enter and the first subsequent display on the terminal's screen.

Sidestream network management is a mature technology for analog communications circuits but a recently emerging technology for digital service. The digital secondary channel must be provided by the communications carrier and is only now being made available in telephone company offices. Approximately 75% of the data circuits terminating at the MTA data center were digital lines. To take advantage of the secondary channel, the MTA would have needed to replace all existing transmission circuits and data service units–channel service units.

Network management recommendations to the MTA included upgrading the new network control center to IBM's NetView network management environment. Although it consists mainly of repackaged existing network management products, NetView is clearly IBM's network management system for the foreseeable future. The data center planned to upgrade to the latest levels of other communications software (e.g., the network control program), and migration to NetView proved a logical extension of this process.

Problem detection is accomplished jointly through NetView and the port-alarm features of the digital matrix switch. NetView's capabilities for detecting hardware and software failure are effectively complemented by the digital interface monitoring in the matrix switch. Up to eight signal pins on the RS-232 or V.35 interfaces, including data carrier detect and data set ready, may be monitored. Any signal transition may be programmed to trigger alarms and issue event reports that are monitored by the network operations staff.

Problem isolation at the data center is accomplished by the combination of NetView and the switching capabilities of the digital matrix and line access switches. When necessary, spare lines, modems, codecs, multiplexers, front-end ports, and the cold standby front-end processor may be quickly connected to replace the suspected failed component. Problems are determined using NetView in conjunction with the digital interface and telephone company line analyzers in the network control center.

Response time analysis is performed using NetView and IBM's Network Performance Monitor software. Response time measurements are also available through the digital interface analyzers in the network control center. This second source of measurements provides an independent verification of the measurements obtained through software. Response time measurements are also becoming available as the MTA migrates to IBM's new 3174 control units. The response time monitor feature is standard on 3174 controllers but optional on the 3274 control units.

RECOMMENDED MODEM AND DATA SERVICE UNIT STRATEGY

The data center decided to continue using AT&T's analog modems and data service units. Except for circuits from the New York City Transit Authority, all communications lines terminating at the data center used AT&T equipment. AT&T devices were located at both the data center and the remote user locations.

Specific recommendations for data circuit terminating equipment located in the network control center included the deployment of the AT&T 2600 Series data service units for all digital service circuits. These compact devices are packaged in multiple mounting racks and feature a shared front panel for a shared diagnostic unit. The 2600 Series data service units were fully compatible with the 2500 Series and the older 500 Series of data service and channel service units, which represented a significant installed base within the MTA. AT&T 2600 Series diagnostics include continuous monitoring of the data service unit, the digital service line, and the shared diagnostic unit; status of the data service unit options; speed; diagnostic tests and commands including self-test, disable-enable, and local and digital loopback; and sending a test pattern.

AT&T 2096A modems were used for analog communications circuits. They represent the low-cost, no-frills model of the 2096 modem series and can operate at 2,400, 4,800, and 9,600 bps. The 2096A analog modem was already in use at the data center. The AT&T 2224 modems were used for dial-up communications connections. This type of modem operates in a synchronous or asynchronous fashion, at 300, 1,200, or 2,400 bps and has an integrated automatic calling feature. This was the data communications equipment device chosen to support personal computer dial-up access to the data center's 3710 network controllers.

Because the MTA rented its modems and data service units from AT&T, the cutover of the data communications lines and data communications equipment devices was made considerably easier by the vendor. All AT&T data communications equipment devices were duplicated and the new units installed in the network control center. After the lines were cut over to the new equipment, the old devices were returned to AT&T.

This was not the case with all the data communications equipment devices at the data center. In particular, the New York City Transit Authority owns standalone modems and codecs supplied by Codex and General Datacomm. Unless duplicate equipment—preferably rack-mounted—were secured, the cutover of the Transit Authority circuits to the network control center would involve more downtime than when the other circuits were moved. It was a good opportunity for the Transit Authority to consolidate and upgrade its data communications equipment devices, so it provided a duplicate T1 multiplexer and a Timeplex Link-1.

DIAGNOSTIC DEVICES

Diagnostic instruments are an important part of the network control center. Four types were recommended: a telephone company voice frequency (analog) line interface analyzer, a digital interface analyzer, a bit error rate tester, and a response time analyzer. Digital interface analyzers are now designed to include bit error rate testing and response time analysis. Therefore, these requirements were incorporated into the performance specifications for the digital interface analyzer.

Two analyzers of each type (i.e., analog and digital) were recommended. Although these devices are very reliable, they can become inoperable. Ownership of two devices protects the data center from being without diagnostic capabilities if one is unavailable (e.g., during repair), and two units permit simultaneous troubleshooting of more than one problem. Two digital interface analyzers are necessary to perform end-to-end diagnostics. To simplify operation and training, the purchase of two identical units of each type was recommended. The budgeted cost of the two analog line analyzers was $15,000. A comparison of selected line analyzers considered by the MTA is summarized in Exhibit IV-3-8. Bradley Telcom's PB-IC/D was the most inexpensive product that fully met the specification requirements.

The two digital interface analyzers had a total budgeted cost of $37,000. A summary comparison of possible vendors for selected digital interface analyzers considered by the MTA is depicted in Exhibit IV-3-9. The Digilog 800 was the most inexpensive product that fully met the MTA's requirements.

ADDITIONAL NETWORK CONTROL CENTER COMPONENTS

The network control center components, including the 3725 front-end processors and 3710 network controllers, are summarized in Exhibit IV-3-10. The digital matrix switch was the important—and most expensive—network management component in the communications control center. The analog line access architecture consists of an analog access switch and analog manual patch fields. The digital matrix switch works in conjunction with the analog access-patch facility for integrated diagnostics and central-site modem substitution.

The remaining components for vendor bidding included modem cabinets and the control center console. Control console terminals, printers, and 3270 coaxial cable multiplexers were provided by IBM. The remaining cables and supplies were specified and competitively bid. All furniture (e.g., desks, chairs, and storage cabinets) was procured from the MTA's existing supply.

Product / Feature	Bradley PB-1C/D	Halcyon 520B	Halcyon 701A	Halcyon 702A	Hekimian 41-01	Hekimian 42-10	Hekimian 3700	Hewlett-Packard 4945A
Frequency Measurement	40 Hz to 4 kHz	200 Hz to 4 kHz	50 Hz to 20 kHz	50 Hz to 20 kHz	40 Hz to 20 kHz	40 Hz to 20 kHz	40 Hz to 4.1 kHz	20 Hz to 110 kHz
Amplitude Level Measurement	Yes	Yes	Yes	Yes	Yes	Yes	Yes	Yes
Noise Measurement	Yes	Yes	Yes	Yes	Yes	Yes	Yes	Yes
Signal to Noise Ratio	Yes	Yes	Yes	Yes	Yes	Yes	Yes	Yes
Fixed or Variable Signal Generator	Yes	Yes	Yes	Yes	Yes	Yes	Yes	Yes
Audio Monitor	Yes	Yes	Yes	Yes	Yes	Yes	Yes	Yes
Portable	Yes	Yes	Yes	No	No	Yes	Yes	Yes
CRT Display	Yes	No	No	No	No	No	No	Yes
Phasor-Domain Measurement Technique	Yes	No	No	No	No	No	No	No
Phase Jitter	Yes	Yes	No	No	No	No	No	Yes
Transient Hits	No	Yes	No	No	No	No	No	Yes
Amplitude-Phase Hits	No	No	No	No	No	No	No	Yes
Envelope Delay	No	No	No	No	No	No	No	Yes
Cost ($)	6,500	8,995	1,695	1,695	1,595	2,505	6,500	14,950

Exhibit IV-3-8. Telephone Company Line Analyzer Feature Comparison Chart

Product / Feature	Atlantic Research International 3600	Atlantic Research International 4600	Atlantic Research Comstate II	DTI Simon 5	Digilog 600	Digilog 800	Digitech 520	Hard Eng. Byte Bug Model 645	Hewlett-Packard 4955A	Hewlett-Packard 4953A	FDACOM IDAXP	Northern Telecom Spectron D-901	Tekelec Chameleon II	Tekelec Chameleon 32
Line Monitor	Yes	Yes	Yes	Yes	Yes	Yes	Yes	Yes	Yes	Yes	Yes	Yes	Yes	Yes
Programmable Simulator	No	Yes	Yes	Yes	Yes	Yes	Yes	Yes	Yes	Yes	Yes	Yes	Yes	Yes
Large 7-in Diagonal CRT	No	No	No	No	No	Color Yes	No	No	Yes	Yes	Yes	Yes	No	Color Yes
Extended Video	Yes	Yes	Yes	Yes	Yes	Yes	Yes	Yes	No	No	No	Yes	No	Yes
Remote Printer	Yes	Yes	Yes	Yes	Yes	Yes	Yes	Yes	Yes	Yes	Yes	Yes	Yes	Yes
Transmission Rate	19.2K bps	19.2K bps	64K bps	64K bps	128K bps	256K bps	256 bps	256K bps	72K bps	72K bps	64K bps	1.54M bps	128K bps	1.54M bps
Bisynchronous	Yes	Yes	Yes	Yes	Yes	Yes	Yes	Yes	Yes	Yes	Yes	Yes	Yes	Yes
Systems Network Architecture	Yes	Yes	Yes	Yes	Yes	Yes	Yes	Yes	Yes	Yes	Yes	Yes	Yes	Yes
Asynchronous	Yes	Yes	Yes	Yes	Yes	Yes	Yes	Yes	Yes	Yes	Yes	Yes	Yes	Yes
X.25	Yes	Yes	Yes	Yes	Yes	Yes	Yes	Yes	Yes	Yes	Yes	Yes	Yes	Yes
Interfaces: RS-232	Yes	Yes	Yes	Yes	Yes	Yes	Yes	Yes	Yes	Yes	Yes	Yes	Yes	Yes
RS-449	No	No	Yes	Yes	Yes	Yes	Yes	No	Yes	Yes	Yes	Yes	Yes	Yes
X.25	Yes	Yes	Yes	Yes	Yes	Yes	Yes	Yes	Yes	Yes	Yes	Yes	Yes	Yes
V.35	Yes	Yes	Yes	Yes	Yes	Yes	No	Yes	Yes	Yes	Yes	Yes	Yes	Yes
Tape Storage	Yes	Yes	Yes	No	No	No	Yes	No	Yes	Yes	No	No	No	No
Diskette Storage	No	No	No	Yes	Yes	Yes	Yes	Yes	No	No	Yes	Yes	Yes	Yes
Hard Disk Storage	No	No	No	Yes	No	Yes	Yes	No	No	No	No	No	No	Yes
Cost ($)	10,000	14,500	19,900	14,995	12,495	19,495	18,500	4,500	18,680	12,000	23,500	15,000	19,500	31,500

Exhibit IV-3-9. Digital Analyzer Feature Comparison Chart

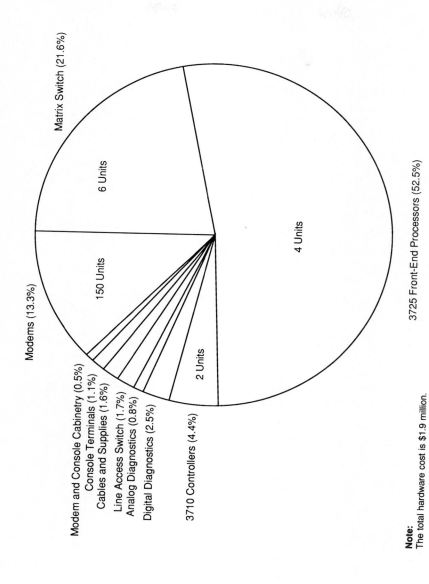

Matrix Switch (21.6%)

6 Units

Modems (13.3%)

150 Units

Modem and Console Cabinetry (0.5%)
Console Terminals (1.1%)
Cables and Supplies (1.6%)
Line Access Switch (1.7%)
Analog Diagnostics (0.8%)
Digital Diagnostics (2.5%)

3710 Controllers (4.4%)

2 Units

4 Units

3725 Front-End Processors (52.5%)

Note:
The total hardware cost is $1.9 million.

Exhibit IV-3-10.   Network Control Center Component Cost Breakdown

SPACE PLANNING

The network control center floor layout is depicted in Exhibit IV-3-11. Space was provided for a maximum of six fully configured IBM 3725 front-end processors, each with the 3726 expansion frames. Equipment racks for the majority of components are located between the front-end processors and the control console.

The control console design provides for growth capacity and flexibility in mounting console components. Although the floor plan indicates specific components in the equipment racks, the electrical and physical site preparations allow changes in equipment location. After the component products were selected, the final control console and equipment cabinet configurations became part of the implementation plan formulation.

The final control console layout is depicted in Exhibit IV-3-12. Four turrets were provided in the center of the console to house analog patch fields, the analog access controller, analog diagnostic devices, and digital diagnostic devices. On each side of the center, additional cabinets were added with angle components to form a horseshoe-shaped control console.

SUMMARY

The Metropolitan Transit Authority data center's network control center implementation plan consisted of four main subtasks. In the first, all new equipment was deployed in the structurally complete network control center. One new IBM 3725 front-end processor, three new IBM 3726 front-end components, and two IBM 3710 network controllers were the first components installed. Analog and digital switches were installed next, along with the equipment cabinets and control console cabinetry. New AT&T modems and codecs were installed with New York Telephone 829 circuit terminating cabinets before beginning the cutover.

The second implementation subtask was installing the proper cabling. The digital interfaces of the modems, codecs, and the front-end ports were connected to the digital matrix switch. The telephone company interfaces for the modems-codecs and the demarcation were cabled to the line access switch. The channel interfaces of the front-end processors were cabled to the IBM channel switch to facilitate testing of the new system generation tables.

The third subtask was migrating the data lines from the 3725 front-end processors in the main computer room to those in the network control center. After all the data lines were moved off of a 3725 front-end processor, the processor could be moved into the new network control center and coupled with a 3726 expansion unit. One offline 3725 front-end processor was moved at a time. The C-Host 3725 front-end processors were migrated

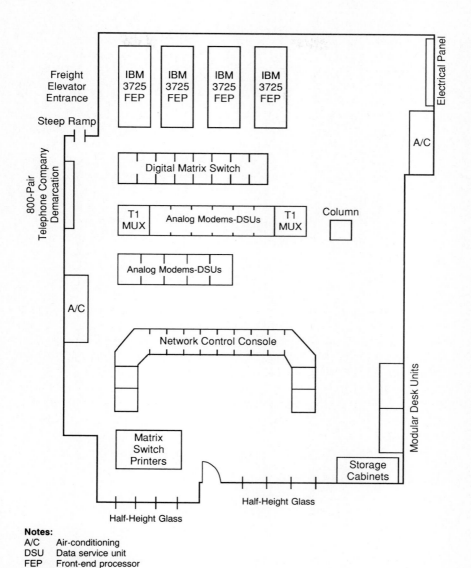

Exhibit IV-3-11. Network Control Center Floor Plan

Notes:

A/C   Air-conditioning
DSU   Data service unit
FEP   Front-end processor
MUX Multiplexer

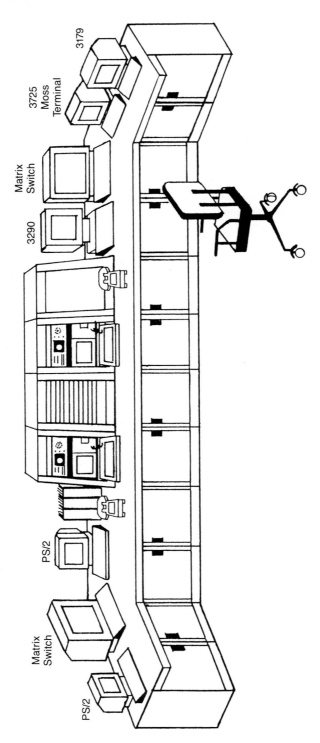

Exhibit IV-3-12. Control Console Layout

first and the Systems Network Interconnection (SNI) 3725/3726 front-end processor followed.

All telephone company data line cutovers were accomplished at the central office. Line cutovers were ordered from the telephone company as outside moves of the station termination at the MTA data center. New York Telephone installed a new dual 600-pair feeder cable from the basement to the 10th-floor network control center. The cutover plan called for New York Telephone to move each line's appearance from the old demarcation point to the new point terminating in the control center. Each circuit was interrupted during the weekend off-hours—for a period of time ranging from five minutes to several hours—while the cutover occurred.

The final subtask completed the cutover of the data lines from the 3705 front-end processors to the 3725 front-end processors in the new network control center. Front-end ports that had duplicate modems in the new network control center were the easiest to migrate. Cutover of the multiplexers was much more difficult. The most difficult facilities to cut over were the T1 multiplexers, because of their numerous ports.

After all the lines were moved off the 3705 front-end processors, the old communications components in the main computer room were dismantled and removed. These components included 3705 front-end processors, modems, data service units, cabinets, line terminating equipment, and cables.

Chapter IV-3-3
Evaluating the Effectiveness of Data Communications Network Controls

James V. Hansen

THERE ARE IN GENERAL FIVE MAIN AREAS for data communications controls: data integrity, data security, backup and recovery, hardware, and administration. Communications technology has brought about the prevalence of flexible distributed network systems and has consequently introduced new areas for control and audit. Primarily, integrity controls, security controls, and hardware controls for these networks should be reevaluated, and the cost-effectiveness of existing controls should be reassessed. Although a distributed data communications network does present some additional risk, there is also opportunity for previously unavailable control techniques, particularly through the use of communications hardware and secure micro-mainframe links. In this chapter, these control areas are discussed in detail, along with systems controls that are not unique to communications networks but are necessary for ensuring effective control of a distributed network.

DATA INTEGRITY CONTROLS

Data integrity controls aim to ensure the quality and reliability of data transmitted over communications links. Without adequate control over data integrity, the most sophisticated system may be rendered inefficient. The sources of data corruption may derive from line noise, erroneous input, deliberate manipulation, and equipment or software failure. In this section, the major controls applied to ensuring data integrity are discussed and guidelines for evaluating those controls are outlined.

Line Error Controls

The effects of uncontrolled or inadequately controlled line errors can be catastrophic. As an example, transactions affecting an online data base arrive with one of every 100,000 bits in error. Each record is updated approximately 100 times per month, and if 1 of 20 five-bit characters is in error, an update error will occur. Approximately 4,500 records will be in error every six months. Moreover, if a transaction-type code is in error, erroneous shipments of inventory may be made or records may be deleted from the data base by mistake.

Line errors are usually controlled by error detection followed by automatic retransmission of the message or by error detection followed by automatic correction of any errors discovered. The message correction method requires that special bits be transmitted with each message, resulting in significant redundancy, and there is a possibility that the error correction may also be in error. Error detection with retransmission is therefore thought to be the more desirable and effective control.

Error detection with retransmission requires a special line protocol to indicate the correct or incorrect receipt of a message. Exhibit IV-3-13 illustrates the four-step retransmission sequence.

The control procedures used at the center of data transmission control should be examined. This can be accomplished by observing the errors occurring in central communications processing. This data is usually gathered from the data control log and from interviewing system operators. The organization's methods for error detection and correction (e.g., parity bits) should be reviewed. Error logs should be examined to verify errors that are discovered and to evaluate their disposition. The error detection and correction system can be tested by transmitting a set of test messages. The extent and pattern of errors generated by the various communications links should also be determined.

User Error Control

Training of employees who use the teleprocessing network should be reviewed. Training documentation should include a description of procedures to assist in the prevention of irregularities that can generate errors. Error recording should be reviewed. All errors in transmission should be logged, with at least the following information recorded:

- Time and date.
- Terminal involved.

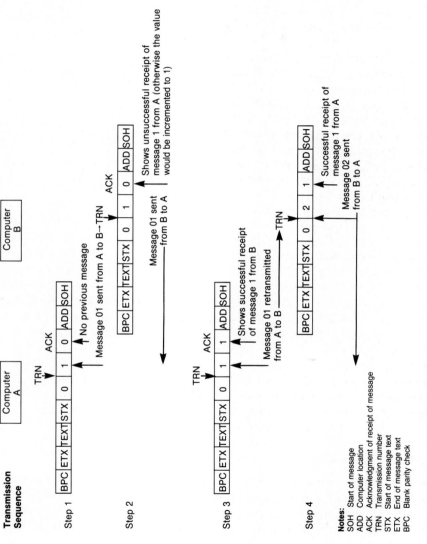

Exhibit IV-3-13. Four-Step Retransmission Sequence

- Number of transaction attempts before message was transmitted correctly.
- Type of error.

An automated error log should be maintained as part of the communications system. The log should include the type of error, the time and date of the error, and the sender's ID. Error correction information should be maintained as well. If the correct message was eventually transmitted successfully, that fact should be recorded along with the number of times the message transmission was attempted. If the error was not corrected, the reason should be indicated. Corrected transmissions should also list the date of correction.

Messages should be stamped with the time and date so that each message can be identified with a unique time frame. That way, when transmission errors are detected, the time, date, and type of error can be logged for later reference. It may be that the incidence of errors is higher from certain locations, over particular lines, or at specific times of the day. Such information can be useful in monitoring system performance and in scheduling processing and maintenance.

Communications Lines

After the communications line is in place, only its reliability can be enhanced. Furthermore, if public lines are used, transmission reliability is governed largely by the quality of the transmission line. Yet even with public lines, the user may have some latitude in specifying the quality of the lines. If a high level of transaction use is expected, a value-added network may provide higher throughput—and a higher level of transmission quality.

The user may also have the option of choosing between analog and digital transmission. Analog transmission lines are widely available; digital transmission is more reliable. In addition, fiber-optic transmission is commercially available. Fiber-optic transmission operates at very high speeds, is reliable, and is difficult for an intruder to tap, so it is ideal for applications in which data integrity is of primary concern.

Concurrent Processing

Multiprocessed online systems are subject to the problem of concurrent updates. For example, the inventory has 100 units on hand. User A and user B simultaneously access record X, which shows 100 units on hand. If user A and user B each concurrently order 75 units on the basis of this information, record X is updated to show 25 units on hand. In fact, 50 units more than are on hand would be sold. To avoid this, the system should have an automatic lockout facility, allowing one user access at a time. With a proper lockout facility, user B would have found record X showing only 25 units on hand.

Micro-mainframe links offer the potential for high-level security leakage, even when data is reasonably well protected on the mainframe. Employees with access to microcomputers must be made aware of security exposures and know how to prevent them. Each employee can be held accountable for security. Security can be more effectively ensured if the company is using virtual system security controls (i.e., security software that provides password identification directories and provides system administrators a way to allocate disk space and file access equitably and efficiently).

After user departments have determined what mainframe data is accessible, a secure communications link should be established. Security clearances should be required to access data in the mainframe files. Specialized file transfer software is typically used. The following procedure should be in place:

- File transfer requests can be initiated from either the mainframe or the microcomputer.
- Upload and download access can be limited to specific fields in a record set—The use of file transfer profiles on the mainframe can provide the basis for implementing this control.
- Each profile defines the transfer record format—The profile should also contain default-record-selection keys. A user may be allowed to override these keys and may also be permitted to select a subset of the accessible fields for transfer.
- Users can initiate a transfer by entering the name of the desired profile into the transfer-handling program—After a profile is established, authorized users can initiate transfers of the necessary information.
- These procedures can be combined with password and information-access protection already existing on the host.
- Additional levels of security can be implemented if desired—For example, additional password controls on transfer access could be effected by attaching passwords to the user codes on the profile list of valid users.
- File creation capability can be limited to particular users or locations.

Distributed Data Controls

In distributed systems, data can be divided into two major categories: local and global. Local data is used only by the local computing device; global data is required by programs that run on at least two computers in the distributed system.

The most serious control difficulties are associated with global data. Global data can be partitioned or centralized. If data is centralized, one of the computers in the network serves as the storage location for all global data. When another computer requires centralized data, it is transmitted to the requesting computer over a communications link. If global data is partitioned, it is distributed among several computers in the network. When a computer requires partitioned data, it must first determine where the data is stored, then initiate a request for the data.

Although maintaining centralized global data is simple, if the central computer fails, applications throughout the network are unable to continue processing. Partitioning global data mitigates this difficulty. There are, however, offsetting control considerations. There are three approaches to the control of partitioned data.

The first approach is to maintain a file copy of all records at each computing location, so that a data change made at one location is transmitted through the network to all other computing locations. A second approach is to use partitioned files, so that only the records that are regularly used by a computing location are stored there. Data that is used intermittently is maintained at the central location or at the location at which it is used most frequently. A third method is to place all necessary detail data at each computing location while maintaining aggregate data in summary form at a central site.

Each of these systems requires an appropriate plan for recovery. For example, replications of centralized global data may be maintained at several locations in the network. These locations need to keep lists of computers that have the replicated data, but they do not require directories that show the locations of particular types of data because all of the data is maintained at each storage location. Exhibit IV-3-14 illustrates the options.

| Storage Strategy | Number of Copies | | Control Concern |
	Replicated	Nonreplicated	
Central-ized	Complete copies of global data are maintained at two or more network locations.	All global data is maintained at a central location.	Data integrity as it is transmitted to and from remote locations.
Parti-tioned	Global data is segmented, but several copies of some or all of the segments are maintained.	Global data is segmented with one copy of each segment maintained. These reside at different locations.	Control over data updates can be complex.

Exhibit IV-3-14.   Data Storage Options for Distributed Networks

Distributed Processing Controls

Distributed processing requires controls in addition to general communications controls. A machine log of all program changes to application programs and other software should be maintained. There should also be a complete log of all restart activity, verifying that files are reconciled after restart. The nature and quality of controls maintained over accumulated transaction backlogs and whether operational statistics are maintained including the logged-on table and the current status of all queues, should be investigated. There should also be a formal plan for covering operations in a degraded mode and a group to furnish assistance to users.

DATA SECURITY CONTROLS

Communications network data have a twofold objective. They must prevent unauthorized access to stored or processed data and protect the communications system from physical damage. The principal controls related to these objectives should be evaluated.

Access Controls

Logical access controls include user authentication and user authorization. Authentication verifies the identity of a user; authorization checks the clearance of a user regarding a particular data item.

Authentication. The most common forms of user authentication are passwords and other dialogue methods. Typically, such methods require that the user supply identifying information—a name, specific date, or location—or an identifying item, such as a card.

Password control concerns the issuance, modification, and deletion of passwords from the system. Password control should be reviewed to ensure that terminated employees are promptly deleted from the security file and that passwords are changed at frequent intervals. Records of password violations should be evaluated to determine the effectiveness of password control procedures.

All denied-access attempts should produce a log entry and a time delay. The system should automatically disconnect users who are unable to supply a valid password within a given time or number of tries. It is also desirable that passwords not be displayed on the screen and that the system not be left unattended after sign-on.

Networks that rely on communications switching equipment should include computer identification of the user device. Switching mechanisms sometimes misconnect. For example, a polling mechanism may connect an incorrect device because of address modification by noise in the communications line. Such errors are often undetected, which degrades reliability and can compromise security. Automatic identification provides a backup.

Undesirable outside intrusion can be prevented by implementing dial-up and callback procedures. With this strategy, anyone attempting to gain system access need only supply the necessary authentication information, after which the system hangs up and calls back to a specified legitimate terminal location. This way, if intrusion occurs, it will have to emanate from a legitimate network microcomputer. Dial-up and callback capabilities are expensive.

Unauthorized system access can be further controlled by allowing only certain types of transactions to be transmitted from a given terminal. When terminals are dedicated to performing a narrow range of tasks, this can be an effective and easy-to-apply control. It is also appropriate in some instances to limit terminals to read-only functions.

Authorization. Authorization procedures identify action privileges for the user. Authorization determines what the user can legitimately do once granted system access—which files or programs can be accessed and what can be done with them.

There are two fundamental approaches to authorization. The first method is to maintain an authorization matrix in a protected area of memory. Such a matrix is simply a listing of all legitimate system users, the files and programs they are allowed to access, and what they are allowed to do with those resources. Evaluating the authorization matrix entails attempting to execute a sample of unallowable functions and record the results.

The second method of controlling resource access is to assign passwords to each restricted-access file and program. There can be more than one password for each resource, one allowing read-only access and another allowing read/write access. For such systems, a sample of resources to be restricted should be selected and appropriate passwords tested.

Auditability of Access Controls. The access control mechanism should provide regular audit trails and reports that record which users accessed what data and record unauthorized entry attempts.

Information provided on the audit trails should be sufficient to provide useful information for managing access control functions. For example, information on unauthorized access attempts should identify that event, when the event occurred, the terminal being used, what data was sought, and any other details that may facilitate clear reconstruction of the attempted intrusion. In addition, the audit trails should be protected from unauthorized access. Audit trail logs should be designed to summarize and highlight important information.

In communications systems, an audit trail must be collected and made available concurrently with application processing. Early warnings of attempts to circumvent access control mechanisms are essential to effective security follow-up. It is unusual for an intruder to break into the system on the first try.

Automatic Shutdown.   An attempt to perform an unauthorized event from a terminal should be made to determine that the terminal does, in fact, shut down. The security log indications of the number of invalid accesses should be compared with the shutdown log to verify that the shutdown process is working.

Encryption

Encryption can be an effective process for protecting data during transmission within distributed computer systems and networks. The degree of protection provided by encryption depends on the encryption algorithm employed, the implementation of the algorithm, and the administrative procedures regulating the use of the algorithm. Additional security requirements of user identification, access authorization, and security auditing can be satisfied by combining encryption technology with a network access control machine in a network security center.

Encryption uses data encodings to hide messages. Although no method has yet been devised that cannot be broken, relatively simple encryption methods are sufficient for business applications. The unencrypted message is called plaintext. Encryption translates plaintext into ciphertext and an algorithm is required to decode the ciphertext at its destination. Three common encryption methods are concealment ciphering, substitution ciphering, and transposition ciphering.

Concealment ciphers are symbols of the plaintext disguised by a predetermined algorithm. Substitution ciphers include monographic ciphers, Caesar ciphers, and transposition ciphers. The Caesar cipher is the simplest, requiring that each letter of the plaintext be shifted by some number. With the transposition cipher, the letters of the plaintext are transposed according to an algorithm. There are other and more sophisticated schemes, but these are easy to understand and use. They can provide adequate security at moderate cost.

Any encryption technique should be evaluated with the particular application in mind. The sensitivity of the encrypted data and the abilities and aims of the typical intruder should be accounted for. The severity of a particular security risk should be met by the complexity of the encryption method implemented. Encryption, used in conjunction with a network access control computer, satisfies the security requirements of typical multiuser multicomponent computer networks.

BACKUP AND RECOVERY CONTROLS

Backup and recovery controls are crucial to ensuring the reliability of a teleprocessing network. There must be procedures, hardware, and software for restoring systems to prior status with limited loss of data and capacity.

Transaction Logs

A complete log must be maintained for all transactions that are transmitted by means of communications media. Some or all of the following items should be included in this log:

- The time-and-date stamp—This is the unique indication of when the transaction updated a date item in the data base.
- The transaction identifier—This is the unique identifier of transactions.
- The user identifier—This is the unique identifier of the user who initiated the transaction.
- The terminal identifier—This is the unique identifier of the terminal from which the transaction emanated.
- The before image—This is the prior value of data item to be updated.
- The after image—This is the new value of updated item.

A variation on the transaction log is the recovery journal, which covers a short period of time. It is intended to facilitate prompt recovery from a minor system failure. In some of the more advanced data management systems, the recovery journal can be accessed automatically by the recovery features of the system software.

Hardware and Systems Failure Controls

In case of hardware failure, recovery equipment, such as a backup computer, should be available to take over critical functions of communications and data recording while the computer is down. For software failure, such contingencies as facilities for restarting in a degraded mode of operation or the use of a previous generation software system should be considered. When a restart is required, each remote location should be sent a priority message including the last message number received and queued from that location as well as the last complete message.

Other types of systems failures can result from unexpected peak volume, which overloads available queue space. System software should include the capability to warn the control terminal operator of a possible overload.

Provision should be made for recovery from program failure. This usually includes checkpoint/restart facilities. Whenever line or terminal failure occurs, users must wait until the equipment is repaired or they must reroute their messages. It is important that microcomputer software restrict alternative routing to prevent transmission of proprietary messages to an unauthorized terminal. In addition, terminal locations should be notified so that they can actively investigate the status of their messages. Physical access to central and remote terminals should be limited (e.g., by securing terminal rooms and alerting personnel to ensure that terminals are accessed by authorized users for legitimate purposes). Procedural controls, such as those that inhibit dial-up access with callback systems, should be considered.

Recovery Procedures. The audit trail log must be sufficient to aid in the reconstruction of transactions and data files. In particular, incoming and outgoing messages should be logged by both sending and recovering stations. Step-by-step recovery procedures should be approved and documented, and all users should be trained in their use. It is desirable to periodically simulate a disaster in order to assess the adequacy of recovery procedures.

A formal system prescribing restart and recovery procedures should be established. Such a system should specify which applications need to be recovered, how quickly recovery must occur, and the resources required to effect recovery. Operating personnel must be adequately trained in recovery procedures. Backup copies should be made at frequent intervals during heavy periods of system use. To ensure efficient system recovery:

- Backup copies of application packages should be maintained.
- Backup copies of all critical data files should be made.
- A log of all changes to data files since the last backup copy was made should be maintained.

With regard to hardware backup, one or more of the following are available:

- Duplicate hardware from the company's hardware vendor.
- Off-hour access to resources from an organization using similar hardware.
- Rented or leased temporary replacement hardware.
- Duplicate hardware used by another department in the company.

HARDWARE CONTROLS

Hardware components can be the most effective means of ensuring adequate communications controls. Although these devices are the result of state-of-the-art electrical engineering concepts, the user need not be an electrical engineer to understand the nature of these controls and achieve proper implementation of them.

Modem Controls

Several secure dial-up modems provide multiple security checks. The most straightforward simply require a password before connection is made to the computer port. At a slightly higher level are modems that require the user to supply a password disconnect and wait for a callback.

Another approach to modem security control embeds registration numbers in authorized modems. A central site modem then verifies whether an authorized modem is being used. The sensitivity of data transmitted by modem must be assessed before the nature of control is determined. In general, modems that provide at least minimal protection from line intrusion by unauthorized users should be used.

Some modems provide fault isolation to identify malfunctioning equipment. Evidence of this can be provided by testing and examining past performance. The maintenance contract for modems should be reviewed to determine how soon maintenance is available in case of failure.

The newer modem models are smarter and more compact than their predecessors. Added controls include voice encryption and password entry. In addition, several modems now screen incoming messages for missing characters and, when such errors are found, request a retransmission of the message.

Vendors are providing modems with memories that store information even when the corresponding computer is turned off. This could lead to an undesirable exposure. If this feature is offered, security features that preclude a third party from accessing the stored data should be investigated.

Multiplexer Controls

Multiplexers should provide store-and-forward capability. That is, if a message is destined for a busy station, it can be stored and then forwarded when the receiving station is not busy. The message logging capabilities should be evaluated for their adequacy in preventing message loss, providing audit trails, and preventing illegal messages. Each message must be appropriately identified to the sender and receiver and adequate physical security for the devices should be provided. Error recording should also be reviewed to determine that all errors in transmission are logged. This log should include the type of error, time and date, terminal, terminal operator, and the number of times the message was retransmitted.

Terminal Security

Terminals should have lockable keyboards, physical locks on the terminal power switches, lockable communications facilities, and when appropriate, should be limited to specific transactions. Terminal failures are usually limited to that terminal, and recovery can be performed using the terminal ID, transaction numbers, sequence numbers, and block numbers. This procedure should be verified.

Front-End Processor Controls

The primary function of a front-end processor is to provide a communications interface between a central mainframe and a network of terminals or other computers. This removes communications-housekeeping functions from the mainframe CPU. Such activities as polling and selecting terminals, handling protocols and transmission speed differences, assembling messages into bits for synchronous or asynchronous transmission, editing and assigning identification data to messages, buffering, queuing, and handling priority schemes can be included in the front-end processor's repertoire. In addition, a front-end processor can handle user detection and control, message logging, and multiplexing.

The vendor should supply a list of controls implemented in the front-end processor and it should be verified that these are in effect. Such controls include:

- Polling of terminals to ensure that only authorized terminals are using the network.
- Error detection.
- Gathering of network traffic statistics.
- An audit trail of all messages sent and received.

ADMINISTRATIVE CONTROLS

These key administrative controls are not unique to communications networks, but without their implementation, other well-planned controls can be compromised.

Personnel Training

Employees should be taught the appropriate skills and kept up to date on technology and systems changes that affect their responsibilities. The frequency of human error should be evaluated, and actual training should be compared with training plans and actual proficiency of personnel with desired levels of proficiency.

Systems Documentation

All of the documentation required to operate successfully and continuously in a communications environment should be reviewed. Such documentation covers security, privacy, access capability and restrictions, error control, and backup and recovery. Documentation standards can be used as the basis for evaluating operations and applications documentation. If these standards are inadequate, the weaknesses should be noted along with a remedial recommendation. This information can then be used to evaluate the existing documentation.

SYSTEM OPERATIONS

A. Uses of application output:
   1.
   2.
   3.

B. What transmission links are used?
   1.
   2.
   3.

C. Dollar risk ($) versus application downtime:

Downtime	Dollar Risk	Reason for Risk	Time	Cost

D. Critical data or files:
   1.
   2.
   3.

E. File backup requirements:
   1. Are the files backed up?
   2. If so, how often?
   3. How are the files backed up?
      a. Magnetic tape and value.
      b. Microfilm.
         1. If microfilm, where stored?
         2. Estimate time to recover.
         3. How often is it updated?
      c. Is listing only backed up partially?

F. Recovery system:
   1. Is a recovery system in place?
      a. If so, how long to put it in place?
      b. If not, cost to put in place.
      c. Estimated running cost of recovery system is _____ per _____.
   2. Time frame when manual recovery system ceases to be feasible.
   3. Estimated loss as a result of interruption:

Duration	Revenue Loss (%)	Revenue Loss ($)	Critical Time	Remarks

G. Access system data:
   1. What type of terminals are used?
   2. Terminals are connected to the computer by means of:
      a. Leased lines.
      b. Dial-up and acoustic coupler.
      c. Other (state) _____
   3. Are the terminals in the same building as the computer?
   4. Are passwords used to sign on the system? If so:
      a. Who determines the password?
      b. How?
      c. How often will passwords be changed and by whom?

Exhibit IV-3-15.   Risk Assessment Checklist

d. Are the passwords changed when an employee terminates employment?
e. How many people know the password?
5. Is the information transmitted or received over the terminal:
   a. Business confidential?
   b. Personally private?
   c. Information whose dissemination should be controlled?
   d. Used in making management decisions?
   e. Information that may be disseminated to anyone within the department without control?
6. Are cryptographic methods employed to protect vital data? If so:
   a. Are software or programmatic techniques used?
   b. Are hardware devices used? If so, name the manufacturer and model number.
   c. The cryptographic methods are used because of:
      1. Pertinent legislation.
      2. User priorities.
      3. Other.
7. Which of the following security measures pertaining to terminals are considered adequate for the organization's needs?
   a. That a nondisplay-screen mode for entering the sign-on parameters and update passwords be implemented.
   b. That the defined terminal access be restricted to time of day.
   c. That the defined terminal be automatically signed off after extended periods of inactivity.
   d. That in the case of attempted violations, the system identify the responsible terminal and user.
   e. That the transaction be able to be entered only from the terminals so authorized.
8. Security audit reports applicable to this system:
   a. Access matrix model.
   b. Security system audit trail.

Exhibit IV-3-15.   *(Cont)*

Error Reporting

Error reports and related procedures should be examined to determine their sufficiency; particular attention should be paid to whether problems are being resolved in a timely fashion. Error correction logs, error suspense files, and the completeness of error detection and correction procedures should all be evaluated.

ACTION PLAN

The implementation of computer networks and distributed processing systems has generated a difficult and challenging environment. The potential for the propagation of data errors and for breaches of security is sufficient motivation for extensive study.

The issue of how to measure the costs and effectiveness of many network controls still needs to be resolved. As a consequence, one of the most difficult decisions the organization must make concerns its investment in controls. Most organizations cannot afford to implement controls that cover every possible contingency, nor do they need to. Moreover, even if such coverage were feasible, it would likely result in processing inefficiencies.

Sound business practices, the Foreign Corrupt Practices Act, and increasing concern over security and control in computer systems argue strongly that close attention should be given to choosing the best combination of controls according to the needs and resources of the individual organization.

One aid is for management to go through a risk analysis exercise, identifying threats and estimating potential losses. Although there are many effective checklists and questionnaires for conducting a risk assessment, the checklist shown in Exhibit IV-3-15 can be used to develop a worksheet tailored to the requirements of a particular system. The worksheet will aid in both risk assessment and cost assessment of existing and alternative control programs.

Chapter IV-3-4
The Third-Party Maintenance Alternative

THE NEED FOR FAST, RELIABLE COMMUNICATIONS EQUIPMENT SERVICE is critical in today's communications environment. Surprisingly, however, many organizations have complex procedures for selecting hardware and software yet put little effort into selecting service vendors. Communications system managers should evaluate other alternatives before accepting the services of the equipment vendor, which often have serious drawbacks.

Vendor-supplied maintenance is often slow because the vendor's work force is not large enough to respond quickly. Service can be even slower if the vendor is located far from the contracting organization.

Many organizations used to lease entire systems from a single vendor. Now it is common practice to assemble a multivendor system. Multivendor systems experience service problems, particularly when one vendor does not provide service on another vendor's equipment. For example, if a problem occurs with a front end and its service representative finds that the problem is in a peripheral, the manager must then call a second service representative to repair the defective equipment. Many managers fail to explore alternative maintenance and servicing until after they have experienced such problems with vendor-supplied service.

Communications managers should consider third-party maintenance if they are dissatisfied with current service arrangements, are planning to acquire new equipment, or are expected to mix vendor equipment. This chapter discusses the benefits of third-party maintenance and helps managers select a reliable third-party maintenance company.

THE THIRD-PARTY MAINTENANCE INDUSTRY

Third-party maintenance is service performed on communications (and computer) equipment by a firm other than the original equipment manufacturer. Maintenance is perhaps the fastest-growing segment of the computer industry today. The overall computer service market includes in-house, manufacturers, local dealers, and third-party maintenance.

The growth of the third-party maintenance industry is attributable to increased sophistication among end users, greater recognition of its value as a service alternative, and the proliferation of new and old communications equipment.

Equipment availability, coupled with increased user knowledge, results in many multivendor systems. As the number of multivendor systems increases, however, so do their maintenance and service problems. Rather than do business with several service companies or manufacturers, many organizations are recognizing the time and cost benefits that can be realized by contracting with a single third-party maintenance organization.

The growing communications equipment population also contributes to the growth of third-party maintenance. The increasing use of minicomputers and microcomputers equipped with communications functions is creating a service market that is expanding faster than many manufacturers' service departments can handle.

Servicing Used Equipment

Much of the communications equipment installed over the past 20 years is still in operation and requires service. Although some equipment is still used by the original purchaser, much of it has been resold. Because it is often difficult for the used equipment buyer to get original manufacturer service contracts, third-party maintenance becomes the only alternative.

In many cases, used equipment must be brought up to specified standards before a manufacturer's service contract can be obtained. Even if the equipment is in working order when purchased, meeting the manufacturer's standards can be prohibitively expensive. Third-party maintenance companies, however, can usually service the equipment without requiring an upgrade. They also can handle used equipment installation.

The manager should note that some third-party service companies are developing working relationships with used-equipment brokers. Through these service relationships, brokers are offering maintenance contracts with third-party firms as incentives to buy used equipment.

Modified Equipment

Third-party maintenance companies are also developing working relationships with value-added distributors (VADs) and value-added resellers

(VARs). VADs and VARs market new equipment that has been modified to perform a specific function or enhanced with an additional feature before sale. Together with a third-party maintenance organization, VADs and VARs can efficiently offer their customers a complete hardware, software, and service package.

ADVANTAGES OF THIRD-PARTY MAINTENANCE

Third party maintenance can offer cost-effective, quality service; for example, third-party maintenance agreements cost substantially less than original manufacturer-provided services. In addition, the faster response times of third-party maintenance firms bring systems back online in less time, lowering the costs incurred by long downtimes.

Some manufacturers regard service as a necessary expense and not as a source of profit, but most third-party maintenance organizations can take advantage of profit margins through service density and geographic efficiencies. Fees are based on the costs of doing business plus a reasonable profit, not on the service fees set by original manufacturer service departments.

Third-party maintenance companies can also provide preventive maintenance to avoid more costly failures and future repairs. Although some managers might consider preventive maintenance a waste of money and time, it can save money in the long run, particularly for users of older equipment. Newer communications systems generally do not require as much preventive maintenance.

Third-party maintenance is specifically advantageous in a multivendor system. A system malfunction can be solved with a single service call, and the system can usually be brought online quickly.

An organization solely in the maintenance business will not compromise service. Because its success depends on consistent, high-quality maintenance, a third-party maintenance firm must have top-quality field engineers and management personnel constantly available to maintain and expand a satisfied customer base. In addition, many third-party organizations are managed by former field engineers.

Third-party maintenance companies must strive to provide quality service because maintenance and service are their only business. Although manufacturers derive most of their income from hardware sales, third-party maintenance companies depend entirely on service fees. As a result, third-party firms generally offer more personalized, focused service than manufacturers. For example, third-party firms offer extended time contracts, which provide greater flexibility to the customer; they can also schedule preventive maintenance around the customer's work routine so that it does not interfere with deadlines.

Customers often think they have more leverage if their systems are serviced by the original manufacturer. If customers are dissatisfied with the manufacturer's service, they usually threaten to buy replacement equipment from another vendor. Such an approach does not foster a positive—or effective—working relationship. Some manufacturers will use pricing to push certain products. A third-party maintenance company, however, should remain impartial at all times and provide excellent service for all customer systems.

DISADVANTAGES OF THIRD-PARTY MAINTENANCE

As the third-party maintenance market grows and more companies enter the field, the chances of receiving substandard service increase. Some companies may try to enter the market without the technical knowledge or state-of-the-art diagnostic equipment needed to provide quality service. Often, such firms use discount prices rather than quality service to attract business. The better organizations resist price wars and operate with reasonable profit margins that allow them to continue to provide service excellence for the customer.

The Third-Party Service Lag

Owners of new equipment may be unable to get third-party maintenance contracts until their systems become more commonplace. Most third-party firms will not service recently introduced equipment unless they have sufficient time to prepare; to do otherwise would be counterproductive for both the third-party firm and the customer. Realistically, third-party service companies need time to thoroughly train their technicians to service the new equipment, build up spare parts inventories and sources, and wait for sufficient numbers of the model to be installed to make repairs.

CONTRACTING FOR SERVICE

When contracting for service, the customer and the third-party maintenance organization should work together to determine the scope of services to be provided. Although fees are an important consideration in any service agreement, they are not the most important factor. Most third-party maintenance companies have standard fees that are based on a combination of factors, including:

- The failure rate of a particular model.
- The average time it takes to repair the system.
- The number of spare parts used per repair.
- The cost of spare parts.
- The technical qualifications the field engineer requires to service the equipment.

- The system's complexity.
- The number of systems to be serviced.

Frequently, a third-party maintenance company will thoroughly inspect a customer's system before quoting a service fee.

If the installation needs extended time coverage, this affects the service cost. Contracts are available to cover several time periods, usually 9, 12, 16, 20, or 24 hours a day, up to seven days a week. Although the extended coverage is more expensive, it may be necessary for communications systems operating for extended periods each day. Some organizations use their communications equipment 24 hours a day, seven days a week.

A manager responsible for providing a specified percentage of uptime to the users must consider a third-party maintenance company's response time. For on-site service, the options range from a permanently stationed technician to a 1-, 4-, 8-, 12-, or 24-hour turnaround on a service call. The manager must choose the type of service response that is compatible with the communications center's needs, the budget, and the user service-level agreement.

The manager should also ask about the number and proximity of service locations operated by the third-party maintenance company and its repair and refurbishment capabilities. Given the importance of quick response time, the manager should choose a service organization with nearby repair facilities. If a company has multiple regional or national locations, the manager should choose a third-party maintenance firm offering wide geographic coverage.

Spare Parts Availability

Spare parts availability is another consideration in the selection of a third-party maintenance firm. Most large companies have spare parts agreements with major manufacturers or have other dependable secondary supply sources to sustain an adequate inventory. In addition, between 75% and 99% of all the parts in a system are produced by a company other than the system's manufacturer and can be obtained from the original parts manufacturer. In either case, before entering into a service agreement, managers should ensure that the third-party service company can quickly obtain needed parts.

Servicing tasks become more difficult as newer, more technologically sophisticated equipment is introduced. Communications system managers should study the third-party maintenance firm's education programs to determine if training is a high-priority, ongoing process and if adequate training equipment is available.

When a communications department uses specialized equipment, equally specialized repair equipment and facilities may be needed. Some equipment must be serviced within a clean room. Clean rooms are rated on a parts-per-million contamination factor and should be fitted with the latest high-technology maintenance equipment, an air filtration system, a wind tunnel, a stereo microscope, and air guns to ensure the proper environment.

Off-Site Service for Microsystems

Unlike large, permanently installed equipment, which requires on-site service because of its physical size, smaller, portable communications equipment does not have to be repaired at the workplace. Instead, microsystems can be serviced at a repair depot or over-the-counter facility.

Most third-party firms will provide on-site service at a higher cost for a 12-month contract. Generally, on-site service offers the fastest repair time, but most companies will not provide on-site microsystem service unless the customer has a service contract.

The least expensive service is obtained by shipping the microsystem or malfunctioning system component to the third-party company's nearest repair facility. This method should be used only if time is not critical, because service can take several weeks or more as a result of the two-way shipping process.

Two major service options are available: carry-in service and pickup and delivery. Although both types of repairs are made at a service center, carry-in service costs less because the customer brings the equipment to the repair facility. This, however, is time consuming and requires that the third-party maintenance firm have a reasonably close repair facility. Third-party organizations will also tailor service options to meet a user's needs.

Contract Versus Per-Incident Service

A preexisting service contract is usually not needed except for on-site microsystem service. Per-incident repair is available, but managers should estimate the number of needed repairs before making a decision. Typically, annual contracts for carry-in service cost between $20 and $40 per month, including parts and labor. In comparison, repairs to a logic board could cost more than $200.

Although some microsystem retailers offer servicing, others are reluctant to invest in the technical personnel, training, and diagnostic equipment needed to service the various systems they sell. Dealers with the same concerns often simply ship the equipment back to the manufacturer for service; however, shipping to and from the manufacturer can take two weeks or more.

Some manufacturers allow customers to ship units directly to the factory for service. Again, the time delays can prove costly if the customer depends on the equipment. Because a manufacturer's major purpose is to produce new equipment, turnaround on repairs may be longer than anticipated.

Some of the larger third-party maintenance companies have developed special working relationships with dealers. Under such arrangements, when a dealer sells a communications system, it also can sell the customer a service agreement from one of the leading third-party maintenance companies in the industry.

Components, Peripherals, and Auxiliary Equipment
- Have the services to be provided under the maintenance contract been agreed on?
- What equipment will be covered by the contract?
- Will service ever be required beyond the usual workday, or will a nine-hour-per-day contract suffice?
- Does the service company provide preventive maintenance? If so, how often is it provided (e.g., monthly or annually)? Is it scheduled around the data center's work routine?
- Does the service company offer carry-in or over-the-counter service for lower-cost maintenance of systems?

Parts Availability
- How and where are spare parts stocked?
- What is the size of the service company's own spare parts inventory?
- How quickly can spare parts be obtained if they are not in stock?
- How quickly can spare parts be located and delivered?

Service
- How large is the service force?
- Are the service installations located close to the data center?
- Are service personnel immediately available during an emergency?
- What is the company's typical response time to a service call?
- How much experience does the average field service engineer have?
- Are backup personnel available to handle problems that may be too complex for the local field engineer?
- Does the company have a national technical information center and offer national technical support?
- How much training does each field engineer receive?
- Is their performance closely monitored?
- Is their training continually updated in the field (e.g., with videotapes from a technical education department on the latest equipment upgrades)?

Company Evaluation
- What is the company's reputation in terms of service? In other areas?
- Can the company document its success claims?
- Does the company have a ready list of customer references?
- How flexible is the service company?
- Does it service a range of equipment or specialize?
- Does the company offer carry-in service for microcomputers and related peripherals as well as on-site service?
- What does service cost? (Costs should be compared only after reviewing all other questions.)
- Does the company have specialized diagnostic and repair equipment and the facilities needed to service special items?

Exhibit IV-3-16.   Third-Party Maintenance Company Checklist

SUMMARY

When choosing a third-party maintenance organization, the communications system manager should first determine exactly what the service needs are. Is the system a mix of equipment types? Is it a multivendor system? Is it new or old equipment? How often will service be needed and at what times of the day? Once the service needs are determined, the manager can begin to search for a third-party maintenance company.

The manager should then compare these needs with the service options offered by third-party maintenance companies. Does the organization offer single-source, national service capabilities? Does it have the experience, diagnostics, and technical knowledge needed to repair the equipment right the first time? Can the third-party company provide quality work at a reasonable price? A checklist is provided in Exhibit IV-3-16 for use in evaluating a third-party maintenance company's service.

Part V
Communications System Issues and Trends

THIS PART OF THE HANDBOOK covers two broad categories: supplementary topics and trends. Section V-1, "Supplementary Topics," cuts across the organization of the handbook; its chapters apply to all phases of the communications system management process. The section includes material on subjects as diverse as documentation, training, and the regulatory environment.

Section V-2, "Future Trends in Communications," looks at a few specific shorter-term technological issues that will have a significant impact on communications systems. Although not covered by separate chapters in this handbook, there are other, more subtle trends of which the communications manager needs to be aware. An example is the increasing number of multivendor installations, which can cause maintenance problems, increase costs for training, and raise the overhead costs for vendor liaison. These drawbacks could, perhaps, be countered with an increase in third-party maintenance contracts.

Another trend is the rapid growth in the demand for bandwidth. To keep the network cost-effective, all forms of information might have to be more tightly integrated into a single network. This, in turn, might require a merging of now separate voice and data departments, which has many ramifications for the communications systems manager.

For the communications systems manager, the next decade will be challenging; it will also be exciting and rewarding. The informed communications systems manager will prove a valuable asset to the organization as the network becomes the core of the information processing and transport system.

Section V-1
Supplementary Topics

IN THIS SECTION, *supplementary* does not mean *unimportant.* On the contrary, the chapters in this section address issues that form an important facet of the communications systems manager's job.

There are, for example, few more frustrating tasks than searching for a technical document when a critical need arises and then, immediately following the euphoria of discovery, finding that the document is partly or entirely out of date. Chapter V-1-1, "Maintaining Accurate Documentation," explains the reasons why accurate technical documentation must be maintained and discusses methods and resources useful in ensuring that the needed document is on hand when the critical need arises.

Chapter V-1-2, "Guidelines for Using Consulting Services," addresses another significant issue: when should the communications systems manager employ outside consulting services, and how is this relationship best managed? The discussion covers roles and expectations, types of services, selection criteria, and economic issues.

Another example of a peripheral or ancillary issue is training, which is discussed in two chapters in this section. The first is Chapter V-1-3, "Training Communications Professionals." This chapter presents observations, ideas, and approaches that will support the development of an ongoing training program. Chapter V-1-4, "Evaluating Training Methods and Vendors," will aid the manager in choosing a cost-effective training method. Advantages and disadvantages of in-house and vendor-supplied training are reviewed. Checklists and sample evaluation focuses are included.

Chapters V-1-5 and V-1-6 focus on the relationship between the communications department and the external environment of standards and regulations. "Weighing the Costs and Benefits of Standards Involvement," Chapter V-1-5, reviews the trade-offs involved in an organization's participation in the communications standards–making process. Benefits of participation are outlined side by side with costs.

Chapter V-1-6, "Protecting the Network Through Regulatory Involvement," examines the opportunities for user participation in the regulatory process. The author argues that corporate users can have an impact on the regulatory decision-making process if they will only get involved. Guidelines for tracking the process and submitting views and positions are included. The chapter briefly discusses two current issues—price caps and line build-out—as examples of the need for user participation in the regulatory process.

Chapter V-1-1
Maintaining Accurate Documentation

MANY ORGANIZATIONS ENFORCE DOCUMENTATION MAINTENANCE only when it is demanded by users who discover that information is inaccurate and obsolete. A more efficient approach is to assume that if the need for such documentation exists, there is an equal need for it to be accurate and current.

Management must provide sufficient staff and appropriate resources to the documentation function as well as institute planning procedures that address the relationships among personnel, resources, changing information, and documentation maintenance problems. For example, the usability of documentation improves if it is not so bulky; the implementation of online documentation to replace paper offers further advantages.

Maintaining accurate technical documentation requires knowing what and when information is changing and modifying the documentation to suit these changes. This calls for the same commitment and planning associated with any important corporate resource.

The term *documentation* in this chapter refers to internal manuals for the administration or description of services and resources. Such manuals include documentation for communications systems development, program users, and policy and procedure systems.

DOCUMENTATION MAINTENANCE PROBLEMS

The maintenance of accurate technical documentation can be made difficult by one or more of the following problems:

- Lack of notification of change.
- Frequent changes.
- New versus original documentation.

- Staff shortages.
- Inaccurate scheduling.
- Unknown documentation distribution.

The following sections analyze each problem and identify realistic solutions.

Lack of Notification of Change

Because they must know exactly when and how information is changing, the documentation staff should be actively involved in the change process and apprised of any variances through formal preliminary, progress, and completion reports. If this process is overlooked or is not well received by the people changing the information, the value of the documentation function decreases. Management must be convinced that documentation staff involvement and change notifications are vital to operations.

Frequent Changes

Although the documentation staff may be well informed of technical changes, documentation may be unavailable if the changes occur too frequently. Accuracy in documenting rapidly changing information requires innovative personnel procedures and sophisticated automated resources. Online documentation can alleviate this problem; because changes are more easily made, documentation is more likely to be kept up to date.

New Versus Original Documentation

It is difficult to determine whether accuracy should be maintained by creating new documentation or by updating previously distributed information. Because the annual production of 10 new user manuals is more impressive (on paper, at least) than the production and maintenance of five, more importance is often assigned to the newest documentation project than to maintaining the accuracy of existing manuals. Most technical writers find it more challenging to develop a new manual than to revise the existing one. Neglecting such revision, however, quickly destroys the confidence in the manuals and reduces the effectiveness of the organization's daily operations.

As mentioned, business not only thrives on but demands (and not unreasonably) timely, accurate information. The scope of technical writing support services that provide this information must be based on the production ability of the documentation staff, the number and type of existing manuals, and the number and subject of proposed documentation projects. The scheduling of new products should reflect the organization's commitment to maintaining accurate published documentation.

Staff Shortages

Many staffs lack the one or two qualified people theoretically required to perform the documentation function within a regular workday. This shortage is often the result of budget constraints or a lack of skilled applicants.

Budget constraints typically limit the number of technical writers on the staff or the salaries offered to attract experienced, qualified documentation specialists.

The lack of qualified specialists stems from the widely varying backgrounds of applicants. Relatively few colleges and universities offer communications degrees that emphasize the sciences—most documentation specialists have journalism, English, or science degrees. Depending on the individual, position, and available training, however, applicants with almost any degree can do well in technical communications if they are expressive and have an aptitude for understanding technical information. Someone with formal training in written communications and with publications experience in the applicable technical area is best suited for supervisory responsibilities.

The consequences of staff shortages can be alleviated by wisely using available skills and enforcing time management. If the quality performance of the current documentation staff justifies the cost of additional or better-qualified staff, budget constraints can be lessened.

To stay within budget constraints, the organization may want to consider online documentation, which can significantly reduce documentation costs.

Inaccurate Scheduling

The scheduling of documentation projects is complicated by both predictable and unexpected setbacks, delays, and interruptions. Maintaining the accuracy of existing publications might therefore seem to be an overwhelming task, regardless of how small an effort is required. Updates can be accommodated, however, through disciplined time management and by scheduling time for unavailable information, delayed reviews, consultations, rewrites, and unproductive writing time.

Unknown Documentation Distribution

The problems associated with documentation distribution can be solved with strong management support and an automated distribution control system that can be purchased or developed in-house. Management support consists of ensuring that manuals are obtained from the organization's official documentation distribution facility (e.g., a library) and encouraging open communication to promote user cooperation and understanding.

ANALYSIS OF DOCUMENTATION MAINTENANCE RESOURCES AND METHODS

The problem of maintaining accurate documentation is similar to a recurring problem faced by communications system personnel: the timely collection, formatting, and distribution of information to those who need it. Both problems can be solved by having adequate resources and using innovative methods. The specific resources and methods used to overcome documentation maintenance problems are discussed in the following sections.

Resources

The most significant resources are:

- A sufficient documentation staff.
- The automatic generation of change notification memos.
- A periodic newsletter to disseminate changed or new information.
- An automated document distribution control system.
- An online documentation formatting system.

Sufficient Documentation Staff.   Because the size and experience level of the staff are basic factors in estimating how much documentation can be generated, the staff's documentation specialists are the primary production resource. Logic and experience should indicate that if new manuals are to be continually developed and the accuracy of the existing ones maintained, the documentation staff's annual work load will increase. Such growth may necessitate staff expansion. Management's commitment to expansion, however, often depends on staff productivity and efficiency in assisting management and technical peers in understanding the role of the documentation function.

Change Notification Memos.   Corporate or departmental procedures for notifying the documentation staff in writing of pending changes and new information enhance the maintenance of accurate documentation. Change notification memos are particularly useful in tracking system alterations. The notification procedure can be streamlined if, during the implementation of a system change, notification is automatically generated through an established routine. This efficient method of prompting unplanned documentation revisions provides a description of the change and identifies its effect on users.

Periodic Newsletter.   A timely method of communicating information to users, a newsletter provides interim documentation until manuals can be updated. The newsletter should be restricted to technical topics, reviewed before publication, and published as necessary (e.g., monthly, quarterly, or when required). Because of its technical content, the newsletter is also a valuable training tool for inexperienced software writers.

Automated Document Distribution Control System. Attempting to manually maintain multiple, current documentation distribution lists can be more difficult than maintaining accurate documentation. An automated distribution control system is invaluable in controlling the distribution, inventory, pending orders, and recovery of technical manuals. The system requires that users obtain manuals from or in cooperation with a distribution facility. Although a distribution control system can be undermined by users unofficially receiving manuals from seminars or co-workers, the system usually receives full user cooperation and management support.

Online Documentation Formatting System. An online formatting system vastly improves the efficiency of the documentation staff by:
- Allowing writers to respond quickly to changes.
- Enabling writers to enter, format, and revise their work.
- Ensuring effective document standardization, formatting, and quality.
- Reducing the time required for manual development.
- Filing documentation masters electronically on disk.
- Producing a complete manual (cover to cover) and thus eliminating the need for many housekeeping and proofreading activities.
- Generating quality review copies quickly and inexpensively.

Methods

The most significant methods for efficient documentation maintenance are the following:
- Scheduling the work load and backup projects realistically.
- Reviewing existing documentation periodically.
- Dividing documentation development into analysis, design, writing, and review phases.
- Instituting internal, on-the-job training for documentation specialists.
- Employing documentation assistants when applicable.
- Conducting two final reviews for editorial quality and completeness of information.
- Using appropriate printing and binding methods.
- Maintaining document control records and receipts when applicable.
- Assigning a writer to software project development teams.
- Appointing a writer specifically to maintain the accuracy of manuals.
- Implementing online documentation.

Realistic Scheduling of Work Load and Backup Projects. Because an estimate of the documentation that can be realistically completed within established schedules forms a basis for performance measurement, excessive work must be identified and priorities set.

For example, writers could be assigned to one high-priority and two lower-priority projects. This practice varies writers' concurrent projects and allows those with particular skills to be assigned specialized responsibilities. Writers have the benefits of managing their time and working on another assignment if the higher-priority project is delayed or becomes tedious. Such work distribution is particularly useful when the priority of a less important project is suddenly increased. Special functional responsibilities are a means of acknowledging and using the superior skills of specific writers, thus encouraging them to seek additional duties. These responsibilities include forms design and control, editing and proofreading, adherence to quality assurance requirements, monitoring the availability and costs of reprographic services, and graphics production.

The techniques described increase documentation staff productivity, which allows the simultaneous development of new manuals and the maintenance of existing ones.

Important scheduling questions to consider are:
- What is the deadline?
- How many people need the information?
- How much work remains before publication?

If possible, special emphasis should be placed on finishing nearly completed manuals or updates, especially if only a day or week of work remains. This scheduling method permits the development of more documentation at one time, encourages above-average performances, and allows managers to focus on project analysis rather than on time management.

Periodic Documentation Review. Prescheduled, periodic reviews help maintain the accuracy of published manuals. Quarterly, semiannual, or annual reviews are adequate for most documentation. The review period should be specified by the managers of technical information and documentation; the reviews should be scheduled only by the documentation manager. Documentation reviews are the most reliable method of ensuring the continued accuracy of printed information.

Documentation Development Phases. Starting a documentation project is the most difficult phase of development. The identification of user requirements, input sources, and the scope of the documentation should be assigned to a writer experienced in documentation analysis. The placement of informational text and graphics within the manual requires less experience but substantial knowledge of manual formatting and information retrieval techniques. Although translating technical input into finished text (the writing process), ensuring accurate entry of information, obtaining and incorporating review comments, and preparing the manual for publication are time-consuming tasks, they require only basic technical writing skills.

If a manual is developed in modular phases, the work can be divided among writers with the appropriate skills or performed entirely by one qualified writer. Inspired by programming development projects, this method greatly increases the productivity and efficiency of a documentation staff with more inexperienced than experienced writers. It also ensures documentation completeness and accuracy through analysis and layout and promotes more frequent maintenance of existing documentation by allowing greater writer productivity.

Internal Training. One method of resolving the disparity between management expectations and applicant availability is to institute a comprehensive in-house training program for documentation specialists. This program can be used to develop the skills of both people within the organization and applicants who appear to possess strong technical communications abilities. In-house development produces effective software documentation specialists, low turnover, and career change and advancement opportunities.

Documentation Assistants. Assistants should be responsible for entering documentation into a word processing system. If a documentation assistant enters handwritten or typed text or changes, the writer is free to work on other tasks. This function can be expanded to include additional responsibilities (e.g., graphics production).

Final Reviews for Quality and Completeness. Documentation reviews have long been acknowledged as a critical element in establishing and maintaining accuracy. In addition to reviews during documentation development, two final reviews for editorial quality and informational completeness should be conducted before the information is published. These reviews—which can be performed separately by an editor responsible for final proofreading and by the documentation manager—ensure that the documentation conforms to the specified editorial style and meets the project's requirements and goals. To achieve overall quality and accuracy, all reports, mock-ups, and any printing instructions prepared by the writer should also be reviewed by the documentation manager before publishing.

Printing and Binding. The timely dissemination and accuracy of documentation depends on an organization's distribution methods. If the documentation is to be distributed in hard copy, it should be printed on a laser printer. The binding—whether ring, spiral, glue, or staple—also affects the accuracy of documentation. Although binding manuals to facilitate updating by page removal and insertion often results in missing pages, three-ring binding is the most common and effective method, unless the document is completely reissued when revision is necessary. Secure binding methods seldom remain accurate; updated pages cannot be readily added to the manual.

Document Control Records and Receipts. Both document control records and receipts describe why and how the documentation is to be revised. The document control record is a permanent part of the user manual. The receipt, however, is signed and returned to the documentation distribution facility after the user has received the manual and updated pages as required; the signature implies that the user understands the information and is therefore responsible for its use.

Periodic audits ensure the accuracy of the manual, its copies, and the distribution list. This control over the accuracy of confidential and safety-related design documentation ensures that documentation users update and understand the information.

A Writer on the Software Development Team. The most efficient method of tracking information generated for a software development project is to assign a writer to the project team. Because the writer is familiar with the subject, issues, and resources, this approach is suitable for large projects requiring volumes of project and user documentation to be created from recently developed information. This method also enhances the systems development effort by allowing the project team and associated corporate advisory committees to work from well-written, accurate, and current information.

A Writer Specifically Assigned to Maintain Accuracy. When the documentation staff and maintenance function are large, one writer may be assigned exclusively to maintain documentation accuracy. This method ensures that published manuals remain accurate.

NEW DOCUMENTATION PRACTICES

Two new documentation practices should also be evaluated for maintaining user documentation.

Less Is More

User documentation should be as simple and to the point as possible, fulfilling its intended function without extraneous information. A smaller manual can be used quickly and easily, promoting greater productivity for corporate employees.

The best documentation should help a user to perform a specific task with as few words as possible and yet be easy to understand. It is far better to have several brief, targeted manuals than a conventional, large guide. Each communications system was probably developed to allow users to perform certain specific tasks. User documentation should correspond to this approach.

To achieve these results, corporations with large manuals should reevaluate both user information requirements and user documentation contents. This is not an easy task, because each format and item of information in previously approved and published user documentation must be reconsidered.

Online Documentation

The second practice that should be evaluated when maintaining user documentation involves moving the user documentation from paper to online. This enables users to access documentation and apply it at one place. Online user documentation may reduce large distributions of paper-based manuals as well as reduce the amount of time spent in the documentation effort.

Effective online user documentation is possible only after the minimal documentation concept described has been achieved. Narrowing down a quantity of available information to a volume that can be displayed on a standard screen or a window within a screen can be difficult. Considerations for online documentation should include the presentation of general and detailed menus, format and size of information screens, style of instruction statements, methods for updating online documentation, and the ability to protect the online documentation from unauthorized changes.

DOCUMENTATION MAINTENANCE COSTS

The ability to identify and justify the cost of changes within user documentation and the maintenance process is difficult but important. The decision to create user manuals and keep them accurate is easily cost justified, but the decision to incur greater developmental costs to enhance or relocate user documentation that is already considered acceptable is more difficult.

The plan for maintaining and enhancing documentation must take into account the need for change and the economics of change. The need for change can be demonstrated by the technical accuracy of the information, the problems that occur from outdated information, the ways in which the information or the documentation is used, or the potential economic benefits of a change. Economic considerations should include the cost of current documentation use, the cost of necessary research for and development of the proposed changes, the associated cost of the proposed changes (e.g., the cost of new hardware or software), and the potential long- and short-term operational savings or payback that is expected to result from changes. For example, in considering whether to convert existing user manuals to online documentation, the cost reduction analysis depends on such factors as whether the organization continues to distribute paper-based documentation to all users, distributes hard-copy documentation by request only, limits paper manuals to sign-out library copies, or eliminates paper manuals completely. In addition, in cases of online and paper publishing, each form of documentation may present the same information, but they cannot be identical; the presentation requirements are simply too different.

Few corporations today question their need for documentation or its timely accuracy, and most are implementing methods to achieve their documentation goals—where the documentation is going and how it will get there.

ACTION PLAN

To maintain accurate documentation, an organization must complete the following steps:

- Assign the responsibility to develop, produce, and maintain documentation—This individual should have extensive knowledge of this function's responsibilities and its related problems as well as of the resources and methods that promote accuracy.
- Ensure that the documentation staff is of sufficient size and experience to perform its function effectively.
- Implement appropriate resources and methods that are vital to maintaining the accuracy of technical documentation.
- Answer documentation challenges with innovative techniques.

Chapter V-1-2
Guidelines for Using Consulting Services

Hugh Harvard

COMMUNICATIONS SYSTEMS MANAGERS must continually face the challenge of building and maintaining the staff and expertise to fulfill their responsibilities. They must ensure that their staff members remain current in the technologies, methodologies, and other tools of their trade. At the same time, managers must often balance and establish priorities for their staff with regard to numerous projects while operating under internal headcount constraints.

Business requirements that affect systems development often cause peaks and valleys in staff requirements or the use of technology that is unfamiliar to the department. Because communications systems are often deployed in an atmosphere of urgency—if not crisis—users will not accept a lack of time or staff as a valid reason for delaying projects.

Consequently, managers are increasingly turning to external consultants to complement their internal staff. Often, data communications departments have developed continuing relationships with consulting or contracting firms to balance resource requirements with project demands.

This chapter assumes that outside services are occasionally sought to supplement the data communications department in some manner. It does not address situations in which major communications functions are permanently offloaded to external services—that is another area of consideration. This chapter focuses on the primary considerations of an arrangement between the communications department and an external consulting service, from the client's perspective. The consulting service's expectations of the data communications department, however, are also addressed. This chapter also describes the advantages and disavantages of using consultants to supplement the in-house staff and the issues that the manager should consider when establishing a relationship with such services.

THE BENEFITS OF USING CONSULTING SERVICES

Almost without exception, it is easier to bring in and release consulting service personnel than internal staff. External consultants enter the agreement knowing that it is temporary and understand that they will leave or be reassigned after specified objectives are achieved.

Most organizations have more freedom to allocate money for a specific project than to increase the number of internal staff—that is, they are willing to spend a certain amount for a product but are often reluctant to hire internal staff to do the work. From the data communications department's perspective, work that is contracted out is viewed as a financial commitment, rather than a commitment by people. Aside from this concern, managers are faced with other challenges as well.

New project opportunities might require technical expertise that the internal staff does not have. Although changing technology is a fact of life in the data communications department, the speed with which an organization can gain expertise is often limited by training requirements and conflicting priorities. Consulting firms offer an alternative source of specialized expertise. In addition to providing instant expertise, consultants possessing the special technical knowledge can provide training by example, or even formal training, to the internal staff. The consultants might be the means for expanding internal expertise.

Although external consulting firms are often thought of as body shops in that they furnish a certain number of people for a certain time, many of them provide complete project management services as well. This concept offers a whole other set of options to the programming manager, which are discussed in a subsequent section.

THE ROLE OF THE CONSULTANT

Consultants provide a spectrum of services to the communications community, ranging from specialized technical programming expertise to the broadest business information analysis. Some concentrate on these extreme ends of the spectrum, others overlap in the middle ranges, while still others cover the entire spectrum.

Therefore, the role to be played or the service rendered by the consulting firm can vary. The role can range from a programmer being assigned to specific tasks to a senior consultant being expected to make broad directional recommendations.

Many clients view these external firms as falling into one of two broad categories: consultants or contractors. Similarly, a certain image of the function and service to be performed is immediately formed.

The consultant often deals with less structured assignments, is expected to offer advice to the client, defines options, and influences decisions. The consultant may also contribute to the client's strategies, plans, and training programs. The consultant also might be expected to challenge decisions, offer alternatives, and participate in the management process.

The contractor, however, might be expected to simply do precisely as instructed—that is, the client is not looking for a discussion or new ideas, but merely to get the work done. For the purposes of this chapter, however, both roles are referred to as consultants.

Consultants who do not clearly understand their roles and functions are at significant risk. Many engagements have failed because of confusion over the boundaries within which the consultant will operate.

Although it is unlikely that misunderstandings will occur in the more specialized areas, it is important for both parties to clearly understand the roles and expectations at the outset. Those assigned to perform according to precise, predefined specifications operate differently than those called in to develop a general system design. Therefore, the communications manager should specify the expected deliverables and the acceptable latitudes within which the consultant may operate. In addition, the consultants should be advised of any constraints and nonnegotiable decisions pertaining to the engagement.

TYPES OF SERVICES AVAILABLE

The type of arrangement and the degree of responsibility that the consultant will be given depends on the specifics of the engagement. One of the most common arrangements is for the consulting service to furnish personnel with specific talents to a project team managed by the data communications department.

In this arrangement, the consulting service assumes no responsibility for the daily assignments or performance of its people, other than to ensure that they are professionals and will fulfill the department's expectations. Consulting service personnel will probably work alongside internal personnel and be an integral part of the project manager's team. This is undoubtedly the simplest arrangement for both parties. The consulting service simply furnishes people and the department manages them, and when the engagement is completed, the consultants move on to something else.

In another type of agreement, the data communications department can delegate total project responsibility to the consulting firm. With this arrangement, the consulting service furnishes management as well as technical expertise to the communications department and has overall responsibility for assignments, schedules, and the quality of the end product.

With this arrangement, the consulting service-to-client interfaces, reporting relationships, and the degree of the consultant's autonomy must be clearly delineated. Several questions must be answered at the outset, such as:

- Will the data communications department be expected to approve

> personnel selection and assignments and have input regarding daily operations?
- If not, how will the communications systems manager be assured that everything is on track?
- Will the consultant be judged entirely on the accomplishment of various milestones and deliverables, on the specified results at the end, or on a combination of these?

The established agreement must be adhered to throughout the engagement.

The arrangement in which the consulting service manages the project exposes both parties to greater risk. The consulting service is more immediately identified with, and responsible for, negative turns in the arrangement. The in-house manager might also be held accountable for delegating responsibility to outsiders if something goes wrong. With planning and understanding from the outset and professional execution of the plan along the way, however, the arrangement can be equally beneficial to both parties. In fact, both parties are becoming increasingly confident with this arrangement because of the mutual benefits—the consulting service reaps economic rewards and status and the internal staff members benefit from the options the arrangement offers them.

PREREQUISITES TO A SUCCESSFUL ENGAGEMENT

The effective use of a consulting service depends not only on the quality and management of its resources but on the degree to which the data communications department is prepared. If the arrangement is such that the consulting service furnishes only personnel, the external personnel will simply conform to the client's mode of operation and emulate the process and procedures of their peers in the data communications department.

If, however, the consultants have project management responsibilities, the data communications department should have a solid infrastructure in place to support the management of the arrangement. The data communications department should have a methodology to specify the process to be followed and a project management system to provide the planning, monitoring, and adjustment mechanism for controlling that process. Otherwise, the data communications department must accept whatever methodology and project management system the consulting service uses—assuming it has them in place. Equally important are the design, implementation, and documentation standards needed to guide the consultant's work. In the absence of standards in the data communications department, the consultants would use their own.

Some vendors have complete and suitable methodologies, standards, and project management systems that are either directly acceptable or adaptable to many environments. If the data communications department is deficient in these areas, the consultant can provide a convenient, economical alternative for closing those gaps. Regardless of the origin, a solid infrastructure places the data communications department in a stronger management position and provides the consultant with a pattern to follow throughout the arrangement.

The most important elements of the relationship are well-understood expectations and clearly defined deliverables for the arrangement. Sufficient time and effort must be spent on the front end to ensure that both parties have the same perspective on the engagement. If both parties agree on exactly what the consultant is expected to do and what is to be delivered, and if the consultant is presented with predefined guidelines, the engagement is more likely to be successful.

Some communications systems managers prefer to delegate all responsibility to the consultant. With this arrangement, the communications systems manager simply specifies the requirements and expectations to the consultant and holds the managers of the consulting service responsible for the performance of the consulting staff. The data communications department might gauge the consulting service's performance only on predefined deliverables and completion of the end product.

Under these circumstances, the consultants should have already proved their process and ability to the client through their past performance. If not, the communications systems manager should have a lot of experience in developing precise specifications on the front end. Many, however, do not have that level of experience or that much confidence in the consulting service. Consequently, they must be more involved with the consultant's activities during the engagement.

CRITERIA FOR SELECTING CONSULTANTS

After the communications systems manager has decided to bring in outside assistance, the decision of which consultant to use is the most important. Many factors must be considered, any one of which might be the most important for a particular arrangement. Certain criteria, however, should be considered in all cases: the consulting service's experience, ability to provide the skill needed, reputation, stability, and adaptability as well as the cost of the service.

Experience. The communications systems manager must know not only how long the consultant has been in business but how much experience the consultant has in the industry and in the particular application being developed. The manager must also examine the maturity and longevity of consulting service personnel in general.

857

Ability to Provide Needed Skills. Every arrangement requires certain and sometimes unique skills. For example, the consulting service must demonstrate the ability to assemble a staff with the specified skill requirements and maintain a prescribed level of expertise throughout the engagement. In this regard, the communications systems manager might want to examine the service's recruiting processes and its ability to maintain a continuity of staff members.

Reputation. The communications systems manager should gauge the integrity, reliability, and performance of the candidates. Previous or current users of the consulting service are the best source for this information. The consulting firm should supply a complete, comprehensive list of customers with specific contacts for references.

In checking these references, the communications systems manager should seek more than just subjective feedback regarding what kind of job the consultants did. It is important to know how the consultants performed against the budget and schedule, the degree of continuity of personnel, the degree of professionalism they demonstrated, and the quality of their work. One of the general issues communications systems managers should cover is whether the organization would use the same vendor again for a comparable engagement. The consultant's past record is more important to the organization about to hire a consultant than a sales pitch.

Stability. The consulting firm's financial status and stability should be an important concern. A financially secure firm can more freely concentrate on serving its clients, and clients can be confident in the firm's longevity and reliability. If the firm's stock is publicly held, the financial information is readily available. With private firms, however, this information is more difficult to obtain.

Adaptability. Although it is an intangible aspect of the selection process, communications systems managers should consider the consulting service's ability to adapt to their environment and culture. Communications systems managers vary in the degree of rigidity in their house rules, dress code, and work styles. Both parties must be compatible in their cultures and value systems.

It is possible that the communications systems manager might prefer a consultant with an entirely different culture to effect change in the organization. In this case, the communications systems manager must be prepared to deal with a unique set of management challenges.

Cost. A manager naturally looks for the best service or product at the lowest cost. If all other factors are equal, the lowest-cost consultant should be selected. The communications systems manager who is driven by the lowest cost at the expense of other factors, however, might be more at risk from the long-term perspective.

The communications systems manager must feel confident that the work will meet quality standards within schedule and that the consultant will perform as advertised and expected. If any of these factors are sacrificed because of cost, the communications systems manager should reconsider the selection criteria and perhaps rethink the decision to use outside services.

From another standpoint, consulting services sometimes offer a low-bid contract to break into a new organization or to use idle resources. Business dynamics notwithstanding, however, both parties have a vested interest in the economic health of the consulting firm. Although the data communications department is not expected to be charitable toward the consulting firm, it is in the communications systems manager's best interest to pay a reasonable rate and deal with financially sound firms.

One of the major causes of disenchantment with consulting services is the fact that lowest cost was the main consideration in the selection process. If expectations are not fulfilled, the engagement is not a success, regardless of the cost.

CLIENT AND CONSULTANT EXPECTATIONS

In any service agreement, the service organization renders the service or product and the client pays the bill. Both parties, however, have certain expectations and requirements of the other.

The Client's Expectations

There are certain gauges that the communications systems manager can apply to assess the consultant's performance. Although it is more subtle than the contractual terms of the agreement, the general conduct of the consultant is important to the client. How the consultant goes about business is a good indicator of the health of the engagement. If the consultant is crisp and methodical in carrying out daily business, the communications systems manager would feel secure that the department is in good hands. A free-spirit consultant who merely wants to be left alone to get the job done, however, should cause concern.

The communications systems manager should expect a high level of professionalism from consulting service personnel. Professionals make reasonable adjustments to the environment and exhibit dedication to producing the specified deliverables. They clearly convey the status of work in progress and express problems objectively and constructively. Professionals not only accept and adhere to standards and guidelines for the performance of their work but even expect them.

Professionals are keenly aware of, and are dedicated to, reasonable and mutually agreed-on schedules and deadlines. If they detect any problems in fulfilling their commitments, they advise their client as early as possible, fully appreciating the communications systems manager's revulsion to surprises in this regard.

The communications systems manager should also expect the consultant to honor the prescribed reporting and management structure and refrain from any political involvement within the organization. In addition, the communications systems manager should expect the consultant to honor any agreement of confidentiality regarding the engagement. Even in the absence of a contractual obligation regarding confidentiality, the consultant should demonstrate discretion in relating experiences about the particular organization.

A correlation exists between the way a consultant conducts business and the results that are achieved. Both parties benefit in remembering this.

The Consulting Service's Expectations

It might seem that because the client is paying the bill, the client's sole obligation to the consultant ends with monetary remuneration. The consultant, however, should indeed expect certain treatment and feedback from the client.

Consultants should be treated as professionals and have the client's respect and reasonable latitude to work. In addition, if it is part of the arrangement, the communications systems manager must furnish adequate facilities, resources, and tools for the consultant to work.

The consultant as well as the data communications department should clearly understand how the consultant's performance will be gauged. Together, they should specifically define intermediate and end deliverables and define and agree on schedules. The communications systems manager is obligated to adhere to the originally defined specifications and deliverables of the engagement or adjust expectations if there are significant changes in the engagement. In addition, consultants should expect periodic feedback on their performance from clients. Consultants dislike surprises as much as communications systems managers do.

Finally, the consultant should receive prompt payment for the services rendered. The terms of payment should be specifically defined and adhered to during the engagement.

It is mutually beneficial when both parties contribute to a healthy, businesslike relationship. Although both parties enter into arrangements with their own vested interests in mind, one common denominator exists—the end product and how it was accomplished. Any ingredient that contributes to a successful conclusion is worth the effort.

THE IMPACT ON THE CLIENT ORGANIZATION

When weighing the advantages and disadvantages of using outside services, the communications systems manager should be aware of certain potential effects on the organization beyond the usual business considerations. Communications systems managers using consulting services for the first time should be particularly sensitive to these effects.

Although consultants can offer many beneficial services, data communications staff members might react negatively to the arrival of outside personnel. They might perceive the consultants as an intrusion to their domain and a threat to their security.

The consultants might have been selected for the unique skills that they can bring to the organization, which might cause the internal staff to act defensively, if not resentfully. The internal staff might even apply other, ulterior reasons for using outsiders. This is a natural reaction, even for professionals. Therefore, the communications systems manager must be sensitive to staff members' reactions and make every effort to accurately communicate the facts and reasons for their actions. Those considering the use of consultants, however, should be encouraged by the fact that most organizations accept the use of consulting services rather quickly.

When a data communications department first uses a consulting firm, some communications systems managers experience an intense interest in consulting services from senior management. Although the decision to use the service was a sound business judgment, an invoice for cash payment often raises questions regarding the arrangement. Unlike expenditures for internal services, the payment to a consulting service probably stands out as a unique transaction. Consequently, the communications systems manager might have to explain the decision to use outsiders and justify their contribution to the organization.

The communications systems manager must prepare for the administration and management of consultants not only for the duration of the engagement but also for their departure. The communications systems manager might not wish to simply release the consultant after obligations are completed. Depending on the engagement, the consultant might have developed a unique knowledge of the application or service for which he or she was responsible. The communications systems manager is then faced with life after the consultant is gone.

Specific preparation must be made for the internal staff to maintain the product after the consultant departs. The procedures regarding departure and the transfer of knowledge should be part of the overall plan for the engagement. Lack of recognition and planning in this regard has been a shortcoming of many communications systems managers. In fact, consultants probably have little motivation to initiate action to prepare for their departure. They are often willing to continue their services, which is beneficial for communications systems managers if it is by their design, but unfortunate if it is by default.

THE ADVANTAGES AND DISADVANTAGES

Some advantages and disadvantages of using outside services have already been discussed in this chapter. This section examines some of the more subtle, less obvious considerations.

The Advantages of Using Consultants

Consultants can bring instant expertise to the data communications department. Special technical skills can be a prime consideration in the selection process. Because the consulting service furnishes the personnel, the communications systems manager is relieved of time-consuming screening and recruiting activities. The communications systems manager can reserve veto rights over the selection and assignment of consulting firm personnel or leave these activities up to the consulting service to decide. Either way, the primary burden of staff selection and recruitment falls on the consulting firm.

Communications systems managers may reserve the authority to remove consulting personnel from the engagement on short notice at their discretion. Most communications systems managers find it much easier to handle this process with outside resources than with their own employees.

Depending on the arrangement, some or all of the administrative burden falls on the consultant, allowing the communications systems manager to focus on planning, the status of the effort, and the consultant's performance. Management of procedures regarding the end of the engagement is also the consulting firm's responsibility. The consultant, rather than the communications systems manager, must deal with the relocation of the people.

The Disadvantages of Using Consultants

Minimal negative aspects should exist when the consulting firm simply furnishes personnel to the data communications department. With this arrangement, the communications systems manager observes the consultant's performance and simply replaces individuals or even terminates the engagement, if warranted—much like the procedure followed when correcting an internal hiring mistake.

When placing greater overall responsibility on the consultant, however, the communications systems manager must face more complex circumstances. For the reasons already discussed, the communications systems manager must perceive consultants differently from internal staff.

In early experiences with consultants, the communications systems manager might wonder if it is worth the trouble. Because it is a new experience for the organization, the engagement itself will receive a high degree of attention. The communications systems manager must be prepared to be second-guessed at various points, especially during the usually predictable difficult periods in the engagement. Communications systems managers are often subject to criticism for delegating important organizational needs to outsiders. This is all the more reason for the data communications department to have a strong infrastructure, a project management system, and well-defined deliverables.

The Cost Factor

Whether the cost factor belongs with the advantages or disadvantages is subject to debate. This chapter has already addressed the cost factor in selecting consultants. This section discusses the relative costs of internal versus consulting services. The initial perception regarding this issue is usually that outside services cost more. When comparing the hourly expense of consulting service personnel to internal staff, the consulting service cost is higher. That, however, is not the end of the argument. Management must consider other cost factors. For example, the consulting firm carries all additional expenses (e.g., insurance, taxes, benefits, vacation, sick leave), recruits the personnel, and handles personnel and administrative functions.

In addition, the consulting firm furnishes the resources for handling peak client requirements, which can be released on short notice at the end of the engagement. The use of a consulting firm for leveling client resources can enable the data communications department to handle projects that otherwise could not be considered.

The communications systems manager might have a project that requires skills that can neither be assembled nor developed in time to meet the organization's business needs. In this case, the use of consultants is not a matter of preference but one of necessity. The justification is from a business perspective rather than an economic one. The communications systems manager might find that using the consultant's instant expertise as a means of developing knowledge in the internal staff can be translated into actual economic benefits.

Therefore, the question of whether consulting services are more costly than internal resources is not answered in this chapter. The answer is less obvious than it might seem. Of course, the use of outside services is a feasible alternative for many data communications departments and should be given serious consideration.

FUTURE TRENDS

Data communications departments are using consulting and contracting services with increasing frequency, and this trend will continue. Communications systems managers are assuming more of an engineering approach in conducting their business—that is, like engineering project managers, communications systems managers are maturing in their ability to assemble technical talents from various sources and manage those resources toward a successful conclusion.

Similarly, data communications departments are gaining strength in instituting methodologies, project management systems, and standards to support the efficient use of outside resources. As they continue to mature in those areas, they will broaden their options for assembling resources for their projects and processes.

The rising cost of systems development will drive the data communications departments increasingly toward the purchase of outside systems, as management's fear of not developing systems in-house rapidly continues to decrease. The purchase price of the systems often includes the consulting services required to install and maintain the products. With this arrangement, purchased systems and consulting services are often virtually inseparable.

As the need increases, many organizations will move faster into the use of consulting services and strengthen their abilities to manage in that environment.

In addition, consulting services are continuing to grow in their professional stature. Many data communications departments routinely use consultants to conduct broad-scale business and systems analysis and contribute to strategic decisions. Consultants are also maturing and expanding in their ability to offer full-scale project management services. Data communications departments have been known to reserve portions of their portfolio of projects for vendor development and maintenance. Many organizations are increasingly concerned with the number of internal staff members and are using outside services as a means of limiting this number. There are some convincing arguments that this mode of operation is less expensive in the long run.

Finally, senior managers are increasingly pressing communications systems managers to consider and weigh the opportunities for use of consulting services. As senior managers understand more about data communications functions and the mystique lessens, they are becoming more aggressive in seeking and evaluating broader alternatives, sometimes literally demanding serious consideration of using consulting services.

All communications systems managers should objectively consider the prudent use of consulting services at all levels of involvement. The trend is in that direction. The communications systems manager who is not equipped to seriously consider the use of consulting services might deprive the organization and its senior managers of a significant, feasible business option.

ACTION PLAN

The communications systems manager should follow a thoughtful, deliberate plan in preparing to use consulting services. The communications systems manager should:

- Specifically define the role the consultant will fulfill and ensure a mutual understanding of this role.
- Define the boundaries of authority and responsibility that the consultant will assume.
- Establish the methodology, standards, and project management system to be used in the engagement.
- Determine the deliverables the consultant must render.
- Establish milestones to help gauge the consultant's progress during the engagement.
- Conduct a thorough analysis of the consultant's past record and ability to perform according to expectations.
- Weigh all selection criteria objectively.
- Prepare the internal staff for the consultant's arrival.
- Ensure that senior managers are informed of the advantages and disadvantages of using consultants.
- Think carefully about the degree to which cost will affect the decision to use a consulting service.

If a great deal of thought and preparation is given before a consultant is selected and arrives, the communications systems manager will be in a strong position to optimize the resources a consulting service can provide.

Chapter V-1-3
Training Communications Professionals

Phillip R. Evans

THE COMMUNICATIONS INDUSTRY HAS EXPERIENCED a technological and organizational revolution during the past few years. Such new solutions as the integrated services digital network (ISDN), T1, local area networks, digital PABXs, electronic mail, and fiber optics have been growing faster than the ability of organizations to exploit them. As a result, the demand for communications professionals has skyrocketed as more business, government, and educational institutions expand their private networks of information systems. Retention of talented personnel in such a market is of utmost importance but difficult to realize. One way for an organization to stay competitive in the field is to offer ongoing training for its communications personnel who will be installing and maintaining a profusion of new equipment and services. This chapter discusses the need for training and sources of training programs, which can attract more qualified candidates to the organization and reduce turnover among valuable personnel.

REASONS FOR TRAINING

The rapid and relentless advance of technology has fueled the transition to an economy that is increasingly information intensive and service oriented. As the economy changes, so do the professional requirements of the personnel who manage and support teleprocessing functions on which business progressively relies. Their training needs are further complicated by the trend toward integrating voice and data networks and by the challenge of managing today's multivendor environments, with different companies providing the local exchange connections, the interexchange portion, and the various equipment that constitutes the network. In addition, these new products and services have new terms, conditions, and prices. For example, long-distance telephone services are available from more than 50 common carriers. Each company has different names for the services they offer, and each service has different pricing schemes and contractual commitments. Without proper training, the staff responsible for billing reconciliation has no chance of performing the job accurately. Because it is not unusual for the communications bills to run in the tens of thousands of dollars per month, the potential for large losses through billing inaccuracies is very real without continual scrutiny by knowledgeable staff.

The communications function is increasing in complexity at a time when most organizations have come to recognize that a strategic deployment of information technology can have a substantial impact on their competitive advantage. To plan, engineer, install, operate, and manage that technology, however, requires training.

Finally, the efforts under way to effect worldwide communications standards that will greatly simplify the current situation are extremely complex and technical, with great political and economic ramifications. Communications personnel must be familiar with the standards and regulations that vendors are complying with as well as the degree of their compliance. The staff cannot keep up in this area without training.

TYPES OF COMMUNICATIONS PROFESSIONALS

Exhibit V-1-1 is an organizational chart for a typical communications department of a large organization. Management usually includes the vice-president or director and managers of the engineering, operations, and planning and administration functions of the department. Generally, the communications department is part of the MIS or information resource division.

Because organizations have given increased attention to communications applications, communications managers are included more often in business planning, and their continued training requirements are usually in business rather than technical subjects. Interpersonal skills are equally important because these managers must first sell senior management on projects and then negotiate contracts to carry out those projects. Responsibilities include managing the department budget, staff, and projects and supervising relations with in-house and outside contacts. Employees at this level usually have a college degree and more than 14 years of professional experience, including technical training from telephone companies or the military.

Supervisors or senior analysts are responsible for overseeing specific projects and selected daily operations as well as lower-level personnel and

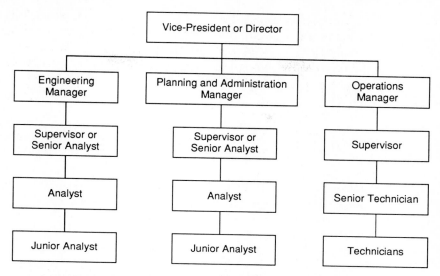

Exhibit V-1-1. Organizational Chart for a Typical Fortune 100 Communications Department

vendor and client contacts. About half hold college degrees; most of the others have some college credits. They usually have 10 years of professional experience, and most of their training has been on the job. The growth in the number of applications as well as the quantity and diversity of communications equipment and services requires that these employees receive continual vendor-specific training to handle the operations and diagnostics of networks and related components. Communication and interpersonal skills are important to the success of these employees as well.

Analysts or senior technicians have daily operational responsibilities and may also be responsible for specific projects or selected parts of specific projects. They usually have about five years of professional experience. Approximately one-third have college degrees, and most of the others have some college credits. Most of the experience of this group is gained through on-the-job training, but additional training is becoming increasingly important as more systems, equipment, and applications are installed. Vendor-specific courses are valuable to employees to facilitate their understanding of the operations, maintenance, and diagnostics procedures associated with the systems for which they are responsible.

Junior analysts or technicians are responsible for certain aspects of daily operations and specific parts of some projects. One-fourth of them have college degrees, and about one-third of the others have some college credits. Most of their training has been on the job, and they have two to three years of professional experience. Their training requirements are also usually vendor-specific. This group can focus on more specific aspects of the product or service, whereas senior levels require a broader perspective.

TYPES OF TRAINING

Training is basically focused on three areas: technical issues, business and management concerns, and interpersonal skills. Although it is desirable for all personnel to be trained in every area, time and cost constraints usually dictate that higher-level management focus on business and management concerns, operations staff (e.g., junior analysts and technicians) focus on technical issues, and supervisors and senior analysts balance their training between technical and business and management areas. As important as interpersonal skills are, training in this area is often neglected.

Technical

Engineering personnel are concerned with specifications and standards applicable to the equipment and services that constitute a system. In addition to a thorough understanding of communications engineering concepts, they must be trained regularly to keep up with new equipment, standards, and specifications. Because they are responsible for developing requests for proposals and authorization for expenditures (documents that describe various system requirements), they must also have technical writing skills. Once a project has received approval, the contracts must be awarded and the project schedule must be managed. Effective project management necessitates competence in the use of such productivity tools as spreadsheet, computer-aided design (CAD), and project-management software.

Operations personnel deal daily with the technical management of communications systems. They must be competent in the operation and maintenance of the network and understand how to perform and analyze the necessary test and diagnostic routines applicable to each of its components. In addition, operations personnel must provide current and accurate documentation on inventories of maintenance spares, operational equipment, jack and cable assignments, circuit numbers, vendor maintenance and escalation procedures, client requirements, and pending work orders. Training on a routine basis is essential. It allows personnel to function effectively and to become familiar with new or modified services or equipment in a timely manner.

Training in interpersonal skills is often overlooked but is of real importance to operations personnel. These employees must interact with both vendors and clients, often in an environment of anxiety and uncertainty. The ability to communicate effectively can be of immense value in allaying concern and focusing attention on the issues at hand.

Business and Management

The communications department basically functions as a utility service company responsible for the efficient transmission of voice, data, and graphic information. The increase in the importance of these functions to the organization means that administrative, motivational, and financial skills are now required of personnel who tend to be technically oriented. Communications managers are involved in corporate planning and must have oral and written communication skills. Often, managers with technical orientation must be trained in the fundamentals of business management. Conversely, those with a business orientation must learn the technical concepts on which the networks and systems are based.

SOURCES OF TRAINING

The success of any training program depends on the selection of the most appropriate type to satisfy staff development needs. The primary sources of training are vendors, professional associations, university programs, and on-the-job training. Each is discussed in the following sections.

Vendors

Equipment and service vendors constitute one of the best sources of training opportunities. Inclusion of vendor-provided training stipulations at the time of contract negotiations can ensure that the staff members who install, operate, and maintain the new or modified equipment and services are qualified to do so. Larger vendors (e.g., AT&T, IBM, and Northern Telecom) have established separate divisions that are devoted solely to training. For example, AT&T publishes a quarterly catalog of the courses that it offers. The catalog describes the curriculum, instructor, duration, location, and cost of each course and whether the course is new or revised. Other companies have comparable services.

Another vendor category comprises professional education groups, including educational institutes, consulting companies, and publishing companies. A sampling of such vendors includes:

- American Institute Inc, Madison NJ.
- BCR Enterprises Inc, Hinsdale IL.
- Data-Tech Institute, Clifton NJ.
- The DMW Group, Ann Arbor MI.
- Lee's ABC of the Telephone, Geneva IL.
- Omnicom Inc, Vienna VA.
- Systems Technology Forum, Burke VA.

A third vendor category comprises companies that provide self-study courses. Audiocassettes, videotapes, and tutorial software for microcomputers have been available for some time, and interactive videodisk systems are becoming increasingly popular. With these systems, personnel can focus on courses matched to specific needs and progress at their own pace. The transportability of the training methods enhances their practicality and desirability.

Professional Associations

Several professional associations offer conferences, expositions, seminars, and short courses that address topics of current interest. The recognized need for continuing education motivated the founders of the International Communications Association (ICA) in 1948; since then, the ICA has been continually enhancing the educational services that it provides. Many of these organizations are international in scope but have local chapters. There are also regional associations that provide training. The training sources are particularly important because of the personal networking that they afford. In fact, the development of acquaintances who have similar professional responsibilities at other organizations can be of immense value. Sharing experiences at association meetings or by telephone can provide valuable insight into how vendors and their products function. Also of value are the formal educational programs. For example, the ICA sponsors an annual conference and exposition that provides an excellent opportunity to learn about numerous topics—vendors, services, and products. More than 90 technical sessions are presented, and more than 300 vendors display their products. Other international and regional groups sponsor similar conferences throughout the year.

In addition to its annual conference and exposition, the ICA offers continuing education through a series of seminars that are continually revised to respond to changes in the communications industry and member needs. Those courses, as well as the annual Winter Program and Summer Program, are available to the public. The Winter Program is a three-day course on specific management topics; the Summer Program is a one-week technical study on a specific area of network management and operation.

A sampling of professional associations includes:
- Canadian Business Telecommunications Alliance, Toronto, Canada.
- Chicago Industrial Communications Association, Chicago.
- Communications Managers Association, Bernardsville NJ.
- Energy Telecommunications & Electrical Association, Plano TX.
- Industrial Telecommunications Users' Group, London UK.
- International Communications Association, Dallas.
- Michigan-Ohio Telecommunications Association, Detroit.
- North Texas Telephone Association, Dallas.
- Southeastern Telecommunications Association, Richmond VA.
- South West Communications Association, Houston.
- TeleCommunications Association, West Covina CA.

University Programs

An increasing number of universities are responding to the growing need for communications educational programs, and many offer undergraduate or graduate degree programs in communications. Recognizing the time constraints faced by employed communications professionals, some universities now provide evening and weekend programs. The programs range from a management to a technical focus. Southern Methodist University has one of the oldest established communications degree programs; it is available through satellite transmission or videotaped classroom sessions. Some of the degree programs that cater more to employed professionals include:

- Christian Brothers University, Memphis.
- City University of New York, New York.
- Golden Gate University, San Francisco.
- New York University, New York.
- St. Mary's University, Minneapolis.
- Southern Methodist University, Dallas.
- State University of New York, Utica NY.
- University of Dallas, Dallas.
- University of Denver, Denver.
- University of San Francisco, San Francisco.

On-the-Job Training

Formal on-the-job training is often neglected. Certain personnel are experts in specific areas, including equipment, operations, diagnostics, and administration. In a formal on-the-job training program, personnel needs are measured against the expertise of current employees, and programs are established to match novices with experts. Because equipment, services, and applications are continually changing, this type of program must be reviewed regularly. A small investment of time to assess personnel needs, plan training, and motivate the staff to participate can provide big dividends in both employee technical knowledge and staff morale.

SUMMARY

Communications is a rapidly evolving area because of radical changes caused by a combination of technological advances and the deregulation of the industry. There has never been a better time for communications professionals to enhance their careers. Because there are more opportunities than professionals to fill them, acquiring and retaining a qualified staff is critical. Training must be used to ensure technical competence and current awareness of alternatives and how to employ them profitably, as well as to satisfy personnel needs and maintain commitment. Although the organization must also provide attractive career opportunities and competitive compensation and benefits, offering a solid, ongoing training program can ensure that qualified candidates will seriously consider employment and, once hired, will display a serious commitment to the communications department and the organization as a whole.

Chapter V-1-4
Evaluating Training Methods and Vendors

Norman H. Carter

THE OBJECTIVE OF A TRAINING PROGRAM is to develop employee potential through the correct mix of training and development activities. The individual training plan shown in Exhibit V-1-2 should be prepared by communications systems managers and their employees to reflect the needs and preferences of both the individual and the organization. The completion and acceptance of such a plan forms the basis of a contract between them.

The information summarized from each training plan becomes the foundation for the planning, scheduling, and course selection tasks of the training administrator. The training administrator determines the training experience and timing necessary for each job task by conferring with the individual's supervisor and examining the appropriate job family/task training matrix. A sample matrix is shown in Exhibit V-1-3. (The employee should not be trained more than three to six months before use of a skill or special knowledge is required.)

CHOOSING A TRAINING PROGRAM

The training administrator must identify the most cost-effective method for addressing the training task to be performed. Exhibit V-1-4 compares characteristics and materials related to five methods, and Exhibit V-1-5 lists some major advantages of each.

The training administrator must also decide whether to fill a given training need by developing courses and workshops in-house or by going outside to a vendor. The diagram provided in Exhibit V-1-6 is helpful in making this decision.

If use of a vendor is preferred, the training administrator must consider the three major categories of training that all employees require and then

Year: _____	
Name	Section
Present Position Title	
General Career Path	
Next Position Title	
Courses to Be Considered Within Next 12 Mo (Job at Hand)	
Courses to Be Considered Within Next 12 Mo (Career Path)	
Employee's Desires with Regard to Training	
Comments: Employee	Manager
Signatures Employee	Manager	Manager

Exhibit V-1-2.   Sample Individual Training Plan

DEPARTMENT
Services Support

JOB FAMILY
Communications System Operations

Course Names / Skills (Skills: Manual Dexterity, Conceptualization, Problem Solving)

Task	Communications System Overview	Communications System Fundamentals	Specific System Fundamentals	Operator Training OS	Basic Operator Training	Console Operation	Communications System Data Entry	Data Security	Effective Verbal Communication	Effective Written Communication	Communications System Management	Coaching/On-the-Job Training Skills	Supervisory Skills	Scheduling Techniques	Notes
OPERATIONS TASKS															
Communications System Operation: Tasks 1–5	X	X	X	X	X		X								
Shift Supervision: Tasks 6–15								X	X	X	X		X		
Operations Management: Tasks 16–19								X	X	X	X	X	X	X	
PLANNING TASKS								X	X	X	X	X	X	X	See also technical supervision.
ADMINISTRATIVE TASKS										X	X	X	X		See also technical supervision.
GENERAL TASKS												X	X		See also orientation courses in Basic Training Core.

Note:
X indicates courses to be taken for specific task levels.

Exhibit V-1-3.   Sample Job Family/Task Training Matrix

	Programmed Instruction Package	Audiovisual Package	Classroom	Small Group Workshop	On-the-Job Training
Characteristics					
Cost to Develop	High	High	High	Average	Low
Unit Cost to Present	Low	Low	High	High	Low
Lead Time to Develop	6 mo	6 mo	60–90 days	20–90 days	20–90 days
Unique Feedback to Student	No	No	Yes	Yes	Yes
Flexibility of Presentation	Low	Low	Average	High	High
Learning Paced to Individual	Yes	Possibly	No	Yes	Yes
Possibility of Boredom	High	High	Average	Low	Low
Trained Leader Required	No	No	Yes	Yes	No
Materials Required					
Leader Manuals	Yes	Yes	Yes	Yes	No
Student Manuals	Yes	Yes	Yes	Workbooks	No
Projection Equipment	Yes	Yes (video)	Yes	No	No
Classroom Aids (e.g., flip charts)	No	No	Yes	Yes	No
Computer Terminals	Yes	Yes	Yes	Possibly	Yes
Most Effective Audience Size	1	1–5	15–25	10–15	1–2

Exhibit V-1-4. Comparison of Training Methods

determine which vendor can best address each category. A professional development program, which includes the three categories, is illustrated in Exhibit V-1-7. In summary, professional training provides knowledge about the organization and the associated industry. Technical training teaches the job skills needed to perform the current job or a future position. Through managerial training, individuals learn to accept greater levels of responsibility as their careers evolve. In both formal and on-the-job training, each category is addressed.

Vendor Selection

Before the training administrator hires a vendor, a checklist should be used to identify and rate vendors consistently and thoroughly. A sample checklist is given in Exhibit V-1-8. When a vendor is selected, a checklist of contract terms and provisions (see Exhibit V-1-9) may be needed to ensure that the organization has considered every factor.

Training Method	Advantages	Disadvantages
Programmed Instruction Package	Content is controlled. Objectives can be set and measured. Overlap with other courses can be prevented. Instruction can be taken at student's convenience. Stress is low. Quality of production is high.	Sequential presentation does not take into account prior knowledge. Package may become partially outdated quickly. Student receives little stimulation. Customization is impossible. Development cost is high for an effective package.
Audiovisual Package	Content is controlled. Objectives can be set and measured. Stress is low. Package can be somewhat customized to an organization's needs. Quality of production is high.	Special equipment is needed. Student receives limited stimulation. Cost of development is high.
Classroom	Content for each class is focused. Objectives can be set and measured. Instruction can be totally customized to an organization's needs. Stress is low. Leaders can bring real-life experience to class.	Material for some classes may be extraneous. Digression into other material occurs easily. Trained, knowledgeable, dynamic leaders are required.
Small Group Workshop	High flexibility of content is allowed within program design. Workshop promotes concentration on individual needs. Situation can be stressful. Interaction is high. Leader need not possess the most knowledge of the subject.	A trained leader is required to facilitate group processes. Digression from subject is common. Training process is slow if many individuals require training.
On-the-Job Training	No time is required away from the job. The same stress is felt as on the real job. A high content of real knowledge and experience is transferred. Training can be done as problems surface. A network for future communication is created.	Digression from content is common. Intensity is low. Coverage of subject may not be complete or effective. Quality varies with the person giving training. Training of supervisor to coach effectively is needed.

Exhibit V-1-5. Advantages and Disadvantages of Training Methods

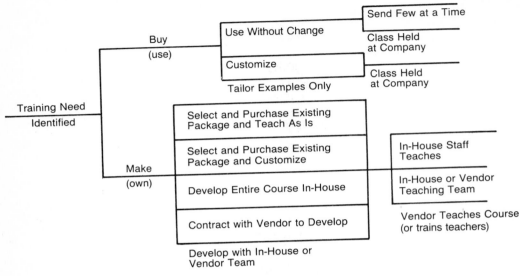

Exhibit V-1-6.   Typical Options for Training Sources

Each participant in a vendor course must evaluate the quality of the training activity. A form for this purpose is provided in Exhibit V-1-10. Changes in course quality may affect future relationships with the vendor or signal the need for reevaluating the vendor. Use of this form also ensures that the attendees examine the training they are receiving and its possible effect on others in the organization.

MAINTAINING RECORDS

The training administrator's final task is to maintain records on completed training. The two required forms are a course completion record for individual courses taken (see Exhibit V-1-11) and a training summary (see Exhibit V-1-12). The results are compared with the requirements listed in the individual training plan so that the cost-effectiveness of the program can be reported to the communications systems manager. If possible, the instructor should fill out the course completion record to provide an independent view of each participant's performance. The participants should know that this report is being prepared and should review the comments with the training administrator and supervisor.

ACTION PLAN

Communications systems managers who are developing a training program for their staffs should evaluate and select training methods that best

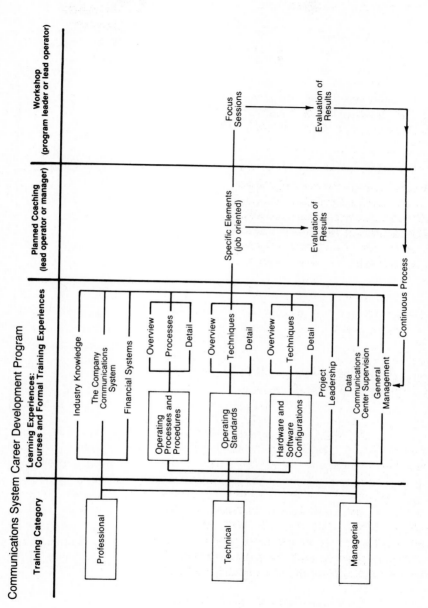

Exhibit V-1-7. Training Categories

Company		Date
Reason for Evaluation		Length of Course

Vendor Organization Size of Training Function Number of Staff Available for Project	Vendor Experience History in Subject Area Previous Clients	Yes ☐ No ☐
Own Staff Used Yes ☐ No ☐ Consultant Used Yes ☐ No ☐ Résumés Submitted Yes ☐ No ☐ Behavioral Approach Yes ☐ No ☐ Used	Willing to Give Reference Previews/Evaluations Available	Yes ☐ No ☐ Yes ☐ No ☐

Evaluation Weaknesses

Other Comments

Terms for Use	Owned by ☐ Vendor ☐ Company

Vendor Provides
☐ Outlines
☐ Behavioral Objectives
☐ Learning Objectives
☐ Case Studies
☐ Workshops
☐ Programmed Instruction
☐ Other
	☐ Instruction Manuals
	☐ Student Manuals
	☐ Tests
	☐ Other
		☐ Audio Aids
		☐ Visual Aids
		  ☐ Slides/Overheads
		  ☐ Film
		  ☐ Videotape
		  ☐ Other

Manager	Have Attended Course ☐ Yes ☐ No	☐ Use ☐ Use with Modification ☐ Do Not Use

Exhibit V-1-8. Vendor Evaluation Checklist

1. Product or Service Contracted
2. Delivery
   - Delivery dates
   - Impact if vendor is late for reasons under its control
   - Date by which contract is null and void if delivery is not made
3. Acceptance and Performance
   - Factors constituting acceptance of product or service and obligation to pay
4. Warranties
   - Extent of vendor obligation and liability for a given period
5. Maintenance Provisions
   - Guaranteed periods of no charge
   - Guaranteed response time to a problem
6. Upgrading and Trade-in Opportunities
   - Obligatory upgrade to newest version at minimal cost
   - Ability to trade in for newer product
7. Payment Plan/Price Protection/Security Interest
   - Guaranteed price protection through contract duration
   - Minimized risk to organization
8. Lease/Rental/Purchase Provisions
   - Rental credit toward purchase
9. Title or Ownership/Proprietary Rights
   - Maintain rights through or beyond contract
10. Contract Termination
    - Date for termination of contract without carry-over liabilities

Exhibit V-1-9.   Sample Checklist of Contract Terms

Return To		By Date	
Name	Position	Date	
Course Title		Length	Grade
Where Taken	Course Rating Value Expected (on a scale of 1 to 10)	Value Received (on a scale of 1 to 10)	
Course Description			
Evaluation of Value for Others			
Manager	Line Manager	Employee	

Exhibit V-1-10.   Sample Training Activity Evaluation

Return To		By Date	
Student Name		Position	Date
Course Number Description		Title	
When Taken		Where Taken	Grade
Instructor's Comments			
Strengths			
Weaknesses			
Added Action			
Manager	Instructor	Student	

Exhibit V-1-11. Sample Course Completion Record

Name		Period	Date
Course Title	Course Number	Date Completed and Grade	Not Completed
Comments			
Manager	Supervisor		

Exhibit V-1-12.   Sample Training Summary Form

meet their employees' needs. To facilitate this evaluation and selection process, the organization's training administrator should perform the following tasks:

- Prepare an individual training plan for each employee.
- Develop a matrix of job families and task training areas for each job family in the data communications department.
- Research the characteristics and materials required for different training methods.
- Weigh the advantages and disadvantages of different training methods.
- Identify potential sources of training.
- Consider different training vendors for different staff levels (e.g., professional, technical, managerial).
- Evaluate training vendors in terms of their staff size and experience and what training materials they provide for use.
- Ensure that the contract made with a training vendor meets all necessary requirements.

In addition to preliminary evaluations, the communications systems manager and training administrator must review completed courses to evaluate the quality of the training provided. Each participant should complete a training summary form to assist in this evaluation process.

Chapter V-1-5
Weighing the Costs and Benefits of Standards Involvement

Michael E. Varrassi

THE COSTS OF PARTICIPATING in the development of industry standards can vary significantly, depending on the standards group, the degree of participation, and the term of the involvement. The benefits can extend to an improved alignment of a company's product growth strategy with market needs and a technology edge over the competition. The purpose of this chapter is to review the costs and benefits of standards work and to offer guidance in getting started.

THE CORPORATE PERSPECTIVE

Direct Costs

The direct costs of involvement in the standards process in the US generally fall into three categories: dues, travel, and personnel. Dues vary widely, depending on the standards group and the participant's status. This cost can range from gratis for observers in the various American National Standards groups to $20,000 a year for membership in the International Telephone and Telegraph Consultative Committee (CCITT).

Travel is usually the largest single overall cost (except for salary in the case of full-time standards personnel). National standards meetings are scheduled in different regions of the country to balance the costs to participants and their sponsoring organizations. Travel expenses generally range from $150 to $200 a day plus airfare for domestic sessions and, on average, perhaps $250 to $400 a day plus airfare for international meetings. Typically, domestic standards organizations meet four times per year for a period of one week each. Consistent participation in domestic standards groups costs approximately $8,000 per organization. Attendance at a CCITT study group, which meets two times per year for two weeks each meeting, runs $15,000 to $17,000.

As the responsibilities of the individual engaged in standards work increase, so too does the cost of involvement. Travel and living expenses for those working as a CCITT special rapporteur can reach $50,000 a year.

Indirect Costs and the Standards Representative

The most significant cost of standards involvement is most likely felt by the participants themselves, who, when committed on a full-time basis, can expect to travel approximately 60% of the time. For those engaged in international standards work, this figure reaches 80%.

Clearly, the standards representative must be committed to standards work. This is particularly true in light of the US involvement in the international environment. Participants should be able to complete at least one four-year period in order to guarantee the consistency and quality of US standards representation.

Certain personal qualities are important for standards work. The standards representative should be a committee-oriented individual skilled at group dynamics who possesses solid technical abilities and is a logical thinker. Because a significant percentage of standards activity touches on political concerns, an understanding of and sensitivity to the various political and social environments of the representative countries is essential. Further personal characteristics of note include self-motivation and self-discipline.

In addition, multilingual skills are an advantage. Fluency in one of the three official languages of the CCITT—English, French, and Spanish—is required; otherwise, the attendee must provide for simultaneous translation. Smaller meetings are generally conducted in English without translation.

Benefits to the Corporation

Corporate participation in standards development paves the way for state-of-the-art product development capabilities. This is particularly true in the case of companies that are involved in research and development or manufacturing. Involvement in the standards development process and consequent knowledge of standards trends can help companies avoid misdirected development and implementation strategies.

Other often-overlooked benefits to the corporation include early identification of the product strategies of competitors and expanded marketing opportunities; standards broaden corporate markets by creating the basis for product compatibility. In addition, tracking the evolution of standards can give manufacturing and user organizations a leg up in the adoption of new technology, thereby enlarging the options for product growth. Combined, these benefits yield a better product at lower cost to the user.

Legal Concerns

Another consideration for the individual and company considering involvement in standards work concerns the legal issues surrounding participation, particularly potential liability. Standards representatives may contribute to situations that can be perceived as restraint of trade under the Sherman Anti-Trust Act. Such was the case in *American Society of Mechanical Engineers (ASME) Inc* v *Hydrolevel Corp* 456 US 556 (1982). Here, two subcommittee members were determined to have led the ASME to issue a standards decision that resulted in a competitive disadvantage for Hydrolevel, which then filed the suit. In a review of the case, the US Supreme Court found the ASME liable for the actions of its members. Since this decision, liability has been determined to extend to persons serving on committees as well as to organizations sponsoring the standards group. As a result, the standards community has enacted bylaws that are aimed at avoiding liability actions. The most significant of these include the establishment of due process and public procedures and the implementation of methods to resolve negative votes and document such actions. The intent is to ensure the opportunity of representation to all who have an interest in or who may be affected by the development of or changes to a particular standard.

ACTION PLAN

Corporate commitment to participation in standards activity is essential to long-term success—both for the company and the standards community. To obtain such commitment, the standards manager should first identify key research and development and manufacturing activities within the company that are of an ongoing nature. The intent is to focus on key technologies as a means of guiding standards group selection. Exhibit V-1-13 lists the major national and international standards organizations and their addresses.

Once the appropriate standards group has been determined, the chairperson of the group should be contacted to determine meeting schedules. Because meetings are open to the public, attending sessions in order to pinpoint technical subcommittees of interest is advisable.

SUPPLEMENTARY TOPICS

National

American National Standards Institute
  (ANSI)
1430 Broadway
New York NY 10018

Electronic Industries Association (EIA)
2001 I Street NW
Washington DC 20006

Institute of Electrical and Electronics
  Engineers (IEEE)
Service Center, Publications Sales
  Department
445 Hoes Lane
Piscataway NJ 08854

National Institute of Standards and
  Technology (NIST)
Technical Information and Publication
  Division
Washington DC 20234

Underwriters' Laboratories (UL)
UL Publications Stock
333 Pfingston Road
Northbrook IL 60062

International

International Electrotechnical
  Commission (IEC)
I rue de Varembe
CH 1211 Geneva 20

International Standards Organization
  (ISO)
1 rue de Varembe
CH 1211 Geneva 20

International Telecommunication Union
  (ITU)
Place des Nations
Ch 1211 Geneva 20

Exhibit V-1-13.   Major Standards Organizations

As noted, participation in US-sponsored international standards work is open to all interested parties. Inquiries for information on international activity can be directed to the US State Department, Bureau of International Communications and Information Policy, Earl Barbely, Director, Office of Telecommunications and Information Standards and Chairman of US CCITT National Committee, 2201 C Street NW, Washington DC 20520.

Chapter V-1-6
Protecting the Network Through Regulatory Involvement

Nathan J. Muller

TO COMMUNICATIONS PROFESSIONALS concerned about the cost of services and equipment, it may appear that the deck is stacked against them. Tariffs for long-distance services and the extent of carriers' participation in other markets are controlled by regulators. The regulators, however, seem to be influenced only by special interest groups, particularly the carriers and the largest communications users.

Communications systems managers therefore believe they have no control over the cost and characteristics of their own networks—the important questions are decided without their input. These managers, consequently, may not be paying serious attention to such issues as whether the seven regional Bell holding companies receive permission to engage in manufacturing, whether there is significant competition between IBM and AT&T, or, most significantly, the shape of the whole competitive landscape during the 1990s.

RESULTS OF LOSS OF USER CONTROL

It still is difficult to achieve efficient and economical transport of voice and data. The reason is that the costs associated with local access and long-haul transport are not based on the carriers' costs. In fact, the rates of return on some digital services are as high as 80% (the traditional rate of return allowed under Federal Communications Commission (FCC) regulations is 12.2%). There is therefore room for aggressive rate cutting, but users are still paying far more for voice and data transport than is strictly necessary, and this deprives them of a great deal of flexibility in configuring their networks.

RATE-OF-RETURN VERSUS PRICE-CAP REGULATION

A recent decision by the FCC illustrates the value of monitoring regulatory decisions at their earliest stages. The FCC had been considering an alternative to rate-of-return regulation; the alternative was generally referred to as *price caps*. The price-cap plan, on which two rounds of public comments were solicited by the FCC, limited the number of changes that could be made in a year to the price of each rate element of a service.

Late in the deliberations, Dennis Patrick, then FCC chairperson, introduced a new scheme that applied price-banding limitations to entire groups of services, called service categories, instead of to each rate element of each service. Within a service category, consequently, a long-distance carrier (in this case, AT&T) can raise or lower rates for services by any amount, provided that the cumulative charges for the bundled service category do not exceed 5% a year. This proposal deviated so radically from previous FCC proposals that further public comment should have been solicited, which would have been more in keeping with the FCC's own procedural rules. Perhaps out of expediency, however, this did not happen.

Nevertheless, the revised price-cap plan devised by Patrick was approved unanimously by the commission. This decision gives AT&T the potential ability to manipulate pricing for its private line services and customer premises equipment in ways that can greatly raise the cost of private networking. AT&T dominates the market for terrestrial private-line services, so it has the incentive—and with price caps, the opportunity—to skew rates to favor its own plans and reduce or eliminate customer premises equipment competition.

With price caps in effect, the burden is now on users to prove that rates are properly cost-based. Regulatory and judicial processes typically move very slowly, so corrective measures could take years. The FCC has extended price caps to the access services offered by the regional Bell operating companies and GTE. The approximately 1,500 smaller telephone companies across the US have the option of adhering to traditional rate-of-return rules or price caps, so users still have some influence on this issue.

LINE BUILD-OUT FUNCTIONS

Another example is the line build-out function embedded in channel service units that terminate T1 lines on customer premises. BellSouth is seeking FCC approval to migrate the line build-out function of 1.544M bps channel service units to a carrier-provided device that would be placed on the network side of the network interface. Such an arrangement would impede the efficient operation of private networks because it would unnecessarily complicate their design, integration, maintenance, and management. It could add substantial costs and prolong the return on investment in customer premises equipment and could even render useless much of the inside wiring currently used for high-speed data communications, especially unshielded twisted-pair wiring used for local area networks (LANs).

The placement of a carrier-provided device on a customer's premises to provide line build-out on the transmission path is contrary to the Computer Inquiry III rulings of 1987, which expressly prohibit carriers from providing any function on customer premises that interferes with the performance of competitively supplied devices. The FCC will consider waiving this prohibition, but only if a carrier demonstrates that such devices provide functions that increase the efficiency or delivery of a particular service and that would be unobtainable by comparable unregulated customer premises equipment.

If BellSouth's proposal is approved, a precedent will have been set that carriers can use to justify the reregulation of still more communications functions, to the carriers' advantage. If, on the other hand, the FCC reaffirms its procompetitive position, the end-user community will benefit from the ever-increasing availability of innovative customer premises equipment, which will result in lower overall costs and more reliable data communications.

THE USERS' CONTRIBUTION

The price cap and line build-out issues (and many others) illustrate the need for users to attend to the FCC rule-making process. Users cannot always count on their industry associations to do this for them. Many associations have limited staff for regulatory affairs, and the technical aspects of a particular issue can easily overwhelm them. Even if the staff is large, the issue may appear too innocuous to merit much attention. Or the association membership may meet too infrequently to respond in a timely manner to carrier initiatives.

Once corporate network users understand the formal and informal actions of regulatory bodies, they can wield considerable influence over the outcome of regulatory decisions. Staying abreast of regulatory developments is not a problem, because the FCC and other federal agencies are required by law to make public their intention to formulate rules and to solicit relevant input. This policy was established through an executive order issued by President Jimmy Carter in 1978. This order requires that federal agencies issue formal invitations to those most affected by proposed regulations. This policy is still in effect.

STAYING INFORMED

Users can monitor the status of regulatory issues through participation in industry associations and user groups as well as through seminars and numerous trade periodicals. The *Federal Register* publishes the agendas of the various regulatory agencies, including the FCC.

An agenda will include action items, proposals under review, requests for comments, texts of proposed regulations, revisions to proposed regulations, cutoff dates for responses, and the text and effective dates of final regulations. All responses to FCC proposals are a matter of public record, and copies may be ordered in their entirety from a commercial service.

CONVEYING THE USERS' OPINIONS

Communications systems users may not believe that their contributions are taken seriously if comments filed with the FCC in response to a proposed action or rule appear not to affect the final decision. This tends to occur if comments and technical information are submitted at the last minute. During the early, conceptual stage of the rule-making process, it is possible to discuss the potential regulations with the agency staffers assigned to research the issues, collect input from the industry and user community, and develop the first drafts of the rules for presentation to the commissioners. It is by making their opinions known during this conceptual stage that users will have the best chances of seeing their recommendations reflected in the final ruling. The dialogue helps the agency become familiar with issues they might not fully understand, which in turn helps them assess the real-world impact of their proposals.

It also is possible to provide input into regulations that have not yet been formally proposed. In this case, it is necessary to identify the various steps in the internal review process as well as the key agency staffers. By making contact with these people, the user can establish itself as a credible source of information. The goal is to help the agency define the problem, define and refine a solution, determine the cost of compliance, consider worthy alternatives, and assess the long-term practical marketplace impact of the proposal.

PRESENTING ALTERNATIVES

One way to prevent the approval of regulations that may be incompatible with users' interests is to present alternatives. The agency's staffers may not know enough about the business issues to suggest—let alone consider—more feasible alternatives. In such cases, they depend on users and other industry representatives for advice. In many cases, they hear only from the carriers that they regulate.

It may be necessary to convince agency staffers that a proposed regulation is or is not needed. Once a credible relationship with those staffers has been developed, a user's arguments will hold much more weight than if they were communicated by means of an angry telephone call to someone the user has never worked with or met before. Without the participation of many users at this stage, carriers and the largest vendors will be in a position to present their cases, which may be one-sided. Views also can be made known by writing to the House Subcommittee on Telecommunications and Finance and the Senate Subcommittee on Communications, which oversee the regulatory activities of the FCC.

During the period between a regulation's conceptual development and its formal proposal, there are many opportunities for users to provide input and feedback. There are usually 60 to 90 days within which to act on an informal basis, through correspondence, telephone calls, and face-to-face meetings. With complex issues, this time period may be extended to a year or more.

The final proposal of a regulation takes place through its publication in the *Federal Register*. After this, ex parte rules apply, which means that well-defined procedures must be followed in all contacts with the agency on that particular matter. The FCC has specific procedures to guide regulatory development from this point on; these are also published in the *Federal Register*. At this point, all communications are formal and a matter of public record.

THE COST OF NONINVOLVEMENT

Through regulatory involvement, communications systems managers may be able to prevent their organizations from paying more than necessary for communications services. If regulations were to permit carriers to reduce network maintenance, transmission quality might deteriorate substantially.

SUMMARY

By tracking regulatory developments and providing relevant input into the rate- and rule-making processes, users can ensure that appropriately priced, usable new products and services reach the marketplace. Moreover, regulatory involvement also ensures that investments in networking technology are not rendered prematurely obsolete. The FCC's mandate is to consider the needs of those most affected by its proposals: the public. Communications systems managers have a lot to gain by voicing their opinions. They also have a lot to lose by remaining silent, because in doing so they invite others to decide critical issues for them.

Section V-2
Future Trends in Communications

THE THREE CHAPTERS in this final section address issues that affect the near future of data communications. Chapter V-2-1, "Managing the Transition to Electronic Data Interchange," discusses the conversion to a new technology that, since its standardization during the 1980s, has followed a bumpy road that now appears to be smoothing out. Users can now see the cost and performance benefits of applying electronic business document interchange to their specific environments. The chapter provides a background on and forecasts for EDI and suggests an approach that can be used to plan for the implementation and growth of an EDI network. Communications systems managers should play a critical role in this process.

There is little doubt that digital transmission services now being deployed under the aegis of the integrated services digital network (ISDN) will have tremendous impact on the future of networking. Successful operation of ISDN requires an associated signaling system. Chapter V-2-2, "ISDN, OSI, and Signaling System #7," provides the communications systems manager with an introduction to the relationship and synergies that exist between these important networking elements.

Chapter V-2-3 examines an entirely different facet of future networking—the internationalization of the communications network. Increasingly, organizations depend on networks that cross national boundaries. The international flow of information is a sensitive issue in many regions. This chapter, "Transborder Data Flow," suggests ways that the least restrictive movement of data that is possible can be maintained.

Chapter V-2-1
Managing the Transition to Electronic Data Interchange

Kenneth A. Nowlan

ELECTRONIC DATA INTERCHANGE (EDI) is the electronic interchange of business documents in standard format between trading partners. EDI will radically change the way an organization conducts business. The amount of interaction with trading partners will increase, though the visibility of that interaction will diminish. It will alter the internal organizational structure, changing the nature and content of many jobs. In addition, procedures that have evolved over many years of trial and error will change drastically. Finally, EDI will necessitate the upgrading of employee skill levels.

Change in any organization can be a slow, painful process. From a technical standpoint, however, EDI is not a complicated challenge. The challenge is to be able to determine the long-term effects that EDI will have on the organization, anticipate the impact of the change, and ensure that a rational, long-term plan for EDI is established and followed to ease the impact.

To help communications systems managers face this challenge, this chapter summarizes eight elements that must be included in the plan if this objective is to be achieved. In addition, it provides a brief description of EDI, outlines the factors that must be considered in assembling a team responsible for the introduction of EDI, and provides a list of the various sources of information and assistance available to the systems development department.

BACKGROUND ON EDI

In most industries, EDI encompasses (or is in the process of including) all business documents regularly exchanged between any two organizations.

Invoices, purchase orders, order status inquiries, shipment advices, shipment information for import or export declarations, payment details, customs manifests, price lists, and more than 150 additional types of documents (i.e., transaction sets) can be exchanged electronically.

The EDI process consists of three major parts:

- The message standards under which the documents are formatted.
- The translation software, which converts the organization's format to or from the standard format.
- The communications networks on which the documents are exchanged.

EDI Forecasts

In 1978 and 1979, the American National Standards Institute (ANSI) formed the X12 Committee, covering eight industry sectors. The grocery industry adopted the Uniform Communications Standard (UCS) Message Standard in 1980, which was later adopted by the drug and retail industries. Currently, message standards are available for 16 industry sectors, which (in addition to the aforementioned) include the automotive, banking, chemical, electrical, mining, office products, textile, and transportation industries.

In Europe, a similar pattern was followed. The resultant standards, known as Tradacoms, were developed by the Article Numbering Association. The International Standards Organization (ISO) recognized the reality of international trade, and after three years of development, released its EDI for Administration, Commerce, and Transport (EDIFACT) in 1988. EDIFACT is intended to replace both X12 and Tradacoms—a prospect that is causing concern in both Europe and North America. There is a reluctance to move away from well-established standards.

Forecasts for EDI are extremely optimistic. Although it seemed as if it were off to a slow start, EDI's momentum is building rapidly. Large organizations have recognized the competitive advantage EDI can offer. This advantage is manifested by substantial cost savings and improved responsiveness to the market. Cost savings are being achieved through:

- The automation of highly labor-intensive functions (e.g., the matching of purchase order and invoice line items).
- A significant reduction in errors caused by poor transition of information from one medium to another (e.g., entering data from a purchase order).
- The introduction of just-in-time (JIT) inventory management.

Improved responsiveness is being achieved by eliminating the delay inherent in using the postal service, working within usual business hours, and manually handling documents.

As more large organizations implement EDI, their smaller (and more numerous) trading partners are often directly or indirectly encouraged to implement EDI. Thus, there is a multiplying effect.

STRUCTURING AND SELECTING AN EDI TEAM

EDI can dramatically alter the manner in which an organization conducts business, even though the technical impact might not be that great. Managers must recognize the impact that EDI can have on the organization and adjust their areas accordingly. The best way to do this is by appointing a project manager who understands the business environment.

The most successful EDI implementation teams are characterized by strong user leadership supported by executive-level approval. If an organization is confronted with an EDI program led by the MIS group, leadership responsibility must be moved to a business area, and the MIS group must focus the sponsor of the EDI program on the potential business impact.

EDI also could alter the structure and size of the accounting department. A JIT inventory process becomes practical in this environment. In addition, requirements for the physical area (i.e., offices and the locations of those offices) could decrease or increase.

Because of the business impact, a key member of the EDI project team is the business analyst. It is the business analyst who points out how the organization can and will change.

The business analyst, however, is only one member of the implementation team. A typical EDI planning and implementation team should include individuals functioning in the following roles:

- A project manager, who is responsible for overall EDI implementation—This should not be someone from the technical area.
- Trading partner coordinators, who provide the key interface into the trading partner or partners involved in the implementation.
- A contract administrator, who is responsible for the coordination of any contracts between trading partners or a trading partner and vendors.
- Business area analysts, who are responsible for the analysis and coordination of changes to the business function or functions.
- A corporate EDI coordinator, who is responsible for administering the EDI program after initial implementation.
- A technical coordinator, who is responsible for the coordination of systems development, data communications, and computer operations.
- Sales representatives, who represent the vendors of EDI products and services.

- A legal representative, who provides legal advice on the ramifications of the EDI program and its contracts—This person might also be called the contract administrator.
- An EDI consultant, who provides advice on all aspects of the EDI program, including the preparation of a business case for EDI and project planning—This position is optional.

KEY COMPONENTS OF THE SYSTEMS DEVELOPMENT PLAN

The following are eight key components in the systems development plan, which, if properly determined, can minimize the traumatic impact of an EDI program:

- Standards selection.
- Hardware platform alternatives.
- Choice of transactions.
- Choice of trading partners.
- Choice of translation software.
- Choice of mailboxes.
- Interface to current business systems.
- Migration strategy.

Each is described in the following sections.

Standards Selection

Order is inherent in EDI because of the concurrence on standards. Standards exist for data element attributes and values, data segments (i.e., logical groups of data elements), transaction sets (i.e., business documents), and functional groups (i.e., logical groups of documents). Operating rules, procedures, and policies are superimposed on the data structures. The available standards include:

- ANSI Accredited Standards Committee (ANSI ASC) X12—Cross-industry.
- Automotive Industry Action Group (AIAG) standards—Automotive.
- Transportation Data Coordinating Committee (TDCC) standards—Transportation.
- Warehouse Information Network Standard (WINS)—Warehousing.
- Voluntary Interindustry Communication Standard (VICS)—Retail.
- Uniform Communications Standard (UCS)—Retail.
- Electronics Industry Data Exchange (EIDX)—Electronics.
- Chemical Industry Data Exchange (CIDX)—Chemical.
- EDIFACT—International.

In North America, the standards are grouped and coordinated by two main bodies—the TDCC and ANSI X12. Examples in this chapter are from TDCC standards.

Each standard data element has a unique number assigned to it. The attributes and values of the data element are specified by the standard. An example of a data element is data element number 8, bank client code, which is an alphanumeric field with length 1 and has the values E (which stands for payee) or R (which stands for payer).

A data segment contains one or more data elements, and each segment is named. Within the segment, the data element may be mandatory, conditional, or optional. In addition, the standards specify relationships between data elements in the segment. Data elements within a segment could be:

- Paired—If one is present, all must be present.
- Required—At least one must be present.
- Exclusive—Only one must be present.
- Conditional—If the first is present, all must be present.
- Conditional paired—If the first one is present, at least one of the others must be present.

An example of a data segment is C2, bank ID, which is used to specify data necessary for electronic payment. It consists of data elements 8 (mandatory bank client code), 66 (mandatory ID code qualifier), 67 (mandatory ID code), 20 (optional client bank number), 7 (optional bank account number), 107 (optional payment method), and 38 (optional currency code).

A transaction set contains one or more segments. Each transaction set is assigned a unique number. Groups of segments may be repeated within the transaction set.

The choice of the appropriate standards, particularly segment selection, should fit the EDI relationships into which the organizations will enter. It should take into account the organization's industry, the industry of the trading partners, the country of the trading partners, and the volume of EDI traffic. For example, it is possible that although the organization is in the chemical industry (which uses CIDX transaction sets), the major trading partners are in the retail industry (which uses UCS transaction sets). If the organization planning to implement EDI is currently trading with partners in Europe or is planning to in the future, EDIFACT should be seriously considered.

As the industries for which the standards are developed evolve, so do the EDI standards that apply to them. Although many of the changes from release to release are minor (e.g., consisting of clarifying notes or additional values for data elements), the systems development plan should include a regular review of the impact of the current standard level.

As the organization's EDI program matures, newer EDI trading partners might use a standard version different from that of the organization. Operational documentation that is intended for aiding in the analysis of problems should note the standard version number as one item on the checklist. Fortunately, one segment in a functional group always contains the version number of the standard.

Hardware Platform Alternatives

A typical step in any systems life cycle is the review of hardware alternatives. For EDI implementation, this step should not be overlooked.

The choice of the initial hardware platform used could be a component in a migration strategy decision (which is discussed in a subsequent section). EDI can be based on an in-house microcomputer, midrange computer, mainframe, or service bureau. It is possible to initially implement on one platform and then migrate to another platform.

Microcomputers provide a relatively inexpensive platform on which to start or base an EDI program. For a smaller organization, a microcomputer, or a network of microcomputers, might be the ideal platform. Microcomputers offer a way to implement an EDI program quickly, with little or no direct impact on systems, other hardware platforms, or other microcomputers. Microcomputer-based translation software (which is discussed in a subsequent section) is readily available from numerous vendors and might be supplied or recommended by larger trading partners. If an organization's commitment to EDI is uncertain or it is reluctant to risk capital expenditures, a microcomputer hardware platform might reduce the exposure.

Numerous pitfalls exist, however, in choosing a microcomputer-based hardware platform. If EDI meets its forecasts, the capacity of the microcomputer might be quickly exhausted. Personal computing has a way of bypassing the disciplines applied to formal IS processes, including programming standards, documentation standards, auditability, and security (both physical and logical) issues.

Midrange computers and mainframes can be viewed in much the same way. The size is a function of the organization's processing requirement and the overall hardware architecture used to meet that requirement. The choice of this type of platform for the EDI program could place EDI under established MIS disciplines. Capacity becomes less of an issue because capacity monitoring and planning tools are usually already in place.

One additional advantage of using midrange computers is departmental or regional computing (i.e., using small computers rather than mainframes in various regions in which the organization operates to support operations). For example, an EDI program that is intended to address only the accounting department (e.g., purchase orders and invoices only) or only one branch might be isolated from other mainframe applications.

The factors that make microcomputer-based solutions attractive might cause problems in mainframe- and midrange-based EDI programs. The lead times to implement an EDI program can be excessive. The capital cost of the required hardware and software might appear prohibitive and impede full corporate commitment to EDI (see subsequent section entitled, "Choice of Translation Software").

Service bureau–based EDI programs offer an interesting alternative. Many mailbox vendors (which are discussed in a subsequent section) are service bureaus, providing EDI mailbox services as one of their products. Often, they offer EDI translation software and consulting services in addition to mailbox services. A service bureau–based EDI program could offer rapid start-up, relatively low initial cost, and implementation assistance. In addition, capacity concerns could also be reduced.

Service bureau revenue is based on the use of processing cycles. The ongoing cost of a service bureau–based EDI program could increase rapidly. At some point, a decision to move to an in-house platform is likely. Thus, the portability of the application and compatibility with in-house hardware and software architectures must be considered.

Choice of Transactions

With more than 150 transaction sets (i.e., business documents) available, planning the introduction of the first and subsequent transactions is critical. In fact, there might be no choice to make. Larger trading partners (e.g., General Motors) often insist that their suppliers have an EDI program. In addition, they often specify the standard and transaction sets that they will accept. Although not mandatory, certain transaction sets might already be in regular use by trading partners, and the organization might receive direct or indirect pressure to implement those transaction sets.

Organizations entering into an EDI trading relationship where there is a choice, however, should clearly understand the impact of transaction set decisions. For example, the transaction with the highest volume might have the largest impact—both positive and negative. A low-volume transaction might not have sufficient benefits to satisfy the project sponsor, though its implementation might be easy to control. During the development of a business case, the systems development and communications managers should consider the timing of the projected benefits.

A thorough comparison of the data element requirements of suggested transaction sets for current business systems should be conducted. Some data elements that do not exist in current business systems might be required in an EDI transaction set. Thus, current systems might have to undergo substantial modification to supply the mandatory data elements.

The implementation of transaction sets should be thoroughly planned. Time must be allotted for the members of the organization to adjust to the new manner in which they must conduct their business. This includes planning for the development of major systems or changes in processing cycles that must be undertaken to implement certain transaction sets. For example, the JIT manufacturing operation that an organization's customer has might require the organization's purchasing system to calculate order quantities and place orders for parts hourly as opposed to the current weekly calculation frequency.

This plan for the introduction of transaction sets can affect the choice of the translation software supplier. Consideration should be given to the ability of the translation software to add transaction sets or remove those that are not required. In the costing stage, the costs of transaction set implementation throughout the EDI program must be considered.

Choice of Trading Partners

An EDI trading relationship is truly a relationship. Two or more businesses must work together in the implementation of a program that might be foreign to them. Cooperation on all parts is required. Experience in implementing EDI by one or more parties is beneficial.

As mentioned previously, the choice of trading partner and the decision to enter into an EDI program might have been forced on the organization by a current trading partner. In the automotive industry, General Motors, Ford, and Chrysler all require that their suppliers use EDI. In the transportation industry, Consolidated Rail Corp (better known as Conrail) has made the use of EDI mandatory, but only for large-scale shippers. Provigo, a multibillion-dollar Canadian retailer, insists that all of its suppliers use EDI, as does Wal-Mart Stores.

Even in cross-industry relationships, EDI may be mandatory. For example, Kimberly-Clark Corp requires Transystems, a Montana-based shipping company, to use EDI. An EDI program might be required in an organization simply to maintain an existing trading relationship or to maintain at least the current level of trade with a partner.

If a choice is available, it might be best to look for a trading partner who is experienced first in EDI but second in the transaction sets that have been chosen for implementation. The organization should take advantage of the advice that can be obtained from a trading partner experienced in EDI and, if possible, should avoid a situation in which both partners are in the pilot stage.

As with transaction sets, the volume of trade can be a determining factor in choosing an EDI trading partner. A trading partner with which there is a high trading transaction volume might return the benefits of EDI faster than a low-volume partner would. If a low-volume partner is well experienced in EDI, however, the benefit of that experience can make subsequent implementation with a higher-volume partner proceed more smoothly.

The choice of initial and subsequent trading partners should be made carefully. Trading partners should be informed of their place in the implementation plan. Goals should be set for numbers of trading partners with which EDI relationships are established. It is important to recognize that, at some point, the organization will be the experienced partner, and it must allow time for working with the partners' systems development teams.

Choice of Translation Software

Translation software is software that implements the EDI data and operating standards. It acts as an interface between the application systems of the organization and those of the trading partners or mailboxes.

Translation software is often purchased or leased from a vendor. Although it could be developed by an organization, it is usually acquired externally. The vendor has the responsibility of maintaining currency with the EDI standards.

There is a long, ever-expanding list of translation software vendors. Some of the more established ones include Supply Tech, DNS Associates, EDI Inc, APL Group Inc, Louis A. Wright & Associates, Sterling Software, Tran-Settlements Inc, and Telecom Canada. In fact, some vendors are distributors for software written by other firms. Current prices for translation software vary according to the hardware platform chosen.

Because all translation software does not support the full range of EDI standards, the standards selection and plan for transaction set implementation is important in choosing translation software. In choosing translation software, the systems development and communications managers should ensure that the transaction sets that are planned to be implemented are supported by the translation software. Because the standards are evolving, the translation software supplier should have a track record of updating the software to match the standards releases.

Some software prices include an initial series of transaction sets and provide subsequent transaction sets at additional cost. In the costing stage, the managers should ensure that the ongoing cost of the translation software is identified.

Microcomputer-based translation software in particular can include a user interface. The interface could consist of fixed-format input screens to manually enter transaction set data or screen painting tools to allow the input to be customized to meet the current document formats. This feature might be important if the EDI migration strategy is to begin with a stand-alone EDI program.

The services that the vendor offers should be considered in the selection of translation software. Local distributors of translation software sometimes supply training and installation assistance for a price. Translation software offered by a mailbox vendor might be reasonably priced and might include installation support but might restrict the organization to the vendor's mailbox only. This is less true now than it was in the early stages of EDI, because most translation software is now certified to operate on most of the major mailboxes.

When the number of EDI trading partners increases, it is inevitable that trading partners will operate on different versions of the standard. Therefore, the translation software that is chosen should be able to accommodate different versions. In addition, it should have adequate error and irregularity detection and logging capabilities.

The translation software that is chosen and the vendor supporting it should fit the migration strategy chosen for the EDI program. If the intent is to use microcomputers as the initial hardware platform and then move onto a mainframe, it should be determined whether the operating environment of the translation software would provide portability. If it does not, the software vendor should have a proven plan for migration to another package, preferably without additional cost or at least with some type of purchase credit.

The translation software may become an integral part of the organization's systems. As such, it should be able to integrate with existing and planned systems. EXITS (or similar operators) should permit two-way links between the translation software and the organization's application systems.

Choice of Mailboxes

In a one-to-one trading relationship, the logistics of coordinating data transfer between the two trading parties is relatively simple. At an agreed time, one partner connects to the other using a specified communications protocol at a specified speed and conducts an EDI session.

As the number of trading partners increases, so does the complexity of scheduling transfer times, ensuring common communications protocols, and matching communications speeds, among other activities. The mailbox allows data from an EDI session to be temporarily stored in a computer center in a manner and at a time suitable to the sending partner, to be retrieved in a manner and at a time acceptable to the receiving partner. One-to-many and many-to-one relationships are facilitated.

As with translation software vendors, numerous firms offer mailbox services. Because of the capital investment required in computer and communications hardware, the list of vendors is not as great as that for translation software. The more popular mailboxes are McDonnell Douglas's EDI*Net, General Electric Information Services' EDI*Central, Sterling Software's Ordernet, Control Data Systems' REDINET, and Telecom Canada's

Traderoute. In addition, Kleinschmidt and IBM have a product currently on the market.

Initial, small EDI programs might not require a mailbox. Point-to-point data communications, magnetic tape, and diskettes are alternatives to mailboxes. Eventually, one mailbox will likely be required. In fact, the choice of mailbox can be driven by the choice of trading partner. Large trading partners often use more than one mailbox, but certain trading groups use only one mailbox.

As EDI trading expands, however, there becomes a need to conduct EDI trading sessions among companies whose major trading groups use different mailboxes, and the organization might be forced into using the services of more than one mailbox. Because interface procedures (e.g., sign-on and security) might be different for each mailbox, a need has arisen to standardize that interaction. The International Telephone and Telegraph Consultative Committee (CCITT) has developed the X.400 standard for electronic-mail interexchange. The presence of, or stated direction toward, an X.400-based mailbox service should be considered in the mailbox choice. This is particularly relevant with the expected transborder trading generated by the Canadian/US Free Trade Agreement and the elimination of trade barriers in Europe in 1992.

The chosen mailbox vendors should provide the organization with training and assistance in EDI implementation. Their operating hours become extremely important as coast-to-coast and international trading relationships move to adopt EDI.

Certain EDI transaction sets (e.g., pricing updates) are intended for wide distribution. The mailbox might have a distribution generation capability or might insist that the sender perform the distribution. In the first case, a distribution charge is usually levied, but in the second case, the sender could be required to incur the overhead of multiple deposits.

Another factor that must be considered in the selection of mailboxes is how well they match the communications strategy chosen by the organization. This strategy includes network architecture, protocols, and speeds.

The use of a mailbox is not expensive. In the selection and costing stages, all of the current cost elements as well as those that will be imposed as the EDI program expands must be considered. A spreadsheet with all of the cost elements included can quickly indicate the impact of various scenarios.

Most mailbox vendors charge a fixed monthly fee, plus fees for use, which are often based on the number of characters deposited and retrieved. Sender fees are typically higher than receiver fees. In addition, a minimum session charge is usually imposed, which is calculated on the basis of 1,000 characters.

Interface to Business Systems

The exercise of mapping the application systems data dictionary to the EDI data dictionary, which was discussed previously, is required to ensure that all mandatory EDI data elements are present. For those data elements not present, it must be determined how they will be made available. It must be decided whether they will be manually entered using the input capabilities of the purchased translation software or whether they will be built into the data capture of existing application systems.

In addition to new data elements being required in the application systems, the manipulation of current data elements might have to allow for truncation, lengthening, concatenation, or a breaking apart of the elements. For those data communications departments not already working with a data dictionary, EDI can act as the initial catalyst.

The identification of the presence of the EDI data elements is only the first part of the exercise. The required availability of the data elements might necessitate significant alteration to current data structures or processing cycles. For example, data elements that will be needed daily might require modifications to six systems that currently make them available weekly.

Business analysis plays an important role in this component of EDI. EDI can eliminate much of the current printed paper. This can have many obvious consequences for the business operating departments that depended on that paper as well as for the application systems that support those functions. For example, manual verification processes (e.g., checking purchase order line items against receiving notice line items against invoice line items) could be replaced by an automated verification process, and the business department will change into a department that mostly deals with exceptions. Thus, the application systems that support exception handling might require modification in order to be more responsive to adjustment transactions.

Other consequences are not as obvious. Because much of the information flow is informal and undocumented, situations such as the vice-president of sales casually scanning through the purchase orders in the in basket of the purchase order clerk, checking for purchase orders from the newest large client, are common. Therefore, the business analyst should identify the requirements for informal access to data that might require formalization as a result of EDI. Restriction of informal access to previously available information might reduce the EDI program's credibility and success.

Two additional items for consideration in the interface to business systems are legal agreements and audit concerns. Current printed business documents might have legal agreements printed on the reverse side of the forms. For example, the reverse of an invoice might cover such issues as late-payment penalty terms, exchange, and restocking terms. Acceptance of the document legally implies acceptance of the printed terms. The list of documents that can be eliminated must be brought to the attention of the contract administrator of the EDI program. Specific legal agreements between trading partners might have to be executed.

Somewhat related to the legal issue are audit concerns. Both the internal and external auditors require adequate audit trails in EDI translation software and in modifications to application systems. This requirement can extend into the audit trails of trading partners.

Migration Strategy

The migration strategy is a combination of the key components, covering the total evolution of the EDI program within the organization. This implies that the plan for EDI is long term. Each step is clearly planned to minimize the impact of the transition from one stage to the next. If the organization's EDI implementation plan is not firmly established, the organization must have alternatives that are flexible enough to accommodate multiple scenarios.

The transition from a manual process to a microcomputer base or to a midrange or mainframe base should be isolated from the business departments. Input screens that were developed using translation software screen painters should resemble the eventual mainframe-based screens. The choice of terminal device should permit the migration to different hardware platforms.

The plan should consider changes to application systems and operating procedures. The types of changes should match both the schedule and approach taken in altering the business functions. For example, if a pilot EDI program is to implement one transaction set for one trading partner, the plan should not force the business function into implementing more than that transaction set. The changes, however, should be scheduled to be in place when additional trading partners begin EDI for that and other transaction sets.

The adjustment of the business function to EDI will be traumatic. There will be a natural desire to revert to current methods of conducting the business. This tendency has two implications. First, parallel operations must be included in the migration strategy to build confidence in the new business operating environment while keeping the current environment available. Second, there should be a contingency plan available if a reversion to current operating methods is required. A corollary to these implications is that the disaster recovery plan for the organization must be amended to accommodate a loss of EDI.

The introduction of EDI might require temporary or permanent changes to the communications department. For example, in establishing the pilot EDI program, a decision could be made to implement a microcomputer-based solution in a remote branch office—the first attempt at distributed processing. The most appropriate location for the system's support might be at the remote location. A plan must be developed for the training of technical staff members in the required skills (e.g., new programming languages), other technical skills (e.g., communications and computer operations skills), and interpersonal skills. The challenge is in deciding which staff members to use in the EDI program.

SOURCES FOR EDI INFORMATION

With the fanfare that EDI is receiving, it is easy to become overwhelmed with the possibilities and alternatives. Fortunately, assistance with and information about EDI are readily available. It is recommended that communications systems managers begin the search for assistance in EDI as early as possible. The experience of others can be used to help to produce a complete picture and, subsequently, to develop a plan.

Associations

There are four categories of associations that can be contacted for EDI information: technical, industrial, national standards, and EDI. Each is discussed in the following sections.

Technical Associations. Technical organizations include the Data Processing Management Association and the Canadian Information Processing Society. These associations have education sessions, special-interest groups, dinner meetings, and libraries, which can be sources of information. An important purpose of these sources is personal networking. Through these sources, the communications systems manager can come in contact with industry peers who have gone through what their organizations are contemplating.

Industrial Associations. Industry associations include trade-related groups such as the National Retail Merchants Association and the Chemical Industry Association. EDI is currently so well established that most industry associations have subcommittees focusing on EDI. They usually provide background information and advise on EDI in their industry.

National Standards Associations. National standards associations have subcommittees involved in the establishment and administration of

EDI standards. They can be contacted for standards documents and some planning advice. Examples of these subcommittees are the Data Interchange Standards Association of ANSI, the Canadian Standards Association, and the European Article Numbering Association.

EDI Associations. EDI associations are now well established in most countries. They are typically cross-industry, nonprofit associations formed for the promotion of, and assistance in, EDI. They provide user forums, education sessions, membership lists, and general information. They act as a lobby group for the promotion of EDI and encourage and organize EDI user groups. Examples of these organizations include the EDI Council of Canada, the Electronic Data Interchange Association, and the EDI Council of Australia.

Similar Organizations

Communications systems managers in companies within the industry that have implemented or are planning to implement EDI can be used as a source. Experience with EDI is discussed with enthusiasm. There should be no difficulty in obtaining advice that can be used in planning. If communications systems managers are not available, most companies involved in an EDI program have assigned an EDI coordinator.

Trading Partners

Because EDI is implemented between two or more organizations, it is likely that one of the companies with which the organization is planning to conduct EDI trade has already been involved in EDI. Extensive assistance is often available, particularly where one of the trading partners is extremely large and is influencing the move to EDI (e.g., General Motors).

Vendors

Numerous vendors—always ready to sell products or services—are pursuing organizations moving into EDI as potential customers. If the organization's knowledge of EDI is at the formative stage, vendors provide valuable introductory information at no cost. Some vendors target products and services at specific industry segments, thus furthering the depth of the information.

Vendors also provide EDI training, from introductory sessions (typically at no cost) to advanced planning and implementation sessions (targeted at both business and technical analysts).

ACTION PLAN

The introduction of EDI into an organization can be traumatic. The business functions will experience the majority of adjustments. The objective of the plan should be to reduce this trauma.

Numerous direct and indirect sources of information and assistance are available to a systems development department that is inexperienced in the implementation of EDI. These sources should be used early in the implementation process so that the alternatives available in systems development are clearly understood. The communications systems manager should use the experience of other communications departments in similar industries and in trading partner organizations to avoid pitfalls.

Although a successful EDI program can provide an organization with a competitive advantage, this should not prevent a high level of cooperation among all parties involved in EDI, which an organization planning to implement EDI should use to its advantage.

The organization must ensure that the business departments are clearly in control of the EDI project. Although the data communications department plays a key part in the EDI team, it should not act in a project management role.

The choice of EDI standards should include a view to the organization's long-range EDI trading patterns. The choice of hardware on which to base the EDI program should allow for timely, flexible implementation while accommodating the forecast explosion of activity in EDI. Transaction set implementation plans should consider the corporate objectives of the EDI program and the degree to which the organization is risk averse. The organization might not be able to choose trading partners in the initial EDI implementation. The plans, however, should consider the order and speed in which new trading partners are to be added, including the level of EDI experience of those trading partners.

The choice of EDI translation software will likely determine a long-term relationship with a software vendor. The communications systems manager should ensure that the long-term requirements of the EDI program are met by both the software and its vendor.

An EDI program mandated by a large trading partner could force a decision to use a specific mailbox. Eventually, multiple mailboxes will be used. The interface to the mailbox should be designed to accommodate multiple mailboxes. Eventually, a CCITT X.400 standard should be considered.

An EDI program has both obvious and subtle effects on existing business application systems. Mapping of the data element requirements can provide an initial picture of the changes required. Business analysis can probably help to identify currently innocuous data flows that, if not accommodated, could have a significant negative impact on the credibility of the

EDI program. The communications systems manager should plan for an orderly upgrading of existing systems.

The migration strategy should combine all of the elements to carry out a smooth transition into the EDI environment. Included in the plan should be a way for the organization to back out if necessary.

The staff involved in EDI should follow the same training and exercises as the business departments. A well-planned EDI program enables the organization to ease into EDI with little or no technical trauma. The eight key components described in this article should be used as a checklist to meet that objective.

Chapter V-2-2

ISDN, OSI, and Signaling System #7

William H. Heflin

THE OBJECTIVE OF WORLDWIDE CONNECTIVITY for the integrated services digital network (ISDN) established by the International Telephone and Telegraph Consultative Committee (CCITT) depends directly on the accuracy and compatibility of international signaling protocols and systems. The CCITT has specified that the seven-layer model open systems interconnection (OSI) of the International Standards Organization (ISO) be used, along with its standard common channel signaling system #7 (SS#7). An understanding of these two ISDN fundamentals is important for the communications manager planning a network. Equipment purchased for private networks that may someday be connected to a public network, a private virtual network, or a Centrex system should contain provisions that ensure compatibility with both the OSI and SS#7 standards to avoid early obsolescence. This chapter describes the OSI reference model protocols and details the SS#7 architecture.

THE OSI SEVEN-LAYER REFERENCE MODEL

The objective of OSI standards is to permit the interconnection and interworking of computer systems that differ in terms of manufacturer, location, or operating system. Interconnection permits file transfer, electronic messaging (e.g., through the X.400 electronic-mail standards), and program-to-program communication. The CCITT and the ISO have worked in close cooperation because of their joint concern with interconnectivity.

The first three layers of the OSI model, shown in Exhibit V-2-1, are network specific and concern themselves with the establishment of end-to-end connections and data transfer between end systems. Layers 4 through 7 apply to end systems communications and are therefore transparent to the network. Layer 4, the transport layer, gives specified qualities of service

Layer 7	Application
Layer 6	Presentation
Layer 5	Session
Layer 4	Transfer
Layer 3	Network
Layer 2	Data Link
Layer 1	Physical

Exhibit V-2-1. OSI Model for ISDN Protocol

(i.e., error rates and throughput). The standards for layers 5 and 6, called the session and presentation layers, provide functions to support the exchange of information between the data systems at each end of the circuit using the network transfer facilities set up by the first four layers of the OSI model. The top layer, the applications layer, provides for the interworking between application processes in end systems—for example, the integration of the summary results in a branch operation with the full report at the head office.

Four application areas have been addressed. The CCITT has already set standards in the first two areas, and standards are close to being finalized in the last two:

- File transfer, access, and management (FTAM)—This standard has been agreed on and allows one end applications system to access and operate on the files in another end system.
- Messaging—The CCITT and ISO have jointly agreed on the X.400 electronic-mail (E-mail) standard for storing and forwarding messages.
- The virtual terminal standard—This standard specifies communications procedures between terminal controllers and host computers for remote access.
- Job transfer and manipulation—This standard specifies procedures to set up and control asynchronous processing activities across a network of end systems.

APPLICATION OF THE OSI MODEL TO ISDN

The physical layer protocol must support all the functions supported by the higher layers because all the information exchanged between higher-layer peer entities flows through the physical layer. The basic rate interface requires a balanced metallic transmission line for each transmission direction capable of passing 192K bps. This layer provides bit streams to support two B channels at 64K bps and a D channel at 16K bps. It also provides the maintenance, timing, and synchronization functions and the on- and off-hook condition for devices connected to the network. The protocol will permit point-to-point or point-to-multipoint operation for multiple device use.

The primary rate interface physical layer, in the North American version, requires a standard 1.544M-bps interface with extended superframe format and bipolar eight zero substitution (B8ZS) line coding format. The extended superframe length is 193 bits of 24 eight-bit time slots and one framing bit. The B8ZS is a coding format for maintaining the density of ones and zeros in the T-carrier primary rate interface link. The signaling channel on the T-carrier must be assigned to time slot 24.

The layer 2 protocol—the D-channel data link layer based on the X.25 link-access protocol, balanced—is called link-access protocol D. This protocol provides the following functions:

- Multiple logical links in the D channel.
- Detection and recovery of transmission errors.
- Flow control.
- Sequence numbering and control.
- Transparency.

The layer 3 signaling protocol, specified as Q.931, provides the following functions:

- Signaling relationship for multiple calls between the user and the network.
- Call connect, disconnect, and maintenance.
- Access protocols for a packet-mode service called user-to-user signaling.

The Q.931 message structure is outlined in Exhibit V-2-2. Protocol discrimination identifies the type of packet that follows. The call reference field in the message identifies the call to which the message pertains. Additional information associated with a call is provided by the information element fields—for example, a request regarding the bearer capability—and includes speech, 64K-bps clear data, packet data service, and the channel identification.

Layer 4, the information transfer protocol, spans all layers of the OSI model; different layers are required for different applications. For example,

Message	Function
Protocol discrimination	Identifies the packet type
Call reference	Identifies the call
Message type	Identifies the message—disconnect
Mandatory information elements	For example, bearer capability in setup
Additional information elements	For example, channel identification

Exhibit V-2-2. The Q.931 Message Structure

a voice application in layer 1 provides a coding sequence as contained in the A-law or mu-law transport carriers; circuit-switched data in layer 1 enables synchronous data transfer at 64K, 384K, or 1.536M bps; packet-switched data in layer 1 provides for synchronous data transfer at 16K bps on the D channel and 64K bps on the B channel. Layer 2 uses the X.25 link access protocol B on the B channel, and the Q.921 protocol specifies the use of the link access protocol D on the D channel. Layer 3 uses the standard X.25 layer-3 protocol.

SIGNALING SYSTEM #7

The common channel signaling system (SS#7) has been specified as the backbone signaling system for ISDN. SS#7 is designed for digital transmission at 56K bps or 64K bps and can support variable message lengths to 272 octets; one octet comprises eight bits for each signal message. SS#7 uses the first three layers of the OSI model: layer 1 includes the signaling link hardware; layer 2 monitors layer 1, ensuring the accuracy and reliability of data transfers; and layer 3 routes the messages received by the signal transfer point to their destination. Exhibit V-2-3 shows the physical links that are involved in the SS#7 architectures.

The service control points contain the data bases that manage the information that allows the system to route messages, provide credit card validation, and provide number translation for 800-number services. Both the service control points and signal transfer points are connected to the signaling point or service switching point by SS#7 links. The SS#7 links are shown by the solid lines. The signaling point would be a central office or tandem (i.e., long-distance) office in the common carrier's network and the service switching point might be the point at which the signal information from a primary rate interface is introduced into the SS#7 network. The dotted lines represent the physical transmission network among the central offices, tandem offices, and the interLATA (local access and transport area) network. The physical network between central and tandem offices would be, most likely, 1.5M-bps (T1) digital carriers in North America and 2.0M-bps carriers under the European plan. The interLATA signaling network would be either on digitally derived fiber-optic circuits or on digital microwave. For reliability, each signal transfer point is duplicated in the SS#7; this pair is referred to as mated signal transfer points.

Signaling points and service switching points use the physical facilities much more efficiently than the older in-band signaling systems, in which the entire voice channel is tied up while the next link in the network is connected and the signaling information is passed along the route. SS#7 is also faster than the older common channel interoffice signaling (CCIS#6) system that was pioneered by AT&T. The CCIS#6, meant to be used on an

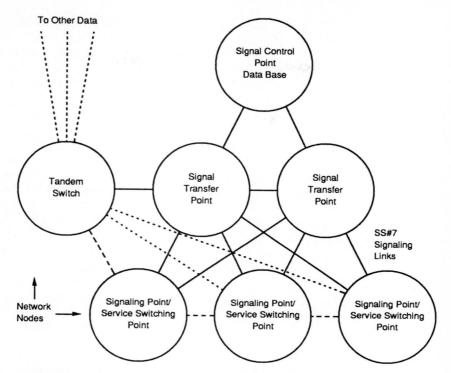

Exhibit V-2-3. Signal Control Points, Signal Transfer Points, and the Physical Links

analog network, had a speed of only 2.4K bps. CCIS#6 was a small advance over the in-band systems in that it used a separate channel for signaling. The SS#7 signal transfer points, transmitting at 56K or 64K bps, handle hundreds of times more traffic than the old CCIS#6.

The service control point is a high-transaction-oriented computer system that centralizes network intelligence regarding call control and allows advanced network services. Call origination information will be passed to the receiving telephone or terminal so that originating call information can be made available for several uses. It will also be possible for the end subscriber to set up conference calls using a point-to-multipoint service. The service control point provides a real-time high-speed inquiry and response function.

The signal transfer point handles two types of signaling messages: circuit-related and data base access messages. Circuit-related messages are used to send connect and disconnect data between two signaling points, located at central or tandem offices. The signaling points are referred to as signaling nodes on the SS#7. The signal transfer point transfers these messages from the originating signaling point to the terminating signaling point. The information in these messages includes the identity of the calling party, the identification of the circuit being used between the two signaling points, the called number, the response, and disconnect and release.

Data base access messages are used to obtain information from the signal control point. An inquiry message is sent from the originating signaling point to the signal control point over the signal transfer point. The response message is returned from the signal control point to the signal transfer point, which, in turn, sends the information back to the originating signaling point. Each message routed by a signal control point contains an address field for determining the destination of the message. If the information in this field contains the required destination data, the signal transfer point will immediately send the signaling message along its way to the destination signaling point. If the destination data is not complete, the signal transfer point must translate the data into an acceptable message using the signal control point if necessary. An example of this type of transaction would be the translation of an 800 call to a regular telephone number for routing.

SUMMARY

SS#7 will be an intelligent network with the intelligence distributed throughout the ISDN. The signal control point need not be physically located within the LATA that it serves. Possibly one of the most important applications of the intelligent network that will be available under ISDN will be the establishment of private virtual networks. These will be less expensive than fully dedicated circuits but will offer many of the same services (e.g., abbreviated dialing, specified classes of service, the network control) through software managed by the communications manager.

One of the other significant possibilities the ISDN intelligent network will bring to the communications manager is that of universal automatic number identification. When the calling party number identification is generally available, a variety of services—automatic callback when busy, selective call forwarding and distribution, call trace, and integrated electronic data interchange—can be used by the imaginative manager to build new cost-effective business systems based on computer and communications technologies.

Chapter V-2-3
Transborder Data Flow

Jan L. Guynes
C. Stephen Guynes
Ron G. Thorn

RAPID TECHNOLOGICAL PROGRESS within the international data communications industry has fostered the development of global electronic information networks. An increasing number of international service transactions have been conducted, causing data communications to shed its once rather anonymous auxiliary role and become something far more important— the infrastructure of international trade. The industrial era policy models that continue to govern global trade relations, however, largely ignore services and lack the flexibility to adapt to rapid technological innovation. Because there had been no adequate international framework, many countries have established national policies designed to promote their domestic service sector by protecting core information resources.

Although these restrictions have not significantly hampered the effectiveness of international data networks, the tendency of governments to enact such protectionist measures creates considerable uncertainty regarding the continued availability and accessibility of communications networks for the conduct of international services trade. Moreover, as electronic data resources continue to gain strategic political importance in a rapidly evolving information society, nations are likely to increase the vigilance and vigor with which they monitor the flow of this resource over their boundaries.

This trend toward the fragmentation of an increasingly interdependent world along information policy lines could seriously undermine the world trade system. Having begun to recognize this, policymakers from industrialized countries have initiated negotiations aimed at establishing an international organization to monitor trade in services and transborder data flow. Because the outcome of these negotiations could significantly influence the business parameters of the data communications industry, the issues being addressed in current international forums are of consequence for today's data communications professional.

This chapter details those issues and describes current attempts to resolve them.

REASONS FOR NATIONAL AVERSION TO UNREGULATED DATA FLOW

In regard to the free flow and access of information, countries are divided into those that support it and those that want to restrict it. Advocates of the free flow of information and trade argue that it benefits all parties, whereas opponents argue that it only increases the advantage of countries that already dominate the information market.

Nations opposed to unregulated transborder data flow cite personal, corporate, and national privacy, national sovereignty, data access, data value, effect on employment, and the protection of home industries and markets. These concerns are described in the following sections.

Personal Privacy. Protection of the rights of private citizens was the original reason for regulating transborder data flow. A current trend is the prohibition of transmitted personal data to nations with less stringent data legislation, forcing some multinational companies, notably banks, to decentralize their information processing activities.

Corporate Privacy. Some countries (e.g., Austria, Denmark, Norway, and Luxembourg) have set up provisions for protecting the privacy of corporations, or "legal persons," inhibiting some companies from creating business data bases for fear that competitors would use privacy laws to examine data stored about them and possibly sue for unintended misuse of data.

National Privacy. Nations want to control the flow of information concerning their national interests (e.g., financial plans, economic trends, defense strategy, emergency plans, and technical data).

National Sovereignty. Many nations fear the dilution of their indigenous cultures by the influx of Western information, programming, and interpretation. Fear of domination has even prompted some countries to restructure external radio frequency allocations.

Data Access. Nations fear that if vital data is stored in other countries, they can lose access to it, especially during times of war or other national emergencies. More important, they are concerned about seizure of their data centers by acts of data piracy or destruction. A relatively free flow of information and advanced communications and computer technology make it virtually impossible to account for the storage and processing of all copies of data.

924

Data Value.   Nations that consider data to be a product support the same import-export controls as for other products. When data is classified in this manner, tariffs or taxation could decrease data flow across national borders if multinational companies find it financially inhibiting. Conversely, if multinational companies are willing to pay for access to levied data, they provide income to these nations. Most developing and underdeveloped countries have considered taxation.

The Effect on Employment.   Many nations (e.g., Canada) fear that information processing outside their borders increases employment opportunities for citizens of other nations and possibly results in unemployment at home. Brazil, Canada, and West Germany have enacted regulations requiring some preprocessing of certain types of data before it can leave the country.

The Protection of Home Industries and Markets.   Many countries fear that unregulated transborder data flow can inhibit the growth of their own communications and information industries, thereby hurting their domestic economies. To protect domestic industries, some countries (e.g., Venezuela, Brazil, West Germany, and Japan) permit the use of only locally manufactured computers, software, and communications equipment—a restriction that often forces multinational companies to rely on inferior products.

Smaller developing nations fear that unrestricted transborder data flow can further strengthen multinational companies' local economic domination. They also are afraid that multinational companies' technological advances can help those companies achieve local monopolies on technology, making host countries further dependent on the companies. This fear is probably well founded, though it appears to be the lesser of two evils (the other being technological ignorance).

INTERNATIONAL TASK FORCES ON TRANSBORDER DATA FLOW

Effective management of international information flow depends on handling legal implications—not an easy task. In addition to regulations stipulated by individual countries, many countries have joined intergovernmental organizations that define and limit transborder data flow. The following sections provide an overview of the major organizations and legislation.

The Council of Europe.   Based in Strasbourg, France, the Council of Europe consists of 21 Western European nations concerned with the protection of personal privacy. In 1980, it adopted the *Convention for the*

Protection of Individuals with Regard to Automatic Processing, which all member nations are required to implement. The council has not established, however, legislation regarding manual files.

The Organization for Economic Cooperation and Development. Members of this intergovernmental organization come from 19 Western European countries as well as Australia, Canada, Japan, the US, and New Zealand. The Organization for Economic Cooperation and Development promotes policies designed to:
- Achieve the highest sustainable economic growth and employment and promote a rising standard of living in member countries while maintaining the financial stability needed to develop the world economy.
- Contribute to economic expansion in member and nonmember countries.
- Help expand world trade on a multilateral, nondiscriminatory basis in accordance with international obligations.

In 1980, the organization adopted *Guidelines Governing the Protection of Privacy and Transborder Flows of Personal Data* to provide a unified approach to protecting private data moving across national borders. These guidelines strive to safeguard unimpeded transborder data flow between nations that maintain adequate domestic privacy legislation while warning member countries of the possibility of imposed sanctions if their privacy legislation is inadequate.

Although similar to the guidelines in the Council of Europe treaty, this organization's guidelines are voluntary. Member countries were asked to send letters to multinational companies urging their acceptance and implementation of the guidelines. More than 180 US companies and trade associations have publicly endorsed these guidelines.

In 1982, the US proposed that the Organization for Economic Cooperation and Development adopt a policy to minimize unjustified barriers to transborder data flow. In 1985, the organization adopted the *Declaration of Transborder Data Flows*, through which member governments agreed to promote free access to data, removal of unjustified barriers, transparency of regulatory policies, common approaches to resolve differences, and sensitivity to effects of transborder data flows on other countries.

The General Agreement on Tariffs and Trade. Endorsed by more than 90 member nations, the General Agreement on Tariffs and Trade (GATT) covers trade in goods, including computer and switching gear equipment. GATT members have unanimously endorsed new multilateral trade negotiations that specifically exclude communications services.

Developing countries oppose extension of GATT rules to include free trade in services, charging that they are at a disadvantage with developed nations. Because GATT includes more than 70 developing nations it is doubtful that the service issue will be resolved to the satisfaction of developed countries.

The Intergovernmental Bureau of Information.   Based in Rome, this organization helps member countries deal with information flow and derive benefits from transborder data flow technology. Fewer than 40 countries are members of the Intergovernmental Bureau of Information; almost all of them are in Africa, Europe, Central and South America, and the Middle East. The two exceptions are Spain and Italy.

Reflecting this constituency, the organization is considered an importer's advocate through which member countries discuss the importation of data goods and services. One of its basic principles is that transborder data flow should not violate national sovereignty or privacy laws of countries in which data is stored.

The International Telecommunications Union.   With more than 160 member nations, the International Telecommunications Union (ITU) is part of the United Nations and has its headquarters in Geneva. It is responsible for regulating international telecommunications by setting standards through the International Telephone and Telegraph Consultative Committee (CCITT) and the International Radio Consultative Committee (CCIR).

Regulations governing international transmission have been in effect for more than a century. As technology advanced and administrative structures changed, these regulations grew to include communications in space and all forms of data transmission.

ITU's mandate has expanded to include assistance of Third World countries establishing and maintaining communications networks.

The United Nations Educational, Scientific, and Cultural Organization.   Although the purpose of the United Nations Educational, Scientific, and Cultural Organization (UNESCO) was to establish a worldwide bibliographic system through which all nations could share and exchange information, Third World countries cited political, legal, cultural, economic, and technological grievances against the free flow of information. This emphasis on restricted information flow resulted in the withdrawal of both the UK and the US from UNESCO; other nations have threatened to withdraw or have reduced contributions, resulting in a bleak outlook for a united information policy.

The European Community.   The European Community, with headquarters in Brussels, promotes economic unification of the following European countries: Belgium, France, Italy, Luxembourg, the Netherlands, West Germany, Greece, Portugal, Spain, Denmark, Ireland, and the UK.

In 1988, the European Community announced its 1992 program, which aims to remove the remaining barriers to the free movement of goods, services, capital, and people among member nations. December 31, 1992, is the target date for a fully integrated European market.

The major concern among non–European Community nations is how their trade with member nations will be affected. The European Community contends that a unified Europe will stimulate the world economy and trade and that by 1992, any service provider approved by any member nation will be permitted the same service agreement as any member. The European Community is also actively involved with GATT, working to liberalize trade and services among GATT members. If the 1992 program progresses as expected, this could be a major benefit for multinational companies.

The International Telecommunications User Group.   This organization, composed of businesses in 13 developed countries, seeks to influence the policies of the ITU, the European Community, CCITT, and the Organization for Economic Cooperation and Development by promoting the free flow of information. Multinational companies doing business in any member nation are encouraged to join.

BARRIERS TO TRANSBORDER DATA FLOW

The goal of most multinational companies is to expeditiously move data across national borders while incurring minimal cost and regulatory hindrances; however, they are often thwarted in their efforts to achieve this objective. The major implications of restricted transborder data flow are trade barriers and loss of flexibility.

Trade Barriers.   Trade barriers can be caused by regulations (e.g., tariffs and taxes) or by unfavorable attitudes toward buying foreign services and software. Whatever the cause, trade barriers usually result in such increased costs as discriminatory pricing of data transmission services, inconsistent tariffs, and border taxes in compensation for loss of local jobs.

Excessive taxation and regulation can cause multinational companies to cease operating in certain countries. As a result, it has been suggested that "data havens" be created in countries that have few or no laws regarding transborder data flow. Data havens would allow multinational companies to avoid trade barriers but could thwart advances in computer technology in highly regulated countries.

Loss of Flexibility.   Information resource management, although critical to corporate operations, can be hampered when a multinational company is forced to rely on often inferior local hardware or software.

The value of information can also be affected by loss of flexible data access. Highly volatile information (e.g., stock prices) quickly loses value. Other flexibility restrictions include duplication of files and data bases, limited leased lines, and requirements that internationally transmitted data be processed locally.

Unrestricted transborder data flow could benefit multinational companies as well as less technologically advanced countries. Multinational companies could expand their markets while effectively managing their data resources. And information-poor nations could benefit through the influx of knowledge and technology. Many countries that restrict data export also have restrictions on importing technology. Although new industries (e.g., data base services, communications services, and computer manufacturing) can significantly spur a country's economic development, multinational companies are unlikely to contract such services unless they are of high quality.

SOME SOLUTIONS

Multinational companies should recognize the legitimate concerns of all nations regarding the flow of information across their borders. The guidelines in this section can help them manage their data resources as well as address host countries' concerns.

Communications managers must be aware of the rules and regulations regarding transborder data flow in each country in which the company does or plans to do business. They should seek to conform to each country's transborder data flow guidelines, allaying fears of privacy violation and data abuse. For example, the company can contact a country's data protection authorities and state and commerce departments to find out which international organizations (e.g., GATT) the country belongs to; these organizations can be contacted to provide current guidelines regarding transborder data flow.

When planning to move hardware between countries, multinational companies must conform to local import-export requirements. For example, US Customs requires an export license before computer hardware can be taken out of the country. An export license from one country, however, does not signify permission to import that hardware into another country. Companies unable to meet these guidelines without significantly altering the manner in which they do business must either change their business practices or cease operating within restricted countries—at least temporarily.

Communications managers should become active in organizations that their multinational companies can join or consult to learn more about transborder data flow restrictions and legislation and to promote data protection regulation and uniform standards.

Because the rules and regulations regarding hardware, software, and communications lines vary among—and even within—countries, communications managers should cultivate working relationships with people familiar with local systems and bureaucracies. These liaisons could save time and money by helping procure compatible equipment and data transfer capability that meets standards established by host countries.

SUMMARY

Negotiations regarding the flow of data across national borders involve divergent interests and objectives. Multinational companies have benefited from the free flow of information and technology between developed countries but are stifled by restrictions in less-developed countries. One of the major reasons that developing nations are primarily data exporters is that they lack the technology to handle the data efficiently within their own borders. This problem is not likely to be resolved until developing countries allow the technology to be imported, enabling them to become technologically developed importers and exporters of data. Underdeveloped nations that continue to prohibit importation of technology will probably never compete effectively in the global information market.

Negotiations within intergovernmental organizations should help resolve these issues. For example, the European Community may lessen restrictions and advance the cause of worldwide economic growth through computerized communications without loss of privacy or national sovereignty.

Multinational companies' efforts to comply with local regulations are also mandatory. Competitive gain should not be achieved at the expense of the basic rights of personal or national privacy.

Index

A

Acceptance testing
  documentation 725
  evaluating test results 728
  network 719–729
  objectives 720–722
  planning 722–725
  preparing test data 725–727
  running the test 727–728
Access controls 35, 200, 231, 821–823
  dial-up 249–256
  NetWare 231
Adaptive or learning algorithm 557
American National Standards Institute,
  see ANSI
American Telephone & Telegraph Corp,
  see AT&T
Amplitude modulation 441
Analog PBX 523–524
ANSI 269–271
  EDI and 900, 902
  T1 270–271
  X9.9 message authentication code
    215–216
  X9.17 key distribution 216
  X9.23 message confidentiality 217
  X9.26 access management 217
  X3.105–1983 209
  X12.42 electronic document
    interchange authentication 217
ANSI/ECSA T1 Committee 425
Application context management
  310–311
Architecture
  MAN 600–602
  OSI security 197
  TCP/IP 319–321

Association control service element 310
Associations 912–913
Asymmetrical modems 451–452
Asynchronous protocols 466–467, 645
Asynchronous transmission
  FDDI 645
Asynchronous versus synchronous
    techniques
  micro-mainframe links 146–147
AT&T 420–421
  metropolitan area networks 608–610
  PBXs 527
Audit trails 231–232
Authentication 35, 199–200, 821–822
  and identification 249–263
  data 258–259
  Kerberos 213
  message 258–259
Authentication mechanisms 207
Automotive Industry Action Group 902

B

Backup and disaster recovery planning
  232
Backup and recovery controls 824–825
Bandwidth allocation 93
Bit-synchronous protocols 468–469
Bit versus byte architectures 461–462
Boot infectors
  computer viruses 239–240
Bottom-up modeling 46–47
Bridges
  Token-Ring Networks 555
Broadband ISDN 426
Broadband modems 755–757

Business organization
  planning 24–26
Buy versus lease decision 129

C

Cabling
  broadband coaxial 737–739
  choices 136–139
  cost comparison 139–141
  fiber-optic 632–634
  splicing fiber optics 498
  twisted-pair station cabling 734–737
Cabling network installation
  baseband coaxial cabling 737
  broadband coaxial cabling 737–739
  building the network control console
    751
  case studies 139–141
  data center construction 747–749
  Digital Equipment Corp connectivity
    755–757
  final configuration 761–762
  management requirements 746–747
  planning methodology 133–135
  relocation plan 731–732
  twisted-pair station cabling 734–737
  Wangnet broadband modems 755–757
CAD/CAM
  centralized control 341–342
  data exchange 342–343
  Ethernet LAN 337
  FDDI and 340
  functional overview 333–336
  international standards: MAP, TOP,
    and STEP 343–346
  remote file access 338–339
  standards for distributed applications
    336–337
  TCP/IP protocol set 337–338
  wide area networks 79–100
Call destination
  X.25 379–383
Call restriction
  X.25 364–369
Call setup and clearing packets
  X.25 401–402
CCITT 274–278, *see also* OSI; X.500;
    X.400; X.25
  T1 424–425
  US participation 276–278

X.400 message handling standard 476
X.25 386–387
Centralized control 341–342
Centralized networks 80–81
  architecture 93–95
Centralized PBXs 524–527
Change control network 28–29
Channel banks 417
Charging-related facilities
  X.25 369–371
Chemical Industry Data Exchange 902
Circuit pricing 82–90
Closed user group facility
  X.25 367–369
Cluster controller emulation
  micro-mainframe links 148
Coaxial cable
  baseband 737
  broadband 737–739
Common management information
    protocol 308
Common management information
    service element 307–308
Communications cabling, *see* Cabling
Communications equipment
  fiber optics 499
  micro-mainframe links 152
Communications facilities 654
Communications professionals
  types of 868–870
Communications protocols 120–123,
    465–473
Communications services
  nonswitched 86–88
  pricing 82–90
  switched 86–88
Communications system acceptance
    testing, *see* Acceptance testing
Communications system failure 180
Compatibility standards
  modems 662–664
Components
  PBX 519–523
Component sparing 792–795
Compression-performing modems 450
Computer-aided design and
    manufacturing, *see* CAD/CAM
Computer-controlled PBXs 528
Computer viruses
  boot infectors 239–240
  counteracting 243–248

executable-file infectors 240–241
mainframe countermeasures 247
microcomputer countermeasures
244–246
network countermeasures 246–247
recovery techniques 247–248
systems infectors 240
Concurrency 232
Connection matrix 104, 114
Connectivity
Digital Equipment Corp 755
Constraints
network planning 27
Consulting services
advantages and disadvantages
862–863
benefits of using 854
criteria for selecting 857–859
future trends 864–865
network 853–865
types 855–856
Control
network 824–825
security 821–824
Control center
construction 747–755, 787–813
Controlled-access security modems 254
Corporation for Open Systems 292–293
Council of Europe 925
Cryptographic techniques 189–191,
204–207
Current situation analysis 50

D

DACS, *see* Digital access and
cross-connect system
Data applications
PBX 528–529
Data base access
micro-mainframe links 148–149
Data compression, *see also* Image
compression
video 501–515
Data confidentiality 200–201
Data encryption 35, *see also*
Cryptographic techniques
Data Encryption Standard
registry 206
technique 204–208

Data flow
barriers 928–929
task forces 925–927
unregulated 924–925
Data integrity
controls 815–821
service 201
Data interchange 41
Data multiplexers 417
Data rate
modems 656–657
Data security, *see* Security
Data service unit
recommended strategy 805
Delay 92
effects 456–459
Delivery
message 177–178
DES, *see* Data Encryption Standard
Design requirements
security 171–180
Diagnostic capabilities
modems 666
Diagnostic devices 806
Dial-up modems 655
Dial-up security 249–256, 259–263
Digital access and cross-connect system
419–420
Digital Equipment Corp connectivity 755
Digital multiplexed interface 426–427
Digital PBX 523–524
Digital signatures 206–207
Digital switching 796
Directory services
security and 218–221
Disaster recovery 128–129
and backup 232
Distributed networks 81–82
architecture 95–98
Distributed PBX 524–527
Distributed processing
data interchange 41
definition 31–32
development 32–33
electronic mail 38
implementation considerations 33–36
load balancing 36
network management 38
remote procedure calls 39–40
standards 36–41, 336–340
transaction processing 40–41

Distributed processing systems
  evolution 32–33
  management issues 34
  planning 31–42
  security of 35–38
Distributed-queue, dual bus protocol
  data types 571–572
  internal structure 581–585
  isochronous data unit structure and
    transmission flow 590–591
  layer access control procedures
    576–581
  logical link control sublayer 591–593
  media access control 586–590
  network topology 570–571
  physical attributes 574–576
  services 572–573
Distributed queuing
  MANs 617–619
Distributed systems
  capacity planning 33
  data control 819–821
  load balancing 36
  management 34
  reliability 33–34
  security 35
Distribution closet 741–743
Documentation
  acceptance testing 725
  maintenance costs 851–852
  maintenance problems 843–845
  network 843–852
  new practices 850–851
  resources and methods 846–850
DS-1 415
DS-0 416
DTE-DCE interface characteristics
  392–394
Dual bus proposal 604–606

E

ECMA, *see* European Computer
  Manufacturers Association
EDI, *see also* Electronic data
  interchange; Electronic document
  interchange
  for Administration, Commerce, and
    Transport 900, 902
EIA, *see* Electronic Industries
  Association

Electromechanical switching 796
Electronic data interchange
  background 899–900
  forecasts 900–901
  information sources 912–913
  just-in-time inventory and 900–901
  mailboxes 908
  migration strategy 911–912
  sources for EDI information 912–913
  standards 902–904
  structuring and selecting an EDI team
    901–902
  systems development plan 902–912
  transition to 899–915
  translation software 907–908
Electronic document interchange
  systems development plan 902–912
Electronic Industries Association 271
Electronic mail, *see also* X.500; X.400
  assessing the need for 478–479
  benefits and disadvantages 476–478
  distributed processing 38
  estimating network costs 482
  future 484
  implementing a network 483
  key management issues 479–480
  needs assessment 478–479
  selecting a network 480–482
  standards 38
  types 475–478
  X.400 message handling standard 476
Electronic mail networks 480–483
Electronic matrix switch 796
Electronics Industry Data Exchange 902
E-mail, *see* Electronic mail
Emanation protection 193
Encryption 189–191, 823
  Data Encryption Standard 204–208
  line 257–258
  link 208–211
  network 212
Encryption algorithm 207–208
Ends plan 48–50
End-to-end encryption 212
Enterprise analysis 12–13
Environment
  network management 776–778
  T1 417–420
Environmental analysis 13–15
Equipment selection and deployment 90
Equipment sparing requirements
  792–796

Error controls
line 816
Error reporting 829
Ethernet 337
European Community 927–928
European Computer Manufacturers
Association
security standards 202–204
European Telecommunications
Standards Institute
composition of 352
government control and 351–352
liaison with 353–354
mandatory standards 350–351
structure 348–350
voluntary standards 350
European Workshop for Open Systems
289
Event logging 250
Executable-file infectors
computer viruses 240–241
Exposures and threats
NetWare 228
Extended superframe format 422–423

F

Facilities, *see* Communications facilities
Facility and campus networks 58
FDDI
asynchronous transmission 645
CAD/CAM and 340
frame stripping 644
frame transmission 643–644
optical bypass 637–638
protocol data units 639–643
ring monitoring 645–647
ring scheduling 644
standards 627–650
synchronous transmission 645
token-ring operation 638–639
Fiber-optic connectors 497–499,
636–637
Fiber optics
benefits 486–489
cable 632–634
communications equipment 499
cost analysis 496
FDDI 627–650
fiber types 485–486, 496–497
LAN applications 493–495
noise immunity 489–490

performance advantages 495–496
point-to-point applications 492–494
repeaters 498
security 490
splicing 498
standards 492
trends 485–499
File transfer, access, and management
918
Financial services
security 215–218
FIPS 140 (Federal Standard-1027) 210
FIPS 139 (Federal Standard-1026)
208–209
Flow control parameter negotiation
facility
X.25 373
4,800-bps modems 448–449
Frame stripping 644
Frame transmission 643–644
Frequency modulation 441
Front-end processors
configuring 790
controls 827
Functional profiles 287–296

G

General Agreement on Tariffs and Trade
926–927
GOSIP
UK 294
US 294
MAP, TOP, and 771–772
Graph theory
network design using 101–117

H

Handshaking
micro-mainframe link 152
Hardware controls 825–827
Hierarchical topology
centralized 708
distributed 708–709
Host port protection devices 253
HR⁴-NET 615–616, *see also* Distributed-
queue, dual-bus protocol;
Metropolitan area networks
Hub networks 58

Hybrid networks 82
  architecture 98–100
Hybrid switches 534–535

I

IBM
  3270 PC 147
  Token-Ring Networks 553–558
Identification and authentication
  security measures 249–263
IEC, *see* International Electrotechnical
  Commission
IEEE 271
IEEE 802.6, *see* Distributed-queue, dual-
  bus protocol; Metropolitan area
  networks
Image compression
  general issues 504–510
  methods 502–504
  statistically based techniques 510–512
  symbolically based techniques
    512–513
Impact of delay
  on data transmission 457–459
  on voice transmission 456–457
Implementation
  network 27–28
Industry standards organizations 271,
  290–293
Information sources
  EDI 912–913
Installation
  LAN requirements 62–64
Institute of Electrical and Electronics
  Engineers, *see* IEEE
Integrated Services Digital Network, *see*
  ISDN
Integration 130–131
  LAN requirements 62–64
Integrators
  systems 693–704
Integrity mechanisms 206–207
Interface protocols
  TCP/IP 320–321
Intergovernmental Bureau of
  Information 927
International Electrotechnical
  Commission 279
International Standards Organization,
  see ISO

International standards organizations
  272–279
International Telecommunications
  Union 164, 272
International Telecommunications User
  Group 928
International Telephone and Telegraph
  Consultative Committee, *see* CCITT
Internet, *see* TCP/IP
IRMA board 152
ISDN
  basic rate interface 918
  broadband 426
  field-trial analysis 434–435
  field trials 431–434
  government use 435
  international implementation 435–436
  investment requirements 430–431
  MANs 606–608
  OSI and 917–922
  PBXs and 535–537
  primary rate interface 919
  Q.931 message structure 919
  signaling system #7 and 917–922
  standards 425–427
  T1 425–426
ISO 278
  standards development 284–287
Isochronous data unit structure
  distributed-queue, dual-bus protocol
    590–591
ISO 9594
  directory systems 218–219
ISO 10021
  message-handling systems 221–222

J

Just-in-time inventory
  and EDI 900–901

K

Kerberos authentication 213

L

LANs, *see* Local area networks
Layer access control procedures
  distributed-queue, dual-bus protocol
    576–581

Layers
  network 57–60
  TCP/IP 326–330
Learning or adaptive algorithm 557
Leased-line modems 655
Lease versus buy decision 129
Limited-distance modems 654–655
Line encryption devices 257–258
Line error controls 816
Line requirements
  switching 797–802
Line routing
  graph theory 109–117
Link encryption 208–211, 257–258
Links
  dial-up 250–251
Load balancing 36
Local area networks
  fiber-optic 492–495
  installation and integration
    requirements 62–64
  long-range planning 43–55, 57–65
  management 782–784
  management requirements 64–65
  planning 43–56
  preparing recommendations 549
  security 225–236
  selecting 543–551
  selecting and evaluating vendors
    547–548
  strategic planning 43–55
  system justification 60–64
Local area network security
  planning 226–230
Logical link control sublayer
  distributed-queue, dual-bus protocol
    591–593
Long-range planning 11, 15

M

Macintosh MacTerminal 148
Mailboxes
  EDI 908
Mainframe
  file server 150
Mainframe countermeasures
  to computer viruses 247
Mainstream network management
  802–803

Maintenance
  documentation 843–845
  third-party 831–838
Management
  electronic mail 479–480
  LANs 64–65, 782–784
  network 776–785
  security 202
Management information base 305
Manhattan Street Network 612–613, *see
  also* Distributed-queue, dual-bus
  protocol; Metropolitan area
  networks
MANs, *see* Metropolitan area networks
Manufacturing Automation Protocol/
  Technical Office Protocol
  and GOSIP 771–772
  CAD/CAM 343–346
  early MAP/TOP 764–768
  goals 763–764
  MAP 3.0 768
  standards for 343–345
  TOP 3.0 769–771
  user groups 290–292
Means plan 54–55
Media access control
  distributed-queue, dual-bus protocol
    586–590
Mesh topology 708–709
Message accountability 179
Message and data authentication
  258–259
Message authentication 176–177
Message authorization 175–176
Message delivery 177–178
Message handling standard 476
Message handling systems 221–222
Messaging
  security 175–179, 221–224
Message-oriented security and reliability
  175
Metropolitan area networks 58
  architecture 600–602
  AT&T and the multiport bridge 608
  distributed queuing 617–619
  dual-bus proposal 604–606
  history of IEEE 802.6 603
  HR⁴-NET 615–616
  initial deployment 600
  ISDN 606–608
  Manhattan Street Network 612–615

Metropolitan area networks (*cont*)
  performance evaluation and
    comparison 619–626
  physical attributes 574–576
  protocol 569–595
  queued packet and synchronous
    switch 616–619
  queuing analysis of QPSX 623–626
  services 572–574
  support 598–599
  use 599
Microcomputer countermeasures
  to computer viruses 244–246
Micro-mainframe links
  accessing data bases 148–149
  advantages 145
  applications 146–151
  approaches to 146–148
  cluster controller emulation 148
  communications devices 152
  difficulties of 153–155
  future 161–164
  handshaking 152–153
  IBM 3270 PC 147
  implementing 155–156
  IRMA board 152
  Macintosh MacTerminal 148
  mainframe as file server 150
  methods 151–153
  organizational issues 155–156
  products 160–161
  protocol conversion 153–155
  security and control 157–160
  software 151–152
  synchronous versus asynchronous
    techniques 146–147
  Systems Network Architecture 153
  terminal emulation 146–148
Microwave equipment manufacturers
  422
Migration strategy
  EDI 911–912
  PBX 527–528
Modeling
  about 129–130
  bottom-up 46–47
  network 129–130
  top-down 47–48
  wide area network performance 91–93
Modems
  advances 439–454

asymmetrical 451–452
broadband 755–757
compatibility and standards 662–664
compression-performing 450
controlled-access security 254–255
controls 825
data rate and throughput
  requirements 656–657
data transmission distance 656
diagnostic capabilities 666
dial-up 655
features 664–666
4,800-bps 448–449
leased-line 655
limited-distance 654–655
management 666–667
networking 659–660
9,600-bps 449
nonstandard 449–450
operating costs 669
packetized-ensemble-protocol 450
physical interfaces 660–661
port contention switch and 757
PSTN and 439
recommended strategy 805
security and 249–256
security systems 825–826
selection 453–454, 653–668
short-haul 757
standards 444–450, 662–664
support issues 667–669
synchronization 657–659
300-bps 444–446
300-bps to 1,800-bps 446–447
2,400-bps 447
user application 653–654
Wangnet broadband 755
Modified minimum spanning tree
  technique 113
Modulation
  amplitude 441
  combined methods 443–444
  frequency 441
  optical signals 635–636
  phase 442–443
  trellis-coded 444
Multidrop configurations 127
Multilink procedures
  X.25 395–398
Multiplexers 455–456
  controls 826

statistical time-division 671–682
versus PBXs 535

N

National Institute for Standards and
    Technology 271–272, 288–289
National standards organizations
    269–272
Needs assessment
    electronic mail 478–479
NetWare
    access control 231
    audit trails 231
    backup and disaster recovery
        planning 232
    concurrency 232
    exposures and threats 228
    physical security 230–231
    restricted commands 228
    security limitations 231
    security review 226–227
    user rights and access 227–228
Network acceptance testing 719–729
Network administration
    TCP/IP 318–319
Network analyzer 806–808
Network architecture
    centralized networks 93–95
    distributed networks 95–98
    hybrid networks 98–100
Network availability 712
Network change control 28–29
Network configuration
    wide area networks 93
Network consulting services 853–865
Network control
    backup and recovery 824–825
Network control center
    additional components 806
    component sparing requirements 792
    configuring front-end processors 790
    construction, case study 787–813
    front-end processors, configuring 790
Network costs
    electronic mail 482
Network countermeasures
    to computer viruses 246–247
Network delay 92

Network design
    establishing objectives 710–713
    fundamentals 707–717
    gathering information 713–715
    generating and optimizing a
        configuration 715–716
    graph theory and 101–117
    implementation and follow-up 716
    optimal 69–71
    optimization 119–132
    options 708–709
    process 709–716
    software 69–100
    tool selection 71–77
    trade-offs 707–708
    user requirements 72–73
Network design software
    centralized networks 93–95
    distributed networks 95–98
    hybrid networks 98–100
    selection 79–100
    system requirements 72–73
    user interface 73–74
Network documentation 843–852
Network encryption 212
Network growth 714
Networking
    modems 659–660
Network interface protocols
    TCP/IP 320–321
Network layers 57–60
Network maintenance, *see also*
    Maintenance
    third-party 831–838
Network management
    components 776–779
    construction of control center 747–
        755, 787–813
    diagnostic devices 806
    distributed processing 38
    elements 778–779
    environment 776–778
    example 780–782
    functions 779–780
    LANs, current status 782–783
    LANs, outlook 783–784
    local area networks 64–65
    mainstream 802–803
    manager's perspective 775–786
    manufacturing local area networks
        341–342

Network management (*cont*)
  OSI 303–314
  recommended strategy 803–804
  reports 124
  sidestream 802
Network management system
  example of 780–782
Network optimization
  disaster recovery 128–129
  lease versus buy decision 129
  modeling 129–130
  multidrop configuration 127
  protocols 120–123
  sensitivity analysis 128
  switching technology 123
  traffic load 124–126
Network planning
  constraints 27
  implementation 27–28
  requirements 27
Network problem tracking 28–29
Networks
  centralized 80–81, 93–95
  electronic mail 480–483
  hub 58
  hybrid 82, 98–100
  implementing 27–30
  long-range planning 57–65
  protocol standards 120–123
  traffic load 124–129
  traffic types 123
Network security, *see also* Security
  concepts and techniques
    186–193
Network topology
  distributed-queue, dual-bus protocol
    570–571
Network types 57–60
9,600-bps modems 449
NIST, *see* National Institute for
  Standards and Technology
NIST Workshop for Implementors of OSI
  288–292
Noise immunity
  fiber optics 489–490
Nonrepudiation 201–202
Nonstandard modems 449–450
Nonswitched communications services
  86–88
NPA-NXX 86
Nyquist limit 440

O

One-end protection 252–254
Open communications architectures
  security 185–194
Open Systems
  European Workshop for 289
Open Systems Interconnection, *see* OSI
Optical bypass
  FDDI 637–638
Optical signals 635–636
Organizational issues
  micro-mainframe links 155–156
Organization for Economic Cooperation
    and Development 926
Organizations for standards 269–282
OSI
  application context management 310
  application to ISDN 918–920
  association control service element
    310
  common management information
    protocol 308
  common management information
    service element 307–308
  development of 284–287
  file transfer, access, and management
    918
  functional profile development
    287–295
  functional profile harmonization
    295–296
  GOSIP (UK) 294
  GOSIP (US) 294
  Implementors' Workshops 288–290
  issues of systems management 304
  management information base 305
  MAP, TOP, and GOSIP 771–772
  network management 303–314
  remote operation service element 309
  security 197, 211–213, 217–224
  security architecture 197–202
  security management 202
  security services and mechanisms
    199–202
  seven-layer reference model 917–918
  specific management functional areas
    311–312
  standards development 284–287
  structure of management information
    312–313

systems management application
   entities 306–311
systems management application
   process 305
X.400 476
X.25 386–387
OSI Asia-Oceania Workshop 289–290

P

Packet format 362–363
Packetized-ensemble protocol modems
   450
Packet-switching networks, *see* X.25
PBX
   analog versus digital 523–524
   AT&T and 527
   centralized versus distributed
      524–527
   components 519–523
   computer-controlled 528
   data applications 528–529
   definition 517–519
   hybrid switches 534–535
   ISDN and 535–537
   memory 521
   migration strategies 527–528
   processor 520–521
   switching matrix 521–522
   T1 418
   versus multiplexers 535
   voice and data integration 532–533
   wide area networks 529–532
Performance evaluation
   MANs 619–626
Personnel
   EDI implementation team 901–902
   identifying LAN project coordinator
      544
   project team 27
   structuring and selecting an EDI team
      901–902
   support 29–30
   training 29, 827, 867–886
   types of communications
      professionals 868–870
Phase modulation 442–443
Physical attributes
   distributed-queue, dual-bus protocol
      574–576

Physical interfaces
   modems 660–661
Physical layer 392–394
Physical security, *see* Security
Planning
   business organization and 24–26
   distributed processing systems 31–42
   elements 15–18
   guidelines 18–20, 23–30
   interviewing users to determine need
      545–546
   LAN security 226–230
   levels 9–10
   local area networks 43–56
   long-range 11, 15, 57–65
   network 26–27
   operational 10
   process of 8–9
   strategic 9–10
   strategic synthesis 15
   tactical 10
   types 9–11
Point-to-point applications
   fiber optics 492–494
Port contention switch 757–759
POSI 293
Presentation encryption 212
Pricing
   communications services 82–90
Private branch exchange, *see* PBX
Problem tracking
   network 28–29
Products
   micro-mainframe links 160–161
Professional associations 872
Professional standards organizations
   271, 290–293
Protocol conversion 153–155
Protocol data units
   FDDI 639–643
Protocols
   advantages and disadvantages
      469–470
   asynchronous 466–467
   bit-synchronous 468–469
   choices available 470–471
   communications 120–123, 465–473
   distributed-queue, dual-bus 569–595
   metropolitan area networks 569–596
   selection criteria 471–473
   synchronous 467–468

Protocols (*cont*)
  TCP/IP 326–330
  wide area networks 91
Protocol standards
  network 120–123
PSTN
  and modems 439

Q

QPSX, *see* Queued packet and
    synchronous switch
Quality of service
  X.25 371–376
Queued packet and synchronous switch
    616–619, *see also* Distributed-
    queue, dual-bus protocol
Queuing analysis
  of QPSX 623–626

R

RBOCs, *see* Regional Bell Operating
    Companies
Recovery
  controls 824–825
  from communications system failure
    180
  from computer viruses 247–248
Reducing equipment and service costs
    683–692
Regional Bell Operating Companies
    421–422
Registry of cryptographic algorithms 206
Reliability planning 33–34
Reliability, windows, and
    acknowledgment facilities 376–379
Remote file access 338–339
Remote operation service element 309–310
Remote procedure calls 39–40
Repeaters
  fiber-optic 498
  Token-Ring Network 555
Resources
  documentation 846–850
Ring monitoring
  FDDI 645–647
Ring scheduling
  FDDI 644
Ring topology 708
Routers 556–558

S

Secure data communications
  design requirements 171–180
Security
  access controls 821–823
  authentication and identification
    249–263
  basic concepts 186–188
  computer viruses 237–248
  controlled-access security modems
    254–255
  cost 263
  cryptographic techniques 189–191
  data security controls 821–824
  design requirements 171–179
  developmental goals 170
  dial-up 259, 262
  dial-up access 249–256
  dial-up links 250–251
  directory services and 218–221
  distributed systems 35
  emanations protection 193
  encryption 189–191, 208–211, 823
  encryption algorithm 207–208
  end-to-end encryption 212
  European Computer Manufacturers
    Association 202–204
  event logging 250
  fiber optics 490
  financial services 215–218
  host port protection devices 253–254
  identification and authentication
    249–263
  Kerberos authentication 213
  line encryption device 257–258
  local area networks 225–236
  management 202
  message accountability 179
  message authorization and
    authentication 175–177, 258–259
  message delivery 177–178
  messaging 175–179, 221–224
  micro-mainframe links 157–160
  modems 249–256, 825–826
  NetWare 226–235
  network encryption 212
  one-end protection 252–255
  open communications architectures
    185–194
  operational requirements 180–183

OSI 197–202, 211–213, 217–224
presentation encryption 212
registry of cryptographic algorithms 206
standards 195–214
techniques 188–193
TEMPEST 193
terminal authentication 256–257
terminals 826
transport encryption 212
two-end protection 255–259
UNIX 213
user authentication tokens 256
user identification and authentication 249–250
verification techniques 192
Security event logging 250
Security services
model 222–223
Selecting
LANs 543–551
modems 453–454, 653–668
protocols 471–473
vendors 547–548
Sensitivity analysis 128
Services
distributed-queue, dual-bus protocol 572–574
Short-haul modems 757
Sidestream network management 802
Signaling system #7 920–922
Single-link procedure 398
SNA, *see* Systems Network Architecture
Software
micro-mainframe links 151–152
network design 69–100
SONET, *see* Synchronous Optical Network
Spanning tree algorithm 557–558
Sparing
component 792–795
Splicing
fiber optics 498
Standards
CAD/CAM 333–346
corporate involvement 887–895
distributed applications 336–340
distributed processing 36–41
electronic data interchange 902–904
electronic mail 38
European Computer Manufacturers Association 202–204
European Telecommunications Standards Institute 350–351
FDDI 627–650
fiber optics 492
ISDN 425–426, 429–436
ISDN primary rate interface 425–426
MAP/TOP 290–292, 343–345
modems 444–450, 662–664
network protocols 120–123
organizations 269–282
OSI 284–287
security 195–214
T1 415–436
X.25 359–387, 389–414
Standards bodies active in security 196–197
Standards organizations
American National Standards Institute 269–271
CCITT 274–278
Corporation for Open Systems 292–293
Electronic Industries Association 271
European Telecommunications Standards Institute 347–355
European Workshop for Open Systems 289
IEEE 271
industry, trade, and professional 271, 290–293
international 272–279
International Electrotechnical Commission 279
International Telecommunications Union 272
ISO 284–287
national 269–272
National Institute for Standards and Technology 271–272, 288
OSI Asia-Oceania Workshop 289–290
POSI 293
security 196–197
standardization process 279–282
transborder data flow 925–928
Underwriter's Laboratory 272
Standards proposals
international 281
national 279–281
Star topology 708

Station cabling user outlet 739–740
Statistical image compression 510–512
Statistical time-division multiplexers
  data source support 675–676
  flow control 674
  operations 672–673
  performance comparisons 673–674
  problems in use 676–679
  service ratio differences 674–675
  switching and port contention
    679–680
STDMs, *see* Statistical time-division
  multiplexers
STEP
  CAD/CAM 343–346
Strategic planning
  business considerations 24–26
  contents 18
  current situation analysis 50–53
  elements 15–18
  ends plan 48–50
  guidelines 18–20, 23–30
  levels 9–10
  local area networks 43–55, 57–65
  long-range 11, 15
  means plan 54–55
  modeling 44–48
  operational 10
  process of 8–9
  strategic synthesis 15
  system comparative analysis 53–54
Support
  MAN 598–599
  modem 667–669
Switched communications services
  86–88
Switching
  digital 796
  electromechanical 796
  electronic matrix 796
  PBX 521–522
  technology 123
  telephone company line requirements
    797–802
Switching and port contention
  multiplexers 679–680
Symbolic image compression 512–515
Synchronization
  modems 657–659
Synchronous Optical Network 424
Synchronous protocols 467–468

Synchronous transmission
  FDDI 645
Synchronous versus asynchronous
  techniques
  micro-mainframe links 146–147
System comparative analysis 53–54
System development plan
  EDI 902–912
System failure 180
System justification
  LANs 60–64
System requirements
  network design software 72–73
Systems infectors
  computer viruses 240
Systems integrators
  description 693–694
  model for evaluating 697–704
  vendors of systems integrator
    services 694–697
Systems management application
  entities 306–311
Systems Network Architecture 153

T

Tactical planning 10
TCP/IP
  architectural design 319–321
  CAD/CAM and 337–338
  history 317–318
  Internet protocols 320
  layers 326–330
  network administration 318–319
  network interface protocols 320–321
  overview of the protocols 326–330
  protocol architecture 319–321
  protocol set 337
  protocol standardization 318
  underlying mechanisms 321–326
Technical Office Protocol, *see*
  Manufacturing Automation Protocol/
  Technical Office Protocol
Telephone company line requirements
  797–802
TEMPEST 193
Terminal authentication 256–257
Terminal emulation
  micro-mainframe links 146–148
Test data
  preparing 725–727

Testing
  planning 722–725
  running the test 727–728
Third-party maintenance 831–838
  advantages 833–834
  contracting 834–837
  disadvantages 834
  industry 832–833
3270 PC 147
Threat analysis and risk evaluation
  188
300-bps modems 444–446
300-bps to 1,800-bps modems 446–447
Throughput requirements
  modems 656–657
Token-passing bus 339
Token-ring networks
  basics 553–554
  bridges 555
  expanding 554–558
  FDDI 638–639
  hard failures 560–562
  operation 559–560
  repeaters 555
  routers 556–558
  soft failures 562–564
  tools for analysis 564–565
T1
  ANSI 270–271
  ANSI/ECSA T1 Committee 425
  applications 427–428
  CCITT 424–425
  channel banks 417
  clear-channel capability 423–424
  committee on telecommunications
    270–271
  data multiplexers 417
  digital access and cross-connect
    system 419–420
  equipment 417–420
  extended superframe format 422–423
  features 422–424
  ISDN 425–426
  PBXs 418
  standards 415–436
  T-carrier hierarchy 416
  transmission resource managers 420
  voice compression 418–419
  zero suppression 423
T1 providers
  AT&T 420–421

private microwave equipment
  manufacturers 422
Regional Bell Operating Companies
  421–422
Tool selection
  network design 71–77
TOP, *see* Manufacturing Automation
  Protocol/Technical Office Protocol
Top-down modeling 47
Topologies
  centralized hierarchical 708
  centralized tree 708
  distributed-queue, dual-bus network
    570–571
  mesh or distributed hierarchical
    708–709
  options 708–709
  ring 708
  star 708
  wide area network 80–82
Trade standards organizations 271,
  290–293
Traffic load 124–126
Traffic types 123
Training
  choosing a program 875–880
  maintaining records 880–886
  network 827
  on-the-job 873
  professional associations 872
  reasons for 867–868
  types 870–871
  university programs 873
Transaction processing 40–41
Transborder data flow
  barriers 928–929
  reasons for regulation 924–925
  regulatory organizations 925–928
Translation software
  EDI 907–908
Transmission Control Protocol and
  Internet Protocol, *see* TCP/IP
Transmission distance
  modems 656
Transmission flow
  distributed-queue, dual-bus protocol
    590–591
Transmission resource managers 420
Transportation Data Coordinating
  Committee 902
Transport encryption 212

Trellis-coded modulation 444
2,400-bps modems 447
Twisted-pair station cabling 734–737
Two-end protection 255–259

U

Underwriter's Laboratory 272
Uniform Communications Standard 902
United Nations Educational, Scientific,
     and Cultural Organization 927
University training programs 873
UNIX
   security 213
User application
   modems 653–654
User authentication tokens 256
User charges
   X.25 369–371
User error controls 816–818
User identification and authentication
     249–250
User interface 73–74
User rights
   Netware 227–228

V

Vendors
   selecting 547–548
   systems integrator services 694–697
Verification techniques 192
Virtual networks 531–532
Voice and data integration 455–463
   bit versus byte architectures 461–462
   hybrid switches 534–535
   impact of delay on data 457–459
   impact of delay on voice 456–457
   implications for network design
     460–461
   PBXs 532–535
Voice compression 418–419
Voluntary Interindustry Communication
     902

W

Wangnet broadband modems 755–757
WANs, *see* Wide area networks

Warehouse Information Network
     Standard 902
Weighted connection matrix 105–106
Wide area networks 58–60
   bandwidth allocation 93
   circuit pricing 82, 94
   computer-aided design 79–100
   equipment selection and deployment
     90
   network configuration 93
   network delay 92
   PBX-based 529–532
   performance 91–93
   protocols 91
   topology 80–82
Wiring, *see* Cabling

X

X.500
   directory systems 218–219
X.400
   message handling standard 476
   message handling systems 221–222
   security services model 222–223
X9.9 message authentication code
     215–216
X9.17 key distribution 216–217
X9.26 access management 217
X9.23 message confidentiality 217
X3.105–1983 209
X12.42 electronic document interchange
     authentication 217
X.25
   call destination management facilities
     379–383
   call restriction facilities 364–369
   call setup and clearing packets
     401–402
   CCITT-specified facilities 386–387
   charging-related facilities 369–371
   closed user group facility 367–369
   DTE-DCE interface characteristics
     392–394
   flow control parameter negotiation
     facility 373–375
   introduction to facilities 364
   multilink procedure 395–398
   network user identification facilities
     370–371

1988–1992 recommendations 389–414
overview 359–363
packet format 362–363
physical layer 392–394
quality of service 371–376
reliability, windows, and
    acknowledgment facilities 376–379

single-link procedure 398
standards 359–387

Z

Zero suppression 423